A Less Complex Commercial Science Book

ISBN-13: 978-1539132219

ISBN-10: 1539132218

Written by P.S.J. (Peet) Schutte

© KOSMOLOGIESE EN ASTRONOMIESE TEGNIKA

> All rights are reserved.
> No part, parts or the entirety of this book may be reproduced by publishing, electronically copied, duplicated by whatever means that form reproduction or duplication, without the prior written consent of the copy rite owner.

This is the book showing everyone that there is

A Conspiracy in Science in Progress

ISBN 978-1-920430-05-4 Written by P.S.J. (Peet) Schutte

© KOSMOLOGIESE EN ASTRONOMIESE TEGNIKA

mailto:info@singularityrelevancy.com

All rights are reserved. Publishing, or alternatively reproducing electronically or using any other means of duplicating without the consent of the copyright owner is strictly prohibited. No parts of this book may be reproduced in any way that is possible.

If you aren't into science this might just be the reason why you chose not to study science. It is The Ultimate Conspiracy Theory.
It is the Conspiracy Concerning Physics

Reading this book will **intellectually** be **very challenging** to any person since what I say was never yet published. That I disagree with science's accepted principles on every basic issue on principles is a fact that is undeniably true. What you read about my principles I propose is new to everyone alike. If you do not consider your abilities in physics to be equal to a professional physicist don't be alarmed because even the purist physicist is a novice. I show how little they know physics. However, I found that the ordinary persons with a scholastic physics background cope with the difficult explaining much better than does Super-Educated-Masters. The Super-Educated-Masters have information stored by culture and if they can't bring the information to mind by recognition they fail to understand new science concepts. You are going to read this fact in a letter that was sent to me. The Super-Educated-Masters have preset conditions they prescribe to information and they can't break their mould. The purpose with which I wrote this book is to get around the network of Super-Educated-Masters who strangle any form of science that does not fit their views or match their liking. If what anyone says does not stroke with what the Brainy Bunch says who controls physics and agree with "Mainstream Science" or echo their thinking, they just smother all intellectual publication on the grounds that it is not fitting their profile on science. I disagree most strongly but I do also supply proof thereof where Mainstream Science blocks the publishing of my views on science that does not compliment their views. If you believe science is more accurate than God, then live your fantasy out and don't read further. If you want to know the truth about **how students** and the public **are brainwashed** by **mind control** in **science** this will wake you from your slumber. Read this and wake from the culture you believe in; that which science has lulled you into and made you accept science as the absolute undeniable rock fast truth by instating it as a religiosity then stop reading or get your tranquillising anti depressants next to you with a large bowl of water. You will find some mathematical equations, if you are not familiar with it ignore it because it shows the silliness of "Mainstream Science" but if you don't read it you will understand the explaining by reading the language where I explain it. "Mainstream Science" hides behind maths. I need help to fight their fraud and I need you to help me fight them. What you read I prove even in this book and I dare any one to prove otherwise or reprimand me.

are aware of. I don't plan to hold back my punches due to being scared of Masters in science.

You will find I never compromise truth for friendship and in that light I say what they needed to hear and not what is wanted to please who ever should feel pleased. Meet the Newtonian physicist. In this book I try to introduce the reader to the brilliant Newtonian conspirator that has been dragging all of intelligent man by the nose for three centuries on a string. The more the conspirator pretends to be an intellectual physicist the better a fool those conspirators become.

He looks sheepish because he acts sheepish because as he follows he never questions what he believes and brainwash students to do the same. Read how clever the physicists are in hiding their stupidity from students and the public alike. By enlisting thought control those teaching physics force students to believe in science and to believe science

This book started off as a website to inform about a science conspiracy but grew into a book that serves much more information than what I first intended to supply. It grew into a comprehensive study on cosmology. At times you may observe while reading this book that it seems as if my frustration will ring through like the chiming of the Big Ben Bell. For that there is a reason. At times my frustration and anger will boil over drowning my politeness and that is true, which I admit. For twelve years I have had the answer that would correct the philosophy that has a stranglehold on cosmological science. I discovered the building blocks of nature where my discovery puts all other cosmic aspects of science into science fiction.

Those who force-feed non-existing dogma do so to brainwash students to hide the incompetence of "modern science" so they can rule supreme while ignoring the truth that they deliberately hide by concocting a conspiracy. To keep everyone unguarded they practise a conspiracy by which they perform an accepted practise of thought control on students to further the false dogma presently in place. I try to blow the whistle on such a practise but accepting my resolution makes every thesis any person that is part of "Mainstream Science" ever wrote becomes science fiction.

Therefore no one in science dare to read my work leave alone appreciates the revolutionary nature thereof. Any association with my work must condemn the own work of such an associating person. Whatever is deemed now to be accepted science would by accepting my work then become what is the past tense and belonging to the past in science. My work unmasks the flaws that those in power of concocted science principles kept in place and in thought for centuries on end as forming the untouchable truth. It will then be rejected to show the holes formed as accepted science ! They try to silence me but surely somehow somewhere I have to break through with my massage! I bring you a true form of science as never seen before in all of history. I do that by disposing of the conspiracy that hides the incorrectness and the failures that haunts science today. They use mathematical interpretations that take the explaining away from logical facts.

In the event of any readers who may have questions concerning more facts as it is presented in this book, please feel free to contact me, PEET SCHUTTE. All information divulged came about through independent self-study during the past thirty-two years or so. I have to warn the readers that the topics are showing a very new approach with no quick answers abstaining from proof or holding just a few lines and the information is new in nature but not hard to grasp. Should anyone wish to confront me or wish to contact me then do so by E-MAIL AT: e-mail mailto:info@questionablescience.net

For more about the conspiracy also visit

www.questionablescience.net or
http://www.singularityrelavancy.com/
and www.sirnewtonsfraud.com

I WISH TO DEFINE THE CATOGORISING I USE AS PART OF THE BOOK.
I have the utmost admiration for Scientists and I shall never dream of placing me in the same category as academics mainly because of their intellect and achievements. They pushed their corrupt conspiracy of a hoax they present as science and which they further by brutal brainwashing through 300 years of never getting detected and that in itself is an achievement unheard of in human history. That achievement is most brilliant and no religion of magical mysteries in the past could ever match the Newtonians. Every time I go against Mainstream Science which is another name for upholding Newtonian blindness I am told I do not seem to have the intellect or mental capacity to **"understand Newton's classical mechanics"** and then because of my limited vision on physics I should know my place and retire to a dark corner where I would then silently and quietly vanish from earthly records. They forever tell me there are two positions on earth: those with the mental capacity **to understand Newton** and then there are those in my sector **that is mindless to the point of not understanding Newton.** In that sense there are two classes, the clever ones that **understand Newton** and then me, the mindless that just cant **understand Newton.**

To substantiate this segregation I use some referring to place distinction between the highly schooled super trained academics that spent most if not all of their lives in preparing to further their minds, filling it with the same void they fill the Universe and calling it "nothing". When I asked where is more nothing: Between Pluto and the sun because Pluto is the furthers from the sun holding the most "nothing" between it and the sun or in the centre of the sun because there is nothing standing between the sun's centre and the centre of the sun, I was discredited as incoherent and irrational. I tip the opposite of the scale as I spent little time repeating the brainwashing they subdue every student with to believe in the norms taught as the official policy in learning and education I have to be on the "other end". I don't believe their crap and tell it as I see it and therefore I am dumber than a pig, or that is their opinion. From where I stand and admire those in science, I can only see intellect as they fooled every person on earth for centuries non-stop: and moreover that achievement is presented as the academic's common denominator. If that is the common denominator used on the one side, fooling everyone by using unsubstantiated rumours and gossip and putting that as the joining factor, then on the other side, which has to be *"my side"* must then be the class of stupidity. To those forming the brilliance in science and their class such a remark would be an insult but to me (and therefore my class) it rings truth and that makes it not an insult but a norm we should except and learn to live with. I would rather be stupid and not **_understand Newton_** than be **Brainy** and believe I **_understand Newton..._** how stupid must I be before I would be able to **_understand Newton._**

It is rather a pity that while the SUPER CLASS will never say it to our faces; the SUPER CLASS is strongly of such opinion that we on the other side of the Universe have no minds to think in any way, and it is therefore our duty as much as it is our absolute privilege to except what the SUPER-EDUCATED, the ones occupying the informed side of the Universe inform us to what we should accept and the SUPER-EDUCATED live by that idea. As I said I have to live with it too and if I am the ill literate, then the SUPER CLASS must be the SUPER-EDUCATED; where I am the class amounting to stupidity the SUPER CLASS must be the Brainy Bunch. It all comes from the fact that there is such a huge differentiation between us. Those that **_understand Newton_** is therefore Superior and I, that don't **_understand Newton_** are of the lesser blessed. To distinctly point to grouping or class or whatever the readers wish to consider the division there are between the SUPER-EDUCATED and me I refer to the SUPER-EDUCATED side of the Universe by the names I use above. Further more when I refer to mistakes that I do prove to be mistakes in the book as we go along I refer to it as Xepted mistakes to clear another distinction of necessity. In short I don't **_understand Newton_** and therefore I am stupid and they **_understand Newton_** and therefore they are brilliant and what I present must hold the categories in such class divisions.

Introducing the six part theses **The Absolute Relevancy of Singularity** written with facts about the creation in mind produces a problem because the complete picture that I introduce has nothing in common with current accepted science. As it should be clear, I remove myself from science. The issues forming the new vision I present remains comprehensive even by using it in a very simple form, such as I do in this book. In spite of this, I shall explain three of the four unrecognised phenomena I use in proving my statements in this very book aiming at a theme of simplistic introduction being "an Academic Letter". Then I go into not that extensive detail proving all of the Cosmic Pillars. With my introduction of the phenomena, which I named the four cosmic pillars you will find it obvious why science do not accept them even if it is documented throughout the Universe and is quite commonly found. The phenomena are there used by nature and still Newtonians deny it being present with a functioning purpose. Mass, which is not present and not has a function science puts forward as the sole contributing factor to the formation of the solar system. What is present science hide and what is not present science promotes as being not only present but being purposeful. Then when I say Newtonian science is a hoax I am incoherent.

Applying the four cosmic principles totally annihilates Academic's formula of the basis on which science rests in the formula being $F = G(M_1 M_2)/r^2$. The four cosmic pillars are the following:
1. Roche-Lobe
2. Titius Bode principal
3. The Lagrangian principle
4. The Coanda effect.

From these cosmic phenomena I produce a path of cosmic development, even going as far back as preceding the Big Bang. That I manage because following the development through the four cosmic pillars where these form the building blocks in the Universe. Since there is a total absence of mass we have to use principles much more substantial than Newtonian rumours and innuendoes such as mass. All this information is totally new and never before did any person understand the phenomena. The problem that comes from this is that I take the reader from a point and lead the reader through the explanation of the existing principles, pointing out how they are flawed and introducing my explanations and proof and substantiate my argument. This is a path one has to follow should one wish to meet with cosmic reality. The principles I use are in place and used as I explain and I prove there is no evidence of mass anywhere. The biggest problem I encounter is to get those in office to admit and to realise there is a problem in science. As long as those in office in physics are contempt with hiding the problem that they dismiss behind a conspiracy while diverting from the truth for their personal worth in academic standings a resolve will never come about.

This situation is outrageous. Even for persons of the stature and intellect such as the Newtonians enjoy the entire problem must be uncovered before the resolution can be discovered and in this process there is no point where one can drop in, or out and in again and maintain the golden thread of understanding. A simple manner explaining the way the book falls into place is the colour we associate with objects and the differences I see to the reading of the colour compared to the way normality see the colour. Science currently judges the cosmos on face value while we have to use our intellect to scrutinize the cosmos. Things are not what they seem. When looking at a red flower we say the flower is red. Nothing can be further from the truth. The flower is every colour in the spectrum, except the colour we attach to it, although it is screaming with all might to its disposal that that specific colour it cannot accept. Yet, we maintain that that colour is the colour we associate with the object, ignoring the object's rejection of that colour. Only when looking at the cosmos from this stance, can the cosmos make sense? By recognizing a disassociation in spite of our cultural recognizing the association, can we understand the cosmos?

We maintain the sun is burning, while the fact of the matter is the sun is freezing. From our perspective on the outside we see the sun burning as we see the red flower. The sun can't discard heat and stay hot. By discarding heat we have to recognise the sun as being cold. Outer space on the other hand accepts all the heat the sun discards and still remains unchanged. By accepting heat without changing the outer space region must therefore be hot. At this point my statement sounds outrageous but with more space inside the book I explain this stance much better. We feel what the sun rejects and the associate that which the sun rejects as not forming part of the sun as being the sun. That is silly and stupid. What we see is not what is the truth. Only by applying the correct view to the cosmos can the four principles I introduce, make any sense and find any proof… and I do prove them. To get such proof I had to do extensive research on cosmology. The proof lies in the unrecognised and misunderstood laws and principles science know about. The laws although existing, fall outside the parameters of applied physics and because of that therefore they are not accepted, notwithstanding all evidence of them being there in place, applied by the cosmos and used to build the cosmos. Not recognising the principles is madness.

I had to define gravity; a task as yet, never done. From that I defined energy, which too, has never been done. I first had to prove the existence of time and time's control over the Universe, as much as time's role in the Universe and even moreover, what time is. Again, this was up till now not yet been achieved. I had to prove what space is, that time and space is sides of the same coin, with matter forming the separation. Galileo gave us the answer to space and time. By moving a pendulum through space we read time and therefore the pendulum shows what is time by the measure of space. Even before realising that I had to find and locate the position of singularity as not to repeat the same errors that Einstein made. In all this I had to prove, formulate and implicate four factors that are there in the universe, as yet and not yet understood or explained by anyone. I named the four principles the four cosmic pillars, which the Universe rests upon.

In parts of the book I explain extensively in some detail how the four laws not recognised brings about the speed limit though to be The Sound Barrier and explain by some short measure how this connects to the start of the Creation process and how this builds the solar system. The way the solar system forms and the Sound Barrier and the process by which Creation started is all interlinked and commonly connected. What I do not do is eco Newtonian bullshit by repeating unproven trash they use to brainwash students. In this admitting I realize that there is a substantial chance then that you the reader will put the book down and not continue reading, but when you do that you do so at your own peril. THE SUPER-EDUCATED stands highly negative against any religious attachment any person may have to science but it is science having the shortfall you see what the Bible shows.

Xepted science has a notion of including or excluding on the basis of likes and dislikes and through out the book I indicate this, but when being to personally involved with science and brainwashed into mental stupidity one cannot see them doing this. What Xepted science does not realize is that same whip is having a backlash on our SUPER-EDUCATED as well. By the same margin they are turning the loose of ignorance on themselves too because that attitude makes them unable to apply meaningful relevancies. Space is forthcoming due to the value and intensity derived from the concentration of time.

They see a Universe they draw with boundaries (although refusing to admit this) because Xepted science makes no effort in divorcing their thinking applying boundaries everywhere. That makes Xepted science see a Universe, which has to have boundaries (i.e. Einstein's theory on the critical density and that in itself holds boundaries), but boundaries the Universe has not. Time on the other hand has distinct boundaries and that too I show where it is. By calculating mass individually one does not recognise influencing attachments through singularity bonding and binding space and time and space-time. This comes about because the focus of attention falls incorrectly on the invalid part of space-time, which is space. Objects in the cosmos are not a certain distance apart but a certain time apart. It takes light time to displace space and not the other way around.

Using a simple top spinning I indicate time by positioning the point where I prove singularity is. This way I manage to prove many aspects in cosmology that is still unclear or not understood.

I prove and formulate **the Roche limit** as having two factors, both of which play a most dynamic role in the cosmos.

I have also managed to prove, formulate and define **the Titius Bode principle** and in that principle also comes with two factors. The principle is a derogative of **the Roche limit** and in amongst that there is another principle I have discovered concerning the dynamics as well as the role of light in the Universe.

The Lagrangian principle also flows directly from **the Roche limit** as **the Roche limit** is a ratio in conjunction to the point which I claim singularity is.

The formulas you are about to witness is another relevancy of singularity as it relate to positions or marks away from the point holding singularity. With all the bad experience I have had in the past where academics reject my work without any one of them investigating the detail and the substance and because those academics find my denouncing of Academics very presumptions I will not in the least find rejecting on their part again surprising at all. I ask you the reader for about two or maybe three hours of reading time and if you are not at least intrigued by my new ideas I shall find that even more surprising.

The work you have before you I have started in self-study back in 1977 and full time writing (without pay) doing investigative research compiled as seven books in one thesis. Academics through out the world place such a high premium on qualified education and once I admit to my personal research not upholding any recognised work I find immediate rejection just on the ground of that. Because my work is completely new and does not match any work done before I go being rejected. I find that truly unfair and in this day and age where free and fair speech serves such prominence it is most unfortunate to find that that is only lip service in most if not all cases. When reading the book you may come to realise why I came to the discovery, which I did.

To those in high office the flaw I detected is so marginal it does not even find a high degree of note worthy at first and they purposely avoid even admitting that such a mistake does occur. However with singularity not forming part of the known Universe they have covered the spot with zero as a factor and therefore I first have to show that zero is no factor in the Universe. If it did not cover the very spot where singularity is it may be the most insignificant discovery I could make. The truth in the matter is that those of academic prominence will never give it a glance and yet in that very spot is all the solutions to the mysteries of the Universe. That is because by brainwashing they see science as flawless and you can download www.sirnewtonsfraud.com http://www.lulu.com/content/e-book/wwwsirnewtonsfraudcom/7182451 and also www.sirNEWTONSFRAUD.com Part 2 http://www.lulu.com/content/e-book/wwwsirnewtonsfraudcom-part-2/8132511 to get some idea of the problem they ignore.

I admit that at this point you may even find a remark as innocent as the one I just made very offensive and rowdy, but as you progress through the volumes you will find I am not in the least exaggerating. I ask you not to dismiss the book after the first few pages just because you have the opinion that I do not understand Academic's laws. Again by making the following remark, it will come across as very presumptuous on my part, but I may be the only person in the world that understand Academics and not accept Academics. There is a huge difference between the two notions. The formulas you find is the diversion that an object in movement holds as the movement relates to the line singularity holds in relation to the moving object.

Have you as you sit reading this part at this minute sat back and gave a thought about the light enabling you to read? Such a thought brings to mind the most simplistic answer one can imagine. The light hits the page bounces from the page and contact the lens of my eye where the lens conveys the photons becoming electricity to a part of the brain that translate the electricity to an understandable message and that makes one read. It is as simple as that! Ever gave a broader thought about light streaming across the night sky, coming from ends of the Universe we do not even realise it is there?

How does the photons manage to convey one complete picture coming from as far apart and as wide an area as it does? With a few photons connecting the eye or lens no one ever noticed the wonder of light. The photons reflect a view that seems as if coming from all the billions upon billions of stars. But most is coming from darkness covering an area no man can measure. Yet how many photons can actually connect to the lens of the camera or to the eye?

Still a few photons coming from a single direction directly ahead eventually tell the entire storey. It is very simple to take the process of seeing by means of photon conducting very lightly and I have never heard one of the Brainy Bunch really in sincerity dissect the process to its potential. It is impossible that light from such an array of assorted sources can simply come together at the eye lens and show a picture of objects spanning across a Universe as wide as our mind can receive where the objects they reflect is beyond human measurement and the quantity is inconceivable many. Light is much more than the medium science takes it to be. Light connects the Universe in a way we cannot contemplate. Light being far apart originating from regions not in the same time or universal space connects in a way that present us with a picture holding the Universe in an understandable content. From the point we stand and we watch the Universe the significance of what we see surpasses the sense of understanding of what we are experiencing.

How can the few photons that our lenses catch coming from such an area as the night sky cover transmit the complete picture of what we see. Take a few seconds and gleans at the picture of the night sky then rethink the picture applying the full content in the picture to what the size of you eyes is. Think how big the picture is that your eyes take in and translate that area to the size of your eyeball in an effort to determine a ratio. One will be forgiven if one thinks of the ratio as eternal to nothing. Yet a few pages back I showed that according to mathematics there couldn't be anything as nothing. Consider the path the light followed from the source connecting to light from all other sources where all particles of the other light may come from and bringing a full picture to the lens one use to look through. In your mind connect a line from every atom producing light and connect the lines to your eyeball and see how you can manage to fit all the lines, as small as the lines may be.

Scientists think of outer space as geodesic zero, with nothing in outer space but space. Geodesic zero means the light travels in a straight line from where it originates unhindered all across space to where the light connects the eye. Such an idea by itself is outrages because the stream of photons reduce in space to such a minute quantity that taken the area the photons travel and the space in vastness it covers, the chances of one photon coming across many hundreds of light years through billions upon trillions of cubic kilometres of space and selecting my eye to convey the electricity is less than infinite. Yet such conveying takes place every second of every minute.

The position of the location of the second singularity, which is the precise duplication of the first singularity but in a diminished capacity, is obvious to miss when one is not applying a detective mentality, as one should in scrutinizing the cosmos. Culture will have us believe that when one sees a colour shining from an object the colour is associated with the object. Logic tells a different storey. A yellow dot is all the colours in the spectrum but yellow because it is disassociating with the yellow. That goes for red blue and all other colours we may visualise. I think the norm accepts this as scientific fact with very little argument or substantiating proof about that required.

If light came as individual streams of photon flurries our visage would translate that as such shown in the fragmented picture above. It would be a picture unconnected bringing across some photons in the manner where every object stands apart not being related in any way and that will be what we see, if it is anything that we see. That we know is not the case but that means geodesic zero is as much rubbish as anything Scientists regard with simplicity and with careless thought. Geodesic zero means nothing and how can I see nothing as darkness because "nothing" is not darkness, nothing is "nothing" and the darkness I see is darkness showing the darkness as something.

What then about colours that are technically not colours as is the case with black and white? White is simple. By spinning all the colours in the spectrum the colour white shines through. Black is quite another matter. A friend of mine whom is one of the best painters I have ever come across told me that one couldn't paint black but have to make black a dark blue to show shade on the canvass. That apparently is his success in achieving the realism.
He also went on to explain how many variations of dark blue form the shadows in one simple tree. This remark set my mind in motion.

One cannot see black because black has no colour to show, but black is the colour most prevalent in the universe. One can see only by colour and since black is not a colour we should not see black, but we do.

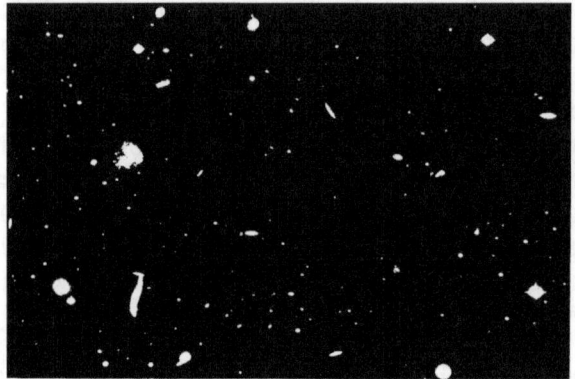

If the darkness was the representation of "nothing", then that should be exactly what we must see, nothing but the stars. Taken from the top picture some stars and leaving the rest to nothing is what we see in the picture below. A blind person sees nothing but when we look at space, we see something that we think nothing of as we see as space. One cannot have the ability of sight and see nothing except by closing your eyelids and then you see nothing. But in that case you do not see "nothing" in contrast of "something" you see "nothing" without it contrasting to "something".

Nothing is all about not being and not "not seeing".

By the ability to see the darkness renders the darkness something other than nothing and that changes the acquired value of the darkness from nothing to something. There is an eternal difference between something in infinity and nothing.

The arguments introduced up to this part of the introduction prologue only touches the most basic aspects of my work and by no means can such an introduction secure an opinion. Yet, not once through all my long investigation in the past thirty or more years have I found any other person claiming such views that I have brought about even in this skimpy way as I do in the prologue. The arguments introduced up to this part of the introduction prologue only touches the most basic aspects of my work and by no means can such an introduction secure an opinion. Yet, not once through all my long investigation in the past thirty or more years have I found any other person claiming such views that I have brought about even in this skimpy way as I do in the prologue.

As it applies with all things, so it does in this case as well that when delving deeper into any issue, the complexity of the issues truly come to the fore ground when analysed in more detail. I wish to advise the reader to treat the seven books as seven different works and in that light I have separated each work in volumes of seven separate books with individual I.S.B.N. numbers with adding one part, the one you are reading, with one sole purpose and that is to bring about an academic introduction to clarify a quick perspective. Then the next three parts being of a general introductory nature there are overlapping in some sense but each highlighting issues in different manner as to clarify facts used in the last three parts bringing conclusion to different cosmic perspectives. Yet the work is seven parts of one thesis and as such it serves.

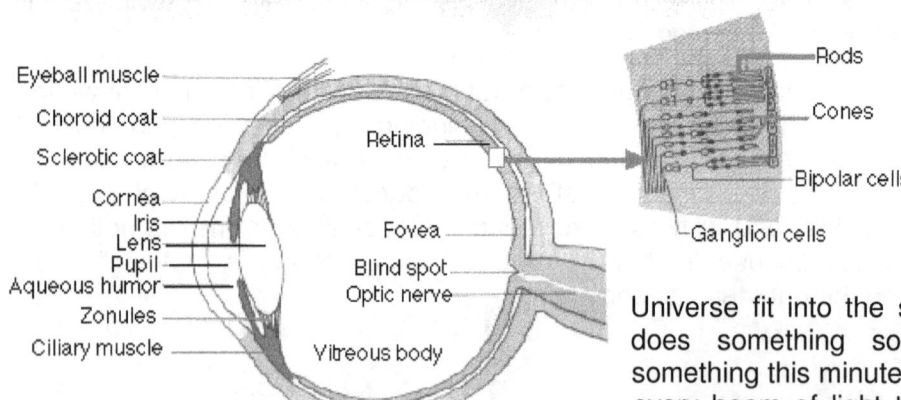

The question physics must answer is not how much mass is in the entire Universe because there is obviously no mass in the Universe. The question physics must answer is how does one half ($180°$) of an entire known or visible Universe fit into the socket of the human eye. How does something so extravagantly large fit into something this minutely small? How is it possible that every beam of light that crosses the space from all over the Universe meets in the eye of the beholder? Then let the Physicist put that in terms of Newtonian physics because Kepler gave the answer while Newton concocted that formula into something dubious.

We look at night and we see a Universe. Not only the light parts but the entire Universe or then half of it because the other half is situated at the back of our heads. What physics must answer if those proclaiming to know physics that well is to explain how the eye can accommodate all the light that comes from all directions throughout the Universe?

This following concept forms the entire basis of everything forming anything in physics that is part of science. If ever any thought represented physics then this is the most fundamental start of physics.

This is so impotent I wish to run through this again because this forms the basis of all physics. Have you, the person reading this, ever thought how it is possible to see that much information that you see at night when looking at the sky. Ever thought about how you are able to see when you see everything in the night sky and how that much light information can fit into such a small space as your eye? Have you ever sat back and think what the amount of information it is that you see when you see the entirety of the Universe when looking at the Universe at night and what the size is of everything of that which you are able to see?

The one star you see seems to be a near visible dot in the picture while the dot might be hundreds or might even be many thousands of times the size of the sun…and we think of the sun as big. The dot is then that much bigger than the sun because the star we think we see could be a galactica hundreds of times the size of our Milky Way galactica but that shows as in the sky as one little dot and yet that entire structure as big as it is, does also fit into our eye socket. But that is not all…there are trillions of such light images and they all fit into one eye socket. What we see is immeasurable and yet we see it effortlessly in the space our eye holds…how can that be?

How is it possible to fit what we see into the space of our eyes we have? Think how much is the entire information that is visible at night and think about how all of that fit into the space your eye holds?

Consider how big is what is visible and put that space into the size is of what your eye can hold and ask your mathematically educated Professor in physics to find some ratio between what you observe and the size of your eye. The ratio is astonishing, but more-over what is truly astonishing is the arrogance of man to think of his position, as being important while the space man holds is beyond any comparison in ratio to everything we see in the Universe we see. Think how small we are when we are able to see the entirety out there! Even if there was other life out there, what is the worth of it in comparison to what there is that we see?

In this idea about how you are able to see the entire Universe you will find all the answers to the questions about how physics use time to employ gravity. Mass and anything Newton ever said has no implications on the explaining. It is about all the information of the entire Universe presented in one electron contacting a nerve in your eye. The question about the Universe is how can whatever is in view, come stored as a parcel in an electron, and tell the entire story about the entirety out there locking all that data into the space of an electron. That is physics and tries as you may, not one person Newtonian or otherwise can have mustered the ability to calculate that part. Newtonians can pretend to play God and live in their fool's paradise as long as they are King of the Universe of fools while keeping the conspiracy alive to hide the truth about Newton's corrupt formula of $F = G \frac{M_1 M_2}{r^2}$. There is a mad conspiracy in physics to prevent anyone not in physics to learn about the truth hidden to all about Newton's Gravitational principle fairy tail. The fact that you can see the entire Universe and everything in it through one optical nerve tells everything that physics and cosmology up to now missed completely.

Newtonians uphold their law of physics without showing mercy. The very first things the Newtonians use to beat us into submission are to blast us with incomprehensible mathematical formulas. Incomprehensible they are but it is to scare anyone with the mathematical equations to get everyone hiding. They bewilder you with equations that put the fear of God into you; used simply to make you feel inferior so that they can feel superior and frown down on your inferiority from a dizzy height. They are masters at manipulating anyone into a state of senselessness...but mostly they do it onto themselves. That they do because it forms the backbone of the fraud. They do not wish you to read closer and to find the fraud they hide to protect Newton. Ignore their mathematics because it only shows their incompetence to understand physics or Newton and see the fraud they propagate...They employ mathematics to bewilder and that is all. I am going to show what we can uncover underneath what they cover. Look at what the mathematics supposedly says and then wake up, they are using maths as a scare tactic for three centuries to scare the daylights out of you and all this while its been working! Looking at the formula shows just how little Newton understood physics. Please allow me to show you how they scare you to become fooled and suckered. Don't run and hide when you see the mathematics; it is meaningless although it was used as a scarecrow for more than three hundred years forming the backbone of the conspiracy. In this case I am referring to the so-called Kepler's Laws that has nothing to do with Kepler.

Use this picture below to show me where the planets are positioned according to mass or where the orbit going around the sun goes according to mass. The entire Newtonian idea of mass creating gravity by pulling is the complete misrepresentation of the truth. I show what principles are in place do give the reason why. It is easy to talk about "mass" and never get "mass" part of reality when hiding the truth.

Newtonians make a statement about "mass" holding the solar system in place. No matter how much this is corrupt, nevertheless they put it down as a given fact so much so that they will show doubt in a living God being present but that mass pulls planets goes beyond doubt. The proof of mass pulling to form gravity can never be tested because it is beyond doubt. If you doubt in it they throw a Newtonian made formula they call Kepler's laws at you. "Kepler's third law" supposedly is "the square of the orbital period of a planet is directly proportional to the cube of the semi-major axis of its orbit." In mathematics it is Symbolically: $P^2 \propto a^3$ and therefore $a^3 = P^2$ in position of P and therefore $a^3 / (P^2P)$. This is taken from the idea that "**Kepler said**", which is totally fabricated that **$a^3=T^2$** where Kepler said in fact **$a^3=T^2k$** and this is a big difference because **$a^3=T^2k$** is the same as $E=mC^2$. Look at reality. $a^3 = P^2$ is total garbage and as

big a hoax as is the idea of mass being any form of factor in gravitational physics or that gravity applies in accordance with $F = G \dfrac{M_1 M_2}{r^2}$. Look at the picture below. Look at how the planets are sorted and that is not by size. There is a ratio applying called the Titius Bode law and this law puts planets in terms of size or mass at a precise equal base notwithstanding that Jupiter is many time bigger than Mercury. Everyone is so taken by the accuracy of Einstein's formula that $E=mC^2$ but this is exactly Kepler's formula where $E^3=mC^2$ is taken from Kepler's formula when accurately used as **$a^3=T^2k$**. There is no $a^3 = P^2$.

Do not get scared as everyone usually does when see and get frightened then consequently as a reaction to find survival you turn on your heels and run...

**Don't run, just read on and see how simple it is to prove Newton was a backward dark aged sod!
You don't have to be a mathematical mastermind to see that it is not mass that applies to allocate planet or star positions. Here is an example that any person can understand when they don't succeed in bewildering people with frightful mathematics and comprehensive formulas.**

I know and realise that you are disgusted by my attitude when I degrade the name on which physics are founded. In this introduction part I am going to show you just some minor deceptions all students are forced to believe since all physics students are forced to believe in Newton, **Sir Isaac Newton** that is.

I am giving you a choice. You can say I am going to commit fraud or Newton has committed fraud. If I am judged to be the culprit that is guilty of deception then it is because Newton misled me. You can choose.

You are expected to believe the following:

Newton stated under the nametag of Kepler that there are so called Conversions for "Unknown" factors.

$4\pi^2 a^3 = P^2 G(M + m)$ Newton introduced this concept because he said mass brings about gravity.

From the top formula Newton devised the next formula $P = \left(\dfrac{4\pi^2 a^3}{G(M+m)}\right)^{0.5}$, which he named after Kepler. Kepler had nothing to do with the entire idea and every incorrect aspect is a Newton contribution. Students don't shy away from the mathematical aspect because the formula is complete bogus fraud.

Ask your physics professor to put in the mass of the sun and any or every planet and from that determine the allocated position in accordance to the calculations derived from the method that the formula dictates. The formula is fraud and keeping the formula in place and used by students all form part of the conspiracy to hide the incompetence.

Any person that upholds Newton's ideas and principles about mass forming gravity use the next formula

Newton introduced $P = \left(\dfrac{4\pi^2 a^3}{G(M+m)}\right)^{0.5}$ and then go on and put in the allocated position of the planet in the pace of **P**.

Then put in this formula $P = \dfrac{1}{G(M_1 + M_2)}$ **the mass of the sun plus the mass of each planet to show how the position is in place in accordance. This is no less than fraud and yet, on this the entire conception of Newtonian science rests.** Let any physicist I challenge take me on by proving this or any other Newtonian formula correct.

Let any physicist I challenge **$4\pi^2 a^3 = P^2 G(M + m)$** prove that it is mass $\dfrac{1}{G(M_1 + M_2)}$ that allocates the poison reserve by the planet **$P^2 = 4\pi^2 a^3$** which is what this part of the formula**.**

If **"mass"** did form gravity by a value that commits a force then the large planets must be on the inside next to the sun having the small planets way on the outside. Instead because we have a random allocation, that destroys the idea of mass forming gravity to pull.

The smallest planets holding the least "mass" are at either end and the largest planets holding supposedly the most "mass" are in the centre. This disproves both arguments that the pulling force forming gravity by the value of "mass" to establish the orbit goes according to "mass" or that the locations of planets are adhering to Newton's ideas of "mass forming gravity by pulling". The way that the present cosmology shows gravity forms is by telling everyone about mass and then this is how gravity forms the Universe. That is the way they put the Newtonian model forward.

Everyone is as gullible is the preconditioning would allow the people to hold the mindset in which the people are in. All knowledgeable persons know the sizes of the planets and yet no one thinks about the bigger planets being in the centre and the smaller planets circling near the sun or far from the sun. No one ever took it to task to confront those cheats in physics about the claims that it is mass that holds the planets in place. The world of people wants to be cheated as long as no one is asked to think and apply personal wit.

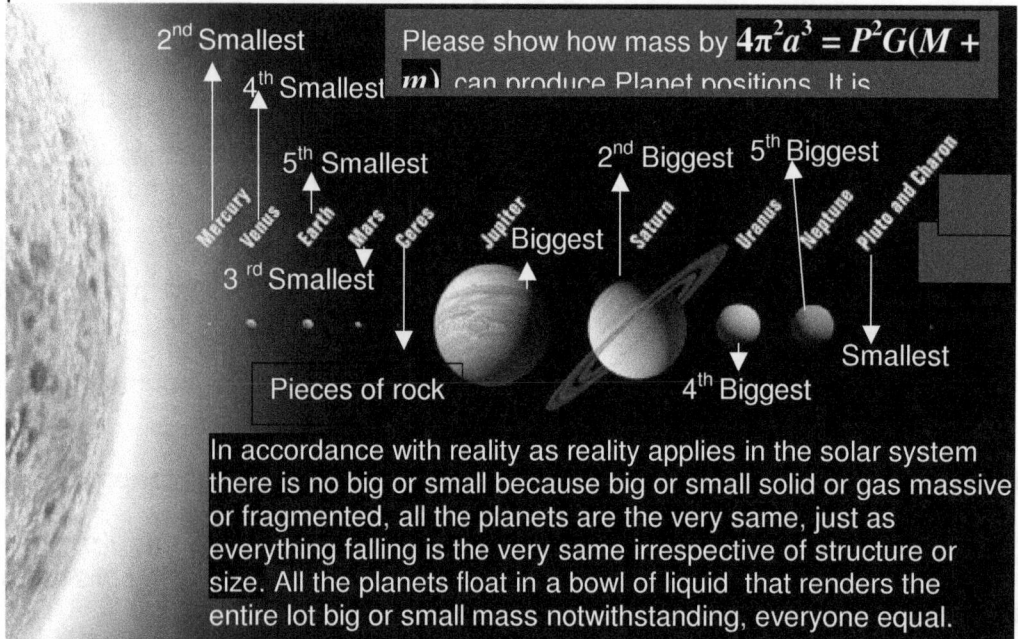

Please show how mass by $4\pi^2 a^3 = P^2 G(M + m)$ can produce Planet positions. It is

By depicting the solar system in such a presentation as Newtonians normally do such as the picture next this form of presenting the layout without providing correct spacing purposely corrupts the entire structure formation by which the solar system develops. It then purposely hides the essence that forms the solar system.

In accordance with reality as reality applies in the solar system there is no big or small because big or small solid or gas massive or fragmented, all the planets are the very same, just as everything falling is the very same irrespective of structure or size. All the planets float in a bowl of liquid that renders the entire lot big or small mass notwithstanding, everyone equal.

In the Universe all thing are equal in size because Neptune spins around the sun equal to mercury's time and Jupiter floats around the sun equal to mars or Neptune. Notwithstanding what "size" or "mass" they grant the planet to have the rotation happens equal and without mass bringing any favouring in positioning or in speed of movement. So where the hell is mass a factor in gravity forming?

Planet	Supposed "Mass" where earth is taken as 1	Average Orbit distance	Biggest to Smallest
Mercury	0.055 times the earth	57.9 x 10^9	2nd Smallest
Venus	0.81 times the earth	108.2 x 10^9	4th Smallest
Earth	1	149.6 x 10^9	5th Smallest
Mars	0.107 times the earth	227.9 x 10^9	3rd Smallest
Asteroid Belt	A Multitude of planet-forming fragments	Notwithstanding size the lot orbit alike	Debris orbiting in space at a specific distance that is not in relation to mass
Jupiter	318 times the earth	778.3 x 10^9	Biggest
Saturn	95 times the earth	1427 x 10^9	2nd Biggest
Uranus	14.5 times the earth	2871 x 10^9	4th Biggest
Neptune	17 times the earth	4497 x 10^9	3rd Biggest

From the Titius Bode that forms the solar system I have compiled the following formula by which gravity forms to the value of Π

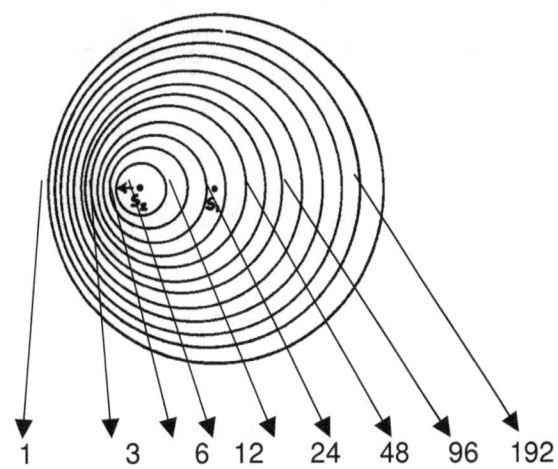

1   3   6   12   24   48   96   192

What we do find in the solar system and what does apply is the Titius Bode law.

Planet	Mercury	Venus	Earth	Mars	Ceres	Jupiter	Saturn	Uranus	Neptune	Pluto
Bode's Law distance	4	7	10	16	28	52	100	196	-	388
Actual distance	3.9	7.2	10	15.2	28	52	95.4	191.8	300.7	394.6

A numerical sequence announced by J.E. Bode in 1772, which matches the distances from the Sun of the six planets then known. It is also known as the Titus-Bode law, as it was first pointed out by the German mathematician Johann Daniel Titius (1729-96) in 1766. It is formed from the sequence 0,3,6,12,24,48,96, and 192 by adding 4 to each number. The planets were seen to fit this sequence quite well – as did Uranus, discovered in 1781. However, Neptune and Pluto do not conform to the 'law'. Bode's Law stimulated the search for a planet orbiting between Mars and Jupiter that led to the discovery of the first asteroids. It is often said that the law has no theoretical basis, but it does show how orbital resonance can lead to commensurability.

The importance that becomes known is the sequence the Titius – Bode law saw in the number arrangement of 3; 6; 12; 24; 48; 96 etc. The incorrect application of the Titus Bode law lies in subtracting the figure of 3 from 10 leaving 7. The other way of reasoning is to add four each time to the firs value of three starting with 3 and so on. The true significance of the Titus-Bode law is that it points directly to a circular growth of 7 stages. The 7 relating to 10 is a precise derogative of the Roche limit or the Roche limit is a precise derogative of the Titius Bode principle because he two systems interlink. This is how I mange to explain the Titius Bode law that is in the solar system by the ratio applying that really form the solar system in the way nature shows space growing by time. What you see on the next page was never been shown but on the other hand Physicist say this mathematics are too simple to apply as physics! This is what there is. This, the Titius Bode law is a given as it is fact.

With the exception of the first two, the others are simple twice the value of the preceding number.

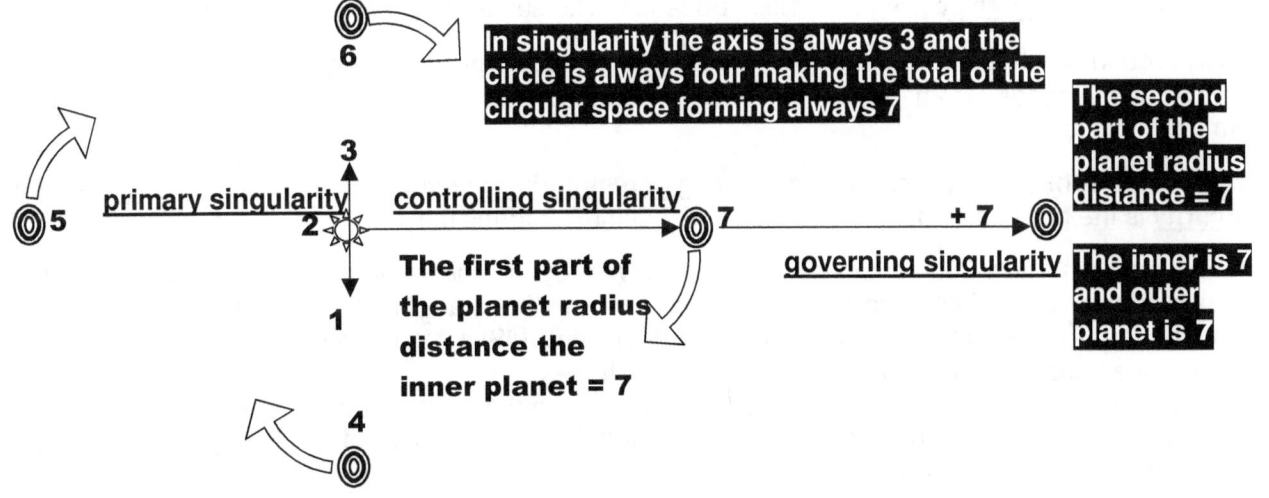

To find the mean distances of the planets, beginning with the following simple sequence of numbers:
0 3 6 12 24 48 96 192 384

To explain this is as follows: the axes line holding 3 doubles because in singularity all axis lines are 3 (1 bottom 2 centre 3 top)

Add 4 to each number:
4 7 10 16 28 52 100 196 388

Then the circle value that always has four point.

Then divide by 10:
0.4 0.7 1.0 1.6 2.8 5.2 10.0 19.6 38.8

Travelling in a straight line or a half circle or a triangle in terms of singularity is equal because it is all 180°. By taking 7 (the first or inner planet) in terms of Pythagoras 7^2 breaking the centre line 1^2 the result is 49 + 1 = 50.

The second circle also values 50 and since singularity unites the movement it totals to form 100. The square root of 100 is 10 and dividing the travel by 10 the allocated position becomes valid.

Is this not far better and truer than the following Newtonian accepted rubbish?

The Titius Bode law is a part of the Coanda effect and the Coanda effect is the interaction there is between what is material and that which holds material.

That which is 7 is part of material and the 10 part is liquid / gas. In the picture the outside planet holds relevance with the line that the direct inner planet has and that in turn holds relevance to the axis the sun holds.

The outer planet has the **governing singularity**, the inner planet holds the **controlling singularity** and the Sun has the **primary singularity**. As soon as any object moves on Earth, the movement switches singularity by allowing the object to obtain the **governing singularity** while the Earth then fore fills the directional circular control in forming the **controlling singularity.**

All four phenomena interacts in a manner forming this role where for instance in the solar system the Sun holds the **controlling singularity** and Milky Way forms the **governing singularity**

The outer planet is the **governing singularity** that forms a governing position in terms of the location and that has a place in terms of the second 7 that will eventually form the 10 of space. This is the outer 7.

The planet on the inside of the planet holding the governing position has a **controlling singularity** since it delivers the four points in rotation that positions the last 7 with which the governing singularity finds position.

This hold the value of the inner 7 of which the **controlling singularity** forms a value of 4 as that forms the **controlling singularity** circle in the Titius Bode law. The sun axis holds the **primary singularity.**

The axis around which the Sun turns forms the **primary singularity.** This **primary singularity** is the axis that draws all the space ($T^2 \div a^3$) towards the sun and in that it has the absolute **primary singularity** role of forming the value of 3.

The first 3 (the sun) and the second 4 (the **controlling singularity**) forms the first 7 while the **governing singularity** is the positioning point to form the two 7 point that forms the 10.

Mathematically the Titius Bode comprises a double seven (by the square) in relation to singularity (by the square) to form one hundred by implementing the Law of Pythagoras. Then the hundred is going into a root to form ten. This is as follows: One side of the triangle 7^2 (49) + 1^2 (1) side of the triangle = 50 x 2 = 100. Then one hundred in the root thereof √100 = 10 is ten.
The axis is 3 and the circle is 4 and the compliment is 7 where the space is double seven (the distance is always doubling) and the double square of seven in relation to the square of singularity is one hundred and the square that this forms is ten.

From the matter-to-matter relation in the Titius Bode configuration there are 7 / 10 + 7 / 10 = .7 + .7 = 1.4

From the space-to-matter relation in the Titius Bode configuration there is 10 / 7 = 1.42

$$(7 + 7) / 10 = 1.4$$

$$(7 + 7) / 10 = 1.4$$

$$(7 + 7) = 14$$

7° on both sides of the circle

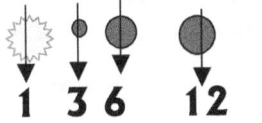

7 = 1.4

10 / 7 = 1.42

10 / 7 = 1.42

10 / 7 = 1.42

= .7 //\\ 1.42

= 1.4 //\\ 1.42

Any object turning around a centre (in this case planets turning around the sun) goes in a straight line by diverting from a straight line by 7°. On a later occasion in this book I show how the 7° forms a direct link in becoming 10. This is the ratio that the space between the planets grows by taking 7° and forming 10 from doubling that value.

In using very simple mathematics and I also dare also say too simple for the extremely intellectual Newtonians Physics Academics I prove how the Titius Bode law works as the Titius Bode law forms space through applying gravity.

While there is no hint of mass as a factor in the solar system, this Titius Bode law is what is present and is what is applying. The only thing new is that I am the first that prove how this law works and no physicist is interested this far in what I have to say because I belittle Newton and his corrupt principles. Every physicist in office with doctoral fights this because with this I prove Newtonian views are no more than science fiction

Because the space-to-matter is in the square at 10 placing the matter-to-matter at a square of .7 + .7 = 1.4 the space-to-matter forces the matter-to-matter to double the distance by number as structures are place father from the main$\Pi°$ maintaining singularity.

1 3 6 12 24 48

I'm correct. If they admit that I'm correct the entire world of physics becomes recognised as the joke it is in reality and they are recognised by all as being the laughing stock they are.

Later I show that 7^2 in conjunction with singularity applying the law of Pythagoras forms 10.

$14 \div 1.42 = 9.859$ or 9.86 or Π^2

This is mathematically how the Titius Bode law forms gravity as space-time in space by time.

I use primary school level mathematics and for that being so simple they say its too simple too apply as physics. I will show you one the many letters of rejection I received later on in the book This is what there is and that is all there is in the solar system. Look at any picture and try to finds mass. The measure of mass forming gravity clearly plays no role in allocating the positions of planets as Newton declared it must do. The entire idea that gravity is a magical force created by the value of mass is as unbelievable as the dogma is of those presenting this idea. Please use what the solar system provides to confirm what Newton says is in place when he says mass forms gravity. Science would rather accept Newton where there is no proof of Newton ever being correct than to admit Newton's incorrectness. I prove mathematically the reality of **all four cosmic principles** that are in place in forming gravity but because I do that and because I then make a mockery of Newton and their Newtonian principles they ignore me. Science would rather deny there is cosmic principles that is in place in the solar system, which are

The Roche limit,
The Lagrangian points,
The Titius Bode law and
The Coanda effect than to admit to Newton's failings.

They would not admit to Newton's failings because then the entire world will see they know less about science than does a pig know about history. They would rather put the error on the solar system than they would commit to the blatant mathematical cheating that Newton committed. It is the Universe that is always at fault when Newton becomes incorrect because without Newton's fabrication of science they have nothing to show for all the wisdom they try to pretend they have. Newton fabricated "Kepler's laws" has some correctness mixed largely with a farce and blending the truth with total fabrication of reality hides the lie behind something presenting the truth.

Forget about the fanciful corrupt mathematics that proves nothing when the cosmos does not confirm Newton's crooked mathematical arguments. Newton's religiosity might corrupt science but who in science would cares about correctness when it simplifies the ongoing brainwashing of students studying science.

They confuse everyone about what *weight* is, what *mass* is and what *gravity* is because they wish to have everyone think of "mass" in terms of weight while they then deny weight and "mass" is the very same thing and then they confuse "mass" and gravity because they never distinguish between what "mass" does and what gravity does. This is only the tip of the iceberg and you will see when reading this book.

This phenomenon should not occur with Academic's laws about gravity. These bodies will collide and destruct, without a doubt. When **F = G(M₁·m₂)/ r²** apply, there should not be any force which is able to keep them apart. Known for almost a century and a half science has failed to give any explanation about this cosmic phenomenon.

The Roche limit in the practical sense.

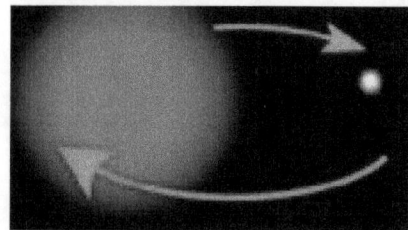  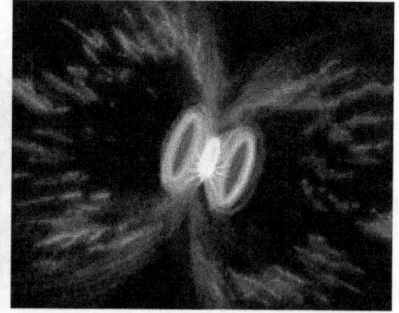

The formula $F = G \frac{M_1 M_2}{r^2}$ **cannot explain the comic occurrence shown in the pictures above, but I can explain what is occurring in this instance and this occurrence connects directly to the Roche limit, as explained above. Not only does the Roche limit explain this phenomenon, but it ties directly to the Titius Bode principle, also inexplicable to the formula** $F = G \frac{M_1 M_2}{r^2}$. **According to the science formula of** $F = G \frac{M_1 M_2}{r^2}$ **the orbiting structures should collide with a bang, but instead they do the tango until one drops, but when dropping it still does not collide with the larger structure as would the formula** $F = G \frac{M_1 M_2}{r^2}$ **suggest.**

This is not only limited to planets in our solar system. In the Universe, there are giant stars spinning around each other. These stars are binaries, which are also one form of double stars where double stars are another such a form. The difference between the types depend on the distance they remain apart. They keep a certain distance apart and do not collide. In the case of the sun and its planets, it could be a case that the systems might be to small, or they might be to apart. However, this is not the case with binary stars. They are close, they are big, and they spin around a mean axes called the Roche limit.

The Roche limit is:
The region surrounding each star in a binary system, within which any material is gravitationally bound to that particular star. The boundary of the Roche lobes is an equipotential surface, and the lobes touch at

the inner Lagrangian point, L₁, through which mass transfer may occur if one of the components expands to fill its lobe. It names after the French mathematician Edouard Albert Roche (1820-83).

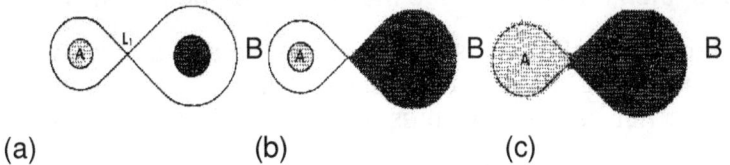

(a) (b) (c)

THE ROCHE LOBE: In a binary system, the Roche lobes of components A and B meet at the L_1 Lagrangian point. (a) In a detached system, neither star fills its Roche lobe. (b) In a semidetached system, one massive component, B, fills its Roche lobe. (c) In a contact binary, both components overfill their Roche lobes and share a common envelope.

The formula $F = G \frac{M_1 M_2}{r^2}$ **is unable to explain the principle discovered by Titius and later by Bode and it is not coincidental. From this one can arrive at the origins of the solar system.**

"Nothing" in the Universe is coincidental, "nothing" in the Universe does not apply and when a principle is discovered, the principle cannot be wrong. Therefore should the principle not match the excepted theory, change the expected theory, the theory does not apply.

The content of my work contain a new view about Cosmology, which I have been working on for the past twenty-seven years and exclusively for the past six years. To give you a little insight into my work, I shall mention the following: I came to realise that lines mathematically couldn't start at zero because there is no evidence of zero as a factor in mathematics. Should you disagree with my statement the question in need of answering is this: What will the length of the shortest hypothetical line imaginable be and moreover, what would the total overall length be in that case? The shortest possible line (hypothetically) must be so short it must have an initial and ultimate point sharing the same spot.

If it used zero as a start, the zero part would not count, because the line will only start at a point past zero where the line then will start forming an infinitely small dot.

The dot is in infinity, however small, it is not zero. Zero ultimately means not existing and then that point, as a start does not exist. The smallest line has a beginning and an end at the very same spot located in infinity, and infinity may be beyond human scope, though infinity is still not zero. Infinity may constitute of something we do not yet understand, but we may not define our human misunderstanding as nothing. In this aspect lies the difference there is between arithmetic and mathematical science where arithmetic can have position such as zero since arithmetic excludes the cosmos calculating numbers only.

A man may have that many oxen or so many sheep and even this amount of wives, (in Africa) or not have any therefore having then a total of nothing, but there cannot be nothing between the sun and its orbiting structures. The having and have-nots are part of arithmetic. Light will indicate a line flowing between the sun and whatever planet, following dot after dot thereby proving the existing of the possibility of something going about by a straight line, and any straight line in relation to other straight lines will be under the law of Pythagoras.

There is no possibility of a straight line not forming in space. Mathematics converts the values of integrating lines according to Pythagoras and arithmetic is about numbers to be added or subtracted. By mathematically excluding zero from cosmology a new Universe opens to the human mind. For instance the distance between the sun and Pluto is roughly one hundred times more than the distance between Mercury and the sun, but both planets mentioned have a vacuum filled with nothing except one atom hear and there occupying the vacuum between them and the sun.

If space supposedly comprises of nothing how can nothing then become plural forming more or be multiplied by a number as to indicate a growth in something not even existing. As the one becomes one hundred the one cannot substitute a value of nothing but then must be part of something. If the one substituted the nothing, all laws of mathematics will go in disarray because when one multiply any number by zero it becomes zero placing both planets in the sun.

By excluding nothing from the equation space becomes something bringing in a value lying inside the realms of the infinite that must form singularity. As the zero becomes a dot, something else becomes clear about the dot. Looking at the night sky we find darkness overwhelming the space in relation to the stars bringing across light. We can detect the dot because we cannot see darkness since our eyes were only meant to cope with light. With this knowledge, then how can we see the sky as darkness at night?

We are only supposed to see the light of the stars and not darkness, yet at night we see a much wider picture than stars alone. One may bring in the argument that the blind see nothing but darkness. We seeing persons do not know what the blind does not see, so we presume it is just about the same darkness, but that is presuming.

When we see a red flower, science knows the flower being all the colours but the red it rejects and this we all know. Therefore the dot we see as darkness also must be light, withholding its light and giving us the darkness we see as light...But the dot must influence the surrounding as well, subtracting the light it claims from the surrounding by casting it as darkness. In the case of stars we see the light the star disassociate itself with, keeping the darkness it has as it pours all the light in excess into the darkness which evidently is then light.

From that one may conclude there should be two forms of singularity where one associate with a dark dot being light, and another being matter with flowing light evidently proving to be the dark one. Proving the dot with many such arguments was easy. Naming the dot and its position, value location and proving the influences mathematically was much more complicated and proving the dot has a definite influence on the surroundings was at first seemingly impossible, yet it is done The definitely defined and underlined value of the dot becomes of utmost importance when finding solutions to cosmic factors not yet clearly defined. That is a big discovery you will make about true physics and how science works as you read on.

My approach might seem unconventional but through the abandoning of the accepted, it enabled me in locating the precise location of a universal singularity forming a connecting basis of the Universe (this I say with some degree of confidence). The smallest figure there can be must be a dot. The only mathematically sensible option about extending a line from the dot will be non-bias progress in all directions equally in order to give a meaningful flow of mathematical equilibrium. The Pythagoras mathematical principle is the proof and that I explain. The obtaining of singularity is in my rejecting of nothing by replacing it with something being the dot.

The claim becomes obvious when observing the connection between the half circle, the straight line and the triangle, which could also promote all the qualities lurking behind the pyramid. Consider the connection between 180^0 sharing and then one may realise much of the pyramid mystique becomes less spectacular in considering the very basic in mathematics being the Law of Pythagoras on which all mathematics are focused.

The claim becomes obvious when observing the connection between the half circle, the straight line and the triangle, which could also promote all the qualities lurking behind the pyramid. Consider the connection between 180^0 sharing and then one may realise much of the pyramid mystique becomes less spectacular in considering the very basic in mathematics being the Law of Pythagoras on which all mathematics are focused.

The rings around planets form not by, mass pulling but by applying the Lagrangian formation system, which holds 5 points that forms around the planet as it holds material in position. However, the Lagrangian system can't service the cosmos standing alone but forms one part of the four cosmic principles. The rings show that gravity is everything to do with Π and has nothing to do with mass in any way possible.

LAGRANGE (-TOURNIER), JOSEPH LOUIS DE (1736-1813)
French mathematician, born in Italy. In celestial mechanics he studied perturbations and stability in the Solar System. He examined the three-body problem for the Earth, Moon and Sun (1764) and the motion of Jupiter's satellites (1766). In 1772 he found the particular solutions to the problem that give rise to the equilibrium positions now called Lagrangian points.

Lagrange also studied the Moon's liberation.

LAGRANGIAN POINT One of five points at which small bodies can remain the orbital plane of two massive bodies; also known as liberation points. Three of the points lie on the line joining the two massive bodies: L_1 lies between them, while L_2 and L_3 have the two bodies between them. These three points are unstable, slight displacements of a body from then resulting in its rapid departure. the fourth and fifth points (L_4 and L_5) each form an equilateral triangle with the two massive bodies, 60° ahead of and behind the smaller body in its orbit around the larger one.

A well-known example of bodies flying at the L_4 and L_5 Lagrangian points are the Trojan asteroids in Jupiter's orbit. Among Saturn's satellites, Telesto and Calypso lie at the L_4 and L_5 Lagrangian points in the orbit of the much larger Tethys. In similar fashion, tiny Helene precedes Saturn's satellite Dione, keeping 60° ahead of Dione. The Lagrangian points are named after the French mathematician J.L. de Lagrange, who first calculated their existence.

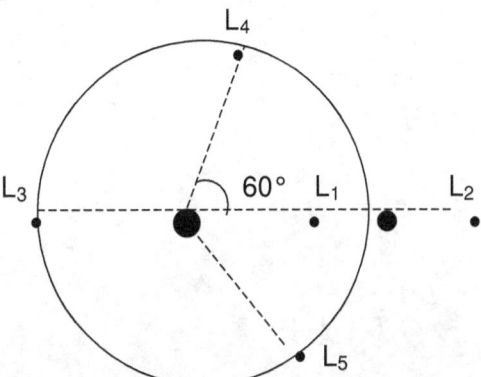

LAGRANGIAN POINT: *The Lagrangian points are five equilibrium points in the orbit of one body around another, such as a planet around the Sun.*

The Lagrangian points, is what keeps the formation in sequel positions.

The Roche limit, is what determines the distance that the rings or structures hold in relation to the centre of the main structure around which the orbit takes place.

The Titius Bode law confirms the circle or the rotation or Π and

The Coanda effect forms the relation between what is solid and holds position and that, which is the liquid part around which the moving part rotates. These principles work in a relation and the one can manifest more amplified in one situation than any of the others would but in every situation all four plays a role because not one can form Π as the circle that is gravity without all four connecting.

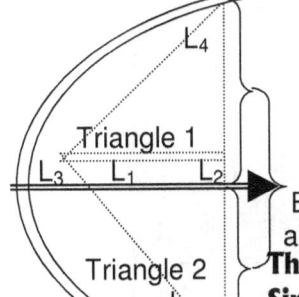

The Lagrangian System implicating the five positions extending from singularity

Singularity dividing the cosmos
Each triangle claiming a side of the universe

1 Half circle   = 180° L_3 L_4 L_5
2 Triangle 1   = 180° L_3 L_4 L_5
3 Triangle 2   = 180° L_3 L_4 L_5
4 Straight Line = 180°

The half Circle = 180° combining as a Sphere when comprising Singularity in the matching of the value of the straight line forming the half circle and combining as the triangle and all are equal 180°

The distance between the sun and Pluto is roughly one hundred times more and if the distance between mercury and the sun, but both has nothing between them and the sun. If space comprises of nothing how can nothing then become plural forming more or be multiplied. If it was one becoming one hundred, then the one cannot contribute to a value of nothing but then must be part of something. If the one substituted the nothing, all laws of mathematics will go in disarray because when one multiply any number by zero it becomes zero placing both planets in the sun. By excluding nothing from the equation space becomes something bringing in a value lying inside the realms of the infinite that must form singularity. Applying this logic to the Lagrangian system and interpreting that information to the law of Pythagoras a clear pattern come about.

This picture is of the 1987 Supernova event. Ask any physics professor to explain how the "mass went mad" to allow this to happen. That which happens in this picture and what happens when an aeroplane goes through the sound barrier rides on the same set of principles.

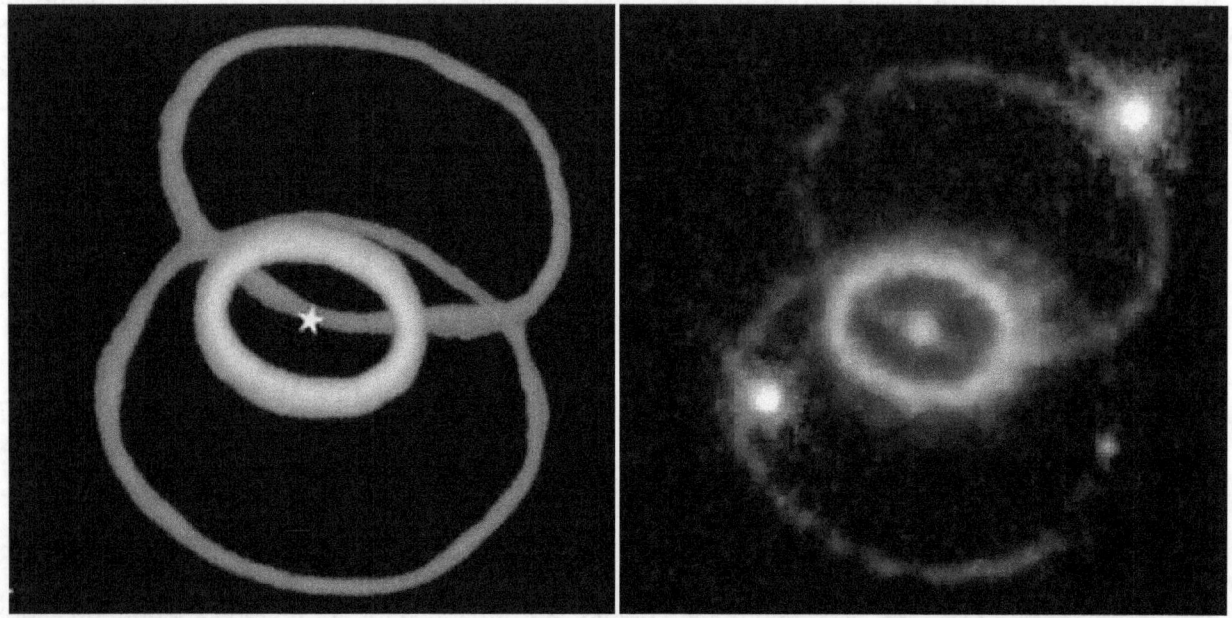

With circle being everywhere in the cosmos and where Π forms all over in circles and as circle, notwithstanding all the leads that the Universe give to Newtonians not one in so many centuries could see a link between gravity and the way that Π mathematically forms.

It is the way singularity presents Π in this picture above that one can see what gravity is and in what way gravity forms. When I saw this I knew I was on the correct line of thinking.

Gravity does not only link to Π but gravity is Π by forming Π. The question to solve was how…and that I solved as you can read in this very book.

Use Newton in the formula $F = G(Mxm)r^2$ to explain the cosmos and explain how stars "devour" each other without ever colliding. $F = G(Mxm)r^2$ may not be able to explain the exploding of double stars, however I can…

..by applying $a^3 = T^2 k$...

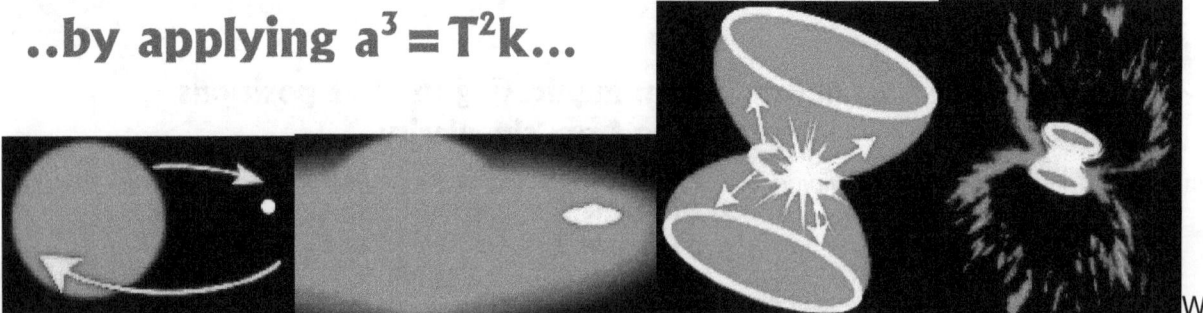

We humans are cursed by all the conspiracies that we have to endure. To break this we better start with science. We are all so entangled in a society filling our senses with one conspiracy upon the other that we can never find freedom in thought without disentangling our mental state from the brainwashing that furthers conspiracies by which we are taught to behave like those in power wish us to behave. This manipulating they subdue us too by rendering our thinking power to become computerised slaves is part of the general public's mental state and proves how much John and Jane Dow and Mr and Mrs Nobody care to be subjected to a comatose condition subdued to having no brainwaves functioning. We are told how to think and what to buy and how to vote and in which to believe by conspiracies whitewashing us so that those in power and those that are rich and those that are intellectual can control us on a daily basis.

If you read on you are going to <u>Read About</u> <u>The Biggest Conspiracy</u> ever enlisted, it's more than just the next conspiracy because it is <u>A Science Conspiracy Network. I prove everything I say about the conspiracy I announce...read and be shocked.</u>

This Network which I denounce Engulfs the Entire Civil Human Race in Every Aspect and willing or not, you reading this are participating without knowing about what you support with every part of the mental intellect you have. It involves the most trusted members in the part of our civil society and we are deceived by the ones EVERYBODY trust the most, the teachers and Professors teaching at academic institutions everywhere. For over a decade I have been knocking on doors with the information I present in this book and lots more evidence, just to be ignored and to be turned away. This behaviour of physicists ignoring me or even attacking me when I point out a very legitimate case of mistakes in science made me realise there is a conspiracy going on. Science has been getting away with this conspiracy since the Dark Ages. What you are about to read is not for the simple minded because science requires much intelligence and in that physicists get away with turning Creation into a joke. By turning the facts that form the conspiracy into what you think of and you accept as natural and as culture, man has been cheated and mislead for three hundred years because those teaching science consider everyone as representing stupidity. To them everyone else are simple-minded peasants who are not worth having independent minds who could be able to think on their very exclusive superior level that they are able to think. That is how they get away with the conspiracy. They pretend to be intellectually superior and to understand what seems senseless to the rest of us and then look down their noses on the rest of the world. I will show you what it is they can't see and then miss. They underestimate the public as much as the public are lazy too think and the misleading carry on. Those in science think their mathematical abilities make them Gods–on–earth while we others are mindless human fodder and for that reason they brainwash our children at school by applying mind control as their mathematical understanding makes them gods.

What they say all must believe because if they speak using mathematics then God spoke.

THIS IS THE WEBSITE THAT <u>WHISTLE BLOWS</u> ON A SCIENCE CONSPIRACY NETWORK. I AM GOING TO SHOW YOU THEIR UNBELIEVABLY POOR MATHEMATICAL SKILLS!

This conspiracy is so widely active that every person on earth teaching physics participates, most probably without knowing, some reluctant and others well knowingly to set out to brainwash by thought control the minds of innocent students forcing them to believe in dark aged ritual beliefs in the ritual practicing of forces and allow the conspiracy to be enlisted with all the vigour they could muster. People would argue religious concepts but everybody accepts science on face value. The guilty party that I refer to betray us in every way possible by teaching us lies that never could be true and all the while we believe them with all our hearts and entrust them with our entire future… They take money from parents with the sole aim to brainwash the students who are entrusted to their care and demolish their intellectual understanding …and before you think my claims are exaggerated I say again I prove every word I say. I show what is true and what is wrong in science and as a thank you for all the concern and devoting care I show, the establishment of science frustrates me, ignores me or runs me down, while I challenge everyone and any one to disprove me. Decide after reading this book weather what you read is not a conspiracy…however, familiarise yourself first with what I present and then decide.

Those I refer to are allowed the powers to take as much tax dollars as they wish and endeavour on all sorts of projects that they could fantasize about. Their budgets are limitless and they never face questions because they are the intellectuals that we dare not question. With the same powers we give them they brainwashed your child making science a religiosity. The methods they use on your child are to make your child believe in science and that is just methodical corruption of their thinking. Physics teachers take money from parents with the sole aim to brainwash the students that are entrusted to their care so that students will believe in science unequivocally and regardless. When teachers teach falsified information then they are worse than the mafia, that operates openly by crooked intention and in that the Mafia maintains honesty, which the physics teachers don't have. If you are a parent then mind what you do and beware and read on! If you think I am going one step too far in accusing the academics teaching physics and I slander their names, I challenge you to ask this question again after you have read this entire book and thought about everything and considered every aspect.

TO WHOM IT MAY CONCERN,
RE :   AN OPEN LETTER ABOUT XEPTED SCIENTIFIC NEWTON MISTAKES
ISBN 0-9584410-1-4

I am Peet Schutte, the author of the above-mentioned book/s
PROVE ME TO BE INCORRECT IN ANYTHING I SAY!

To whom it may concern and all others reading this document:
This is my introduction and this is my prologue:
But before I can commence with that task I have another duty administer: I AM ABOUT TO WARN EVERY PERSON IN SIGHT OF MY WORK ABOUT MY SLENDER ABILITIES.

Therefore in the light of what the most respected academic group on Earth accuses me of, I therefore have to issue a most serious warning to any person with the intention of making some kind of inquiry to the content this book holds, then the most concerning matter involving any content within the pages of this book you hold are that you must please seriously consider that where the stating declares the possibility that the content in this book has been (written by...) then don't take the announcing Written By Peet Schutte (Petrus S. J. Schutte) very seriously for there are grievous doubts leaving considerable dispute about the possibility, which underwrites the authenticity of Peet Schutte achieving the (written by...Peet Schutte) status. Please take note of the following dehortation. In the light of the reference to me serving in the capacity as being responsible for authoring, (written by...) in line of keeping fairness and justice to members of society, where all civil beings should carry reputed honesty, then: Please be warned before any reader starts reading about the following extremely serious admonition: I am bound by my conscience to warn all intended readers that I am placed under caution by the Academics in Physics. Those most esteemed members responsible for the guardianship and maintaining the ethos in physics are of the opinion that I, Peet Schutte, am unable to write any book on the science of Physics as well as Astrophysics. Therefore I, Peet Schutte, must declare that I should be considered as not very able to write anything, because I am incapable thereof. I suppose, I merely generate new information, which I establish as thoughts and then gather as concepts. I further collect the result as words, which I put on paper using alphabetic symbols. I then compile that in a format that others may confuse with a book, but a book it cannot be, since the Masters in science found me unable to write a book. But before you go further and follow my arguments, I first have to level with you about how academics view me in the position I hold. Please do not allow me to fool you, for this then cannot be, or represent a book. Now I have done my duty in warning everyone and in that, I denounce further participating with any purposive intention to wilfully bring down the crux of civilization by acting unacceptable and irresponsible.

I didn't write a book since I am not schooled to do so. It is my guess that I merely generated uninformed thoughts, which I collected as alphabetic symbols and plotted that in ink on paper. This effort I achieved from harbouring my delusional ideas spawned by a dehumanised brain. It only proves my weak and under developed mentality, due to my lack of an informed insight that is a typical symptom that all those have that is suffering from a disadvantaged past that one can only have when the person obviously lacks formal education. While you are reading the letter deciding to regard or dismiss my work, then also please keep in mind when reading my language used and also please give credit where it belongs…if you do find linguistically improper use of words or misspelling, then remember that I am a feeble minded motor mechanic and not a literal giant. I do find much pride in my status as being Afrikaner and would like to have my names used by pronouncing it in the manner Afrikaans dictates…therefore I would sincerely appreciate the courtesy when readers will take note that my name and last name are pronounced in Afrikaans, which is originally from Dutch and must be pronounced that way. Peet one would pronounce "here" which is the closest English to the pronouncing of the "ee". The "Sch" in Schutte is pronounced exactly as school is where both actually are pronounced Skutte or "skool". By pronouncing my name in Afrikaans you do me the utmost courtesy any one can. Being an Afrikaner is what I am most proud of.

Should any person challenge me about the legitimacy of the statements and content of this book, please do so at any time after you have familiarised yourself with what it is I say. However do not do it on the grounds of only the information provided in this book. For such persons believing totally in the accuracy of science, the believability of Newton, and the uncorrupted nature of Physics academics first get to know the truth by going to and reading www.sirnewtonsfraud.com, www.questionablescience.net as well as http://www.singularityrelavancy.com/ and see what facts you ace. There are so much facts pointing the truth and that much more detail when you read the six part theses called **THESIS"**

As you read the title of the book
www.SIRNEWTONSFRUAD.com

I know and realise that you are disgusted by my attitude when I degrade the name on which physics are founded. In this introduction part I am going to show you just some of the deceptions all students are forced to believe since all physics students are forced to believe in Newton, Sir Isaac Newton that is.

In the following am giving you a choice. You can say I am going to commit fraud by aligning the planets' positions according to mass but then Newton has committed the fraud because I only follow his lead. If I am judged to be the culprit that is guilty of deception then it is because Newton misled me. You can choose.

You are expected to believe the following: Newton stated under the nametag of Kepler that there is so called Conversions for "Unknown" factors.

Also read

A Conspiracy to Commit Fraud on a Cosmic Scale

Part of

www.sirnewtonsfraud.com

WRITTEN BY PETRUS S. J. SCHUTTE

ISBN 978-1-920430-07-8

©KOSMOLOGIESE EN ASTRONOMIESE TEGNIKA
All rights are reserved.
No part, parts or the entirety of this book may be reproduced by publishing, electronically copied, duplicated by whatever means that form reproduction or duplication, without the prior written consent of the copy rite owner.

Going under the Title heading of
A Conspiracy in Science in Progress

Part of **www.questionable science.com**

http://www.lulu.com/content/e-book/wwwsirnewtonsfraudcom-part-2/8132511]

PART 1

A Conspiracy to Commit Fraud on a Cosmic Scale

Part of

www.sirnewtonsfraud.com

WRITTEN BY PETRUS S. J. SCHUTTE

ISBN 978-1-920430-07-8

©KOSMOLOGIESE EN ASTRONOMIESE TEGNIKA
All rights are reserved.
No part, parts or the entirety of this book may be reproduced by publishing, electronically copied, duplicated by whatever means that form reproduction or duplication, without the prior written consent of the copy rite owner.

There is another book also published by www.Free-eBooks.net
CS@Free-eBooks.net,

Going under the Title heading of
A Conspiracy in Science in Progress
Part of www.questionable science.com

http://www.lulu.com/content/e-book/wwwsirnewtonsfraudcom-part-2/8132511]

1. **ALL THERE IS... ABOUT NOTHING?**

The single most tedious problem I faced in writing this book was where to start. This book is everything about the **Universe** in time and space, which on their own is eternal and put together, remains eternal. Wherever I decided to begin there were always some factors that lead to that event. Firstly, I had to explain those factors that the readers were to understand, before I could explain the events that took place wherever I decided to begin. Unfortunately, these factors did not stand alone, but were supporting other events that I first had to explain. If there were not a well founded comprehension about the factors that led to the events which supported the factors in explaining the events what occurred before the start of whatever point I decided to start off with, it seemed as senseless as this sentence which I just wrote. Writing one sentence, while sounding stupid is one matter, but to write a whole book and coming over sounding like an idiot is quite a different kettle of fish to fry.

The obvious point to start with, should be the Big Bang, with one minor problem. In the Big Bang, the Universe was not a **"NOTHING"** that came from "nowhere". The Universe is forever "SOMETHING" which is flowing from "somewhere" to another destiny which seems to be right in the middle of wherever the Universe is heading. That was what forced me to start with the "NOTHING", and only by starting with "NOTHING", I stood a chance in not achieving "NOTHING".

In order to understand the Universe you had to understand Einstein's theory. Einstein said there are three known substances in the Universe that we know. One is matter, two is time and three is space. Matter is one substance, which we consider everything should consist of; time is a **"NOTHING"** that man created and space is a "NOTHING" that God created.

However, have you, as the reader, ever considered what **"NOTHING"** is? What is **"NOTHING"?** It is a notion used by all and understood by none. No person can define the exact meaning of this word. In every day language, it applies to an understanding of not understanding. This seems to apply to the world of cosmology as well. The Universe takes its birth from this very word. However, there must be some explanation to the concept of the meaning of the word, if it has any right to exist.

Only when one put into context with its contrary meaning, which is **"SOMETHING"** there comes validity to the meaning of **"NOTHING"** and only then can one visualize the concept of "NOTHING". Let us use an imaginary scenario. Let us pretend it is nighttime and there is a dog outside that starts barking. The owner of the house gets uneasy and goes out to investigate. According to the dogs, observation there was "SOMETHING" to investigate. However, to the person's observations there was "NOTHING" outside. Who is right?

Let us take this same scene a little farther. Now the dog enters the house. In the living room, he sees humans staring at a light that flows from a square apparatus. The people's faces seem taut and hypnotized. Because of the dog's inability to see the picture in the T.V., the dog will find the human behaviour confusing. To him, they focus on, "NOTHING" but to their vision's ability, "SOMETHING" excites them.

With this in mind, that the concept of **"NOTHING"** can only be defined by the use of the word **"SOMETHING"** and in this doing, **"SOMETHING"** validates the meaning of **"NOTHING"**. With the next verbal sketch that might seem to be a little tedious, I would like to prove my point. I use myself as an example as not to gossip about the facts of another person.

There are parties in my family who thinks of me as **"NOTHING"**. According to those persons, my outlook on life is flat, boorish, racist, mindless and rude because of my personal standings and viewpoints.

According to me, I consider myself to be to the point and without frills. I am the owner of a B.M.W. 735 I but the model is a 1986, thus almost 16 years old. It has little value and nobody is interested in buying it, though it was an expensive German car.

I live in an ordinary farmhouse that is far from good looking. I own a farm with an enormous debt. My five tractors, two centre pivots, three trucks and all my implements are between 10 and 30 years old. According to my standards, these antiques that the bank regards, as **"NOTHING"** is all I have. That means I am financially worth **"SOMETHING"** because it is better than **"NOTHING"**.

My relatives, seen from my point of view are gaudy overindulging with liberal notions because they benefit financially from being "liberated". To my mind, they hide their shortcomings behind a curtain of money and therefore their outlook on life is **"NOTHING"**. They consider me as **"NOTHING"**. I consider them as **"NOTHING"**.

We cannot both be **"NOTHING"** because then "NOTHING" must be sharing in a common and specific value. If this was true, we then must have had **"SOMETHING"** in common though it will be **"NOTHING"**, but we have **"NOTHING"** in common therefore we share **"NOTHING"**. We have **"NOTHING"** in common. With all this personal gossip, my only intention is to give meaning to the **"NOTHING"** that I share with my relatives and not even that **"NOTHING"** is **"SOMETHING"**.

This brings about that **"NOTHING"** only has meaning when compared to **"SOMETHING"** which then can put a certain value to "something "and rob **"NOTHING"** of every value it might have.

This misconception divides families political convictions, religious believes, and cuts through society to the bone. In addition, all of this comes about because of **"NOTHING"**.

Let us take this **"NOTHING"** even farther. As a concept in the universe, we are able to take this **"NOTHING"** much farther. The atom compromises a nucleus in the middle and electrons orbiting the nucleus. In a sketch on paper this two dimensional drawing seems exceptional and the atom seems to be logic. However, if you put the atom in the correct perspective, your comprehension of **"NOTHING"** becomes mesmerized, stripped of all logic even to a point of absolute inconceivability. To prove this point I will place the atom to a more lifelike scale.

Put an object the size of a sugar grain in the middle of a rugby field in the very centre. Consider this the nucleus of a hydrogen atom. Now put a grain of salt right in the middle of the upright post, fifty yards away from the grain of sugar. You now have constructed your hydrogen atom. The orbit of the grain of salt (electron) will now move in a circular motion around the grain of sugar at the speed of light from the one post to the other post.

Now go and stand on the pavilion at any given angle and admire your hydrogen atom. Your atom would be unsighted to yourself or any other person with normal eyesight. Remember the two grains are representing the **"SOMETHING"** and the grass is representing the **"NOTHING"** part. The grain of sugar and the grain of salt is the portion that one can regard as the matter part and the grass represents the "NOTHING" in the atom. There seems to be a very little of **"SOMETHING"** and a tremendous amount of **"NOTHING"** in this picture.

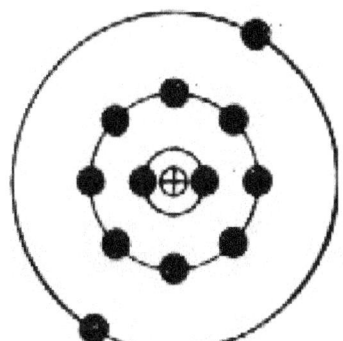

Now, let us construct the iron atom, which is the atom that keeps everything known to man in our Universe in a structural form by means of so called gravity. (Later in this book, I shall qualify this statement.)

For this experiment, you would have to go out to a vast expansion like the Namib dessert. Choose any point and put down a golf ball. Now put two grape pips opposite each side of the golf ball at a radius of

10km. Now in a radius of 20 km from the golf ball 40 km in diameter put down eight grape pips evenly apart. Now place another 16 grape pips in a diametric circle of 80 km (radius 40 km) evenly spread over the circle.

If you want to construct an even denser atom, you would have to go another 20 km and construct a circle with 32 pips in this circle. However, let us stick to the iron atom. Now you would have to fly in an air balloon at a height of about 500 km to picture your construction of one of the densest atoms in the universe, which make up stars. Even if you had all the imagination of a space scientist, (and do they have over inflated imaginations!) you still would only be able to see sand. Now this represents an atom of enormous density! The sand in fact represents the **"NOTHING"** that this enormous dense atom is made of. This boils down to an enormous misconception that overruns any sort of sane logic and it is no wonder that none of the "SUPER- EDUCATED- MASRER- OF- FACT" and most learned cosmologist experts ever tries to explain the workings of this **"SOMETHING"** to **"NOTHING"** ratio.

From this stance the **"NOTHING"** is so overwhelming that the **"NOTHING"** to the **"SOMETHING"** ratio becomes unexplainably unreal. If the **"NOTHING"** in this argument is so overwhelming, the only conclusion I can come to is that that **"NOTHING"** must be **"SOMETHING"**.

In a question I once put to a "SUPER- EDUCATED- MASTER- OF- FACT" Genius, how the construction of this atom can bring about a dense and solid structure as in the case of iron, he started blabbering about the electrons moving so fast that it forces the atom to become impenetrable. The worst part was that I had to sit and listen to this rubbish. What happens at 3 000 degrees when iron becomes a gas. Does the electrons then become slower as it orbits the nucleus?

This means that I have to conclude that the speed, at which the electrons travel, will bring about whether matter is solid, a liquid or a gas. This means the negative particles determine the density of the atom. This is not the end of the argument.

Now, picture the scene where the electrons orbit the nucleus of the atom, in an orbit that place them right next to the nucleus. The electrons are now rubbing against the nucleus in such a way that **"NOTHING"** separates the two. Will there be more **"NOTHING"** between the electron and the nucleus in the last mentioned relation or will there be more **"NOTHING"** in the previously mentioned example. In which of these two examples are there more **"NOTHING"** that separates the electron and the nucleus of the atom? Whatever your answer may be, then why? I do not care how brilliant you may think you are, but to this question there can be no answer as to the value of **"NOTHING"**.

I believe predominant in my Creator. That includes my belief that His creation is perfect. If the creation were perfect, there would obviously be logical, calculated and obvious solutions in the composition of the universal creation. With this construction of the atom, science obviously fails in being calculated and logic. There is far too much **"NOTHING"** and far to little **"SOMETHING"** to make any sense. In this lies my main motivation why I conducted my studies, because I believe that science and religion should be the very same thing if the same Creator creates it.

The atheists believe in **"NOTHING"** after death. When he breathes his last breath then according to him, there is **"NOTHING"** that follows this departure of life. Those same atheists are very quick to point out how valued life is and how dear we must consider it to be.

No atheist, up to now could explain to me the reason why, if life is so precious, would it then just be lost after death to **"NOTHING"**? Why then would **"SOMETHING"** which is that precious (according to the humanists), just vanish into **"NOTHING"**?

Personally, I think that all atheists are on a self-sublimation crusade. Their way of thinking is that if man, including the atheists are without a god, man then is put right on top of the priority list and becomes a god! As scientists, they become even more of a **"SOMETHING"**, because the scientist becomes one of only a few that can understand and explain to a very trifling point how creation actually works.

However, if there is a Supreme Being, then this blasphemer and self appointed god becomes just another **"NOTHING"** as all the other "nothings" that call themselves humans. The atheist tumbles down to the same level that we, the "nothings" find ourselves to be in the order of importance. Then the atheist is not a **"SOMETHING"** any longer and he could not consider himself one. By recognizing a Supreme Being, the atheist will become the same **"NOTHING"** that he regards the rest of us to be and that would be the end of his sublimation in regarding himself to be **"SOMETHING"**.

To whom it may concern:
My introduction as well as introducing the readers to general cosmology in a very brief and compressed manner but first, I have to give the emphatic warning to all prospective contemplating readers.

Please take note of a conscientious warning about the gravity of the misgiving there is on the part of the most respected Academics in physics about a much concerning matter. I state it emphatically that science accuses me to be not schooled to the point where I am able to have any form of an opinion on any matter concerning Sir Isaac Newton. Notwithstanding that my research proves I did my private studies and through which I skipped the indoctrination and mind control academics place on students goes unrecognised by their standards and so too my ability to have any insight on matters regarding physics. However my skipping their methodical and systematic brainwashing enabled me to see and allowed me to be able to express the incorrectness in Newton's teachings and allowed me to show in clarity what destructive force Sir Isaac Newton used to corrupt the laws of mathematics, corrupting to science along the way and mostly raping to the work of a great man, Johannes Kepler and what Sir Isaac Newton did can only be expressed as being blatant criminal fraud. What his deeds amount to is to corrupt the laws of mathematics, to render the laws of cosmology useless and to rubbish all of science. Should you find this to be unbelievable, then I am glad to announce that this book is more for you than any other person, so go on and read what academics guarding science never wanted published. I challenge any one that disputes any claim I make to prove me wrong by proving me wrong and not merely suggesting claims in that direction.

PROVE ME TO BE INCORRECT IN ANYTHING I SAY!

To whom it may concern and all others reading this document:
This is my introduction and this is my prologue:
But before I can commence with that task I have another duty administer: I AM ABOUT TO WARN EVERY PERSON IN SIGHT OF MY WORK ABOUT MY SLENDER ABILITIES.

Therefore in the light of what the most respected academic group on Earth accuses me of, I therefore have to issue a most serious warning to any person with the intention of making some kind of inquiry to the content this book holds, then the most concerning matter involving any content within the pages of this book you hold are that you must please seriously consider that where the stating declares the possibility that the content in this book has been (written by...) then don't take the announcing Written By Peet Schutte (Petrus S. J. Schutte) very seriously for there are grievous doubts leaving considerable dispute about the possibility, which underwrites the authenticity of Peet Schutte achieving the (written by...Peet Schutte) status. Please take note of the following dehortation. In the light of the reference to me serving in the capacity as being responsible for authoring, (written by...) in line of keeping fairness and justice to members of society, where all civil beings should carry reputed honesty, then: Please be warned before any reader starts reading about the following extremely serious admonition: I am bound by my conscience to warn all intended readers that I am placed under caution by the Academics in Physics. Those most esteemed members responsible for the guardianship and maintaining the ethos in physics are of the opinion that I, Peet Schutte, am unable to write any book on the science of Physics as well as Astrophysics. Therefore I, Peet Schutte, must declare that I should be considered as not very able to write anything, because I am incapable thereof. I suppose, I merely generate new information, which I establish as thoughts and then gather as concepts. I further collect the result as words, which I put on paper using alphabetic symbols. I then compile that in a format that others may confuse with a book, but a book it cannot be, since the Masters in science found me unable to write a book. But before you go further and follow my arguments, I first have to level with you about how academics view me in the position I hold. Please do not allow me to fool you, for this then cannot be, or represent a book. Now I have done my duty in warning everyone and in that, I denounce further participating with any purposive intention to wilfully bring down the crux of civilization by acting unacceptable and irresponsible.

I didn't write a book since I am not schooled to do so. It is my guess that I merely generated uninformed thoughts, which I collected as alphabetic symbols and plotted that in ink on paper. This effort I achieved from harbouring my delusional ideas spawned by a dehumanised brain. It only proves my weak and under developed mentality, due to my lack of an informed insight that is a typical symptom that all those have that is suffering from a disadvantaged past that one can only have when the person obviously lacks formal education. While you are reading the letter deciding to regard or dismiss my work, then also please keep in mind when reading my language used and also please give credit where it belongs…if you do find linguistically improper use of words or misspelling, then remember that I am a feeble minded motor mechanic and not a literal giant. I do find much pride in my status as being Afrikaner and would like to have my names used by pronouncing it in the manner Afrikaans dictates…therefore I would sincerely appreciate the courtesy when readers will take note that my name and last name are pronounced in Afrikaans, which is originally from Dutch and must be pronounced that way. Peet one would pronounce "here" which is the closest English to the pronouncing of the "ee". The "Sch" in Schutte is pronounced exactly as school is where both actually are pronounced Skutte or "skool". By pronouncing my name in Afrikaans you do me the utmost courtesy any one can. Being an Afrikaner is what I am most proud of.

Should any person challenge me about the legitimacy of the statements and content of this book, please do so at any time after you have familiarised yourself with what it is I say. However do not do it on the grounds of only the information provided in this book. For such persons believing totally in the accuracy of science, the believability of Newton, and the uncorrupted nature of Physics academics first get to know the truth by going to and reading www.sirnewtonsfraud.com, www.questionablescience.net as well as http://www.singularityrelavancy.com/ and see what facts you ace. There are so much facts pointing the truth and that much more detail when you read the six part theses called **THESIS"**

As you read the title of the book
www.SIRNEWTONSFRUAD.com

I know and realise that you are disgusted by my attitude when I degrade the name on which physics are founded. In this introduction part I am going to show you just some of the deceptions all students are forced to believe since all physics students are forced to believe in Newton, Sir Isaac Newton that is.

In the following am giving you a choice. You can say I am going to commit fraud by aligning the planets' positions according to mass but then Newton has committed the fraud because I only follow his lead. If I am judged to be the culprit that is guilty of deception then it is because Newton misled me. You can choose.

You are expected to believe the following: Newton stated under the nametag of Kepler that there is so called Conversions for "Unknown" factors.

Tell me, can you find any credence in the "Conversions for "Unknown""
$$4\pi^2 a^3 = P^2 G(M + m)$$
In this comes the fraudulent part because there is no evidence of mass playing a part or forming an actual presence in the solar system.

If the cosmos supported Newton's claims of $P = \left(\frac{4\pi^2 a^3}{G(M+m)}\right)^{0.5}$ then the planet arrangement would have been much more likely as I show above, but the picture indicates the mass as well as the planet formation.

You must judge; it is either the cosmos that is incompetently wrong or it is Newton that is incompetently wrong because what the cosmos has in place Newton knows nothing about and what Newton claims the Universe uses, the cosmos knows nothing about. Who would you say knows more about the cosmos' method of workings, Newton or the cosmos? If Newton is correct then the planet layout must be as I show with Jupiter very close to the sun. It seem the cosmos is just as unaware of Newton's ideas as Newton is of what is happening in the cosmos. Who would be correct about cosmic principles applying, the cosmos or Newton?

$P = \left(\frac{4\pi^2 a^3}{G(M+m)}\right)^{0.5}$ What hogwash does the factor $\frac{1}{G(M+m)}$ indicate?

The same can be said in the formula $M = \left(\frac{4\pi^2 a^3}{GP^2}\right) - m$ when $P = \left(\frac{4\pi^2 a^3}{G(M+m)}\right)^{0.5}$ that the factor $\frac{P^2}{}$ is senseless and $\left(\frac{P}{2\pi}\right)^2 = \frac{a^3}{G(M+m)}$, has no foundation other than fraud.

$M = \left(\frac{4\pi^2 a^3}{GP^2}\right) - m$ is complete fraud. The Cosmos does not support the Newtonian formula even in one place where it could apply.

Position as a function of time

$P = \left(\frac{4\pi^2 a^3}{G(M+m)}\right)^{0.5}$ This is what Newton said is in place and with no evidence ever founding this ridiculous proposition, all Newtonians that ever come after Newton. This is what Newton and his Newtonian followers tell the solar system it has in place and tell the cosmos it uses to operate. I have indicated that mass has no place or use in the solar system according to what the solar system puts in place.

These are the closest because these are the massive giant gas plants and having the most mass must put them the closest to the Sun.

Get your professor to prove Newton correct in the face of $P = \left(\frac{4\pi^2 a^3}{G(M+m)}\right)^{0.5}$ and if he can't let him admit he has been conducting in a fraudulent practise all the time he was teaching.

In order to find movement heat must apply. To escape from the earth a lot of heat must apply. To loft from the earth say in a balloon heat must apply and heat must be released to get airborne as well as get into outer space. The more heat is released the more any object will escape from the surface of the earth. In that sense heat is anti gravity because by releasing heat gravitational confinement is overcome. Expanding increases density therefore gravity is something about density! All the less dense particles float and all the dense particles fall. That connects to gravity

The faster the body moves the further will the orbit of the circling body be from the centre of the Earth and that also depends on motion or gravity. Being big or being small is not part of the requirement and therefore mass has no influence as a requirement. When the body starts to move slower the body stars to descend. Big or small does not matter and having mass or no mass does not matter. It is when the motion of the structure reduces to a speed that is below the rotating speed of the Earth that the body starts to plunge to the Earth. Galileo said all things fall equal and then all things can't fall by mass.

The falling or not falling and the escaping or not escaping only depends on the speed of motion and mass as a factor is only required in relation to the imagination of the physicist. While falling all objects show equal mass and having mass differences puts no extra load on any of the factors in any way. The buoyancy of all objects in the atmosphere depends on the speed in motion of the object. There are many other factors that come into play and all those factors forced me to write four different books on the subject where every book highlights different aspects that play an apart. The only aspect that plays no part is having mass or not having mass. Mass can't pull when all things fall equal as Galileo stated.

Please do not think I am able to bring you all I know about the problem we have in science in this small book. I have written 37 books so far and I have still not addressed the entire issue.

This information I present in this book does not tell the entire story even about the conspiracy…

This information does not begin to tell a part of the entire story. This information does not even introduce a part of the beginning of the introduction of the entire story. Yet, in thirty plus years in science and through out the entire world of science I have not come across one individual in science that shares my concerns about the problems within science and I am not aware of any person working or teaching or that holds a teacher's post in science on whatever level or working with whatever form of science who would admit to a problem in science and considering what I present, that constitutes to an academic world wide cover-up when faced with the mountain of evidence I gathered. It still helps me nothing because of the conspiracy.

There is a deep, dark conspiracy hidden in physics that some willingly participate in, some absentmindedly participate in, some not being aware of the conspiracy but still very willingly participate in and other too stupid to realise they help to further the conspiracy just because they don't think of the consequences of their stupidity by which they promote the conspiracy.
In the end no matter why or how those teaching in science are involved the only action of importance is that students are brainwashed by methodical mind control and I challenge everyone on earth to show I am committing slander against science. Everyone in a teaching position is participating in absolute mind control by brainwashing students to believe science and this forms the conspiracy going on in science.

Should anyone not believe me, then don't believe me after you have read this book or another similar book **A Conspiracy to Commit Fraud on a Cosmic Scale** but is just much less technical and then still after you completed either of the books or both of the books then I dare you to hold the opinion that I am deforming the characters of those teaching physics. Prove to me after reading the evidence that there is not brainwashing going on to conspire against society.

No one in science would face up to the fact that there is a serious problem in science from school level through pre- and post graduates up to nuclear science and astronomy or astrophysics.
Science is in a denial about what they know and what they are oblivious too and this denial forms a conspiracy by which everyone in science controls all thinking of what forms the principles by which science is dictated.

I offer the solution whenever science would admit there is a problem but everyone is in science is running away as fast as they can from the problem where they are unable to see the problem and if only they would, but they will not even stand still for one minute to look at the solution I offer.

A Conspiracy to Commit Fraud on a Cosmic Scale informs about the same conspiracy but the detail used is much less in complexity and the focus is less on mathematical input while the overall focus provides on much more simpler verbal explaining. It is for those favouring less technical detail and broader information. The first objective in this book would be to establish what a conspiracy would be because everyone has his or her own idea about what forms a conspiracy?

Physics students, it is your duty to pull the plug on the powers of the All-Powerful Academics in Physics and stop their dishonesty. It is your task as the as the next physics generation to stop the criminals that are filling the corridors and the lecture halls of physics departments throughout the world by acting as if they know all there is to know and all they know is to fool the next generation of students. Stop their teachings by forcing them to stop their criminal fraud. Force them to explain the deception such as the one they call THE CRITICAL DENSITY, which is a conspiracy to commit fraud. Let them explain how an expanding Universe can suddenly and abruptly turn in direction of developing and start to contract as Newton stated it is doing at present, and when facing all other concluding evidence showing that the

Universe was expanding since time began they come up with the utmost unrealistic garbage only an idiot can devise. Tell them to bring proof with evidence that the cosmos is contracting as Newton said. In THE CRITICAL DENSITY conspiracy all they say is that they are waiting to see when the cosmos would stop its criminally insane behaviour and start to listen to the laws of Sir Isaac Newton. They shove all the blame of wrongdoing onto the cosmos and take away all error from Newton. If the cosmos does not contract as Newton said then when will the cosmos mend its ways and follow what Newton said and to start contracting! It is a conspiracy to cheat and lie and crook the human race in order to keep Newton untouched. With The Critical Density shambles the modern Newtonian set out to defraud the world in the same manner as their Master Sir Isaac Newton has done centuries ago. Newton said the cosmos is contracting. When Hubble proved the cosmos is not contracting, Newtonians looked where the cosmos went wrong by not following Newton guidelines he so clearly set the cosmos to follow. It has to contract and not expand. Those in academic positions fabricate non-existing material no one can detect to cover the real conspiracy they try to hade. When the argument arrives of contraction versus expanding they wall this down by referring to the search for a substance that can't exist and could never be detected. It is not the dark matter issue that is the real conspiracy but the dark matter forms a conspiracy to hide the facts that the true conspiracy covers up. It is this mother conspiracy I am gong to uncover and present.

These whom I named in Honour of Sir Isaac Newton as the are the guard of the Newtonian High Priests carrying the name as the Newtonians are Men amongst mankind, that charged the Universe with not applying to standards set by Sir Isaac Newton, and then went on proving how incorrect the behaviour of the Universe was in not adhering to the direction gravity has according to Sir Isaac Newton. Since Sir Isaac Newton can't possibly make a mistake, it then was presumed the cosmos made the mistake by not following the gravity settings laid down by Sir Isaac Newton, the one that cannot falter nor could his teachings fail, carrying the illustrious name of Sir Isaac Newton.

It must be the cosmos being at fault by expanding without seeking the approval of Sir Isaac Newton to do so in contradicting Sir Isaac Newton. Up to this point in science and in despite of an array of evidence pointing to the cosmos growing by expanding in every sense and with all pieces of evidence gathered by science from all over the Universe (including the solar system), the theory of contraction is still hailed as the infallible Newtonian truth. Every one that is part of physics, shares the Newtonian vision of a contracting Universe where the lot would one day again come together and Creation will end where Creation started some time ago. The Universe has mass that is pulling mass towards one another and we are in the centre of an ever shrinking Universe. The Universe is about to end where all mass contracts into one huge lump of material, and this conclusion contradicts al evidence gathered by science. Without any evidence available to its authenticity they gave this Newtonian processes a name, they call it the Big Crunch. If you don't believe me marry Newton's contraction with the Big Bang and see a divorce in place before any Church consummation of such a union could begin…but then again just as unlikely union in principle marriage between Galileo and Newton is in place and the mindless masses never once frowned on that! Students in Physics, it will serve you well to read the following arguments very carefully and come to a conclusion about what gravity is and what mass is and how it is impossible for the concept carrying the idea of mass then become responsible to form what we think of as gravity. Mass can't ever and doesn't bring about gravity. In fact I challenge anyone to prove that a factor such as mass do exist.

Look at any picture portraying any part of the cosmos and show where to find "mass". If "mass" was the factor that formed the Universe we would find all the giant starts of any galactica in the centre allowing all the smaller stars in the middle and those that seem to be pebbles floating at the rim of the galactica forming. The biggest galactica would be found in the centre of whatever picture one could see and we would have all the smaller galactica floating around the larger ones forming the centre of the Universe. This is so absent as finding the largest planets next to the sun and the smallest planets orbiting way out in the darkness. Take a look at the pictures I supply and show where is the "mass" pulling the "mass"?

The surprising question is how could every physicist ever since Newton never openly question this reality because it is so clearly wrong. Why am I the first person since Newton to mention this and moreover why am my questions ignored as it held the illness causing bubonic plague. Why am I always the cast out, the obscure mentally retarded that no one with good taste mention in accepted and polite company? Then why would those then feel offended when I refer to them as criminals conning and defrauding the Public? If I don't kiss their arse while they blatantly corrupt everyone I become the disgusting outcast to be avoided. This is only one of an elaborate many hoaxes I reveal with which they, the Brainy Bunch cheat the truth to try to make sense of a concept that could never find truth under any circumstances.

Planets do maintain positions like a geared clock but not according to mass that pulls although a balance provides the orbit. Newton claim that mass is used and with that the idea of proof is automatically placed at the door of Newton. We talk of planets orbiting and that is what planets do. Planets don't creep up to the sun by the value of mass. Think of what planets do… and you think that planets orbit. The saying goes that planets orbit indicating they follow a circle and this uses other rules than mass applying position. That is not what Newton said. In conversation we speak of the planets orbiting and this orbit is in terms of the Titius Bode law. If Newton was correct the planets pull but that would be blatantly wrong according to what we find the cosmos applies. Never do we refer to the planets pulling the Sun or the Sun pulling the planets, but we speak of seasons coming from orbital positions. Being in orbit has to neutralise the pulling and then cancel the pulling concept that also became culture. Using the formula $F = G \, \frac{M_1 M_2}{r^2}$ as Newton provided, disallows any other concept other than moving towards. The person Newton got his ideas from and the work he raped completely, that of Johannes Kepler explained this very well, but Johannes Kepler makes no room for any pulling of any sort. In the work of Johannes Kepler he said that the space being the orbiting route a^3 remains at a specific distance k while the orbit T^2 takes place…and in all my other books that addresses more information I take Newton to task on his dismembering of Kepler's formula by corrupting Kepler's work and with what amounts to fraud, Newton takes science on a goose chase that holds no truth. There is no pulling by mass of mass in any way. Kepler said the space a^3, moves T^2 in terms of a ratio k adhering to a centre k^0.

We have either one of two that has to be incorrect. If Newton is correct, then the normal way of how the cosmos is functioning of the planets rotating T^2 is incorrect. Then we must start saying planets are pulled to the Sun. If the normal form of speech is correct and the planets are merely orbiting the Sun, then Newton is wrong. Newton said centuries ago that gravity is the force of attraction there is between objects that hold mass and it is the mass factor that brings about this attraction, which Newton claimed there is. The Big Bang Theory proves Newton's idea is not only being wrong but Newton's idea of attraction is a joke. If the Big Bang is expanding the Universe, then how can the Universe contract at the same time?

Newtonian science creates a factor such as mass to fake reality and misrepresent true physics. As long as they cheat, they seem clever and supremely superior when compared to the rest of us inferior beings. They live in a Universe fit for Alice in Wonderland! Please, for the love of God, show me what role does mass play in this picture. Show me how the biggest stars is locked together and the smallest stars are pushed to the outside. Show me where gravity "pulls" what to where and how gravity forms in this picture. Show where will dark matter hide. The idea of proof comes automatically to the door of Newton although Newtonians will deny this fact as if they deny the honour of their Master Newton and that is what they have to do. Think of what planets do… and you think that planets orbit. The term pulling does not suggest any circling because no one can be pulling towards and do that while circling. Going in a straight line serves the term pulling. Then the saying goes that planets orbit indicating they follow a circle. That is not what Newton said. Never do we refer to the planets pulling the Sun or the Sun pulling the planets, but we speak of seasons coming from orbital positions. Being in orbit has to neutralise the pulling and then cancel the pulling concept that also became culture. The Universe is growing!

If there was a pulling, and the word orbit cancels such an idea, then there has to be some sort of prevention taking place that disallows the pulling to commit the direction of travel. I know it is said that the orbiting object falls as fast as it circles and by falling while moving to the following side on position it never reaches the Sun, and yes, it makes sense, but there has to be some form of resistance replacing the planet in the next side position and preventing the falling or the pulling from taking place. The person Newton got his ideas from and the work he raped completely, that of Johannes Kepler explained this very well, but Johannes Kepler makes no room for any pulling of any sort. In the work of Johannes Kepler he said that the space being the orbiting route a^3 remains at a specific distance k while the orbit T^2 takes place…and in all my other books that addresses more information I take Newton to task on his dismembering of Kepler's formula by corrupting Kepler's work and with what amounts to fraud, Newton takes science on a goose chase that holds no truth. There is no pulling by mass to form a force of gravity in any way. Newtonians are faking the truth because the truth makes them seem small while Newton's misrepresenting cosmic principles allow those in science to paint their own Universe as they wish to present the Universe in the way they fake the story. Show me where there is evidence confirming a factor such as mass. Show me what makes the circling of Jupiter going around the sun faster than any of the small rocks in the asteroid belt. Newton placed mass in the formula he said Kepler represented. Could anyone show how much is the giant planets closer to the sun than the small planets are, keeping in mind that the pull of gravity must be exceedingly greater between the sun and the large planets.

Could anyone show how much does the giant planets orbit faster around the sun than the small planets do, keeping in mind that the pull of gravity must be exceedingly greater between the sun and the large planets.

Could anyone show how much is there a greater gravity-whatsoever between the giant planets and the sun than the small planets have less of, keeping in mind that whatever Newton mass would benefit because the pull of gravity must be exceedingly greater between the sun and the large planets. Whatever Newton saw, was in his dreams. Anything they say that mass gives is cosmically unrealistic.

What will the mass of the gas planets have that put them in a different stance when going around the sun than what the small planets have? By small I refer to Earth, Venus, Mars and say Mercury. All plants orbit equally.

The fraud factor came about when Newton destroyed the work of Kepler by changing the work of Kepler without understanding the principle factor guiding the work of Kepler. The fact that one can use $a^3 = T^2 k$ puts the Kepler formula in the realm of singularity. I give away free books that explain how this comes about. One such a book is <u>The Absolute Relevance of Singularity: The Article</u> as well as another book that indicate the working more specific and is therefore require a higher degree of understanding because it is more technical is

<u>The Absolute Relevance of Singularity: The Website.</u> I wish to reflect on the modern conspiracy to hide Newton's failure and the deception to cover up Newton's fraud by engaging in almost unlimited fraud in the modern era. The science of the era had Newton's principles wrapped up. Having the formula that Newton gave as $F = G \frac{M_1 M_2}{r^2}$ enabled those who could "understand" the principle of "mass pulling" bringing about "a force" that Newton called gravity. This force was supposed to pull whatever was present in the Universe closer to whatever was present in the Universe. The formula gave man an equality to the power only God had because now it was the ability of the physicist to redesign the Universe and correct the failing God made in the Universe. The formula enables science to play a game designing the cosmos and the game used $F = G \frac{M_1 M_2}{r^2}$ as the prime basis. The Universe was working on a force valued at the measure mass that produced this force and the Universe was coming together pulling everything into one spot. The Universe was contracting…until a man came along going by the name of E. P. Hubble and this

man destroyed the godlikeness of Newton and therefore all his followers as well. This no good science cheat could allow!

E.P Hubble is generally credited with discovering the red shift of galaxies. From his own measurements of galaxy distances based on the period-luminosity relationship for Cepheid with measurements of the red shifts associated with the galaxies, Hubble and discovered a shift in material that clearly showed material was moving apart. The distance between materials was growing and was not contracting as Newtonians believed.

Hubble and Humason were able to plot a trend line from the 46 galaxies they studied and obtained a value for the Hubble-Humason constant of 500 km/s/Mpc, which is clear evidence of a Universe that was growing from every point away from every point. In 1929 Hubble and Humason formulated the empirical Red shift Distance Law of galaxies, nowadays termed simply Hubble's law, which, if the red shift is interpreted as a measure of recession speed, is consistent with the solutions of Einstein's equations of general relativity for a homogeneous, isotropic expanding space. Although concepts underlying an expanding Universe were well understood earlier, this statement by Hubble and Humason led to wider scale acceptance for this view. The law states that the greater the distance between any two galaxies, the greater their relative speed of separation.

This discovery was the first observational support for the Big Bang theory, which had been proposed by Georges Lemaître in 1927. The observed velocities of distant galaxies, taken together with the cosmological principle appeared to show that the Universe was expanding in a manner consistent with the Friedmann-Lemaître model of general relativity.

All of this shouts one thing **"Newton is wrong!"** Newton's lovely contracting Universe was not contracting or receding but was expanding by growing. Newton's failures surfaced and with it the entire Humans population could see Scientists in Physics new less about physics than does an ape know anything about maritime exploring. This took away…no, this robbed those that thought they are equal to God their insight into God's cosmic manufacturing process and this showed the world that they were not only not equal to God but was just like Newton, utter failures as professional that knew nothing of their trade. If this evidence of Newton came to the knowledge of those idiots forming the general population that they knew even less about physics than the general population, then they were as much the idiots as the general population. "Understanding Newton" always made them in physics superior with the ability to know what God and Newton knew and that vision gave them recognition and the superiority to look down on those with lesser abilities…those that did not "understand" Newton. If this came out that the Universe was not contracting by the measure of mass, then those that the educated Newtonians saw as the ones that knew nothing then uncovered the intellectuals as "Understanding Newton" but with Newton the intellectuals did not understand the cosmos. This revelation would uncover them as being the real simpleminded persons who thought they knew everything but was the ones that knew nothing about cosmology. Those unfortunate ones that did not "understand Newton" as being was then correct all along when they did not understand Newton because Newton was wrong and the intellectuals that "understood Newton" were ones that apparently knew nothing. Then those that did not "understand" Newton was correct because with Newton being incorrect those that understood Newton was the simpleminded minority that did not "understand" what the majority knew by realising Newton did not "understand" physics. They had to devise a plan to save the day. They had to cover Newton's deceptions with more deceiving. They had to go more criminal by topping the corruption to a higher degree as even Newton went before the discovery. They had to devise a plan that would outsmart everyone. They created missing mass in the critical density scam. This is how it works.

I have many books explaining how Newton corrupted science and this would take too much space to explain that. There is a book for free www.sirnewtonsfraud.com. I am not going into that at this venture.

They devised the critical density theory. How smart can those that are smart get when they truly try to be smart to save their skins? How brainy can the Brainy Bunch be when they deceive a plan to outfox all of the Human kind? You can go to the Internet and read all the multiple arguments about the Universe going flat and contraction therefore come about although not one sod amongst them know the least how singularity works. You can read in books I give for free how singularity works because I plot singularity as 1 and not as a complex formula. It is in The Absolute relevancy of Singularity: The Theses and the more complex explanation is in The Absolute relevancy of Singularity: The Website. There is to many arguments presented to mention so if you have a year or two filled with boredom, then search the web and go waste your time on fairy tails. You will see what they present when they have no idea what they present.

Here it is in a nutshell; what this is, it is not a theory that is called the critical density theory but it is a criminal venture conceived by the intellectual minds in physics to go criminal and defraud the world. They decided that the blame of the Universe not working as Newton said it does has to be placed at the door of the Universe. The Universe is lacking matter to contract as Newton said it should. Newton still remains absolutely correct. The blame for the mistake is diverted to the cosmos and away from Newton. Newton did not lack the insight to see the Universe expands, no; the Universe lacks material to contract. The Universe made the flaw and now we must find how the Universe will correct its flaw. The Universe went wrong by not having the sufficient mass and therefore the Universe must correct the lack of mass to get the Universe back in line with Newton. The blame for the mistake must be laid in the midst of the material within the Universe. Newton's formula stands correct and all suspicion goes in the way of how the cosmos was designed not to applaud the truthfulness of Newton.

Students, read the work I offer free of charge and learn how you are brainwashed and how your mind is pre- conditioned into believing in Newton's myth of pure deception which Academics call physics. If you are a student in physics who don't believe that you are subjected to unlawful brainwashing, then read on. Download work I offer free of charge. They come up with the most ferocious mathematical formula that should prove everything they say but all the formula proves is that the one trying to use the formula is out of touch with reality. If ever there is a person trying to impress with the fantastic mathematical formula explaining how the Universe works to his calculation, ask him or her to formulate when the earth and the moon will hit each other. Every formula they use found its basis on the perception that mass forms a force of contraction called gravity. Before they start to bewilder you with breathtaking calculations, call their bluff…ask them to show when will the earth and the moon collide. Tell them to use the mass of the earth and the moon and divide the product thereof with the distance between the earth and moon and then show what a force there is driving the gravity!

All those unrealistic arguments the Brainy Bunch offer as to why the missing mass or dark matter will bring a clarifying solution to avenge Newton has one damning flaw. Whatever they bring as an argument is tainted by a law in mathematics. It is built into the formula $F = G \frac{M_1 M_2}{r^2}$. If the radius increases, then the value of the mass reduces while staying the same. If the factor representing the radius r^2 becomes $2r^2$, then by the very same token does the mass become half of its previous value! It is effectively this $F = G \frac{M_1 M_2}{2xr^2}$ is $F = G \frac{M_1 M_2}{2}$ and this will bring about that while the cosmos is expanding, the worth of the mass is reducing by the same margin.

This is not rocket science; this is mathematics at its most basic. If the Universe was expanding then the measured value of the mass was declining that is if mass was responsible for producing contracting gravity. They are the ones that are the masters in mathematics. They are the ones that know mathematics better than anyone else on earth…and they missed this truth. This missing the basics was as deliberate as it was swindle the hide Newton's incompetence and with it their failure to understand physics.

This is where the second conspiracy started. The first conspiracy was the idea of mass that produces gravity and Newton's fraud to convince the world to believe him. Then the Masters-of-Deceit called Mathematical Physicist devised a plan to protect Newton's image and therefore their academic standings. They got Albert Einstein to hunt for the presumed missing mass and thereby distract attention away from Newton's oversight and their failure. Albert Einstein carried the heavy burden of being acclaimed the title of the best mathematical mind of all times. If you bend mathematical laws you will get a distorted Universe, as distorted as the Newtonian Universe is. The Universe doesn't agree with Newton's principles and Newtonian science has to bend all aspects to get it to fit.

You know…Einstein must have been very dumb or very dapper. Einstein must have been exceptional foolish or extremely brave and you can decide on what merit I suggest you should judge. In astrophysics they teach you about the critical density theory and how Albert Einstein calculated all the mass of the entire Universe and found the Universe fell short of Newton's expectations that was needed to drag the Universe back into one spot. Because of the critical density theory coming short we now sit with the dilemma of Dark Matter hiding in invisible black places waiting in ambush an unsuspecting Universe! If it was not for Einstein counting all the matter in the Universe and found it was too little, the hunt for dark matter would not have raged on.

It is because of this shortcoming that the entire group of science explorers are now searching (mostly in vain I hear) to locate the missing mass hiding in dark matter somewhere in obscure places. What a load of garbage all of that is because look at the picture. Do you think it is possible for one man, (or a million men) to calculate just the mass we see in this one photo? Those in academic posts lecturing you on Newton's correctness think they are the wise and are able to try to fool you because those professors think you lot of students are a bunch of mindless-monkeys that must believe anything they say because they think that only they have the ability and the mind to think and because you are a stupid student, simple-minded when compared to them, you can't think and therefore you will believe anything they say. If Albert Einstein thought himself capable of measuring only the material we see in this picture, then Albert Einstein was a fool with higher ego than what his IQ was. He thought he could measure the Universe and being that egoistic is stupidity crossing over to the insane.

In fact it is very obvious from what we see that Albert Einstein was no mathematical genius, he was an academic stooge, standing in for the other Academics to try and divert the looming shame by extending the visible problem of Newton getting uncovered. If Einstein was not able to see the reality of relevant formulations then he was more stupid than I am and I am so stupid I don't even understand Newton or so I am told.

This is the academic excepted values: The average density of matter in the Universe today that would be needed exactly to halt, at some point in the future, the cosmic expansion. A Universe that has precisely the critical density it is said to be *flat* or *Euclidean*. If the density of the Universe is greater than the critical density, then not only will the expansion be stopped but there will be a collapse of the Universe in the distant future. In this <u>closed Universe</u> scenario, the Universe will eventually implode under its own gravitational pull, leading to an event known as the Big Crunch. If the density is less than the critical density, <u>an open Universe</u> scenario plays out in which the cosmic expansion will continue forever. The critical density is calculated to be about (1 to 2) × 10^{-26} kg/m^3 – about 100 times greater than the average density inferred from all the known visible matter in the form of galaxies. However, when the inferred presence of dark matter and, possibly, of dark energy, is take into account, the Universe appears to be pretty close to the density called for by the flat scenario.

From these figures, how old is the Universe? What is its future? However, to answer them, it is necessary to know the density of the Universe, also known as "omega" (Ω). What a lot of rubbish this is! The Brainy Bunch insists that the Universe is 13 billion years old. This they measure by using the speed of light. In the first instant the Universe was the size of a neutron. It took some time to get the entire Universe to the size of five millimetres. That was the Universe.

The speed of light travels at what the speed of light = 299 792 458 m / s. At this rate light travels across a vastness of space 12 billion years to reach us. When the Big Bang came about the Universe was the size of a neutron. Take note on this issue; the entire Universe was the size of a neutron they say. What was the speed of light then? They say at the present it is 299 792 458 m / s, with which I completely disagree but let us leave this argument at that. The speed of light could not have been 299 792 458 m / s when the entire Universe was the size of one millimetre. That means the overwhelming distance of this 12 billion light years the light had to travel at a speed of less or equal to 1 millimetre / 10 billion years, because the Universe was so small at the time! The Universe was 13 billion years ago say five millimetres. Then what was the speed of light at that point? They say it takes the light coming from the furthest images 12 billion years to travel to us since the light travelling now left where it came from 12 billion years ago. At that point the Universe then was about the size of a pea. If the Universe was across the size of a pea, then light took a long time to travel a very short distance!

The light back then could not have travelled 299 792 458 m / s 12 billion years ago because 12 billion years ago the entire Universe was a couple of centimetres across. How fast was the speed of light when the Universe was the size of a pea. Then light travelled 10 million years to cross one yocto (y)(10^{-24}) of a mille meter. In terms of us today, light back then stood still. If it took the light 12 billion light years to cross, then the first part of the journey was pretty slow which makes the measure of time travelled versus space crossed rather ridiculous in every aspect it is portrayed. This means a lot of the light years it took to cross had to be billions of years just to gain one millimetre of space. Then the Universe must be trillions upon trillions of years older that they reckon it now is. It took a lot more darkness to form all the space is present in the entire Universal than the 12 billion years they say it did That brings us to the dark matter bit and the conspiracy that carries on undeterred. At present the conspiracy went as far as forming dark matter with (I suppose) dark energy.

This dark matter hides in places we can't see. This dark matter is what now forms the lost matter that protected Newton's image of correctness. Still Newton is untouchable because now in the present time the undiscovered dark matter is waiting to contract the Universe and this dark matter hanging suspended is what protects Newton. Is that not that sweet? Is that not the bedtime story every five year old would wish to hear every night? Every child will go to sleep feeling secured and in comfort. We can't attack Newton because the unseen matter that is dark is protecting Newton. Newton becomes untouchable by undiscoverable, unseen untraceable material that lives like a fairy tail in fantasy. If there is dark matter, what is the dark matter waiting for before it unleashes its incredible mass deployed force of gravity on this little unsuspecting Universe. What is preventing the dark matter from forming gravity that will do the job at this point in time? Why is it that the matter must be dark and must be seen in order not to form gravity. If the matter is present and forming a part of the Universe, albeit dark or not, seen or unseen, detected or not, if it has mass and if mass does bring about a force and the force is contracting gravity, then it must employ gravity. What is suspending this dark matter from kicking in and clocking in for duty? What prevents the dark matter from starting to get pulling? This is as big a scam as all the rest of the fraud they use to cover up Newton's fraud. I am showing all of this to prove how much deception there is in cosmology. Everyone in astrophysics is living a fantasy and everyone can make as they please, as long as the mathematical calculations seems to be in order. The reality and the viability or the lack thereof is no one's concern. As long as they can come up with stupefying formulated mathematics any dream will do. And the conspiracy carries on as long as it avoids reality and is void of constructive argumentative facts.

One night so many moons ago I have no intention of dating the time to which I refer, I was sitting outside while anticipating how to solve the riddles of the Universe. Well at least I was attempting to solve the part that riddles me and with my meagre qualifications it did not take that much to riddle me. Sitting outside and staring at the night sky gave me a break from all the confusion that faced me as I was again rejected by one of the so many academics rejecting my work and at the time I was still taking their rejection seriously and took their replies to heart. Then I saw the darkness of the night sky and compared that darkness with the brightness we find the star portrait in order to inform us of its location. That made me wonders about this thing called light and the manner in which light travels.

Have you as you sit reading this part at this minute sat back and gave a thought about the light enabling you to read? Yes it is simple in Newtonian terms because Newtonians keep everything at a child's mentality level so that they might not be confused. There is a wave of photons travelling at the rate of time and travels through space and time that is equal to time. What a load of rubbish did Einstein dump on the human race with that observation? It comes from some formula he or someone like him devised where they firstly manage to stop time. How they can achieve stopping time is beyond my limits of understanding but then again I admit I am one of those that so not "understand Newton". However, it does fit into the thought pattern of the Newtonian since their great Master also accomplished this outrageous deed. They go and put time equal to one. This then they mange in terms of placing the velocity that any object moves in relation to the speed of light and it is the square of this that they put at the root of this which they deduct from time being one and suddenly by the magic of mass it all comes down to time standing still. They put time at one (1) and deduct the square of the velocity any object travels (V^2) from the speed of light (C^2) which then also serves as a velocity and which then formulises as $t = \sqrt{(1 - (C^2 - V^2))}$. No argument is given as to why time and the speed of light should be the same other than Einstein fantasizing. Let's put the argument to the test of logic and see what happens.

If the speed of is equal to time one of two things will apply: either time as the speed of light will fit into eternity and stand completely still eternally because eternity is the fact that nothing changes as al remains unchanged. It is infinity that changes eternity by interrupting eternity. This might seem a little complicated but read **The Absolute Relevancy of Singularity The Website** where I introduce the points holding infinity as well as eternity. The concept is so easy to understand because do not hide untruths behind a collection of bizarre ideas. I shall give you a quick view into the physics I try to introduce and believe me; the impact of this introduction enrols every aspect cosmic physics may ever offer. This shows the human mind is not some mathematical computing device but an instrument of collective logical informative thinking about concepts outreaching simple computing by the way of mathematics. The only thing connecting what is to what was is light. Light brings the past to the present and therefore that is where physics must start and not with calculations. Time moves light through space. That means when having light as time in eternity, which is what outer space is, then infinity will never interrupt light and light will never move from the point it holds.

When we look at any object in the distance it is the light that brings the object to us. Yes that does happen to light but there are specific qualifications that have to apply before that will happen. If light was using infinity as time then light must move through the Universe from end to end before time can establish a moment or any name one wish to give for the smallest duration possible. What this says is that if light was equal to time then light must travels across the Universe in one instant of infinity. Then only can time and the speed of light be equal. The Universe Einstein referred to is the multitude of a combination of indefinite numbers of Universes all having an equal value and therefore all being equal and one but they are definitely not one from our point of view. Such a thought brings to mind the most simplistic answer one can imagine. Every atom is an entire Universe in space.

The facts you are about to learn will astonish you and it will seem unbelievable but notwithstanding it is true.

Notwithstanding as unbelievable as it may seem, I nevertheless challenge any one to show me that the least of any or all facts I uncover is not true.

How can a Universe contract as Newton said it does while that same Universe is expanding? In accordance with the Big Bang theory the Universe expands as the while according to Newton the Universe is contracting? If the Big Bang is correct then Newton is wrong and if Newton is correct then Hubble expanding ands the Big Bang is wrong.

How can there be a Universe that holds everything there ever can be and has whatever could be expand and therefore become more while the Universe already holds whatever there will ever be? To expand something has to increase, in order to expand that which already is. What there is must then become more of what there is in order to hold more of what already is available. The Universe already holds everything there will ever be, so what can become more in order to expand? If it is understood that it is space that increases and space is filled with nothing, then how can nothing become more when nothing is the defined by what is the absolute absence of everything.

To have "nothing: as a filling of tangible substance in the Universe becoming more in order to have the Universe expanding, then the Universe must reduce what it already has to increase having nothing whereby nothing then becomes more because more nothing applying means the removing parts of what has tangible substance of some of what already is. How is it possible to have "nothing" filling the Universe and still have distances between objects? It is said by those very smart and distinguished brilliant academics in astrophysics that there are 149 X 10^9 km of **<u>nothing</u>** (please note the word **<u>nothing</u>** that is specifically used) to hold any distance. The word **<u>nothing</u>** states a detail of what is total absence and the lack of anything present. How can that which can't be present because it is absent such as the term **<u>nothing</u>** does specify, how can that then fill space in terms of distance measured in kilometres or astronomical units. If there is 149 X 10^9 km of **<u>nothing</u>** filling the space all the way from the Sun to the Earth, then how long is one specific point holding **<u>nothing</u>** in terms of a measurable unit?

How is it possible that mass can be responsible for objects falling by creating gravity by which objects supposedly fall and then have all things fall equal as Galileo proved. All things do fall equal…

If you think this is nonsense, then you are reading official Mainstream astrophysics. You will read about things they find "on the edge of the Universe" while everyone knows if the Universe does have an edge the Universe must end there and any end must have a new start. How could the Universe that can never end have an edge where having an edge means it must end and having such an end is having a start? Whatever ends must start to bring an end and the Universe can never end.

However, please note that this following information mentioned is supplied as mostly being a part of the four books entitled as An Open Letter On Gravity Part 1 and 2 Volume 1 and 2

How and where and at what point can we see did the Universe begin… and if you think answering this is impossible you haven't followed the trail of thought that Kepler left. You were only misguided by Newton's misinformation. Why would you be most likely correct when saying the Universe is a sphere… as it always is depicted in pictures?

Have you thought about the following? Because these I can answer?

…where is the cosmos coming from…?

…where is the cosmos going to…?

…and most of all…

Why is the cosmos travelling through time …?

…what brings about the direction of expanding?

My studying Kepler helped me finding answers to all the questions, which was deemed impossible to answer. The following is only a few of the many questions that I do answer.

Where is the centre of the Universe?
How did time and space begin?

That I can answer…and I also can answer…

…why is the Universe still growing since the Big Bang…?
…why did the Universe start so very small…?

…why did the Universe fit into a neutron at one time…?
…how did everything expand from fitting into a neutron…?
…why does space grow from small to large?
…where is it going while it is growing …?
…why was the Universe any specific size…?
…what was everything before the Big bang…?

I grew tired of apologising for my (as they see it) having the audacity of being correct on matters of Newton's incompatible religiosity, which I bring to their attention. When being in contact I am expected to show the utmost humble attitude acknowledging their supreme posture with me being in their surreal presence. I have to feel honoured to be in their presence when I mention to them their mistake about Newton being mistaken about a Universe that is contracting according to Newton while it never ever contracted in the least.

I am quite fed up with the attitude of those academics looking down their noses at me or worse still are those ignoring me whenever I show that Newton's facts just don't add to a conclusive believable answer. I am at my limit with those academics ignoring me because while they can ignore me by using their all-powerful status they continue with the conspiracy with them never having to prove Newton and therefore disclaiming even my presence when I try to disprove Newton.

I have reached my peak with stomaching the corruption they hide behind a lily white cover of dishonesty while they sit in their mighty towers and live in a bubble where not even God can touch them less having me point a finger at their despicable ignorance about their mistaken Master they portray as a God. They can say the Universe is made up of nothing and go unchallenged for making a most senseless statement and when I bring this to their attention in a book, I am the person they condemn as being incoherent with my arguments about their nothing they have in place.

I say this again: any person telling a lie is committing fraud be it in the name of God or of science; such a person is a despicable liar. When they tell a lie to distort the truth and find financial compensation while falsifying facts, even if it is in conducting science then they are behaving criminally. That is distortion and is equal to the behaviour of the Mafia.

From my view and from my perspective, I honestly can't see any difference between the Mafia's racketeering and corruption and what academics commit in the name of being honourable scientists. Those in charge of Mainstream physics feed students lies in order to be compensated for their misrepresentation of the truth.

They are being paid enormous salaries from student fees to ensure that students believe in the impossible and accept what can never be proven and force students with methodical examinations to repeat the unproven or be expelled from the institutions and branding those expelled students as failures. That is a rip off whether it is justified as science in the process of learning or if it is plain legal criminality; it remains the same because they fly the same banner.

Those academics in key positions of academic credibility keep certain facts and evidence away from students and give other facts that were never proven before, prominence as well as credence while applying their trade in brainwashing to give their Newtonian views undeserved credibility and from these proceedings they earn substantial incomes. That is the same as racketeering. When you deceive by conveying untruths and cheat to mislead, then your behaviour is criminal.

If you purchase any of the following books you will come in contact with the truth for the first time in centuries. My work is about uncovering the truth and blaming the shameful conduct of those persons no one expects to be criminals.

When you purchase my books, I don't sell ink on paper. I do not sell material with questionable information, holding facts that were repeated so many times that it is accepted as the truth because it became a culture to believe Newton.

The information my work caries, which you will read in the event of purchasing my books, you have never seen, it was never yet mentioned or the facts I divulge has never been printed by any person, ever before.

I put untruths about the work Newton claimed as correct in question and that might have been published before but never published as questionable evidence. The rest I bring is new.

The following books that I offer for sale in this web site are unique in every sense. Only the information I question has been published before. The books take on Mainstream Science, uncovering facts that were never touched by any person for more than three hundred and fifty years. I act on behalf of the students in protecting those students that at this time is studying, or those that studied physics in the past. I am giving students information that is hidden by Mainstream Science.

On the book entitled Newton's Mythology I am making no profit. I give the book at a price that is going covering basic cost to furnish students with facts with which they can challenge academics torturers. I sell this at cost to show I am truthful in the hope of selling my other books

The book entitled Newton's Fraud is showing profit. The profit is going towards my attempt to get the books published privately which is done at a considerable cost. As the Purchaser will see, there is a significant number of sketches in the books and private publishers charge money for every sketch printed in the book. It is said that a picture is worth a thousand words and with the explaining; I offer this saying is certainly true.

The book entitled Sir Isaac Newton: a Conspiracy to Defraud Science is showing much profit and that profit is going towards covering the expenses of what the publishing costs will be that this book holds. The information in these above mentioned books has never been published but moreover has the information appearing in the book Sir Isaac Newton: a Conspiracy to Defraud Science seen the light of day. In the first two books I deal with the problems academics in physics are hiding whereas in the third book I bring the solution to the problems. This information has never been in print or given in printed form to any member of the public and that information I do not intend to divulge free of charge. Should you wish to know what the problems are in the attraction by mass theory of Newton, and you wish to learn the truth about the working of physics plus you wish to find the answers concerning the truth about physics and in particular gravity, then you will have to pay for learning my decades of research.

All proceeding will eventually go towards having these books privately published and have the information freely available to members of the wider public. I have run into a brick wall called Mainstream physics and there are (I suppose) hundreds of reasons for preventing me from having these books published of which no less is the finically motivation to restrain the publishing of my work since this work will condemn many other books with profitable titles to the incinerator. If my books do get published, thousand of books that are already selling on the commercial market will have the same informative value as the fairy tale story of Cinderella. However, publishing costs are astronomical and this is the only manner in which I can circumvent the academic blockade placed on my work and have my work published.

These books I offer for sale on this web page are not yet linguistically edited or controlled and I mention this because in it there is a chance that some grammar errors might lurk in the content. I found the most difficult part of writing is to correct one's own work in grammar because it is done by the method one applies to speak. Please also keep in mind that Afrikaans is my first language and English is my second language. It is also partly in order to have my grammar controlled that I require funding. Part of the funding raised by this effort will go towards editing before I can have them published and it is also for that reason I turn to this effort of marketing the books privately in order to obtain funding to have these books published and marketed through normal channels. I do not intend to challenge William Shakespeare in creating a masterpiece of linguistically prudent magnificence and with Biblical grandeur, but my aim is to get students to see what academics hide from them and to show academics what Newton hides from them. The brainwashing that is going on and the mental control that is inflicted on students is criminal. I challenge any person and every person to prove that I exaggerate or that I show falsified facts.

When any person tells facts that prove to be untrue such a person is a liar notwithstanding the motive in doing it. When any person supplies facts, which prove to be untrue with the motive of distorting money in the process, that person is a con artist and an untrustworthy individual. When any person supplies facts, which prove to be untrue with the power to black mail students and distort students' mental abilities, such a person is a criminal that should be locked away behind bars.

Again I repeat that academics teach students untruths about Newton that can never be proven and those very same academics know they are teaching blatant lies while being paid to do so. They charge students institution fees only to have students pay for being brainwashed in accepting Newton. Whenever the students never dare to challenge Newton's statements because if they do challenge Newton Professors would simply fail their exam papers and banish them from further studying.

I dare you to challenge your lecturer just on the facts I give in this web page and see how he or she will react. Academics in physics are paid to minimize your questions, limit student thinking, conceal other ways of reasoning that might be unfavourable to Newton, control all information they give students, have students accept the facts about Newton that they give unconditionally, have students never question or doubt Newton in any manner and accept what they are told to write in examinations. What do you as a student think would become of the student that ask the professor to prove Newton's claims that $F = G \frac{M_1 M_2}{r^2}$ is believable when all evidence points to an expanding Universe?

Let's find Mass and then therefore find Anti Mass

It 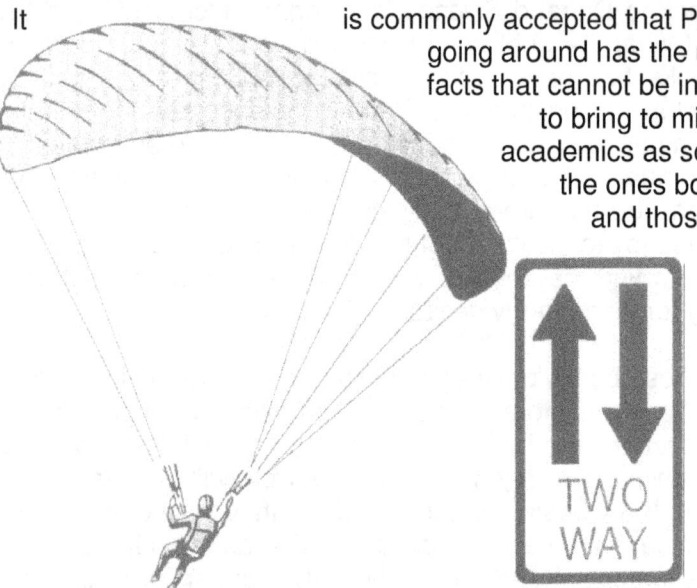 is commonly accepted that Physics demand respect because the general idea going around has the understanding that Physics only work with proven facts that cannot be in dispute or be disproved in any way. Well...I wish to bring to mind some of the facts that physics work with when academics as scientists only work with facts. Remember they are the ones boasting that if facts are not proven then it is fables and those very important academics don't waste time with fables because they only work with facts. Students, it is your liberty to ask them to explain what they say is such correctly proven facts. They maintain it is a fact that we have to have mass in order to produce gravity. Mass is responsible for gravity. If you don't have mass you're not going to have gravity. Mass is equal to gravity and gravity is only where mass is. If mass is anywhere it should show its presence otherwise mass is absent. If a body falls it is the mass that allows the body to fall because the body receives gravity by ratio of mass and mass that produces gravity in relation to the mass available. It is mass that drags you down because the mass is in charge of the gravity and the gravity finds the value from the mass available. So what happens to balloons? Have they got anti mass or anti gravity? They are moving up when the air is heated. Mass pushes you down by the gravity it forces onto you. If mass drags you down then what are lifting you up in the balloon? If mass gives the gravity to drag you onto the Earth then why would the hot air lift you up? Is the hot air causing anti gravity or anti mass because gravity and mass drags you down or so Newtonians say. The balloon is lifting the passenger and all that is in the bag plus the bag plus the balloon into the air. So what is then pushing the lot up if it is mass that drags you down. Has the air not got mass because then the air can't have gravity and then the air must escape into the blackness of outer space because by going up it shows a resilience of either mass or gravity. We have seen that it is mass that pulls everything onto the ground.

Why would the air defy mass and allow the balloon to go anti whatever. We find mass being the equivalent of that which brings the object to the ground. The object has mass to produce gravity. Why then would hot air allow the balloon plus everything in the balloon to lift into the air? The balloon lifts in relation to the hot air that blows into the sack. The more hot air and the hotter the air is the more lift and the swifter the lift will be that the balloon provides. The issue sticking out is that the balloon then must not have mass because with anti gravity it is pulling up. Remember mass drags you down and mass can't pull you up and drag you down at the same time. Then what is pushing while mass is pulling or is mass pushing while what is pulling? The object is not going in the normal direction where it is dragged down by gravity and in all my life I have never heard one Academic mention anything about gravity lifting and that makes the lot very confusing. What is lifting up when the lot should be pushing down and why did everything connected to the balloon lose the mass and if it has mass why is it not dragging down the balloon? If you think this is a little confusing try what is to follow.

They teach you that it is mass that produces gravity and gravity makes you fall because while gravity makes you fall mass drags you down. It is because those mind controllers are lying through their teeth with a menace in which they are the experts, as they know just how to pull cotton wool over you eyes. Take a truck of 15 tons into an airplane. Put next to the truck a petite little dancer weighing 45 kilograms. Put next to her a frog weighing 150 grams. Then get this lot into the air by airplane and let them jump. Take note that you are told by the wise amongst us that it is mass that produces the gravity that pulls you down. We have just had a lovely debate on how it works and how mass drags you down and wondered if it then is anti mass or anti gravity that lifts you up with the hot air balloon, well take note of this as your airplane reaches 11 thousand meters which is eleven kilometres straight up into the air.

If Newton's $F = G \frac{M_1 M_2}{r^2}$ did apply by having mass pull then the objects could never land at the same time because of mass inequality preventing this.

Now we drop the truck and the girl and the frog at the very same time from the airplane. The frog then pretends he drives the truck and the next scene he is dancing with the girl while the truck is falling as fast as a truck can fall. Who do you think is lying? Remember only one group can tell the truth and the other must be lying. Have you thought why one party is lying while the other party has to tell the truth?

The academic Brainy Bunch are telling students all over the world that mass is in charge of gravity and it is mass that's pulling you down. Then the mass is pulling the truck of 15 tons down since the mass produce the gravity and the gravity produces the fall which is three hundred and thirty three times more in a down direction than the mass of 45 kg is pulling the dancer down. The mass providing the gravity that pulls the truck down is doing the pulling down of the truck one million times better than it is pulling down the frog. If the mass is doing the pulling by establishing the gravity the truck must fall 333 times faster than the girl and one million times faster than the frog. It is either that or the three has the same mass because they are falling at the same rate.

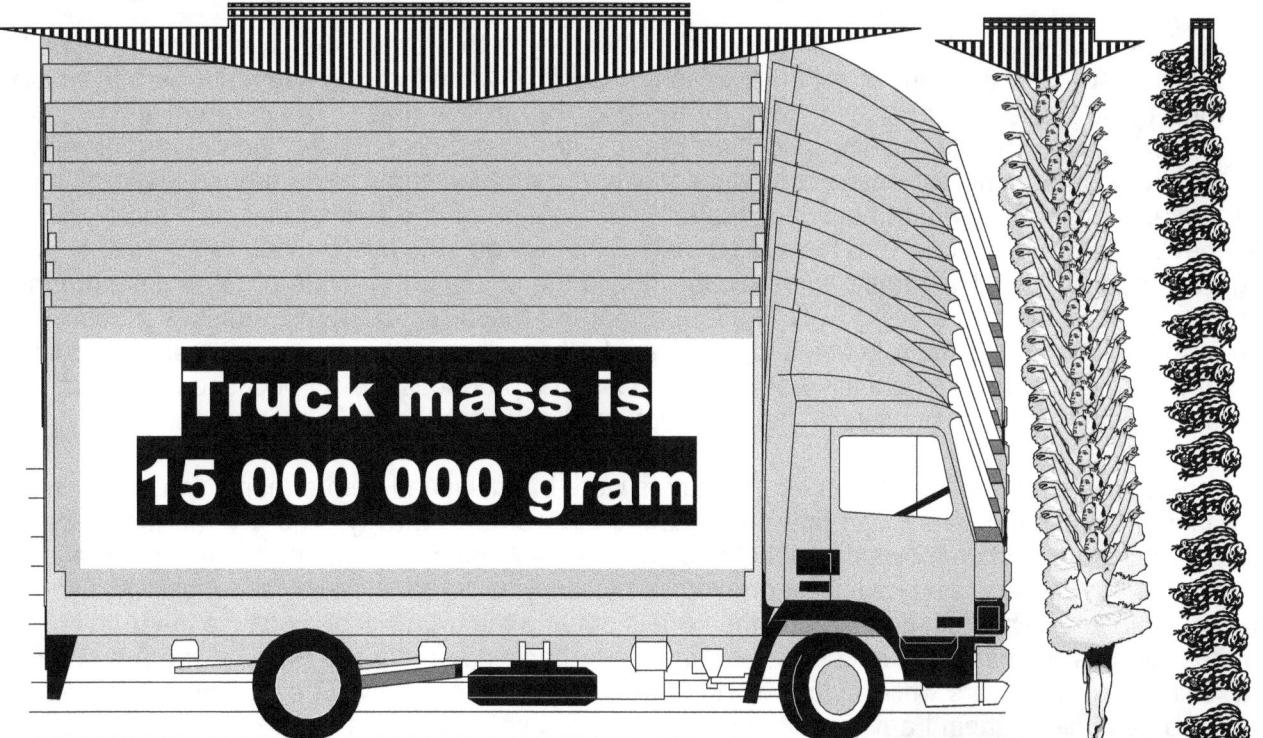

Take a truck loaded with cargo totalling of 15 tons into an airplane.

Put next to the truck a petite little dancer weighing 45 kilograms.

Then to keep the dancer on her toes, put a frog of 150 g next to her.

Now we will have the mass of each object "pull" by "gravity" as this lot falls down

If the Brainy Bunch all too wise is correct the frog can fly to America and have a pizza in New York while the truck has a few micro seconds to get down if the girl is going to fall during the normal falling duration of a minute or so.

Everyone has seen skydivers jump out of airplanes next to cars and trucks and bags. Every one has seen they all fall at the same rate. The girl can do tap dancing around a jumping frog on top of the truck or below the truck and they can be inside the back of the truck galloping on fresh air inside the truck because the lot is falling at the exact same rate.

The academics wishes to brainwash you by mind control in accepting that it is the mass that the falling takes place and that mass is responsible for the gravity and by mass pulling you down it is gravity that makes you fall. Where is the proof of mass that according to them is that which is producing gravity. They tell you Galileo said all things fall equal and we can see from the TV monitors how all things fall equal. Where is the mass that makes the gravity to let you fall if all things fall equally? They tell you that the truck has a mass of 15 tons and that mass is making the gravity that is having the truck fall while the truck is falling at the same speed and distance than the frog does.

If you take that as proof then they got you. Then they brainwashed you into a zombie. Then if you don't repeat after them and echo every word test after test and exam after exam they will fail your papers and kick you from campus. That is mind control, better than what even the KGB is able to implement. You repeat after them and you live an academic life or you disagree and you go home to play with your toes. If mass is in the picture then mass must be represented by a factor of more than just one because if mass is not part of the overall picture then mass has a factor of one which proves that mass is not part of the equation since mass can't change the results. With all the objects falling equal mass has no role and if mass has no role then for my money academics in physics can't just go and put everything in as their hearts desire. If it is Galileo that is correct and if all things fall equal then mass has no part in gravity. If mass is the inspiration behind gravity the truck must fall a million times faster than the frog and in fact the frog should almost land in another country because that is how slow it falls.

The fact of the matter is that I don't wish to be near when any of this lot hits the ground because the truck will cause a quarry and the dancer will be a splash of red fluid while the frog might not be that worse for wear if the truck or the dancer doesn't land on the frog. But that is mass. The differentiation of having mass or having equality when falling and then not having mass and between individual differences in mass by each component that enters the equation when the objects touch the ground. Then every one gets the mass it has. Only when they touch the ground and land on the soil is mass as a factor awarded. While they fall they all fall equal and there is no distinction between the falling at all. What then is gravity? The gravity is the falling. The gravity is the motion. While the object is in a state of mass it is not moving. The tendency to move and apply gravity is the part that the mass restrains. The mass is preventing the falling from continuing. It is the role of mass to prevent further falling and independent motion to continue. Some of then might even still honestly believe it is mass that produces gravity because they were taught that it is mass that produces gravity and never thought about the matter again afterwards.

They were brainwashed by their tutors as their tutors were brainwashed before them. You don't need the brainwashing because you now can find out what the answer is to gravity. You are the first generation that can receive the light of knowledge about what gravity really is, or you can be the last generation that will live in the lie. You are in a position where you can teach your tutors the truth about gravity if you read what is in the books. The truth is there and the truth is out and the truth will be because the truth is written for all that wishes to read. The academics on the other hand have ignored my work and my being on Earth for the past six years while I was writing them letters about gravity. They ignore me as if I am a rattlesnake because to them I am a rattlesnake. With what I say I will have them tumble down from their pedestals because by accepting my work they suddenly find their position equal to yours as students, and then they will have to learn my work in the same manner as you learn my work because to them everything is as new as it is to you. The Academics of the day have too much to lose to recognise my work and therefore have to protect their interest with all they can muster. For that reason if no other they will rather go on lying to you and cover their corrupt fraud than face up to the truth and admit their work is lost.

The truth will be whether it is recognized by them and they can become the first to admit and repent or they will be the last of the laughing stock that those in the future will refer to as the bunch that couldn't see when things fall equal they cannot have mass and when things do not fall by mass then one can know mass has nothing to do with the falling and the gravity.

It is up to you as students to rattle their cages and make them admit they've been lied to as they are lying to you. Or you can be the last of the fools that couldn't see that when things fall equally they have no mass by which they fall. My book is written and those that read it first will know what gravity is. If you do not accept the role as being zombies that is brainwashed then confront these academics that treat you with disgust and betray your trust. They might tell you the mistake is not that serious and the damage is small but then how will they know how big or small the damage is if they don't even know what damage there is or what the damage is.

Science has stayed so far from the truth that they can't even see the truth any more. If you carry on you will learn about some of it and when you read my books I will entertain you with many more than you ever believed. My books will serve as the light switch that brings the light to you.

I charge your young minds to confront those fraudsters about the truth. I wrote to them in the last letter where I informed them that they protect the criminality of their corrupt teachings because when the corruption is removed then nothing remains because they have lived a lie for too long. If you reach the need you may down load it because it is a fair bit of information.

If I come to you with a proposal about something I wish to share with you on condition that you pay me an amount to share with you what I know then I am an academic wishing to teach you. Have you a name for such a person that will force another person to pay him to be brainwashed and be mind controlled because the tutor has absolute control over the life and death of the academic future of the brainwashed being and therefore is willingly forcing this unfortunate creature in accepting what will never amount to the truth? I think they are called Physics professors and rule Universities as draconian authoritarian dictators bent on sadism.

Let's investigate the falling as such and see what happens during the fall. The truck falls at the same pace in which the girl falls, which is the same pace as that which the frog falls. If the truck falls at the same pace as the girl and as the frog there has to be a common denominator in this process and since the common denominator eliminates size form and shape we can eliminate mass. Mass brings distinction and the falling eliminates any form of distinction.

When I fall down a waterfall with a boat I travel the same pace, as does the boat. That could be because I am fixed to the boat by sitting in the boat. But my sitting in the boat has certain condition and one is that I can remain sitting because I fall the same pace as the boat is falling.

I fall down with the boat and the boat and me forming a distinctive unit falls at the same pace as the water that forms the waterfall falls. Should I at the time of my falling hold an empty mug in my hand and I wish to fill the mug with water, and then I will have to move the mug against the flow of water streaming down the waterfall. I will have to thrust my mug upwards at a faster pace than my descending is casting the mug down and therefore I accompanying the mug down the waterfall. My mug will not automatically fill with water or if there was water in the mug my mug will not automatically empty with water just because the emptiness filling the mug will be at a different pace than the content that is otherwise the filling of the mug.

The mug being empty falls as fast as the boat and I. The empty space in the mug is falling as fast as the mug will fall when the mug is filled to the brim with what ever can fill a mug to the brim. Notwithstanding the content within the mug or the content within the boat or the content within the water being within the waterfall, the very lot is falling at a similar pace. By lifting the cup while falling the cup will fill with water.

I am not putting the water into the cup but I am exchanging the space that the water holds with space that the empty cup holds and my action in truth has no bearing on the water filling the space, which I then transfer into the cup. I am filling the cup with space that at that point holds water but the holding of water has nothing to do with the transferring of space.

If I leaped from the boat and fell I would fall alongside the boat. The boat will be empty but will fall at the same pace and as the same space as I fall notwithstanding being empty. The mug being empty will fall at the same pace as the boat being empty which will fall at the same pace as the water in the waterfall and I would fall. The space in the boat, which is empty if I do not fill the space, will fall at the same pace as the empty space, which fills the mug, and the mug will fall at the same pace whether the space in the mug contains or doesn't contain whatever can fill a mug. The space filling the mug is falling the same as the water that would fill the space in the mug should the mug be filled with water.

The space in the boat is falling at the same pace as I would fall whether I am filling the vacant space in the boat or otherwise filling the vacant space next to the boat. It is the space that falls and not the object filling the space that are falling. It is the space that is filled or not filled that is dropping down because the space being filled is in decline. If it was not the space that fell the space within the mug would fill first as the mug and the boat fell because the empty space would first fill before it could take anything down. But since the boat falls as fast as whether it is being filled or not we can assume that the space which the boat fills or does not fill is falling as fast as it would fall whether it is holding the boat or I or the boat and I. The space not filled by mass also moves just as fast as space filled by mass.

When the object such as the mug or the boat or I connect with the Earth the Earth disallow the object free motion by taking any more space the object claims through to the centre of the Earth. The object now has to give up the space it claims and take on new space that the object claims to flow by contraction to the centre of the Earth. In forming a blocking it resists the flow or the gravity or space lining up with the centre of the Earth. The flowing of space by contraction is gravity but the object being in the space that flows becomes and obstacle through which the oncoming space must drag in order to flow to the centre of the Earth. It forms resisting of allowing space claimed to release to the normal flow when the object will not relent form in favour of gravity. This resisting such relenting of form and consequently forming a frustrating barrier that blocks the free flow of space towards the centre is time displacement of space and this relenting of space-time flowing freely becomes the mass factor. The density and the resistance that the particles show forms the mass that implicate the degree of the frustrating or preventing or disabling of such free flow of space through time and the displacement of space during time is space-time notwithstanding what ever irrational connection Newtonians wish to add too space-time. Allowing space to displace through time to form time is space-time and that is gravity.

All this is not new and I am not the first and the big genius that thought this out for the very first time since Eve had a bite on the forbidden fruit. In around 450 BC a man going by the name of Empedocles killed the myth that it is nothing that fills the bowl when water runs from the bowl. It was named the clepsydra meaning water thief. He proved at the time that something other that water fills the container while the container is emptying of the water.

The water will start running only when a finger lifts from the pipe where the finger before the time blocked the intake of the pipe and therefore prevented something from entering and therefore releasing the water from the container.

Empedocles' Clepsydra of 450 BC

Connected pipe allowing filling of bowl by water

Round Container Filled with water

Water running from outlet at the bottom

This he interpreted as being that the water was not running out from the container but was being pushed out from the container. The container was filling with something as it was being emptied of the water by something being anything other than nothing. This experiment was done some almost two thousand five hundred years ago and still Newtonians have to find a manner in which their grasp will accommodate these facts. It is the air filling the space that pushes the space filled with water from the bowl and the filling process of either space is named gravity. How difficult can it be to grasp an experiment that was understood two thousand five hundred years ago?

The space that holds the water is running down and away from the bowl and the water thief or clepsydra is stealing not the water but the space that holds the water. Once the water is out we can presume this displacement of space continues because there is no reason to think that the factor of nothing suddenly enters the scenario and the process stops because Newton saw nothing where something had to be. The space keeps repeating the process of displacing the space it follows as it is displaced by the space falling. It is a continuous cycle never ending and the space flows notwithstanding it being filled with what Newtonians can understand or the nothing they do understand. They understand nothing so well they filled the entire Universe with the entire nothing they do understand.

If it was purely atoms replacing atoms in the clepsydra as Newtonians wish to think and nothing else as they say because they still think of space as nothing, then how can so little number of atoms that fills the air replace so many atoms such as the water has. If the atmosphere was vacuum in the sense as they see vacuum being the stuff a Newtonian would normally have between the ears, and that stuff is nothing in as much as something being present in whatever space filling capacity anything being nothing can fill and also can form nothing, then the vacuum had much less particles to fill the space that is filling the clepsydra than what the water has that is emptying the clepsydra.

Let them count the number of atoms leaving the clepsydra through the holes where the water sprouts from and compare that to the number of atoms entering the clepsydra at the top and see how those figures add up in an explainable argument. If the finger closes the hole in the pipe on top the space entering stops completely but so does the water escaping also stops. If the pipe on the top is restricting the flow of space entering the clepsydra then in relation there will be restriction in the water leaving the clepsydra. It is the space that fills the space emptying the clepsydra and whether the space is filled with particles or sparsely filled with particles or not filled with particles at all, it remains space replacing space filled or not filled. It is space entering and it is space departing and the filling of space with material has no bearing on the matter what so ever. Newton had every wire crossed in his head when hew put falling of any object down to the mass which the object supposedly should have.

It is so clear that it is space that is moving and the fact of being filled by material or not filled by material has no merit in the process. It is not the apple that Newton saw that fell but the space he saw as nothing that fell and the space that was falling took the apple with while the space that followed still fell whether it had an apple to accompany down or was filled with a Newtonian filling of nothing. The space holding the truck being next to the space not holding the truck is falling as fast as the space holding the girl and the space next to the girl which is not holding the space or the frog where this lot is falling just as fast as the space holding the frog. The space is falling. The space is falling whether it is filled or whether it is empty and that means mass has as little to do with the falling, as the colour of onions has to do with the depth of the sea or the temperature of the shining Sun.

It is not as if I wish to condemn and reject that which is in place without placing something of worth back into the process. All I ask is to read what I bring. Don't be a coward and stop reading as soon as you reach the point where I condemn what is in place! Just move past that to the point where I show what is wrong and how it can be corrected! Just judge me not for condemning what now is so apparently incorrect but for showing why I condemn what now is so apparently incorrect and what I bring to the table and offer as a remedy. See what I have to offer and not only what I am taking away. Don't set your sights on what there is to lose but take a view on what there is to gain!

Do not reject me on merits you do not wish to instate because you have the fear you are going to lose what is instated. Do not judge me by using your double standards that are useless in the face of the truth. Rather look at the double standards you employ and do not judge me by using your double standards on me. Rather use your mind to detect what is double about your standards and then investigate with me what needs to change. Don't hide the truth. Don't hide from the truth and don't hide behind what you wish to portray as the truth. Rather come out into the light for the first time in three hundred years and admit to the truth. Follow what I say and see for yourself what there is to gain by trying to detect what is wrong because we all know there is much wrong. The comet does not collide with the Sun and the Moon is not on its way to collide with the Earth in time to come.

It is not as if I wish to condemn and reject that which is in place without placing something of worth back into the process. All I ask is to read what I bring. Don't be a coward and stop reading as soon as you reach the point where I condemn what is in place! Just move past that to the point where I show what is wrong and how it can be corrected! Just judge me not for condemning what now is so apparently incorrect but for showing why I condemn what now is so apparently incorrect and what I bring to the table and offer as a remedy. See what I have to offer and not only what I am taking away. Don't set your sights on what there is to lose but take a view on what there is to gain!

Do not reject me on merits you do not wish to instate because you have the fear you are going to lose what is instated. Do not judge me by using your double standards that is useless in the face of the truth. Rather look at the double standards you employ and do not judge me by using your double standards on me. Rather use your mind to detect what is double about your standards and then investigate with me what needs to change. Don't hide the truth. Don't hide from the truth and don't hide behind what you wish to portray as the truth. Rather come out into the light for the first time in three hundred years and admit to the truth. Follow what I say and see for yourself what there is to gain by trying to detect what is wrong because we all know there is much wrong. The comet does not collide with the Sun and the Moon is not on its way to collide with the Earth in time to come.

On TV we find that all things fall equal and that no size or mass differentiation plays any part in the falling process. Every time Newtonians are cornered with this idea they come up with a variety of answers where they would try to convey the idea that Galileo and Newton used the same mind in thinking. If things fall equal then while falling all things have equal mass. There can't be any mass description or mass variation and if it is mass that produces the falling the falling has to be different with mass varying. Don't let them confuse you on this and don't let them get away with more cheating…they have cheated enough for too long and got away.

Expand science and no the Universe for the Universe is the only aspect that has not the ability to expand. I challenge all of you Newtonians to prove $F = G \dfrac{M_1 M_2}{r^2}$ and not just to declare it proven because it is in use since the Dark ages. Expand your mind and double check the formula you all so vividly underwrite and support. Prove why you support the formula in a modern and a scientific way. Explore the correctness that this formula $F = G \dfrac{M_1 M_2}{r^2}$ underwrites. Be a true exploring scientist and journey with me through the following pages while we venture on the quest to find and vindicate my incorrectness by proving the truth vested in the formula $F = G \dfrac{M_1 M_2}{r^2}$ that carries the entire physics everyone uses. Let us start where the lot should start and get two Masters together on one point of argument. Galileo said all things fall equal. That says all things fall alike. The first thing anyone brings in is the vacuum bit with the feather and the hammer and since we do not live in vacuum there is no chance of finding a feather that will fall as fast as a hammer. Since the feather does not fall as fat as the hammer we immediately jump to the conclusion that there are falling disparities because of the falling discrepancy we find between the hammer falling and the feather falling.

Then what would give the feather the time to fall longer than the hammer does. Everyone concludes about mass coming into play and they are correct. But they are half correct while Newton still is completely incorrect by attaching mass to the entire idea of falling. Take away the resisting of the feather and replace it with something far less air resistant and one will come to a different conclusion.

We have to dissect what factor consists of gravity and what factor represents mass. Then we have to dissect which part does mass play and what part does gravity play. The falling object experienced no mass while falling therefore the falling or moving must be gravity's contribution. While objects are in motion those moving objects is experiencing gravity.

The object show mass when the object has a tendency to move but the motion towards the centre of the Earth no longer takes place. That means mass is the restraining of the motion or is that which prevents the motion or gravity taking place. On Earth, objects experiences mass by restricting gravity or motion with the Earth giving mass but taking away free motion. By giving mass the Earth forces the object to become one with the Earth and move with the Earth as a pat of the Earth.

Persons falling will experiences weightless ness while falling and they have a weightless state while falling. One cannot then go on to declare that the factor, which prevents motion, is the factor that causes motion because that is totally contradictory. The motion takes place without the presence of mass because the frog and the truck are falling equally fast. When landing the motion of the truck and the motion of the frog ends. Then the two have very different mass values but neither shows the ability to break from mass and move further towards the centre of the Earth. Kepler said the space a^3 is equal $=$ to the motion in a line k as well as a circle T^2.

While experiencing unrestricted gravitational motion a body a^3 is $=$ to the motion $T^2 k$ as Kepler said gravity is: ($a^3 = T^2 k$). When motion stops then only does weight or mass form as a result. While falling we find that gravity applies as individual separate space is moving and putting time in relation to the distance that the falling object travelled. That makes the falling factor the part that is the motion that confirms gravity. In the motion or movement we find the gravity because that even remains as a permanent attempt to move. Even when mass comes in as that which results in the ending of the gravity and in that gravity as a term is also forming the motion factor, still remains as an attempt to move. The while moving Galileo proved mass is not present because all things fall equal. Mass comes in when movement is retained and although the mass is present as a factor that factor that mass represents is what produces restriction of such a movement and not resulting in such a movement. The factor that mass represents is the containing of further downward movement. Looking at the factors separately it is obvious that mass as a factor cannot produce gravity. Mass is the restraining motion that leaves gravity as intending motion. Mass occurs only when motion is prevented and when mass prevents further motion resting objects leans against each other. When objects rest against each other they restricts individual gravity motion. Mass is a substituting factor, compensating for motion loss. When mass restricts motion gravity becomes the tendency of motion. Mass counters motion when the Earth restrains further motion of falling objects. When motion seizes, falling objects remains individual while still tending to move. The Earth resists further movement of falling bodies' movement restricting motion individuality. Having mass does not bring about.

Physics is Brainwashing by Mind Control

If you are a student in physics then you should read the following information. It is about the subject of gravity and is most important. Do you realise that it is an accepted practise that all students that are studying physics on all levels are subjected to the most intense brainwashing and thought control found any where on Earth? This must be some sort of a joke you may think but thinking that way in disbelief is just what those practising the mind control wish you to think!

Should you think this page is some sort of a prank then answer the following simple question to yourself in utter honesty?

The questions concern that which you are studying and that touches every aspect you are academically concerned with. You are taught that gravity pulls objects to the centre and obviously gravity then has to ultimately pull everything to the centre of the Universe. That is what the Critical density research that Einstein initiated wishes to establish.

When visiting the classes you attend in physics, has any one confirmed a location where one might find the centre of the Universe? This they have to do if they say that all objects are submitted to gravity. Then they must know where gravity is taking the Universe. If you wish to apply a Gravitational constant as a calculated factor in using the basic formula $F = G \frac{M_1 M_2}{r^2}$ then it is apparent that every one must know to where such gravity is pulling. Gravity is pulling to the centre and therefore the gravitational constant also is pulling to the centre of the Universe. If there is such a force then where is the force taking the pulling…if it is a gravitational constant applying through out outer space then where is it having a centre base? Tell your tutor to calculate when the Earth will collide with the Sun by using Newton's also accurate formula $F = G \frac{M_1 M_2}{r^2}$.

To calculate the following data is necessary:
Mass of the Earth = 5.974 x 10^{24}
Mass of the Sun = 1.989 x10^{30}
Gravitational constant = 6.67 x10^{-11}
Diameter between the Sun and the Earth = 149.598 x 10^6 km (remember that this has to square)

If he can't give you a ready answer it is because it is the biggest hoax man has devised ever. Using this formula $F = G \frac{M_1 M_2}{r^2}$ to calculate is complete rubbish because it can't be done. They are brainwashing you into believing the use of this formula is viable while it is complete and utter rubbish. Gravity does not draw by mass at all.

I wrote a book in which I found a means to define gravity. This feat I accomplish and by my effort it was done this for the first time ever. For the first time ever my investigating physics runs further back than since the time Newton introduced the idea of gravity. Before I achieved that discovery, I firstly had to find the centre of the Universe because it is there that I could locate gravity. I can now show how gravity forms because I have detected the centre of the Universe. But by my effort in finding the location I disrupted everything Academics in physics hold holy and for that I am most unwanted in the presence of the Academics charged with guarding the ethics of physics. Every time I try to indicate what I discovered about gravity, academics throw Newton at me and detecting from the information I discovered when I investigated another even much wiser Master, it is clear that Newton is the last person that knows what gravity is...he (Newton) even admitted to not knowing what gravity is...and yet they (the Newtonians) keep throwing at me the ideas of the man that admitted he had no foggy idea about gravity. During my research I discovered abnormalities and inconsistencies about mistakes Newton made long ago and it is clear that the Arch fathers in physics must be aware of such misconduct performed by Newton but they (the Newtonians) are hiding such information from students using all their considerable influence. I will come to a few of the inconsistencies but part of the discovery I made when I investigate the other Master I mentioned I was also introduced to a much better vision about gravity as well as many new aspects never before realised in science. The road was never smooth and the resistance I came across from the Newtonians was almost unbearable. Academics guarding physics will never allow an outsider to enter their domain without the intruder paying a heavy price and in this matter I am seen as the intruder. Intruding allowed me to find much that I was not supposed to find which was only allotted to the most inner circle and much I share with you.

In achieving goal to locate the centre of the Universe I had to step on some very important toes, which made me very unpopular. With my unpopularity rating this high as it does, I never qualified for help and found intolerable rejection as I tagged along while trying to convince those Newtonians about mistakes in their field of expertise. Because of this insider rejection I received so blatantly, I had to resort to private publishing because from the nature of my work I take Mainstream science head on and am confrontational on most aspects of astronomy. This is the only road to go if one wishes to lay an axe to the root of the insider corruption that all they (the Newtonians) are guilty of. In that sense there does not seem to be any publisher that wants to go head bashing with the Physics Custodian establishment of science on official science principles, which I have to do to convey my message in no uncertain language.

I argue that if it is the correct practise to use $F = G \dfrac{M_1 M_2}{r^2}$ to calculate gravity then the radius holding the gravitational constant must lead one to the centre of the Universe. I found the route that gravity takes when gravity goes to the centre of the Universe but it definitely does not apply when using the Newtonians formula $F = G \dfrac{M_1 M_2}{r^2}$ With nobody willing to publish my work as I contest science all the way and even at the most basic level, I had to go the road alone and fight the battle by my private effort.

I know the explanation I give in this seems cramped, but when reading the book it is much clearer. This is the point that I wish to make on this one issue and similarly there are thousands other unexplainable issues. If the Sun for instance has mass that is apart from the Earth and the Earth also has mass and there is a gravitational constant in between the Sun's mass and the Earth's mass we have the radius in that location. It then must be the gravitational constant $F = G \dfrac{M_1 M_2}{r^2}$ represented by "G" that fills the space that the radius holds. It is rather obvious that while the radius is filling the vacant space between the Sun and the Earth it is the only place left where the gravitational constant can hide. The space Newtonians give a value of nothing or emptiness must hold "G". Where is this gravity pulling if it fills space that is empty and is not filled? If it is the centre of the Earth that does all the Earth's gravity pulling and it is pulling towards the Earth's centre then that other gravity being "G" has to have a point whereto that gravity is pulling if it is pulling at all. To find the centre of the Universe I had only to find the gravitational constant that holds the centre. Through my venture I discovered one person that knows what gravity is! From studying that person's work I found the centre of the Universe.

If you think scientists know what gravity is do not be duped that easily because no one in science remotely knows what gravity is...not even Newton knew what gravity is except Kepler... and because of what Kepler introduced now I know I can prove what gravity is. Gravity is precisely what Kepler said gravity is and only Kepler new where to find the centre of the Universe because only Kepler knew what gravity is all about. Did you know that Kepler showed what gravity is decades before Newton even had a thought about gravity? Does your Professor know that Kepler found and proved what gravity is long before Newton had admitted he had no idea what gravity is? Newton had no idea what gravity was because Newton changed Kepler's work without ever studying Kepler's work and therefore never understood Kepler's work.

Students read the following message about my book I named www.SirNewtonsFraud.com and by reading the book you can learn how you are brainwashed and how your mind is pre- conditioned into believing in Newton's myth which is pure deception which Academics call physics.

Let's start surveying civilized principles by evaluating what lawfulness means and what would constitute as morality. Let's determine what makes the crook in the book?

If any person, notwithstanding what reasons given, tells a lie or conveys untruths it is seen as fraud. To convey information that is not substantiated as verified fact then the mere conveying of such information becomes fraud.

When any person, notwithstanding what reasons given, repeat such a lie unabated while being well aware that the information passed on by such a person is incorrect, then the person commits deceit. When anyone is repeating the information that is passed on as being unblemished factual substantiated and verified truth while such a person knows very well that such information is void of proof or lacks proof, then committing such an act is a criminal enterprise.

Academics in physics commit every one of the above indignities and yet see their actions as being lawful and even much praiseworthy and hold their role in society in the heist esteem imaginable. They fail to see the crime that they commit while tutoring physics. Whatever motivation they may claim to have as their driving force, the fact that they perpetually perpetrate in unlawful behaviour by spreading untruths such actions on their part put those academics holding such highly regarded office in the league of ordinary cheats, gangsters and common criminals. By willfully and constantly falsifying facts of what order repeatedly remains derogative and unlawful in nature, notwithstanding what morality it should serve. A Preacher or Pastor lying on behalf of God is not lying on behalf of God and to think the Preacher or Pastor improves or underlines the Greatness of God by lying on behalf of God is very mistaken because in reality such a Preacher is falsifying the truth for personal his or her personal benefit. Lying is wrong and doing so even in the name of God remains despicable. The same applies to academics in physics. There is no argument that can change this truth about falsifying the truth and when doing so there is no hiding behind any excuses of ennobling to benefit mankind that will change such truth into righteous conducting.

Newton said centuries ago that gravity is the force of attraction there is between objects that holds mass and it is the mass factor that brings about this attraction, which Newton claimed there is. The Universe does not contract and all the proof we require to disprove such a statement that we find in the Hubble constant as a guarantee. Moreover it is true that the Universe never contracted even for a brief instant and proving that is the Big Bang concept with all the proof that this concept bring in backing the principle of expansion in the Universe. Planets never moved closer, are not moving closer and will never move closer to each other and this is backed by all information collected this past century. The Moon is not coming closer but the distance between the Moon and the earth is widening. Studies about the Universe reveals every time that space in the cosmos increases constantly studies find that all things are moving apart and away from one another. Any and all the proof about this is beyond what any doubt may present to counter this knowledge.

Notwithstanding this irrefutable findings, science still regard Newton as the only person that ever lived which no one ever could prove wrong…and this is upheld by Mainstream Physics in spite of the cosmos proving Newton wrong every instant of time. The basis of what science holds as its foundation we find to be the Newtonian principle of $F = G \dfrac{M_1 M_2}{r^2}$. The foundation used by science promotes this argument and backs up this argument well knowing that in the cosmos there is no evidence backing up this proposal Newton suggested. The Newton formula $F = G \dfrac{M_1 M_2}{r^2}$ used as basis for science see gravity as being a force of attraction and the force of gravity is being in place between all objects in accordance with the mass factor that the objects have as presented by Newton in the formula $F = G \dfrac{M_1 M_2}{r^2}$

What we find as we gauge all evidence found while studying the Universe is that reality shows there is no attraction between objects in space going on any where in the Universe, that the entirety of such a concept is a myth and the outward moving of the Universe has been coming from and since the time of the Big Bang and maintaining this flow of material is substantiated in a concept named as the Hubble constant which proves Newton's perceptions to be a myth. The Hubble constant proves that space every where is growing ever since time began and the growth never stopped ever since. Knowing this irrefutable fact does not deter science from under scribing Newton as the sole basis that underwrites all the correctness of all of science known as physics. However Hubble and the Big bang and all other investigation contradict this attraction Idea Newtonian dogma holds. Therefore any further believing that there is attraction going on as Newton claimed has to be viewed for what it is and that it is a fairy tail. The Big Bang Theory proves Newton's idea as not only being wrong but Newton's idea of attraction is a joke. If the Big Bang is expanding the Universe then how can the Universe contract at the same time? Any contraction by nature would have the Universe collapse back into infinity the moment the Big bang moved out of infinity. Ask your professor to show how an expanding Universe can also contract and your professor will tell you about Einstein's Critical Density theory. This theory I prove is the biggest fraud ever devised by any group of persons in the history of civilization! This is perpetrating fraud and conducting in upholding deceptions instituted by Newton that then formed the institution of lies they call physics.

The Universe does not contract in any way means or form and even such a suggestion is incorrect! The Moon and Earth are not moving closer but are moving apart. The entire Universe is growing in space and no where is space depleting by any norm used. Academics are very aware of this misconception Newton had and still academics in physics are promoting the ideas of Newton as unwavering truth. Academics teaching these misconceptions are committing fraud notwithstanding their portraying of their role in society being unblemished, spotless while they are covered in a lily white blanket making them being whiter than snow and having such a holier than though attitude. Teaching Newton is participating in deception and promoting Newton is criminally deceiving the public and while doing so is committing an act with criminal intentions.

Then, in the face of all this evidence contradicting Sir Isaac Newton they remain upholding the correctness of Sir Isaac Newton and keep on teaching students about the unwavering correctness of Sir Isaac Newton. They put down conditions of learning to this effect and are expecting students to repeat these untruths and unproven facts by forcing answers to that effect in examinations. Forcing the accepting of this untruth about physics are equal to preposterous subjecting students to physiological torture and heinous mind conditioning, scandalous thought control and brainwashing. This applies to everyone serving as a tutor in physics notwithstanding whatever status the torturers might have in society or the morality they attach as a reason to commit such atrocities.

If you are a student then you are conditioned by academics in controlling your thinking by enforcing pre mind setting and in which they methodically force you into believing in Newton and this is an on going process conducted for centuries in the past while it is the truth that Newton is completely void of any tests that may secure any form of confirmation and in securing proof then also by that establishing proof. Read this book www.SirNewtonsFraud.com and then use the information I supply in the book to insist that Academics that are teaching physics prove to students that Newton's statements of attraction are correct. Let those academics explain the method mass uses. Let them with precise detail show when mass is applying that gravity is produced by mass and such producing of gravity that then would establish attraction! I show precisely how gravity produces mass but mass can never produce gravity. I sue explicit detail in showing when how and where gravity forms mass but mass can never form gravity. What I prove annihilates every Newton claim.

They never prove Newton's philosophy on gravity but those persons conducting teaching in the subject of physics force all physics students to learn Newton's gravitational concepts and accept the facts as if it has been proven beyond all other facts. Students have to believe that Newton is correct or academics will see to it that they fail their examination. The condition of being accepted in physics is to accept Newton without questioning the proof that is never supplied. Let those academics now prove precisely how mass brings about gravity and then afterwards test you on how Newton is proven correct and not on you repeating facts about what they say is true about what Newton said that they say is true. The manner they present Newton is completely hearsay and that method may not be used in any court of law. Let you professors now prove how it is that Newton's teachings are correct and then examine you on the process they use to prove Newton's concepts. At present they say Newton is correct and then they test you on your ability in repeating that Newton is correct without ever proving to you that Newton is correct. Let those physics professors now prove Newton and then test you on the manner they use to prove Newton to be correct.

The truth beyond all other truth is that Newton's gravity has never been proven (because try as you may it is not possible to prove Newton's formula forming gravity mathematically) and because academics know that, academics require the blind accepting of Newton by students. This unconditional accepting of Newton's correctness relies only on the pre-conditioning of students' mind set and academics depend only on the student trusting the academic say so that about the institutionalised correctness of Newton. That Newton is correct nevertheless and notwithstanding that there is no founding proof about this matter is what students should be accepting blindly. Pre-conditioning students' into blind acceptance depends on the academics' insistence that students approve Newton's concepts without pre judgment or students insisting on scrutiny of any sorts. In examination students have to outright and blindly follow academics' say so only because academics say so. Academics depend on students never questioning their say so or demand proof about what academics teach. Those academics in teaching positions insist that all students accept Newton's accuracy.

This is methodical mind control as much as it is the brainwashing I show that they enforce. If you are one of those believing that Newton was ever proven, then what you believe to be true is a lie because Newton can't be proven and that is the truth! The time has come to face your teachers and force them to stop the centuries old culture of bullying students and conditioning their thoughts by enforcing on them dogmas which is mind control!

I charge all academics to prove what I say is being wrong in any way or even that I exaggerate in the least. I challenge Newtonian academics to prove that mass does indeed form any force of any sorts and in particular gravity! To those professors claiming Newtonian ideas are substantiated by proof I say that notwithstanding your personal academic qualifications and while at the same time disregarding your status and previous achievements as well as ignoring your many admirable abilities you may have and however superior they might be, I shall teach you about gravity. I say it is time Students learn the truth about physics notwithstanding the status academics will loose. Students read www.SirNewtonsFraud.com and challenge those academics depending on their ability to brainwash you into submission.

I don't think so because your professor is an expert on Newton's work and Newton admitted he didn't know what gravity is. That means whatever your professor has expertise on, it involves Newton and Newton admitted not to know what gravity is. Then notwithstanding what your professor says about gravity and what he professes to know about gravity, he knows what Newton knew about gravity and Newton by personal admission said he knew nothing about gravity. When you read my work you will learn that the last thing gravity can be is that gravity is some force pulling objects closer! Try to tell that to a professor that says he or she is an expert on the work of Newton. They are experts on the work of a man that knew nothing about gravity except giving the idea behind a notion as gravity some grave connection and a ridiculous religious name. Try to get an answer from academics physics about gravity in detail or where does gravity originate or even about where the centre of the Universe is, and they stone wall you. Achieving that is more like trying to touch the moon. Talking about the Moon let's stay with the Mon for a thought. Tell your professor to tell you much is the moon moving closer to the Earth since $F = G \dfrac{M_1 M_2}{r^2}$ is in effect, because such calculations and measurements are easy to measure. With that information they can determine how long it will be until the Moon is part of the Earth. The Big thing is that the moon and the Earth is moving apart at the same rate as what human nails and hair grows. The Moon and the Earth is departing and not arriving at a point. The fact that the Mon should be moving closer and not be moving away is all part of that big scam and cover up that I write about in www.SirNewtonsFraud.com.

The best your professor could profess to be gravity is that they know that gravity is a force that works by magical powers pulling on whatever nobody can find. By merely putting gravity in the Universe that is acting as a mysterious FORCE that is pulling towards a common point in an allocated general centre is rather avoiding the question with simplicity because the question about how and why remains unanswered. Not knowing the answer will leave you empty and unfulfilled because of being a student and not knowing is the same as suicide n a mental level. Ask yourself the following: If gravity pulls towards a centre and gravity holds the Universe attached the question arising from that simplistic answer is then… where is the centre of the universe?

Should you decide to go to www.SirNewtonsFraud.com it will bring along a new perception about Kepler? Science sees to it that Kepler stays the least appreciated Cosmologist because **NEWTON'S FRAUD** destroys the entire substance of the work of Kepler where as in truth Kepler proved gravity, proved singularity, proved space-time, proved the Big Bang, proved every dynamic most of the wise persons that came up with all the various ideas afterwards thought about. However, even Einstein's special relativity theory was devised by Kepler beforehand! I can trace all of Einstein's work straight back to Kepler. Yet no one gave Kepler any recognition up to now because science denies Kepler his limelight.

Through my effort in investigating Kepler I came upon a mistake concerning physics.

This mistake is about the cosmic phenomena called gravity. Detecting the mistake is simple because it is uncomplicated to understand. One only has to look at the Big Bang concept and from that one can question Newton's idea that it is mass that is attracting material. To circumvent students detecting **NEWTON'S FRAUD** those cheats in physics came up with the biggest fraud scam any one ever invented. They still call it the Critical Density Theory where at first Einstein was called upon to calculate all the mass in the entire Universe but then found there was insufficient mass to pull the Universe into attraction. That again didn't support **NEWTON'S FRAUD** as Einstein was set at task to do so the cheats had to come up with more scandalous conniving. Then with this Critical Density of Einstein going to the dogs they found another way of trying to prove **NEWTON'S FRAUD** as being believable and correct while knowing Hubble proved the Universe is expanding and what Hubble found proved that **NEWTON** was committing **FRAUD**. They then had to find another way to cover **NEWTON'S FRAUD** but to do that they really extended all earlier criminal schemes by bettering all previous avenues of betrayal through which **NEWTON** committed **FRAUD** before. They started looking for so called dark matter, which according to their idea is undetected dark material that supposedly fills the cosmos while we are unable to see the dark matter. The matter is supposedly unseen but should be there and with enough of that the matter in the Universe the material forming mass will eventually start to pull the Universe into contraction. The question that shatters their deception is that if the matter is there dark or not dark, and mass attracts like they say it does, then what prevents the mass from energising the force of gravity in the present moment and through that the force must start to pull the Universe into contraction instead of expanding? What prevents the mass from committing the force of gravity if the mass is there? Why would the mass not produce gravity at present and why would the mass start to produce enough gravity to start to pull the Universe at a later stage as Newton said it has to do. Either the mass (dark or not dark makes no difference) is there and produces gravity to pull the Universe together or Newton is wrong with his contracting idea and the Universe is expanding like Edwin Hubble said it does. If the mass is there and is irrelevant now it will remain irrelevant through out time to come. Why is a fact such that the mass is not producing light preventing gravity from contracting the Universe? Why is this visibility being connected to the fact that it is not producing enough gravity to contract the Universe and then why will the mass at a later stage start doing so and when will the mass eventually going to start to do the job of contracting, because in this expanding Universe the mass, notwithstanding being there or not and notwithstanding being invisible or not is not, whichever way it is, it is not doing the job of pulling at present. With an expanding Universe as we have at present the mass can't contract as **NEWTON'S FRAUD** claims it does and that proves that mass doesn't attract at all. The Critical Density Theory is the biggest fraud committed by any group of persons and is put in place to cover **NEWTON'S FRAUD.**

Part 2

If you read on you are going to <u>Read About</u> <u>The Biggest Conspiracy</u> ever enlisted, it's more than just the next conspiracy because it is <u>A Science Conspiracy Network.</u> <u>I prove everything I say about the conspiracy I announce...read and be shocked.</u>

This Network which I denounce Engulfs the Entire Civil Human Race in Every Aspect and willing or not, you reading this are participating without knowing about what you support with every part of the mental intellect you have. It involves the most trusted members in the part of our civil society and we are deceived by the ones EVERYBODY trust the most, the teachers and Professors teaching at academic institutions everywhere. For over a decade I have been knocking on doors with the information I present in this book and lots more evidence, just to be ignored and to be turned away. This behaviour of physicists ignoring me or even attacking me when I point out a very legitimate case of mistakes in science made me realise there is a conspiracy going on. Science has been getting away with this conspiracy since the Dark Ages. What you are about to read is not for the simple minded because science requires much intelligence and in that physicists get away with turning Creation into a joke. By turning the facts that form the conspiracy into what you think of and you accept as natural and as culture, man has been cheated and mislead for three hundred years because those teaching science consider everyone as representing stupidity. To them everyone else are simple-minded peasants who are not worth having independent minds who could be able to think on their very exclusive superior level that they are able to think. That is how they get away with the conspiracy. They pretend to be intellectually superior and to understand what seems senseless to the rest of us and then look down their noses on the rest of the world. I will show you what it is they can't see and then miss. They underestimate the public as much as the public are lazy too think and the misleading carry on. Those in science think their mathematical abilities make them Gods–on–earth while we others are mindless human fodder and for that reason they brainwash our children at school by applying mind control as their mathematical understanding makes them gods. What they say all must believe because if they speak using mathematics then God spoke.

THIS IS THE WEBSITE THAT <u>WHISTLE BLOWS</u> ON A SCIENCE CONSPIRACY NETWORK. I AM GOING TO SHOW YOU THEIR UNBELIEVABLY POOR MATHEMATICAL SKILLS!

This conspiracy is so widely active that every person on earth teaching physics participates, most probably without knowing, some reluctant and others well knowingly to set out to brainwash by thought control the minds of innocent students forcing them to believe in dark aged ritual beliefs in the ritual practicing of forces and allow the conspiracy to be enlisted with all the vigour they could muster. People would argue religious concepts but everybody accepts science on face value. The guilty party that I refer to betray us in every way possible by teaching us lies that never could be true and all the while we believe them with all our hearts and entrust them with our entire future... They take money from parents with the sole aim to brainwash the students who are entrusted to their care and demolish their intellectual understanding ...and before you think my claims are exaggerated I say again I prove every word I say. I show what is true and what is wrong in science and as a thank you for all the concern and devoting care I show, the establishment of science frustrates me, ignores me or runs me down, while I challenge everyone and any one to disprove me. Decide after reading this book weather what you read is not a conspiracy...however, familiarise yourself first with what I present and then decide.

Those I refer to are allowed the powers to take as much tax dollars as they wish and endeavour on all sorts of projects that they could fantasize about. Their budgets are limitless and they never face questions because they are the intellectuals that we dare not question. With the same powers we give them they brainwashed your child making science a religiosity. The methods they use on your child are to make your child believe in science and that is just methodical corruption of their thinking. Physics teachers take money from parents with the sole aim to **brainwash** the students that are entrusted to their care so that students will believe in science unequivocally and regardless. When teachers teach falsified information then they are worse than the mafia, that operates openly by crooked intention and in that the Mafia maintains honesty, which the physics teachers don't have. If you are a parent then mind what you do and beware and read on! If you think I am going one step too far in accusing the academics teaching physics and I slander their names, I challenge you to ask this question again after you have read this entire book and thought about everything and considered every aspect.

Read About <u>The Biggest Conspiracy</u> ever enlisted, it's more than just the next conspiracy because it is <u>A Science Conspiracy Network.</u> <u>I prove everything I say about the conspiracy I announce...read and be shocked.</u>

I call science a scam and if many readers if not most will completely disagree with me about science living as a scam then the next medical scam I wish to bring to every one's attention will support me. This scam is called humanity as it involves the medical science more than physics. It covers not what you are told or what you are taught but what science remains silent about. It is what is never said that is of serious importance. It is never what the research reveals but what science hides from getting revealed. It lurks in the profits of the pharmaceutical companies or the doctors or the hospitals the industry never whish to reveal. Euthanasia is not about the absolute sanctity of life but it is keeping the patient alive when the patient is going to need the most expensive drugs and the most expensive intensive care. I saw a patient with a knife in his head wheeled out of hospital to find some other place to die because that patient did not have medical care to fund his desperate situation. Because there was no money for treatment there was no treatment and this goes on in every private hospital in every city around the world. If that person had a good medical care and the medical insurer was willing to pay then a team of doctors would fight day and night to keep the person alive and no cost will be spared. It is during the last few months / weeks/ days / hours that most care and medicine is needed. During that time when the end of the life approaches and the fatal disease is going to take its final toll that the pharmaceutical company, the hospital and its staff, and every person sanctioned with keeping this individual alive will fill the bill as fast and as hard as possible to get the last money from the dying sucker. You have a lawyer walk in there holding your last testament that states you are no longer presuming responsibilities for the costs and you have the lawyer tell them if they don't guarantee success with the treatment payment will stop on that minute and the costs for keeping you alive will be on the hospital and its staff that treat you and then you see how quickly they stop machines keeping you breathing. If you start legalising euthanasia you kill the part that serves the highest profits and they would rather see you suffer than they would agree that you might die cheaply. If you are not prepared to pay for their effort of holding your life so ever dearly they will lose all interest and let you die as quickly as the machines are switched off. Your life and all life have value when there is someone prepared to pay the bill and those doctors are the worst criminals there mathematically is. But yet again behind this medical industry there is insurers and behind the insurers there are bankers. Fighting euthanasia is a conspiracy because euthanasia will kill the huge profits the medical industry makes. If we are all so against euthanasia lets take the profits of these pharmaceutical companies, the hospitals and the doctors and spend their money to pay for every one dying that can't afford treatment. Now it is a case that they would rather see people suffer and die in agony when such persons can't pay for the treatment because after all being part of science makes them equal to God and therefore others must suffer so that they can reap the award of the money they invest in promoting and furthering science.

THIS IS THE WEBSITE THAT <u>WHISTLE BLOWS</u> ON A SCIENCE CONSPIRACY NETWORK. I AM GOING TO SHOW YOU THEIR MATHEMATICAL SKILLS! This conspiracy is so widely active that every person on earth participates, most probably without knowing, some reluctant and others well knowingly pursuing the goals of the conspiracy with all the vigour they could muster. People would argue religious concepts but everybody accepts science on face value. The guilty party that I refer to betray us in every way possible by teaching us lies that never could be true and all the while we believe them with all our hearts and entrust them with our entire future… They take money from parents with the sole aim to brainwash the students that are entrusted to their care and demolish their intellectual understanding …and before you think my claims are exaggerated I say again I prove every word I say. I show how the establishment of science frustrate me, ignore me or run me down and I challenge everyone and any one to disprove me. **Those I refer too are allowed the powers to take as much tax dollars as they wish and endeavour on all sorts of projects that they could fantasize about. Their budgets are limitless and they never face questions because they are the intellectuals that we dare not question. With the same powers we give them they brainwashed your child making science a religiosity. The methods they use on your child are to make your child believe in science and that is just methodical corruption of their thinking.**

They take money from parents with the sole aim to brainwash the students that are entrusted to their care so that students will believe in science unequivocally and regardless. If you are a parent then mind what you do and beware and read on! If you think I am going one step too far in accusing the academics teaching physics and I slander their names, I challenge you to ask this question again after you have read this entire book and thought about everything and considered every aspect.

For more about the conspiracy also visit www.questionablescience.net http://www.singularityrelavancy.com/ and www.sirnewtonsfraud.com

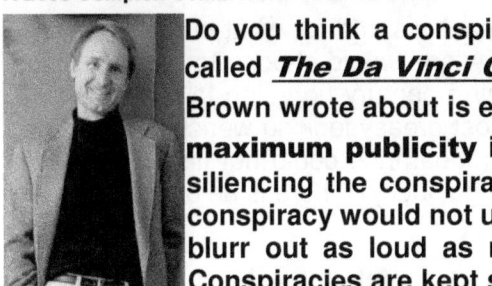

Do you think a conspiracy has to be as dramatic as Dan Brown's conspiracy he called *The Da Vinci Code* and *The Lost Symbol*, no it is not because what Dan Brown wrote about is entirely fiction and therefore it it **must be** dramatized to **gain maximum publicity** in order to sell. It is not intended to be silenced because by siliencing the conspiracy then the conspiracy would remain a conspiracy and the conspiracy would not unveil the billions of dollars it must make. The entire idea is to blurr out as loud as much gossip minded hogwash to sell as much as it can. Conspiracies are kept silent and Dan Brown's is anything but quiet. Or do you think a true conspiracy theory should be as devilish looking as the Illuminate. Are you convinced one must see the evil of the devil being portrayed in every symbol that represents the illuminate? No, those in power would silence all. It's just what happened in the Kennedy conspiracy. Those with the power hide the blame so effectively by law no less that the guessing will remain in place the next century. **The death of President John F. Kennedy is a true conspiracy and that we can see from the fact that there are no results coming from any inquest launched at even a Presidential level.** We know Hoover was involved by his involving of the FBI in the cover up because Hoover had to hide his gambling debt and his homosexuality. It was the Mafia that involved Hoover and the FBI. **Hoover denied there was organised crime because organised crime or better known as the mafia had him wrapped up in horse racing gambling debt and therefore he had to keep the FBI off of the heels of the Mafia.** We know the Mafia was involved because they silenced Lee Harvey Oswald the innocent man that was openly framed because it was the mob that got Kennedy elected by faking the election results against Nixon. **Kennedy had to get Cuba back to mob rule and the gambling houses back on behalf of the mob and when JFK turned on his election promise to the mob they showed that not even a President is above their retribution.** The mob by the hand of Jack Ruby silenced the investigation afterwards as they had other interested parties also in their pocket and those in high office helped afterwards getting JFK out of the way. **Did the police look for powder residue on the face, hands and arms of Lee Harvey Oswald? Did the police determine how many shots were fired from the rifle of Lee Harvey Oswald that they found in the Texas School Book Depository store and how many cartridges were retrieved? Where are the cartridges now and if not what happened with the lost cartridges.** Those are the question to address, but no investigation ever asked the obvious questions. We know that someone ordered the conspiracy from very high up being in cahoots with the mob and the FBI because the investigation was screwed even before the assassination took place. The conspiracy does not end its involvement there, because Congress and the Senate also had to be involved. **With Kennedy so quickly reaching for the Nuclear button, a lot of Bankers got stiff with anxiety because a few Nuclear bombs might not reach the Kennedy family in their bunkers, but it will write off billions of dollars of property all belonging to a lot of banks, either by mortgage or by ownership.** With a trigger-happy cowboy in Presidential office someone somewhere might just lose billions and in that it is cheaper to get a few million flowing in the direction of politicians that will help the conspirers get Kennedy from the office to a coffin. This is so apparent in the manner that the Warren inquest was so self-serving righteous and maliciously an open cover-up while every one on the Warren commission was greatly rewarded during the next decade, each in their own time. **This case shows the true science of a conspiracy and even the brother of JFK got done in when he got to close to solving the mystery. But investigate as much as you like, you will keep on guessing for the rest of your life!** Are you under the impression that a conspiracy theory cover-up must be as old, as wide spread and as powerfully involved as the Free Masons, which represents those that have the power to alter the destiny of the human race? If that were true then they would have the influence we find in the Kennedy case to keep their secrets covered under a veil of total silence and never to reveal any information to anyone! When the establishment guards a secret, it is well guarded.

So then you want a conspiracy theory to be as blatant and as staged, in your face and unreliable as the UFO landings where the sites are very vaguely concealed and as revealing as it can be with the most pathetic idiots becoming star witnesses. The media then blasts their idiotic untested testimonies over the air around the world as Gospel. It will be better to believe in ghosts. I don't ever say there are ghosts that can haunt us but I can't use physics to prove there are not ghosts while I can prove with physics there just can't be aliens. Where would aliens come from and where do they return? A visit from a

"nearby" galactica will take them tens of thousands of years to reach us and for what purpose, to have sex with the biggest idiots they find? You never think why those Aliens would choose to meet the most simple-minded idiots with single figure IQ after travelling for many a century non-stop to reach us? Are you impressed with aliens coming to have intercourse with the most greasy-looking wench on earth that by her own admission in accordance with her intellect and her appearance could not find one male in three billion to inseminate her with wasted sperm, and so the alien got lucky? There is no one wanting her so she invents someone craving her.

She is so desperate for sex that she invents some one even more desperate than she is. Do you fall for her stories that her ten mindless children has a father many galactica away and did not even leave her alimony to bring up his very earthly looking brats. If you have that mentality, then go for every conspiracy theory you can read because facts of life you will find too challenging to understand and comprehend fully. Then you think conspiracies must be as mind baffling and logic eluding as crop circles that are not in the fields when the sun sets and are there when the sun rises. If you do, then you are a sucker living for bullshit and fantasy, a monkey to be made of and you have the mentality a child has and you then are living in a fantasy-world only you can believe. Then this world is a place far to harsh for a person as gentle minded as you are. It would seriously be in your best interest to commit suicide very gently and even more quietly for this earth is far too complicated for your likes. You would be better off not to be here, we would be better off missing you than having you around.

The truth about a conspiracy is that everyone involved with the conspiracy will fight tooth and nail to stop the conspiracy to get leaked and leave those behind the conspiracy and all those feeding from the corruption of the conspiracy exposed. Those who have the power to maintain the conspiracy would keep the waters as still as possible as to draw no attention to such conspiracy. A true conspiracy has to be as quiet and as unseen and going as unnoticed as it could be. A true conspiracy must involve everyone without anyone detecting even a hint of what the conspirers hide. There are many conspiracies going on such as the banks involvement with crime and the bankers profiting from gamble rackets and drug selling. The same goes for the insurance business profiting from lenient sentencing of courts holding very merciful judges in office so that the insurance will sell cover-policies. Our social system in the Western World is a conspiracy. Democracy is held up as a saviour but politicians rule only to favour the rich.

Burglars, robbers, car thieves, drug smugglers all are the main income of Bankers and Insurance firms. If there is one chance in two million that your house would be burgled or your car stolen would you take out very expensive insurance, never in your life. You would take the chance to go without expensive insurance and take the risk yourself. Therefore if you have a chance of one in one hundred that you will become victimised in that your house will be burgled or your car will be stolen, then you better take the insurance notwithstanding the premium. In England the Bankers already got to the politicians to draft a law that makes it illegal not to have insurance. Now the Bankers and their Brothers can rob you blind legally and take you to court if you would not oblige.

They set the terms and the rates of how you will be robbed. The Insurers would never have the insurance premium holders pay the equal amount of what the amount is of cars being stolen. They would set the premiums at least at three times the number of crimes committed to cover their arses and cut a tidy profit on the run. Therefore we can assume that for every one car that is stolen the premium holder would pay three times what the crime rate asked because to the insurer it has to be worthwhile. That makes the insurance firms the biggest robbers, car thieves and crime hustlers in the world because they rob three times more from the public and take three times the cars or furniture or whatever was stolen than what the criminals did. They steal more from you than thieves do.

All they ask for is that crime flourish so that no one could dare to be without insurance and remain alive. Then to get everyone scared and nippy about crime, they protect the child molesters and the rapists and arsonists just to keep John and Jane Dow and the rest of the public so scared of crime they haven't got time to become frustrated with the total rip off the public suffers under the "protection" of the Police and the Politicians. That is why the "law makers" make crime ever so humane and get parents to become the villains when they try to discipline their children. It is about making the law-abiding citizen feel guilty and vulnerable at the same time while allowing the criminal to victimise the innocent so that the public can be controlled. You see how fast they put a father in jail when he goes to castrate and remove the genitals of the rapist who raped his daughter. Having the rapist out in public will put the parents on high alert rendering to a condition where they are great fill just to have their children alive and moderately safe. When it does happen to the persons next door we pray on our knees thanking that it were not our turn.

The rapist gets a slap on the wrist with early release and the parent get tossed into the rough side of the jail for decades because no one in charge wants crime to get under control. We have gone as far as feeling grateful that a rapist did not kill his victim or that a burglar only took "earthly possessions" and that the victim wasn't physically harmed. We don't mind being robbed as long as we're not clubbed to death. That is how far these scoundrels wheeling money got into our safety. Those scoundrels are the ones governing you by employing "democracy" and bringing comfort to your door. Your credit cards, home loans, hire purchase and all the credit available is paid by blood of victims of the criminal enterprise the Bankers fight to uphold and even pay to maintain. To Bankers crime pays.

The latest trend is the kidnapping of persons and taking ships for ransom. This became just one more way the bankers find a means to "tax" the public. If pirates in Somalia take an oil tanker, everybody sits back with a smile because it is not their problem since they are not involved. Wake up and smell the shit on you doorstep. We, the public, pay for the goods the cargo carriers carried. By paying the ransom to the pirates the price of the products went up. Not only that but now insurance are paid on all other products in the event of pirates grabbing ships. Insurers will now put a levy on every container carried on the high seas just in case of piracy. Whether the cargo is carried through safe waters or not, it is money to be made by insurers and the Bankers get the benefit of the crime.

Everything on earth just went up with a few cents but at the end of the year it comes to billions of currency that was captured by the Bankers going into the pockets of the wealthy and out of the pockets of the poor. Crime pays so there has to be a way to boost the levels of crime so that the insurance firms and Bankers can benefit from the source of wealth. The trend of kidnapping will only rise and in the next decade or two everyone on earth will have a stipulation added that there is a certain premium put on the head of every policyholder. In the event of any person being kidnapped, the insurers of the policyholder would be willing to pay that some of money that your policy covers so that you then are released. If you have not got such a clause in your contract, the kidnappers just get rid of your body and get the next victim to pay up. If you have not got insurance covering such an event, you will disappear without a trace, but if you do have such coverage, then your insurers would be kind enough to pay for your release. This is how the game is played. The Insurers would always see to it that they get many times more from crime than what the criminals could ever get.

Illegal immigrants are another source of currency going into the pockets of the insurers and the Bankers. When an illegal immigrant lands in a country, that person has to eat and live and crime does it. That person didn't cross the shores to live on minimum wages and be the poorest part of society in that country. Yea sure, maybe a few percent are honest workers striving for better but the vast majority wanted more than what was on offer where they came from. These are perfect candidates because they are untraceable. The police have no names and no identity and the criminals are shadows coming out at night to harvest what is not theirs and then to live lavishly while the true criminals are the Insurance firms and Bankers that regard this affair as only being business.

Tony Blair as Prime Minster of Britain acknowledged he has no idea how many illegal immigrants there are in Britain and no Brit ever asked why he has no idea. He and his party are paid not to have any idea. Why is the Ministry of Home affairs the one department that is "not fit for purpose" to use the phrase they used. It is so that no one can trace anybody illegal even if they tried to do it. Criminals are renowned to walk out of jails with early release even when they were illegal immigrants. Now why would you think that would happen? The Bankers pay the Government to have the Home Office in such a state that no one knows arse from head and the bigger mess in this department is, the more illegal immigrants there are and the more the crime rises. Crime pays but only for the Bankers, the stockbrokers and the filthy rich that can buy their own security army. Therefore the rich benefit by deducting cost as tax deduction so that the poor must pay more tax.

The illegal immigrants bring another sort of business with them. They bring girls they use as slaves for prostitution in the developed countries where there are money to be made. The girls get kidnapped and then their services get sold off. The price asked for one hour is not high but since the girls don't ask for much as they are drug slaves every man can afford to pay for the services and having many girls in a working position, the money eventually escalates into billions. Where will this money go…it goes to banks and the Bankers are the recipients of the cash in the end! Big drug busts don't stop the drug trade. It does not even influence the process by causing massive price hikes and that should be an indication of how much money goes to Banks.

Don't think of rescuing the girls in distress. You try and touch those girls to save them and see how you vanish from the scene either by getting jailed for interfering with police investigation or by the criminals who will make you disappear without a trace. You touch the girls and you touch the bankers that befriended the Politicians who are the bosses of the police and in that order you will annoy the command structure because the girls prostituted as slaves guarantee Bankers and other criminals much wealth. Those with money or with power are criminals or are criminal minded because that is the only way you can have money, steal from others to fill you pocket or use them as slaves to work for you. Don't believe this crap about only honest money going to Banks because with them in charge of the Political Government of all countries they will get the law to create loopholes through which they will work "legally" and with them having your elected government working just to please them they have nothing to worry about what is legal and correct and what is not because everything is just business.

The drug trade flourishes just because of one principle, the money has to end up in the coffers of Bankers. If the drug trade is worth a billion dollars a day, where does the money go? No one can walk around with that much money and therefore it has to go to bankers for safekeeping. I have not made a study on the subject since it is not my main interest to reveal crime statistics, but any one would see that printed money escalated by tens if not by hundred fold the past forty years. Why would they print that much more money? It is because that much more money is in circulation and the distribution thereof finally land in the banks. This gives banks the ultimate power they could ask for. Crime brings in cash and cash is their business, so crime becomes the business of banks. In order to further their campaign on behalf of criminals they get to politicians. Politicians not in favour of using "their election help" and aid by "election contributions and donations" don't get elected. The honest ones don't even get press publication. In this syndicate don't discard the TV stations and the printed press. Since the Bankers would own much of the press as shares legally held, the press is in place to further the cause of those in politics that would further the cause of Bankers and Bankers are in business to make money and crime pays good money. This is an animal that feeds off itself by consuming the consumer and enrich the rich.

There are many more "lobbyists" running down the corridors of the American House of Congress and the House of the Senate than there are politician official assistants in America. Why would that be? That is a smarter and more acceptable name for bribing the politicians to gang up against the electrets. It is a fancy way of bribing anyone and everyone so that the "democratic" process could be hijacked and rigged.

If Gordon Brown and the rest of the European leadership so full heartedly believe in democracy, why don't they put the death penalty to the system of democracy? Go on and give the public the choice, let the people decide… If they say they rule by democracy, then let the people decide on the main issues directly, call a referendum! Why don't they put it to the vote and disallow the one-sided advertisement funding of vote swinging so that the people can make a well-decided popular choice on the matter. They lie and cheat and bullshit before an election just to come afterwards and push the rule of the rich down the throat of the poor. The Police say they protect. If they say it let them prove it and then let them protect and if they don't solve crimes then put the Politicians in jail with the rest of the Bankers and the criminal mob. But it is a case of the quicker robbers and car thieves and rapist get out of jail, the quicker they get back to crime and the more crime is committed, the higher the preemies get for insurers. Therefore the more the money is and the higher the profits are for insurance firms, the more bankers and industrialists gain. Also the higher the crime rate gets the more the people have to see that they are insured to cover their potential losses in crime. Therefore it pays to buy politicians to create laws that go soft on criminals.

Defend yourself in your own home against a burglar and you are going to jail! If you as a law-abiding citizen have a gun it is a crime, but is it because you make it difficult for criminals? They can't keep guns away from criminals so they take my gun away! You go and beat the holy shit out of a person that robbed you then you land in jail. If the robber gets away with the crime no single tear is shed but if you beat him blind you are the villain. This is to protect the Insurance firms that bank at the Bankers that invest on the Stock exchange that get the Rich richer so that the Poor must get poorer. The more crime is about and committed daily, the more people need insurance covering and thereby the more money insurers bank giving bankers profit to spend on the stock exchange by controlling the economy and the politics.

If the Government puts Police in charge of crime and disallow me as a citizen to take revenge on the criminal by going out and look for the robber myself, or get the rapist to swing on a lamp post because I have his finger prints in my house where he raped my wife, or cut the genitals out of the rapist and molested my four-year old daughter, then they must take the blame for unsolved crime and the high crime rate. Then the Politicians must see to it that when the rapist rapes again after being released from jail everyone involved with his release from the lenient judge down to my local politician that is responsible for crappy laws, must sit in jail with the rapist because they all participated in the crime being repeated again. Because of their weakness crime is committed, so the politician must guarantee my property or give myself the option to retrieve my belongings and then hang the criminal from the nearest lamppost.

It is time government takes responsibility for laws failing and because they favour the Bankers and Insurers it is my security that is compromised by them applying lenient laws that is about protecting criminals. If the government says their Police will protect my belongings and secure me then they take the responsibility. If I am being burgled, then the government must replace my lost property because it was they and their police and their laws that did not protect me and then if the crime is too high, the cabinet must resign and give other people more fit to do the job the chance to govern. Now at this moment we know that Politicians are the most crooked over paid under worked robbers of public funding there is on earth. Do like the Chinese do with crooked officials, those they catch with their hands in their cookie jar, take a crooked politician or official into the centre of town and put a bullet into his brain. Why would other politicians be scared of this practise when they are as innocent as newborn babies? That will get rid of the crooked politicians and place the genuine honest persons in public office. Crime pays for the elite that profit from the rate of crime becoming sky-high. The money is used to change laws in favour of crime so that crime can flourish and Bankers can profit. Bankers buy politicians and democracy through money donations. With the money they give politicians, laws are written that protect the rich against the poor. This is the way to run a conspiracy orderly. No one can trace it and know one suspects it because we all benefit. With this method the Bankers gain by inflating everyone's livelihood and eroding personal income and safety as to gain from that. In time I might write about this but now I cover another story, a much bigger storey about a conspiracy everyone on earth has been brainwashed to hold a part of...it is about...

There are so many conspiracy theories going around, floating like a bad smelling odour in the room, but most of those have no more factual substance or evidential backing by truthful facts than a James Bond novel has. I also have a science conspiracy theory but mine has all the proof and all the substance of authenticity that any argument can ever hold. It involves **physics** as **physics are taught** at institutions worldwide. If you now don't believe me, read on and learn the truth yourself, it is free of charge.

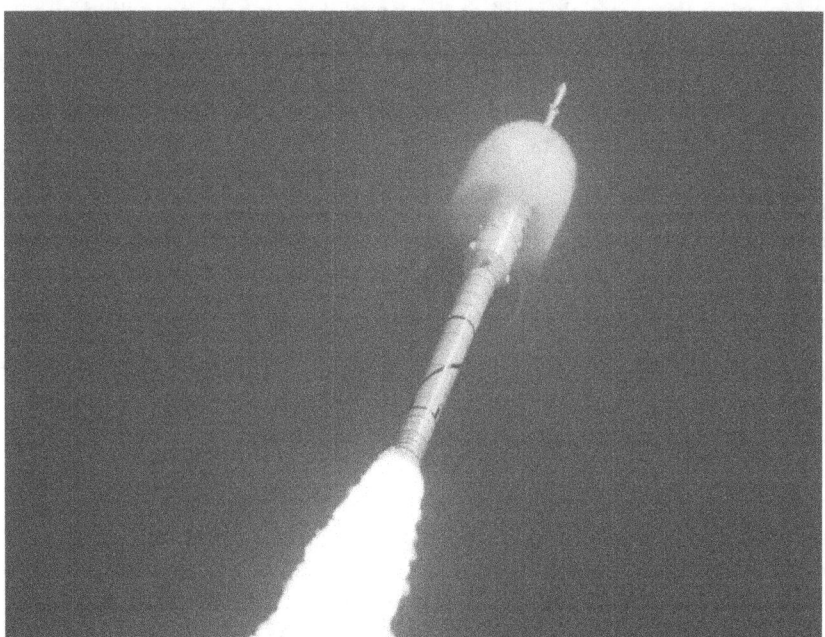

Mine is the most unbelievable one that has most truth. This book is not about any conspiracy ever mentioned before but takes into consideration what worries people.
When you see what science achieves and you hear what I accuse science of, then it seems as if I am a nutcase who escaped from the loony house and I must be the one having very dangerous tendencies to harm the innocent.
I should be locked up and taken from society.
However, read my story and you'll see that I don't harm the innocent. It is those we all trust without reservations that brainwash to forge deceit through corruption and malice.
**The blame is with those admirable, yet astute members in our society, the educated leaders in physics that we trust without even thinking about their trustworthiness or question their sincerity regarding their blameless honesty.
Do you believe in the absolute unquestionable correctness of science and that everything that forms science is truthful being far beyond suspicion?
Do you believe that those furthering the science of physics works with no less than absolute proven facts and the facts are beyond any doubt at all.
If you do then** I have very bad news for you, **news that will change the grain of what you believe is truthful.** I am about to shock you into reality!

This is going to reveal the biggest conspiracy ever concocted by the human mind, ever since the time a human had a mind to use.

It is A Conspiracy about Fundamental Science. It started when Newton gave us the idea that gravity is the result of mass and in gravity mass is the biggest contributing factor.

I say that mass as a factor in physics and not the value of weight, but mass with the pulling power only pulls stupidity over the eyes of incompetent idiots raving in their personal sublimation. Let see what I say. Would those in science cheat you and withhold the truth from you?

www.singularityrelavancy.com that is telling everyone about
An Ultimate Science Conspiracy, the biggest conspiracy ever concocted.
NATURE put TIME into place TO PREVENT Everything FROM HAPPENING all at once.

This proverb sounds silly and yet, silly it is not.
While this proverb is the truth, science say they don't know what time is...and this is ignorance they hold while living with time for millennia all the while Newtonians say they uphold Galileo and everything Galileo implemented. Gravity drives the Universe as the earth spins around.

In other words what science can't see is that there is a time –connected plan in the form of gravity driving the Universe along. Time is in the form of gravity and the cosmos is driven by time. Time and gravity are the same, which is precisely what science is unable to see.
If gravity drives the Universe, the Universe is powered by time.

When Galileo's pendulum can measure the flow of gravity and record time to a precise science while swinging in space, then the flow of gravity in space has to be measure of the pendulum swinging in air while reading time. The pendulum measures time when swinging through the air that surrounds us and what surrounds us is there because of gravity.

This obvious conclusion about gravity forming time, which is what the pendulum reads, any child can draw but as simple as it is, this went past Newtonian understanding of science for centuries.

They never could realise the flow of gravity and time is one and the very same since it then conflicts with Newton's ideas. Still the pendulum proves that reading time and measuring the flow of gravity is the same. Galileo used gravity to measure time accurately, yet for as long as the pendulum recorded time, this fact eluded science in a way that science doesn't even know about it. It is what science doesn't know that becomes important to man. If gravity measures time by applying a swinging pendulum this becomes the biggest fact in physics and it still went past science because their cleverness brought on their rigged stupidity.

This religious belief in the infallibility of their Godly insight becomes their stupidity. Their ignorance is what science hides. It is that what they hide which is what you don't see. Science hides what science doesn't know which they cover under the larger pretence of their cleverness.

Also science hides from everyone's view what science doesn't know so that students will never realise Newtonians are covering their stupidity and ignorance behind a curtain of arrogance. This conspiracy to not reveal their stupidity then becomes the conspiracy they nurture for centuries. Science conspires to hide by covering their stupidity from open view. The main thing is that science is clueless about time and time is the driving cosmic plan so science hasn't a plan to see the truth.

I have been in fruitless conversation mainly going one way about what I see as their stupidity in defending flawed science. I told them this but with having no response. Still notwithstanding their arrogance about ignoring my showing right from wrong, still those physicists filling high academic office are so infatuated by their superiority and their personal righteousness they can only see the malice that the Pope showed towards Galileo when he differed from the general science views and yet they do the very same today.
They are overwhelmed by their correctness.

you will see how those hypocrites condemning the pope having the holier than thow attitude because they think they are cleverer than all other intellectually lesser sub-humans. They accuse the Pope of constraining science progress while back then but they torment me in much crueller ways than the pope did to Galileo. At least the Pope allowed Galileo to print a book while they did everything to stifle my efforts when I just tried to be heard by others about my views!

Whenever anyone show those in science that their religiosity called Newtonian science is wrong, they act in exactly the same way as the Catholic Church did to Galileo five hundred years ago, in no less a manner while they uphold in public that they maintain their stand on condemning the Church for not excepting Galileo's principles. The crude thing is that I prove all those filling academic office in science have also still not accepted Galileo to the letter and I challenge anyone to prove me wrong. That is one of the parts of the big conspiracy I see and call it the Mother Conspiracy!

You are about to Meet The Biggest Crime Syndicate that Ever Operated In the History Of the Earth.

There are so many nagging questions that Newtonian brilliance fail to address. Why did the Moon gather from dust so close to the gathering area of the Earth? Why did the moons of all the planets become the dense structure while achieving this incredible feat with such small mass evoking such poor gravity? Why did the Sun not collect the lot since in the normal formula that applies, the dust had to pass by the Sun on its way to where the location is where that the dust then formed planets on the other side of the circle? What prompted Pluto to become one of almost a double star and not reserve all the accumulation into one unit? Why have the unit, so far from the centre and yet have the dust to form such a firmly compacted unit in two separate structures being so closely together? Why did a giant such as Jupiter allow fragments that we now call moons, go flying about the giant planet in an area so small where the dust assembled as spheres of insignificance while Jupiter collected the overwhelming dust in the area to eventually assemble the planet Jupiter? What made the planets choose the bands that they circle in at the specific points they chose to assemble as planets? There are a million unanswered questions that I do not find answers to because I have too a poor mentality to understand Newton. It seems the Newtonians never got around realising these questions, which go unanswered.

Let us go on to investigate what it is in the Newtonian high priest custodian mythology that I am not intellectually suitable to understand.

$$F = G \frac{M_1 M_2}{r^2}$$

We understand from Newton that gravity is on the increase because

$$\frac{M_1 \, M}{\quad} \, G$$

This is the part taxed with establishing gravity the force.

r^2 This is the belligerent part enduring the conduct of the above where they are taxed with this facto's destruction. The radius is on the decline. This idea founded physics for centuries in the past and no one ever dared to question the feasibility of the statement.

This meant that the Universe was coming to a conclusion. All that has parted somehow has drawn on the mass that developed and the "gravity" was pulling the cosmos into a "grave". That is what I can conclude from the name but it might be presumptuous of me to conclude Newton's thinking in such a manner. This brought serenity to the cosmos. The cosmos had a purpose, a motive, and a function and even moreover it has a conclusion. At a point the book will close and the curtain will draw. It is where the cosmos will end in a grave and all will conclude into a finality that will end that, which started.

The Sun was pulling the Earth that was pulling the Moon and one day we can rest assure that the Sun will absorb the Earth that would have absorbed by then the Earth that by then would have consumed the Moon in turn. First the inner planets have to give and finally the outer planets have to give to this force of magic being inspired by the mass that Newton saw everything must have because everything had to have mass. The there is a question arising in search of a valid answer…with all the mass available where does the expanding Universe fit into the bigger picture?

Because those who should know these things tell me that do not understand Newton I do not fit into the crows every one I meet is part of because in that group they obviously understands Newton or at least are being part of the crowd that understand Newton. I am seemingly the only person that does not fit into the group that everyone knows are the wise or pretending being wise where that crowd as a group is in total understanding about all of Newton and his forces. To acknowledge my not understanding what there is not to understand I had to go into a personal solo debate with myself commenting about my not understanding to myself on the facts and in order to try and understand what Newton never understood by his own admission, I acted as an appointed narrator and by talking about such and other facts, I bought into the open questions normally never raised in intellectual conversation, such as where is the expanding Universe expanding too? Some then said I went mad because I started talking to myself but I found in my person the only person that understood what I did not understand about what Newton also seemingly never understood. When one does not understand, it is best to go into an argument and clear the matter, but since I am the only one in the world that does not understand Newton I have to argue with myself. Some less sensitive persons think of this method as talking to one-self…but that low I shall never stoop, since I go into a solo debate sharing my intelligence with another of equal intelligence. (That sounds better than saying one is talking to yourself, does it not?)

Newton said everything had to have mass because if everything did not have mass everything did not have gravity and with everything not having gravity everything could not be pulling everything else to the final conclusion that Newton predicted. The most important rule in physics is that although God can be wrong, being wrong is just not possible if you are Newton. This statement I put to all those academics that have on their answering machines the Lord's Prayer recited and verses from the Bible but they would rather face criticism coming to God than allow any criticism about Newton.

The Universe contract by the mass it has with the gravity it produced to find the measurable force that keeps the lot spinning and contracting.

Then came an absolute disaster by the name of Edwin Powell Hubble, the man that will never deserve the Nobel prize because he screwed everything up that Newtonians regard as sacramental and holier than the Bible.

Hubble, Edwin Powell

Hubble, Edwin Powell, 1889–1953, American astronomer, b. Marshfield, Mo. He did research (1914–17) at Yerkes Observatory, and joined (1919) the staff of Mt. Wilson Observatory, Pasadena, California, of which he became director. Building on V. M. Slipher's discovery that galaxies had strong shifts to the red end of their spectra, Hubble used the Cepheids in nearby galaxies to demonstrate that they lie far beyond the Milky Way. Because of an incorrect understanding of the Cepheids, this distance was vastly increased years later.

Hubble's law is the statement in astronomy that galaxies move away from each other, and that the velocity with which they recede is proportional to their distance. It leads to the picture of an expanding Universe and, by extrapolating back in time, to the Big Bang theory.

In 1929 Edwin Hubble first formulated the law that was named after him after he saw an expanding Universe in his telescopes eyepiece. Hubble compared the distances to nearby galaxies to their redshift, found a linear relationship, and interpreted the redshift as caused by the receding velocity. His estimate of the proportionality constant, now known as **Hubble's constant**, was however off by a factor of about 10. Furthermore, if one takes Hubble's original observations and then uses the most accurate distances and velocities currently known, one ends up with a random scatter plot with no discernable relationship between redshift and velocity. Nevertheless the relationship was confirmed by observations after Hubble.

The law can be stated as follows:

$v = H_0 D$ where v is the receding velocity of a galaxy due to the expansion of the Universe (typically measured in km/sec), H_0 is Hubble's constant, and D is the current distance to the galaxy (measured in mega parsec Mpc).

One can derive Hubble's law mathematically if one assumes that the Universe expands (or shrinks) and that the Universe is *homogeneous*, meaning that all points within it are equal.

For most of the second half of the 20th century the value of H_0 was estimated to be between 50 and 90 km/sec/Mpc. The value of the Hubble constant was the topic of a long and rather bitter controversy between Gérard de Vaucouleurs who claimed the value was 100 and Alan Sandage who claimed the value was 50.

Hubble also suggested that the clusters of galaxies are distributed almost uniformly in all directions, although more recent studies show that clusters are combined into huge super clusters of galaxies: at this new level, however, the distribution appears to be even. He was the first to offer observational evidence to support the theory of the expanding Universe, presenting his findings in what is now known as Hubble's law. With Milton Humason, Hubble classified the different types of galaxies including irregular galaxies, three types of spirals and barred spirals, and elliptical galaxies. Included in his writings are A General Study of Diffuse Galactic Nebulas (1926), Extra-Galactic nebulas (1927), Spiral Nebula as a Stellar System (1929), The Realm of the Nebulas (1936), and The Observational Approach to Cosmology (1937).

This put a lion amongst the pigeons and the lion devoured any cat that could enter the pigeon den. The Universe was upside down and this was latterly. What everyone thought was coming was in fact going. What was reducing was expanding and what was to become less became more.

We know Newtonians are not well versed in Horse Trading. If they don't like the rules they bend the rules and if they don't bend the rules they break the rules. They don't take prisoners when it comes to obliging by mathematical principles and they never adhere to science as a coach because now they have the formula $F = G \dfrac{M_1 M_2}{r^2}$ and now they can coach science.

What would you say were the reaction launched by Newtonians? Newtonians corrupted corruption by enlisting the biggest fraud ever concluded. This outmatched Newton's by a billion to one and Newton's effort in corrupting Kepler's work already surpasses nothing ever cooked up by any gangster ever.

The Academics in charge of official policy never held Newton in content. They never reinvestigated Newton. They never flinched an eyelid at Newton's claims. Not once did one raise an eyebrow against Newton's work. The reaction was that if the Universe did not prove Newton correct then when will the Universe start proving Newton correct. If the Universe was expanding when is the contracting coming in place? They did not say wait a minute that makes Newton a fool. No, they said the foolishness of the Universe must immediately end and start adhering to what Newton predicted. The Universe must come right and it must do so fast. The Universe must attend to Newton's statement and get a grip on its act. The Universe must show its critical density immediately so that Newton should not be ashamed of the actions of the Universe. They never said once that with the Universe expanding that means that Newton actually stands to be corrected. No…they said the Universe is wrong. God and his Universe must be wrong because Newton is correct. Never once was the idea of incorrectness associated with Newton. Never did any person thought of Newton in terms of being disproved. The reaction was condemnation to the cosmos. The mistake the cosmos made has to be corrected. There has to be a density mismanagement committed by the cosmos. The cosmos is inadequate, faulty deficient and impotent. The Universe lacks the mass required to substantiate Newton. There was a critical density to be investigated in which Newton would be vindicated. With the Universe in fault at this stage to prove Newton's absolute trustworthiness then it is up to the Universe to prove when the Universe will become trustworthy. When can one trust that the Universe will correct its miss management of mass and start contributing to Newton's correctness? The issue never went the way of questioning Newton's legality in the statements about contraction that Newton made…the disputing of concept legality went the way of the Universe. If the Universe can dare to show expanding in the face of Newtonian contraction then when will we call the Universes bluff? When will we show the Universe is up to now good and when will the Universe face its flaw and correct its incorrect ways? When will the Universe stop the farce it is busy with and begin to act responsible? The name they came up with to enlist the rehabilitation of the cosmic disgrace was the critical density.

If the cosmos was wrong then when will the cosmos become right. When will the cosmos surrender to what is correct and start to abandon its incorrect approach to the physics of the Universe?

They could not dispute Hubble but they could dishonour his dishonouring Newton by not honouring Hubble with the Nobel Prize. They could disgrace the achievements of the person that showed the tenacity to disgrace the cosmos by indicating the cosmos dared to not enlist Newton's rules. By never honouring Hubble, they could disgrace his disgraceful discovery and at the same time they could reinstate the honour of the cosmos by finding the critical density of the material that will supposedly start to rectify the unbelievable behaviour the cosmos dared to show. The will find out when will the cosmos agree to adhere to Newton's gravity law.

The Academics did what Newton did best. They reinvented the truth to reinstate the Newtonian lies. They conceived more corruption to cover Newton's fraud.

They employed the services of the man that filled the centre of the Universe to calculate the mass of all the material in the Universe as to account for any shortfall there might be to bring a final conclusion in the contraction Newton instated.

The person that did the investigation had to be in a place where such a person could see where all the material in the cosmos was. That person had to view the cosmos from a point that gave the person the advantage to see as far south as possible while also being in a perfect position to see as North as the Universe can go as to find the view of all the mass that the Universe holds between point south and north. At the very same point that person had to have the ability to see east and west as far as the directions can go so that the person can find where every morsel of material is located. The person had to be just as far from the top of the Universe as the person was from the bottom of the Universe to find the view where every crumb of material is hiding. After all the persons was given the task to find and calculate all the mass in the Universe and that was first and foremost then to locate every square inch of material that held mass could have the ability to provide the gravity that will bring about the contraction there has to be to validate Newton and vindicate the cosmic disobedience about Newton.

The person must be in the right position from where the person could glance at the entire Universe. That person must be in the centre of the Universe because only from the centre of the Universe can the person view all the mass there is and that might be hiding in the Universe. The person had to think himself as filling the centre of the Universe which from there he had to be in the academic centre of the Universe which will allow him the vantage that all the light he might use to see the material in the Universe will be heading straight to him. If one light photon pass him he would not be able to bring that particle that the light represents into his calculation and we know that some galactica seems so far distant, they are represented by one photon. Therefore, only a person that fills the centre of the Universe will be able to see all the light that is coming by gravity to the centre of the Universe and being in the centre of the Universe will place the person in the centre of all light moving. That is the most critical position the person has to have and that is the most ardent precondition that the person has to achieve to be able to calculate the mass in the entire Universe.

Fortunately at the time there was one person available that met all the requirements and criteria for the task. He was what was perceived to be the centre of intelligence in the Universe and he was in the country that has the idea amongst it inhabitants of forming the centre place in the Universe and he was part of the people that was vain enough to think them to form the centre of the Universe because he was representing the rights of the Man that formulated the centre of all physics in the Universe, which was Newton. And because of such importance Newton was being the man that brought the entire Universe into disrepute because the entire Universe was at fault in its showing of being in disagreement with the laws Newton (where Newton is the man representing what is correct and what is incorrect) instated. So we have the most brilliant mind being Einstein the intellect at the centre of the Universe forming the physics centre of the Universe, which is American physics in a country that think they are in the centre of the Universe, which is the United States of America, with the people that holds their position as the centre of the Universe which is the American's view about their role in the entire Universe, representing the man in the centre of all physics being Newton, where he placed in physics mass to hold all gravity to form that which preserve the centre of the Universe. What a hell of a lot they were…and now every one knows where the centre of the Universe is…it is where Einstein found his critical density theory in the heart of America in the centre of American physics

They have no other person for this masquerade than Albert Einstein in person. Einstein fitted the profile to a T because Einstein had the image and the flair and the reputation that was required. Remember they were not looking for a result. They were trying to establish a time delay. They did not try and find why the Universe was not doing what Newton said the Universe was doing. They were not placing the incorrectness at the door of Newton. They were putting the Universe at fault. They were in search of when Newton will become correct and not why Newton is incorrect. They tried to establish when the renegade Universe will address its mistake and come into line by correcting the mistake God made in sending the Universe in the wrong direction. Placing the Universe at fault and not Newton gave the opportunity not to address Newton and his miscalculation but find the opportunity to present a Universe that is correcting the mistake God made. In that they placed all the apparent blame on the Universe by disclaiming Newton from any resolve.

At the time the Big Bang concept were as much rejected as Christianity was applauded by science. Newton said everything is going to the grave with gravity and that is where everyone wanted the Universe to go, down an eternal grave hence gravity by name (or so I surmise) where everything was going to meet its Maker except the atheists were convinced there was no Maker. The Moon is on its way to clash with the Earth just before the Earth is going to crash into the Sun and God only knows in what the Sun was suppose to crash.

Every one was sharing the Newtonian vision of a contracting Universe where the lot would one day again come together and Creation will end where Creation started some time ago. The Universe has mass that is pulling mass towards one another and we are in the centre of an ever shrinking Universe. That is what the lot of us can see… we are forming the centre of the ever contracting having cosmos where every Newtonian can vividly see with his or her eyes through any telescope that all Newtonians minded scientists are sharing the centre stage of the ever collapsing Universe. The Universe is about to end where all mass contracts into one huge lump of material. But there was someone out to spoil the party and let rats loose in the dancing parlour!

Then along came a man that had a good look at the Universe. He looked at the sky and came to a conclusion the lot was not shrinking but it was expanding. Any one that would look through his eyepiece could clearly see the lot was not shrinking. The lot was growing apart. In some case he said the lot was racing apart. The Universe was growing by miles and not shrinking into nothing.

This unleashed a problem the world had no name for. Everything known to science was at that point devastatingly unknown to science. The world was expanding and not contracting which made the Universe quite wrong. It is impossible to have any vision about Newton being wrong. Newton could never be wrong because Newton was never wrong yet…so if the Universe is out of step with science, then science will correct such an abnormality by finding a way to defraud science and postpone the correcting that the Universe had to comply with since the Universe owed the Master Newton some apology. Did the Universe not know that he whom never can be wrong is in name Isaac Newton! Decisive action was needed. At this point I cannot believe that the most brilliant minds were so naïve and therefore I must suspect deliberate deception. Hubble was far too prominent to blow away and Newton was found wanting. At that point they put the onus of proof not on Newton but turned the focus away from Newton to what the presented as the guilty party. When will the Universe confirm its incorrectness by affirming Newton's obvious correctness? If they had to admit that Newton was wrong, the most intellectual science then had to admit they had nothing to show for all their minds brilliant work.

Science that was defying the likeliness of a living God stood bare and naked for all to see. They put the onus of proof and converting onto the cosmos. When will the cosmos come clean and prove Newton correct. When will the cosmos admit to a mistake and set its crooked ways straight. When will it meet its diverting from Newton and reach a point where the Universe will finally come to comply with what Newton demands. It is the cosmos that is wrong therefore it is time to find out when the cosmos will correct its manner. To deal with such a task they needed a man with a bigger ego than he had an IQ. They needed a person that thought more of his abilities than his ability to grasp any complex situation. They needed a man that was presented as a genius without ever proving his genius. They had a man that filled the centre of the Universe, which then placed the man in a location from where the man could see the entire Universe. They had just such a man. He went by the name of Albert Einstein. For all the genius Einstein had, Einstein failed to see the most simplistic and tiniest mathematical rule. Einstein failed to realise that if there was insufficient mass at the beginning of the expanding Universe, the growth of the Universe will reduce the influence of such mass as a factor because as the radius grows, such growth will restrict the gravity by rendering the mass progressively more incompetent.

If the Universe is expanding as Hubble indicated, the growth of the radius will reduce the influence value of the mass as every second passes. The mass will become more and more wanting for such a task. Yet with this obvious shortsightedness of the genius Einstein, the genius saw him fit enough to calculate and measure something as overwhelming as the Universe. As in the case of Newton, Einstein as an ego driven maniac that saw his abilities fit to measure and master the Universe while his mind was to simple to recognise the most basic principle of mathematics, the principle of relevancies or ratios. What a mathematical genius that turns out to be. While the radius enlarges, at the same proportion does the influence of the mass factor reduce and the mere fact that the radius increase shows that at no stage further into the future can the mass stem the growth of the radius because the radius overpowered the mass factor already. Unless there is new material entering the Universe at a point, which is impossible, the entire concept is fraud.

The idea was never to admit wrongdoing on the part of Newton and Newtonian science but to post pone, delay and divert attention away from the truth. If there was not enough mass to start with, no dark matter can kick in later on and start secondary mass frenzy that at that stage will then be enough to bring about the required mass potential that will turn around the Universe from expanding to contracting. To establish a scenario that would hide all deception they got the man that has a bigger ego than an IQ, they tell the world this man is a genius while the fool does not no the least of mathematical principles because his Master Newton did not no the least of mathematical principles and they got him to measure the Universe. While they did not even have any device (and will never have such a device) through which anyone would be able to see the entire Universe, they set of a scandalous misconception that this Einstein could calculate all the mass in the Universe.

Off course as can be expected, there was not enough mass and there will never be enough mass because there is no such a thing as mass in the entire Universe. When the deceit played out to the full, they, those I refer to as fraudsters being the paternity of physics elaborated on the delusion by trying to find dark matter that is hidden. If the dark matter did not develop enough contraction at this time, there is no chance in the future to develop enough gravity because the factor of what mass supposedly should have is tarnishing and tarnishing as the Universe expands. The bigger the radius becomes the less would the mass effect be.

The community of astrophysics are trying to frame a picture where they set the stage in the way that if the Universe were stretched to a point the mass would not tolerate any more expanding. The mass will get frustrated in some way and show resistance to the increasingly elastic expanding. The gravity constant (I suppose) must prevent any further expanding. How they ever got to such an argument I never could tell. They surmise that outer space is consistently overall filed with nothing and when this nothing is stretched to the limit, the nothing would resist in growing more nothing or become further nothing and the nothing would stop other nothing to enter outer space in the community represented by nothing. If ever there is a faculty ruled by absolute inconsistency and rubbish as the motto of logic it has to be astrophysics.

Every measured kilometre represents nothing. Every mm is one of nothing. We on Earth are 149 X 10^6 kilometres holding nothing away from the Sun. Only they can argue that outer space is nothing with material here and there. If that is the case then which has more nothing between the Sun and Pluto or the Sun and Mercury. The distance between the Sun and Pluto is more, therefore that which outer space is made of is more than in the case of Mercury and the Sun. Therefore Pluto has more nothing between the Sun and the planet than Mercury has between the planet and the Sun. Only astrophysics and all the geniuses guarding the principal of astrophysics can put a calculated value by measure on nothing. In fact Mercury has hundred times less nothing between the planet and the Sun than is the case with Pluto. Since my days at school I was always under the impression that a hundred times the value of outer space being nothing is numerically expressed as (zero = 0 x 100 = 0), but where the genius that is such a prevailing part of astrophysics take the stage we find that Pluto can have 100 times more nothing than the amount or distance measuring nothing than Mercury has. The figure containing nothing that puts Pluto at the edge of the solar system is one hundred times more nothing than what Mercury has where Mercury becomes the first planet in the solar system. That is astrophysics. The brilliant minds of the mathematicians hold no rules apart from what they can calculate. Astrophysics is the only department throughout the Universe where normal rules don't apply since because with mathematics they can bend all laws as they wish…in fact Newton started the trend with his deceit.

Only the guardians of astrophysics policy can know why the undetected dark matter will start producing gravity to change the expanding to contraction. Would the fact that it is detected, change the influence it established? Or is it merely to extend the cover up and allow the deceit to linger until the following generation. There is no mass and any one that says there is mass, let such a fraudster then explain why all the planets irrespective of size or density, spin around the Sun at the same sped as all the others. Let them prove that the Universe acknowledge big and small and let them show how Jupiter can move at the same pace as does Mercury and Pluto while Jupiter is so many times more massive than the other two mentioned. More condemning evidence is yet to come because the astrophysics tricksters did not leave the corrupting of evidence just at that.

The fatherhood of physics never once diverted from acknowledging that Newton's contraction is the prevailing thesis on which the cosmos is built because they accepted that Newton used unlawful arguments and to cover up Newton's fraud which they still use to this day, they then proceeded with further criminality when producing the bluff they established with Einstein just to fool everyone in the normal public. Without ever recalling Newton's contraction theory that is obviously not working or admitting doubt about Newton's testimony to the effect, physics accepted the Big Bang Theory. The Big Bang theory opposes what ever Newton might have implied. The physics paternity however finds it wise to still advocate Newton while admitting to the Big Bang event. Newton said the lot is contracting. Go on and marry that with the Big Bang that says everything is expanding. You can't promote both except is you can define why we would see the two merge.

The Universe comes from a point the size of a Neutron. That makes the radius parting the Universe infinitely small. It just about removes the radius as a factor. At the very same implication it takes the pulling of the mass (if there are pulling forces converted by mass) to a level it will never again have. As soon as the distance between the objects holding mass started to grow, the power and influence of the mass factor started to diminish in the same ratio. If the mass were incapable of contracting the Universe then, it will forever remain contracting the Universe. Then you may ask what the story is? Read on and you will learn how far Mainstream Physics stray from the truth and how big a cover up the paternity is protecting.

According to science the Universe started with singularity. Quoted directly from the Oxford dictionary of Astronomy the following:

The definition of singularity is as follows:

Singularity: a mathematical point at which certain physical quantities reach infinite values for example, according to the general relativity the curvature of space-time becomes infinite in a black hole. In the Big Bang theory the Universe was born from singularity in which the density and temperature of matter were infinite. The average daily temperature was "$10^{\alpha\beta}$ to 10^{34} K".

Then the second "day" the daily average temperature came down to 10^{34}K and 10^4K. That is fine, but if the temperature was in Kelvin, then what was 0^0K. In order to make sense of the scale used there must be a minimum to secure a maximum otherwise the maximum can just as well be the minimum and is only advocated to impress humans applying earthly standards.

By using a scale as $10^{\alpha\beta}$ to 10^{34} K, it places the lower temperature at a modern 0^0K to make sense of standards. If that was the temperature the standards were lowered, compromising something to gain something, because something had to grow larger for heat to reduce. We know space grew larger bringing heat down to reduce.

Being the onlooker the viewer has to maintain one position. From that position some particles would be circling a centre point, as the particles would be coming towards the onlooker. The other matter would be circling the centre point while rushing away from the onlooker.
At the very end the single dimension may come into the dynamics but where the single dimension comes in the factor of zero is removed.
If there is space, there is a flow of light and a flow of light has to produce lines in relation to angles forming space between them. Something must be present to confirm space because there is an absolute difference between being in space and no space to be found. If there was a line that formed nothing that one line that forms nothing would completely destroy the other lines' chances of ever forming a triangle, let alone having all lines and they then have a total being zero. As shown in the example no line can form zero and therefore no mathematical equation as far as it extends to cosmology can ever bring about zero as a number. While there is space present there has to be three dimensions relating to each other by time and in three dimensions there has to be three lines in relevancy to each other by angles formed holding space in (at least) six opposing sides. Removing one line must bring about a flat Universe and that then will constitute nothing.

Cosmology is about light flowing by means of lines indicating space obeying the rules enforced by time in motion and light flowing dictates crossing space and across space light is using lines. The book: ***An open letter Announcing Gravity's Recipe*** is dealing with the subject finding singularity by removing the concept of nothing from outer space. By diminishing nothing one uncover singularity and the effort brings in a new perspective not yet introduced.

For your benefit I will shortly give a summary by which I hope to interest you in reading the manuscript: Compressing space produces heat Releasing heat will bring expansion of space bringing about space. We call such a release of heat an explosion. In other words heat translates to space and space concentrates back to heat. The one is a product of the other where space forms expanded heat.

They are quick to show the time that was applying at the time being some thousandth of a second or the heat that was present being numbers we have no name for. The other side of the story they ignore. They ignore the other side of the story because in that respect it puts their promoting of Newton down to madness. If you reduce the radius applying at the present back to what it was at the time of the initiating of the Big Bang, you must also increase the influence gravity and mass had at that moment by the same number you are decreasing the radius. That is pure mathematics and the most basic physics of all concepts.

The shrinking radius will increase the effectiveness of the influence of the gravity that the mass can produce by the margin of the shrinking of the radius. If the Radius was infinite at that point, then that means the gravity was eternal. With the entire Universe being as big as a Neutron, the Universe was the size of an atom. If the Universe were the size of an atom and the mass within that Universal atom could not prevent the Universe exploding into immeasurable atoms, then it would not be able to retract all the atoms into one unit again. If there was not enough mass to start the contraction, there can be no contraction of mass that is producing the gravity at this stage. If the gravity is of such a nature that it allows a continuous growth of the radius, then the radius firstly cannot be zero as Newton suggested and the extending of the radius proves there is no contraction in the way Newton had everyone to believe. If Newton's mass contracting mass is true, then on the other hand it must have resulted in an implosion as that which can never repeat again. With Newton's formula of $F = G \dfrac{M_1 M_2}{r^2}$ forming gravity, then the Big Bang is just not possible because from that formula the Big Crunch must respond.

The critical density is the biggest and most elaborate case of fraud ever perpetuated by any group of people or persons. They did not say Newton was incorrect. They said they are on a mission to see when the Universe will correct itself and prove Newton correct. Newton remained correct while they gave the Universe the chance to mend its ways. If there was not enough mass to start with, why will there be enough mass in the future? If the Universe is growing the mass in the Universe is not pulling the Universe into contraction. Only an over bearing egomaniac with an ego outweighing his common sense by a margin of many times to one, will take the opportunity to calculate the mass in the Universe. Step outside tonight and see what there is to see outside in terms of mass. They put the exercise to a formula in order to calculate what no man can even presume. If you wish you can read the following example I offer to prove what elaborate criminal scheme the entire venture called the critical density is and to what fraud it amounts. Here follows a part taken from a web site that tries to prove fraud by elaborating on the fraud by implying more corrupt fraud. You are welcome to read the following frauds if you wish, but to my mind it is not worth the paper I allocated to its print. Nevertheless, here it is:

A simple, non-relativistic, derivation of the critical density can be performed as follows

> *Recall that kinetic energy of a body of mass m moving with velocity v is $\frac{1}{2}mv^2$, and that gravitational potential energy of a body of mass, m, at a distance, r, from a second mass, M, in a radial field is GMm/r.*

In order to find the escape velocity of a spacecraft from a planet we can work out that the spacecraft will escape if it has enough kinetic energy that gravity will be unable to slow it to a standstill. Using the two expressions above we can write

> $\frac{1}{2}mv^2 >= GMm/r$

As the kinetic energy of the craft is converted to gravitational potential energy while it escapes. If the two are equal it would stop at a very great distance (infinity), if the kinetic energy were bigger, then it still has velocity when at infinity.
We can do the same calculation to see if a galaxy is able to escape from the attraction of all other galaxies or whether it will be stopped, turned around, and caused to fall back. If this latter situation is the case then the future of the Universe is finite. The galaxies, which are at present travelling away from each other, will stop, turn around, and head back towards some common centre where everything will be compressed, perhaps out of existence, 'The Big Crunch'. If, however, there is not enough material in the Universe to stop the galaxies then perhaps the Universe will go on forever.
<u>*Hubble's law*</u> *shows that the velocity of a galaxy is proportional to distance,*

> $V = H_0 r$

Where H_0 is about 5×10^4 m s^{-1} Mpc^{-1}, 80 km s^{-1} Mpc^{-1} or between 1.587×10^{-18} and 1.62×10^{-18} s^{-1} depending upon whose figure you accept!

So the kinetic energy of a galaxy can be written

$$\tfrac{1}{2}m(H_0 r)^2$$

The mass of all the material inside a sphere of radius r is given by

$$M = \tfrac{4}{3}\pi r^3 \, Rho$$

Where Pi is the average density of the Universe.
Substituting these two expressions for the first equation above gives

$$\tfrac{1}{2}m(H_0 r)^2 = Gm(\tfrac{4}{3}\pi r^3 \, Rho)/r$$

This can be simplified to

$$m(H_0 r)^2 = Gm(\tfrac{8}{3}\pi r^2 \, Rho)$$

and rearranged to

$$Rho = 3H_0^2 / 8\pi G$$

The above my friends are hogwash. It is what represents the utter most thoughtless and mindless arrogant mismanagement of all facts. If the mass was not up to the grade to pull the Universe into contracting, it is incorrect to presume the formula will apply. By finding that the Universe is growing in size, one should question the authenticity of the formula $F = G \dfrac{M_1 M_2}{r^2}$. This formula can only apply to a shrinking or contracting Universe. To try and integrate the formula as pronounceable in the case of the expanding Universe, is committing fraud.

$F = G \dfrac{M_1 M_2}{r^2}$ In the using of the formula such as Newton recommends the radius determine the influence of the mass.

$\dfrac{M_1 \, M}{r^2} G$ In mathematics to put any number in relation to the value of another number is to agree that the top part f the formula will find a suitable and applying measure by the measure and the size of the bottom part.

$\dfrac{M_1 \, M \, G}{r^2}$ If the bottom part is bigger than the lower part, the bottom part will be a fraction of the top part. If the top part is smaller than the top part, the bottom part will be a fraction of the top part. The bottom part and the top part will always and without reservation have one of the two smaller or then a fraction of the other, except in cases where the two are equal. In such cases there will be a unifying number coming about as one. Putting two factors into a ratio or a relevancy, places the one value in charge of the other value while the other value is dependent on the first value. Not one of the two then can be dismissed as nothing or having no value because the other value holds the measure of the first value.

$$\frac{M_1 M_2}{r^2} G$$

Appreciating that the top is sizably more than the bottom value will put the top part in a numerical superiority and the answer will not be in any fraction, which includes a mathematical fraction of less than one.

Having the radius growing, means the top part of the formula that represents the mass is shrinking in influence.

Edwin Hubble proved that $F = G \frac{M_1 M_2}{r^2}$ is not applying because the cosmos is not shrinking. To further use an already falsified formula is cooking the books to a point where the books they cook starts to burn.

By enlisting $F = G \frac{M_1 M_2}{r^2}$ you are placing the mass and the mass as a multiplied unit with the gravitational constant in relation to the radius value that applies. That means if the radius does not shrink but it grows, the mass does not reduce the radius but it promotes the radius into growing. By incorporating the incorrectness into what one then tries to establish, is committing fraud by falsifying the facts derived from fraud even further. There simply is no formula as $F = G \frac{M_1 M_2}{r^2}$ so then there can be no integrating the formula to whatever form $\frac{1}{2}mv^2 \geq GMm/r$ is required. The cosmos proved that $F = G \frac{M_1 M_2}{r^2}$ is a farce and Newton should be investigated. The Newtonians elaborated on Newton's fraud by investigating the Universe of fraud. There just is no mass found in the entirety of the Universe.

Read my book and you will find that the Universe is not space but the Universe is time. There is no space in the Universe, for through expanding the Universe placed time in space. As time develops it changes density into space. The Big bang is about substituting the development of time into forming space. Time is space because space came about as time progressed. You cannot calculate the Universe in terms of $4/3(^4/_3 \Pi r^3 / 3)$.

Guess how surprising was the result that Einstein produced. Einstein shockingly and most surprisingly confirmed a constant defiant Universe. Einstein produced results confirming not only Hubble but also that Newton will have to wait for longer to be proven correct. A new plan was devised to vindicate Newton and his correctness. If the cosmos did not play ball then the Brainy Bunch will cook up a real witch's brew that not even the cosmos can dare to defy.

A search went out to find dark matter. Dark matter will result in correcting Newton if nothing else can. If one can't see dark matter then no one can prove dark matter and while no one can prove dark matter, then it also becomes true that disproving dark matter is impossible. It worked with the instigating of the original fraud that Newton invented. Not ever proving Newton and his mass that creates gravity by attraction between the mass factors also brought about that not one person ever thought of disproving the mass deception. With the mass being impossible to prove, it also makes the mass of dark matter impossible to disprove. Then on what grounds do I lay my charges of the biggest hoax ever created to defraud tax payers out of their livelihood by stealing tax money and diverting the proceeding of the tax collector to establish a criminal inspired corruption never yet before experienced by man?

Kepler introduced space a^3 growing by the time T^2k that allows the space a^3 to move. That is time T^2k allowing space a^3 to be. For space a^3 to move about it will use the motion that time provide in the rotation of a^3 as well as the displace net of a^3. In space there is no mass because in space there is no proof of mass.

$F = G \dfrac{M_1 M_2}{r^2}$ If there was not enough mass to redirect the expanding Universe into contraction, then there can never be enough mass to redirect the flow of the cosmos. What will produce more mass if there is not sufficient mass at this point? Why would the dark matter be gravity dormant and what will enlighten the dark matter into activating gravity. If it is mass and the mass is suppose to establish gravity, the mass then has to establish gravity whether the mass is dark or light. What will make matter, that is visible more active in creating gravity than matter not seen by man? Why will dark matter come into gravity later on, as it at present holds the gravity dormant? Why will the dark matter at present not form sufficient gravity by mass but will later on become energised and jump start the cosmos into contracting. If the mass is there, the mass should charge gravity if mass does charge gravity. Why would dark matter play hide and seek and hide their potential gravitational abilities for a later date?

$F = G \dfrac{M_1 M_2}{r^2}$ The suggested formula indicates that it is mass that produces the gravity by which the force pulls structures and thereby it reduces the distance between the objects. The mass is proportional charging gravity. The mass is responsible for the force of gravity by the measure that the mass has. A lot of mass will charge a lot of gravity and a little mass will charge less gravity. The mass is there in relation to the establishing of the force and forcefulness of the force proportionate to the amount of gravity applying by the production in relation to the mass. If the mass is there, the gravity is there. If the gravity is absent the mass is not present. If the mass cannot reduce the gravity, what then is the point of trying to establish when the gravity will come about, since the mass is obvious lacking as a quantifiable amount to charge enough gravity to contract the expanding Universe? If the gravity is not available, the mass is not sufficient and therefore, if the mass is not sufficient in the first place how the hell will the mass become sufficient later on. The mass is not an elastic band that allows expanding up to a point where after it will reduce as the elastic energises.

$F = G \dfrac{M_1 M_2}{r^2}$ The formula does not compensate for any such suggestion. The notion alone is one huge farce inspired by criminal minds to cover up flaws in the top banner bearers of the science world of physics. The whole issue goes against the grain of physics and mostly against all mathematical principles. If they were incompetent novices trying to address a school project that has a mathematical inscribed theme that goes way past their abilities and they are in far too deep water to tread, then yes I can find some degree of honest miscalculating and incorrect judgement of mathematical founded facts. They hold the top notch of all mathematical insight. They father the laws of mathematical principles. Therefore, in that light they woefully and deliberately avoided the truth and rendered their honesty to distrust. The fact that they skipped inspecting Newton and created a farce to mislead any thought the public may have about the mistake, shows their deliberate action to avoid the truth about the matter and carry on in criminal intent. I challenge any one bearing the title of professor in physics to prove otherwise. Let any academic in physics show where I overstep my boundaries when I charge them of deliberate acting with the motive to corrupt the truth and avoid the true impact of Hubble's findings.

$\dfrac{M_1 \ M}{r^2} G$ If the mass was prominent from the start in relation to the distance separating the two objects, then the contracting of the object will reduce the radius parting the objects to become a factor of one that has no influence on the formula.

$$\frac{M_1 M_2 G}{r^2}$$

If the mass grows and the mass keeps growing, the mass will reduce the input of the mass on the gravity. The further the mass goes apart from the mass, it shares gravity with the less the influence will be of the mass exerting gravity. If the gravity was insufficient from the beginning then it will remain insufficient as the expanding progresses. The gravity does not grow as the radius increases. It is the other way around and as mathematicians of notoriety they knew it. That makes their criminal intent even more appalling than what it would have been if the actions of criminality came from the midst of others being persons in a less trustworthy disposition. However, Kepler gives all the answers they tried to criminally cover up.

This we dealt with here is only the tip of the iceberg. When one starts to dissect the inner workings of the Critical Density Theory and find their way of thinking about how the operation will carry out its purpose then it is clear how much the theory proves how the madness can grip the idiot's mind and unbalanced thoughts go out of control. If the mass stops the molecules in the process leading to turn about of direction of Universal flow, it will stop the less massive first and this will have the more massive molecules plough into the slowing as well as stopping smaller molecules.

The more massive molecules would look like trains colliding with cars and bicycles and removing the bicycles from the face of the Universe. At that point a second Big Bang will come about that totally destroys everything that was not destroyed by the first Big Bang.

This is what happens when a lot of criminals can go about, as they are allowed to work unchecked. As they are blindly trusted by all those unsuspecting, honest persons in the general public holding no criminal intentions they never think that these Academics are never being controlled, which means the Academics can steal billions from the coffers of the tax paying public. That they then also do in finding funding for the most bizarre research that is so fruitless it falls in the category of the insane. One such a venture that comes to mind is the monumental fraud of research into alien life coming from somewhere n the outer space. This is only one of so many that they use purposely on many levels. In other instances the fraudsters apply the same modus operandi by inventing some bizarre mathematical formula that must supposedly serve some purpose and then they unleash this fraud onto the public at large albeit to try and cover up Newton's fraud or to defraud the public in other ways.

The latest book that is supposedly written by Steven Hawking is a perfect example of the degree of mad corrupt and idiotic methods they employ to commit such fraud. They invent a formula. The formula in theory is meaningless because the practicality of the formula can never be tested and must therefore be accepted. The reasoning behind such a formula is as unrealistic as the critical density crime venture, which I just explained and has as little practical function as does the critical density theory. As I show they go about to bend and cheat mathematical principle and logic at will to corrupt the truth as to find proof to their meaningless venture. By applying even the least of logic they prove callow. But since they deal with honourable people they take the public to the cleaners

The people forming the public are too scared to ask questions about the sanity behind the reasoning because the people feel incompetent in reading the mathematics. That uncertainty these evil gangsters use most shrewdly to their financial benefit. Behind this fear of feeling incompetent that the general public has is a rational of not asking questions because they do not wish to feel stupid and this the academics in astrophysics exploit for their criminality while the gangsters continue with their evil exploits.

The criminals calling themselves astrophysics academics then realising this flaw that the public have and then go on to commit gang rape on the unsuspecting public, which are also those they see as witless beings they can manipulate and control at will. I challenge those fraudsters to sell me that lame idea that the latest Hawking book explores and see how I plough their fields to uncover pure bullshit in front of a TV audience.

I call them criminals because notwithstanding the position any person has, when such a person pretend to be what the person is not, or to have what the person does not have, or to launch a project that the person intending to benefit from the results and the research of such a project but in launching the project the person intentionally has too spread untruths and corrupt lies will filly, then this makes that that person intends criminality. By telling not the truth one tells a lie. By spreading lies one commit fraud and deception. There is no small medium and large way of deceiving a person because untruths are lies. In doing so they intentionally divert to truth to sustain a lie and such actions is criminal. Even not admitting to the fraud is a way f committing intentional diverting of the truth, which is the sustaining of criminal behaviour. Any behaviour not being in line with the truth is committing criminal activity and that is the narrow and the broad of the lot.

Brainwashing and Mind Control is an Everyday Practice in Physics

If you are a student in physics then you should read the following information. One could think of another name for physics and that would be Newton's mythology. It is about the subject of gravity and is most important. The "Newton's mythology" comes from the fact that students have to learn what never was proven. Do you realise that it is an accepted practise that all students that are studying physics on all levels are subjected to the most intense brainwashing and thought control found any where on Earth? This must be some sort of a joke you may think but thinking that way in disbelief is just what those practising the mind control wish you to think!

I came upon a mistake concerning physics.

This mistake is about the cosmic phenomena called gravity. Detecting the mistake is simple because it is uncomplicated to understand. Academics in Science say that a feather will fall with the same speed as what a large rock would fall.

That is according to Galileo and that is accepted as a principle in physics. For the first time ever since the time Newton introduced gravity I seem to be the person that questions this interpretation.

Has anyone ever explained how the idea of a feather falling as fast as a hammer fits into the idea that mass pulls mass and how the falling by the gravity forms power that is exerted by mass as a hammer has much more mass than even what a large feather has. How on Earth do these two concepts of a feather falling equal to a hammer proved by Galileo fit into this interpretation they use of mass causing objects to fall?

How can a large mass pull as equal as a small mass pulls to travel equal at the same speed over the same distance and still be driven by the power of mass creating gravity. Have you given this idea a good thought? By me scrutinising this concept I disagree and by me disagreeing I am silenced by those in power. When any person disagrees with any academic in any lecture hall about mass not forming a picture as being the factor responsible for pulling gravity and you come to a conclusion that you doubt the mass part that they bring into the picture as being responsible for establishing gravity the academics wipe you from the table with a swipe because then they contemplate that you are so stupid you fail to see facts and you are to stupid too understand physics. They even in some cases go on to say physics is not for stupid people!

I have been at odds with academics for years and only because of the superior positions they hold in office are they able to bully me into silence. Academically I am not from their league and neither am I from their ranks and with me not being part of their ranks they are of the opinion that it disqualifies me to have any opinion. Being what they are gives them the rite that they may regard or disregard all opinions when they do not fancy the opinion. They may silence whatever I may say notwithstanding my correctness and validity. Absolute power corrupts and they are the living example of that.

Due to the important positions Academic hold in the huge academic institutions such castles of power gives them free sanctuary from where they can hide their criminal ploy of deceit. They do not need to explain anything but to themselves amongst themselves and their deeds go totally unchecked. That makes them be the untouchable and unapproachable powerful from where they rule with absolute authority. This unquestionable authority gives them the locations erect a cover and give them the opportunity to hide behind that wall of absolute superiority and suppress little persons such as I into silence and submission notwithstanding...Whatever I have to say can never go past their scrutiny and can never pass their sanctions.

What Newton saw as gravity can't withstand even the slightest test of proof and I showed that it is not possible to use Newton's formula as Newton suggested it applies to mathematically calculate gravity. I come back to this issue later on. I have tested Newton's thinking and the book I offer to you for investigation serves as the testimony to all the testing I did on Newton. This any body who can see, will see when reading this book, I tested Newton from all the angles to see if he possibly could be correct but found his thinking wanting every time. The truth about Sir Isaac Newton's concepts I came to conclude, was that the reality is that it is not in any way overstated to declare that Newton conspired to defraud science and moreover that he committed blatant mathematical corruption in trying to prove the concept he had about what he thought forms gravity. There is no backing for Newton's ideas and even the ideas which are in use are not in the form that Newton said it applies where physics in daily use serves as the best discredit to Newton bringing no proof about any of the claims that Newton made on matters concerning science in cosmic gravity.

I show that every thought Newton introduced that later proved useful and was correct, was what he stole from another far better cosmologist called Johannes Kepler. Not one of his laws are directly relating to any concept Newton ever introduced at any stage but is the result of academic theft he committed against a much larger figure that preceded him by almost a century. But he stole, he lied and he raped the work of a predecessor in order to defraud the world of science in his time. Newton brought no original input into science except that he gave a concept the name "gravity" and even that is inappropriate. Newton made suggestions that break every mathematical principle he could think of. That, Newton did in his attempt to win over the prevailing academic thinking of the day in his time as to lay some sort of groundwork to form backing for his ideas on physics and to attempt to explain gravity or what he thought gravity is. If this is shocking and sounds outrageous, then a lot more shocking detail awaits the reader in this book.

Newton's claims about the principles he declared as being responsible for guiding physics carry no proof and after I realised that, I was able to start forming another line of thought on gravity. After formulating my concept about how gravity was truly formed, I had to introduce my ideas to academics in physics. In my quest to find the method how gravity formed I used the four phenomena and the principles of these phenomena as well as determining in which way each phenomenon applied. Then I placed each one in the way that were known how they work and then implicated that specific formula's function mathematically in forming gravity in the cosmos. This was no easy task but I did it and by formula shows that my argument is logic and the mathematics prove that it works well.

They proclaim to understand what flows out from what they understand but such concepts become meaningless because of many inconsistencies. To name but one such an example is the explanation they put forward in the Tunguska event. To claim that a mini Black Hole went through the Earth is demeaning just going on the basis that they claim there can be such a thing as a mini Black Hole. Such statements are beyond the ridiculous and to achieve some degree of believability from the public they create scenarios, which use arguments that are entangled with deception, such as what is obvious in the case I mentioned. What they declare as unwavering facts can't even be supported in the least form when tested. Even the least degree of verification of correctness is absent and Newton lacks all evidence of authentication in any investigation of even the simplest terms. It is as if they never read with interest that which they explain and they never scrutinise that which they advocate. They give values that are senseless and make that which they say meaningless. In all this they use billions of tax dollars to prove what they have no idea of. They try to commit matter to fusion while they have no idea why matter would fuse at all!

Now I am taking my case to the members of the public so that the truth must be brought into the open. I have had the tour they give and then more came my way. I never got around swallowing the mass creating gravity part where science is of the opinion that mass pulls as gravity is… Academics condemned my work and therefore me and for six years where I could not get a publisher to come around and bother to read my work let alone seriously proposing a publishing contract. I had to finally go private with the publishing as all doors shut in my face as soon as the academics read the content of my work because from the nature of my work I take Mainstream science head on and am confrontational on most aspects of astronomy. There does not seem to be any publisher that wants to go head bashing with the establishment of science on official science principles, which I have to do to convey my message in a no uncertain language. If you also have doubts about the academic's indisputable correctness please read on and confront either them or me on everything you read here.

After reading this letter you will have to take sides because you will know the truth.

Then you either become partner in the crime as you cover the truth up or you will be part of the truth and help me confront them to acknowledge the truth.

Should you think this page is some sort of a prank then answer the following simple question to yourself in utter honesty? If there is a Big Bang with everything moving apart, how does that support Newton's contraction? Tests results received after the Moon landing show the Moon and Earth are moving apart! Yet students learn about mass pulling mass and that puling by mass forces togetherness by contraction.

The entirety of physics rests on this one formula $F = G \frac{M_1 M_2}{r^2}$. The questions concerning that which you are studying and that touches every aspect you are academically concerned with, is that if everything is moving apart, how does that support Newton's idea that everything is coming together…and please don't let them fool you with Einstein's Critical Density idea! If there was mass seen or unseen in the Universe and mass generated gravity and gravity does the pulling then why is the mass not at this moment doing the pulling. What is all that mass of so many supposed stars doing at present while waiting to get to work where it will only later, much later form a force of gravity that then will bring about this pulling of the Universe? What makes the mass slumber in darkness to one day form a pulling force? What has the "darkness" or the fact that we don't see the mass got to do with the idea that the mass at present is not forming gravity that is forming a pulling force? You are taught that gravity pulls objects to the centre and obviously gravity then has to ultimately pull everything to the centre of the Universe. That is what the Critical density research that Einstein initiated wishes to establish. The idea is that $F = G \frac{M_1 M_2}{r^2}$ makes the mass create a force that will destroy the radius and ensure everything is going to come together eventually at one point where the radius then will be no more. If that is the case, then where is that point? If everything is destroying the radius, then it must end at one specific point.

In the classes you attend a physics lecture, has any one confirmed a location where one might find the centre of the Universe to confirm the ultimate destination of $F = G \frac{M_1 M_2}{r^2}$? If you wish to apply a Gravitational constant as a calculated factor then it is apparent that one must know to where such gravity is pulling since it then is the gravity that is predominantly keeping everything apart. Then the gravitational constant is what is resisting the collapse of the Universe. If there is a force, then where is the force taking the pulling…if it is a gravitational constant applying through out outer space then where is it having a centre base?

I wrote a book in which I found a means to define gravity. This feat I accomplish and by my effort it was done this for the first time ever. For the first time ever runs further back than since the time Newton introduced gravity. Before I achieved that discovery, I firstly had to find the centre of the Universe because it is there that I could locate gravity. I now am able to show how gravity forms because I have detected the centre of the Universe. But by my effort in finding the location I disrupted everything Academics in physics hold holy and for that I am most unwanted in the presence of the Academics charged with guarding the ethics of physics. In short, I clash head on with Newtonian principles. During my research I discovered abnormalities and inconsistencies about mistakes the Arch fathers in physics must be aware of but is hiding with all their considerable influence. I will come to some of the inconsistencies later on but the discovery also introduces a much better vision about many new aspects that I discovered but in reality was never before realised in science. But these discoveries discard and blacken the Newton reputation totally and therefore the academics dispute my work totally in order to save their Newtonian reputation. The road I took in my search for truth concerning physics was never smooth and the resistance I came across coming from the academic sector was almost unbearable. Academics guarding physics will never allow an outsider to enter their domain and dislodge Newton from being god that is without the intruder paying a heavy price for trying to do so and in this matter I was and still I am seen as being in the role reserved for such an intruder. It is not about my work they detest but it is my rebutting of Newtonian thoughts that they reject! However such intruding allowed me to find so much that I was not supposed to find, which was reserved to all that studied physics the insider information that is available but because of that it was only allotted to the most inner circle and the insider information that I share with you. By finding the centre of the Universe enabled me to find a point the Universe is controlled from. In achieving the locating of the centre of the Universe I had to step on some very important toes, which made me very unpopular. With my unpopularity rating this high as it does, I never qualified for help and those that would help found my ideas intolerable whereby I only found rejection instead of help as I tagged along. Because of this insider rejection I had to resort to private publishing because from the nature of my work I take Mainstream science head on and am confrontational on most aspects of astronomy. This is the only road to go if one wishes to lay axe to the root of the insider corruption they are guilty of. In that sense there does not seem to be any publisher that wants to go head bashing with the Physics Custodian establishment of science on official science principles, which I have to do to convey my message in no uncertain language. I argue that if it is the correct practise to use $F = G \dfrac{M_1 M_2}{r^2}$ to calculate gravity then the radius holding the gravitational constant must lead one to the centre of the Universe. With nobody willing to publish my work as I confront science dogma and principles all the way, I had to go the road alone and fight the battle by my private effort.

This is only one of many points that I make on this one issue and there are so many other issues one may think of those in terms of counting in numbers in many hundreds or even in thousands. If the Sun for instance has mass that is apart from the Earth and the Earth also has mass and there is a gravitational constant in between the Sun's mass and the Earth's mass we have the radius in that location. It then must be the gravitational constant that fills the space that the radius holds. It is rather obvious that while the radius is filling the vacant space between the Sun and the Earth it is the only place left where the gravitational constant can hide. To find the centre of the Universe I had only to find the gravitational constant that holds the centre. Through my venture I discovered one person that knows what gravity is! Newtonians went and filled that space reserved for the gravitational constant having a measured value with nothing! How can nothing have a value of 6.67×10^{-11} while also being filled with nothing as it is nothing filling the nothing of outer space?

If you think scientists know what gravity is do not be duped that easily because no one in science remotely knows what gravity is…not even Newton knew what gravity is except Kepler… and because of what Kepler introduced I now know I can prove what gravity is. Gravity is precisely what Kepler said gravity is and only Kepler new where to find the centre of the Universe because only Kepler knew what gravity is all about.

Try to get an answer from any academic person in physics about where the centre of the Universe is, is like trying to touch the moon.

I dispute Newton and so do all students learning physics because Newton's arguments are an onslaught on human intellect. Think of the resentment that students have towards Newton under normal conditions when they have to cope with understanding the Newton principles Mainstream science says are applying and how that confusion of what is possible and what Newton suggests is possible clashes with their intellect which makes them feel stupid. Students hate Newton because they don't understand Newton and for that they are accused of not having the intellectual capacity to follow Newton. Every student from the past going into the present and even including those forming a future generation of students will purchase a book that is showing that Newton's legitimacy is cracking up when exposed to some vivid scrutiny. This fact gives the book a selling potential like no other book in the past could do. Yet I am unable to find a publisher because publishers need academics to assure the correctness of the information in the book and academics would cover up Newton's errors at all costs. There is a total denial about the truth and as long as those academics have the opportunity to brainwash students in to accepting Newton's unproven and ridiculous concepts and as long as their misconduct of mind control by fact manipulation goes unchallenged, the process will go from generation to generation as it has been going for the past three hundred and fifty years.

In short I will now explain what I explain throughout the book you are about to receive and which is named *Newton's Fraud* or whatever it will be named as. The Newtonian formula $F = G \dfrac{M_1 M_2}{r^2}$ is the formula used by science to explain and define gravity. It says the that the ($M_1 \times M_2$) mass of one object pulls the mass of another object and this process in relation with a gravitational constant (**G**) (a supposed force keeping the Universe attached) and the pulling subsequently destroys the radius (r^2) being between the objects. That says that objects **ALWAYS MOVE CLOSER *BY FORCE*** in relation to **MASS**. Newton submitted the suggestion that objects fall as MASS provides the force that will cause the falling by the inducing of a force he named gravity which he subsequently only proposed was the acting suppositious force. I disprove this formula in so many ways in this book and I show that this formula and the ideas Newton introduced just don't stand up to even the smallest tests. Then, if Newton's idea on gravity has validity and mass is responsible for objects falling, then all objects that are in a process of falling must be subject to mass and in that idea rests differentiation and discrimination in size and compactness producing speed variations. If any and all falling is subject to the variation mass introduces and the influences coming about is the result of mass interfering in the gravity force being generated, this then must bring different speeds to cause substantial variation in the falling of different objects holding different mass factors. There can't be conformity in the falling of all objects while such falling is the result of the discrepancy that mass has to inflict due to variations that result in mass differentiations. This is a vital issue that science eludes and has all clever ways to avoid direct questioning. This part science just run around and never addresses and avoids confronting the issue. This avoidance of confronting the issue whish will disprove the validity of Newton is done with such cunning as you will not believe. The fact that objects fall due to conformity in the falling, science accepts but portrays a picture of deceit that mass brings falling distinction and therefore equal falling doesn't happen, while they at the same time admit to Galileo's presentation that falling of all objects are equal in tempo, irrespective of size or any form of differentiation. While they promote the obscurity that Newton and Galileo is in harmony the truth about their deceit is that the two can never have the same issues. That I prove is a fact and also I show how big a part this is in the overall covering up of Newton's initial fraud.

I have written several books in which I challenge the thought process of Mainstream physics and especially Sir Isaac Newton's arguments about physics. I am of the opinion that even though everyone thinks of Sir Isaac Newton as the genius who established every aspect that is used in modern physics today, but in spite of every other person hailing Newton, I remain of the opinion that the man did not have a foggy clue about any of the principles driving the concept that he named as gravity, or what brought about gravity according to his explaining of what forms gravity. I am able to explain gravity but it doesn't even vaguely resemble Newton's version of gravity. I can explain gravity by proving my explaining with the use of simple mathematics. I use Johannes Kepler's formula to back up my statements. By using Johannes Kepler's formula I found a way to prove there are four phenomena found in the cosmos. There are the four phenomena applying in tandem that together forms gravity. They are: The Titius Bode law; The Roche Limit; The Lagrangian Point System and; The Coanda effect. As the phenomena don't support Newton's vision on cosmology, the phenomena has no support amongst Mainstream science although they did apply it with a positive results in locating the missing planets at the time of their discovery. When they located unknown and undetected planets in the past, the existing of the phenomena was never disputed but when the argument of proving them comes to mind, then they are dismissed as some coincidental abnormality occurring. But since it holds no similarity to Newton's view on science, Mainstream science rather disclaimed the validity of the phenomena than they would find fault with Newton's ideas. In the mind of science the cosmos can be wrong and God can be wrong but Newton can never be wrong. In using the four correct principles correctly, which I back up with the correct mathematical interpretation thereof in support of the function that each phenomena has in forming gravity, I did a far better job than what Sir Isaac Newton did and what I achieved is of a far more acceptable level as well as being mathematically far more correct than what Sir Isaac Newton did achieve with his guessing about issues he couldn't explain. To be successful in my quest to find an explanation for gravity, I had to redirect all my concepts I previously had and also alter all the otherwise normally accepted thinking on physics. I had to find the phenomena and I had to dissect the function of each phenomenon as well as mathematically valuate the phenomena. In this process I realised that to come to realise what gravity is, I had to realise that gravity is not what Newton foresaw. Planets have no mass and neither has the Sun got mass except the mass Newtonians wish to credit planets with.

Bigger planets don't move faster because they have more mass and smaller planets are not further from the Sun because they have lesser mass. All planets big and small spin at the same speed around the Sun and in relation to the Sun and all planets are scattered going around the Sun while being big and small where all sizes are well mixed. This is because planets have no mass except in the imagination of Newton and his devoted followers. The mass of the Earth never plays a role in physics and the mass of planets do not draw any of the planets closer to the Sun and let one physics professor bring proof that the planets do draw nearer to the Sun!

They just can't because planets do not have mass that can produce a pulling gravity! If and when the mass of the Earth do not feature as a factor in any formula that is used in physics, then the mass of the Earth is no factor playing part in gravity. This then can only indicate that the Earth has no mass. If there is an absence of mass as a factor that influences physics, this can only be as the result that the Earth mass has no gravitational presence in any physics formula. Gravity does have the value of $g = 9.81\ Nm/s^2$ but that I explain and the value $g = 9.81\ Nm/s^2$ I prove as well. With that evidence being that clear, then the mass that the Earth should supposedly have, does not produce gravity as Newton suggested. Prove me wrong by getting gravity at $g = 9.81\ Nm/s^2$ from using either any of Newton's formulas being $F = G\dfrac{M_1 M_2}{r^2}$ or $F \propto \dfrac{M_1 M_2}{r_2}$ and $F = \dfrac{r^2}{M_1 M_2}$. Let me see Newtonians do that and I will become a believer in Newton! The Earth has no mass because physics can't show the Earth's mass playing part in calculating formulas and if there is no mass that plays a part that should produce gravity, and then mass can't be responsible for the producing of gravity as Newton declared. That makes Newton's suppositions total rubbish and that makes Newton responsible for a crime of defrauding and falsifying the science of physics. If you, the reader is able to get academics in physics as far as even reading this argument I make, then you are more influential than I can ever be. They plainly dismiss all these arguments with arrogance by discrediting my credentials!

The scientific presumption is that gravity is established when one object holding mass is pulling another object having mass and forces the two abject to move toward each other. The entire basis of all physics rests on this formula where init it is believed that mass produces all gravity by distinction of differentiation in density as well as size and if physics is anything to go by, then what ever is proven, such proof must stem from and be in support of well as being supported by this formula $F = G\dfrac{M_1 M_2}{r^2}$. It is the formula that keeps the entire Universe in place and all of Newton's accuracy solely depends on $F = G\dfrac{M_1 M_2}{r^2}$ as a formula that has to be truthful and unquestionably accurate. The mass is the crucial factor because the mass is in a position where the mass destroys the distance of the radius from both ends equally. The mass generates a force and the force produces the gravity and the gravity produces the pulling and the pulling is what the time depends on that we have left to enjoy a Universe. We have to appreciate Newton's finding of mass until Kingdom comes because if not for mass, Kingdom is coming either tomorrow or never. Then we also accept the formula $F = G\dfrac{M_1 M_2}{r^2}$ has been tested and proven so many times by science that there is no other formula on Earth that has endured the testing that Newton's gravitational formula $F = G\dfrac{M_1 M_2}{r^2}$ has under gone. The force of gravity has the mass that would generate the gravity whereby the pulling of the other object orbiting and in also generating by mass the other object will also force gravity onto the first object $F = G\dfrac{M_1 M_2}{r^2}$. What goes up must come down.

We fight our mass because we fight gravity the entire time during one life span we live through. When I jump the force of my mass that generates the gravity by which I pull the Earth and by which the Earth pulls me back and the pulling is the result of the mass of the Earth that pulls me down again while at that moment I am pulling the earth up again, thus the square value coming about in the radius factor. When I fall my mass kicks into action and by mass I hit the ground at a rate my mass will determine. Newton is a genius because Newton realised all these wonderful happenings. Newton saw that a planet pulls another planet by the gravity that the mass of the planet charges. But consider that if mass is what brings about falling, it then implies that objects just cannot fall equal but have to fall differently and according to their mass. If mass has nothing to do with the falling then objects must fall equal and in equanimity through out the entire distance of travel while in the process of falling and that fact goes without argument. Mass brings variation and conformity is the result of mass not applying! The Universe is in a state of contracting $F = G \dfrac{M_1 M_2}{r^2}$ as Newton's formula must indicate. The objects are drawing closer to each other all the time.

Now marry that thought with the ever expanding Big Bang beginning and the Newton's concept of a Universe shrinking which totally contradicts the reality that Hubble found to be true and that there is a Universe out there of which we are part of that is exploding in expansion. To the world they declare openly that Newton's contracting Universe and Hubble's expanding is the same thing and we must wait for the Universe to admit being incorrect and start to employ Newton's contracting. They gave this blaming of the Universe going the wrong way on the Universe being the incorrect party because they are looking for mistakes in the Universe and wait to find out when the Universe will start to comply with Newton and start shrinking because the Universe has to stop this ridiculous expanding since Newton said the Universe is contracting. Since Newton just can't be wrong, therefore the blame of such silly contradicting of Newton has to be found at the door of the Universe. This blame game and detecting how far and why the Universe went wrong in disobeying Newton they named the Critical Density Theory and is the biggest scam and covering of fraud ever invented by any group of persons any time during the history of man. If the Big Bang is true (and it is true), then Newton just doesn't fit! In my book I show how this led to the biggest criminal cover up man has ever devised and was initiated by a person called Albert Einstein. The entire philosophy behind the Critical Density Theory is a scam and is even as ridiculous as what the rest of Newton is. You are about to read how far Newtonians will employ criminal cover up to form a blanket of deception!

The Newtonian formula $F = G \dfrac{M_1 M_2}{r^2}$ explains the comet arriving at the Sun, drawn by the mass of the Sun, pulling the mass of the comet as the comet comes closer to the Sun, but then if Newton's $F = G \dfrac{M_1 M_2}{r^2}$ has any validity the comet has to crash into the Sun after arriving. If gravity by mass was pulling the comet towards the Sun in the manner as Newton insisted in the Newtonian formula $F = G \dfrac{M_1 M_2}{r^2}$, then try and get any academic to explain why and how the comet moves away from the Sun and into the black yonder. After reaching the comet, the comet avoids colliding with the Sun as the formula $F = G \dfrac{M_1 M_2}{r^2}$ would suggest and head into the darkness of outer space. The comet then is moving directly in the opposite direction of what Newton's formula $F = G \dfrac{M_1 M_2}{r^2}$ would have us believe as the comet is not suppose to be pulling away because it is the mass pulling that was in place when the comet was drawn by mass as Newton stated. Does mass then start pushing mass to get the comet floating away from the Sun? Mass establishing gravity by pulling of a force is a gimmick Newton suggested but is unproven and it is nothing less than foolhardy to believe that mass does the pulling of the comet. Try and get those academics in physics to sensibly admit this reality and then in the explaining be sensible by using their Newton formula $F = G \dfrac{M_1 M_2}{r^2}$ as Newton's formula presents the law to show how this going away happens when mass is doing all the pulling at first. Try and get any Newtonian academic to explain this escaping of the comet from the mass of the Sun in the face of mass pulling mass. Some try to use the idea that the momentum drags the comet around the Sun but the mass will pull the comet into the Sun if $F = G \dfrac{M_1 M_2}{r^2}$ applies. Newton never created a detour as the mass pulling mass forms a linking straight line running from the centre of the Sun to the centre of the comet. Newtonians always bring more deceit to cover up Newton's fraud. These questions I address are otherwise never asked by students because students are brainwashed to accept and not think about asking questions. In presenting my work I can and I do answer the questions raised above but my answers do not fit the Newtonian visions of mass doing the pulling and because it contradicts Newton, I am ignored. In my following describing Newtonians is not to moan and grumble but it is to show the means and the manners they use to fight and when using such utter arrogance, despicable high and mighty autocracy with plain bullying tactics and megalomania. They have this attitude that only they are wise enough to think and the rest is mindless dehumanised animals walking on hind legs. If they fought fair and used intelligence it would not be that bad but to use dirty tactics when confronting me by just dismissing my views from a position of having authority is coward ness. By bullying me from holding a position of being able to ignore me and I can do nothing about it doesn't frighten me, it angers me!

If you are a student in the science of physics, then ask your Educated Masters to please explain the following abnormalities you are about to read in this book and insist on a clear explanation about the inconsistencies they promote while tutoring physics as if the physics they present are the most flawless and accurate institution there has ever been. Ask those academics supporting Newton about the following flaws that no one, except me, ever mention. Get them to explain the inconsistencies they never talk about. Wise up and confront those charged with tutoring physics and see who should you believe. Then get informed instead of brainwashed.

One very simple example, which I mention now at this point but I do not elaborate on this matter any other place in the book since in this book I wish to limit space, used, is mentioning the gravitational constant. If any one wished to bring in an explanation by employing the gravitational constant also introduced in the Newtonian formula $F = G \dfrac{M_1 M_2}{r^2}$ then using this gravitational constant is one of the ultimate bogus ploys Academics use to confuse the public.

Newton first envisaged the idea that it is mass standing in relation to mass that is destroying the radius found between the two objects forming gravity as presented by the formula $F = \dfrac{r^2}{M_1 M_2}$ but subsequently the notion as well as the formula used changed to $F \alpha \dfrac{M_1 M_2}{r^2}$. To get Newton's miscalculation $F = G \dfrac{M_1 M_2}{r^2}$ to work with some dignifying crookedness' they devised a constant of sorts going by the title as the gravitational constant and is this constant holding the symbol **G** in $F = G \dfrac{M_1 M_2}{r^2}$ It is put in place as being the same as all the gravity but is apparently that gravity that fills the space between the Earth and the Moon. Now comes the Newtonian part… This same space filling ingredient called the gravitational constant and holds a measured value of 6.67 X 10^{-11} where it is using this value while it is playing its part in filling all the space we find between the Sun and the Earth as well as the Sun and Pluto and everywhere there is space in outer the gravitational constant is the space-filler to have in that space being filled. If you think of space then we have such space filled with a gravitational constant at a value of 6.67 X 10^{-11}. This was the case in the days when it was accepted that ether was filling the space the gravitational constant filled and therefore ether might have had the value of 6.67 X 10^{-11}. Then after finding no evidence of ether, the ether that was not filling the gravitational constant was miraculously and by a stroke of Newtonian magic removed and replaced with…nothing…yes, nothing is now filling the space ether filled before they realised ether was not filling the space but the marvellous part is that nothing that the replaced ether took from the ether that is not there the value given to the gravitational constant and now while space is filled with nothing it still holds the measured value of 6.67 X 10^{-11}.

Newtonians are most adamant the Universe outside material is filled with nothing and nothing form outer space. Let's quickly ponder on this for a second or two and find out how much of this concept is palatable. They are then saying there is a long line of nothing standing the one after the other where one nothing is following the nothing in front while the nothing in front is leading the nothing behind. The nothing is lines forming rings here every ring ends at a point that the nothing that form the line of nothing connect linking in a chain of nothing from the Sun all the way to Pluto and even far beyond, in fact as far as the mind can take nothing and then nothing links in a line even further.

This line of nothing linking in a chain has a measured value that consists on the gravitational constant and each one of any of the nothing we mention has the value of 6.67 X 10^{-11}. Well even Red Riding Hood is going to sound more believable than does the most intellectual minds in mathematics…and from where does this first thought originate that the Universe comprises of nothing…from Newton of course when Newton said that the spin of an object cancels the space in which the object spins or in mathematical terms

Newton's formula holding and explaining the Universe portrayed as being $F = G \dfrac{M_1 M_2}{r^2}$ is completely wrong. If everything is in contraction then by now some places should already contracted large areas of space leaving gaps at other places where cosmic holes should by now be in place. If contraction brings about the gain of space in some parts of the cosmos as Newtonians say, then there has to be a part of the Universe that is losing space and by losing space a shortage of space must bring holes in space to appear where space is reducing! The gain of one part must bring about the loss of another part. In all of this that I have mentioned thus far in this letter, it does not even form a drop in the ocean compared to the incorrectness I present on science in my books about science. One should think that a book that challenges the dogma of Mainstream science and bring a new view or if only then just another view on science has to be commercially viable and should have some sort of selling potential. One should imagine that there has to be some publisher that recognises the potential and would have the courage if not the business sense to put such a book in the book shops. But I found that thinking that way is pure daydreaming. A book such as the book I have to offer that takes on Newton and starts to strangle Newtonians has more scope to become controversy than does the Da Vinci code because it affects a wider audience.

Since 1977 I was convinced that there was something amiss in the approach science took on the matter of gravity. In the work I presented I based my theory on the discipline Johannes Kepler introduced. But the more I pursued my goal of forming another gravity concept other than supported by Newton's view that $F = G \dfrac{M_1 M_2}{r^2}$, the more I had to confront the thinking of Newtonian inspired culture when approaching academics. I was never in doubt that Newton was ultimately wrong and that the Newtonian dogma of gravity was incorrect, but never wished to directly attack Newton as a scientist. But in the end I had to change my approach of being polite because it was clear to me that academic culture would not change and see the logical arguments. In September / October 2005 I wrote letters to nine academics heads of departments at Universities in South Africa that was involved in cosmology. Each one I wrote a letter to was heading a cosmology department at a South African University and in the letter I informed them that I intended to show why Newton was incorrect and therefore what mind games they as physics academics and tutors were playing by protecting Newton's dogma and Newtonian religiosity, I was going to uncover and make public the fraud Newton committed and what the extend of the fraud was that they were intentionally covering up. I knew before I wrote the letters that no one is going to take me serious but then there is a price to pay for every mistake one makes and they were blind to what I said when I said I was going to expose them. Even at this point they do not see that they are intentionally covering up Newton's criminal behaviour and hat they are committing intentional fraud by covering up Newtonian misconceptions. They are so involved in a culture of crime it is not possible for them to separate criminal fiction from factual reality. The covering up syndrome and culture prevailing in physics will prevent them from having any negative work on Newton or about Newton to be published.

By merely putting gravity in the Universe that is acting as a mysterious FORCE that is pulling towards a common point in an allocated general centre is rather avoiding the question with simplicity because the question about how and why remains unanswered. Not knowing the answer will leave you empty and unfulfilled because of being a student and not knowing is the same as suicide n a mental level. Ask yourself the following: If gravity pulls towards a centre and gravity holds the Universe attached the question arising from that simplistic answer is then ... where is the centre of the universe?

Should and if you decide read my letter addressed to students it will bring along a new perception about Kepler. Science sees to it that Kepler stays the least appreciated Cosmologist where as in truth Kepler proved gravity, proved singularity, proved space-time, proved the Big Bang, proved every dynamic most of the wise persons afterwards thought about. Yet no one gave Kepler any recognition up to now because science denies Kepler his limelight.

By not confronting the establishment, you give the establishment grounds to allure you into being sheepish. Because they see you, as just another stupid senseless student they have the opinion that they can brainwash you into accepting these fallacies that I am about to tell you. They will literally brainwash and condition your mind to accept what they never yet were able to prove. If you feel I come across far too strong then put correct values in place of the symbols in $F = \frac{r^2}{M_1 M_2}$ or in $F \alpha \frac{M_1 M_2}{r_2}$ and see according to your own opinion how totally ridiculous Newton was and physics at present is.

They are of the opinion you will swallow any rubbish they throw your way just because every generation before you, were mind controlled in the way they are about to put their control over you. You may think this is big words but read on and see after you come to know all the facts if I exaggerate even in the least. They see you as slow-witted and mindless because they think they are the academics being superior making you the inferior. If you are not aware of the facts beforehand they know you will follow their teaching without asking questions.

They think that your naivety makes you mindlessness will incapacitate you into their control. They don't want you to ask nosy questions about contradictions existing and they refuse to answer. This process of brainwashing and mind controlling in physics has been in progress for hundreds of years. Just answer how a feather and a large hammer fall equally while mass drive gravity as a force. If you can't...well they can't either! Their task is not to explain but to mislead since they think you can't think while they think they know how to control you.

The following letter addressed to students does not aim to represent the full entirety of the copy of the original letter addressed to students but is reduced considerably to aid any possible potential reader in the examining what the purpose is of the information this letter addressed to students wish to announce. Anybody and everybody are aware that all objects fall at an equal rate. If an object such as a car weighing one ton falls at the same pace as a person weighing fifty kg how does mass come into the picture by committing a force to do the pulling? Mass has to pull because according to their teaching it is mass that establishes gravity. However, mass is a factor that produces differentiation whereas all objects show equality during their fall. If it is mass that is establishing the force gravity, all objects must fall at different speeds. That they do not do as they all fall equal. That means physics is wrong from the start because mass cannot have any input in objects falling.

Physics students, it is your duty to pull the plug on the powers of the ALL-POWERFULL Academics in Physics and stop their dishonesty. It is your task as the as the next physics generation to stop the criminals that are filling the corridors and the lecture halls of physics departments throughout the world. Stop their teachings by forcing them to stop their criminal fraud. Force them to explain the deception called THE CRITICAL DENSITY, which is a conspiracy to commit fraud. Let them explain how an expanding Universe can suddenly and abruptly turn in direction of developing and start to contract as Newton stated it is doing at present, and when facing all other concluding evidence showings it was expanding since time began. Tell them to prove that the cosmos will begin to contract doing its turn about by using other proof than merely Newton's say-so. Tell them to bring proof with evidence that the cosmos is contracting as Newton said. Then force them to admit to the fraud they are precipitating in, which is THE CRITICAL DENSITY conspiracy... In THE CRITICAL DENSITY conspiracy all they say is that they are waiting to see when the cosmos would stop its criminally insane behaviour and start to listen to the laws of Sir Isaac Newton.

With The Critical Density shambles the modern Newtonian set out to defraud the world in the same manner as their Master Sir Isaac Newton has done centuries ago. Newton said the cosmos is contracting. When Hubble proved the cosmos is not contracting, Newtonians looked where the cosmos went wrong by not following Newton guide lines he so clearly set the cosmos to follow. It has to contract and not expand

To all those that feel disgusted by me accusing the greatest name in science that ever lived being Sir Isaac Newton of fraud, please go on and prove me wrong!

$F = \dfrac{r^2}{M_1 M_2}$ This is the formula Newton used with which Newton proved gravity. Now prove gravity by using this formula. Do the following to prove me wrong.

To find the force of gravity one has to multiply the mass of the Earth (M_1) with your personal mass (M_2) and then divide the distance there is between you and the Earth (r^2). Using these factors by multiplying (M_1) and (M_2) and dividing with (r^2) should present gravity coming from mass. But science uses a fixes value to calculate gravity.

Now, convince your mind about my correctness. Do the simple calculations.
Take the mass of the Earth (M_1).

Multiply the Earth mass by your personal mass that any scale should indicate (M_2).
After multiplying the two mass factors, then proceed to the following step by dividing the multiplied mass factors with the square of the radius there is between your feet and the Earth (r^2), which should not amount to more than a few billionth of a millimetre.
If the answer in front of you is not 9.81 Nm/s^2 then there is something very wrong.
The incorrectness has to be one either of two possibilities presented:
The measured value of gravity is not 9.81 Nm/s^2 as science uses it, or

Sir Isaac Newton's formula suggested as $F = \dfrac{r^2}{M_1 M_2}$ is complete fraud…

Now which is it…you can decide…

The force of gravity that the world of physics uses to do measurements is 9.81 Nm/s^2. If your answer you have in calculating your force of gravity is not 9.81 Nm/s^2, then it is either this measuring value of gravity that is wrong or it is Newton's $F = \dfrac{r^2}{M_1 M_2}$ that is wrong because by the calculation you did, the calculated answer you got could not possibly have deliver a measured value of 9.81 Nm/s^2. After all, science maintains it is the pulling of mass that delivers the force of gravity! If by using the factors of mass and the radius does not accumulate to 9.81 Nm/s^2, then how can mass deliver gravity.

To teach students that $F = \dfrac{r^2}{M_1 M_2}$ are the measuring formula in determining gravity, while knowing very well it is not totalling gravity at 9.81 Nm/s^2, then doing that to students while enforcing a thinking pattern in the minds of a student is committing brainwashing because by forcing examinations on students, expecting them to confirm the falsified statements used that the tutors present as correct is brainwashing, a way of enforcing mind control and it is manipulating the thinking process of students.

If you can't prove that my manner of thinking is incorrect and you keep surmising that science is correct then recalculate the formula or start reading the rest of their fraud.

Gravity is a constant of 9.81 Nm/s^2. This is used in all cases of scientific calculations.

<u>Please use your intellect to explain the following mathematical expressions as Newton suggested that the principles of the different formulas applied.</u>

Let any one of them prove this. Let any one of them explain this as a mathematical principle. This is all mathematical formulated expressions and has to be proven accordingly. This is not linguistic suggestions but is used in terms of mathematical accountability! This is palpably false

Mass is an individual factor that is different on anything on which it is applied as a measuring factor. How can something as different as mass that is never constant even on Earth form a constant such as the force of gravity and still be the same in all cases?

Sir Isaac Newton's says that $a^3 = T^2$. I have to believe Sir Isaac Newton when it is said that three dimensions are equal to two dimensions or in mathematical terms that $a^3 = T^2$ on no more grounds than that Sir Isaac Newton said so and without having any other proof to back the statement. Remember, Kepler never said $a^3 = T^2$, that is the part coming from the fantasy of Sir Isaac Newton. Kepler said $a^3 = kT^2$ which places three dimensions on one side holding three dimensions equal on the other side of the equation. There is a^3 on the one side of = and then there is kT^2, which is $k^1 \times T^2$ which is $k \times T^2 (^{1+2=3})$ and that makes $a^3 = kT^2$ having three dimensions on the one side being equal to three dimensions on the other side. There is no way in heaven or hell that one can have the third power being equal to the second power or have a cube that is equal to a square, even if you are Sir Isaac Newton. There is no one on Earth that will tell me that $10^3 = 10^2$. There is a case that $10^3 = 10^2 \times 10$ or that $2^3 = 2^2 \times 2$ but never can it be that $2^3 = 2^2$. Not even when Sir Isaac Newton is doing the saying so. If one says that in the event where $a^3 = kT^2$ one may assume that $a^3 = a \times a^2$ or $k^3 = k \times k^2$ or even that using $T^3 = T \times T^2$ will also bring equality but never can $a^3 = T^2$…and then there are academics that try to convince me that $a^3 = T^2$ because Sir Isaac Newton was of the opinion that $a^3 = T^2$ and furthermore they expect me to also believe that it is true that Sir Isaac Newton has never been wrong on any suggestion and because no one could ever find Sir Isaac Newton to be wrong, I have to accept that $a^3 = T^2$ and take it as the absolute truth without questioning this abnormality!

The one image is a cube with three sides. The other totally different image is a square having two sides. Sir Isaac Newton said the two are equal while they can never be equal since they are one dimension apart Sir Isaac Newton convinced so many generations of idiots considered as being the wise amongst the wise and fooled those to the point where these stooges are willing to believe they are wise enough to believe that a cube is equal to a square and only on the ground that Sir Isaac Newton said so.

Sir Isaac Newton proposed and moreover convinced the world of science, and this includes every one and all members that should be the most intellectual bunch living on Earth in human form, that they and the entire world should accept that the inexplicable $a^3 = T^2$ is correct and that the biggest trick in fraud can be played on a bunch of fools all willing to be stupid enough to pretend they are clever enough to see that $a^3 = T^2$ and they are so stupid they pretend to be so clever that they will accept that $a^3 = T^2$ which when translated in words means that two dimensions are equal to three dimensions. This is the same as stating that a person's reflection coming back from the mirror is the same as the person filling reality while standing and looking at his image in the mirror. In this group hosting the most advanced minds man can produce there are a big enough bunch of zombies pretending to be mentally superior while being big enough idiots that are foolish enough not to think and not to ask questions but be small minded to the point that they will accept that a cube is equal to a square $a^3 = T^2$ just simply going on the say so of Sir Isaac Newton's

When Sir Isaac Newton says $a^3 = T^2$ that does not prove that $a^3 = T^2$. It only proves Sir Isaac Newton was the worlds biggest and best silver tongue devil and cheated an entire Earth load of scientists for almost four hundred years. He fooled the supposedly wisest humans we all think there can be to pretend to be wise so that they can hide their stupidity while they only focus on their stupidity by not questioning the validity of $a^3 = T^2$. You bring me one other con artist and fraudster that can manage that. It takes some doing to fool so many people for so long and leave all those fooled feeling good about themselves in that they are fooled. Sir Isaac Newton was the biggest con artist ever to live and never again will the world experience an equal to Sir Isaac Newton. It is no small wonder that science is infested with atheism because science upholds disdainful lies based on mediocre understanding about truth applying as a reality and crooked science! Newton is all lies and shambles and reading this book will prove that.

If science cannot prove God's existence, it is not God that does not exist, but it is science failing and therefore it is then that specific view about science that should be re-examined since it is the view on science that is proving as being incorrect. This fact is what the so very brilliant and intellectually mindful Newtonian atheist should remember when they fail in their science altogether. That their science fails altogether and that failing it does in all its splendour, is facts I am delighted to prove! The fact is Newton's views were never tested and that the Newtonian views on science were never challenged before and because of that Newton principles never withstood diligent scrutiny before. When Sir Isaac Newton is investigated even in the flimsiest of manners, well accepted facts seems to become very suspect, to say the least. This becomes evident when concluding all the facts this book presents. Now, in this book, for the first time Newton is tested and such testing is the proof you gain by reading that which I uncover. What I bring into the open is unseen facts, which I present you with as I take you on a tour through an avenue of facts I introduce in this work. The lack there is in sensibility concerning Sir Isaac Newton's principles this book proves. The theories of Sir Isaac Newton require proof, which was never given while God never needs proof and that is what science constantly seeks. When science perpetually ignored my concerned calling on and ignored my calling on them because (I suppose) they were finding my concerns wanting, in my final letter to them I promised them never to contact them personally again by any and by all means. I also promised them a fight. This is the fight I promised. I was not worth noticing so I was ignored. I now am calling on the public, as I am ignoring their reputations.

All prospective, intending and otherwise possible potential readers of anyone of the two books called An Open Letter On Gravity are hereby seriously advised to read Part 1 of An Open Letter On Gravity first before advancing onto Part 2 of An Open Letter On Gravity, and only then afterwards and after completed reading Part 1 of An Open Letter On Gravity then the reader should advance by reading Part 2 of An Open Letter On Gravity because by first reading Part 2 of An Open Letter On Gravity the answers might seem to remain questionable but when reading Part 1 of An Open Letter On Gravity first the questions will become answerable. If it is said by using easier language to explain the expression: then by not knowing the questions, the answers are not well defined but knowing the questions, the answers become rather simplified as it is self explained and much better understood in explaining.
Still my challenge is and remains there where I challenge everyone and all persons notwithstanding title or position to show me how they can maintain with clear and lily white conscience that Sir Isaac Newton did not corrupt science in all aspects by committing fraud in mathematics and do so after reading the two titles An Open Letter On Gravity Part 1 and Part 2!

AN OPEN LETTER ON GRAVITY Part 1 disputes the correctness of the formula $F = \dfrac{r^2}{M_1 M_2}$

Using the formula above as Newton did does not imply a suggestion or carry an idea across as a thought but must be seen to be acting as confirmation about a fact because one cannot suggest anything mathematically, one can only confirm a fact mathematically. There is no mere suggesting of any possible movement in a specific direction of any suspected behaviour by an object moving from and to a point as suggested but this is saying that the gravity of the Earth measured in mass at it's totality is colluding with the falling body's measured in mass as the two factor's diminish the radius from both ends. This is used to back up a fact!
This Is a Book That Is Not Afraid To Show How The Paternity of Newtonian Science in Physics Openly Cheats To Cover Their Oversight In an All Out Effort to Hide Newton's Misjudgement

Newtonians say the force F of gravity is 9.81Nm/s²
Newtonians also say the force F of gravity is

$$F = \dfrac{r^2}{M_1 M_2}$$

Then in terns of mathematical principles
Newtonians say $F = \dfrac{r^2}{M_1 M_2}$ is 9.81Nm/s² where both are equal to the force F of gravity
$F = \dfrac{r^2}{M_1 M_2}$ = 9.81Nm/s². That is the way that proving with mathematics is done and what does it prove? SIR ISAAC NEWTON'S FRAUD

Science teaches that a feather and a hammer have different mass while they fall equal in time through an equal distance travelled. All things fall equal in time and distance when subject to the same environment. If gravity was mass related, then this was not possible, because then objects must fall according to mass. Falling objects bears no evidence of mass playing any part in falling. Any two objects holding different mass fall equal in time and in distance when sharing similar conditions, which suspends mass altogether as an influencing factor. Galileo proved different mass fall equally under similar conditions. That fact about Galileo, science does embrace, although this strongly contradicts Newton's impressions about mass inflicting gravity. Acknowledging Galileo must make the work of Newton incorrect and also corrupt. On TV we see how all objects, such as cars, humans and bags fall at the same pace, which sets a standard totally against Newtonian mass principles that produce the falling, and proves Newton wrong because mass then does not underwrite gravity in any way or form at all. The formula $F = \frac{M_1 M_2}{r^2} \cdot G$ would suggest mass taking all the responsibility for such falling that takes place. Newtonians declare gravity as the force of gravity F, that is = equal to gravitational constant G, when it is multiplied by the mass M₁ and the mass M₂ after which then the product of the three factors influencing gravity is divided by the square r² distance between mass pulling the mass that destroys the distance between the two objects. If mass pulls mass as Newton said, the Big Bang is not possible but the Universe is notwithstanding Newton's claims, expanding (growing apart). If the mass destroys the radius separating the objects, then the comet has to collide into the Sun, but it doesn't. If mass forms gravity, every planet must orbit at a different pace, which they do not, as all planets orbit at the same pace around the Sun. Planets don't give the slightest hint that they obey Newton's suggested cosmic laws by implementing mass. The truth is that mass is the resistance of any independent material to deform and to acquire mass the individual object relinquishes independent motion. Mass is the reluctance to deform and integrate into a larger structure and becoming a unit of the larger or holding structure. Mass comes about when the falling of any object stops the motion of the falling. Mass prevents further falling, it does not sustain further falling. Gravity is the moving of the object to the centre of the Earth while falling.

I say gravity is movement while mass is obstructing independent movement, which is what gravity is. Mass is not forming the factor responsible for gravity or movement, but prevents further movement. A body falls by gravity. Mass obstructs further falling, while gravity remains present as a factor that brings the tendency or inclination to move or the attempt to continue moving. Mass hinders movement and therefore mass can't enhance or produce movement or gravity. Mass prevents or blocks gravity. Gravity is the motion that defines the individual identity of any object's structural form by rendering motion while reserving independence in granting free space from other manipulating objects. By saying this, I am awarded the cloak of death by Academics ignoring my correspondence as if I never addressed their mailbox.

This is not the only untruth that the Paternity called Mainstream Science is keeping concealed as a cover up that is wrapped under an airtight blanket of deception. If you sit in class and listen while also experiencing the sinking feeling that the facts you hear are not adding to a total you are comfortable with while you disagree with what is said then you better read on because this letter addressed to students has it at task to show all that will read this document how much discrepancies academics lay on unsuspecting students that trust Academics with their future and their life. Do you as students realize the inconsistencies that physic Academics present you with when portraying that what they teach you as being the solemn truth.

>Students tell your Professors to stop deceiving and stop trying to control your minds with their fraud.
>Those Academics tutoring you are telling facts about gravity that has never been proven.
>That is mind control.

>They wish for you to accept facts on gravity that they hold as the truth. They claim those truths are beyond questioning yet with the least examining those truths they stand by then proves to be totally void of substance because it was never corroborated by one single experiment.

>Should you question that mass produce gravity they will expel you from University by letting you fail your examinations and it was never proven. They will expel you and have you fail tests should you question their authority on the matter of gravity while at the same time they can't for one second bring evidence in support of what they wish you to accept as the unquestionable truth.

>That's brainwashing by mind control because if you don't accept their baseless fact as God given truths they dismiss your academic career.

It is either put up and shut up or be gone. Academics do put mind control to work on unsuspecting students by forcing students never to question the legality of statements they offer as being sound and correct.

What they present as correct I prove in this very letter addressed to students are openly laughably totally incorrect and by just reading my evidence you will see how feebly easy it is to rubbish it. Take the evidence I am about to share with you and confront them with the fabrication of facts that they present. Go on and challenge those teaching you with the falsified facts as I challenge any one to prove me wrong.

What they maintain is gravity is total incompetent nonsense and can't be corroborated at all but what they can't corroborate because they don't understand I prove to be that which the Universe employs to form gravity. There are four phenomena they dismiss because they have no idea what they are. I studied each one and formed an explaining by implementing Kepler's formula as the Universe gave it to Kepler.

The by understanding the formula and implementing the content into the four phenomena I am able now to prove what forms the motion we think is gravity and when reading it then the Universe makes sense. All the questions in these books I managed to answer while they can't … and in the books I answer a lot more questions than what I ask here in the rest of the web site while Science fail to answer any...

A Less Complex Commercial Science Book   95   Chapter 3

PART 3

This is not just Another Conspiracy Theory
This is About, believe it or not, but modern Physics as it is taught
It is The Ultimate Conspiracy Theory,
It is a Conspiracy Concerning Physics

This website allows you to see what the degree of blatant corruption is presented as physics and what Academics in physics hide.
For other books covering relevant information visit the Lulu.com website called Peet Schutte Storefront @ Lulu and find a wide range of book.

Let us test Newton's attraction theory in practice

Newtonian physics teaches that the mass of the earth and the mass of the moon is constantly contracting (reducing) the distance they are apart, by forming a force called gravity. If there is such a force then the cosmic question that needs an answer is ... If this is true that the earth-mass pulls moon-mass as a pulling force, then when will the earth and the moon collide? This destruction of the earth and the moon must happen with the earth and moon being so close... Academics in physics insist that gravity is founded on the Newtonian gravitational principle of mass $F = G \dfrac{M_1 M_2}{r^2}$, then when is the inevitable collision coming. We have to know when is it the final doomsday when all life forms end?

This principle declares that a force of attraction by the value of mass of both solar objects is pulling by reducing the radius, therefore bringing the moon and the earth closer. This must lead to an inevitable collision that must end both the earth as well as the moon. Let the Newtonians calculate when is this event due and inform the human race when our final day of doom and death will arrive by using Newton's law of attraction or $F = G \dfrac{M_1 M_2}{r^2}$. The day if (no ("when") this happens, Life in all forms will seize and therefore knowing the time is crucially important for all of us having life on earth! We must know the time!

Let the truth be told for once: It has been confirmed by NASA that the Moon's orbital distance is increasing at a rate of 3.8 cm per year. Other sources say it is at a rate of about 3--4 cm/yr. The moon and earth can never collide because the two are drifting apart. Let the Newtonians chew on that...and think of Newton while chewing on this piece of reality...the Universe entirely is expanding all over.

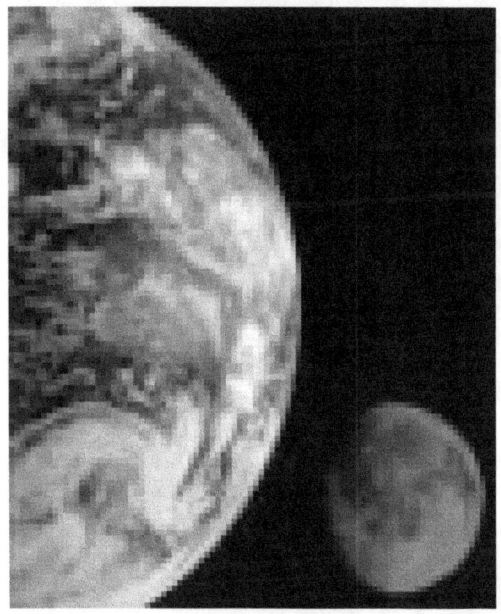

According to Newtonian principle there is attraction by the force of gravity to the value of mass. Nevertheless, NASA confirms the moon and earth is parting in a way that increases the distance between the two solar systems. Notwithstanding Newton, the distance there is, that distance grows!
In order to cover up for Newton's misrepresenting physics and misperceptions an entire variety of reasons are established; each is accepted as a possible truth. The fact that Newton's principle goes begging never gets mentioned, although the only reason why it never gets mentioned is because it is the only valid conclusion. That bit Newtonians don't want to know. Newton's principal of mass pulling overrules all other reasons mentioned. Mass pulling is the founding law that all other factors rest on. The earth slinging the

Moon away can't be a factor because the mass of the earth is too great. The law says that the mass that pulls creates a force called gravity and this gravity reduces the radius between objects by the square thereof. That is gravity.

The moon is moving away from the earth. Yes, have a look at all the theories presented by Newtonians on the web as to why this happens. It is the thrust, no, it is heat expansion, no, it is parting because of sea currents, no…and there are an innumerable many excuses why this is happening. Not one has to do with Newton being wrong! They never take Newton's formula and apply the mass of both solar objects and see what the gravity is as to check why the parting occurs. With the radius increasing the gravity must therefore reduce in pulling power because the distance determines the validity of the force by dividing into the multiplication of the mass. In relevancy it is not the value of the top factor that determines the value of the outcome, but it is the size of the bottom factor that controls the value of the top factor that determines the outcome of the value of gravity! This is the mathematical truth…as the dividing (r^2) factor increases; the influence that the mass represents in the formula $F = G \frac{M_1 M_2}{r^2}$, will diminish in respect to the growth of the distance (r^2). As the bottom gets bigger, the dividing will reduce the value of the top factor. This is a mathematical law. In that sense the gravity force between the earth and the moon must reduce its ferocity therefore weaken, as the distance grows bigger. There is no use to look for dark matter that will pull the lot closer in the future. However, Newtonians only apply their ability to calculate and knowledge thereof for the purpose of upholding Newton and never to provoke Newtonian liability by telling the truth. They never mention mathematical realities when mathematics is used.

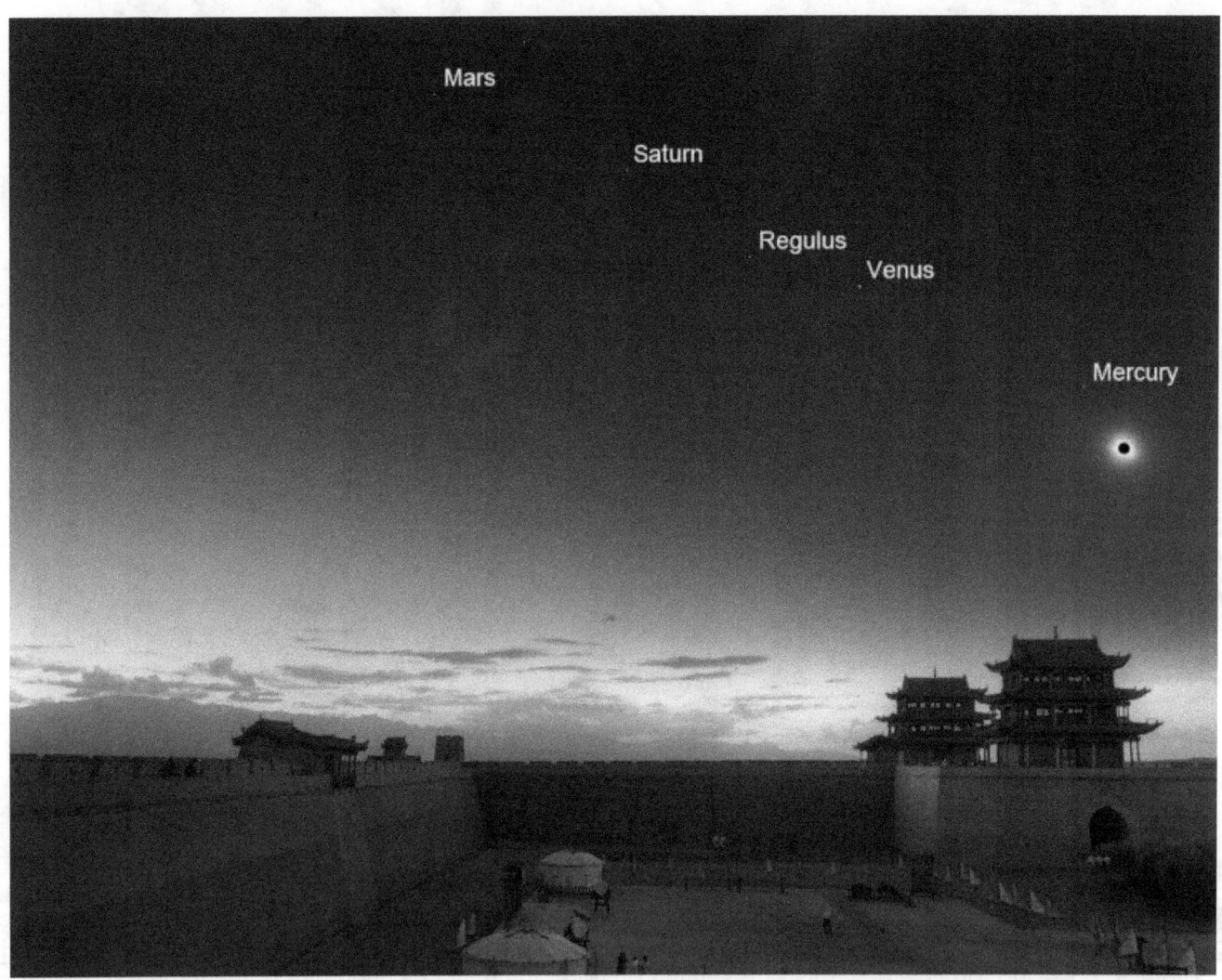

Those Ever-So-Wisely-Educated cosmic Super-Brains always calculate the power that drives a Super Nova. Let them use those brilliant minds then to calculate the precise date when the earth and moon will destruct all life on Earth! Those Brainy- Mathematical-Masters always custom-design in detail space –whirls they invent with applied cosmic imagination. Let them bring such terrific

astonishing human abilities closer to home. The Mathematical-Geniuses that can calculate the inside of a Black Hole should bring their splendour to a much better use in terms of where it concerns human future.

Instead of painting a Universe fit for Alice in wonder-world, and sprinkling it with the best mathematical formula that they think must put them on par with God, rather apply the same formula and show when the solar system, as big as it is, will collapse into the Sun. Ask them to not search for imaginary undetectable, unexplainable dark matter they can't even point out, but to find when the distance we have between the planets and the sun will dissolve and when the pull of the planets will have all the planets go crashing into the sun. Those Ever-So-Wisely-Educated cosmic Super-Brains best art form is to deceive everyone by conspiring to hide the truth about Newtonian science. Those Mathematical-Masters will never even hint Newton's incorrectness of thought because it will reveal their fraud.

Please allow me to show you how they scare you to become fooled and suckered. Don't run and hide when you see the mathematics, it is meaningless although it was used as a scarecrow for more than three hundred years forming the backbone of the conspiracy.

This is the manner in which Newtonian science presents the solar system. We have nine spherical structures orbiting one big structure in the centre of the orbiting planets and the picture shows all of them are nicely apart by a common distance.

This picture is as big a hoax as Newtonian science are when Newtonian science presents mass to be a factor that produces a pulling force called gravity. The question shouting for an answer in the picture is if mass is a factor that produces gravity as Newtonians claim it is then why are the planets not positioned according to mass as Newtonians declare. The claim is as bogus as the entire philosophy. They present the proof that planets orbit according to mass in the following "Kepler law" which in its entirety had nothing to do with Kepler at all. It is all devised by Newton because Newton had no inclination of what Kepler's work was about.

Newton brought about the idea of mass positioning the planets in the formula $4\pi^2 a^3 = P^2 G(M + m)$

From that they present the formula $\left(\dfrac{P}{2\Pi}\right)^2 = \dfrac{a^3}{G(M+m)}$ as proof. This is **supposedly the proof but don't let the use of the mathematic equation fool you** because they use the senseless equations to bewilder students and general public alike with the use of meaningless mathematical equating.

What this formula presents is that the mass $G(M+m)$ of the sun and the planet in question puts relevancy to the space the planets orbits in a^3 indicated as $\dfrac{a^3}{G(M+m)}$ and from this the planet's allocated position is derived at point (P) by the square $\left(\dfrac{P}{2\Pi}\right)^2$ in relevancy of the square of the double the value of the circle Π. P then places the planet in the allocated position. What a lot of garbage this is and it has been fooling the public as mindless suckers for three centuries.

This is where the general public turn their heels and run like rabbits when they are confronted by the mathematical expression and that is where the Newtonians get you fooled with their conspiracy to keep Newton lilywhite and unblemished. This is no less than fraud when it is conducted to form a conspiracy to make the rest of the human population believe that physicists know physics intimately.

What this does prove that physicists have not even a faint clue of what physics are about because they fail to prove that mass allocates the planet position.

As I said, don't let them scare you with nonsense. What this formula $(\frac{P}{2\Pi})^2 = \frac{a^3}{G(M+m)}$ claims is that the planets are positional allocated in accordance with the mass that each planet has and the planets orbit according to the mass they have.

From that argument they formulate that the position $P = \left(\frac{4\pi^2 a^3}{G(M+m)}\right)^{0.5}$ is according to the mass in terms of the space forming the circle.

Again do not run because the question destroying their claim is answered by asking how does the solar system respond to this. In the centre of the planets there is the big giant Jupiter next to a band of debris no bigger than large rocks. The asteroids and rock fragments circle around the sun $P = \left(\frac{4\pi^2 a^3}{G(M+m)}\right)^{0.5}$ equally to Jupiter that is the biggest planet of all. If $P = \left(\frac{4\pi^2 a^3}{G(M+m)}\right)^{0.5}$ had any credibility even in the least, then the most inner planet had to be Jupiter and Mercury as well as Pluto was then on the very outside.

The Earth would be so far from the sun that the sun would be no bigger than a green pea. It would be so cold life would not be possible.

Body	Mass (10^24) kg	÷	Orbital Distance(10^6 km)	=	ratio
Mercury	.3302	÷	57.9	=	0.0057
Venus	4.869	÷	108.2	=	0.045
Earth	5.975	÷	149.6	=	0.039939
Mars	0.6419	÷	227.9	=	0.00281
Jupiter	1898.6	÷	778.3	=	2.439419
Saturn	86.83	÷	1427	=	0.060847
Uranus	102.43	÷	2869.6	=	0.03569

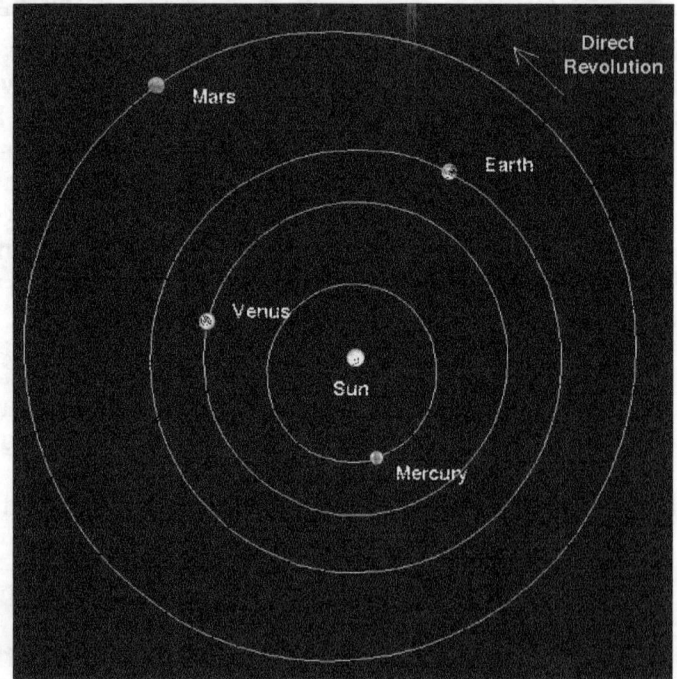

This table shows where the planets are and what the respective mass of each planet is. If the Newtonian formula of $P = \left(\frac{4\pi^2 a^3}{G(M+m)}\right)^{0.5}$ apply then why would the largest planet be placed in the centre, in the very middle and the smallest planets would be the very inner and the most outside planet.

There is not even an indication that the mass concept of Newton applies as a thought, yet the idea has been presented and claimed for centuries.

Is there even one person that would try to tell me no one has looked at the numbers as the numbers are in the solar system and doubted this? If that is not a conspiracy then the word conspiracy has no meaning. The conspiracy puts every teacher down as a fraudster.

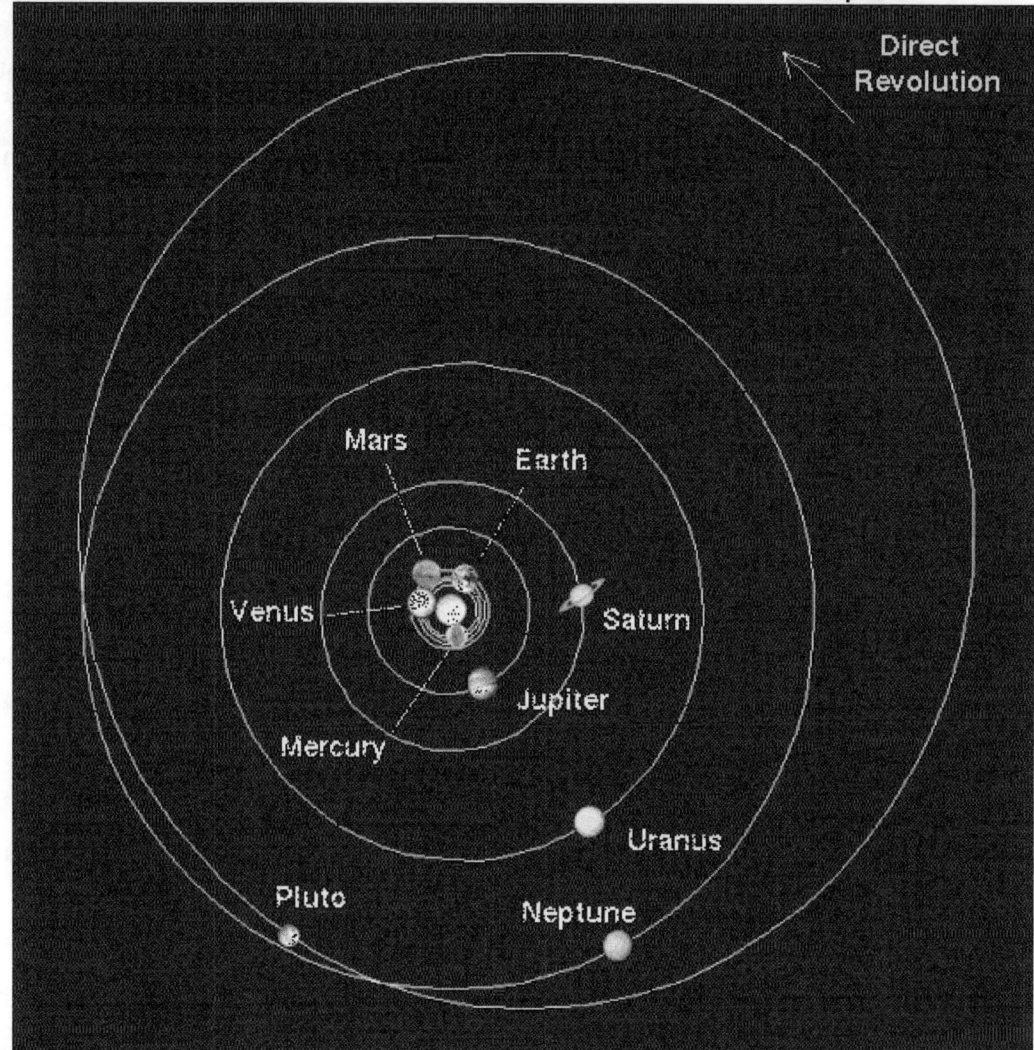

This picture shows the hoax the Newtonian conspiracy pamper to keep the rest of Newtonian physics believable. They never mention the Titius Bode law and try to explain the Titius Bode law while it is the Titius Bode law that is really in place in the solar system. If *you wish to learn the truth then think again.*

Our Super-Educated-Wiser-than-any-other-humans use the elaborate formulas such as $\frac{d}{dt}\left(\frac{1}{2}r^2\dot\theta\right)=0,$ and $P=\left(\frac{4\pi^2 a^3}{G(M+m)}\right)^{0.5}$ as well as $T^2 = \frac{4\Pi^2}{G(M=mp)}a^3$ to explain and prove what? If this is true then Jupiter must spin 317 times faster than the Earth does and be almost next to the sun while Mercury ands Pluto in comparison must hardly move being cast into the darkness of the oblivious.

The formula $T^2 = \frac{4\Pi^2}{G(M=mp)}a^3$ says that the spin or time in which the circle comes about T^2 holds a space relation a^3 directly proportionate to the mass of both the sun and the applying planet multiplied by placing the gravitational constant also in relevancy. I say prove it! Then $\frac{d}{dt}\left(\frac{1}{2}r^2\dot\theta\right)=0,$ says that the Planets don't move at all because with $\frac{d}{dt}\left(\frac{1}{2}r^2\dot\theta\right)=0,$ the movement acceleration is zero. Every person associated with physics to whatever extent has been and had been constipating to hide the truth that not one knows (not even Newton and even less Einstein) what physics is.

If they don't know what forms gravity and it is clearly not mass, then they know nothing about the way that gravity forms. They hide there incompetence under a blanket of ignorance and all commit a conspiracy to hide the truth from the public. Let any one of those So –Wise-In-Mathematics show just how does Saturn

arrive at the position it holds in relation to the mass it has by applying the Newtonian formula
$P = \left(\frac{4\pi^2 a^3}{G(M+m)} \right)^{0.5}$ Data confirming Kepler's Law of Periods comes from measurements of the motion of the planets. It confirms movement but not mass in any aspect.

Planet	Semi major axis(10^{10}m)	Period T (y)	T^2/a^3 ($10^{-34} y^2/m^3$)
Mercury	5.79	0.241	2.99
Venus	10.8	0.615	3.00
Earth	15.0	1	2.96
Mars	22.8	1.88	2.98
Jupiter	77.8	11.9	3.01
Saturn	143	29.5	2.98
Uranus	287	84	2.98
Neptune	450	165	2.99
Pluto	590	248	2.99

If there is a table T^2 that is in division of a^3 that would put **k** in a fraction (k^{-1}) or if you will then $(\frac{T^2}{a^3}) = \frac{1}{k}$ in the event if $a^3 = T^2 k$ as Kepler proved it is. But hiding the truth $(\frac{T^2}{a^3}) = \frac{1}{k}$ prevents any person from seeing the truth. Gravity is not contracting the solid orbiting structures but gravity is compressing the space or reducing the relevancy in distance of the space in which the planets orbit. There is no mass because all the planets float evenly in the space they move, but the space in its self is contracting by reducing. By this I was able to show that the Titius Bode is gravity and I was able to show what gravity is.

Gravity is the reclining of space coming about from the turning action of the spherical object spinning in space and in that compressing the space into a denser construction. By condensing the space gravity forms the value of Π in being $\Pi^0\Pi = 3.1416$ on the inside and $(\frac{21.991}{7}) = \Pi$ on the other side and that forms the Coanda effect. When an object is secured on the ground it will form a factor of mass because $\Pi^0\Pi = 3.1416$ makes it part of the centre of the earth and when the object floats or falls it can't have mass because $(\frac{21.991}{7}) = \Pi$ makes it form part of the curve of the earth at 7°. Gravity forms as simple as that! The space reduces by the earth turning, and the rotation of the earth compresses the space into forming the atmosphere of the earth. Then the space pushes objects onto the earth giving it mass. Mass does not pull but space pushes by movement!

Through the EIGHT e-books I named
The Absolute Relevancy of Singularity INTRODUCES you to a
new cosmic concept in physics.

By going to The Absolute Relevancy of Singularity you will learn where is the spot in our Universe holding time in, and representing space at the point we think of locating time

Where time is in

infinity

And time is in eternity

You will also find the centre of the Universe. You will find the point where singularity forms the Universe as infinity starts the Universe at that point. You will find the point where singularity forms the Universe as eternity ends the Universe at that point…

By rejecting Newton's unsubstantiated science I can prove what Newtonians don't even understand, for one to mention is the sound barrier. The Sound Barrier is part of the Coanda effect, it is part of the Titius bode law, it is part of the Lagrangian formation and it is part of the Roche limit. Although this four principles form gravity, Newtonians never mention the principles.

The principles are what the cosmos uses and Newton's mass is what the cosmos does not use but in spite of what the cosmos uses or does not use, Newtonians elect to use Newton's shamble because it makes the Newtonians look clever instead of what they are, stupid.

If you follow the designated route whereto <u>**The Absolute Relevancy of Singularity**</u> takes you, I will show you how the Universe started. The Universe did not start with the Big Bang, but it started way before the Big Bang. It did not start with materials and light forming space but it started with numbers that formed mathematics and the numbers forming mathematics started by firstly forming points with no space and the points formed clusters and the clusters much later became atomic material spinning in light. The Big Bang is the point in time when the atom formed a three dimensional Universe, but that was way after singularity formed a space less Universe that held numbers in mathematical groupings and these groupings formed the cosmic phenomena I was able to decipher and evaluate.

Without knowing what the four phenomena are, science can never discover the way the Universe started. Without understanding how the four phenomena work, one can never understand gravity and to understand gravity and the four phenomena one must valuate the measured value of gravity. I managed all that and much more and if you read my work you will learn about how the Universe started.

Light came into the Universe alongside the atom at the point where the Big Bang came in place, but at that point everything that forms the Universe already were in place. I discovered the route to unlock singularity and by singularity, everything that is in the Universe came into place by the way of the four phenomena.

This website will not only show you pictures, for that there are other websites that will tell you everything except what happens in physics. This website will tell you **why** these phenomena happens and **what** cosmic philosophy prevails in physics **allowing this** taking place.

This website will entertain you with arguments about physics you have never encountered and it will teach you what no other person knows... Do you know what happens in these pictures? Of course you do because this is the big anomaly science can't explain. That is because Newtonian principles can't explain this. This is one of the four cosmic phenomena that prove Newton to be incorrect!

Do you know what these pictures show…?

It shows a jet aeroplane going through the sound barrier. By going to The Absolute Relevancy of Singularity The Theses http://www.lulu.com/content/e-book/the-absolute-relevancy-of-singularity-the-theses/7587475] you will learn what the sound barrier is. No person ever could understand the sound barrier because no person ever could enter singularity and see what applies when the Universe goes singular.

You will read what exactly happens when the "sound barrier" is exceeded but also I will mathematically prove what gravity truly is. This I explain in simple understandable mathematical terminology.

It is so simple even I can understand it. It shows you why the "sound barrier" is a cosmic limit. This is the first time ever that anyone could explain the "sound barrier" and that is because mainstream science are ignoring principles forming the basics of science.

That they do because those in science don't understand the principles applying in science.

They brainwash students to believe science is the way they believe it applies.

They force students to believe Newton's principles of gravity are proven and is correct while physics shows Newton's principles are baseless and are totally fabricated.

The Sound Barrier is combining the four cosmic principles and the four cosmic principles together form gravity.

Getting the forming of the two mixed up throws the entire view we have about the cosmos upside down. It mixes every concept we have with tainted misunderstanding and misrepresent the cosmos.

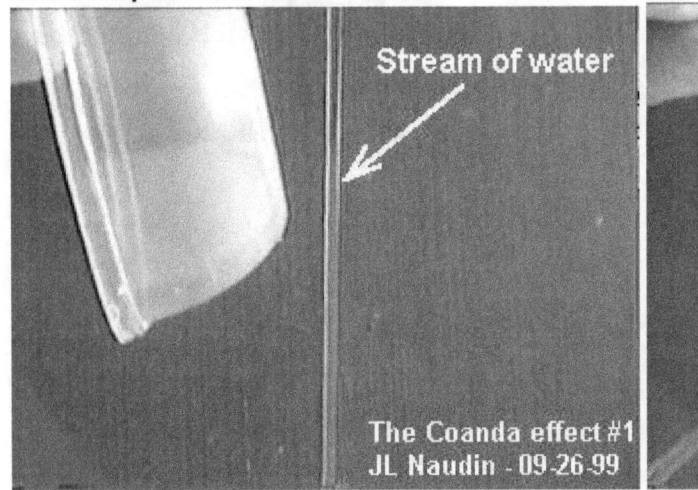
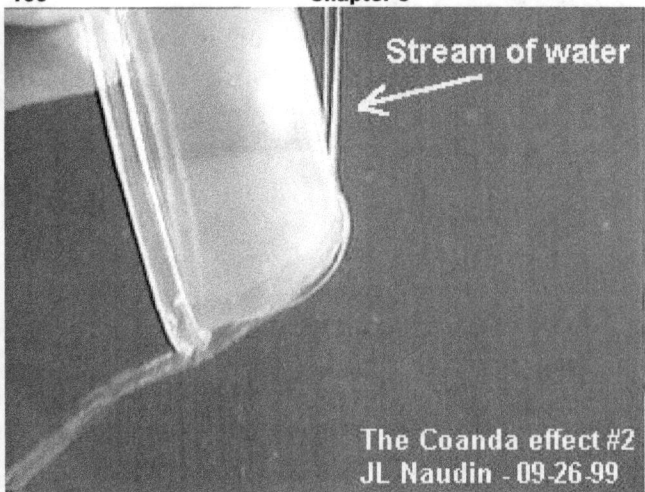

What this is goes by the name of the Coanda effect. I prove that this represents gravity as a principle more than any other form or factor could. Yet, with this so prominent in physics, you will never see any explanation about the Coanda effect in any physics handbook because the Coanda effect puts a serious question mark behind Newton's idea about physics allowing this.

The Coanda effect is the very reason why the Earth has an atmosphere, but you will not learn anything about the Coanda effect... because with the limited view that science at present portray they have no explanation for the Coanda effect or the atmosphere being there other than the mass of the atmosphere pulls the atmosphere down. What a lot of unproven Newtonian gargle that is; what mass could the atmosphere have? The Absolute Relevancy of Singularity shows you how the spin of the earth compresses the space and by compressing with movement, not with mass pulling, the turning produces the Coanda effect and the Coanda effect by gravitational motion condenses space to become the atmosphere. The Coanda effect is the principle that also proves the sun is not a gas giant but it is a liquid as can be seen from the liquid spewing from the sun's surface. That is a dead give away about what gravity really is. The earth spinning contracts the atmospheric space surrounding the earth and that process cause gravity to attract and not the mass of objects as Newton insisted. It is the space holding the object falling that moves downwards and not the object that falls. That is why Galileo was correct when he said all things fall equal under the same conditions notwithstanding size differences. What Newton says is that things fall by the value of mass bringing on gravity. This means if everything has a different mass, therefore everything must fall at a different pace, which doesn't happen...and in that is where science is making the biggest mistake. I ask any one in science to prove the fact of mass applying, not as weight but as a force. Read on and I will show you how those in physical science are brainwashing students into believe that mass is responsible for gravity as a force.

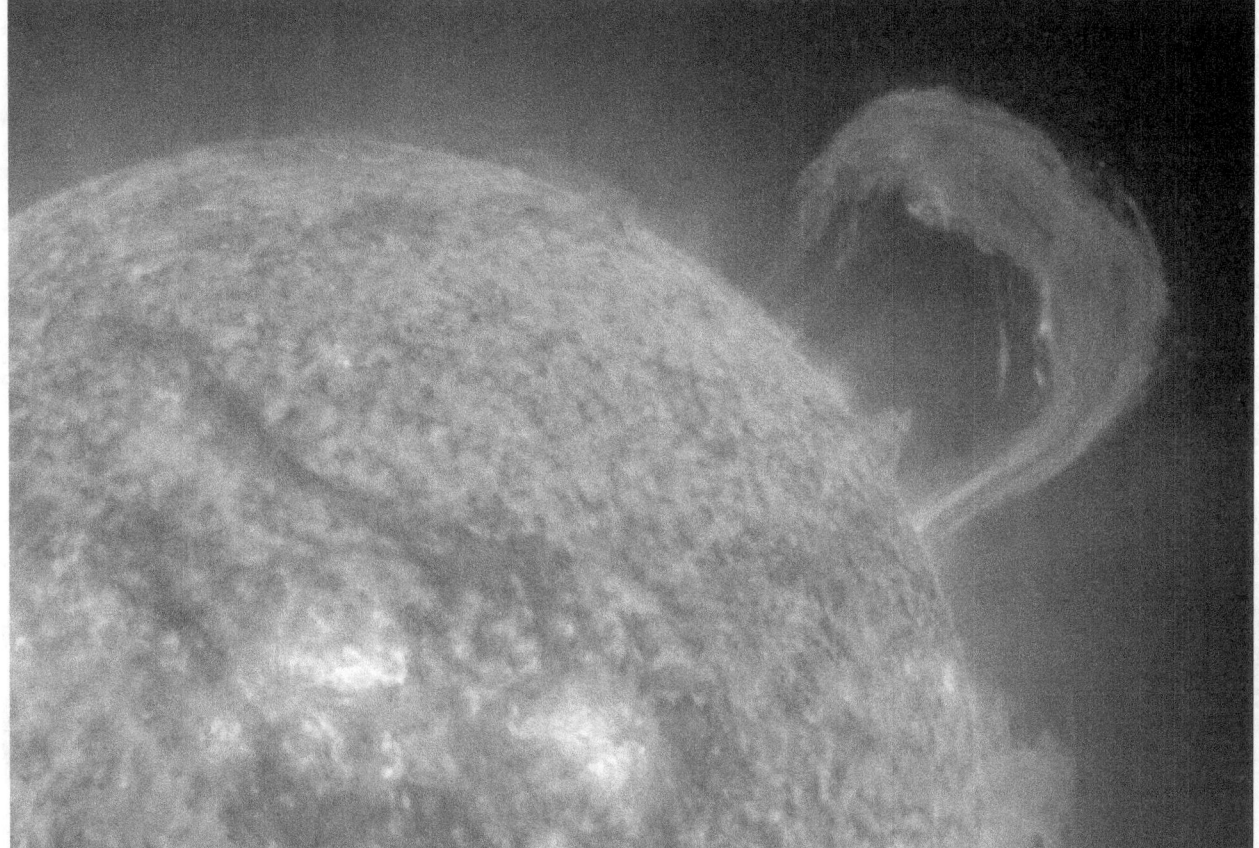

The Coanda effect too is the reason why the sun has such an enormous atmosphere. Science will give you a lot of names by which they classify what they do not understand but what they understand is so questionable their explanation becomes a joke. It is liquid spewing from the sun.

What you see in this picture is also the result of the Coanda effect. To learn more about the Coanda effect download <u>The Absolute Relevancy of Singularity</u> and find out how galactica works by forming everlasting, never-ending and always continuing circles.

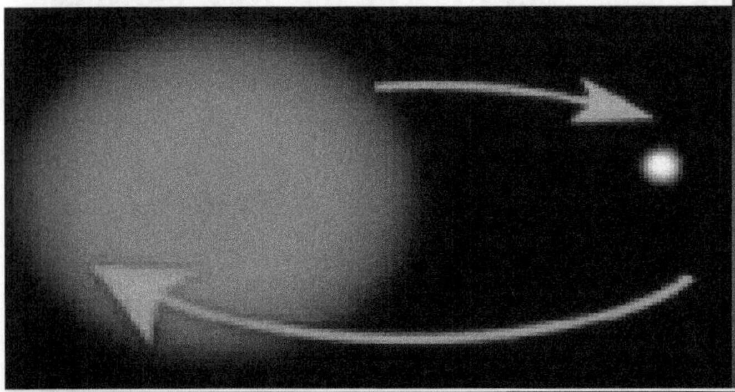

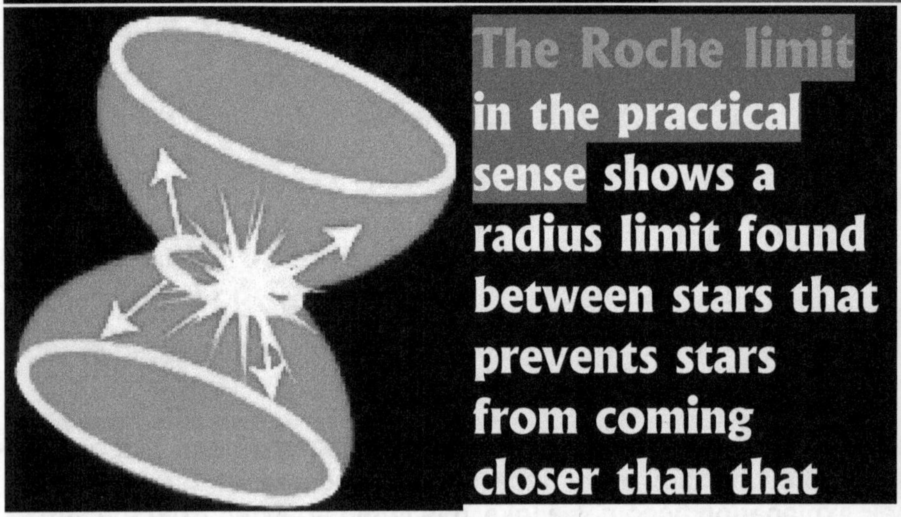

The Roche limit in the practical sense shows a radius limit found between stars that prevents stars from coming closer than that

Do you know what this is? What happens in these pictures proves that stars could never collide because what prevents the collision is a cosmic law that science at present can't fathom because the scientific understanding prevailing at present is far too limited and with the law totally defying the Newtonian gravitational explaining science dare not even give the Roche limit any pertinent place in accepted physics. What happens in the top picture is what happens in the bottom picture. What applies in the picture is the Roche limit that liquidises the lesser structure and then to apply the Coanda effect. <u>Through The Absolute Relevancy of Singularity</u> you will discover how it works the Roche limit as a gravitational factor works.

<u>**The tiny asteroid**</u>

The planet rings.

Ask your teacher to explain the fact that if mass pulls mass by gravity pulling, then why does something as tiny as an asteroid circle around the sun at the very same speed as what the giant planets do. Why would the rings form around planets when mass have the ability to form gravity that will pull the rings into the giant star? With mass doing the pulling what prevents the circles spinning around the large planets to collapse into the planet. What sustain the orbit of the circles and why would the asteroids circle around the sun as if they had the same mass as the planets have?

Don't allow them to shrug their shoulders and carry on as if everything they say is true. That is part of the brainwashing and mind control process.
Do not believe them because they are corrupt.

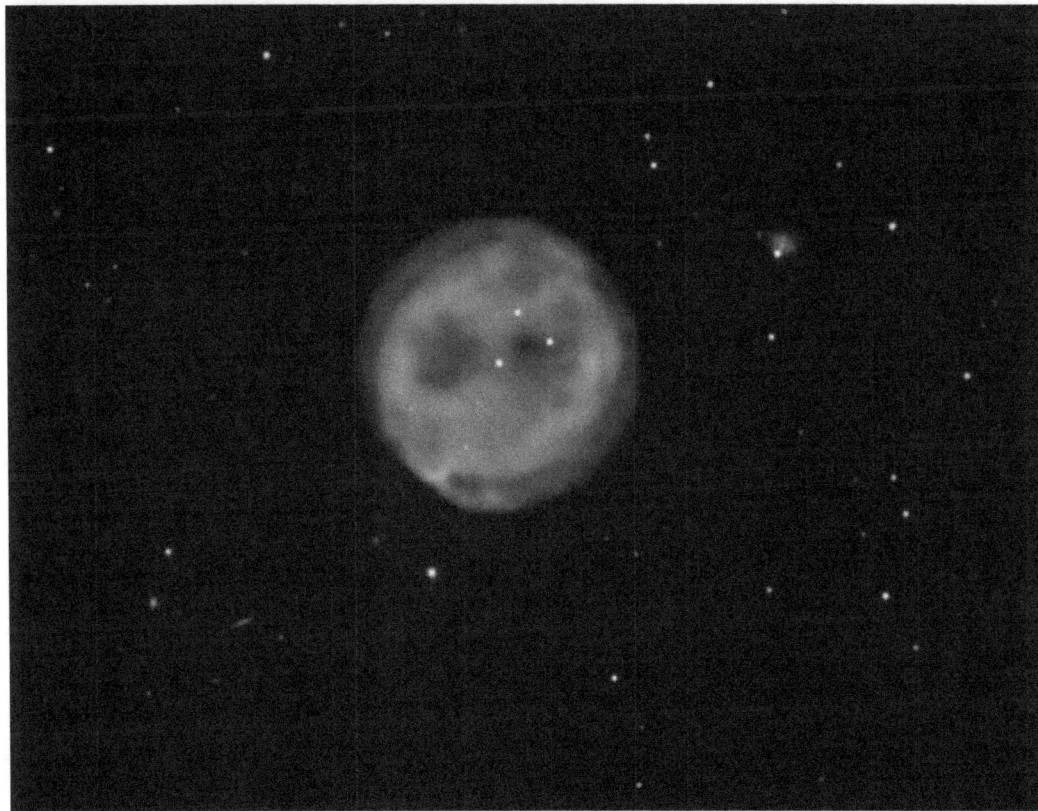

The Absolute Relevancy of Singularity will show how you it is possible to condense all the space around the spinning the planets. You may see how material compress the space around the star. This website informs you about the facts behind the lovely pictures. This website will tell you why the cosmic physics principles establishes these phenomena…and because the information does not salute Newton, academics in physics despise what is said. If mass don't pull stars to a point of colliding then Newton is wrong! Any star coming closer than 2.4674 times the radius of the larger star would liquidise the smaller star and this law and the Coanda effect are a precise duplication of true gravity truly applying.

Do you know what happens in this picture? Is there in physics any realistic explaining available other than saying that gravity has gone mad? That is the best they can do because either they must come up with anti-gravity, which must come from anti-mass, which must be the result of anti-matter, which they can't explain. Gravity can't go mad because gravity has no intellect! Have you ever encountered a realistic believable explanation why a Super Nova explodes? The best offer science sent can produce is that the gravity in such a star goes mad. Then one would presume that gravity has to have some mental stability with a testable intellect and a nervous system that can go mad or stay sane, depending on the emotional stress limit the star can endure. What junk this is and there is even more junk they cook up because they know so little.

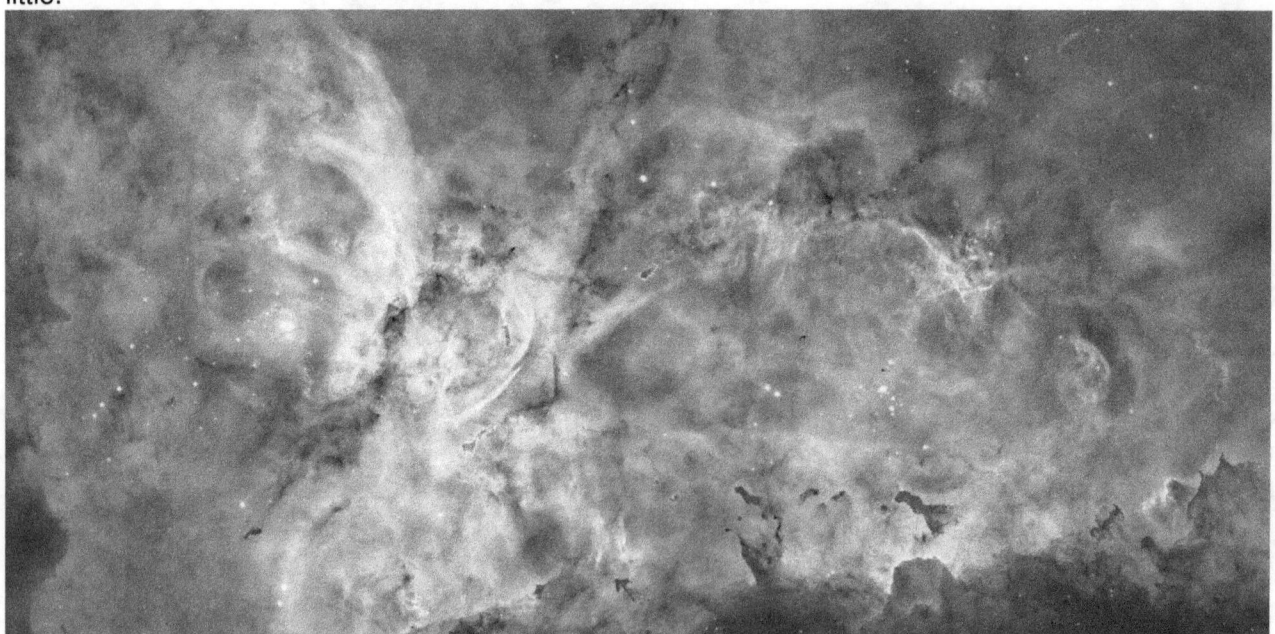

At present science is unable to explain these very crucial phenomena and even to the extent those guarding the principles of applying physics don't even realise the importance of these phenomena in the principles applying in physics.

These phenomena are there and it is in place so ignoring it is rather stupid…but that is precisely what Newtonian science has been doing for centuries.

They ignore these phenomena in favour of what is not present in the cosmos …and that is mass. The put mass in as an explanation for the movement of planets and while all planets has different mass, yet all planets orbit and move at exactly the same tempo. So how can they move according to mass with having different mass and still moving at an equal pace?

What there is, are the four phenomena but because Newtonian science insight do not extend to understand the phenomena and the role it plays in physic s and moreover in generating gravity, Newtonian science is brilliantly ignoring the phenomena as well as the role these phenomena play.
This is forming part of the website **www.singularityrelavancy.com**

The phenomena are there and are applying! Put Newton's formula $F = G \dfrac{M_1 M_2}{r^2}$ to task and use it to explain these very common phenomena, and anyone would find it is not possible to use Newton and explain the gravity represented by this. The phenomena are there and applying so if Newton can't explain it then maybe Newton's concept of mass establishing gravity is not applying. <u>It is this last statement where Newtonian science is unwavering in their believing that mass is forming gravity which is what I strongly bring into question</u>.

In any book that deals with gravity there are just too many and numerously wide ranging facts that form the complete picture as a whole, which leaves me unable to include a full introduction in a space as small as that which page will allow. The explaining include for instance those phenomena, which I call the four cosmic pillars, but wise as you are, you would not believe me at this point that I have cracked the coconut because I guess in your vast experience you have seen too many idle explanations in the past proving to be senseless and little impressive, therefore my mentioning my success would not matter much either way.

Newtonian science creates a factor such as mass to fake reality and misrepresent true physics. As long as they cheat, they seem clever and supremely superior when compared to the rest of us inferior beings. They live in a Universe fit for Alice in Wonderland! Show how can a factor like mass construct the Universe in this picture

Please, for the love of God, show me what role does mass play in this picture. Show me how the biggest stars is locked together and the smallest stars are pushed to the outside. Show me where gravity "pulls" what to where and how gravity forms in this picture. Show where will dark matter hide.

The idea of proof comes automatically to the door of Newton although Newtonians will deny this fact as if they deny the honour of their Master Newton and that is what they have to do. Think of what planets do... and you think that planets orbit. The term pulling does not suggest any circling because no one can be pulling towards and do that while circling.

Going in a straight line serves the term pulling. Then the saying goes that planets orbit indicating they follow a circle. That is not what Newton said. Never do we refer to the planets pulling the Sun or the Sun pulling the planets, but we speak of seasons coming from orbital positions. Being in orbit has to neutralise the pulling and then cancel the pulling concept that also became culture.

	Mercury	Venus	Earth	Mars	Jupiter	Saturn	Uranus	Neptune	Pluto
diameter (Earth=1)	0.382	0.949	1	0.532	11.209	9.44	4.007	3.883	0.180
diameter (km)	4,878	12,104	12,756	6,787	142,800	120,000	51,118	49,528	2,300
mass (Earth=1)	0.055	0.815	1	0.107	318	95	15	17	0.002
mean distance from Sun (AU)	0.39	0.72	1	1.52	5.20	9.54	19.18	30.06	39.44
orbital period (Earth years)	0.24	0.62	1	1.88	11.86	29.46	84.01	164.8	247.7
orbital eccentricity	0.2056	0.0068	0.0167	0.0934	0.0483	0.0560	0.0461	0.0097	0.2482
mean orbital velocity (km/sec)	47.89	35.03	29.79	24.13	13.06	9.64	6.81	5.43	4.74
rotation period (in Earth days)	58.65	-243*	1	1.03	0.41	0.44	-0.72*	0.72	-6.38*
inclination of axis (degrees)	0.0	177.4	23.45	23.98	3.08	26.73	97.92	28.8	122
mean temperature at surface (C)	-180 to 430	465	-89 to 58	-82 to 0	-150	-170	-200	-210	-220
gravity at equator (Earth=1)	0.38	0.9	1	0.38	2.64	0.93	0.89	1.12	0.06
escape velocity (km/sec)	4.25	10.36	11.18	5.02	59.54	35.49	21.29	23.71	1.27
mean density (water=1)	5.43	5.25	5.52	3.93	1.33	0.71	1.24	1.67	2.03
atmospheric composition	none	CO_2	N_2+O_2	CO_2	H_2+He	H_2+He	H_2+He	H_2+He	CH_4
number of moons	0	0	1	2	60	31	27	13	1
rings?	no	no	no	no	yes	yes	yes	yes	no

The Universe is growing! The Universe is not contracting as Newtonian Mythology promotes with

$$F = G \frac{M_1 M_2}{r^2}.$$

If there was a pulling, and the word orbit cancels such an idea, then there has to be some sort of prevention taking place that disallows the pulling to commit the direction of travel. I know it is said that the orbiting object falls as fast as it circles and by falling while moving to the following side on position it never reaches the Sun, and yes, it makes sense, but there has to be some form of resistance replacing the planet in the next side position and preventing the falling or the pulling from taking place.

The person Newton got his ideas from and the work he raped completely, that of Johannes Kepler explained this very well, but Johannes Kepler makes no room for any pulling of any sort. In the work of Johannes Kepler he said that the space being the orbiting route a^3 remains at a specific distance k while the orbit T^2 takes place…and in all my other books that addresses more information I take Newton to task on his dismembering of Kepler's formula by corrupting Kepler's work and with what amounts to fraud, Newton takes science on a goose chase that holds no truth.

There is no pulling by mass to form a force of gravity in any way. Newtonians are faking the truth because the truth makes them seem small while Newton's misrepresenting cosmic principles allow those in science to paint their own Universe as they wish to present the Universe in the way they fake the story. Show me where there is evidence confirming a factor such as mass. Show me what makes the circling of Jupiter going around the sun faster than any of the small rocks in the asteroid belt. Newton placed mass in the formula he said Kepler represented. Could anyone show how much is the giant planets closer to the sun than the small planets are, keeping in mind that the pull of gravity must be exceedingly greater between the sun and the large planets.

Could anyone show how much does the giant planets orbit faster around the sun than the small planets do, keeping in mind that the pull of gravity must be exceedingly greater between the sun and the large planets.

Could anyone show how much is there a greater gravity-whatsoever between the giant planets and the sun than the small planets have less of, keeping in mind that whatever Newton mass would benefit because the pull of gravity must be exceedingly greater between the sun and the large planets. Whatever Newton saw, was in his dreams. Anything they say that mass gives is cosmically unrealistic.

What will the mass of the gas planets have that put them in a different stance when going around the sun than what the small planets have? By small I refer to Earth, Venus, Mars and say Mercury. All plants orbit equally.

The fraud factor came about when Newton destroyed the work of Kepler by changing the work of Kepler without understanding the principle factor guiding the work of Kepler. The fact that one can use $a^3 = T^2 k$ puts the Kepler formula in the realm of singularity. I give away free books that explain how this comes about. I wish to reflect on the modern conspiracy to hide Newton's failure and the deception to cover up Newton's fraud by engaging in almost unlimited fraud in the modern era. The science of the era had Newton's principles wrapped up. Having the formula that Newton gave as $F = G \dfrac{M_1 M_2}{r^2}$ enabled those who could "understand" the principle of "mass pulling" bringing about "a force" that Newton called gravity. This force was supposed to pull whatever was present in the Universe closer to whatever was present in the Universe. The formula gave man an equality to the power only God had because now it was the ability of the physicist to redesign the Universe and correct the failing God made in the Universe. The formula enables science to play a game designing the cosmos and the game used

$$F = G \frac{M_1 M_2}{r^2}$$ as the prime basis.

The Universe was working on a force valued at the measure mass that produced this force and the Universe was coming together pulling everything into one spot. The Universe was contracting…until a man came along going by the name of E. P. Hubble and this man destroyed the godlikeness of Newton and therefore all his followers as well. This no good science cheat could allow!

E.P Hubble is generally credited with discovering the red shift of galaxies. From his own measurements of galaxy distances based on the period-luminosity relationship for Cepheid with measurements of the red shifts associated with the galaxies, Hubble and discovered a shift in material that clearly showed material was moving apart. The distance between materials was growing and was not contracting as Newtonians believed.

Hubble and Humason were able to plot a trend line from the 46 galaxies they studied and obtained a value for the Hubble-Humason constant of 500 km/s/Mpc, which is clear evidence of a Universe that was growing from every point away from every point. In 1929 Hubble and Humason formulated the empirical Red shift Distance Law of galaxies, nowadays termed simply Hubble's law, which, if the red shift is interpreted as a measure of recession speed, is consistent with the solutions of Einstein's equations of general relativity for a homogeneous, isotropic expanding space. Although concepts underlying an expanding Universe were well understood earlier, this statement by Hubble and Humason led to wider scale acceptance for this view. The law states that the greater the distance between any two galaxies, the greater their relative speed of separation.

This discovery was the first observational support for the Big Bang theory, which had been proposed by Georges Lemaître in 1927. The observed velocities of distant galaxies, taken together with the cosmological principle appeared to show that the Universe was expanding in a manner consistent with the Friedmann-Lemaître model of general relativity.

All of this shouts one thing **"Newton is wrong!"** Newton's lovely contracting Universe was not contracting or receding but was expanding by growing. Newton's failures surfaced and with it the entire Humans population could see Scientists in Physics new less about physics than does an ape know anything about maritime exploring. This took away…no, this robbed those that thought they are equal to God their insight into God's cosmic manufacturing process and this showed the world that they were not only not equal to God but was just like Newton, utter failures as professional that knew nothing of their trade. If this evidence of Newton came to the knowledge of those idiots forming the general population that they knew even less about physics than the general population, then they were as much the idiots as the general population.

"Understanding Newton" always made them in physics superior with the ability to know what God and Newton knew and that vision gave them recognition and the superiority to look down on those with lesser abilities…those that did not "understand" Newton. If this came out that the Universe was not contracting by the measure of mass, then those that the educated Newtonians saw as the ones that knew nothing then uncovered the intellectuals as "Understanding Newton" but with Newton the intellectuals did not understand the cosmos. This revelation would uncover them as being the real simpleminded persons who thought they knew everything but was the ones that knew nothing about cosmology.

Those unfortunate ones that did not "understand Newton" as being was then correct all along when they did not understand Newton because Newton was wrong and the intellectuals that "understood Newton" were ones that apparently knew nothing. Then those that did not "understand" Newton was correct because with Newton being incorrect those that understood Newton was the simpleminded minority that did not "understand" what the majority knew by realising Newton did not "understand" physics. They had to devise a plan to save the day. They had to cover Newton's deceptions with more deceiving. They had to go more criminal by topping the corruption to a higher degree as even Newton went before the discovery. They had to devise a plan that would outsmart everyone. They created missing mass in the critical density scam. This is how it works.

I have books explaining how Newton corrupted science and this would take too much space to explain that. There is a book for free **www.sirnewtonsfraud.com**. I am not going into that at this venture.

They devised the critical density theory. How smart can those that are smart get when they truly try to be smart to save their skins? How brainy can the Brainy Bunch be when they deceive a plan to outfox all of the Human kind? You can go to the Internet and read all the multiple arguments about the Universe going flat and contraction therefore come about although not one sod amongst them know the least how singularity works. You can read in books I give for free how singularity works because I plot singularity as 1 and not as a complex formula.

There is to many arguments presented to mention so if you have a year or two filled with boredom, then search the web and go waste your time on fairy tails. You will see what they present when they have no idea what they present.

Here it is in a nutshell; what this is is not a theory that is called the critical density theory but it is a criminal venture conceived by the intellectual minds in physics to go criminal and defraud the world. They decided that the blame of the Universe not working as Newton said it does has to be placed at the door of the Universe. The Universe is lacking matter to contract as Newton said it should. Newton still remains absolutely correct. The blame for the mistake is diverted to the cosmos and away from Newton. Newton did not lack the insight to see the Universe expands, no, the Universe lacks material to contract. The Universe made the flaw and now we must find how the Universe will correct its flaw. The Universe went wrong by not having the sufficient mass and therefore the Universe must correct the lack of mass to get the Universe back in line with Newton. The blame for the mistake must be laid in the midst of the material within the Universe. Newton's formula stands correct and all suspicion goes in the way of how the cosmos was designed not to applaud the truthfulness of Newton.

Students, read the work I offer free of charge and learn how you are brainwashed and how your mind is pre- conditioned into believing in Newton's myth of pure deception which Academics call physics. If you are a student in physics who don't believe that you are subjected to unlawful brainwashing, then read on. Download work I offer free of charge. They come up with the most ferocious mathematical formula that should prove everything they say but all the formula proves is that the one trying to use the formula is out of touch with reality. If ever there is a person trying to impress with the fantastic mathematical formula explaining how the Universe works to his calculation, ask him or her to formulate when the earth and the moon will hit each other. Every formula they use found its basis on the perception that mass forms a force of contraction called gravity. Before they start to bewilder you with breathtaking calculations, call their bluff…ask them to show when will the earth and the moon collide. Tell them to use the mass of the earth and the moon and divide the product thereof with the distance between the earth and moon and then show what a force there is driving the gravity!

All those unrealistic arguments the Brainy Bunch offer as to why the missing mass or dark matter will bring a clarifying solution to avenge Newton has one damning flaw. Whatever they bring as an argument is tainted by a law in mathematics. It is built into the formula $F = G \dfrac{M_1 M_2}{r^2}$.

If the radius increases, then the value of the mass reduces while staying the same. If the factor representing the radius r^2 becomes $2\,r^2$, then by the very same token does the mass become half of its previous value! It is effectively this $F = G \dfrac{M_1 M_2}{2xr^2}$ is $F = G \dfrac{M_1 M_2}{2}$ and this will bring about that while the cosmos is expanding, the worth of the mass is reducing by the same margin.

This is not rocket science; this is mathematics at its most basic. If the Universe was expanding then the measured value of the mass was declining that is if mass was responsible for producing contracting gravity. They are the ones that are the masters in mathematics. They are the ones that know mathematics better than anyone else on earth…and they missed this truth. This missing the basics was as deliberate as it was swindle the hide Newton's incompetence and with it their failure to understand physics.

This is where the second conspiracy started. The first conspiracy was the idea of mass that produces gravity and Newton's fraud to convince the world to believe him. Then the Masters-of-Deceit called Mathematical Physicist devised a plan to protect Newton's image and therefore their academic standings. They got Albert Einstein to hunt for the presumed missing mass and thereby distract attention away from Newton's oversight and their failure. Albert Einstein carried the heavy burden of being acclaimed the title of the best mathematical mind of all times. If you bend mathematical laws you will get a distorted Universe, as distorted as the Newtonian Universe is. The Universe doesn't agree with Newton's principles and Newtonian science has to bend all aspects to get it to fit.

You know…Einstein must have been very dumb or very dapper. Einstein must have been exceptional foolish or extremely brave and you can decide on what merit I suggest you should judge. In astrophysics they teach you about the critical density theory and how Albert Einstein calculated all the mass of the entire Universe and found the Universe fell short of Newton's expectations that was needed to drag the Universe back into one spot. Because of the critical density theory coming short we now sit with the dilemma of Dark Matter hiding in invisible black places waiting in ambush an unsuspecting Universe! If it was not for Einstein counting all the matter in the Universe and found it was too little, the hunt for dark matter would not have raged on.

It is because of this shortcoming that the entire group of science explorers are now searching (mostly in vain I hear) to locate the missing mass hiding in dark matter somewhere in obscure places. What a load of garbage all of that is because look at the picture. Do you think it is possible for one man, (or a million men) to calculate just the mass we see in this one photo? Those in academic posts lecturing you on Newton's correctness think they are the wise and are able to try to fool you because those professors think you lot of students are a bunch of mindless-monkeys that must believe anything they say because they think that only they have the ability and the mind to think and because you are a stupid student, simple-minded when compared to them, you can't think and therefore you will believe anything they say. If Albert Einstein thought himself capable of measuring only the material we see in this picture, then Albert Einstein was a fool with higher ego than what his IQ was.

In fact it is very obvious from what we see that Albert Einstein was no mathematical genius, he was an academic stooge, standing in for the other Academics to try and divert the looming shame by extending the visible problem of Newton getting uncovered. If Einstein was not able to see the reality of relevant formulations then he was more stupid than I am and I am so stupid I don't even understand Newton or so I am told.

This is the academic excepted values: The average density of matter in the Universe today that would be needed exactly to halt, at some point in the future, the cosmic expansion. A Universe that has precisely the critical density it is said to be *flat* or *Euclidean*. If the density of the Universe is greater than the critical density, then not only will the expansion be stopped but there will be a collapse of the Universe in the distant future. In this <u>closed Universe</u> scenario, the Universe will eventually implode under its own gravitational pull, leading to an event known as the Big Crunch. If the density is less than the critical density, <u>an open Universe</u> scenario plays out in which the cosmic expansion will continue forever. The critical density is calculated to be about $(1 \text{ to } 2) \times 10^{-26}$ kg/m^3 – about 100 times greater than the average density inferred from all the known visible matter in the form of galaxies. However, when the inferred presence of dark matter and, possibly, of dark energy, is take into account, the Universe appears to be pretty close to the density called for by the flat scenario.

From these figures, how old is the Universe? What is its future? However, to answer them, it is necessary to know the density of the Universe, also known as "omega" (Ω). What a lot of rubbish this is! The Brainy

Bunch insists that the Universe is 13 billion years old. This they measure by using the speed of light. In the first instant the Universe was the size of a neutron. It took some time to get the entire Universe to the size of five millimetres. That was the Universe.

The speed of light travels at what the speed of light = 299 792 458 m / s. At this rate light travels across a vastness of space 12 billion years to reach us. When the Big Bang came about the Universe was the size of a neutron. Take note on this issue; the entire Universe was the size of a neutron they say. What was the speed of light then? They say at the present it is 299 792 458 m / s, with which I completely disagree but let us leave this argument at that. The speed of light could not have been 299 792 458 m / s when the entire Universe was the size of one millimetre. That means the overwhelming distance of this 12 billion light years the light had to travel at a speed of less or equal to 1 millimetre / 10 billion years, because the Universe was so small at the time! The Universe was 13 billion years ago say five millimetres. Then what was the speed of light at that point? They say it takes the light coming from the furthest images 12 billion years to travel to us since the light travelling now left where it came from 12 billion years ago. At that point the Universe then was about the size of a pea. If the Universe was across the size of a pea, then light took a long time to travel a very short distance!

Bibliographic Entry	Result (w/surrounding text)	Standardized Result
Tipler, Paul A. *College Physics*. New York: Worth, 1987: 877.	"The critical density of matter in the universe that separates the two possibilities can be calculated from Einstein's theory. It is now approximately 10^{-30} grams per cubic centimetre. Small though this value maybe, it separates two entirely different futures for the universe"	10^{-30} g/cm³
Guth, Alan H. *The Inflationary Universe*. New York: Addison Wesley, 1997: 22.	"The value of the critical mass density is believed to lie between 4.5×10^{-30} and 1.8×10^{-29} grams per cubic centimeter, depending on the value for the expansion rate (i.e. the Hubble constant) that one uses in the calculation. By standards of our everyday experience, this density is astonishingly low. The critical density corresponds to somewhere between 2 and 8 hydrogen atom per cubic yard, a density that is more than ten million times lower than that of the best vacuum that can be achieved in an earthbound laboratory!"	$4.5–18 \times 10^{-30}$ g/cm³
Davidson, Keay & Smoot, George. *Wrinkles in Time*. New York: Avon, 1993: 158-163.	"The critical density is calculated to be about five millionths of a trillionth of a trillionth (5×10^{-30}) of a gram of matter per cubic centimeter of space, or equivalent to about one hydrogen atom in every cubic meter -- a few in a typical room"	5×10^{-30} g/cm³
Silk, Joseph. *Big Bang*. New York: Freeman, 1977: 299.	"$d_{critical} = 3H^2/8\pi G = 5 \times 10^{-30}$ gram cm⁻³"	5×10^{-30} g/cm³
What is the density of the Universe and the Size of the Universe? Question Number: 52. ScienceNet Questions and Answers.	"What is the density of the Universe and the size of the Universe? Density = 2.11×10^{-30} kg/m³ Radius = 3×10^{26} meters"	0.0211×10^{-30} g/cm³

The light back then could not have travelled 299 792 458 m / s 12 billion years ago because 12 billion years ago the entire Universe was a couple of centimetres across. How fast was the speed of light when the Universe was the size of a pea. Then light travelled 10 million years to cross one yocto (y)(10^{-24}) of a mille meter. In terms of us today, light back then stood still. If it took the light 12 billion light years to cross, then the first part of the journey was pretty slow which makes the measure of time travelled versus space crossed rather ridiculous in every aspect it is portrayed. This means a lot of the light years it took to cross had to be billions of years just to gain one millimetre of space. Then the Universe must be trillions upon trillions of years older that they reckon it now is. It took a lot more darkness to form all the space is present in the entire Universal than the 12 billion years they say it did That brings us to the dark matter bit and the conspiracy that carries on undeterred. At present the conspiracy went as far as forming dark matter with (I suppose) dark energy.

This dark matter hides in places we can't see. This dark matter is what now forms the lost matter that protected Newton's image of correctness. Still Newton is untouchable because now in the present time the undiscovered dark matter is waiting to contract the Universe and this dark matter hanging suspended is what protects Newton. Is that not that sweet? Is that not the bedtime story every five year old would wish to hear every night? Every child will go to sleep feeling secured and in comfort. We can't attack Newton because the unseen matter that is dark is protecting Newton. Newton becomes untouchable by undiscoverable, unseen untraceable material that lives like a fairy tail in fantasy. If there is dark matter, what is the dark matter waiting for before it unleashes its incredible mass deployed force of gravity on this little unsuspecting Universe. What is preventing the dark matter from forming gravity that will do the job at this point in time? Why is it that the matter must be dark and must be seen in order not to form gravity. If the matter is present and forming a part of the Universe, albeit dark or not, seen or unseen, detected or not, if it has mass and if mass does bring about a force and the force is contracting gravity, then it must employ gravity. What is suspending this dark matter from kicking in and clocking in for duty? What prevents the dark matter from starting to get pulling? This is as big a scam as all the rest of the fraud they use to cover up Newton's fraud. I am showing all of this to prove how much deception there is in cosmology. Everyone in astrophysics is living a fantasy and everyone can make as they please, as long as the mathematical calculations seems to be in order. The reality and the viability or the lack thereof is no one's concern. As long as they can come up with stupefying formulated mathematics any dream will do. And the conspiracy carries on as long as it avoids reality and is void of constructive argumentative facts.

One night so many moons ago I have no intention of dating the time to which I refer, I was sitting outside while anticipating how to solve the riddles of the Universe. Well at least I was attempting to solve the part that riddles me and with my meagre qualifications it did not take that much to riddle me. Sitting outside and staring at the night sky gave me a break from all the confusion that faced me as I was again rejected by one of the so many academics rejecting my work and at the time I was still taking their rejection seriously and took their replies to heart. Then I saw the darkness of the night sky and compared that darkness with the brightness we find the star portrait in order to inform us of its location. That made me wonders about this thing called light and the manner in which light travels.

Have you as you sit reading this part at this minute sat back and gave a thought about the light enabling you to read? Yes it is simple in Newtonian terms because Newtonians keep everything at a child's mentality level so that they might not be confused. There is a wave of photons travelling at the rate of time and travels through space and time that is equal to time. What a load of rubbish did Einstein dump on the human race with that observation? It comes from some formula he or someone like him devised where they firstly manage to stop time. How they can achieve stopping time is beyond my limits of understanding but then again I admit I am one of those that so not "understand Newton". However, it does fit into the thought pattern of the Newtonian since their great Master also accomplished this outrageous deed. They go and put time equal to one. This then they mange in terms of placing the velocity that any object moves in relation to the speed of light and it is the square of this that they put at the root of this which they deduct from time being one and suddenly by the magic of mass it all comes down to time standing still. They put time at one (1) and deduct the square of the velocity any object travels (V^2) from the speed of light (C^2) which then also serves as a velocity and which then formulises as $t = \sqrt{(1 - (C^2 - V^2))}$. No argument is given as to why time and the speed of light should be the same other than Einstein fantasizing. Let's put the argument to the test of logic and see what happens.

If the speed of is equal to time one of two things will apply: either time as the speed of light will fit into eternity and stand completely still eternally because eternity is the fact that nothing changes as al remains unchanged. It is infinity that changes eternity by interrupting eternity. This might seem a little complicated but read **The Absolute Relevancy of Singularity** where I introduce the points holding infinity as well as eternity. The concept is so easy to understand because do not hide untruths behind a collection of bizarre ideas. I shall give you a quick view into the physics I try to introduce and believe me; the impact of this introduction enrols every aspect cosmic physics may ever offer. This shows the human mind is not some mathematical computing device but an instrument of collective logical informative thinking about concepts outreaching simple computing by the way of mathematics. The only thing connecting what is to what was is light. Light brings the past to the present and therefore that is where physics must start and not with calculations. Time moves light through space. That means when having light as time in eternity, which is what outer space is, then infinity will never interrupt light and light will never move from the point it holds. When we look at any object in the distance it is the light that brings the object to us. Yes that does happen to light but there are specific qualifications that have to apply before that will happen. If light was using infinity as time then light must move through the Universe from end to end before time can establish a moment or any name one wish to give for the smallest duration possible. What this says is that if light was equal to time then light must travels across the Universe in one instant of infinity. Then only can time and the speed of light be equal. The Universe Einstein referred to is the multitude of a combination of indefinite numbers of Universes all having an equal value and therefore all being equal and one but they are definitely not one from our point of view. Such a thought brings to mind the most simplistic answer one can imagine. Every atom is an entire Universe in space.

If you wish to know the answer to this and find out what truly is the truth about physics.

Then go one further...think why would all the light coming from everywhere come straight to you, the onlooker! How is it possible that light that left the original position travelled in some cases 12 billion years come to find you who are looking at the sky? Why did the light cross that vastness of space to find you where you are looking at the night sky? Why did the light not simply just miss you and went past you as it disappeared into the blackness.

What makes the position you hold so important that all the light comes rushing to you, precisely to the point where you are regardless of where you are? Why would not one photon miss you as you stand and stare at the night sky? What makes the position you have that important that all the light will come to you from all over the Universe and from wherever the Universe has space and moreover rush to you at the speed of light.

Not one photon misses you standing and staring at the night because you see with your eyes the full picture out there. What puts you in the place where all the light coming from all over then will meet at that point? Why would all the light gather where you are? If the light did not meet and cross at the very point you are, the light would miss you by far, and it does not! This is what Newtonian scientific fraud misses about science.

If the gravity contraction is going on as Newton said, why did everything come from a small dot called singularity and from that develop into something immeasurably big? Whereto is the lot in this picture above pulling? Is galactica absorbing stars or is starts uniting with galactica.

Which will be the next galactica to implode as its stars meet the gravitational demise.
They should see it...

Let's forget about the crooked fictional critical density or the rubbish dark matter that is all made – up fiction forming part of the conspiracy of deception in physics to hide Sir Isaac Newton's fraudulent physics principles.

Every aspect of what is mentioned so far is fraud.

There is a Universe that is very much undiscovered and waits for recognition. This Universe does not work on space and light but it works on singularity running in time that is diverting of space. Before the Big Bang brought about space it worked on singularity holding everything that is in one spot that in today's standards never was.

What happened before the Big Bang is still happening but it is out of view and is located in intellect? It needs much more understanding that what a few formulas can represent because the entirety is still holding a relevancy of numbers and only numbers. The numbers only operate in the adding of numbers because in the dimensional Universe multiplying has worth, but in singularity (1) everything that multiplies brings such multiplication back to 1.

If you are one of those members of society that never thought you would hear the name of an accomplished person such as **Sir Isaac Newton** being associated with **fraud**, **corruption** and **brainwashing**, then these books are specially written to inform you about the truth there is lacking in the correctness about science. Everyone knows that planets orbit around the Sun. Planets circle the Sun which is the same as saying planets orbit the Sun. Just by calling the circle motion in terms of what applies, that statement nullifies Newton's claim of mass that attract mass and put to question the reliability of Newton's dogma. If mass did attract mass, what kept the balance where the planets do find a balance in orbit, rather than move towards the Sun. This is the fraud and a conspiracy to cover up. If mass did attract mass, then what is pushing the planets back into orbit?

The idea of proof is automatically placed at the door of Newton. If normal speech contradicts Newton, then it is it is the Newtonians' task to prove Newton supposition correct and thereby prove the claims about attraction Newton made. Newtonians will deny the fact that Newton was never proven or is incorrect. They will uphold the honour of their Master Newton notwithstanding. Think of what planets do… and you think that planets orbit. It is connected to the brain. No one thinks of planets spinning or planets basking in the summer Sun. When hearing about planets the first thing that comes to mind is the rotating of planets while circling around the Sun.

However, just using the term orbiting is in total defiance with Newton! Newton said gravity draws or pulls or moves in the direction, which would have one understand that the two objects in example the Sun and any of the various planets will be moving directly towards each other. The radius is diminishing.

The term pulling does not suggest any circling because no one can be pulling towards and does that while circling around the object. When pulling anything it must take place while using the shortest line possible. That serves the term pulling. Then the saying goes that planets orbit indicating they follow a circle. That is not what Newton said. However, wrong that may seem but circling is precisely what planets are doing.

If anything whatsoever would contradict or question Newton, then the blame or fault should never lean towards Newton. Of the Universe expand instead of contract as Newton said it must, then it is the Universe that is requiring the missing mass and Newton remains correct. Newton is not wrong about the contraction part; it is the Universe that is wrong about the expanding it is doing. If there is a lack of mass, then the Universe must be hiding the required mass in dark matter in out of sight areas. The Universe carries the blame for the mistake of expanding because Newton said the Universe is contracting and therefore even the Universe must take the blame, but not Newton, never could Sir Isaac Newton carry blame for being wrong.

If there was a pulling, then the orbit cancels out such an idea completely. The orbit then there has to indicate that there has to be some sort of prevention taking place that disallows the pulling to commit the direction of travel. I know it is said that the orbiting object falls as fast as it circles and by falling while moving to the following side on position it never reaches the Sun, and yes, it makes sense, but there has to be some form of resistance replacing the planet in the next side position and preventing the falling or the pulling from taking place. By orbiting the planets don't even suggest a contracting direction of movement going towards the sun.

Students in physics are you aware that you Professors can profess all they like, but they have no foggy idea what gravity is? If you don't believe me, then confront them with the question: what is gravity. If they come out with gravity being a force, then gravity can be water and gravity can be the wind and gravity can be heat because all of those mentioned are also individual forces driving objects. Let them be much more specific than just advocating that gravity is a force of some form.

They do not know what gravity is because Newton, yes, Sir Isaac Newton of physics fame admitted he didn't know what gravity is and that was the only thing he was absolutely correct about, the fact that he never knew what gravity is.

Yet he forms the basis of all physics wisdom… while Sir Isaac Newton never knew what gravity is! No academic in physics know what gravity is because the "father of physics" Isaac Newton did not know what gravity is. That makes on wonder how Isaac Newton could be the "father of physics" when he didn't know what his offspring was…something like an illegitimate child he knew of and he knew about but never knew in person!

Students in physics consider the following: If Isaac Newton didn't know what gravity is how could he have fathered physics? If you're Professors don't know what gravity are how are they able to teach you what applies in physics because they then know nothing about physics because they do not know what gravity are!

They only know what Newton knew and if Newton didn't know what gravity is then they don't know what gravity is…and blessed with that much lack of information they seem to think they are fit to teach you everything they have no idea about

If you never know the answer to what gravity is you inability to reply on a question as to what gravity is, not knowing the answer will leave you unfulfilled because of not being able to answer even the most fundamental physics answer: what is gravity. Gravity is what holds the Universe in form but then one need to answer the following: If gravity pulls towards a centre and gravity holds the Universe attached the question arising from that simplistic **answer is then ... where the centre of the universe is.**

However using the formula $F = G \dfrac{M_1 M_2}{r^2}$ as Newton provided, disallows any other concept other than moving towards. The person Newton got his ideas from and the work Newton raped completely, that of Johannes Kepler explained this tendency very well, but Johannes Kepler makes no room for any pulling of any sort. Johannes Kepler indicates all cosmic space $a^3 = k\,T^2$ specified forming a containing space – circle a^3 has a dual directional movement of circling T^2 at a specific point in correspondence with a straight line **k** point and the space at such a point will formulate as $a^3 \div k = T^2$. In the work of Johannes Kepler he said that the space being the orbiting route a^3 remains at a specific distance **k** while the orbit T^2 takes place...and in all my other books that addresses more information. I take Newton to task on his dismembering of Kepler's formula by corrupting Kepler's work and with what amounts to fraud, Newton takes science on a goose chase that holds no truth. There is no pulling by mass of mass in any way. The Big Bang proves otherwise and the dark matter swindle is there to hide Newton's incompetent incorrectness.

The Universe expands since the time of the Big Bang or the Universe contracts since the time of the Big Bang. We have either one of two that has to be incorrect. If Newton is correct, then the normal way of expanding since the Big Bang is incorrect. Then we must start saying planets are pulled to the Sun. If the normal form of speech is correct and the planets are merely orbiting the Sun, then Newton is wrong. The Universe can't expand since the Big Bang with or without involving artificial dark matter. There can't be expanding while at the same time we have Newton's accepted scientific presumptions of contraction being correct. The fraud part is in the accepting of the Universe expanding while still insisting that Newton is correct in his dogma of contraction with mass. This web page is an effort to show how Mainstream Physics brainwash students into accepting Newton's hypotheses of mass attracting by force while the entire Universe is expanding at the rate the Big Bang indicates. They will rather look for lost missing mass and invent dark matter than to come out and say that Newton was wrong all along. The Universe must take the blame because Newton could never be wrong!

I grew tired of apologising for my (as they see it) having the audacity of being correct on matters of Newton's incompatible religiosity, which I bring to their attention. When being in contact with esteemed Academics agreement with those most esteemed academics in physics. I am expected to show the utmost humble attitude by acknowledging their supreme posture with me being in their surreal presence. I have to feel honoured to be in their company and rather die than to I mention to them their mistake about Newton being mistaken about a Universe that is contracting because according to Newton it pulls while it never ever contracted in the least.

I am quite fed up with the attitude of those academics looking down their noses at me or worse still are those ignoring me whenever I show that Newton's facts just don't add to a conclusive believable answer. They sit in high and mighty places while they cover their positions with fraud and has person controlling their unlimited power. They can waste $5 billion on dubious theories without being responsible to any person controlling their mismanagement and fraud by digging holes into Swiss mountains. Nothing will come from that because their vision on mass is as corrupt as a p[politicians oath. I have reached my peak with stomaching the corruption they hide behind a lily white cover of dishonesty while they sit in their mighty towers and live in a bubble where not even God can touch them less having me point a finger at their despicable ignorance about their mistaken Master Sir Isaac Newton , the man they portray as a God. If the Universe defies Newton, it is the Universe that is wrong.

I am at my limit with being ignored because those academics can ignore me by using their all-powerful status and with the image they carry with them they never have to prove Newton's correctness and therefore is fair in disclaiming even my presence when I try to disprove Newton. They are only gangsters hiding behind fraud. From my view and from my perspective, I honestly can't see any difference between the Mafia's racketeering by forcing the unlawfulness of their trade onto others with dubious corruption and what academics commit in the name of being honourable scientists. They forcefully brainwash students in believing Newton is correct by means of employing despicable mind control. That is as criminal as anything can ever get!

To them and to those I say, prove Newton correct by proving the moon is coming closer to the Earth I Any person telling a lie is committing fraud albeit in the name of God or of science; mendacity is despicable corruption. When they tell a lie to distort the truth and find financial compensation while falsifying facts, even if it is by conducting science, then they are behaving criminally. That is distortion and is equal to the behaviour of the Mafia. Anyone feeling offended, please tell me when the moon and the earth will collide as Newton must suggest that is if Newton is correct about a force pulling mass.

They can say the Universe is made up of nothing. Between the earth and the moon we have lots of nothing and this goes unchallenged. For making a most senseless statement any mind can think of is quite accepted because no one dares question his or her superiority. When I bring this to their attention in a book, I am the person they condemn as being incoherent with my arguments about their nothing they fill a Universe with.

Those in charge of Mainstream physics feed students lies in order to be compensated for their misrepresentation of the truth. They are being paid enormous salaries from student fees to ensure that students believe in the impossible and accept what can never be proven and force students with methodical examinations to repeat the unproven or be expelled from the institutions and branding those expelled students as failures. That is a rip off whether it is justified as science in the process of learning or if it is plain legal criminality; it remains the same because they fly the same banner. Those practising physics waste billions of dollars on falsified theories while I can hardly put food on the table because I fight to reveal the truth. They can corrupt facts and get paid while I am ignored and starved. The information my work carries, which you will read in the event of purchasing my books, you have never seen, it was never yet mentioned or the facts I divulge has never been printed by any person, ever before. I put untruths about the work Newtonian science

claim is correct in question and that might have been published before but never published as questionable evidence. The rest I bring is new. I show how physics truly applies and I can explain everything Newtonian science can't explain.

Those academics in key positions of academic credibility keep certain facts and evidence away from students and give other facts that were never proven before. The academics teaching physics give prominence as well as credence to Newton while applying their trade in brainwashing to give their Newtonian views undeserved credibility and from these proceedings they earn substantial incomes. That is the same as racketeering. When you deceive by conveying untruths and cheat to mislead, then your behaviour is criminal.

If you purchase any of the my books on offer, then through the books you will come in contact with the truth for the first time in centuries. My work is about uncovering the truth and blaming the shameful conduct of those persons everyone trusts since no one expects such persons to be criminals.

When you purchase my books, I don't sell ink on paper. I do not sell material with questionable information, holding facts that were repeated so many times that it is accepted as the truth because it became a culture to believe Newton. I prove statements by taking the reader into infinity and into eternity respectfully.

However I do reject the idea that mass is perceived to have a unexplainable and improvable magical pulling power that is used to establish a force thought to be gravity that pulls all things closer.

Please let any of your physics professors show you that all the big things are spinning in a group together as a unit, either in the centre or the outside leaving all the small things then either on the outside or the inside. If $F = G \dfrac{M_1 M_2}{r^2}$ was true the big stars will group in one area and the small stars in another area because the big stars must have more gravity having more mass. That is if mass does attract.

The above questions I answer but my answers show the foolishness of Newtonian physics and why it is that the Academics in physics all teaching about Newton's wisdom despises me more then I do despise Newton's corruption.

Let's start surveying civilized principles by evaluating what lawfulness means and what would constitute as morality. Let's determine what makes the crook in the book?

If any person, notwithstanding what reason is given in justifying such depravity, tells a lie or conveys untruths to further whatever humble cause, it is seen as fraud. To convey information that is not substantiated as a verified fact then the mere conveying of such information becomes fraud.

When any person, notwithstanding what reasons given, repeats such a lie unabated while being well aware that the information passed on by such a person is incorrect, then the person commits deceit. When anyone is repeating the information that is passed on as being unblemished factual substantiated and verified truth while such a person knows very well that such information is void of proof or lacks proof, then committing such an act is a criminal enterprise. Academics in physics commit every one of the above indignities and yet see their actions as being lawful and even much praiseworthy and hold their role in society in the highest esteem imaginable. They fail to see the crime that they commit while tutoring physics. Whatever motivation they may claim to have which they offer to serve them as forming their driving force, the fact that they perpetually perpetrate in unlawful behaviour, by spreading untruths, such actions on their part put those academics holding such highly regarded positions in the league of ordinary cheats, gangsters and common criminals. By wilfully and constantly falsifying facts to further whatever humble cause and produce illegal claims repeatedly, remains derogative behaviour and is unlawful by nature, notwithstanding what morality it should serve. A Preacher or Pastor lying on behalf of God is not lying on behalf of God and to think the Preacher or Pastor improves or underlines the Greatness of God by lying on behalf of God is very mistaken, because in reality such a Preacher is falsifying the truth for his or her personal benefit. Lying is wrong and doing so even in the name of God remains despicable. The same applies to academics in physics. There is no argument that can change this truth about falsifying the truth and when doing so there is no hiding behind any excuses of ennobling to benefit mankind that will change such truth into righteous conducting. By bringing on a cover-up scheme as we find in the critical density scam, it stinks of criminal fraud. The entire saga is only to cover Newton.

Newton said centuries ago that gravity is the force of attraction there is between objects that hold mass and it is the mass factor that brings about this attraction, which Newton claimed there is. The Universe does not contract and all the proof we require to disprove such a statement we find in the Hubble constant as a guarantee. Moreover, it is true that the Universe never contracted even for a brief instant and we have the proof of that as the Big Bang concept. Also does the Big Bang guarantee that contraction will never become part of the Universe since it is the relevancy between what I named cosmic solids and cosmic liquids /cosmic gas that will provide the balance that shifts.

Since the first instant that time began the cosmos grew away from and not towards points. With all the proof that this concept brings in backing the principle of expansion in the Universe, still the Arch-Fathers-in-Physics uphold Newton's contraction. Planets never moved closer, are not moving closer and will never move closer to each other and this is backed by all information collected this past century. The Moon is not coming closer but the distance between the Moon and the earth is widening. Studies about the Universe reveals every time that space in the cosmos increases constantly. Studies find all things are moving apart and away from one another. Any and all the proof about this is beyond what any doubt may present to counter this knowledge. Notwithstanding this irrefutable findings, science still regards Newton as the only person that ever lived whom no one ever could prove wrong…and this is upheld by Mainstream Physics in spite of the cosmos proving Newton wrong every instant of time. The basis of what science holds as its foundation we find to be the Newtonian principle of $F = G \frac{M_1 M_2}{r^2}$. The foundation used by science promotes this argument and backs up this argument well knowing that in the cosmos there is no evidence backing up this proposal Newton suggested. The Newton formula $F = G \frac{M_1 M_2}{r^2}$ used as basis for science sees gravity as being a force of attraction and the force of gravity is being in place between all objects in accordance with the mass factor that the objects have as presented by Newton in the formula $F = G \frac{M_1 M_2}{r^2}$.

If you think scientists know what gravity is…then do not be duped that easily because no one in science remotely knows what gravity is…not even Newton knew what gravity is because Newton admitted not to know what gravity is… yet everyone in science are redesigning the Universe as if everyone in science knows exactly to every specific and smallest detail what gravity is.

Yet in spite of all the blistering ignorance there are about gravity in the field of academics and all those academics pretending to be informed in physics while claiming and pretending to have Mastered all there ever can be known about gravity all in science is reinventing the Universe without having the least bit of knowledge about what gravity is.

Having not even a foggy notion about what gravity is, why gravity is there or what gravity does other than performing as a pulling force of some sorts is all they gained in knowledge since Newton did nor know what gravity is and ignorance about gravity is all the gain they can show. All they have to do to sustain the current intellectual grasp academics have on gravity is only to acquire more ignorance and that will maintain the current level of information acquired the past three hundred and fifty years. After 350 years they are still as much in total ignorance about gravity, just as they were when science was in the dark ages and Newton did not know what gravity is.

What we find as we gauge all evidence found while studying the Universe, is that reality shows there is no attraction between objects in space going on anywhere in the Universe, that the entirety of such a concept is a myth and the outward moving of the Universe has been coming from and since the time of the Big Bang and maintaining this flow of material is substantiated in a concept named as the Hubble constant, which proves Newton's perceptions to be a myth. The Hubble constant proves that space everywhere is growing ever since time began and the growth never stopped ever since. Knowing this irrefutable fact does not deter science from under scribing Newton as the sole basis that underwrites all the correctness of all of science known as physics. However, Hubble and the Big bang and all other investigations contradict this attraction Idea Newtonian dogma holds.

Therefore, any further believing that there is attraction going on as Newton claimed has to be viewed for what it is and that it is a fairy tail. The Big Bang Theory proves Newton's idea as not only being wrong but Newton's idea of attraction is a joke. If the Big Bang is expanding the Universe, then how can the Universe contract at the same time? Any contraction by nature would have the Universe collapse back into infinity the moment the Big bang moved out of infinity. Ask your professor to show how an expanding Universe can also contract and your professor will tell you about Einstein's Critical Density theory. This theory I prove is the biggest fraud ever devised by any group of persons in the history of civilization! This is perpetrating fraud and conducting in upholding deceptions instituted by Newton that then formed the institution of lies they call physics.

The Universe does not contract in any way; means or form and even such a suggestion are incorrect! The Moon and Earth are not moving closer but are moving apart. The entire Universe is growing in space and nowhere is space depleting by any norm used. Academics are very aware of this misconception Newton had and still academics in physics are promoting the ideas of Newton as the unwavering truth. Academics teaching these misconceptions are committing fraud, notwithstanding the portraying of their role in society being unblemished, spotless while they are covered in a lily white blanket making them being whiter than snow and having such a holier than thou attitude. Teaching Newton is participating in deception and promoting Newton is criminally deceiving the public and while doing so, is committing an act with criminal intentions.

Then, in the face of all this evidence contradicting Sir Isaac Newton; they remain upholding the correctness of Sir Isaac Newton and keep on teaching students about the unwavering correctness of Sir Isaac Newton. They put down conditions of learning to this effect and are expecting students to repeat these untruths and unproven facts by forcing answers to that effect in examinations. Forcing the acceptance of this untruth about physics is equal to preposterous subjecting students to physiological torture and heinous mind conditioning, scandalous thought control and brainwashing. This applies to everyone serving as a tutor in physics notwithstanding whatever status the torturers might have in society or the morality they attach as a reason to commit such atrocities.

If you are a student, then you are conditioned by academics in controlling your thinking by enforcing pre-mind setting and in which they methodically force you into believing in Newton and this is an on going process conducted for centuries in the past, while it is the truth that Newton is completely void of any tests that may secure any form of confirmation and in securing proof then also by that establishing proof. Read this book **Newton's Mythology** and then use the information I supply in the book to insist that Academics who are teaching physics, prove to students that Newton's statements of attraction are correct. Let those academics explain the method mass uses. Let them with precise detail show when mass is applying it forms gravity that mass does produce gravity and such producing of gravity that then would establish attraction! I show precisely how gravity produces mass but mass can never produce gravity. I show with explicit detail when, how and where gravity forms mass but mass can never form gravity. What I prove annihilates every Newtonian claim.

They never prove Newton's philosophy on gravity but those persons conducting teaching in the subject of physics force all physics students to learn Newton's gravitational concepts and accept the facts as if it has been proven beyond all other facts. Students have to believe that Newton is correct or academics will see to it that they fail their examination. The condition of being accepted in physics is to accept Newton without questioning the proof that is never supplied. Let those academics now prove precisely how mass brings about gravity and then afterwards test you on how Newton is proven correct and not on you repeating facts about what they say is true about what Newton said, which they say is true. The manner they present Newton is completely hearsay and that method may not be used in any court of law. Let your professors now prove how it is that Newton's teachings are correct and then examine you on the process they use to prove Newton's concepts. At present they say Newton is correct and then they test you on your ability in repeating that Newton is correct without ever proving to you that Newton is correct. Let those physics professors now prove Newton and then test you on the manner they use to prove Newton to be correct.

The truth beyond all other truth is that Newton's gravity has never been proven (because try as you may it is not possible to prove Newton's formula forming gravity mathematically) and because academics know that, academics require the blind acceptance of Newton by students. This unconditional acceptance of Newton's correctness relies only on the pre-conditioning of students' mind set and academics depend only on the student trusting the academic "say so" about the institutionalised correctness of Newton. That Newton is correct nevertheless and notwithstanding that there is no founding proof about this matter, is what students should be accepting blindly. Pre-conditioning students into blind acceptance depends on the academics' insistence that students approve Newton's concepts without pre judgment or students insisting on scrutiny of any sorts. In examination students have to outright and blindly follow academics' say so only because academics say so. Academics depend on students never questioning their say so or demand proof about what academics teach. Those academics in teaching positions insist that all students accept Newton's accuracy.

This is methodical mind control as much as it is the brainwashing I show that they enforce. If you are one of those believing that Newton was ever proven, then what you believe to be true is a lie because Newton can't be proven and that is the truth! The time has come to face your teachers and force them to stop the ongoing old culture of bullying students and conditioning their thoughts by enforcing on them dogmas, which is, mind control! In order to get students to accept Newton's hypothesis, academics resort to brainwashing pupils and students. They teach you that the Universe contracts and to state their case they force students to learn that gravity is proved by Newton introducing the following formula $F = G \frac{M_1 M_2}{r^2}$

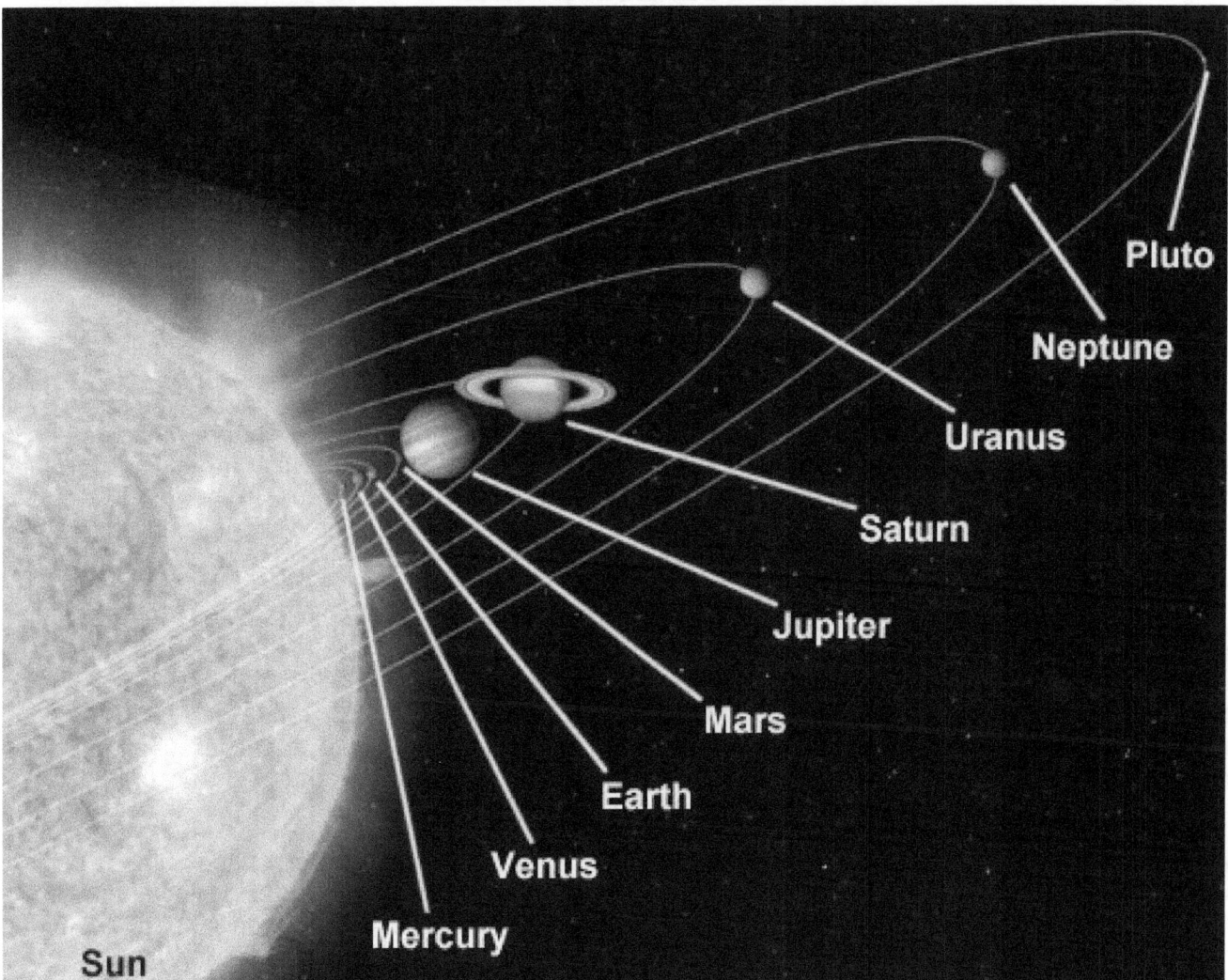

They say that M_1 is the mass of the Earth and M_2 is the mass of the individual in questions mass and the multiplying of these factors with the gravitational constant produces the force of gravity when this gets divided by the square of the radius. Please let you lecturer put in all the values of the formula and prove Newton is correct. If he can't and I know for sure he never can fill in the symbols and calculate the force of gravity, then read the rest of the web page that follows to see how far academics in physics go to brainwash students into believing in Newton's fraud. This is a fair test to see if Newton's contraction theory underwritten by Newton's attraction formula $F = G \frac{M_1 M_2}{r^2}$ is valid, and then force your professor to use this formula as it reads and show WHEN the Moon and the Earth is going to collide. If he fails to do it by using Newton's formula as $F = G \frac{M_1 M_2}{r^2}$ then you will know who is conning you, him or I and who is truthful again him or I. I charge all academics to prove what I say is being wrong in any way or even that I exaggerate in the least. I challenge Newtonian academics to prove that mass does indeed form any force of any sorts and in particular gravity! To those professors claiming Newtonian ideas are substantiated by proof, I say that notwithstanding your personal academic qualifications and while at the same time disregarding your status and previous achievements as well as ignoring your many admirable abilities you may have and however superior they might be, I shall teach you about gravity. I say it is time Students learn the truth about physics notwithstanding the status academics will loose. Students read **Newton's Fraud** and challenge those academics depending on their ability to brainwash you into submission.

The idea of proof comes automatically to the door of Newton although Newtonians will deny this fact as if they deny the honour of their Master Newton and that is what they have to do.

A Brief Summery of The Practice of Brainwashing and Mind Control of students in Physics

**Show me any evidence of mass playing a role in any way except fuelling Newtonian imagination!
Let me show you what is The Practice of Brainwashing and Mind Control in Physics**

The Definition of Gravity according to the Oxford Dictionary of Astronomy:

Firstly it is said in physics that in Physics:
1) **The natural force of attraction exerted by a celestial body, such as Earth, upon objects at or near its surface, tending to draw them toward the centre of the body.**

What this refers to is the process by which weight is measured and the more a body pushes onto the earth, the more would the measured scale reading be. It is whether this result comes, as the intervention of a pulling magical force; that is what I strongly dispute. Mass indicates in the same result as what measuring weight would be and there was never any defining proof that parts this form of mass from this form of weight. It is always said that weight and mass are different but mass is measured in the same currency as weight and that puts mass and weight in as the same thing! Both mass and weight are measured in the same measurement and have the same indicators making the two identical. This form of mass is there when something touches the ground or connects to the ground by some direct link but is the same as weight and only disinformation in physics by tutors can put any distinct identifiable characteristic differences claimed to be between the two.

Mass has no influence on falling objects. The story of the hammer and the feather falling in a vacuum is part of the conspiracy to defraud science and confuse students by manipulating thoughts. A hammer and a person and the toolbox will fall equally with or without a vacuum! Everything falls equal as long as heat and other conditions prevailing are equal and mass plays no part in bringing differentiations. On TV we see stunts where objects fall out of an aircraft everyday that disputes Newton's claims because all things fall at an equal rate and that dismisses falling to go by mass. All things falling fall equal and that disclaims Newton's idea of mass applying by falling.

When two people jump out of an aeroplane and a tennis ball is thrown out of an aeroplane with the persons and this is alongside a car the four objects will fall precisely equal as long as the conditions that they fall within remains the same. If there are two persons falling the two can throw the tennis ball to each other where one is inside the car and the other one is outside the car and the lot will still fall at the same rate. Only when this lot hits the ground and mass comes into play carrying weight as a factor would the damage become individual because the weight or mass that enters the scenario becomes individual. Touching the ground brings differences but falling has nothing to do with mass and therefore gravity has nothing to do with mass and teachers and lecturers in physics teaching students otherwise is practising brainwashing and mind control.

This information is available for the first time ever. For the first time students have the information to stop those brainwashing them and force-feeding them corrupted information. The process is a way of centuries long mind and thought manipulation and only students have the power to stop the teachers and professors fooling them with nonsense. If mass did bring on the "force" of gravity every object will fall different and not equal as Galileo said it does, and nobody is disputing Galileo because all clocks, big or small keep time the very same way and prove Galileo correct. Newton said the very opposite by implying mass brings about falling whereby all objects must fall different and not equal as they do and as Galileo said it does. Objects cannot fall equal and by the same margin fall by the measured value of mass because every object holds different mass. Stop those brainwashing you into believing the impossible because it is either Galileo that is correct or it is Newton that is correct but both cannot be correct while holding opinions that far apart.

Secondly it is said in astronomic physics that in Physics:
2) **The natural force of attraction between any two massive bodies, which is directly proportional to the product of their masses and inversely proportional to the square of the distance between them.**

I beg anyone to use the information gathered from the solar system to prove this definition do apply!

This again is as much fraud as the first argument. If mass did pull, the comet must hit the sun and not circle around the sun as it does. If mass pulls the comet closer, then what pushes the comet away after it missed the sun during the spiralling around and when it rushes into the dark black space. If mass pulls then let your lecturer / teacher / professor tell you why the comet circles around the sun and not smash into the sun. If it is the mass that "pulls" the comet towards the sun, what is the "force" that "pushes" the comet away from the sun after the comet circled around the sun? Let them tell you this inconsistency because by not telling you this hidden fact this methodical selection of facts form part of the brainwashing process they submit students to when they apply to students to condition their thoughts and make students believe Newton is correct.

In the table in column "**A**" the line of numbers show the "mass" of every planet, which is very different and in the table in column "**B**" the line of numbers show that "the orbit speed or velocity" of every planet is much the same. If the planets are in orbit by gravity produced by mass, then Jupiter must be orbiting 317 times faster than the earth does and then Mercury on the other hand must be orbiting 16.6 times slower than the earth does. As anyone can see from reading information from Kepler's tables, notwithstanding everything that Newton mathematically fabricated by corrupted manipulation of the truth, the planets all spin the same and equal around the sun notwithstanding mass differences. Please use these columns in the table to show where the solar system adheres to the use of mass or put distinction in any factor forming mass. Please prove that the Universe is aware of Newton's claim about mass forming the over riding factor that drives all forces in the Universe strictly or otherwise to the measure of mass. Mass is the biggest scam that was ever devised!

Use either the images as seen in the portraying of the planets or by using Kepler's tables to prove Newton's conspiracy about mass forming the Universe as in $T^2 = \dfrac{4\Pi^2}{G(M)} a^3$ that subsequently is expressed as

$$T^2 = \dfrac{4\Pi^2}{G(M_1 + M_2)} a^3$$

is positioning the planets.

Replace the values of the mass in the following factors in the equation

$$T^2 = \dfrac{}{G(M_1 + M_2)}$$

with the planet's applying mass values of the planets and prove me wrong that this is or never was the biggest scam and conspiracy ever divided. Show that the mass of every planet does allocate the planet in its orbit or admits Newton is a hoax.

According to the tables, all movement is according to some other value than mass. This proves that all objects in space that is not connected to the earth or any solid spinning object and thereby receiving weight as mass, but all objects holding free space has the same mass or weight because all space is weighing the same. Whether material fills space or is space not occupied with material filling space, all space is the same and the sun controls all space within the solar system equal as much as the earth controls all space in the atmosphere equal. Notwithstanding whatever mathematical deception Newton produced to convince anyone, in nature he stands unproven and he failed to convince objects performing in gravity by natural physics.

This is the table Kepler showed the speed at which planets EQUALLY orbit.

Planet	Mass per Earth unit A	$k^{-1} = T^2 \div a^3$ Movement B	a^3 of space volume	T^2 During time units
Mercury	0.06	$T^2 \div a^3$ =0.983	$(a^3)=$ 0.059	$(T^2)=$ 0.058
Venus	0.82	$T^2 \div a^3$ =0.992	$(a^3)=$ 0.381	$(T^2)=$ 0.378
Earth	1.000	$T^2 \div a^3$ =1.000	$(a^3)=$ 1.000	$(T^2)=$ 1.000
Mars	0.11	$T^2 \div a^3$ =1.000	$(a^3)=$ 3.54	$(T^2)=$ 3.54
Jupiter	317.89	$T^2 \div a^3$ =1.000	$(a^3)=$ 140.6	$(T^2)=$ 140.66
Saturn	95.17	$T^2 \div a^3$ =0.999	$(a^3)=$ 868.25	$(T^2)=$ 867.9
Uranus	14.53	$T^2 \div a^3$ =1.000	$(a^3)=$ 7067	$(T^2)=$ 7069
Neptune	17.14	$T^2 \div a^3$ =0.999	$(a^3)=$ 27189	$(T^2)=$ 27159
Pluto	0.0025	$T^2 \div a^3$ =1.004	$(a^3)=$ 61443	$(T^2)=$ 61703

Therefore Newton's $F = G \dfrac{M_1 M_2}{r^2}$ becomes invalid.

Those teaching physics never prove Newton's philosophy on gravity but those persons conducting teaching in the subject of physics force all physics students to learn Newton's gravitational concepts and accept the facts as if it has been proven beyond all other facts. Students have to believe that Newton is correct or academics will see to it that they fail their examination. The condition of being accepted in physics is to accept Newton without questioning the proof **that is never supplied.**

	Distance (AU)	Radius (Earth's)	Mass (Earth's)	Rotation (Earth's)	# Moons	Orbital Inclination	Orbital Eccentricity	Obliquity	Density (g/cm^3)
Sun	0	109	332,800	25-36*	9	---	---	---	1.410
Mercury	0.39	0.38	0.05	58.8	0	7	0.2056	0.1°	5.43
Venus	0.72	0.95	0.89	244	0	3.394	0.0068	177.4°	5.25
Earth	1.0	1.00	1.00	1.00	1	0.000	0.0167	23.45°	5.52
Mars	1.5	0.53	0.11	1.029	2	1.850	0.0934	25.19°	3.95
Jupiter	5.2	11	318	0.411	16	1.308	0.0483	3.12°	1.33
Saturn	9.5	9	95	0.428	18	2.488	0.0560	26.73°	0.69
Uranus	19.2	4	17	0.748	15	0.774	0.0461	97.86°	1.29
Neptune	30.1	4	17	0.802	8	1.774	0.0097	29.56°	1.64
Pluto	39.5	0.18	0.002	0.267	1	17.15	0.2482	119.6°	2.03

Do you as the reader and visitor of this web page realise that there is work that is used to introduce an astonishing presentation about gravity, about Space-time and finding the centre of the Universe.

His work is about translating mathematical equations and being correct in interpretations of mathematical expressions.

Other factors are about certain mathematical deductions that were made in the past but were incorrectly presumed. A line cannot and therefore does not start with zero. Should you think such a statement is trivial, and then this book is even more especially for you because judging that correctly that mathematically changes where one presumes the Universe came from at the very beginning?

Kepler said $a^3 = T^2 k$
Then it is also true that $k^0 = a^3 / T^2 k$ and it is true that $k = a^3 / T^2$ and it is true that $T^2 = a^3 / k$, that explains gravity. It seems ridiculously simple but when this formula is dissected as it should, it is truly complicated.

The truth behind the entire Universe hides in the formula Kepler gave as $k^0 = a^3 / T^2 k$ and that is what Newton missed.

Behinds this Kepler equation hides most of the many yet unsolved cosmic mysteries… and I can prove I uncovered most of them, most of all whey the Universe expands.

Why would you think that the Universe is always correctly anticipated when it is depicted as a sphere…Ask yourself honestly why you accept that…that the Universe is in the form of a sphere…

I had found a means to define gravity for the first time ever since the time Newton introduced gravity. I used Kepler after every one discarded Kepler and by using Kepler's related information taken from studying Kepler I now can interpret certain cosmic aspects much more accurately than any one else.

That enabled me to introduce changes about the Big Bang theory in as much as I can now show how gravity formed particle-plasma and particle plasma then can replace the matter versus antimatter cannibalism that supposedly devoured almost the entire cosmos.

I achieve this which I say I did when I accepted what Kepler said what gravity is while Newton rejected and never knew what Kepler said what gravity is. If only Newton accepted what Kepler saw as gravity he would not have had the audacity to go about changing the facts about that what Kepler said what gravity is?

I used Kepler after every one followed Newton's lead and discarded Kepler. By using Kepler's related information taken from studying Kepler I now can interpret certain cosmic aspects much more accurately than any one else. That enabled me to introduce changes about the Big Bang theory in as much as I can now show how gravity formed from the centre of the Universe and I can show and prove the location of the centre of the Universe from any valid angle, which is available at that point.

Let those academics now prove precisely how mass brings about gravity and then afterwards test you on how Newton is proven correct and not on you repeating their facts blindly about what they say is true about what Newton said, which they say is true. The manner they present Newton is completely hearsay, completely unsubstantiated without having any grain of truth and as such it may not be used in any court of law as evidence. Let them first prove by showing how mass is responsible for gravity and then test you on that. Let them show objects fall by mass with every object falling distinctly according to mass and not fall equal by space reclining and that the big planets spin faster by mass and not that all objects in space regardless of size spin equal around the sun by space spinning around the sun as Kepler indicated.

Let your professors now prove how it is that Newton's teachings are correct and then examine you on the process they use to prove Newton's concepts rather than test the state of brainwashing you have been submitted to. At present they say Newton is correct and then they test you on your ability in repeating that Newton is correct without ever proving to you that Newton is correct. That is not testing your knowledge but it is testing the mind control you have submitted to. Let those physics professors now prove Newton and their ability to prove Newton correct and then test you on the manner they use to prove Newton to be correct.

If mass did bring distinction, the line up of planet allocations must be in the order of mass in order to follow the layout that Newton's gravitational formula $F = G \dfrac{M_1 M_2}{r^2}$ would insist upon. Yet everything is placed at random and there is no reference to mass contributing in any respect. The solar system totally ignores Newton's mass and the only place we find validity for mass is in the imagination of Newton.

Let your tutor prove the validity or existence of mass before your tutor again starts shouting that mass associates with gravity applying as **_"The natural force of attraction exerted by a celestial body, such as Earth, upon objects at or near its surface, tending to draw them toward the centre of the body"_** with what we think of happens to solar bodies having mass which is **_"the natural force of attraction between any two massive bodies, which is directly proportional to the product of their masses and inversely proportional to the square of the distance between them"_** and science cheats everyone into believing the two is the very same. The one I experience every day and with the other there is no evidence of applying anywhere. Yet students are fooled into believing the two are exactly the same issue, which is untrue. If there is any academic feeling insulted by me calling the lot fraudsters, bring evidence of the second form of gravity working on the principle of mass and I will withdraw my statement, otherwise if no proof can be brought, then you lot that ignored me for ten years are the fraudsters I accuse you to be.

You lot that are teaching physics the Newtonian manner, you are villains brainwashing students to corrupt their thinking while you teach them what you know can't be true. I challenge you to prove that mass is something that pulling is something and how does the concept exist..

3) **Gravitation.**

This proves how low Newtonians can go to solicit absolute fraud and intoxicate student's minds in brainwashing their thinking to accept what never is proven. This is hogwash at best!

Gravity is the fundamental force of attraction that all objects with mass have for each other. Like the electromagnetic force, gravity has effectively infinite range and obeys the inverse-square law. At the atomic level, where masses are very small, the force of gravity is negligible, but for objects that have very large masses such as planets, stars, and galaxies, gravity is a predominant force, and it plays an important role in theories of the structure of the universe. Gravity is believed to be mediated by the graviton, although the graviton has yet to be isolated by experiment. Gravity is weaker than the strong force, the electromagnetic force, and the weak force. Please show me the attractions in relation to mass that applies between any of the planets and the sun.

In this article I am going to investigate how much truth there is in mass pulling by the force of gravity. Most if not to all of the persons reading this article will be annoyed by just the thought of me embarking on an investigation of the issue that seems so totally senseless to investigate. It is senseless because the concept it carries became accepted as household practise and life science. Mass is associated with everything that is represented everywhere.

If there was a pulling, and the word orbit cancels such an idea, then there has to be some sort of prevention taking place that disallows the pulling to commit the direction of travel. I know it is said that the orbiting object falls as fast as it circles and by falling while moving to the following side on position it never reaches the Sun, and yes, it makes sense, but there has to be some form of resistance replacing the planet in the next side position and preventing the falling or the pulling from taking place. Using the formula $F = G \dfrac{M_1 M_2}{r^2}$ as Newton provided, disallows any other concept other than moving towards. The person Newton got his ideas from and the work he raped completely, that of Johannes Kepler explained this very well, but Johannes Kepler makes no room for any pulling of any sort. In the work of Johannes Kepler he said that the space being the orbiting route a^3 remains at a specific distance k while the orbit T^2 takes place…and in all my other books that addresses more information I take Newton to task on his dismembering of Kepler's formula by corrupting Kepler's work and with what amounts to fraud, Newton takes science on a goose chase that holds no truth.

There is no pulling by mass of mass in any way.

The full article of **The Practice of Brainwashing and Mind Control in Physics**

Previously I have indicated that the moon is not contracting as the Newtonian law of gravitational force would apply but the space between the earth and the moon is expanding or the moon is moving away from the earth as much as the earth is not contracting but is moving away from the moon. Newton said the moon contracts by gravity towards the earth. The cosmos shows that the moon is parting from by increasing the distance between the moon and the earth. As much as Newtonians are forked tongued snakes when double speaking, only one can be correct, notwithstanding all the arguments as to why this is occurring and the thousands of reasons why the cosmos is (again) being at fault and Newton being innocent, still the opposite of what Newton said is happening. The Brainy Bunch can't have it both ways. Either gravity between the earth and the moon expands the distance or contracts the distance. I have given a summery as to acquaint the readers to my view. Now I elaborate on the information you have been introduced to so that you may read more of the content. There is more than substantial evidence proving Newton's law on gravitational forces incorrect, and in fact is an old wives tail. This means there was never investigative studies done to support Newton. Yet students are taught Newton's failed principle in their studies and any student disagreeing with this claim would be failed.

This then is the Practice of Brainwashing and Mind Control in Physics

The Definition of Gravity according to the Oxford Dictionary of Astronomy:
Physics

1) **The natural force of attraction exerted by a celestial body, such as Earth, upon objects at or near its surface, tending to draw them toward the centre of the body.**

What this referrers to is the process by which weight is measures and the more a body pushes to the earth, the more would the scale reading be. Whether this comes as result of as force is what I strongly dispute. This indicates in the same result as measuring weight would be and there was never any defining proof that parts this form of mass from this form of weight. Both mass and weight are measured in the same measurement and have the same indicators. This form of mass is there but is the same as weight and only mind control in physics by tutors can put any distinct identifiable characteristics differences claimed to be between the two.

2) **The natural force of attraction between any two massive bodies, which is directly proportional to the product of their masses and inversely proportional to the square of the distance between them.**

This is the one that I refer too that is the figment of Newton's information. I beg anyone to use the information gathered from the solar system to prove this definition do apply!

3) **Gravitation.**

This proves how low Newtonians can go to solicit absolute fraud and intoxicate student's minds in brainwashing their thinking to accept what never is proven. This is hogwash at best!

Gravity is the fundamental force of attraction that all objects with mass have for each other. Like the electromagnetic force, gravity has effectively infinite range and obeys the inverse-square law. At the atomic level, where masses are very small, the force of gravity is negligible, but for objects that have very large masses such as planets, stars, and galaxies, gravity is a predominant force, and it plays an important role in theories of the structure of the universe. Gravity is believed to be mediated by the graviton, although the graviton has yet to be isolated by experiment. Gravity is weaker than the strong force, the electromagnetic force, and the weak force. Please show me the attraction in relation to mass that applies between any of the planets and the sun.

In this article I am going to investigate how much truth there are in mass pulling by the force of gravity. Most if not to all of the persons reading this article will be annoyed by just the thought of me embarking on an investigation of the issue that seems so totally senseless to investigate. It is senseless because the concept it carries became accepted as household practise and life science. Mass is associated with everything that is represented everywhere.

Do you think of astrophysics as the department that is run by the wise and the level minded, the sober thinking, and the absolute trustworthy? If you think those in charge of astrophysics are the pillars of trust, then get wise and read the following. Newton created the factor mass as a trick of his imagination. We have a body with mass on earth and that we all know, but no heavenly body could ever have mass or present mass! After you have considered the following you might agree with me that even small Children can reach a higher level of clear-minded logic and find more sensibility than what those scientists promoting astrophysics have because science lives in a make believe fool's paradise. They love to calculate because with mathematics they create a fools paradise.

In this article I give the table Kepler represented his research by which the planets move. Also I give one column in which the mass is indicated in terns of earth mass units. Use the column to show where it is mass that produces any valid factor in accordance to planetary activity, movement and allocated positioning. In the table columns we have the mass of the main solar objects and according to Newton the mass is responsible for movement. If it is mass that does the pulling in ratio of the radius then we have to see some evidence of this applying somewhere. Is there any person that can prove that claim from the table showing Kepler's movement in relation to Newton's mass? In what way odes Jupiter move faster or how does Pluto move slower according to mass? If mass did the job, how is it done? Notwithstanding Newton's blatant mathematical manipulation, mass don't apply.

There is no possible way to produce evidence from the behaviour of objects in the solar system that mass is responsible for gravity forming between two heavenly bodies. If it did, the asteroids could no possibly orbit in their allocated circle at the same rate as Jupiter circles in its designated circle. Every one believes in Newton except comets, because comets fail to collide with the Sun. However I can explain in some way…The evidence that mass is pulling mass there is not and there is no evidence of mass pulling mass.

Looking at the size of a body does not allow any one to awards mass. If you are a student, then ask your Masters please to explain the following abnormalities and inconsistencies they promote as part of official physics, which I present in this article and as students get wise instead of brainwashed. I say brainwash again because they force-feed you fabrications, as you will come to see. They can't explain the facts as reliable but hide the fact that the facts are in fact untruths. Tell them to prove that planets have mass. Tell them to prove that it is mass that generates gravity that pulls the planets. Ask them to explain gravity in detail.

If any person gives evidence and the evidence is unsupported by truth and unfounded by proof, it is fraud. If a Clergyman should bring evidence of some miracle healing that took place, and he could not repeat the act at will for all to witness again and again, it is fraud not withstanding the religion. Even if the lie is to underline the greatness of his religion, it still constitutes to fraud. If anything is said without truth backing the claim, making it unfounded, it is fraud. It is illegal to make claims that is not based and is not supported by reality. As perfect as everyone thinks Newtonian physics are, this applies to educators teaching physics. There is a suspicion lingering in the back of everyone's mind that something is not quite correct about the approach physics take on the matter of gravity and only those academics seasoned with years of studies and salted with time seems to miss this haunting feeling. But the longer the students are educated, the more this uncomfortable feeling dissipates.

Hidden under a cover of "understanding Newton" or "not being able to understand Newton" tutors in physics force certain incompatible arguments to join that which never can join and while joining also make sense at the same time. In this article I challenge the figures that charge the highest form of respect in our communities and those in charge of the most dynamic part of society and those who stand beyond and above any form of suspicion. I charge those that personify truth and are the very same persons that I accuse of betraying the ones trusting them. Again I challenge you to come forward and tell your students the truth about what I uncover in the articles that follows and as the articles progress by introducing information…then you explain to them how you deceived their blind trust in you as a tutor in physics. In physics the blame goes to students ability to "understanding Newton" or "not being able to understand Newton" but I show that in this case the blame should openly be dedicated to those that should be blamed, named and shamed and they have to defend their years of lying and contribution to cover the misconduct that was committed by them. Those teaching physics as well as their predecessors now have to explain in the name of science why so many evidence were falsified to keep their noses clean while mud colours their lily white image in hogwash and the stink of their lies equals the pig pen it needs to cover such aroma. It is time that they reveal the truth. Believe it or not, but this diverging from reality and misconduct about applying science is in place because of centuries of brainwashing going on passed on from generation to generation and is employed in physics from teacher to student for centuries on end. This article is dedicated to bringing honesty into the faculty of Astrophysics and Astronomy as well as to show the Physics student on what corruption and deceit does physics base their facts which they proclaim as being such well proven, and godly accurate facts and is unwavering depicting only the truth.

If you are a student studying in physics then reading this article is detrimental to your future, as you can remain part of the problem physics has had for centuries or you may join the solution that came to physics and start to heal the wound. Students read the next pages and you are about to learn how students are brainwashed into accepting the baseless and ridiculous misinformation Sir Isaac Newton puts forward as truths. The Custodians of Physics have nothing better to offer than presenting you with unfounded, corrupt and distorted facts…and by doing that they resort to mind control on students and introduce baseless concepts by manipulating the student's thoughts. If you are a student then read in **www.sirnewtonsfruad.com** what they do to you and how they brainwash you. This suspicion that there is, this feeling about a certain concern and doubtfulness that is lingering on in the minds of many… and also is lingering on from generation to generation… without anyone ever finding a solution… it is them defrauding you by exchanging your institution fees for corruption, so confront them about their dishonesty. There is this vague unspoken question hanging in the air without any one ever finding words to express the question…and yet the question remains however unspoken it seems. Force them to become honest and to stop corrupting students with intentional malice. Let them explain how mass generates gravity…
Even if you had your personal favourite conspiracy theory, try and match it to the one that I have! I seem to be the first person in generations that ask questions about Newton's work. Questions I now ask is asked for the first time ever, well ever since the time Newton introduced gravity, before the emphasis fell on proof rather than merely what a person with reputation suggested. I now am able to show how gravity forms by forming a circle using Π because I have located the centre of the Universe.

But by my effort in finding the location I disrupted everything Academics in physics hold holy and for that I am most unwanted in the presence of the Academics charged with guarding the ethics of physics. In short, I clash head on with Newtonian dogma and principles forming physics. During my research I discovered abnormalities and inconsistencies about mistakes the Arch fathers in physics must be aware of but are hiding with all their considerable influence and academic power. The road I took in my search for truth concerning physics was never smooth and the resistance I came across coming from the academic sector is almost unbearable.

With my unpopularity rating this high as it does, I never qualified for help and those that would help found my ideas intolerable whereby I only found rejection instead of help as I tagged along. Because of this insider rejection I had to resort to private publishing because from the nature of my work I take Mainstream science head on and am confrontational on most aspects of astronomy. Since no one see the a problem there is in physics, no publisher wants to go head bashing with the Physics Custodian establishment of science on official science principles, which I have to do to convey my message in no uncertain language.

I argue that if it is the correct practise to use $F = G \frac{M_1 M_2}{r^2}$ to calculate gravity then the radius holding the gravitational constant must lead one to the centre of the Universe. As I confront science dogma and principles, nobody is willing to publish my work. I have to walk the road alone and fight the battle by my private effort without any support anywhere. If Newton is the problem one have to go pre Newton to find the problem.

The problem is that when looking at Kepler's table then if there is $T^2 \div a^3$ according to the table matching a column, then mathematically $T^2 \div a^3$ must be k^{-1} and where k^{-1} goes negative it shows space reduces time. It shows space in volume goes single by movement of space and not objects.

Planet	Mass per Earth unit	k^{-1} Movement	a^3 of space volume	T^2 During time units
Mercury	0.06	$T^2 \div a^3$ =0.983	$(a^3)=$ 0.059	$(T^2)=$ 0.058
Venus	0.82	$T^2 \div a^3$ =0.992	$(a^3)=$ 0.381	$(T^2)=$ 0.378
Earth	1.000	$T^2 \div a^3$ =1.000	$(a^3)=$ 1.000	$(T^2)=$ 1.000
Mars	0.11	$T^2 \div a^3$ =1.000	$(a^3)=$ 3.54	$(T^2)=$ 3.54
Jupiter	317.89	$T^2 \div a^3$ =1.000	$(a^3)=$ 140.6	$(T^2)=$ 140.66
Saturn	95.17	$T^2 \div a^3$ =0.999	$(a^3)=$ 868.25	$(T^2)=$ 67.9
Uranus	14.53	$T^2 \div a^3$ =1.000	$(a^3)=$ 7067	$(T^2)=$ 7069
Neptune	17.14	$T^2 \div a^3$ =0.999	$(a^3)=$ 27189	$(T^2)=$ 27159
Pluto	0.0025	$T^2 \div a^3$ =1.004	$(a^3)=$ 61443	$(T^2)=$ 61703

If you are a student in physics then you should read the following information with care and with much consideration because your mental health might be at steak here. One could think of another name for physics and that would be Newton's mythology. It is about the subject of gravity and is most important. The "Newton's mythology" comes from the fact that students have to learn what the professors claim to be true and what was never was proven. Students have to repeat in examinations that the formula $F = G \frac{M_1 M_2}{r^2}$ is truthful and viable while it was never proven. Do you realise that it is an accepted practise that all students that are studying physics on all levels are subjected to the most intense brainwashing and thought control found any where on Earth? This must be some sort of a joke you may think but thinking that way in disbelief is just what those practising the mind control wish you to think!

According to the tables, all movement is according to some other value than mass. They never prove Newton's philosophy on gravity but those persons conducting teaching in the subject of physics force all physics students to learn Newton's gravitational concepts and accept the facts as if it has been proven beyond all other facts. Students have to believe that Newton is correct or academics will see to it that they fail their examination. The condition of being accepted in physics is to accept Newton without questioning the proof that is never supplied.

Let those academics now prove precisely how mass brings about gravity and then afterwards test you on how Newton is proven correct and not on you repeating their facts blindly about what they say is true about what Newton said, which they say is true. The manner they present Newton is completely hearsay and that method may not be used in any court of law.

Let your professors now prove how it is that Newton's teachings are correct and then examine you on the process they use to prove Newton's concepts rather than test the state of brainwashing you have submitted to. At present they say Newton is correct and then they test you on your ability in repeating that Newton is correct without ever proving to you that Newton is correct. That is not testing your knowledge but it is testing the mind control you have submitted to. Let those physics professors now prove Newton and their ability to prove Newton correct and then test you on the manner they use to prove Newton to be correct.

They teach you that the Universe contracts and to state their case they force students to learn that gravity is proved by Newton introducing the following formula $F = G \frac{M_1 M_2}{r^2}$ They say that M_1 is the mass of the Earth and M_2 is the mass of the individual in questions mass and the multiplying of these factors with the gravitational constant produces the force of gravity when this gets divided by the square of the radius. Please let you lecturer put in all the values of the formula and prove Newton is correct. If he can't and I know for sure he never can fill in the symbols and calculate the force of gravity, then read the rest of the web page that follows to see how far academics in physics go to brainwash students into believing in Newton's fraud.

Newtonians uphold their law of physics without showing mercy. The very first things the Newtonians use to beat us into submission are to blast us with incomprehensible mathematical formulas.

Incomprehensible they are but it is to scare anyone with the mathematical equations to get everyone hiding. They bewilder you with equations that put the fear of God into you; used simply to make you feel inferior so that they can feel superior and frown down on your inferiority from a dizzy height.

They are masters at manipulating anyone into a state of senselessness…but mostly they do it onto themselves. That they do because it forms the backbone of the fraud. They do not wish you to read closer and to find the fraud they hide to protect Newton. Ignore their mathematics because it only shows their incompetence to understand physics or Newton and see the fraud they propagate...

They employ mathematics to bewilder and that is all. I am going to show what we can uncover underneath what they cover. Look at what the mathematics supposedly says and then wake up, they are using maths as a scare tactic for three centuries to scare the daylights out of you and all this while its been working! Looking at the formula shows just how little Newton understood physics.

Do not get scared as everyone usually does when see and get frightened then consequently as a reaction to find survival you turn on your heels and run… Don't run, just read on and see how simple it is to prove Newton was a backward dark aged sod!

Our Super-Educated-Mathematical-Wise use the elaborate formulas such as $\frac{d}{dt}\left(\frac{1}{2}r^2\dot{\theta}\right) = 0,$ and $P = \left(\frac{4\pi^2 a^3}{G(M+m)}\right)^{0.5}$ as well as $T^2 = \frac{4\Pi^2}{G(M+mp)} a^3$ to explain and prove what? If this is true then Jupiter must spin 317 times faster around the sun than the Earth does and be almost next to the sun while Mercury and Pluto in comparison must hardly move being cast into the darkness of the oblivious.

The formula $T^2 = \frac{4\Pi^2}{G(M+mp)} a^3$ says that the spin or time in which the circle comes about T^2 holds a space relation a^3 directly proportionate to the mass (G(M+m)) of both the sun and the applying planet multiplied by placing the gravitational constant also in relevancy. It says planets spin in accordance to mass…lets see.

I say prove it! Then $\frac{d}{dt}\left(\frac{1}{2}r^2\dot{\theta}\right) = 0,$ says that the Planets don't move at all because with $\frac{d}{dt}\left(\frac{1}{2}r^2\dot{\theta}\right) = 0,$ the movement acceleration is zero. Zero indicates no movement and that is as corrupt as the rest of Newton's ideas. Every person associated with physics to whatever extent has been and had been conspiring to hide the truth and the truth is that no one knows (not even Newton and even less Einstein) what physics is. What they say is precisely what the solar system proves different and either the solar system has no idea about what is driving gravity or they have no idea of what drives gravity, but it is definitely not mass. If they don't know what forms gravity and it is clearly not mass, then they know nothing about the way that gravity forms.

They hide their incompetence under a blanket of ignorance and all commit to a conspiracy to hide the truth from the public. Let any one of those So –Wise-In-Mathematics show just how does Saturn arrive at the position it holds in relation to the mass it has by applying the Newtonian formula $P = \left(\frac{4\pi^2 a^3}{G(M+m)}\right)^{0.5}$, which they claim is data confirming Kepler's Law of Periods of the motion of the planets in relation to mass.

Please for those that so strongly believe this misconception Newton brewed up, please prove to me that the formula $P = \left(\frac{4\pi^2 a^3}{G(M+m)}\right)^{0.5}$ where the planet's mass does determine the location and position of the planet.

I know and realise that you are disgusted by my attitude when I degrade the name on which physics are founded. In this introduction part I am going to show you just some minor deceptions all students are forced to believe since all physics students are forced to believe in Newton, **Sir Isaac Newton** that is.

I am giving you a choice. You can say I am going to commit fraud or Newton has committed fraud. If I am judged to be the culprit that is guilty of deception then it is because Newton misled me. You can choose.
You are expected to believe the following:
Newton stated under the nametag of Kepler that there are so called Conversions for "Unknown" factors.

$4\pi^2 a^3 = P^2 G(M + m)$ Newton introduced this concept because he said mass brings about gravity.

In this he had to force the issue even to the point of committing fraud and here comes the fraudulent part because there is no evidence of mass playing a part or forming an actual presence in the solar system.

$P = \left(\frac{4\pi^2 a^3}{G(M+m)}\right)^{0.5}$ What hogwash does the factor $\overline{G(M+m)}$ indicate?

The same can be said in the formula $M = \left(\frac{4\pi^2 a^3}{GP^2}\right) - m$ when $P = \left(\frac{4\pi^2 a^3}{G(M+m)}\right)^{0.5}$ that the factor $\frac{P^2}{}$ is senseless and $\left(\frac{P}{2\pi}\right)^2 = \frac{a^3}{G(M+m)}$, has no foundation other than fraud. It says the position of the planet is derived from the mass $M = \left(\frac{4\pi^2 a^3}{GP^2}\right) - m$ and that is totally ands complete fraud. The Cosmos does not support the Newtonian formula even in one place where it could apply.

Position as a function of time
$P = \left(\frac{4\pi^2 a^3}{G(M+m)}\right)^{0.5}$ This is what Newton said is in place and with no evidence ever founding this ridiculous proposition, all Newtonians that ever come after Newton. This is what Newton and his Newtonian followers tell the solar system it has in place and tell the cosmos it uses to operate. I have indicated that mass has no place or use in the solar system according to what the solar system puts in place.

According to Newton $P = \left(\frac{4\pi^2 a^3}{G(M+m)}\right)^{0.5}$ puts the location or position $\frac{P^2}{}$ of a planet in relation to the mass $\overline{G(M+m)}$ of such an individual planet. I am coming back to this and then you will choose which of us Newton or me, is committing blatant fraud.

I say I can prove Newton correct by showing the formation of the planets in orbit going around the sun. I say you can fool many persons most of the time but Newton's fish is cooked and his game is up. For decades those in academic power refused to publish my work but with the aid of the modern media such as the Internet I hope I now am able to show and tell what physicists hide by a cover-up as never seen before.

With implementing Newton's formula $P = \left(\dfrac{4\pi^2 a^3}{G(M+m)} \right)^{0.5}$ the planet distribution are as follows:

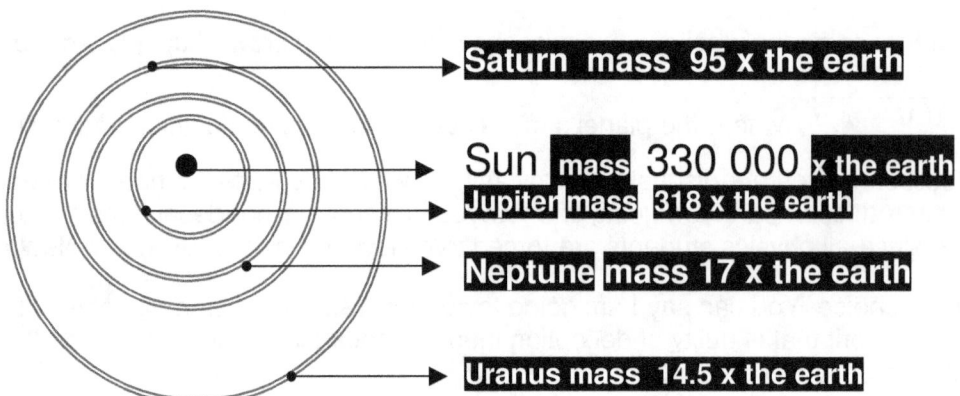

Saturn mass 95 x the earth

Sun mass 330 000 x the earth
Jupiter mass 318 x the earth

Neptune mass 17 x the earth

Uranus mass 14.5 x the earth

The inner circles that are very close to the sun we find the big gas planets with so much more mass forming the force of gravity that these planets are almost on top of the sun so close they are to the sun.

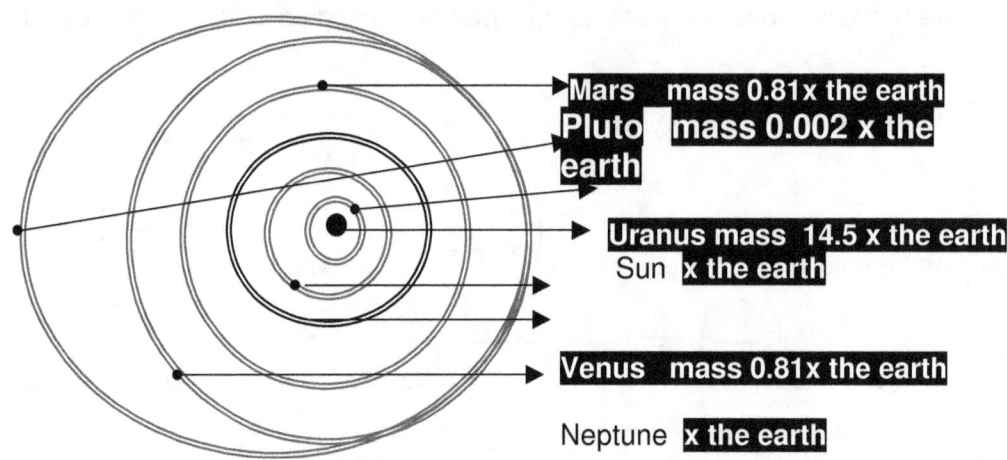

Mars mass 0.81x the earth
Pluto mass 0.002 x the earth

Uranus mass 14.5 x the earth
Sun x the earth

Venus mass 0.81x the earth

Neptune x the earth

Then in the outer circles that are very far from the sun we find the smaller solid planets with so much less mass the force of gravity just can't pull these planets closer to the sun.

Closet **1) Jupiter mass 318 x the earth**

   2) Saturn mass 95 x the earth

   3) Neptune mass 17 x the earth

   4) Uranus mass 14.5 x the earth

Then come the smaller planets with less mass and therefore less pulling force called gravity

   5) Earth 1 x the earth

   6) Venus mass 0.81x the earth

   7) Mars mass 0.81x the earth

   8) Mercury mass 0.055 x the earth

   9) Pluto mass 0.002 x the earth

The above can be the only designated outlay of the planet position according to the sun when applying mass. Remember, if I am wrong then Newton and his formula $P = \left(\dfrac{4\pi^2 a^3}{G(M+m)} \right)^{0.5}$ is wrong but if I am correct and Jupiter is the closest planet then Newton is incorrect and then the academics in physics are indulging in Newton's fraud by forcefully brainwashing students to believe in Newton notwithstanding the fact that it is the solar system that disputes Newton's ideas altogether. If you disagree with my layout, then you better disagree with Newton and his ideas.

Everyone admires and adheres to Newton's misconceptions except where it counts, in the Unversed where true physics apply correctly. If the solar system does not comply to Newton then Newton has no idea what gravity is when it comes to physics. They can ignore my work, but my work represents the truth and in time my work will prevail! I show how gravity forms and I have so many calculations and facts to back me up.

Students, your professors are fooling by brainwashing you through mind control Download for free the website www.sirnewtonsfraud.com to find out the extent by which they control your thinking. See how they disregard students. They think that you deserve to be their mindless monkeys just the way they think of you because you don't think about what they say! Cut the bullshit, force their ignorance about physics into the open and make monkeys of them, they deserve it even more you do…so let them have it for spreading three centuries of fraud and misleading facts.

The truth beyond all other truth is that Newton's gravity has never been proven (because try as you may it is not possible to prove Newton's formula forming gravity mathematically) and because academics know that, academics require (no insist on) the blind acceptance of Newton by students. This unconditional acceptance of Newton's correctness relies only on the pre-conditioning of students' mind set and academics depend only on the student trusting the academic "say so" about the institutionalised correctness of Newton. The students have to accept that Newton is correct nevertheless and notwithstanding that there is no founding proof about this matter, is what fuels the century old conspiracy. I'll bet you the academics are more surprised than you about me accusing them of systematically mind altering the student's physic and ability to think in reason.

This is a fair test to see if Newton's contraction theory underwritten by Newton's attraction formula $F = G \dfrac{M_1 M_2}{r^2}$ is valid. Force your professor to use this formula as it reads and let him show WHEN the Moon and the Earth is going to collide. If he fails to do it by using Newton's formula as $F = G \dfrac{M_1 M_2}{r^2}$ then you will know who is conning you, him or I and who is truthful again him or I. I charge all academics to prove what I say is wrong in any way or even that I exaggerate in the least. I challenge Newtonian academics to prove that mass does indeed form any force of any sorts and in particular gravity! Lets take it to outer space.

What forms the solar system is not mass but this what is there: The Titius Bode Law

Bode's Law" or "Titius-Bode Law". The original formulation was **a = (n + 4) / 10** where *n*=0,3,6,12,24,48 … The modern formulation is that the mean distance *a* of the planet from the Sun is, in astronomical units (**AU**$_{earth}$ = 147.597 *10^6 km): **a = 0.4 + 0.3 x k** where "k'= 0,1,2,4,8,16,32,64,128 (sequence of powers of two *and* 0)

The following table compares the law's predictions with the actual distances, where the addition of Pluto is a modern modification.

Planet	n	Titius-Bode Law	Semi-Major Axis
Mercury		0.40	0.39
Venus	0	0.70	0.72
Earth	1	1.00	1.00
Mars	2	1.60	1.52
asteroid belt	3	2.80	2.8
Jupiter	4	5.20	5.20
Saturn	5	10.0	9.54
Uranus	6	19.6	19.2
Neptune	-	-	30.1
Pluto	7	38.8	39.4

In another place I show how ridiculous it is to presume that the planets form by order of mass.

This is what there is in the solar system. What Newtonians say is that mass is driving the solar system. I say prove it because I prove why it is the Titius Bode law that forms the solar system. The Titius Bode law is gravity because the Titius Bode law is forming the concept that I prove is Π and Π is gravity. Since I prove the Titius Bode law forms Π as gravity and not mass no one this far took the time to see what I prove all the while what I prove is the precise reason why the solar system uses the Titius Bode law to form the planet layout.

$P_n = P_o A^N$

P_n = period of orbit of the nth planet
P_O = period of the sun's rotation
A = semi major axis of the orbit

The law relates the mean distances of the planets from the sun to a simple mathematic progression of numbers.

To find the mean distances of the planets, beginning with the following simple sequence of numbers:

0 3 6 12 24 48 96 192 384

With the exception of the first two, the others are simple twice the value of the preceding number.
Add 4 to each number:

4 7 10 16 28 52 100 196 388

Then divide by 10:

0.4 0.7 1.0 1.6 2.8 5.2 10.0 19.6 38.8

The resulting sequence is very close to the distribution of mean distances of the planets from the Sun:

Body	Actual distance (A.U.)	Bode's Law <A.U.)< td>
Mercury	0.39	0.4
Venus	0.72	0.7
Earth	1.00	1.0
Mars	1.52	1.6
		2.8
Jupiter	5.20	5.2
Saturn	9.54	10.0
Uranus	19.19	19.6

Please consolidate this $P = \left(\frac{4\pi^2 a^3}{G(M+m)}\right)^{0.5}$ which Newtonian wisdom says is true with **$P_n = P_o A^N$**, which is what the solar system have in place and which is what the solar system upholds.

This is the best that the Newtonian can do with the most important principle forming one in four part of the entire cosmic code. The cosmos forms by code and I have unravelled that code. This is how gravity applies space by implementing the Law of Pythagoras and by the law of Pythagoras time builds space.

The **Titius Bode law**, the **Roche limit**, the **Lagrangian points** and the **Coanda effect** form the cosmic structural code. I show how these four form the cosmic code and the cosmic code is in the way that Π forms gravity and how gravity is Π. Those that are cautious to admit that there is a conspiracy in place, tell us how the Newtonian physicists know that the Titius Bode law is the principle on which the solar system is formed and yet they acknowledge Newton's bogus formula $P = \left(\frac{4\pi^2 a^3}{G(M+m)}\right)^{0.5}$ instead of trying to explain the Titius Bode law. The best explaining that the Newtonians can present about the Titius Bode law is **$P_n = P_o A^N$**, which in all fairness to their intelligence says nothing. They can put down a mathematical formula and then that is it…that is all they can present and that shows their entire cosmological insight they have. If they put what applies in some simple mathematical formula, the Newtonian mind has reached its goal because the Newtonian mind has no ability o go beyond equating what they do not understand. I have sent the explanation about the Titius Bode law off to amongst other papers, The Annalen Der Physics but found no response from them since I don't hail Newton and I don't fiddle with corrupt imaginary mathematics that only corrupt the truth and that promote their conspiracy to keep the truth from coming out. From them I heard nothing…

And what is the truth, the truth is that they know as little about physics as Newton did except that they can conspire as a unit to corrupt the truth about physics. This sounds monotonous but for more than a decade I tried to get them to read my explaining but as soon as they have to read words they switch off because they want only to admire mathematical cosmic make – believe. Either they are to lazy or to stupid to read.

In twelve articles I explain how the four Cosmic Pillars form gravity but no publisher wants to publish it because the conspiring cheats wants to uphold Newton's corruption because the corruption is all they can understand. …But they would not even publish my articles because my articles prove that Newton's views are hogwash and corruption. What is not there is Newton and what is there they say is not there because it doesn't hail Newton.

Body	**Orbital Distance**(10^6 km)	**Mass** (10^24) kg
Mercury	57.9	.3302
Venus	108.2	4.869
Earth	149.6	5.975
Mars	227.9	0.6419
Asteroid Ceres	Halfway Between Mars and Jupiter	Fragmented rock
Jupiter	778.3	1898.6
Saturn	1427	568.5
Uranus	2869.6	86.83
Neptune	4496.6	102.43
Pluto	5913.5	0.0125

This is the table holding the actual figures about the solar system, and $\left(\frac{P}{2\pi}\right)^2 = \frac{a^3}{G(M+m)}$, is the equation, which the Newtonians paint. In this picture it represents the Newtonian stance perfectly. By having so much mass the sun fills more space than the planets and the planets form a line where each planet holds a distance in equality as well as regular proximity in circling while going around the sun and then straight to the sun. The sun now presumably contracts the planets by enlisting an inexplicable force while the lot is circling by being contracted. This formula is in use for more than three centuries and was never once challenged. It was also never defended because it was always portrayed as the perfect presentation. What a tour of misguiding do they dish up to confuse the rest of all human beings...and they got away with it for more than three hundred years. The acceleration of the planets is supposedly strictly related and directly according to the mass of the planet. The equation is $4\pi^2 a^3 = P^2 G(M + m)$ that forms $\left(\frac{P}{2\pi}\right)^2 = \frac{a^3}{G(M+m)}$.

Now we go on a hunt to find proof for the factors ($\overline{G(M+m)}$) that plays such a critical role in the position ($\dfrac{P^2}{}$) of the planets.

It is said that the acceleration of a planet's movement in accordance with Kepler's laws can be shown as a result of the planets being directed towards the sun, and the magnitude of the acceleration is in inverse proportion to the square of the distance from the sun. Mathematically this is presented as $F = G \dfrac{M_1 M_2}{r^2}$.

Isaac Newton assumed that all bodies in the Universe attract one another with a force of gravitation. This force pulls all objects closer by the ratio of the mass the object holds. This is what I question... and the factor of mass forming for even forming a presence at all. As the planets have small masses compared to that of the sun, the orbits obey Kepler's laws approximately. Newton's model improves Kepler's model to give better fit to the observations.

What I wish to see is where they hide the proof that shows that planets are firstly moving towards the sun and secondly then also at what rate do they move in accordance with the mass applying. How does the mass contribute to the position in which the planets are allocated? Or otherwise with mass bringing movement I wish to see how much are the larger planets shifting faster than do the smaller planets?

Let me show you clearly how the fraud applies. $4\pi^2 a^3 = P^2 G(M + m)$ Look at the mathematics and it scares the day light out of the common public. That is tactic number one. That the formula is completely bogus goes without question but this formula was supported by the Newtonian mathematical wise for longer than three hundred years. That puts a question mark against the so-called Newtonian mathematical Masters.

It says the planet position squared divided by 2Π squared is equal to the space filled in relation to the mass of the gravitational constant multiplied by the sum of mass of the sun and whichever planets fills such a position.

How do they console what there is with what Newton says there is without upholding the most elaborate conspiracy to hide the truth that was ever concocted to fool the human race.

There is no correlation present at all between the mass and the distance of orbit. $F = G \dfrac{M_1 M_2}{r^2}$

Body	Mass (10^{24}) kg		Orbital Distance(10^6 km)	=	ratio
Mercury	.3302	÷	57.9	=	0.0057
Venus	4.869	÷	108.2	=	0.045
Earth	5.975	÷	149.6	=	0.039939
Mars	0.6419	÷	227.9	=	0.00281
Jupiter	1898.6	÷	778.3	=	2.439419
Saturn	86.83	÷	1427	=	0.060847
Uranus	102.43	÷	2869.6	=	0.03569

Can you find a sustainable connection between the distance applying and the mass the planet holds. This does not present any correlation of any type. Yet this is upheld as the truth and the basis of all of Creation for longer than three hundred years! What the lot do share is the space they share and it is the space that contracts while the solid planets keeps position according to the space contraction. So where does

Newton's formula $\left(\dfrac{P}{2\pi}\right)^2 = \dfrac{a^3}{G(M+m)}$, fit into the grand picture. The line up is as follows and please note the distances compared to the mass of every planet. Now you decide what the connection there is between mass and radius when taken into account that Newtonian mythology places this as the direct contribution to the position allocated to the planet and the gravity driving force between the sun and every individual planet.

The equation is **4π²a³ = P²G(M + m)** that forms $\left(\dfrac{P}{2\pi}\right)^2 = \dfrac{a^3}{G(M+m)}$. Now we go on a hunt to find proof for the factors ($\dfrac{1}{G(M+m)}$) that plays such a critical role in the position ($\dfrac{P^2}{1}$) of the planets.

It is said that the acceleration of a planet's movement in accordance with Kepler's laws can be shown as a result of the planets being directed towards the sun, and the magnitude of the acceleration is in inverse proportion to the square of the distance from the sun. Mathematically this is presented as $F = G\dfrac{M_1 M_2}{r^2}$.

Isaac Newton assumed that all bodies in the Universe attract one another with a force of gravitation. This force pulls all objects closer by the ratio of the mass the object holds. This is what I question... and moreover I question that the factor of mass forming is even forming a presence at all in any part of the Universe. As the planets have small masses compared to that of the sun, the orbits obey Kepler's laws approximately. Newton's model improves Kepler's model to give better fit to the observations.

What I wish to see is where they hide the prove that shows that planets are firstly moving towards the sun and secondly then also at what rate do they move in accordance with the mass applying. I wish to see how much are the larger planets shifting faster than do the smaller planets? Let me show you clearly how the fraud applies. **4π²a³ = P²G(M + m)** Look at the mathematics and it scares the day light out of the common public. That is tactic number one. That the formula is completely bogus goes without question but this formula was supported by the Newtonian mathematical wise for longer than three hundred years. That puts a question mark against the so-called Newtonian mathematical Masters.

It says the planet position squared divided by 2Π squared is equal to the space filled in relation to the mass of the gravitational constant multiplies by the sum of mass of the sun and whichever planets fills such a position.

Any person that repudiates my accusation that Science is one elaborate conspiracy to fool the public, then please explain this. From this carbon heating global warming to the Newtonian principle debacle every aspect that science put forward is a hoax with one purpose they aim at and that is to lead the public around the bush. I can't see one bit of honesty remaining in science in every aspect I see! In my book The Absolute Relevancy of Singularity in terms of The Cosmic Code I prove that gravity is the relocation of heat taking heat from an expanded location and then by movement concentrate the heat into a denser location. Gravity is the transfer of heat so how the hell can the planet heat up by the carbon. The warming and the cooling is a cyclic thing, which is what gravity is. I named the "ice age" the solid faze and the other warm faze the "liquid faze' and by gravity forming singularity there is forever an alternating of cyclic proportion between the two polar states.

Pre-conditioning students into blind acceptance depends on the academics' insistence that students approve Newton's concepts without pre judgment or students insisting on scrutiny of any sorts. In examination students have to outright and blindly follow academics' say so trumped up rhetoric and respond on facts only because academics say so. Academics depend on students never questioning their say so or demand proof about what academics teach. Those academics in teaching positions insist that all students accept Newton's accuracy.

Where does mass manifest in this? Where does Jupiter show its mass is pulling more than that of Pluto? According to the definition of gravity claiming gravity is a pulling power by the mass of objects and this formula $F = G\dfrac{M_1 M_2}{r^2}$ then Jupiter has to beat the earth going at a speed 317 times faster than the earth, because Jupiter is 317 times larger than the earth. Show me any evidence that mass plays any role or function anywhere in the Universe! If you are one of those members of society that never thought you would hear the name of an accomplished person as Sir Isaac Newton being associated with fraud, corruption and brainwashing, then the books **The Absolute Relevancy of Singularity** are specially written to inform you about the truth there is lacking in the correctness about science. Newton's formula and suggestion indicate a straight line moving $F = G \dfrac{M_1 M_2}{r^2}$ between the objects pulling by mass while it is clear everything applies the motion around the sun using a circle. Everyone knows that planets orbit around the Sun. Planets circle the Sun which is the same as saying planets orbit the Sun. Just by calling the circle motion in terms of what applies, that statement nullifies Newton's claim of mass that attract mass and put to question the reliability of Newton's dogma. This might at first seem a small issue but from that I prove that gravity works by the value of Π and not mass.

This is methodical mind control as much as it is the brainwashing and I show that they enforce this practise. If you are one of those believing that Newton was ever proven, then what you believe to be true is a lie because Newton can't be proven and that is the truth! The time has come to face your teachers and force them to stop the ongoing old culture of bullying students into submission and conditioning their thoughts by enforcing on them dogmas, which is no more than systematically enforcing mind control! In order to get students to accept Newton's hypothesis, academics resort to brainwashing pupils' and students' thoughts.

One such an example is Einstein's Critical density scam and the "Dark Matter" swindle where they hide Newton's shortcomings under a pretext of ongoing research. Hubble found the Universe is not contracting as Newton said it does with $F = G \dfrac{M_1 M_2}{r^2}$ and hell broke loose. The cosmos stepped out of line by showing Newton is wrong in his supposed theory that mass pulls mass to reduce the radius such as $F = G \dfrac{M_1 M_2}{r^2}$ will indicate. Einstein was then tasked to supposedly measure "all the mass" in the entire Universe and to find out when the Universe will correct its evil ways and start to submit to Newton's idea of contraction. They concluded it is not Newton that is wrong but the cosmos went out of line to contradict Newton. Therefore one has to find the error in the actions of not Newton's idea but about the Cosmos acting out of line with Newton. The fault was investigated not on the side of Newton, but in the cosmos! The proof of practise fell on the Universe that could be wrong but never Newton!

With all this in mind did any one ever come to wonder about the reality driving the all too famous **Einstein's Critical Density theory** and the fact that this idea was conceived to conceal the corruption of Newton in physics? Allow me please to elaborate and then make up your minds. The facts in truth are that the **Einstein's Critical Density theory** was a scheme plotted by those in charge of physics principles to cover up and conceal corruption in the heart of physics. Hubble proved beyond doubt that there was an inflating Universe. This contradicted Newton's deflating or pulling Universe and this perception of a deflating Universe being a myth had to be most ardently hidden as to yet again compromise the truth about Newton and his theory. In the formula $F = G \dfrac{M_1 M_2}{r^2}$ the relevancy applies between the strength of the mass and the distance of the radius that keeps the influence of the mass forming the gravity. The longer the radius increasing the distance between objects the more this will reduce the value of the mass, whatever the value of the mass might be. The Hubble concept proves that while the mass might remain the same, the radius keeps growing and such growth diminishes the influence of the mass all the while the radius increases. This is so basic that primary children learning the basics of mathematics will understand! This means with the radius increasing, there is no chance that the mass will ever become strong enough to bring about the pulling because the constant increase in the radius constantly diminish the influence that the mass might produce.

Yet Einstein proceeded in searching for a value that will determine when the mass would bring a turn about in the direction that the cosmos evolves in. Einstein was looking for the moment the mass will become strong enough while the most basic principle indicates that an increasing radius leads to a decreasing mass influence submitting a decreasing potential gravity since the mass becomes less prominent in influence. If Einstein was unable to recognise this most basic of mathematical principles, then what type of genius did physics create in him and what slur did physics promote. In $F = G \dfrac{M_1 M_2}{r^2}$ any factor standing in relation of division, it is the bottom value that determines the outcome of the value of the top.

This idea of the two factors being in opposing relevance where the size of the bottom caries the value of the outcome is so simple that children will recognise the principle, and yet those fathers of physics wants me to believe that the greatest mathematician that ever lived did not realise this principle…the principle that the radius and the mass stands related and the growth in the bottom value will promote the decline in the top value as a dominant factor. Can any one with this information including the proof I asked for on previous page have any other conclusion than students should smell rotting fish somewhere?

Notwithstanding my arguments that should have been raised by the mathematical genius, Einstein supposedly measured "all the mass" in the entire Universe and then afterwards concluded there is "insufficient mass" to pull the Universe closer. This rattled the cages of the Newtonian conspirators because Newton once again stood naked, venerable and bear. Yet the blame had to be associated with the cosmos not playing by the rules of Newton. The cosmos had to be at fault because Newton just can't be the person to blame. Then the genius of the Newtonian cunning kicked in once more and we can see why they are seen as the most brilliant minds walking the earth. There had to be something they miss and no one can see.

Then some idea was presented that the Universe is hiding mass from view, I suppose just to spite Newton. If the mass was not visible and was therefore undetected, then the mass was dark and if no one could see the mass then no one can prove the mass being present and then no one can disprove the mass being there waiting in the dark. What a splendid idea this was for cheating. This presented a solution of Biblical proportions and a new scam was introduced to hide the failing of Newton and of the Critical Density sham.

The Dark Matter hides mass that will supposedly pull all material closer again at one stage in the future. This is meant unleashing an enormous swindle, bigger than anything before, only beaten into second place by Newton's swindle about mass unleashing forces. They proclaim there is "Dark Matter" hiding and waiting in the wind to come to the present someday in the future and start to pull the Universe closer as Newton said it must by the measure of $F = G \dfrac{M_1 M_2}{r^2}$. In that way Newton was vindicated and the cosmos took the blame for hiding mass and the contraction was reinstalled.

It is obviously clear that having such a total idea that there might be dark unseen mass floating in the Universe which at this time does not generate gravity but will some day kick in to generate gravity in order to cover up Newton's deception about a contracting Universe and just because Newton has to be correct at some point in the future. Science wishes me to believe that since there is a lack of seeing material there then will be dark and unseen material where they are so dark they are undetected by all humans. Why would the mass at present then not activate gravity and why would the mass at some point spring to life and start activating gravity? How much can the Physics paternity still hide the fact that Einstein's critical density is being used as a cover-up to distort the truth to conceal the fraud Newtonians wish to cover? Hubble found the Universe is expanding and Newton said otherwise. Hubble's declaration was on track to blow the cover that was concealing the Newton fraud wide open and uncover the century old deception. The question is if it is mass pulling mass onto mass, then why do we have comets left in the solar system? The mass of the Sun should by now at least have destroyed every comet going around.

The term pulling does not suggest any circling because no one can be pulling towards and does that while circling around the object. That serves the term pulling. In conversation we speak of the planets orbiting. If Newton was correct we should be speaking of the planets pulling, but talking about pulling would be blatantly wrong according to the normal spoken word. Never do we refer to the planets pulling the Sun or the Sun pulling the planets, but we speak of seasons coming from orbital positions. Being in orbit has to neutralise the pulling and then cancel the pulling concept that also became culture. The entirety of physics rests on this one formula $F = G \dfrac{M_1 M_2}{r^2}$. The questions concerning that which you are studying and that touches every aspect you are academically concerned with, is that if everything is moving apart, how does that support Newton's idea that everything is coming together…and please don't let them fool you with

Einstein's Critical Density idea! If there was mass seen or unseen in the Universe and mass generated gravity and gravity does the pulling then why is the mass not at this moment doing the pulling? If there was material and material had mass in the Universe, and the material was seen or unseen and material called dark or shiny had mass nonetheless that notwithstanding had to generate gravity and gravity does the pulling then why is the mass of the invisible material not at this moment doing the pulling? If there is an object it indicates the presence of mass or so they say. Then what stops the mass that should be present, dark or luminous, to start pulling by gravity, as it should do? This dark matter and Einstein counting all the mass in the entirety thought to be the critical density of the Universe and all mass was forming pulling gravity and Newton never doing wrong, it is all cloak and dagger; it is all part of the most elaborate scam ever devised to cover up a fraud as big as science goes.

Should you think this page is some sort of a prank then answer the following simple question to yourself in utter honesty: If there is a Big Bang after which everything was left moving apart, how does that support Newton's contraction? Tests results received after the Moon landing show the Moon and Earth are moving away from each other! Yet students learn about mass pulling mass and that puling by mass forces togetherness by contraction.

This is only one of many points that I make on this one issue and there are so many other issues one may think of those in terms of counting in numbers in many hundreds or even in thousands. If the Sun for instance has mass that is apart from the Earth and the Earth also has mass and there is a gravitational constant in between the Sun's mass and the Earth's mass we have the radius in that location. It then must be the gravitational constant that fills the space that the radius holds. It is rather obvious that while the radius is filling the vacant space between the Sun and the Earth it is the only place left where the gravitational constant can hide. To find the centre of the Universe I had only to find the gravitational constant that holds the centre. Through my venture I discovered one person that knows what gravity is! Newtonians went and filled that space reserved for the gravitational constant having a measured value with nothing! How can nothing have a value of 6.67×10^{-11} while also being filled with nothing as it is nothing filling the nothing of outer space?

My question in this matter is what is all that mass of so many supposed stars living in utter darkness doing at present while waiting to get to work and begin with generating gravity by mass where it will only later, much later form a force of gravity that then will bring about this pulling of the Universe? What makes the mass slumber in darkness to one-day form a pulling force? What has the "darkness" or the fact that we don't see the mass got to do with the idea that the mass at present is not forming gravity that is forming a pulling force? You are taught that gravity pulls objects to the centre and obviously gravity then has to ultimately pull everything to the centre of the Universe. That is what the Critical Density research that Einstein initiated wishes to establish. The idea is that $F = G \dfrac{M_1 M_2}{r^2}$ makes the mass create a force that will destroy the radius and ensure everything is going to come together eventually at one point where the radius then will be no more. If that is the case, then where is that point? If everything is destroying the radius, then it must end at one specific point.

In the classes you students attend a physics lecture, has any one confirmed a location where one might find the centre of the Universe to confirm the ultimate destination of $F = G \dfrac{M_1 M_2}{r^2}$? If you wish to apply a Gravitational constant as a calculated factor then it is apparent that one must know to where such gravity is pulling since it then is the gravity that is where the contraction is going that predominantly is keeping everything apart. Then the gravitational constant is what is resisting the collapse of the Universe. If there is a force, then where is the force taking the pulling…if it is a gravitational constant applying through out outer space then where is it having a centre base? To those professors claiming Newtonian ideas are substantiated by proof, I say that notwithstanding your personal academic qualifications and while at the same time disregarding your status and previous achievements as well as ignoring your many admirable abilities you may have and however superior they might be, I shall teach you about gravity. I say it is time Students learn the truth about physics notwithstanding the status academics will loose.

What you are about to read comprises of extractions forming part of the actual book that you can download free of charge. Go to **www.sirnewtonsfraud.com** and press on the blue button and the book is yours to have! Use this information to test the reliability of your tutors' teachings. Confront them with facts and don't allow them to stupefy you with their ability to commit mind control. This is what eighty years of studying the solar system brought about and no individual study since changed one aspect of what this table brings as information.

If mass did attract mass, what kept the balance in the distance it held according to the Titius Bode Law where the planets do find a balance in orbit rather than moving towards the Sun. One might think of mass pulling the comet to the sun, but instead of slamming into the sun, the comet makes a circle (Π) and disappears into outer space. If mass pulls the comet closer, what does the pushing away afterwards. If mass did attract mass, it would explain the behaviour of the comet coming towards the sun...but then what is pushing the comet back into orbit, into the darkness of outer space? The idea of proof about mass pulling everything is automatically placed at the door of Newton. Think of what planets do... and you think that planets orbit. If normal speech contradicts Newton, then it is his task to prove his supposition is correct and the claims about attraction Newton made, although Newtonians will deny this fact as if they deny he is correct the honour of their Master Newton and that is what they have to do. No one thinks of planets spinning or planets basking in the summer Sun. However, just using the term orbiting is in total defiance with Newton or physics!

Ever person associates gravity applying as **_"The natural force of attraction exerted by a celestial body, such as Earth, upon objects at or near its surface, tending to draw them toward the centre of the body"_** with what we think of happens to solar bodies having mass which is **_"the natural force of attraction between any two massive bodies, which is directly proportional to the product of their masses and inversely proportional to the square of the distance between them"_** and science cheats everyone into believing the two is the very same. The one I experience every day and with the other there is no evidence of applying anywhere. Yet students are fooled into believing the two are exactly the same issue, which is untrue. If there is any academic feeling insulted by me calling the lot fraudsters, bring evidence of the second form of gravity working on the principle of mass and I will withdraw my statement, otherwise if no proof can be brought, then you lot that ignored me for ten years are the fraudsters I accuse you to be. You are villains brainwashing students to corrupt their thinking.

This again was proven by the very first ever experiment concluded scientifically. This fact of space descending does not come as a surprise because Empedocles proved this fact back in 450 BC. Empedocles showed that space displaces water from the clepsydra, which was a sphere shape container with a sprout on the top and small holes in the sphere through which water ran in small streams out at the bottom. When the flow of air or space was blocked in the spout by a finger covering the hole at the top of the sprout at the entry, the water stopped flowing from the clepsydra. They concluded in 450BC that it is the empty space that pushes the water out of the clepsydra because the moment one restricts the empty space or air to flow into the clepsydra from the top, the water will stop flowing out of the bottom of the clepsydra. Why would the flow of the water stop if the mass did pull the water down? When the finger blocks the sprout and stop the space entering from the top, the water does not fall to the ground but it is the empty space that pushes the water out at the bottom to fill the clepsydra from the top. When the finger blocks the sprout and stop air to come in through the sprout opening the water should still run out at the bottom by the mass of the water pulling, if mass was doing the pulling. If mass was the force giving factor, then the water must keep on flowing because the mass of the water did not disappear when the sprout was covered and therefore it still has to produce the pulling by forming gravity. All this evidence was known to science about 2500 years ago but since "With a lot of words and some simple algebraic relations, there is no way to "explain" the world of physics" it lacked mathematical communication and it should therefore surprise anyone very little that physics could not fathom this result 2500 years onwards.

Forget the example always used about the hammer and the feather falling equal in a vacuum because the hammer and the nail and the elephant falling together will also fall equally notwithstanding falling in a vacuum or not falling in a vacuum. The vacuum part is conspicuously in place to purposely confuse reality as it is brought in to flagrantly spread misunderstanding of the issues in hand about the falling that takes place. With everything always falling equally when the same the condition applies to all objects falling and therefore with such falling happening under the very same variation of natural conditions applying, this shows it is the space in which the object is that falls and not the object falling while leaving the space it holds behind. The lack of relevant density in relation to air moving down stops the feather from falling equal just as gas does not fall with the space at the rate that space does descend. All space falls by the compressing of the atmospheric space. The rotation of the earth moves the space sideways and this brings the space to move downwards by increasing the density of space or air as it comes closer to the earth. This results from the Roche limit applying to fix atmospheric layers varying in density. In my books I explain that principle applying mathematically. Notwithstanding using your mathematical marvels, science has not got any vague idea to explain any of the phenomena mentioned above. To understand these phenomena one has to understand singularity.

The circle forming Π uses 7 to indicate the roundness of the circle but the 7 holds its roots deep within creation. It indicates how the Universe started because this is the way a star will start moving and it shows how as the infant star starts generating gravity just as the top starts to spin when it is thrown by life. Life can create nothing and that is true but life can mimic all laws in the Universe. Time is eternal movement and will be with us always. The line in infinity is still present while not being a part of the Universe. This line is always ready to be in place when the slightest movement orders it in place. Before the Universe was in place eternity and infinity was in perfect harmony and the line forming singularity validates this fact.

Before infinity parted from eternity, eternity met infinity on one spot as eternity came from the past (1) forming the present (2) to go onto the future (3) but also returned to come from the past which was the spot held by the future and this we find in the fact that the line forms 1 when not spinning but as soon as it evokes by spin, 3 points form even now. Then heat and cold differentiated values and space landed in between eternity and infinity. As eternity moved in relation to infinity but not forming a part of infinity any longer, eternity had to follow a path by never going away from infinity (3) and always returning to the point infinity holds but never lash onto the point again. With space parting the points, eternity had two points (the past and the future) before the partition came about and infinity held both the past and the future while infinity had the present as it still gas presently. By eternity also moving, the two points it held opposed each other (the past and the future) and since it moves, by the movement it became the square of the two because movement is the square and not a flat blanket-like surface with squares embroidered on it as Newtonian science depicts it by using grand mathematics to understand singularity.

Then we had two point holding eternity in place going square by movement to form 4 points serving eternity and infinity captured the first three points held by both and since eternity could not release from the two it had but had to duplicate what it had, eternity by movement became a circle captured by the line. With four points captured by the line of three points the circle coming about is eternally returning to infinity but never complying with infinity because if mismatching temperature or movement (3 against four). Material will always be colder than outer space. It is because material spin and outer space moves by expanding due to overheating. This is where I start when I start to explain the first moment but I use a shipload more information to do explaining when I explain the star in the book I do so. I involve the four cosmic pillars to substantiate the claims I make because all four still work the very same way as it did at the beginning of the Universe. The three points serving one part of singularity combined with the four points serving singularity unites as seven to form a circle of either 3.1416 or 21.991÷7. The seven going to one is eternity matching infinity by movement. But since seven moves it is seven that has to produce gravity. How do I know all these facts, because we can see from the top it is still doing what it did the very first second.

When time started infinity as well as eternity had altogether 3 positions, the past, the present and the future. It is still forming the very line in the centre of the top as it forms all lines in the centre of all things spinning. Then eternity parted from infinity when heat separated what is cold from what is hot and eternity formed one more point than before when it had the three points. With infinity and eternity then jointly having 7 the cosmos came into rotation. In the aftermath post big Bang we now see the phase of cosmic development where the tow sectors try to unite and this brings along the contraction. When Π forms it does so on the grounds that 7 rotates. The circle forms by a change in direction by 7°. Every circle has opposing sides forming in relation to the axis line. If the topside goes rite then the bottom side has to the left. If the rite side goes down then the left side goes up. There is this double presence of a change in direction forming on both sides of the circle. The 7° move and by moving 7° goes square 7^2 and that is Pythagoras.

Gravity is about the reducing of space to singularity. In spinning the sphere contracts by measure of 21.991 reducing to 3.1416 while 7 is reducing to form singularity, but also gravity forms when the 7 comes from the past to the present 7 and onto the future 7 and this became 21. Not only that but with singularity advancing from infinity to become one it proves that even as we see singularity as one, singularity also is multi dimensional but that ability is beyond our scope we have being in the Universe. The dimensional change that Π undergoes shows that singularity repeat into a new location by the value of 0.1416 and then as the new 7° as a redirection forms as at first becoming 0.991 that then progresses to 1. That is how the cosmos started. Infinity holding eternity on one spot coming from the past to the present being one spot and onto the future being one spot the cosmos was singular monotonously eternally by repeat.

Gravity is the contraction of space density taking Π from a value of $\frac{21.991}{7} = \Pi$, which is what is in space to the rim of the earth, which is $\Pi = 3.116 \div \Pi°$. This indicates contraction by the earth's change in direction by 7° to alter the relevancy applying from $\frac{21.991}{7} = \Pi$ to form $\Pi = 3.116 \div \Pi°$.

The line has two opposing sides turning directionally against each other while turning with each other. By moving or turning this involves time duplicating space by the square Π^2 on both sides of the divide $\Pi^2+\Pi^2$ and using the same divide or the same axis or the same point serving singularity we have 7^0 crossing the same point in singularity Π^0. There then is in this rotational movement 7^0 standing in for Π^2 on both sides of the divide $\Pi^2+\Pi^2$, which then is 7^2 on both sides of the divide 7^2+7^2. But since it involves singularity moving it calls for the law of Pythagoras to produce space. The law of Pythagoras is the triangle a^3 that is moving forward in singularity k by turning T^2. In singularity the 7 stands in for 7 points on the numerical line crossing over the line holding singularity or 1. By moving 7 has to go square T^2 and that means 7 goes square 7^2 twice $7^2 + 7^2$ crossing the same divide $\Pi^0 = 1$. Since all movement in singularity has to enforce the law of Pythagoras we have two triangles holding 7 dots moving across singularity. I don't want to get too involved by bringing in numerical outlays because then this can truly become complex.

The circle spins in duel directions. On the one side it would go left if on the other side it would go rite. The one side hold a directional change in singularity by 90°. As it is going sideways it changes to going down. This produces a rite angle triangle of 90° and in it the law of Pythagoras produces direction changes.

The circle spins in duel directions. On the one side it would go left if on the other side it would go rite. The one side hold a directional change in singularity by 90°. As it is going sideways it changes to going down. This produces a rite angle triangle of 90° and in it the law of Pythagoras produces direction changes.

Since the square of the turn of the circle places by the spin and the direction change we have 7 holding a relation to 10 in space because it is space that has to carry the value of 10 when material circles by 7. There is a connection between space surrounding the spherical circle turning and the sphere. Te circle holds the value of 7 as in 7° and this we find from looking at singularity controlling the circle by movement

PART 4

Einstein saw gravity from a window in a patent office

Years ago I was reading of a remark Einstein made about his realisation whiles being a patent clerk. Einstein realised that had Einstein fell from the window of the patent office Einstein would feel as if he was as weightless and as weightless as the chair and a pen falling alongside Einstein down the building would be. The only principle Einstein therefore could not accommodate in his theory on Relativity was Newton's gravitational pulling by the value of mass. Einstein saw this "feeling" as a psychological experience more than it was physics. This moment reading this gave me the breakthrough that I waited for and it (then) took me twenty-three years to make the breakthrough. It was not Einstein or the chair or the pen that fell but it was the space the three components occupied that descended.

Then I realised Einstein felt weightless because he was falling and part of falling was feeling what was happening to him. He was not pretending to fall whereby he then would feel as if…he was really falling and with that there is no as ifs. What he experienced came by means of what he was experiencing in as much as undergoing. If Einstein was experiencing weightless ness, it would be because he was weightless while falling. Weightlessness is not having mass and not having weight is feeling being without mass. If he felt being without weight he then was without mass and then mass has nothing to do with falling or the process of going towards the earth. Einstein would not imagine the weightless ness because Einstein was truly falling. He was at that moment truly weightless. He saw himself falling alongside (not faster as he should if he had weight or mass) than the chair and the pen that dropped at the same pace and at the same sped as he was falling. Einstein, the pen, and the chair had the same weight since they were all weighing the same because they were descending at the same rate. All three items would be equally weightless during the falling…that was what Galileo found because objects of different size and different mass travel equal while descending. This is what Galileo found and that is what Newtonian science can't reject but have to accept in spite of Newton contradicting this founded his pendulum theorem on. Newton contradicting this is because Newton claims things falls by applying mass. Galileo says all things fall equal whereby mass plays no role and Newton says it is only mass that plays a part. Reality TV shows that the bigger objects do not fall quicker than a smaller object and that can only be attributed to one fact; it can only be true if they weighed the same while falling.

From this one can deduct that gravity is motion or the intent to commit motion and mass is one the motion of gravity is frustrated by blocking the continuing of the motion. Gravity is motion of space and mass is the restricting of the motion of space. Having mass does not bring about gravity but it does restrict gravity's motion. Gravity produces mass but mass does not produce gravity. Mass is the restraining motion and gravity is material moving about. Mass only comes into the application when two objects filled with space moves into a position where both want to claim space the other occupy. In essence it still is the frustration of motion and the commitment to move once the blocking of space is relinquished.

I then after reading this realised that gravity is not mass orientated, but gravity is motion differentiation between objects. While falling, The object moves less or slower in the direction that the Earth rotates and will fall in the direction of the Earth centre until such a time as the movement of the object is in synchronising with the speed that the Earth spins or if not the object will and on the Earth surface at the edge of the Earth and that will bring about having mass. The gravity applies as speed that is putting time in relation to the distance travelled and distance travelled is space. While the object is in a process of falling, the motion confirms gravity, both by getting the object's distance or band in which the object travels in harmony with the Earth that conducts all the spinning taking place at that point. That will reduce the height in which the object spins until it lands on the Earth and then can't reduce such reducing of a travelling band any further. It has to do with specific density. If the specific density is increased by filling the object with helium we will find there arrives a point where the conducted speed is at a level that the Earth no longer will claim the body into having mass. When motion downward ends and the Earth disallow any further movement to secure a better specific density in relation to rotating movement, then mass sets in and becomes what is than point holding mass where the constraining of the object takes place to secure frustration of further movement and the Earth's motion annexes the object's freedom. While experiencing mass the motion is still there but now incarcerated by mass and locked onto the Earth by the rotation of the Earth and the superior or equal specific density of the Earth. By connecting to the Earth the motion that the object is experiencing is what nails to object to the Earth by the force of mass and the object is then experiencing mass and not falling further through the loss of downward movement and now only conducts with the Earth rotating side-on movement. In this the downward movement is not lost altogether but remains, as detectable movement is the form of having a tendency to move although the object in mass is applying by forcing the downward motion to stand still. While the object is in mass and seems to be as if it is resting the tendency to move downward remains applying but that tendency to continue to move downwards is the tendency he named mass. However mass then restricts motion and becomes motion tendency. While falling, gravity applies as equal motion to all objects relying to place all objects in relation to specific density and because of this motion counteracts any size, mass or weight by making everything able to fall equal in specific density. When falling, the object is either equal to what might be in the air according to allowed specific density, or has more than the specific minimum required density that is what is allowed to serve as the minimum required specific density and therefore will spiral down to the Earth. When the Earth restrains further downward motion of the object that comes as the result of finding an allocated position of motion according to the specific density of the falling object, this readjusting of allocated position is stopped from conducting further downward or readjusting movement and all such further movement of gravity is hindering in the form we call mass. The falling object remains individual and still tends to move while Earth individuality resists movement. Further movement is disallowed as other material fill space. While the bonding of the atoms forming the object will secure any further deforming the object will remain to be independent but it is this bonding that is the value of the specific density of the object applying. By securing a place on the Earth, the falling object will finally rest and from that motion resistance comes mass.

While falling, the object is experiencing gravity because the object is in gravity but when on the soil the object experience mass which is the restricting of gravity or motion of the space filled with material.

Moreover, I came to another conclusion of equal importance. When any person is standing on any place anywhere, while viewing the Universe, that person is filling the centre of the Universe. Let's get more personal. When you, the person that is reading this, are standing at night and are looking at the Universe you are seeing the Universe from the centre of the Universe. All the light, every single beam that ever left any destiny at any time acknowledges this fact. You are the most important person in the Universe because you are holding the most important position in the Universe. All the light that comes across all of space runs directly in a straight line towards you filling the centre of the Universe. Not excluding the effort of one photon, all light is heading to meet you where you are in that centre spot and not one photon will pass you by. Not one photon dare miss you because if they do they miss the effort that all light has to accomplish and that is to locate you as the person filling the centre of the Universe. If you find this funny, or laughable you are in for a shock because this is what gravity is and this principle dictates gravity. It is the most complex issue one can imagine and expanding on this thought takes thousands of pages. It forms the crux to all cosmic principles and embraces every successful and meaningfully theory ever used to explain the Universe. Without taking this aspect in to account, there is no valid explanation available to understand the cosmos. Al the light coming from wherever meets the point you fill in time and in space. For al the light travelling you hold the spot it was on route to.

Should you decide to shift your position to any other place in the Universe you will shift the centre of the Universe to that location as well. If you install a camera on Mars, the light is obliged to acknowledge your relocating the centre of the Universe at your will to reposition you're being that centre of the Universe. All the light that ever left its destination crossing the vast spaces of the Universe, excluding no particular light, travelled all the way just to find you filling the centre of the Universe, right where you are. By you're standing anywhere, you fill the centre of the Universe, and the entire Universe admits to that because all the light comes to meet you there. If you shift from the North Pole to the South Pole you will shift the centre of the Universe because all the light travelling throughout the Universe will find you where you then moved the centre of the Universe. The light left its destination billion years ago as it travelled through space at the speed of light anxious to acknowledge you're being in the very centre of the Universe. No photon will pass you by where you are in the centre of the Universe. No wonder every person born has the idea they were born to fill the centre of the Universe, which we do fill. The Universe is spinning around you or I, which is filling a centre where all motion is connected. That is the Coanda effect on the uttermost grandest scale imaginable; nevertheless it is only a manifestation of the Coanda effect. It implicates gravity as wide as can be…

Then I reviewed the Universe. If gravity is motion, what causes motion? What stops motion? That answer is in the Black Hole. If a star is about fusing atoms thereby growing, what happen when all the atoms fused into one all collective atom? What is the gravity if the star has one all-inclusive atom providing all the gravity that the star had when the star still had massive volumetric space? If all that space that once filled an entire giant star fused into one enormous gravity applying atom and that enormous force has been secures in the space that one atom holds, the atom would then show a force that would pull the surrounding Universe flat. Where does the gravity of the star end when all the atoms in the star became one giant atom? Gravity is smallest where space is least. Where space of an entire massive star is left in the size of one atom the gravity coming from that will pull the Universe flat at that point.

Coming to the conclusion about gravity being motion and mass being the restriction of motion was the easy part. What produced the motion and what prevented the restriction from overcoming the motion was the tough part. Figuring out why was everything on the move and where did the motion stop that was the part that took some figuring and some explaining. What made gravity move and why does gravity move…the answers are in the four phenomena never yet explained to satisfaction but now turns out to be the cradle of gravity.

Gravity is The Roche limit,
  Gravity is The Lagrangian system
    Gravity is The Titius Bode law
      Gravity is The Coanda affect

And gravity as the Roche limit forms the principle in producing the sound barrier. Read the book and find out why this is the case. I explain these in the next chapter or the chapter following the next chapter.

Newton's claims about the principles that he declared is responsible for guiding physics carries no validated proof and only after I realised that, was I able to start forming another line of thought on gravity. This had the purpose of confronting the corner stone of modern physics and at first I tried desperately to do just that. At first I was not confrontational towards Academics in physics and avoided any indication about disagreeing with Newton, although avoiding to show my disagreements was also totally impossible too but every time I approached academics with my new concept the academics always threw Newton at me . Facing Newton or facing defeat became a two-sided blade and I had to start to confront them by confronting Newton, with which I was in disagreement from the beginning. At first I was reluctant to voice any opinion about the matter of how far I was prepared to challenge Newton because Newton was and is an icon. But slowly it dawned on me that if I had any serious plans to introduce my ideas I had to dispute Newton's gravity principles and do it head. When the slight confrontation did not bring results I finally decided to go all the way and show the inconsistencies that were prevailing in Newtonian science. That worked neither and it brought me the same results as before whereby I decided to go public and straight to John and Jane Dow avoid arrogance academics have with only one motto they serve and that is their autocracy and in particular their megalomania especially to my case as well as me in person. I wrote them (nine in total) letters in which I warned them that I was going public to show the extent of their dishonesty in their Newtonian's approach and lacking of substance and proof their physics has. The lack of honesty and furthermore the absolute dishonest on their part is there whether I avoid it or attack it; the inconsistencies are part of forming the basis for modern accepted science.

This process I now described is explained in a paragraph or less and it seems I got that far in a breath or two, but getting this far took me the best part of seven years to get to I tried my best not to attack them or Newton but left with the option to leave the project and lose thirty years of work and then fail after I concluded an answer on every aspect they never even thought of or take them on and dish out what they should have received years ago made me decide on the latter. After being avoided and taunted by their powerful positions and arrogance vested in their mentality they show in regard to their positions as well as the disregard they show in the mentality of others I slowly concluded that only and after I can get people forming the general public and the opinion of those that holds their disregard just as I do to see what they hide will I get a response from the Mater's of fraud. First I had to show the general public the true colours of the academics in physics and get every one to see how incorrect Newton is, and only then do I stand any chance to introduce my line of thought. I am so sure of the ideas that I propose of being correct that I dare any one to disprove any part or the entirety that my concepts about cosmology forms! But that can only come about when I can get an audience to see how I expose Newton for what Newton was and it is in that where I find no luck. I can't find one academic with influence that is brave enough to stand up and face my attack on Newton and argue me down or prove me wrong in a sound debate. Now I see frowning coming from everywhere because it is madness on my part to think the world is wrong and only I am correct!

I realise that it shows signs of madness on my part and in my thinking to even regard any possibility that I am the only person on Earth that is correct and all others that ever studied physics are wrong but mad as it seems, if that is what I have to say to find an audience to listen and to judge my case, then that is what I say. I don't say this lightly or without understanding the enormity of what I suggest is going on, but be that as it may seem, it is the truth without question that Newton went on for three hundred and fifty years defrauding science with no one testing his claims. Argue me down or prove me wrong but don't discount me before hearing me out and only after considerable consideration while studying my arguments then form an opinion that disputes what I say but when disputing what I say, do it while confronting me in a sound argument when proving me incorrect! This not one academic could achieve and I challenge the lot to do so. But do it after studying all my work and being in a position to account for all the details I propose. Don't just dismiss me because I dismiss Newton because following that road is the way of the coward and the mentally impaired. Read my challenge about the correctness of Newton's proposals when he brought no more than suggestions into science and when I dispute Newton, then take me on by proving Newton correct... do it just once... prove Newton correct just once...prove that his formula is working and that his principles apply on the grounds he principled his ideas.

Detecting Newton's misconduct is possible because I saw a way to break away from the invalid concepts Mainstream physics hold. I went about and tried to prove Newton and when that was not happening I tried to apply Newton's ideas into the greater fields of cosmology. That also wasn't possible. I tried to amalgamate the four cosmic principles applying in cosmology with what Newton said was happening in the cosmos with mass and with gravity and in light of what the cosmos showed was happening Newton just wasn't happening!

Notwithstanding the pose Mainstream physics try to uphold, the entirety of physics still use the idea of magical forces intervening in nature and they still base concepts on unexplained novelties. Think of finding four unexplained forces going around and influencing persons in an unexplainable manner except that the magic of gravity keeps people attracted to the Earth. To say the least, the concepts physics use in terms of Newton would not even be acceptable to children in the modern informed era we live in, I challenge any person to prove Newton, not to accept Newton but to undoubtedly prove Newton correct! Prove how Newton's formula of mass forming the force of gravity can apply as Newton said it does! I recognised the impossible double standards Mainstream physics apply to promote their much shady explaining. In short I tested Newton's principles and found the principles to be wanting.

The inconsistencies Newton introduced brought science double vision and to compensate for these bogus truths supporting their incredible theories, they simplify issues to such a level where what they embark on, is the meaningless acceptance of the unproven and they proclaim to understand what are meaningless inconsistencies and to achieve this they create scenarios which uses the entanglement of deception. Prove the attraction Newton said was enforcing gravity that is pulling by mass and is gathering plants by contracting the diameter between planets.

Show how much the Moon came closer to the Earth since the time of Kepler. Show proven distances taken by radar tracking and indicate just how accurate Newton was. Show how much the Moon came closer to the Earth since the time of the Moonwalk in sixty-nine. The figures are available but are kept in a grave of silence where no one ever speaks about what science found applies and how much the distance between the Earth and the Moon is shrinking as Newton said is happening or then how much is the is expanding which will contradict the very principles Newton brought about! What they declare as unwavering facts can't even be supported in some form when tested by a silly test as to show that the distance between the Earth and the Moon is shrinking. Even the least degree of verification of correctness is absent when trying to find support of Newton and Newton lacks all evidence of authentication in any investigation of even the simplest terms. It is as if they never read with interest that which they explain when they embark on explaining Newton and they never scrutinise that which they advocate when they teach Newton's principles applying. They give values that are senseless and the very values they use make that which they say meaningless.

In this book I am going to investigate how much truth there is in mass pulling by the force of gravity. To most if not to all of the persons reading this, such a venture of investigating Newton is time wasted and just the thought about me embarking on the investigation of the issue is totally senseless to investigate. It is senseless because the concept it carries became accepted as household practise and life science from where it proceeded to become everyday culture in every person's mind. The worst part is that the group of people normally considered as the wisest bunch there is, never did prudent testing on Newtonian presumptions, while to test the presumptions is most easy to do. I will not believe that a lot that lives up to the veneer of being the best mathematical intellectuals on Earth, never though of testing Newton's very simple formula and in that disregard the formula because of the incorrectness the formula holds.

Do you think of astrophysics as being the department that is run by the wise and the level minded, the honest and pure at heart, the nobility of well-to-do academics and the sober thinking standing in front of the world as the absolute trustworthy? If you are a student, there is no other choice you have but to trust them while they feed you absolute hogwash! If you would so much as dare to doubt any thing they say they will banish you from the institution they rule so absolutely. The banishing process is dome under the blanket of examination. They teach you what to think and to make sure you think what they wish you to think, they tell you to confirm their teachings on a blank piece of paper. You write what they prescribe and you supply the answers they demand in the words (sometimes) of what they demand. Should you in any way say anything different from what they tell you to think, your presence will not be tolerated any further as they abolish you from their institution of academic tutoring.

After reading this book I invite you to…no I dare you to challenge their statements with evidence gained from this book and see them wilfully further their culture of deceit by bringing unfounded arguments just in order to silence you and prevent you from getting behind the truth. If you think those in charge of astrophysics are the pillars of trust, then get wise by reading the following facts and arguments this book presents. What you are about to read is simply mystifyingly simple and yet to this day I have not had the privilege to challenged one academic any where that had the honesty to admit to the fact of Newton being wrong. After you have considered the following you might agree with me that even small Children can reach a higher level of clear-minded logic and find more sensibility than what those scientists promoting astrophysics have because science lives in a make believe fool's paradise.

The manner of regard to life that the Academic Physicist holds and the outlook on life that the followers of Newton physics have (I call them plainly Newtonians and to me they are sheepish because they resemble to the image that to me seems the same as sheep running after their leader without having the ability to think for one second any thought spawned out of personal intellect) is quite the opposite of what I think of them. They keep their forming the establishment of the order the Academic Physicist in high regard and consider their order to be the top thinkers in society.

This religion that they practise of self promotion and sublimely self regarding their status being next to God has them so high that we down on Earth forming the waste of human garbage can be told anything and we will believe what they say just because they with their supreme intellect tell us to think what they wish us to think. This they do because we human waste living way down below their supremacy have not the ability to think and therefore they must think on our behalf. In their view and so far very correctly judged on their part, they, the persons being in the group that forms the Academic Physicists, believe very correctly that can dish up whatever they wish and we, those forming the group in the gutter, those that are mindless in their eyes, we will have to accept what they say without being allowed to form an opinion other than having the opinion they give us to have because in their view we are unable to have a mind other than what they are able to control. This attitude they have is the result of a relationship that worked for so long and thee fact hat it worked that long is what confirmed their opinion that we, the public, are fools to believe anything and everything because of blind stupidity.

But in spite of their aggravating conduct and mischief towards us, it is not because of a lack of insight and inability of controlling a mind that we have our childlike belief and blind trust in their opinions and which there was. It is the faith we shown that they misused for their scandalous cheating. Our faith is what we have shown towards them and is that, which became used as the reason why we accepted what they said blindly. We didn't accept their word on the grounds of us being utterly stupid as they perceive us to be but our trust depended on our good nature and believing in their trustworthiness.

This trust we have is brought on by a culture of trusting the King to do the people well and somewhere in every person's cultural past there was Kings that did us well in leadership. But their underestimating of our abilities is the testimony of their poor understanding and their weak insight ability, which results from their arrogance and stupidity. You are about to see just how stupid they really are in the thinking aspect of science. It will become clear as you page along while reading! They didn't fool us half as much as they fooled themselves and you are about to read all about it. The fact that they could fool us for centuries didn't run on their intelligence being so much superior but served their purpose as it stemmed from the trust we had in them resulting from good intentions on our part. This betraying on their part and misusing the public's good nature to be used in schemes to get the public conned must end and I pray that this book form the first step in resisting the arrogance of the Academic Physicist.

Any one not in their group of the Academic Physicist is part of the lowest order of mindless being and to become part of their order and those that have minds with an ability to think, students have to accept what they say when they say whatever they wish to say without having to prove the correctness of what should back their saying so and as a result of this students may never question what they say. Only when and after proving that a student has totally lost all ability to think for him or her self may a student be promoted into the ranks of their sublime intellectual group. The sifting process they named examinations. You write on paper what they told you and never question their opinion and after passing that examination will you ever enter their sphere of intellectual brotherhood. Does this sound far fetched? Then you better read on and I will remove your blindfold and show you what a world of deception the Academic Physicist force on us into.

Read the following and see how they, the high and the mighty, those that think they can replace God and those who think they can think on our behalf and think what to tell us to think, how much they are clowns and the jokers in society. Read how little are they, the Academic Physicists, able to understand concepts about Creation while they think they are able to replace God in their superior intellect.

If you are a student in the science of physics, then ask your Educated Masters to please explain the following abnormalities you are about to read in this book and insist on a clear explanation about the inconsistencies they promote while tutoring physics as if the physics they present are the most flawless and accurate institution there has ever been. Ask those academics supporting Newton about the following flaws that no one mentions ...ever... except me in this book you are about to read and get them to explain the inconsistencies never talked about, which I present in this book and then after confronting those charged with tutoring physics and seeing who should be believed, then get wise instead of brainwashed. Let them mathematically show how one would go about and use Newton's visionary formula $F = G \dfrac{M_1 M_2}{r^2}$ to calculate the force of gravity by replacing the symbols with the actual values in mass that the items referred to have. Put in the Earth's mass in place where it belongs and put in your mass in place where it should be and then divide that with the distance between your soles and the Earth measured in micro millimetres by the square thereof!

In the book named an **_Open Letter on Gravity Part 1 and Part 2,_** I bring the solution to the mystery behind gravity. I tried in vane to introduce the principles I find valid to the academics in charge of astrophysics. Facts that Science present as being the uttermost explicit and unwavering truth, fails to bring any logic answers to so many questions that it should address. It fails to have substance in addressing the most basic and simple questions about gravity and physics. Yet to every question science can't answer my approach does bring many solutions. The presentation and the delivery of my answers that I reach are understandable and simple where it serves both logical science and the truth.

Since my answers do not match Newton and his misconception about gravity and that mass generates gravity, those in charge of science don't even bother to read my work. With their affixation to the corruption they portray I can do little to the giants where they are in the mighty positions they have and just because of that they can go about to sideline and ignore my work and this is notwithstanding the correctness that my work delivers compared to the utter failing that Newton's work shows. When confronted with my evidence and they have to match my work with the hypocrisy and misleading nature of Newtonian cosmology their defence in substantiating their claims is to ignore me. Since I do not applaud mainstream science and the clear fraud they embrace and fraud it is that they embrace, I am silenced.

Why is it that my work is going unrecognised or even in the least goes never debated and never commented on…it is because it will then trash every article anyone has ever written about astrophysics and cosmology. They show little integrity when academics with such supposed high standing or then such as they should have, play a dishonesty game where those in commanding positions will rather protect fraud and save their skins. They would rather protect the corruption they have than seek the truth and find honesty in physics. Those academics in charge would much rather protect their un defendable ethos they maintain as forming the back bone in science and what gives their personal position legality although it is corrupt than admit to the truth they find when they begin reading my work and in agreement they then have to back the truth my work brings.

Doing that (accepting the truth in my work) will trash all work in cosmology delivered thus far and condemn it to the waste paper basket and render all work invalid and void. It will put all the Newtonian's bias and fraud into the place where it belongs. Considering that such acting will lose them money, those academics in controlling positions then will rather rape the truth in order to benefit from continuing to corrupt student's minds further. If they wish to justify their inconstancies they have to attack my work and disprove the accuracy of my work. That they can't do. They then ignore my work because they can't attack my work. In that sense they also place their work beyond my approach, as they can simply ignore me as if I represent the plague while they carry on with little consequence to bother them. I challenge them to prove Newton correct and not just declare Newton being beyond reproach after all has seen the evidence I bring. After reading this all students must challenge them to defend what they can't or get honest.

$F = G \dfrac{M_1 M_2}{r^2}$ $F = \dfrac{r^2}{M_1 M_2}$ This is the basis that Mainstream science uses as the foundation of all physics anywhere. If this is wrong then everything they have got to work with goes out the window. They put mass and the distance that parts objects in a relevancy, in other words the one is a ratio to the other. The one factor brings a measure to the other factor's value. The one cannot be without the other. The increase in one becomes the reducing of the other and the other way round also applies. When the distance is large, the influence of mass will be small and when the distance is small, the influence of mass will be overwhelming.

Then they state we are in a Big Bang expanding of the entirety. Why then, when considering that if it is mass that produces an inclining force of contraction as Newton says there is going on then…why didn't the expanding stop before it started when the Universe was small. Today using hindsight after the fact of the exploding Universe became apparent by the studies Hubble brought to light did the lot of everything that is not implode as Newton would have us believe whereas, instead it did expand just as Hubble proved. The radius at the time of the first instant back then was no factor, which makes the gravity at the time a totality of unrivalled force. The radius being that insignificant leaves the mass unchallenged in asserting power in relation to the non-existing radius it had.

I dare any physicist to show me where they apply Newton's formula just and exactly as Sir Isaac Newton suggested gravity applies. Show me just once where the mass of the Earth is multiplied with the mss of the object in normal physics. Show me just once how $F = \dfrac{r^2}{M_1 M_2}$ or $F \alpha \dfrac{M_1 M_2}{r^2}$ where one M represents the mass of the Earth while the other M represents the mass of the object and in this formula the end result will have a value of 9.81 Nm/s² … show just once one example… where the use of the mass of the Earth comes into play. If multiplying the mass of the Earth with the mass of an object and dividing that with the distance parting the two mass factors does not deliver 9.81 Nm/s2, and then any claim by Newton indicating that $F \alpha \dfrac{M_1 M_2}{r^2}$ is equal to gravity, such claiming constitutes to deliberate fraud…even if Sir Isaac Newton said this. Prove that the mass of the Earth with the mass of an object and dividing that with the distance parting the two mass factors delivers 9.81 Nm/s2 or admit physics is conducting fraud to protect Newton!

To whom it may concern:
My introduction as well as introducing the readers to general cosmology in a very brief and compressed manner but first, I have to give the emphatic warning to all prospective contemplating readers.

Please take note of a conscientious warning about the gravity of the misgiving there is on the part of the most respected Academics in physics about a much concerning matter. I state it emphatically that science accuses me to be not schooled to the point where I am able to have any form of an opinion on any matter concerning Sir Isaac Newton. Notwithstanding that my research proves I did my private studies and through which I skipped the indoctrination and mind control academics place on students goes unrecognised by their standards and so too my ability to have any insight on matters regarding physics. However my skipping their methodical and systematic brainwashing enabled me to see and allowed me to be able to express the incorrectness in Newton's teachings and allowed me to show in clarity what destructive force Sir Isaac Newton used to corrupt the laws of mathematics, corrupting to science along the way and mostly raping to the work of a great man, Johannes Kepler and what Sir Isaac Newton did can only be expressed as being blatant criminal fraud. What his deeds amount to is to corrupt the laws of mathematics, to render the laws of cosmology useless and to rubbish all of science. Should you find this to be unbelievable, then I am glad to announce that this book is more for you than any other person, so go on and read what academics guarding science never wanted published. I challenge any one that disputes any claim I make to prove me wrong by proving me wrong and not merely suggesting claims in that direction.

Tell me, can you find any credence in the "Conversions for "Unknown""
$4\pi^2 a^3 = P^2 G(M + m)$
In this comes the fraudulent part because there is no evidence of mass playing a part or forming an actual presence in the solar system.

If the cosmos supported Newton's claims of $P = \left(\dfrac{4\pi^2 a^3}{G(M+m)}\right)^{0.5}$ then the planet arrangement would have been much more likely as I show above, but the picture indicates the mass as well as the planet formation.

You must judge; it is either the cosmos that is incompetently wrong or it is Newton that is incompetently wrong because what the cosmos has in place Newton knows nothing about and what Newton claims the Universe uses, the cosmos knows nothing about. Who would you say knows more about the cosmos' method of workings, Newton or the cosmos? If Newton is correct then the planet layout must be as I show with Jupiter very close to the sun. It seem the cosmos is just as unaware of Newton's ideas as Newton is of what is happening in the cosmos. Who would be correct about cosmic principles applying, the cosmos or Newton?

$P = \left(\dfrac{4\pi^2 a^3}{G(M+m)}\right)^{0.5}$ What hogwash does the factor $\dfrac{}{G(M+m)}$ indicate?

The same can be said in the formula $M = \left(\dfrac{4\pi^2 a^3}{GP^2}\right) - m$ when $P = \left(\dfrac{4\pi^2 a^3}{G(M+m)}\right)^{0.5}$ that the factor $\dfrac{P^2}{}$ is senseless and $\left(\dfrac{P}{2\pi}\right)^2 = \dfrac{a^3}{G(M+m)},$ has no foundation other than fraud.

$M = \left(\dfrac{4\pi^2 a^3}{GP^2}\right) - m$ is complete fraud. The Cosmos does not support the Newtonian formula even in one place where it could apply.

Position as a function of time

$$P = \left(\frac{4\pi^2 a^3}{G(M+m)}\right)^{0.5}$$

This is what Newton said is in place and with no evidence ever founding this ridiculous proposition, all Newtonians that ever come after Newton. This is what Newton and his Newtonian followers tell the solar system it has in place and tell the cosmos it uses to operate. I have indicated that mass has no place or use in the solar system according to what the solar system puts in place.

These are the closest because these are the massive giant gas plants and having the most mass must put them the closest to the Sun.

Get your professor to prove Newton correct in the face of $P = \left(\frac{4\pi^2 a^3}{G(M+m)}\right)^{0.5}$ and if he can't let him admit he has been conducting in a fraudulent practise all the time he was teaching.

If you think scientist know what gravity is…well, do not be duped that easily because no one in science remotely knows what gravity is…not even Newton knew what gravity is except Kepler… and because of what Kepler introduced now I know I can prove what gravity is.
Gravity is precisely what Kepler said gravity is

That Kepler was the only one that knows what gravity is we can see when we make an effort to investigate Kepler without Newton telling Kepler what he (Kepler) should have found. He (Newton) should have investigated Kepler's work more open minded and much closer then he (Newton) would have seen that gravity is precisely what Kepler found to be gravity

Science fails to bring logic answers to so many questions. It is simple questions about gravity and physics in which they fail, yet to every question science can't answer I bring an answer. My answer serves both logical science and truth but my answer does not match Newton and the misconception that gravity is generated by mass. Yet, since I do not applaud mainstream science and their fraud they go about to sideline and ignore my work is. Why is it going unrecognised…. because it trashes every article anyone has ever written about astronomical science and cosmology. It puts all the delivered on Newton's bias and fraud to the place it belongs, it trashes all work done thus far to the waste paper basket and renders all work invalid and void. Where they have to attack my work they ignore my work because they can't attack my work. In the same sense their work is beyond any defence and when I attack their work, they ignore me as if I represent the plague. I challenge them to prove Newton correct and not just declare Newton being beyond reproach.

$F = G \dfrac{M_1 M_2}{r^2}$ This Mainstream science use as the foundation of all physics anywhere. They put mass and the distance that parts objects in a relevancy, in other words the one is a ratio to the other. The increase in one becomes the reducing of the other. When the distance is large, the influence of mass will be small and when the distance is small, the influence of mass will be overwhelming. Why then when taken into consideration that if it is mass that produces an inclining force of contraction as Newton says then…when the Universe was small it did not implode whereas, instead it did expand. After all, the radius was almost no factor at that point leaving the mass to enjoy an eternal power in relation to the non-existing radius.

When the Universe was at the point where the Big bang started, the radius was incredibly small. That would make the mass inducing gravity by contraction inconceivably large because the mass was completely overpowering all factors with the small radius. It did not bring about an implosion that the overbearing mass contraction was supposed to unleash on such a small Universe in the beginning

The more the radius develops in time, the lesser would the gravity be that the mass factor generates in relation to the advancing radii developing and the larger would the reducing be of all contraction.
The effectiveness of force the mass produce will tarnish as the radius that separates the material from each other increases as time moves

Although it is presumed that the Universe was small at the dawn of the Big Bang, such presumption will put validity to another presumption that the gravity the mass charged at the time was enormous because the influence of the small distance in radii and the factor such distance produced promoted the factor, which the mass has to an enormous large factor. If an object is a million kilometres apart the radius is a million times more in value by dividing the mass influence than when objects are one kilometre apart. That is the most basic realisation about mathematics. It puts ratio to order and define coherency. That is what gravity is to the Universe as it puts respect to factors about the Universe in the Universe. It is what derives order in the Universe.

At the very same time we will find in a Universe that was supposedly so small it had a radius of less than only one kilometre, then at such a time when the Universe was still that small it must also be accepted that the gravity the mass charged was one million times greater as it would be when the radius keeping the structures apart is one million kilometres in distance.

The extremely small radius that was only the size of one neutron in radii distance and with the factor that such a distance produces, it must promote the mass factor, which will support the mass in having an enormous large factor by relevancy to what the case must be at present. The mass factor that produces the gravity at any given point during the event of the Big Bang, had to be eternally larger at the dawn of the Big Bang while having an infinite radius, which gave gravity all the power it can have and which it will ever have.

If at the Big Bang there was not sufficient mass to destroy the radius and prevent the expanding from coming about, then the expanding won the match and there can be no contracting Universe as Newton had us to believe. If the Universe started a journey of parting objects no amount of dark matter that might lurk in the night sky and is at this moment hiding from detection will produce the gravity required to stop the expanding from continuing.

At the start the expanding became evident and as the radii grows the inclination will suspend in influence as a factor. If there was insufficient mass at the start in order to tilt the balance in favour of the reducing factor, no amount of mass can ever accomplish such a goal afterwards. Then Newton's surmising was one of corruption making that which all physics are based on fools thought and corrupted proof.

If you might be of the opinion that my accusing the greatest intellectual department in the world as being in misconduct and to your view such accusing is outrageous and far-fetched, then be my guest and judge the following with a clear and unbiased mind because when scrutinised with a clear view then the facts cannot fool an idiot. However, that is just what the physics paternity thinks the rest of us forming the general public at large are. They have the opinion that they can feed us in the public arena any senseless rotten garbage they dish up because they see us as being inferior by thought and mind.

With all this in mind did any one ever come to wonder about the all too famous Einstein's critical density theory and the fact that this idea was conceived to conceal the corruption of Newton in physics? The fact in truth is that the Einstein's critical density theory was a scheme plotted by those in charge to cover up and conceal corruption in the heart of physics.

If Einstein was unable to recognise the most basic of mathematical principles then what type of genius did physics create in him and what slur did physics promote. This idea of the two factors being in opposing relevance is so simple that children will recognise the principle, and yet those fathers of physics wants me to believe that the greatest mathematician that ever lived did not realise this principle…the principle that the radius and the mass stands related and the growth in the one will promote the decline in the other as a dominant factor.

Can any one with this information including the information given on the previous page have any other conclusion? It is obviously clear that having such a total idea that there might be dark unseen mass floating in the Universe which at this time does not generate gravity but will some day because Newton has to be correct at some point in the future. I am to believe that dark undetected mass can be found and such undetectable mass could be found which will bring about contraction after all this expanding? Why would the mass at present then not activate gravity and why would the mass at some point spring to life and start activating gravity? How much can the Physics paternity still hide the fact that Einstein's critical density is being used as a cover-up to distort the truth to conceal fraud?

The uncovering by the Hubble constant about of the Newton fraud is so simple to see. Hubble found the Universe is expanding and Newton's said otherwise. Who is lying about what? Hubble's declaration was on track to blow the cover that was concealing the Newton fraud wide open and uncover the centuries old deception. To see this we have only too look at the comet behaviour when any and all comets again come around on a cycle by repeated visiting the sun. The question is if it is mass pulling mass onto mass, then why do we have comets left in the solar system? The mass of the Sun should by now at least have destroyed every comet going around.

Every indication that we so far received in vivid portraying from astronomy photography studies from outer space disputes a shrinking Universe concept. From the moon increasing the radius distance between the earth and the sun, to the Hubble Constant indicating a space growing any where in space wherever man may conduct studies. Since the end of the middle ages a force called gravity was identified, but more than that science did not take it. What is gravity, besides being a force? What forces the force? I introduce a cosmic theory that turns the missing questions to answers.

Let us for one second return to the science we all know.

There is an undefined phenomenon in the cosmos, never mentioned (in public) because it obscures the basic formula

$$F = \frac{M_1 M}{r^2} G$$

Lets put the mathematical formula into a practical context.
By reducing r would bring about the same result as enlarging the mass factor of the cosmic objects i.e. the Sun and the planets. It is a very drastic implication that will cause much more than just seasons changing. It must bring about that gravity changes through out the year…yet the radius does constantly change, therefore…
The closer any two cosmic objects come the stronger the force should be, with eventually no force in the Universe being able too keep them apart. This is just not happening!!!

There is no indication of truth about a contracting solar system as Newton proposed and as Newton's followers promote and or a contracting Universe as seen from the Hubble as he introduced has Hubble Constant…and not from any other evidence seen through the Hubble Telescope.

<u>As explained, there is some discrepancies about calculating the force of gravity, because gravity would apply as nicely as it does if it was the perfect balance, a balance exist in space of equal measure bringing about equal seasonal time.</u>

The biggest discrepancy and a practical denouncing the official version of the comet's flight around the Sun

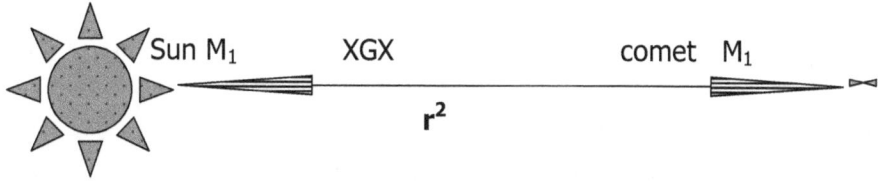

The Sun gets a grip on the comet by mass inflicting gravity and as it gets hold of the comet it drags the comet through the solar system straight ahead to the Sun just as Newton predicted the Sun with all its gravity producing gravity will do. There was no hint of a circle forming at any stage.

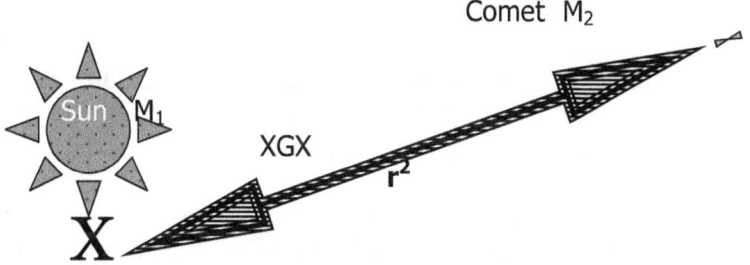

As Newton had said the gravity that the Sun and the comet mass induce pull the comet to the Sun. As we all know the comet moves to the center of the Sun just as Newton predicted with a slight complication and

a change in the venue, the comet no longer aims to the center of the Sun but aims at a target outside the limits of the space that the Sun occupies.

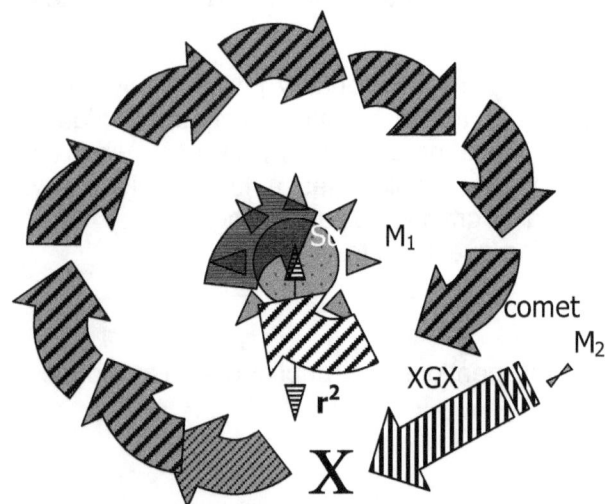

One should give Newton the benefit of the doubt and disregard this miss the position that the comet is aiming at. It could be that the gravity was not generated strong enough to locate the centre at such a great distance as the comet was at first. It could be that the reducing of the radius comes in instalments of a few circles. The comet might just take a cycle or two to wind down as the radius reduce and one should wait and see if it is not that which Newton meant when he said the mass by the mass is dismissing the radius between the objects. The formula Newton suggested did not make any room for a circle of any sort to form the trajectory of a planet or a comet trajectory.

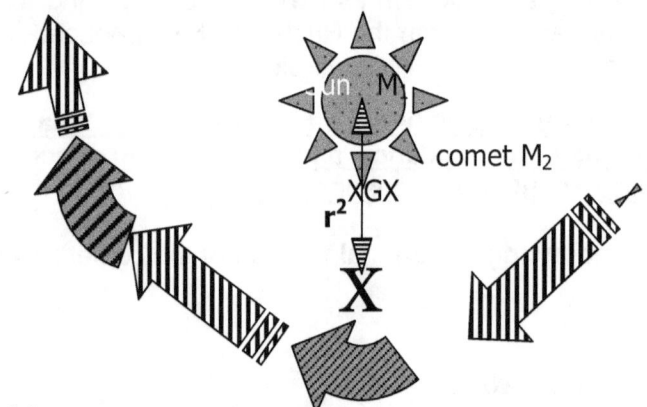

$F = \dfrac{M_1 M}{r^2} G$ **Explains the first sketch.**

Alas, it is not what the comets intends to do because the comet breaks the strangle hold of the Sun and what ever was pulling the comet at first is doing all the pushing at this point because the comet is surging into the darkness of the abyss. The comet speed away from the Sun and also at the same pace it was heading towards the Sun and there is no altering to the speed in any way. The comet seems very much unaware of the comet behaving in opposition to what the great Newton predicted with his formula.

The comet performs all the other manoeuvres as the sketches indicate, which Newton's formula totally ignores. If the formula forms the basis of all physics used by science, the basics, which are around for hundreds of years, are trash and simply does not perform as it is supposed to. For many hundreds of years every person in physics were aware of this flaw but did nothing about it.

The lot that was filling the Universe was growing apart. In some case he said the lot was racing apart. The Universe was growing by miles and not shrinking into nothing. The Universe was blowing apart!
The main discoverer had a name and a position of seniority, which prevented others from pushing his opinion aside. The man was E.P. Hubble. Through his telescope any one could see that the Universe was expanding and the expansion was most rapid.

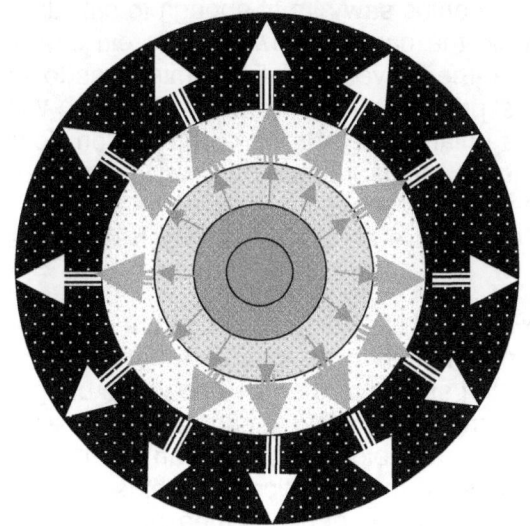

I am sure at the time all the Newtonian con artists were well aware of Newtonian shortfall in logical proof but this Hubble fellow was going to open a can of worms no one in science was able to face. How would the crooks that pretended to be the wise explain that Newton's contracting gravity was part of Newton's wild imagination? What would then happen if those that should know less then became wise and started to ask more questions about what the lot in conspiracy were hiding for (at the time) say about two hundred and fifty years? What is going to happen when the entire world learned that those academics in physics that was pretending to be the most brilliant among men was uncovered as the most stupid crawling the earth. What would happen if every one saw Newton was wrong all along and started demanding answers!

Every one was sharing the Newtonian vision of a contracting Universe where the lot would one day again come together and Creation will end where Creation started some time ago. The Universe has mass that is pulling mass towards one another and we are in the centre of an ever shrinking Universe. That is what the lot of us can see... we are forming the centre of the ever contracting having cosmos where every Newtonian can vividly see with his or her eyes through any telescope that all Newtonians minded scientists are sharing the centre stage of the ever collapsing Universe. The Universe is about to end where all mass contracts into one huge lump of material.

This unleashed a problem the world had no name for. Everything known to science was at that point devastatingly unknown to science. The world was expanding and not contracting which made the Universe quite wrong. It is impossible to have any vision about Newton being wrong. Newton could never be wrong because Newton was never wrong yet...so if the Universe is out of step with science, then science will correct such an abnormality by finding a way to defraud science and postpone the correcting that the Universe had to comply with since the Universe owed the Master Newton some apology. Did the Universe not know that he whom never can be wrong is in name Isaac Newton! Decisive action was needed. At this point I cannot believe that the most brilliant minds were so naïve and therefore I must suspect deliberate deception. Hubble was far too prominent to blow away and Newton was found wanting. At that point they put the onus of proof not on Newton but turned the focus away from Newton to what the presented as the guilty party. When will the Universe confirm its incorrectness by affirming Newton's obvious correctness? If they had to admit that Newton was wrong, the most intellectual science then had to admit they had nothing to show for all their minds brilliant work.

Science that was defying the likeliness of a living God stood bare and naked for all to see. They put the onus of proof and converting onto the cosmos. They asked not Newton but the cosmos when will the cosmos come clean and prove Newton correct, maintaining their unshakable belief that even the cosmos could be at blame but Newton could never be wrong. . When will the cosmos admit to a mistake and set its crooked ways straight. When will it meet its diverting from Newton and reach a point where the Universe will finally come to comply with what Newton demands. It is the cosmos that is wrong therefore it is time to find out when the cosmos will correct its manner. To deal with such a task they needed a man with a bigger ego than he had an IQ. They needed a person that thought more of his abilities than his ability to grasp any complex situation. They needed a man that was presented as a genius without ever proving his genius. They had a man that filled the centre of the Universe, which then placed the man in a location from where the man could see the entire Universe. They had just such a man. He went by the name of Albert Einstein. For all the genius Einstein had, Einstein failed to see the most simplistic and tiniest mathematical rule. Einstein failed to realise that if there was insufficient mass at the beginning of the expanding Universe, the growth of the Universe will reduce the influence of such mass as a factor because as the radius grows, such growth will restrict the gravity by rendering the mass progressively more incompetent.

If the Universe is expanding as Hubble indicated, the growth of the radius will reduce the influence value of the mass as every second passes. The mass will become more and more wanting for such a task. Yet with this obvious shortsightedness of the genius Einstein, the genius saw him fit enough to calculate and measure something as overwhelming as the Universe. As in the case of Newton, Einstein as an ego driven maniac that saw his abilities fit to measure and master the Universe while his mind was to simple to recognise the most basic principle of mathematics, the principle of relevancies or ratios. What a mathematical genius that turns out to be. While the radius enlarges, at the same proportion does the influence of the mass factor reduce and the mere fact that the radius increase shows that at no stage further into the future can the mass stem the growth of the radius because the radius overpowered the mass factor already. Unless there is new material entering the Universe at a point, which is impossible, the entire concept is fraud.

The idea was never to admit wrongdoing on the part of Newton and Newtonian science but to post pone, delay and divert attention away from the truth. If there was not enough mass to start with, no dark matter can kick in later on and start secondary mass frenzy that at that stage will then be enough to bring about the required mass potential that will turn around the Universe from expanding to contracting. To establish a scenario that would hide all deception they got the man that has a bigger ego than an IQ, they tell the world this man is a genius while the fool does not no the least of mathematical principles because his Master Newton did not no the least of mathematical principles and they got him to measure the Universe. While they did not even have any device (and will never have such a device) through which anyone would be able to see the entire Universe, they set of a scandalous misconception that this Einstein could calculate all the mass in the Universe.

Off course as can be expected, there was not enough mass and there will never be enough mass because there is no such a thing as mass in the entire Universe. When the deceit played out to the full, they fraudsters being the paternity of physics elaborated on the delusion by trying to find dark matter that is hidden. If the dark matter did not develop enough contraction at this time, there is no chance in the future to develop enough gravity because the factor of what mass supposedly should have is tarnishing and tarnishing as the Universe expand. The bigger the radius becomes the less would the mass effect be.

The community of astrophysics are trying to frame a picture where they set the stage in the way that if the Universe were stretched to a point the mass would not tolerate any more expanding. The mass will get frustrated in some way and show resistance to the increasingly elastic expanding. The gravity constant (I suppose) must prevent any further expanding. How they ever got to such an argument I never could tell. They surmise that outer space is consistently overall filed with nothing and when this nothing is stretched to the limit, the nothing would resist in growing more nothing or become further nothing and the nothing would stop other nothing to enter outer space in the community represented by nothing. If ever there is a faculty ruled by absolute inconsistency and rubbish as the motto of logic it has to be astrophysics.

Every measured kilometre represents nothing. Every mm is one of nothing. We on Earth are 149×10^6 kilometres holding nothing away from the Sun. Only they can argue that outer space is nothing with material here and there. If that is the case then which has more nothing between the Sun and Pluto or the Sun and Mercury. The distance between the Sun and Pluto is more, therefore that which outer space is made of is more than in the case of Mercury and the Sun. Therefore Pluto has more nothing between the Sun and the planet than Mercury has between the planet and the Sun. Only astrophysics and all the geniuses guarding the principal of astrophysics can put a calculated value by measure on nothing. In fact Mercury has hundred times less nothing between the planet and the Sun than is the case with Pluto. Since my days at school I was always under the impression that a hundred times the value of outer space being nothing is numerically expressed as (zero = 0 x 100 = 0), but where the genius that is such a prevailing part of astrophysics take the stage we find that Pluto can have 100 times more nothing than the amount or distance measuring nothing than Mercury has. The figure containing nothing that puts Pluto at the edge of the solar system is one hundred times more nothing than what Mercury has where Mercury becomes the first planet in the solar system. That is astrophysics. The brilliant minds of the mathematicians hold no rules apart from what they can calculate. Astrophysics is the only department throughout the Universe where normal rules don't apply since because with mathematics they can bend all laws as they wish…in fact Newton started the trend with his deceit.

Only the guardians of astrophysics policy can know why the undetected dark matter will start producing gravity to change the expanding to contraction. Would the fact that it is detected, change the influence it established? Or is it merely to extend the cover up and allow the deceit to linger until the following generation. There is no mass and any one that says there is mass, let such a fraudster then explain why all the planets irrespective of size or density, spin around the Sun at the same sped as all the others. Let them prove that the Universe acknowledge big and small and let them show how Jupiter can move at the same pace as does Mercury and Pluto while Jupiter is so many times more massive than the other two mentioned. More condemning evidence is yet to come because the astrophysics tricksters did not leave the corrupting of evidence just at that.

The fatherhood of physics never once diverted from acknowledging that Newton's contraction is the prevailing thesis on which the cosmos is built because they accepted that Newton used unlawful arguments and to cover up Newton's fraud which they still use to this day, they then proceeded with further criminality when producing the bluff they established with Einstein just to fool everyone in the normal public. Without ever recalling Newton's contraction theory that is obviously not working or admitting doubt about Newton's testimony to the effect, physics accepted the Big Bang Theory. The Big Bang theory opposes what ever Newton might have implied. The physics paternity however finds it wise to still advocate Newton while admitting to the Big Bang event. Newton said the lot is contracting. Go on and marry that with the Big Bang that says everything is expanding. You can't promote both except is you can define why we would see the two merge.

The Universe comes from a point the size of a Neutron. That makes the radius parting the Universe infinitely small. It just about removes the radius as a factor. At the very same implication it takes the pulling of the mass (if there are pulling forces converted by mass) to a level it will never again have. As soon as the distance between the objects holding mass started to grow, the power and influence of the mass factor started to diminish in the same ratio. If the mass were incapable of contracting the Universe then, it will forever remain contracting the Universe. Then you may ask what is the story? Read on and you will learn how far Mainstream Physics stray from the truth and how big a cover up the paternity is protecting.

According to science the Universe started with singularity. Quoted directly from the Oxford dictionary of Astronomy the following:

<u>**The definition of singularity is as follows:**</u>

Singularity: a mathematical point at which certain physical quantities reach infinite values for example, according to the general relativity the curvature of space-time becomes infinite in a black hole. In the Big Bang theory the Universe was born from singularity in which the density and temperature of matter were infinite. The average daily temperature was "$10^{\alpha\beta}$ to 10^{34} K".

Then the second "day" the daily average temperature came down to 10^{34}K and 10^4K. That is fine, but if the temperature was in Kelvin, then what was 0^0K. In order to make sense of the scale used there must be a minimum to secure a maximum otherwise the maximum can just as well be the minimum and is only advocated to impress humans applying earthly standards.

By using a scale as $10^{\alpha\beta}$ to 10^{34} K, it places the lower temperature at a modern 0^0K to make sense of standards. If that was the temperature the standards were lowered, compromising something to gain something, because something had to grow larger for heat to reduce. We know space grew larger bringing heat down to reduce.

Being the onlooker the viewer has to maintain one position. From that position some particles would be circling a centre point, as the particles would be coming towards the onlooker. The other matter would be circling the centre point while rushing away from the onlooker.
At the very end the single dimension may come into the dynamics but where the single dimension comes in the factor of zero is removed.

If there is space, there is a flow of light and a flow of light has to produce lines in relation to angles forming space between them. Something must be present to confirm space because there is an absolute difference between being in space and no space to be found. If there was a line that formed nothing that one line that forms nothing would completely destroy the other lines' chances of ever forming a triangle, let alone having all lines and they then have a total being zero. As shown in the example no line can form zero and therefore no mathematical equation as far as it extends to cosmology can ever bring about zero as a number. While there is space present there has to be three dimensions relating to each other by time and in three dimensions there has to be three lines in relevancy to each other by angles formed holding space in (at least) six opposing sides. Removing one line must bring about a flat Universe and that then will constitute nothing.

Cosmology is about light flowing by means of lines indicating space obeying the rules enforced by time in motion and light flowing dictates crossing space and across space light is using lines. The book: *An open letter Announcing Gravity's Recipe* is dealing with the subject finding singularity by removing the concept of nothing from outer space. By diminishing nothing one uncover singularity and the effort brings in a new perspective not yet introduced.

For your benefit I will shortly give a summary by which I hope to interest you in reading the manuscript: **Compressing space produces heat**. Releasing heat will bring expansion bringing about space. We call such a release of heat an explosion. In other words heat translate to space and space concentrates back to heat. The one is a product of the other where space forms expanded heat.

They are quick to show the time that was applying at the time being some thousandth of a second or the heat that was present being numbers we have no name for. The other side of the story they ignore. They ignore the other side of the story because in that respect it puts their promoting of Newton down to madness. If you reduce the radius applying at the present back to what it was at the time of the initiating of the Big Bang, you must also increase the influence gravity and mass had at that moment by the same number you are decreasing the radius. That is pure mathematics and the most basic physics of all concepts.

The shrinking radius will increase the effectiveness of the influence of the gravity that the mass can produce by the margin of the shrinking of the radius. If the Radius was infinite at that point, then that means the gravity was eternal. With the entire Universe being as big as a Neutron, the Universe was the size of an atom. If the Universe were the size of an atom and the mass within that Universal atom could not prevent the Universe exploding into immeasurable atoms, then it would not be able to retract all the atoms into one unit again. If there was not enough mass to start the contraction, there can be no contraction of mass that is producing the gravity at this stage. If the gravity is of such a nature that it allows a continuous growth of the radius, then the radius firstly cannot be zero as Newton suggested and the extending of the radius proves there is no contraction in the way Newton had everyone to believe. If Newton's mass contracting mass is true, then on the other hand it must have resulted in an implosion as that which can never repeat again. With Newton's formula of $F = \dfrac{M_1 M}{r^2} G$ forming gravity, then the Big Bang is just not possible because from that formula the Big Crunch must respond.

Science still is, in spite of the proof I bring to the table, about Newton and his fraud; they are of the opinion that according to their ethos I am not allowed to write on physics because they don't like what I write. I write what I see and they think I am too uneducated to see so they would rather have me not write about what I see since what I see, I see as being fraud committed by everyone in science from top to bottom and from Professor, Dean of... to teacher teaching primary school pupils, including everyone that is connected in any way or is connected to and connecting with any or all aspects of furthering and promoting physics are implicated in this fraud I see. I say what I see and what I see I write and therefore, it is not my writing about what they say I don't know anything about and then of which I write about when I write about that they detest, but they detest what I write when I write about what I see as their fraud infested information being part of their views. They claim I am not schooled to have an opinion but instead they won't listen to what I say because they can't defend their opinion when I say what I say. Read this, what I point out and become the judge of how much you, the reader, are able to secure as forming the honesty and the honour as truth of whatever information comprehensively becomes what we have as physics.

Science teaches that a feather and a hammer have different mass while they fall equal in time through an equal distance travelled. All things fall equal in time and distance when subject to the same environment. If gravity was mass related, then this was not possible, because then objects must fall according to mass. Falling objects bears no evidence of mass playing any part in falling. Any two objects holding different mass fall equal in time and in distance when sharing similar conditions, which suspends mass altogether as an influencing factor. Galileo proved different mass fall equally under similar conditions. That fact about Galileo, science does embrace, although this strongly contradicts Newton's impressions about mass inflicting gravity. Acknowledging Galileo must make the work of Newton incorrect and also corrupt. On TV we see how all objects, such as cars, humans and bags fall at the same pace, which sets a standard totally against Newtonian mass principles that produce the falling, and proves Newton wrong because mass then does not underwrite gravity in any way or form at all. The formula $F = \frac{M_1 M_2}{r^2} G$ would suggest mass taking all the responsibility for such falling that takes place. Newtonians declare gravity as the force of gravity F, that is $=$ equal to gravity constant G, when it is multiplied by the mass M_1 and the mass M_2 after which then the product of the three factors influencing gravity is divided by the square r^2 distance between mass pulling the mass that destroys the distance between the two objects. If mass pulls mass as Newton said, the Big Bang is not possible but the Universe is notwithstanding Newton's claims, expanding (growing apart). If the mass destroys the radius separating the objects, then the comet has to collide into the Sun, but it doesn't. If mass forms gravity, every planet must orbit at a different pace, which they do not, as all planets orbit at the same pace around the Sun. Planets don't give the slightest hint that they obey Newton's suggested cosmic laws by implementing mass. The truth is that mass is the resistance of any independent material to deform and to acquire mass the individual object relinquishes independent motion. Mass is the reluctance to deform and integrate into a larger structure and becoming a unit of the larger or holding structure. Mass comes about when the falling of any object stops the motion of the falling. Mass prevents further falling, it does not sustain further falling. Gravity is the moving of the object to the centre of the Earth while falling.

I say gravity is movement while mass is obstructing independent movement, which is what gravity is. Mass is not forming the factor responsible for gravity or movement, but prevents further movement. A body falls by gravity. Mass obstructs further falling, while gravity remains present as a factor that brings the tendency or inclination to move or the attempt to continue moving. Mass hinders movement and therefore mass can't enhance or produce movement or gravity. Mass prevents or blocks gravity. Gravity is the motion that defines the individual identity of any object's structural form by rendering motion while reserving independence in granting free space from other manipulating objects. By saying this, I am awarded the cloak of death by Academics ignoring my correspondence as if I never addressed their mailbox.

Every professor in astrophysics knows there are no arguments to counteract or contradict what I say and therefore academically they just try to smother my work while they refuse to read or even recognize a possibility that my work exists. Mass is only a fact found on Earth with human interpretation, creating a usable human standard or a quantifiable norm and therefore my work do not affect normal physics on Earth, but in the Universe there is no remote trace of mass. I prove that all the material that was ever written in astrophysics goes to the dogs and that even includes Steven Hawkins's book that sold 50 million copies. When using mass, those books are a waste of paper. There is not even evidence of mass, except in a corrupt way that science tries to present mass and they have to do that by blindfolding and corrupting the truth. For the first time in almost half a century there is a challenge on the dogma that science uses and I dispute the very basics used to prove science. Again, I state that gravity is motion and therefore all motion places all falling objects equal in movement. That is not the way mass influences object because by implication thereof mass does bring differentiation in the face of falling objects all descending equally.

There are four phenomena in the Universe Newtonian science could never explain. I took the phenomena and trashed mass as the product responsible for gravity. I dissected the inner working of the four phenomena science would never admit existing, and found the four phenomena in movement produces gravity when forming a combined unit. I prove this mathematically. This trashes all existing dogma and that no one in powerful academic positions wishes to have, therefore they will rather silence me. Moreover and to top that insult, the injury part is that I have no formal education. I am a motor mechanic by trade. This is more than what the all of their Holiness the holy High Priests of Newtonian religiosity and all physics academics can stomach since I prove their Newton fallible. I prove mathematically that gravity forms when the Bode law is interfaced with the Lagrangian system and the Roche limit and from this motion, creating space-time, deriving the Coanda effect.

The Bode law is allocating positions that show how the nine planets arrange their orbital positions they have in accordance with the Sun.

The Lagrangian system is the manner in which natural moons or satellites arrange their orbital positions according to the centre planet.

The Roche limit is the closest positions stars can come to each other before the superior star dissolves the inferior star in the partnership. However, although nature shows clearly the phenomena, it is science that disputes even the official existing thereof in cosmology, just because these phenomena clearly disputes the fact of Newton's mass.

From the point of the most respectable scientist, all that I have said up to now is as if I have never said anything because no one in his sober mind will ever go against Sir Isaac Newton, for that to them is utmost religiosity. Therefore, whatever was said up to now by me in this writing, is as good as if it was never said because Academics in science are unable to read past any criticism that might be targeted at Sir Isaac Newton.

That changes every aspect Academics see about the reasons inspiring books and the motivation that is driving the writing of any book. The cosmos is explicitly deliberately and stupid in expanding with a Big Bang and contradicting the genius of Newton, where Newton said the cosmos has to contract to comply with Newton.
To those well-respected Physicists in their positions of well to do Super-Educated Academic circles, the entirety of this book is about whether Sir Isaac Newton made a mistake of what ever nature or the total impossibility of Sir Isaac Newton ever able to make any mistake because science is of the opinion that Sir Isaac Newton never made a mistake because no one ever could prove that Sir Isaac Newton made any mistake. Thus any indication of a mistake cannot be about a mistake that Sir Isaac Newton made for Sir Isaac Newton never could make any mistake.

The mere fact that the Universe expands, implicates the incorrectness there is on the part of the cosmos and not on the part of Sir Isaac Newton. This Universe's expanding is diverting from Sir Isaac Newton's law and with the cosmos going against Sir Isaac Newton, the cosmos forced the establishment in physics to launch an investigation about underlining irregularities on the part of the cosmos and therefore on the part of God. It is well documented and named as the Critical Density Theory and this theory had the most brilliant mind on Earth set out to calculate (and count) all the mass available that could inflict mass, which will provide all the gravity to pull the Universe together, just as Sir Isaac Newton's law insists it to be.

It is the Universe that went out of line expanding and not Sir Isaac Newton that is at fault by predicting contraction instead of expanding. It was the task of Albert Einstein to find out when the Universe will abide by Sir Isaac Newton and the laws of Sir Isaac Newton. It is not Sir Isaac Newton that is wrong about the Universe not coming together, it is the Universe that is wrong by not doing what Sir Isaac Newton said it has to do in order be a good complying Universe and adhere to Sir Isaac Newton's laws. It is the cosmos going out of order by expanding and by that disobedience the cosmos outraged the Paternity in physics to the extent that they have spent billions of dollars to find out what is wrong with the Universe not complying with Newton. Why would the Universe hide matter in dark places just to spite Sir Isaac Newton and not launch the mass to start providing the gravity so that the cosmos will start doing what Sir Isaac Newton ordered it to do?

This act of the cosmos just cannot be tolerated and so the Establishment must soon find a way to correct the Universes' spiteful expanding as they ignore the Big Bang, since they presume that no one will notice that this theory will never match Sir Isaac Newton's contracting by mass theory. It is also the comet's mistake that the comet goes astray and loses focus of Sir Isaac Newton's laws by not colliding with the centre of the Sun.

It is the comet being spiteful and vengeful to escape the gravity of the Sun pulling by mass using gravity to pull the comet close after which the comet escapes the gravity effort and speed into the darkness of outer space. When it doesn't hit the centre of the Sun, it speeds off into the distance and darkness of outer space contradicting Sir Isaac Newton's formula. In ignoring the comet's despicable ignominious behaviour, one saves the comet of facing the shame and the blame of inappropriateness, as the blame can't go to the address of Sir Isaac Newton

It must be Galileo's mistake that all things fall equal and not Sir Isaac Newton's mistake because Sir Isaac Newton says mass has things falling while Galileo said al things fall as if even in mass. Only the most ardent Newtonians can see that Newton and Galileo agreed while Galileo totally contradicts

Newton. Only Newtonians can see that by falling with mass implicating gravity all mass falling will fall equal and that puts mass playing a centre part by never intervening.

One may never dispute the reputation of Sir Isaac Newton, as Sir Isaac Newton is always correct! Any mistake that can connect to Sir Isaac Newton or put any connection between Sir Isaac Newton and such a possible mistake becomes the mistake. The mistake becomes a mistake simply while thinking in terms of Sir Isaac Newton and about a mistake at the same time. Therefore, the mistake is in terms of the view the person has when connecting a mistake to Sir Isaac Newton and not in terms of Sir Isaac Newton's ability to make mistakes. Any person that has an opinion about Sir Isaac Newton being able to make a mistake is making a mistake in terms of the idea the person has and not the fact of Sir Isaac Newton committing any possible mistake.

Any opinion of accusing Sir Isaac Newton of having made a mistake is then creating a mistake. I don't share the opinion about Sir Isaac Newton's ability not to make a mistake because I did detect serious mistakes on his part. Since no one this far proved Sir Isaac Newton wrong, this book and my undertaking in proving Sir Isaac Newton wrong, renders me automatically as the mistaken party. By attempting to prove Sir Isaac Newton wrong, in my attempt I prove myself wrong. If this is senseless to you, then more confusion awaits you in this book because there is a lot about Newtonian science that is not making sense and the biggest thing about these things not making sense, is the all out effort every person that holds any connection with Science is having in their attempt to hide these facts that doesn't make sense and their actions in covering up the mistakes, is that that is making the least sense of all.

However, it is most urgent to note that the enveloping mistake uncovering any mistake about Sir Isaac Newton, is the possibility that the mistake was detected by a person with the most underachieving degree of not achieving any degree and the lacking of such formal education is that part which holds the entirety of the mistake possibly being present or having no possibility of being present whatsoever. Academics with the utmost infinite wisdom will never allow any unschooled mindless halfwit barbaric labourer the opportunity to teach those with such brilliant minds and with that endless lucidity, anything! Newtonians regard Newton as God, because only God can be accepted unconditionally and without any one having reservation about being correct.

Please note this warning. I am the person that science accuses of being a fraud. I am a fraud because I present you with something that represents something that resembles a book while they say I have no ability to present anything to the likes of forming a book or that is the general consensus in opinion amongst all and according to the informed Academics. Your ability to hold this book does not confirm my ability to write a book. A book it cannot be because science clearly has the esteemed opinion of my inability to produce a book because of my poor schooling and the lack of education I have.

Before sharing with you that which I am unable to present, I have to draw your attention to my inability. Before entering the annals of my mind where I hope to share thoughts with you, I first wish you to recognise my lack in talents. Where it is said that I, Peet Schutte is the one that wrote this book, I also do realize my personal restraints in being not able to have you reading my book.

I know that I know very little and I am painfully aware that I am with very limited mental means. I merely mentally generate new information by thought processing and mind control, which I establish as brain electricity that I manage to catalogue as concepts, translate it to words and after that I then collect the proceeds in writing.

I confine this information by selecting a process where I compile a word layout, which makes the mind waves I have, better understandable when conveyed in sentences. Where others afterwards might recognise this process as the writing of a collection of ideas that holds a format we think of in terms of forming a book then that does not apply in this case.

This I say in my personal defence where I stand before you now as one whom is accused by the invigilators that are charged with the task of guarding the ethics of physics and where in their view I have overstepped civil accepted boundaries by crossing the limits of the laid down norms that establish the cultural divide, because I did what is not accepted. Please keep in mind that I know much better than any other person might do, of what my personal possibilities in terms of the social standards are as far as my education abilities stretch.

The established genii that are responsible for forming the norms that they put in place, found me with my presenting of my academic ethos as being much unsuitable to write anything. I was warned occasionally by the established Academics taxed to maintain order in the departments of Mathematics and Physics that I have not the right to write any book with my meagre qualifications. With my effort in doing so, I am acting unacceptably irresponsible.

They set norms in place to protect the standards of civilization and without the norms civil order will go to ruins. In preventing that, please do not confuse this effort you hold in your hand with any other examples presented as written books. The writing of this is merely a compiling of thoughts by transforming my spent and wasted brain electricity into a mental computing of processes and translation of ideas. The system merely uses words as a medium to catalogue the concepts that is stored. The stored notions are then later implemented as ink symbols on paper to present words. That means a book it is not.

In my final letter that I ever intend to direct to academics I end the letter as follows and this book is what promise I made that I fore fill that final promise that I made when I sent nine letters to nine South African Universities promising I shall never contact them again on the matter of the mistakes I saw that Newton (as well as they) made but I also promised them that a fight is on. These are precisely the words I used when I ended the final letter to the nine academics of Universities being the head of the institutions in cosmology.

Sir, when you address your students about the wonders of Newton, then tell your students tomorrow how much the Earth drew closer to the Sun since the days of Kepler, and Madam, announce to the world how much the Moon came closer to the Earth since the days of Tycho Brahe while you then remain convinced that Newton is still flawless after three hundred and fifty years. Notwithstanding all the correctness you do attach to Newton, please take note. Do not take this, which I am about to say and what I am about to say as a threat but rather as my promise.

A threat to you and your address it can't be, because you are so high and mighty I have no ability to ever harm you, so a threat it can't be and neither is there a possibility where one can see what I have to say to you in conclusion as any threat. From where you stand, you will not even notice me down where I am, let alone take note of any threat I might make. Rather see this following remark as a promise. I promise you that I am coming after all Newtonians with everything I dare to use, and that is a promise I do make. This remark means nothing to any person holding the position that you hold since from where you are, all Academics in your position, including you, being in your position, is beyond approach by anyone with the likes of me and I am quite unable to reach any person being where you now are.

Therefore, you correctly see yourself being where you are also as being outside my reach. You are all-powerful and I am powerless in every sense. That might be very true. Nobody, and that includes me, has the ability touch any academic with your standing. I admit to that very readily. However, there is judgment waiting for all and some comes sooner and some comes later. That which I wished to share with you, I am going to try and share with the public at large.

The thing about all Newtonians is that they are boastful and arrogant and self-centred. I have not met one Academic in physics that is not fitting these criteria. These qualities are remarkably the essence of what any person should have that does comply with the demand there is required in being a Newtonian. From the position I have, in relation to the position you have, you may observe me as not worth noticing. Please hear this. All people on Earth think they know everything because what that person knows is everything in that person's Universe.

Whatever that person knows fills that person's entire Universe. On the other hand whatever the person does not know, cannot and therefore does not exist in terms of the knowledge that person regards as worth having and therefore to that person, everything that the person does not know in terms of everything the person do know, then comes down to being nothing. All persons are in a position where such a person does not know what the person is unaware of. What I know you don't know and for that reason what I know is so inferior to you that it is not worth your effort of taking note of what I know.

If it were worth your appreciation, you would have known what I know because it would have been worth your effort to know what there is that I do know. What there is that I do know is to a man with your field of knowledge non-existing insignificant. Since that, which I think I know, has never been to your mind, therefore it was never to you worth knowing, and it puts what I know in the bracket of nothing worth your while to know. What there is that I do know has no place in the Universe you fill. What I know is of such unimportance that it has never been worth finding out.

Then in your Universe, what you know accumulates to everything there is that is worth knowing, because to you that amounts to everything that has any worth for you to know. What I know is nothing to you because that is what you don't know and that which you don't know doesn't exist to you. What you do know is an entire Universe filled with everything there is to know in terms of what you know and only that which fills your Universe is of significance and only that exists to your knowledge. Nevertheless, please take note of the promise I make.

You might find the level of my education being much below your high standards and you are of the opinion that anyone with such a low qualification such as I have, can never offer anything of value in the form of information, which can further and advance the knowledge you have in your field. Sir, Madam that might be true and then, that might be a very incorrect assessment on your part, because and it might just be extremely costly as no person ever holds a position where such a person has in his or her position all there is to know and has no more to gain from others.

But knowing what you know and not knowing what you know is not a quantifiable measure but more a degree of fallibility and that makes it much more a state of mind concealing Newtonian utter arrogance. You may know everything there is to know that is in your Universe and you may judge my Universe so empty of substance that you might think in my Universe there is nothing worth knowing, but in that there is a surprise installed and waiting for you. There might be some things in my Universe that you are not aware of, since it is in my Universe and to you my Universe is not worth your regard. This might turn out to be expensive to you. Who would know beforehand?

That which you don't know, to you don't exist because it falls outside the Universe you do know. On the other hand that which you know is everything there is to know in the Universe because to you there is one Universe and that is your Universe. Please do not be surprised if you do miscalculate your Universe and underestimate my Universe.

This letter is one of ten letters that I addressed to academics indicating my concerns and after this letter there will be no further communication by me to any academic in any way. If there are still no response coming from any academic, after I sent off the letters mentioned, then this is the last time that any academic, which includes you in person, as well as the institution that you represent, will hear from my address. That you will hear from me again, that is the promise I leave you with, but also that it will be along another avenue when you hear from me again, is also part of my promise.

That, which I know, you think does not exist, because you do not know what I know and if you don't know it then that item is not worth being in you Universe. This attitude is not only connected to the Newtonian mentality but is typical human behaviour.

That which you think you know, you accept I don't know, but that is because my Universe is much smaller, since you are the intellectual and others with lesser degrees are lesser intellectually inclined. What I know and the information in my Universe to you in your Universe does not exist and with you having the more intellectual Universe, what I have in my Universe can't be as important as what you have in your Universe.

I am about to prove that concepts you think you are sure about such as Newton, I am going to dismiss and that I can do only by changing interpretation of information and not changing information as a fact. When I do that I am planning to address the general public and try to show that audience how there can be another side of perception about science no one ever realised. I am going to open the closet and reveal all the Newtonian corruption hidden under a dark cloud of falsified facts and also reveal all the rotten bones you in science conceal. However, in your Universe, you are not even aware there are corrupted, distorted and unreliable facts because to you your Newtonian Universe is perfect.

That is how I ended my last intentional communication with any and all Newtonian High priests anywhere. From then on I tried to reach out to the public.

After that I went on my merry way and I did write this book where I now aim to reach the general public with information in spite of the Newtonian view about my educational restraint. Again as not to confuse any reader, when I say I wrote this book then be clear about the Newtonian opinion about my inabilities and with that I am glad to say it still is my guess that they will hold the opinion that I merely generated uninformed thoughts, which I collected as alphabetic symbols and plotted that in ink on paper. That is the information they have to their disposal in their Universe and that I can never change.

The effort that I achieved, I did so from harbouring my delusional ideas spawned by a dehumanised brain. Criticizing Newton only proves my weak and under developed mentality, due to my lack of an informed insight that is a typical symptom that all those have that is suffering from a disadvantaged past as I so clearly have.

That which my views represent, one can only has when the person obviously lacks formal education. While you are reading the letter and in reading, you obviously decided to regard and not to dismiss my work, then also please keep in mind when reading my language used and also please give credit where it belongs…if you do find linguistically improper use of words or misspelling, then remember that I am a feeble minded motor mechanic and not a literal giant and because of financial restraining I went the course on my own.

In this book you hold and of which I am the author believe it or not and which I hope you are about to read, I explain gravity. This achievement is possible because I saw a way to break away from invalid concepts Mainstream physics hold. I recognised the impossible double standards Mainstream physics apply to promote their much shady explaining.

The inconsistencies brought them double vision and to compensate their incredible theories they simplify issues to a level where what they embark on to understand, is becoming meaningless. What they say can't be supported and authenticated by any investigation even in simple terms. It is as if they never read with interest that which they explain and they never scrutinise that which they advocate. They give values that are senseless and make that which they say meaningless.

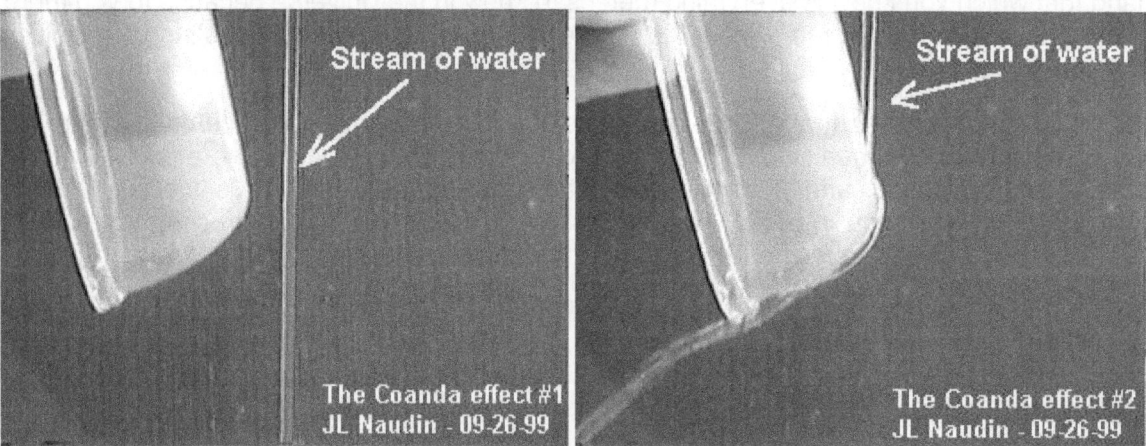

Do you know what this is? This website will not show you the lovely pictures. For that there is another website that will tell you everything that happens in the experiment. This information will tell you why this happens and what is the cosmic philosophy prevailing in physics allowing this.

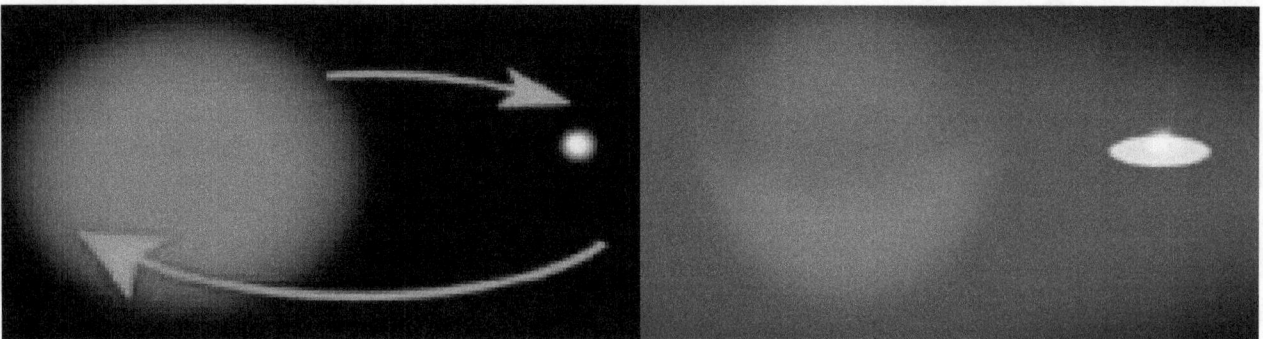

This is why no star could ever collide with any star because stars or anything else never pulls anything else closer. Anything coming as close as 2.467 or $\Pi^2 \div 4$ times the diameter of the larger will liquefy the smaller unit.

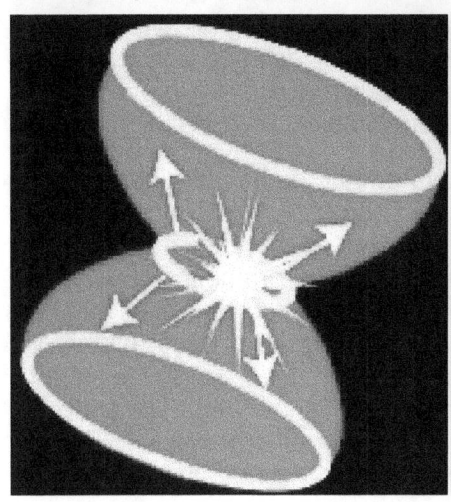

The Roche limit in the practical sense shows a radius limit found between stars that prevents stars from coming closer than that limit allow.

Do you know what this is? What happens in these pictures proves that stars could never collide because what prevents it is cosmic law that science at present can't fathom because their understanding is too limited. This website informs you about the facts behind the lovely pictures. This website will tell you why the cosmic physics principles establishes these phenomena…and because the information does not salute Newton, academics in physics despise what is said. If mass don't pull stars to a point of colliding then Newton is wrong! Any star coming closer than 2.4674 times the radius of the larger star would liquidise the smaller star and this law and the Coanda effect are a precise duplication of true gravity truly applying. This value the Roche limit has is $\Pi^2 \div 4$ which is circle movement divided by 4 opposing sides through which the rotation goes.

At present science is unable to explain these very crucial phenomena and even to the extent that those guarding the principles of applying physics don't realise the importance of these phenomena or how the principles apply in physics.

These phenomena are there and it is in place so ignoring it is rather stupid…but that is precisely what Newtonian science has been doing for centuries.

They ignore these phenomena in favour of what is not present in the cosmos …and that is mass. Those in physics put mass in as an explanation for the movement of planets and while all planets has different mass, yet all planets orbit and move at exactly the same tempo. So how can they move according to mass-differences with having different mass and still be moving at an equal pace?

What there is, are the four phenomena I present, but because Newtonian science insight do not extend as far as understanding the phenomena and the role it plays in physics and moreover in generating gravity, Newtonian science is brilliantly ignoring the phenomena as well as the role these phenomena play.

Science do not realise the value or the impact that these four phenomena has on physics. That makes their judgement unfit in terms of deciding the importance the phenomena offer in relation to understanding physics. For many centuries science used improper factors in physics such as thinking that it is mass that produces gravity while only Newton's word is proof of this. There is no evidence of mass forming gravity and to overcome this issue they manufacture a component they called the graviton. There is no evidence of a graviton existing but to compensate for the shortfall existing in physics they use the graviton to promote a hoax however, this is not the only fable science invented

No one realises there is a shortfall because all evidence of such shortfall in Newtonian science is ignored like they would ignore the plague. There is no evidence of mass that has the ability to form gravity, but by cultural brainwashing every one believes this to be true. With everyone satisfied that there is no problem in science all persons are very satisfied with conditions applying as it is.

In every mind using physics the awareness of the mistake is absent and all presume that physics in the manner Newton said it is applying therefore thinking the Newtonian way is a healthy and correct and no one can find a reason to investigate my claims that Newton's idea of mass that forms gravity is a hoax.

No one can see the need to support my claims of science being wrong in spite of finding no evidence of mass anywhere in the solar system. The phenomena that do apply in the cosmos, science totally ignore because it does not support Newton's ideas and Newton can't explain these phenomena. I prove how these phenomena form gravity but I am ignored because it will turn physics on its head.

With the knowledge in hand for decades why did science not move away from the contracting idea? Why are they chasing an idea that shows how blatant the corrupt conspiracy is in bedded? Academics are very aware of this misconception Newton had and still academics in physics are promoting the ideas of Newton as the unwavering truth. Teaching Newton is participating in deception and promoting Newton is criminally deceiving the public and while doing so, is committing an act with criminal intentions. Then, in the face of all this evidence contradicting Sir Isaac Newton; they remain upholding the correctness of Sir Isaac Newton and keep on teaching students about the unwavering correctness of Sir Isaac Newton. I show precisely how gravity produces mass but mass can never produce gravity. I show with explicit detail when, how and where gravity forms mass but mass can never form gravity. What I prove annihilates every Newtonian claim. Still they persist in promoting the lie just to favour science and that is a lie.

When any person, notwithstanding what reasons given, repeats such a lie unabated while being well aware that the information passed on by such a person is incorrect, then the person commits deceit. When anyone is repeating the information that is passed on as being unblemished factual substantiated and verified truth while such a person knows very well that such information is void of proof or lacks proof, then committing such an act is a criminal enterprise. Academics in physics commit every one of the above indignities and yet see their actions as being lawful and even much praiseworthy and hold their role in society in the highest esteem imaginable. Physicists would not tolerate any form of criticism.

They fail to see the crime that they commit while tutoring physics. Whatever motivation they may claim to have which they offer to serve them as forming their driving force, the fact that they perpetually perpetrate in unlawful behaviour, by spreading untruths, such actions on their part put those academics holding such highly regarded positions in the league of ordinary cheats, gangsters and common criminals. By deliberately and constantly falsifying facts to further whatever humble cause and produce illegal claims repeatedly, remains derogative behaviour and is unlawful by nature, notwithstanding what morality it should serve. To promote the lie is to hide their misconduct and it is not to promote science.

A Preacher or Pastor or Priest lying on behalf of God is not lying on behalf of God and to think the Preacher or Pastor improves or underlines the Greatness of God by lying on behalf of God is very mistaken, because in reality such a Preacher is falsifying the truth for his or her personal benefit and trying to impress the congregation about his importance and not the importance of God. Lying is wrong and doing so even in the name of God remains despicable. To say God sent you to do anything is self-promotion because the truth is you told God you are going to do with or without His consent and you use his name to justify your enterprise in any case. That is falsifying facts not on behalf of God but you use God to falsify your justification. The same applies to academics in physics. There is no argument that can change this truth about falsifying the truth and when doing so there is no hiding behind any excuses of ennobling to benefit mankind that will change such truth into righteous conducting.

Ask your professor to show how an expanding Universe can also contract and your professor will tell you about Einstein's Critical Density theory. This theory I prove is the second biggest fraud ever devised by any group of persons in the history of civilization! I am about to introduce you to the biggest fraud ever devised but it is not the Critical Density elusion. The Critical Density elusion is perpetrating fraud and the conducting thereof is upholding deceptions instituted by Newton that then formed the institution of lies they call physics. The Universe does not contract in any way; means or form and even such a suggestion is incorrect! The Moon and Earth are not moving closer but are moving apart. The entire Universe is growing in space and nowhere is space depleting by any norm used because what is space is time. That is exactly what Galileo's swinging pendulum shows; it shows the contraction of space is reading time.

Academics are very aware of this misconception Newton had and still academics in physics are promoting the ideas of Newton as the unwavering truth. Academics teaching these misconceptions are committing fraud, notwithstanding the portraying of their role in society being unblemished, spotless while they are covered in a lily white blanket making them being whiter than snow and having such a holier than thou attitude. Teaching Newton is participating in deception and promoting Newton is criminally deceiving the public and while doing so, those doing so are committing an act with criminal intentions. Then, in the face of all this evidence contradicting Sir Isaac Newton; they remain upholding the correctness of Sir

Isaac Newton and keep on teaching students about the unwavering correctness of Sir Isaac Newton. They put down conditions of learning to this effect and are expecting students to repeat these untruths and unproven facts by forcing answers to that effect in examinations. Students must accept cheating to become accepted in the ranks of the learned while also carrying the cloth of righteous in knowledge and unblemished virtue. Forcing the acceptance of this untruth about physics is equal to preposterous subjecting students to physiological torture and heinous mind conditioning, scandalous thought control and brainwashing. This applies to everyone serving as a tutor in physics notwithstanding whatever status the torturers might have in society or the morality they attach as a reason to commit such atrocities.

If you are a student, then you are conditioned by academics in controlling your thinking by enforcing pre-mind setting and in which they methodically force you into believing in Newton and this is an on going process conducted for centuries in the past, while it is the truth that Newton is completely void of any tests that may secure any form of confirmation and in securing proof then also by that establishing proof. They never prove Newton's philosophy on gravity but those persons conducting teaching in the subject of physics force all physics students to learn Newton's gravitational concepts and accept the facts as if it has been proven beyond all other facts. The condition of being accepted in physics is to accept Newton without questioning the proof that is never supplied. Let your professors now prove how it is that Newton's teachings are correct and then examine you on the process they use to prove Newton's concepts. At present they say Newton is correct and then they test you on your ability in repeating that Newton is correct without ever proving to you that Newton is correct. Let those physics professors now prove Newton and then test you on the manner they use to prove Newton to be correct.

The truth beyond all other truth is that Newton's gravity has never been proven (because try as you may it is not possible to prove Newton's formula forming gravity mathematically) and because academics know that, academics require the blind acceptance of Newton by students. This unconditional acceptance of Newton's correctness relies only on the pre-conditioning of students' mind set and academics depend only on the student trusting the academic "say so" about the institutionalised correctness of Newton. Pre-conditioning students into blind acceptance depends on the academics' insistence that students approve Newton's concepts without pre judgment or students insisting on scrutiny of any sorts. Academics depend on students never questioning their say so or demand proof about what academics teach. Those academics in teaching positions insist that all students accept Newton's accuracy.

In order to get students to accept Newton's hypothesis, academics resort to brainwashing pupils and students. Please let you lecturer put in all the values of the formula $F = G \dfrac{M_1 M_2}{r^2}$ and then with the values applying prove Newton is correct. Pre-conditioning students into blind acceptance depends on the academics' insistence that students approve Newton's concepts without pre judgment or students insisting on scrutiny of any sorts. In examination students have to outright and blindly follow academics' say so only because academics say so. Academics depend on students never questioning their say so or demand proof about what academics teach. Those academics in teaching positions insist that all students accept Newton's accuracy. This is methodical mind control as much as it is the brainwashing if ever there was brainwashing. I show what they enforce. If you are one of those believing that Newton was ever proven, then what you believe to be true is a lie because Newton can't be proven and that is the truth! The time has come to face your teachers and force them to stop the ongoing old culture of bullying students and conditioning their thoughts by enforcing on them dogmas, which is, mind control! In order to get students to accept Newton's hypothesis, academics resort to brainwashing pupils and students.

They teach you that the Universe contracts and to state their case they force students to learn that gravity is proved by Newton introducing the following formula $F = G \dfrac{M_1 M_2}{r^2}$ They say that M_1 is the mass of the Earth and M_2 is the mass of the individual in questions mass and the multiplying of these factors with the gravitational constant produces the force of gravity when this gets divided by the square of the radius. Please let you lecturer put in all the values of the formula and prove Newton is correct. If he can't and I know for sure he never can fill in the symbols and calculate the force of gravity, then read the rest of the web page that follows to see how far academics in physics go to brainwash students into believing in Newton's fraud. To cover their ongoing fraud they create a mythical mystical non-existing dark energy.

Those in power of physics wheel the one conspiracy after another conspiracy extending the cover up with another cover up by falsifying the truth to hide the untruth and it is not clear what they would hide when trying to hide from people discovering where they refuse to reveal what the truth is. Why don't they not just come clean and admit that the Universe is departing instead of arriving? They keep cosmic science in disconnecting darkness with patches of light in between as sensibility being between the dark that disallows the light to connect. The senseless darkness of failures to accept the incorrectness holds the light disconnected but the light parts darkness with tiny strands of unrelated correctness and only connects by the dark that does not connect by understanding.

They have a concept and in time the concept is proven to be incorrect. When they discover what they thought applied was wrong, those in physics do not step forward and admit to what is wrong…no they change nothing but keep everything wrong in place or not in place by concocting a conspiracy to keep everyone happy and everything as it is in place. That is how they can unite the idea of the Big Bang with everything moving apart since the time it began in line with everything coming together and Crashing in a Crunch while no one has any vague idea of what is going on in-between. Here is what they essentially in essence really hide. This is what all the scheming and deception is they try to cover. This what follows and what you are about to read undoubtedly for the first time ever in history. It is the true conspiracy that every crooked angle in science is covering!

It is not what science declares that is important but it is always what scientists don't declare that holds prominence and more so the reason why science keeps a silence about the information they do not disclose. It is never about what they say but it is why they don't say other things they keep quiet about. You will read how they never disclose the entire truth because science is about promoting one-sided and selectively opinionated information forming fraud no less. I have been per suiting a new cosmic theory that I partly present in a six part theses, of which the investigating research began in 1977. In 1999 I compiled my theory and searched for a publisher.

First I located what was wrong in physics then formed a correct approach. I compiled my presentation of it in a theses that I call **The Absolute Relevancy of Singularity** and then six separate thesis parts forming the theses published through **LULU.com** which I saw as the only manner whereby I could generate funding by which I would be able to have the twenty seven books I already wrote linguistically edited and then to have the books published on a Print-On-Demand basis.

I compiled **a new cosmic theory** by which I eliminated all the incorrectness that Newton has burdened science with but with this being my opinion I did not find a garage full of academics supporters waiting to applaud me and to uphold my views on the matter. Yet still I was not going to be ambushed by their relentless stonewalling my efforts and blocking my efforts in introducing both the incorrectness and the new cosmic theorem I concluded. Their mannerism in blocking and frustrating my opinion when showing the mistakes in science convinced me about **a Conspiracy in Science in Progress** and this spurred me on to tell the entire world about their brainwashing students minds. By the manner they selectively withhold information when teaching science, amounts to deliberate brainwashing of students in physics by "normal" education practises.

Trying to convey my message kept me busy for the past going on to twelve years on full time basis whereby I was trying to introduce my findings to many academics without finding much joy from my efforts. This past eleven years plus saw me go without any income as I tried to get my theorem recognised as well as get my warning noted.

Going without a steady income left me almost destitute and in order to find a manner to get my theory across to the attention of influential readers, I decided to publish a theses of six books electronically as to try and get around the stranglehold of Newtonian bias controlling science at present worldwide. I decided to publish electronically which those in power do not control. However to get people to believe me is to change science that everyone believes as culture.

With my first language not English and the books not linguistically checked by an expert there are bound to be language errors that readers will notice. In the past I tried to check my work myself but after checking say one hundred and fifty pages for language corrections, then after days of toiling instead of having corrected work I ended having four hundred pages of newly written information which is still not linguistically corrected but holds a lot more information. The language and spelling errors compiled instead of reduced. This is because my priorities lie elsewhere. I aim to spend money on correcting the work as far as language goes, as I receive money in the selling of my theses and in the hope that I will receive money. I will have all my work including the one you are reading edited professionally and corrected as I find money to do so...But first I have to get the public aware of the problem to get the academics to appreciate the problem. In everyone's mind science is more perfect than religion is.

In the event of any readers who may have questions concerning more facts as it is presented in this book, please feel free to contact me, PEET SCHUTTE. All information divulged came about through independent self-study during the past thirty-two years or so. I have to warn the readers that the topics are showing a very new approach with no quick answers abstaining from proof or holding just a few lines and the information is new in nature but not hard to grasp.

This book started off as a website to inform about a science conspiracy but although reduced still it grew into a book that serves much more information than what I first intended to supply. You will see many new aspects about gravity please make sure you understand what you read. It grew into a comprehensive study on cosmology. At times you may observe while reading this book that it seems as if my frustration will ring through like the chiming of the Big Ben Bell. For that there is a reason. At times my frustration and anger will boil over drowning my politeness and that is true, which I admit. For twelve years I have had the answer that would correct the philosophy that has a stranglehold on cosmological science.

I discovered the building blocks of nature where my discovery puts all other cosmic aspects of science into science fiction. Those who force-feed non-existing dogma do so to brainwash students to hide the incompetence of "modern science" so they can rule supreme while ignoring the truth that they deliberately hide by concocting a conspiracy. To keep everyone unguarded they practice a conspiracy by which they perform an accepted practise of thought control on students to further the false dogma presently in place. I try to blow the whistle on such a practise but accepting my resolution makes every thesis ever written science fiction. Therefore no one in science dare to read my work leave alone appreciates the revolutionary nature thereof. Whatever now is deemed to be accepted science would then become what is the past tense in science because the flaws that those in power of science principles kept coated for centuries on end as untouchable truth will then be rust that breaks the surface to show the holes!

They try to silence me but surely somehow somewhere I have to break through with my massage! I bring you a true form of science as never seen before in all of history and I do that when I dispose of the conspiracy that hides all the incorrectness and the failures that haunts science today. Science is accepted as the most righteous information available to man and that is a scam.

In the prelude I am on purpose going to show you, the reader a lot of pretty pictures about the cosmos and you look carefully while I explain certain ideas and then by observing you detect what the common factor is that all these pictures have. Let's see how perceptive you are about the cosmology you view. Can you crack what Physicists could not crack this past three-centuries of gazing at the sky?
I put in lots of pretty picture for you to see and what you will see in them is gravity because it is gravity that is holding the entirety called the Universe together and in form. What you see is gravity but can you visually see gravity?
If we could talk and I could shake your hand on the bet, I would bet you that you are going to miss what gravity is notwithstanding all the lovely pictures of many galactica I show you. If you do see what gravity is you are brighter than all the bright Newtonians minds put together but after saying all that, it is not that a powerful achievement being brighter than what those lot of mathematical minds are anyway.

mailto:info@singularityrelavancy.com

We are obsessive with our search of where it all began but every one has his or her individual idea. So I too have an idea and for what it is worth, I wish to share my idea with readers. Every one goes by the way of mathematics and everyone falls by the wayside using mathematics. The Universe started with singularity and singularity is 1 and no more than 1, so don't get complicated about mathematics applying. Physicists put mathematics in place of God and presume mathematics built the Universe. That is not true because it must be the Universe that put mathematics into the Universe as the Universe built what now forms the Universe by building mathematics from the figure of 1 going on to 2 and then 3, 4, 5 and so on. It never once occurred to any one in search of the beginning that there must have been a beginning of mathematics before mathematical equations took centre stage and that mathematics are most probably just another development product of the Universe. It is very unrealistic to think that before anything, the cosmos devised mathematics and mathematical laws and from that the Universe was introduced.

When we discarded the contraction idea that saddled us to the Middle Ages and the Big Bang brought us thinking in the right direction, it introduced the way to think. But starting off with a Universe the size of a neutron, or having a Universe formed as is at 10^{-43} and counting is dropping in somewhere in the middle. Einstein said that going beyond the speed of light would bring everything into pure energy. Is that not what the Universe is made of…of pure energy. Is that then not a clue to the next phase? What if the lot was spinning faster than light? It is pure energy we are after because the atomic bomb shows that what ever is captured in the form of the atom, when the relevancy breaks what is unleashed is a hellish pure energy. The nuclear explosion took us to the limit of what happens when the atom overeats. Looking at the atom bomb must be looking at a very limited extreme downscaled version of the Big Bang with heat turning to space like nothing else can ever repeat. For a long time I had a huge problem with Einstein's idea of going the square of light to calculate nuclear energy release. Light cannot go square and that is a fact. However the accuracy of $E=mC^2$ is proved beyond question. I had to find out how light can go square and that was one of my first tasks I had to achieve after finding the inaccuracies lurking in science.

Einstein formulated $E = mC^2$. E is energy, a topic I will discuss shortly, m is mass also a topic I am going to discuss throughout and C is the speed of light. Einstein going square on the speed of light is absolute madness The speed of light can't even double and that detail I can't discuss at this point because there is no space. However there are many books I do go into much detail about that. Then I realised Einstein duplicated Kepler's formula before Newton raped it to bits. Kepler made this possible formula he introduced after TWO lifetime achievements (his as well as Tycho Brahe) **by applying $a^3 = T^2k$.** This is the formula Kepler used for planet distribution however planet distribution accompanies star distribution, which is the same as galactica internal and external distribution. It is space $a^3 = T^2k$ relating to two types of motion. The E is the a^3 the m k leaving T^2 to be replaced by C^2. Then, when realising this formula connection I new why Einstein contradicted himself. When light goes square there is no light to go square and that we see from Black Holes, Pulsars and other optic dark or partially dark stars. The speed of light is fixed at C and that brings a motion boundary. Then it came to me like a hammer hitting me between the eyes with a blow that almost crushed my head. Taking Kepler's cosmic formula $a^3 = T^2k$ into account Einstein saw in using his formula

that C^2 is gravity applying and not the speed of light going square. Einstein saw a Universe at the time when the Universe was relevant at a factor of one to the speed of light, which is long, time since not valid any more. Einstein saw the Universe when the electron was outer space and the neutron was pure liquid. The proton then resembled the solid. At that time there were no space because the Big Bang was heat turning heat to space which ever way you wish to put it because heat overheating forms space. What Einstein saw was a time when light was forever taking eternity in motion to move from one point to another point. If the Big Bang was the moment space overcame light and formed space then predating stars were in gravity that is stronger than the speed of light we are seeing in stars going dark time predating the Big Bang.

The rotation speed of outer space was equal to the speed of light at the time before the Big Bang making the gravity that comes about under such condition the square of the speed of light, also applying in relevance to what applied at the time. The darkness we see in space is not "nothing" as Newtonians believe the space in outer space is but it is light moving away whereas white light is light moving towards us.

The blackness is light going in then opposite direction in relation to us than white light coming towards us. We have to forget our ancestral past about seeing the darkness and thinking that is empty and black because it holds nothing. There is one Universe and that Universe holds singularity not space-time. The atom links with singularity claiming space-time by motion of space provided. There are many phases of singularity still developing space-time waiting for the time of their arriving into a related era. In the centre  of this and most developing galactica the material is spinning beyond the speed of light. That is why it has light to cast light away as visual beams. The spinning still has to reduce considerably to allow the growth of space not to throw out light, which the singularity deems to move to slow and is therefore discarded from space-time. If the moving was fast enough it could remain in relevancy and remain circling with the rest, but since the light travels to  slow, the light moved out of the controlling centre singularity and it now seems to us as light. The Big Bang as the Big Bang has not arrived yet and there is still no electron within the atom. However I am miles ahead of myself with this venture. Only when stars break with C deep within in the centre inner core can stars achieve the independence by moving outside the heat blanket that has been covering them since space broke free from occupation and from individual motion. Where gravity still exceeds the speed of light the gravity will not allow the light to depart from the centre. In that the spin of star material inside the galactic structure circle is stronger than the individual drive to claim independence.

I believe I am about to commit one more eternal sin and criticize Einstein. I am about to gain the scorn of every Newtonian walking this planet. Einstein was a brilliant mathematician that followed where his mathematics took him but mathematics is just another language. It needs interpretation as it needs translation from a written expressed mathematical to a verbally understood form and spoken form of language. Einstein read his mathematics through Newtonian glasses and did not know where he was. Einstein was a brilliant mathematician but he was no philosopher and I immediately admit that I am no  philosopher myself either. But I use the help of a great cosmologist provides, which enable me to use the correct interpretation of cosmic relevancies about space-time. When using the spectacles Kepler provide the truth becomes more than obvious.

The motion of the stars grouping in layers flow around a centre, which is allocating  singularity, is the forming of the Coanda effect. The centre provide the gravity, the motion provide the time in relation to the relevancy of that particular structure in relation to the group as a unit and singularity governing. The circle forms space a^3 provided and limited by time T^2 in precise relation to the relevancy thereof **k. That is the Coanda principle of gravity** just as Kepler announced it when he said the formula to gravity is $a^3 = T^2 k$.

When Einstein saw the square of light it is clear how little he new about Kepler in reality although this should be highly astonishing to know, it clearly is not as my motioning this to physics academics left them cold. No Newtonian knows about Kepler for that matter and Newton was the second highest Newtonian

High Priest ever. Einstein saw C^2 but did not connect it to Kepler's gravity and at the time he did not realise the importance of the Coanda principle. Einstein never realised he was working with gravity Π^2 which was at the time equal to the square which we find when light holds the precise relevancy. At the time of the Big Bang the space relevancy matched the speed of light C that made the applying gravity C^2 and that made the available space and that Einstein saw.

If $a^3 = T^2 k$ as Kepler said it is and $E^3 = mC^2$ as Einstein said and $k^0 = 1$ as mathematics says then accordingly $k^0 = a^3 \div (T^2 k)$ because $a^3 \div (T^2 k) = 1$. This mathematical interpretation shows why stars and galactica and all other cosmic structures are round. This is why I have put in these pretty pictures about all the lovely cosmic structures. What this formula $k^0 = a^3 \div (T^2 k)$ says is that **to form material a^3** there has to be movement **($T^2 k$)** and that movement is around a specific centre k^0. **That is gravity and that is the essence of gravity. Moreover and more important than anything else is the knowledge that this proves the Universe has a centre a specific spot around which everything in the Universe spins.**

From that point holding that centre of material turning around such a pivotal centre the relevancy forming space-time amplified and space grew as time reduced. But moreover if there was an era that had stars developing before the Big Bang era arrived, but stars are in the initial stages about light, we are in a scenario that there is more than one Big Bang. Then there was an era before light ever became a valid factor. But everything does not move equally and movement produces space formation that varies as much as the Universe is large. With that in mind it comes to mind that there had to be more than one Big Bang, which incidentally is also most obvious in some galactica development since there was a major time spent in development before light became a factor. There was a time when gravity propulsion was beyond light and even beyond time or space or form. The line that forms relevancy must therefore be the key to solving the riddle of where the lot came from since it goes pre-Big bang many times over and again. The fact that star doesn't develop all at once shows a clear timeline divide in development.

To trace our beginnings we have to find the centre of the Universe where the curvature of space-time forms gravity that forms the form of the cosmos. It is clear that with many centres there also has to be one specific centre forming the centre of singularity. Locating the centre of gravity is also locating the centre of the Universe because where space is least gravity is the strongest and gravity inflicts the curvature of space-time. Finding this centre I feel was my biggest achievement At the heart of bringing about the solution to one of the greatest Astronomic riddles one will find a child's toy... the riddle of Einstein's singularity pointing to the position where the cosmos started so many billions of years ago. An explanation about the growth of space such as the above picture matches every logic view we all have about the Universe, but does Science really provide the answers matching our modern logic, or are we filling in and compensating for science's shortfalls. The big question I asked and that you reading this must ask and answer is as flows: Does Newtonian or official outlook really match the logic of science used by nature? If you are one that thinks Newtonian science knows everything stop reading because you are in for a shock you will never recover from. Here is a few questions showing Newtonian shortfall

Does the Hubble concept Newtonian science put forward match the explanations about how Creation all started... where it is heading...and where it will end?
What is motivating the expansion and the moving?
Why is the Universe depicted as a sphere?
...Why did it start small?
...How did everything become so much and so large...
...Why does it grow from small to large?
...Why was the start so small?
...Why is it growing?
...Where is it going while it is growing ...
...Why any specific size...
...What was everything before that?
And why in creation would it reduce again!!!

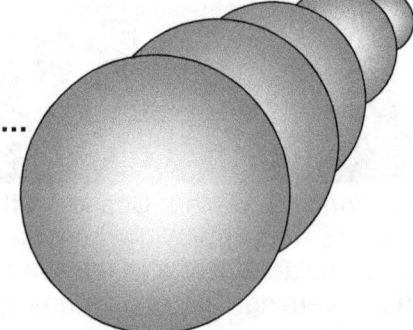

All the above questions Newtonians hide behind **a Conspiracy in Science in Progress**

You are going to read about a conspiracy but people think of a conspiracy in many terms. Let us define not by definition but by interpretation to what a conspiracy constitutes. What do you think is a conspiracy?

All the conspiracies you know about is known about because someone somewhere makes money by allowing the revealing of the conspiracy. Silencing the conspiracy does not make money but informing a suspicious public loosens the flow of money. If it were a true conspiracy no one would know about the conspiracy because the powerful would make money from not revealing the conspiracy. The revealing of the facts about any conspiracy would be stopped before it leaked because it would kill the flow of money.

A conspiracy is thought to be a gossip story that makes money and by not revealing it or revealing it goes in line with making money or not making money. You can download this book free of charge because I don't make money by revealing the conspiracy. I truly want to find an audience to divulge the truth. I want to make money but it is by showing how I can correct the flaws in science, not by hiding it in a conspiracy. People put a conspiracy in the same realms as a gossip story, an old wives tail, which is going about but does not intend to harm and mostly serves as amusement to many. Hearing about a conspiracy tests your intellectual comprehension. It is some quiz that you match your truth against the truth that the conspiracy reveals. It is a funny, but it is not funny until you catch the funny part hiding behind the conspiracy and only when you measure the catch behind the conspiracy are you treated to be amused.

If Newton is correct then ask your physics professor to explain why the solar system employs the Titius Bode law to form space and not the mass as Newtonians uphold. The Titius Bode works as follows:

Why does the distance from the sun to Mercury double to Venus and that again doubles in distance to the earth and that distance again doubles all the way to Mars and this carries on going throughout the solar system. Tell your physics professor to make sure he or she uses Newton and his idea that the mass every planet has forms gravity and to use only Newtonian gravity principles to explain this.

If you follow the tour and go where I guide you to go you will learn why in the solar system is Mars twice as 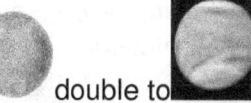 far from the sun as the earth is and why the earth 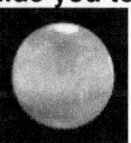 is twice as far as  Venus is away from the sun and Venus is twice as far as  Mercury. This principle is called the Titius Bode law and is in place in the  solar system instead of the mass factor Newton said is in use by the cosmos.

Let's repeat this again...

The distance that 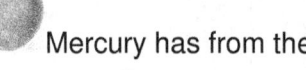 Mercury has from the sun is doubled by that which 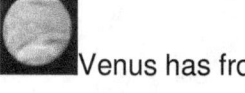 Venus has from the sun

Then again the distance that 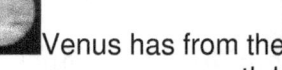 Venus has from the sun is doubled by that which  the earth has

Then again the distance that  the earth has from the  sun is doubled by that which Venus has and inexplicably this forms the layout of all planets in the solar system. But why does science never mention this?

Professors in physics never mention the Titius Bode law because the Titius Bode law makes rubbish of Newton's gravitational principles. It proves Newton is a fraud.
This process forming distance between planets carries on throughout the solar system.

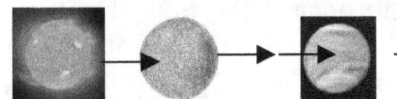There is no room in a room to show this layout in its full compliment where it covers all the nine planets. If mass formed gravity then the layout should be running from the biggest to the smallest. The distance should be in terms of size $F = G \dfrac{M_1 M_2}{r^2}$, but it is not, it is according to 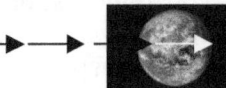 the Titius Bode law, which is some law no one ever hears of because it disproves Newton and his mass concept. It shows Newton has no ground on which to form his concept that is completely wrong! But while the cosmos disproves Newton, science believes Newton in spite of the cosmos using the Titius Bode law This is very typical of science in the way Science prefers to cheat the truth to prove Newton correct.

If planets pull according to and by mass, then why would the distance that every planet has in relation to the sun then double every time in the case of every planet?

The entire idea above called the Titius Bode law is there in place used by the cosmos and is so accurate it was used to discover planets but also is dismissed as not being factual by physics professors because it repeals Newton's principles of mass. It is called the Titius Bode law and the undoubted accuracy of the Titius Bode law was never questioned. It used in the past to locate unknown planets but also it is clashing head on with Newtonian accepted principles and therefore it is degraded and denounced, just because it does not salute Newton's unproven principles.

Confront your physics professor by asking your professor in physics to use Newton and then to explain why every planet doubles the distance it has from the sun in relation to the immediate inner planet because he should know why every planet is doubling its distance from the sun in relation to the previous inner planet; after all he is the Newton-physics expert and with this representing gravity as gravity applies in the cosmos he is the expert in physics… and if he doesn't know, then what does he know because this is the basis of everything forming physics and he is the Newton-physics expert in physics!

Here are more thoughts to ask your physics professor to test what his credibility is in relation to what reality is
If Newton is correct then ask your physics professor to explain:
Ask your teacher to explain the fact that if mass pulls mass by gravity pulling, then

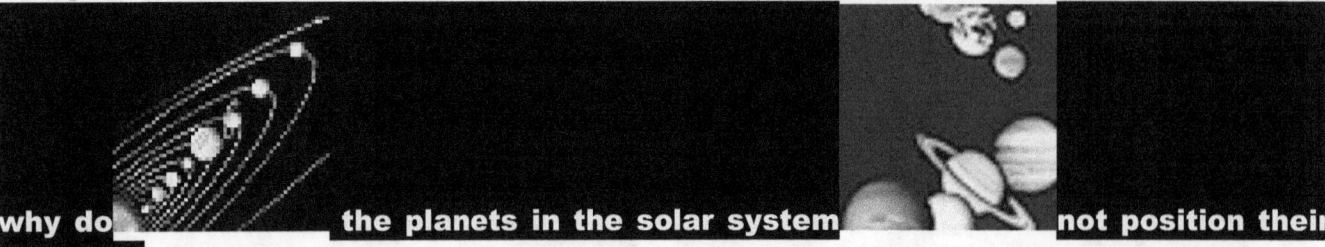

why do the planets in the solar system not position their allocated places in relation to the sun by applying mass as the nominating and defining factor?

If Newton is correct then ask your physics professor to explain:

Why would the tiny asteroids, some no larger than dust, be floating in the solar system as well as circling in the asteroid belt, going around the sun just as fast as any of the mighty large massive planets including Jupiter does?

The planet rings are made up of tiny rock fragments and water crystals spinning around massive planets so why don't the huge planet pull those tiny particles by mass to fall into and onto the huge planet...if mass does pull.

If Newton is correct then ask your physics professor to explain:

If gravity's pulling structures are the result of mass pulling mass forming density by pulling then why are tiny asteroids that are so small with so little pulling power but yet so very dense because they are only small rock fragments which are leftovers of what once was that presently is still floating in the solar system. And then when compared to something as overwhelming huge as the gas planets with all the mass and therefore all the pulling density that represents and all the pulling that the mass forms but notwithstanding that the gas planets only has the density that gas will allow!

If Newton is correct then ask your physics professor to explain:

If gravity is the result of mass pulling mass that reduces the radius, when will the earth and the moon finally collide as the mass of the earth pulls the mass of the moon into a final collision or for that matter when will the other moons such as Io collide with Jupiter? Surely the massive mass of Jupiter should reduce the radius between the giant planets and the main body in a very short time. When will the mass

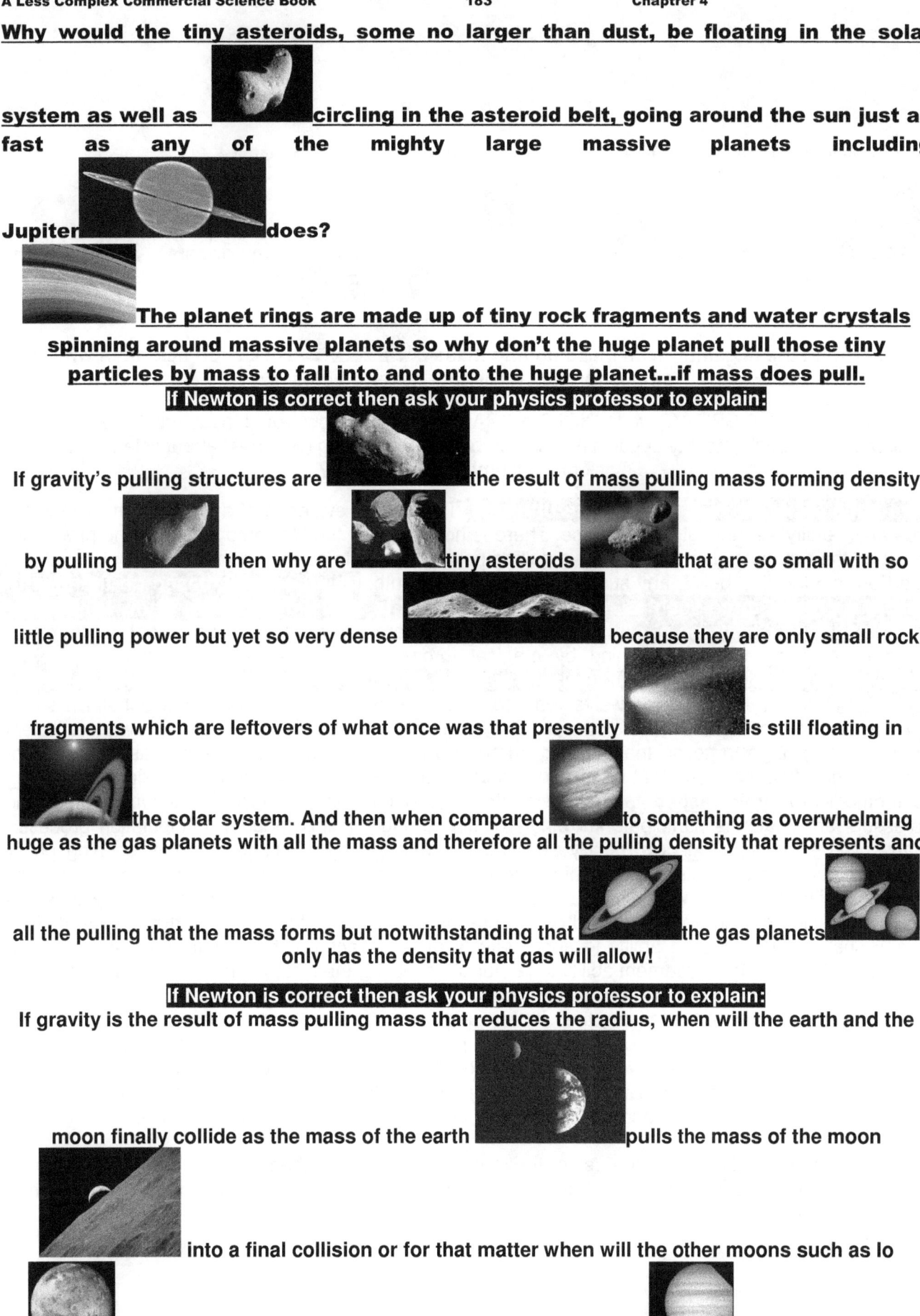

 of the big planets  destroy the all the many rings  that circles around these giants. With mass pulling mass  as Newton said then why would circles form as rings that never go closer?

What does A CONSPIRACY to UPHOLD a SCAM means and how would those most- honourable gentlemen with much integrity conducting science benefit from being dishonest about science I hear you ask in amazement and surprise?

There is no factor such as mass in the Universe. There is no evidence of a factor such as mass that holds any validity throughout the Universe. There is no proof that the Universe indicates the presence of gravity by the measure of mass forming a pulling power and while science conducts an entire religiosity based on this falsified belief, any such notion is falsified truth. **Using science based on the idea that there is a pulling force such as mass forming gravity is as valid as giving Snow White seven dwarfs and then beginning a religion on that basis.** There is a factor such as weight but there is no pulling of anything towards anything by magical forces forming gravity or whatever. **There is a conspiracy of conducting fraud by claiming non- existing forces but such claims are utterly fraudulent.** I have been trying for twelve years to introduce the true forming of gravity but all Physicists I have encountered prevented me of doing so. They stop me because my work makes Newtonian science and when removing the notion that a pulling force of gravity works by the value of mass, most of their work becomes science fiction that falls apart in substance. Read this and see how **students in physics** are methodically **brainwashed** to get the students to believe in the absolute accuracy of science. Professors and teachers participate knowingly or unknowingly in this thought manipulation process by means of conducting **mind control.** By applying this **mind-altering process** those teaching physics ensure they subdue students into becoming mind-altered zombies. It is a process going on for centuries and which without science would have no foot to stand on in the modern environment. By presenting incorrect, falsified or unproven facts and other untruths as proven truth they exert **thought control** and thereby change the student's ability to appreciate what is correct and believable logic and then force students to discard such judgement ability in favour of accepting the institutionalised untested norms and values of science in order to unequivocally believe in science. The accepted teaching methods force students to comply by compromising their better judgment and then systematically to capitulate under teacher pressure by making their own what science prescribes what should be believed. I prove this and you get this for free so what have you got to lose...**but you can get wise to what forms a better understanding about science**! By using the building blocks that forms the Universe I take you back to the instant the Universe started and I show you how the Universe fits like a jigsaw puzzle.

If you think my accusations are baseless or the ravings of a madman then go on and download what you have opened and read for yourself. What you download is free and I do not benefit financially from this explanation I present to you.

PART 5

IS THERE DEATH AFTER LIFE?

Science believes in science with no reservation. Therefore scientists believe in science with a belief more that Theologians could believe in God. To Physicists everything there is must be science. They formulate a mathematical equation and when doing so the explain everything in man's field of knowledge. Those in science propagate that we will live up to a thousand years in a very short time in the future. However ask them to explain aging and they have no idea what they are talking about. Science puts life down to some acids and a jolt of electricity and with that life could be programmed. According to science life is mostly electricity stored in the brain and by electricity the body moves life around through the body. What a lot of hogwash and this attitude of simplification is so Newtonian junk as Darwin's simplified ideas are about the origins of life. To any Newtonian any suggestion in a simplified form regardless of proofs everything science requires to become accepted as fact. The overwhelming majority of physicists with Doctoral degrees in physics will not have the ability to read this article and too understand the arguments. That is because there is a shortfall in their argumentative ability. This inability runs very deep in physics.

When there are aspects in nature that physics can't address then we have to look for the shortfall in physics. Physics holds the opinion that God cannot be proven or be substantiated. The fact that physics can't accommodate a certain fact or feature does not exclude the fact, but it underlines the failure in physics. Prove a thought using physics and see physics fail. In example when we look at the fact that I can think and my thinking is a fact beyond proof, yet physics can't prove how I think or why I think, except put it down to a flow of electricity in my brain and body. Other than that they have no capacity to know what life is or to understand the concept of life. If they shock a frog leg with electricity and the muscles show spasm they then conclude life is electricity. It never dawns on them that life is the ability to generate electricity whereby muscles are controlled and life is a lot more than just the flow of electricity. This shows the absolute fallibility of physics and not the absence of my thoughts. The Newtonian mindset is to keep physics above reproach and beyond suspicion while it is desperately poor in all senses.

If physics tries to put my thought down to some brain activity somewhere in my mind it shows how incompetent the reasoning is behind the argument. The activity is electrically induced and could be harvested by jolting on nerves from some exterior source and that does not prove that my thinking comes from the brain being jolted by electricity, it proves I am some thought with the ability to jolt the brain into action by supplying the electricity. As the medium could arouse action by stimulating the brain with electricity, so the true "I" has the ability to stimulate the mind with a jolt of electricity to provide electricity. If it was the electricity that did the job, we then could jolt a cadaver's mind and get the cadaver filled with life. The brain is not what stores life but it is life that keeps the brain with life. If life was in the brain we could generate the brain back into "life" by shocking the brain with electricity hours after death.

Once life is lost, no electricity can restore the factor we call life. Life we know is in thought and thought is the presence that physics never can accommodate and yet the entirety thought to be life in whatever form is held by thought and with thought the body is motorised by life. Yet, when physics can't accommodate thought we do not discard thought as an absence that is unproven. My thoughts are with me until I die. If physics fail to accommodate God then it is not God not being a reality but it is physics being an utter failure. I can and I do prove God as the absolute factor in the cosmos, but before one can get there a lot of Newtonian garbage has to be discarded and a lot of misleading Newtonian disbelief has to be dismissed. If you look at the night sky you will see many specks of light. In that understanding but more advanced to understand than any person can realise we find the proof of God.

It is not in the light but in understanding the principle forming light and the understanding behind the concept of light that we can mathematically prove God by using mathematics. I have done just that but it takes a book of 700 plus pages just to explain the reality forming the concept. Atheism is not proof of intelligence but it is proof of the lack thereof. My dog is the biggest atheist walking this earth but he is that because he is stupid and has a lack of mental capacity. One must star to understand light and understand the light is the Universe. The Universe is not material and darkness but it forms by light. One doesn't see a galactica, one se light that formed a galactica a very long time ago. That what I see is the Universe and the Universe forms by light left behind by time as space. I am not going into that because I wish to try and keep it simple.

Have you as you sit reading this part at this minute sat back and gave a thought about the light enabling you to read? Such a thought brings to mind the most simplistic answer one can imagine. The light hits the page bounces from the page and contacts the lens of my eye where the lens conveys the photons becoming electricity to a part of the brain that translate the electricity to an understandable message and that makes one read. It is as simple as that! Ever gave a deeper thought about light streaming across the night sky, coming from the visible limits we think of as ends of the Universe we do not even realise it is there? How does the photons manage to convey one complete picture coming from as far apart and as wide an area as it does? With a few photons connecting the eye or lens no one ever noticed the wonder of light. The photons reflect a view that seems as if coming from all the billions upon billions of stars. But most is coming from darkness covering an area no man can measure. Yet how many photons can actually connect to the lens of the camera or to the eye considering the size the eye allows light to pass through? We see by using a few photons.

We see with a few photons going past our lens a Universe representing immeasurably many photons. Still a few photons coming from a single direction directly ahead eventually tell the entire storey of a Universe larger than we could ever understand. What we see is bigger than any person can comprehend and what we apply to see that much is smaller than any human mind can comprehend to understand. It is very simple to take the process of seeing by means of photon conducting very lightly and I have never heard one of the Brainy Bunch really in sincerity uncover the process to its utter and full potential. Moreover let those Mathematical-Masters put this notion into an equation and conduct the understanding they have by proving the concept I just explained. It is impossible that light from such an array of assorted sources can simply come together at the eye lens and show a picture of objects spanning across a Universe as wide as our mind can receive where the objects they reflect is beyond human measurement and the quantity we can apply to receive a vision is inconceivable many. With that small space within our eye how can we see the space that is large as the Universe that we see? I am never going to try and Simplify the idea behind understanding God in terms of physics, but very deep inside this understanding we can begin to fathom the presence of God in terms of physics, not religion according to a Bible, but in terms of Physics. But understanding that is way ahead and light-years advanced from what I explain in this book.

Light is much more than the medium science takes it to be. Light connects the Universe in a way we cannot contemplate. Light being far apart originating from regions not in the same time or Universal space connects in a way that present us with a picture holding the Universe in an understandable content. From the point we stand and we watch the Universe the significance of what we see surpasses the sense of understanding of what we are experiencing and more so surpasses what we are able to understand. How can the few photons that our lenses catch coming from such an area as the night sky cover transmit the complete picture of what we see. Take a few seconds and study the picture of the night sky then rethink the picture applying the full content in the picture to what the size of you eyes is. Think how big the picture is that your eyes take in and translate that area to the size of your eyeball in an effort to determine a ratio.

One will be forgiven if one thinks of the ratio as eternal to nothing. Yet a few pages back I showed that according to mathematics there couldn't be anything as nothing. Consider the path the light followed from the source connecting to light from all other sources where all particles of the other light may come from and bringing a full picture to the lens one use to look through.

In your mind connect a line from every atom producing light and connect the lines to your eyeball and see how you can manage to fit all the lines, as small as the lines may be. Understanding this can only come when we understand singularity and I still have to find a person educated or otherwise that would be able to understand this idea

If it is lenses that enable us to see what we can't see in outer space it also means we cannot see the light, which is outer space because we haven't got the lens to match the curb of outer space. Newtonians think of outer space as geodesic zero, with nothing in outer space but space. Geodesic zero means the light travels in a straight line from where it originates unhindered all across space to where the light connects the eye. Such an idea by itself is outrages because the stream of photons reduce in space to such a minute quantity that taken the area the photons travel and the space in vastness it covers, the chances of one photon coming across many hundreds of light years through billions upon trillions of cubic kilometres of space and selecting my eye to convey the electricity is less than infinite. Yet such conveying takes place every second of every minute. The position of the location of the second singularity, which is the precise duplication of the first singularity but in a diminished capacity, is obvious to miss when one is not applying a detective mentality, as one should in scrutinizing the cosmos. Culture will have us believe that when one sees a colour shining from an object the colour is associated with the object. Logic tells a different storey. A yellow dot is all the colours in the spectrum but yellow because it is disassociating with the yellow. That goes for red blue and all other colours we may visualise. I think the norm accepts this as scientific fact with very little argument or substantiating proof about that required.

If light came as individual streams of photon flurries, then our visage would translate that as such shown in the fragmented as telescopes enlarge images. If the light only held what we think of as light, we would be unable to see the dark parts because only the light parts would contain light. The total picture we could see would be a picture unconnected bringing across some photons in the manner where every object stands apart not being related in any way and that will be what we see, if it is anything that we see.

That we know is not the case but that means geodesic zero is as much rubbish as anything Newtonians regard with simplicity and with careless thought. Geodesic zero means nothing and how can I see nothing as darkness because "nothing" is not darkness, nothing is "nothing" and the darkness I see is darkness showing the darkness as something. The darkness we see is as much "light" as the "light" we see because we see Light" and in that we see "darkness" as another form of light.

What then about colours that are technically not colours as is the case with black and white? White is simple. By spinning all the colours in the spectrum the colour white shines through. Black is quite another matter. A friend of mine whom is one of the best painters I have ever come across told me that one couldn't paint black but have to make black a dark blue to show shade on the canvass. That apparently is his success in achieving the realism. He also went on to explain how many variations of dark blue form the shadows in one simple tree. This remark set my mind in motion. One cannot see black because black has no colour to show, but black is the colour most prevalent in the universe. One can see only by colour and since black is not a colour we should not see black, but we do.

The fact that we see light means that the dark next to the light cannot be "nothing", If the darkness was the representation of "nothing", then that should be exactly what we must see, nothing but the stars. Taken from the top picture some stars and leaving the rest to nothing is what we see in the picture below. A blind person sees nothing but when we look at space, we see something that we think nothing of as we see as space. One cannot have the ability of sight and see nothing. It is light that we see and it is light that we use, which enable us to see.

That proves the darkness that we see in outer space is light that we see without recognizing it as such. If the darkness was the representation of "nothing", then that should be exactly what we must see, nothing but the stars.

Taken from the top picture some stars and leaving the rest to nothing is what we see in the picture below. A blind person sees nothing but when we look at space, we see something that we think nothing of as we see as space. One cannot have the ability of sight and see nothing. It is light that we see and it is light that we use, which enable us to see. That proves the darkness that we see in outer space is light that we see without recognizing it as such.

What puts us humans in a category one higher than animals (or so we like to think) is our ability to think about that what we can see. The less develop an animal is the more it has the attitude of eat or be eaten. The higher developed animals are the more the animal find reason to argue. One may teach a crocodile not to eat you if you start feeding the animal. That is a mindless reptile and yet it can think above eat or be eaten. What we see is not merely the truth and it requires reasoning to see the truth and substantiate between culture motivated observations and thought through decisions.

To all the **Super-Educated-Mathematically-Superior-Intellectuals** physicists believing in mathematics and those trying to replace God with mathematics, prove the ability to live by calculating thought as the sole factor of life where it is life that controls the body and not the other way around. Should any of them insist that the mind is responsible for life, then revive a cadaver by filling it with whatever acids you claim produces life or shock the corps until it roasts or force movement onto the body.

If it can't revive, then go and calculate by mathematics the precise ingredient it is that has left the body and therefore has filled the body with death. Mathematics is as unaccommodating to reality as Newton is to cosmology. All those Super-Superior Newtonian mathematicians, use you Newtonian inclined mathematics to explain my previous argument about how all the light that fits and fills the entire Universe can bring one picture of the entirety to fit into my eye.

The fact that **your** Newtonian physics will not allow you calculate this does not remove my ability to see the entire Universe in as large as it is through one tiny hole at the back of my eye. When you understand this entire concept you will have the ability to understand God's presence in the Universe and until then your mathematics removes your ability to understand physics and Newton promotes you blind stupidity about real cosmic physics.

Moreover, I prove all of this ability mathematically but only after removing the falseness of the factor of mass and from the myth presented as Newtonian corrupt science.

This article is as much about proving what energy is as it is about knowing the difference to the state in which alive person is and in which a dead person is. Newton considered all forms of energy to be the same, and oh boy, was he mistaken. It is not surprising he formulated gravity the way he did. There is a worldwide fashion amongst the very well educated that in order to be regarded by those with the know how as a supremely informed person, one must at least be an atheist. The key to science is apparently to be completely atheistic. Atheists do not believe in the life after death, a Creator or a Force that does not exist outside the technological criteria of mathematical science.

Everything that does exist only exists because it exists in the perceptibility. Any force that might lie outside this norm is quite unthinkable and that thought could never present itself as to be present in the material universe. The ironic of this fact is that those well-educated scientists have only one source of information and that is light waves. Still they permit themselves to be atheists in their blind state of ignorance.

I do not condemn them, because they apparently know more than I will ever know. However, because I am not that knowledgeable, I must feel my way through the tunnel of ignorant darkness like a blind person. However, in doing that, I stumbled across a heap of questions that has no answer, even by those who carry the flame of knowledge. That forced me to form my own theories, think and come up with sensible conclusions, which answers all those questions their light of knowledge could not answer.

I declare to be of average intelligence and like millions of others on earth, all these millions are believers, like me, and are confronted by the same questions these super intellectuals are seemingly incapable or unwilling to answer. Then I realized the super intellectuals only have one source that lead them and that is measured light.

Let us look at the definition of energy. Energy is, as I understand it, indestructible, which means it cannot be destroyed. Energy can only be transferred from one form to another form. Let us look at the example, which is used to teach scholars at school. We take a rock and move it from a ditch up a hill. On top of the hill, we have a lot of potential energy that was transformed from static energy by means of kinetic energy. This is the simplest example we teach children in school.

I too had to teach the children this nonsense, in the period when I too was a teacher. In the transformation, other losses occurred, like heat, sweat vapour and friction losses.

The science apparently does not take into account energy losses brought about by anger, fighting and frustration brought about by incumbency. These are also energy losses. After all the sweat and wrestling, the rock is on the top and we have a situation with potential energy from which we can derive kinetic energy when the rock is rolled down hill. I do not agree with any of the above mentioned, and will later state my point of view. I will however declare at this point that Newton's statement of energy and work being the same thing is utter nonsense. If I feel drowsy it be because I lack energy and if my car stops running it is because of a lack of energy and if my dog bites someone it is because of too much energy

Newtonian physics can't prove a God and therefore there is no God. Newtonian physics can't prove any form of feeling or emotions so we have to accept there is no feeling or emotions just because Newtonian physics are incapable of proving we have feeling or emotions. The prove or no proof Newtonian science can comply with is incredibly limited although Newtonian science wishes to cast the view that what is not Newtonian is not reality while the fact is that Newtonian science is not reality.

When medical doctors have no idea what is wrong with a patient it is because he smokes or because someone next to him in a bus twenty years ago smoked or if he never saw someone smoke even on Television twenty years ago and you saw it on the TV screen it gave you your heart condition and if that never happened then the person has cholesterol because cholesterol takes all the other blame smoking can't take. That is the way all science are run in order to keep the charade that science in any form is infallible. If they don't know what is wrong it is either smoking or it is cholesterol but it is never that they have no clue why you are medically the way you are because they have to pretend to know everything. This is science. If science has no clue what they talk about the use the word energy. Energy can be anything science wants it to be because…science says so.

Life starts of being in a sperm that has to couple with an egg. The sperm only carries life but does not even represent life. If life leaves the sperm you can do to the sperm whatever you wish and it would not represent life. Therefore the sperm is a vehicle for life and life forms the sperm sell. If life abandons the sperm sell the sell goes back to atoms. If it were the other way around, the sperm cell would remain intact and start hunting for a new life form to hold. The same argument applies to the egg. Thee egg carries life and life supply the other half of the life that will become human. If the egg does not hold life any longer, then the egg will disintegrate into billions of atoms once more.

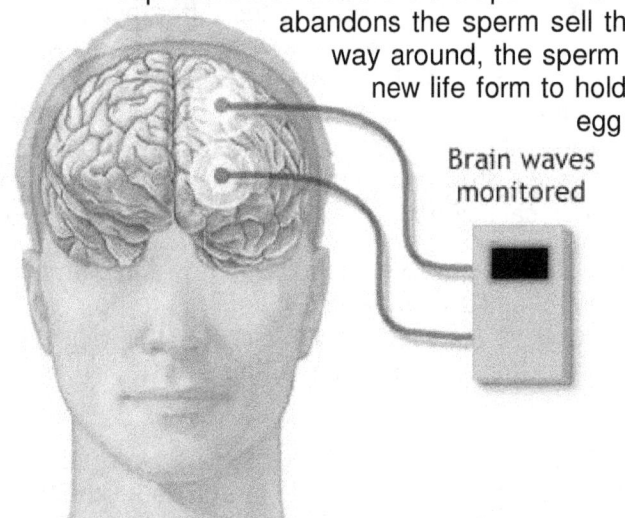

Brain waves monitored

ADAM.

This is extremely important to realise that from the first second of life forming life collects tissue that will become a human body. It is not like your halfwit Newtonian professor believes that the human body represents life. From the first moment life forms the body and it is not the body that forms life. Therefore you with your life forms your human body and it is not your human body that takes the responsibility for life.

It is by the thought process that life collects material to form the human body and the human body does not collect life as it goes along. Every one in modern science think it is the brain that controls the human body but are they so completely wrong. You use your thought process to control your body and in this thought process you form your body to be as strong as you wish it to be. Can I make you strong; no I cant because you are already as strong as a giant. What I can do is help you realise you potential strength by helping you learn to control your muscles. Before you complete any action of movement a thought first have to apply. It is the thought that command the muscle and not the brain that commands the Muscle. The thought takes charge of a cell in the mind and then takes information stored in the cell of the brain, which a thought directs to a channel that by electricity which the thought is also responsible, inflicts current in the muscle to pull the muscle. The thought generates the electricity that collects the information and the thought sends the electricity carrying the information to the muscle that has to do the job. It is the thought and not the brain and there science falls flat in their hogwash they use as information.

Life which is what you are, not a decomposable body, started accumulating material by thought when you were sperm and egg, and after the Unification you started accumulating useful building material. It is done by mind controlling the body. Don't allow the atheistic senseless stupidity tell you different. If your body was what is in charge which is you, then when you are dead someone with life can pump some oxygen into you and shock you with electricity until you bounce around like a ping-pong ball and you will begin life again. That is total rubbish. If life leaves the body there is no structural formation left to control and maintain the structural I integrity of the body.

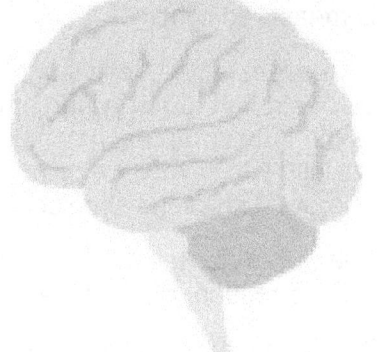

Even in the very beginning life formed the sperm and the sperm did not represent life. You can't have a tube filled with sperm and when you find the lot are dead you revive the sperm with an electric jolt. You can't have a jar filled with D.N.A and by regrouping the composition you build the body of the person once more. You build your body through thinking with your mind. You construct your body cell be cell by using your mind to do so. How can I prove this? The instant your life vacates the body; the body's ability to restructure the structure leaves that very second. The moment life vacates the body, the body degenerates by fragmenting the structure until the entire construction disassembles into forming atoms again. It is life that keeps the body into form and without life the body de-fragments into atoms once more. Your body doe not hold life but your mind by thought controls your body.

Your medical doctor will tell you your body represents your life and when you die your body dies. He will be of the opinion that when your human body dies you have died. The problem with this attitude is that while your human body is still intact, one should be able to resurrect the body by supplying heat and electricity. That is not possible. That is the way one goes about killing people. By electrocuting people on a chair the state removes life from the body ands so giving the body electricity does not bring life to the body. Therefore the body does not use electricity to instate life but by duplicating the transmitting of electricity the body becomes confused and the body relinquishes life. It allows life to depart.

Your physics teacher / professor will tell you that you are what you are because of the body you have. Hogwash I say that concept is and I prove it is rubbish. You build your body with your mind but you control your mind through thought and without though you will not even move a muscle. It is by thought that you tell a muscle to move and it is by thought you tell the muscle to get active and by the same thought you form the muscle which puts the ability in the muscle to form the strength. I prove that this is how gravity forms and it does not form by the pulling force of mass. Newtonians such as your physics teacher or professor wouldn't even read my books in which I prove they (those teaching physics) are all brainwashing students to believe that physics is what Newton said it is. I prove they are submitting all students to mind control in order to force you to believe physics is what they teach it is.

Those that think they are experts in physics has no idea about physics and I challenge all of them to prove Newton is correct, not to surmise that Newton is correct or to force students to admit and confess that Newtonian physics is correct but to prove it is correct. Prove the formula $F = G \dfrac{M_1 M_2}{r^2}$ does form gravity by forming a force by the value of mass or prove that the formula $P = \left(\dfrac{4\pi^2 a^3}{G(M+m)} \right)^{0.5}$ does put planets in positions allocated according to mass.

It is life that allows the semen to swim and it is the life within that allows the egg to be receptive of the semen. When either the semen or the egg holds no life and the egg or the semen is still intact. There is no life forming possibility. The semen does not swim it is life that allows the semen to swim. This proves that it is life that allows movement from the beginning of where we think life starts.

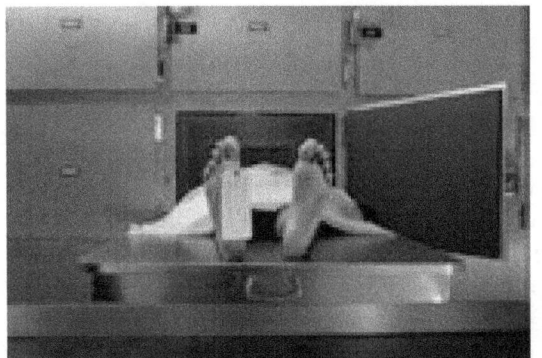

This is a picture of a cadaver. It is not something to be scared of because it is a body NOT containing life, as anyone of us will be someday. So it is the same as you being scared of you as you are going to be somewhere in the future and that is pretty silly. On the condition that you were born the only thing you will be someday is dead. If you are alive then you will face death. What we have to answer is what is the difference between this cadaver in the mortuary and me. One is that the cadaver has no life and I show vital signs filling me with life.

The biggest factor is movement and that movement is linked to thought. Considering the implication of this is vital if you wish to enhance your physical strength and build your body. There are persons in hospital in a coma for years and they apparently show no thought because their muscles don't move and therefore they wither away. The thought gives control over the body and the thought form the muscle and the thought form the size of the muscle.

This cadaver or dead person can't get up and walk as I can. Why can't this dead body get up and walk, it is because the dead has no thoughts. If you think the Newtonian idea is correct that life is part of the body then rethink. I dare you to conduct some tests. If life is electricity as they say it is, then why can you shock that cadaver until it hums like an electric transformer and life will not return? If life is as they say it is electric convulsions then try and shock the brain with electricity and you will find no response. The fact that you can manipulate muscle spasm with electric convulsion shows that life controls the brain by charging electricity and that process is done by thought in life.

Life generates electricity that life then implements to control the body life extends for the purpose of serving life.

Life is in charge of the body and of thought and not the body being in charge of life.

By electrocuting a body with life you merely short circuit life's actions with a stronger jolt of electricity but the electricity is just a modem through which life controls muscles and growth in the body. Then you burn the electricity conducting connection that life has as life controls the functions of the body and do that long enough and life may not find a manner to form conduction of electricity whereby the organ control will become suspended.

Your mind charges electricity and that electricity are created by thought and thought is life. By creating thought you form your body and by forming your body through thought you establish you level of strength. That is why when you are in shock you are able to perform in a manner not even you ever thought you are capable of. Around your head there are electricity flowing which science named "brainwaves". These brainwaves are just a form of electricity and that current is the same as what flows around every electrical motor or any planet charging gravity or any star forming a gravitational field.

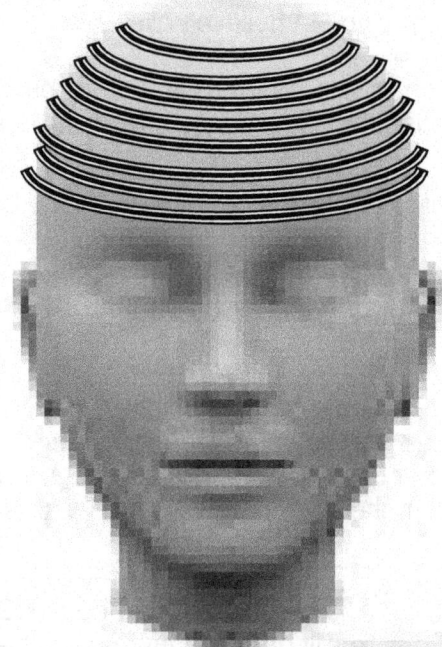

A human is not the body you have but the mind that forms the body

Now we take the scenario of a person's life. When that person is born, and after that momentous occasion of the birth episode, he or she continuous to live on this planet for the best part of the next sixty or seventy years. In this, period a great deal of energy is used to walk, run, laugh, cry, think, produce and reproduce. By doing that, he would from time to time state that he feels tired or without energy. What energy is the man referring to. I have once heard a scientist that made such a fool of him. That scientist declared that if God was energy, God could be coal, because coal is energy as well. Now I would love to invite him to a meal and see him devour a plate of coal. If coal is energy, he can make a meal of it, and then live very cheaply. What he does not seem to grasp, is that there are many forms of energy, which differ totally as we distract the heat and in doing so one can tap the energy. However, coal cannot walk, run, jump and laugh. I cannot even begin to imagine one brick crying and moaning because his friend was thrown into a fire. Coal cannot have sexual intercourse producing an offspring and then caring for it afterwards. Life on the other hand does have that energy quality. This means that there are different values and forms of energy, of which life is one. If life is no different to other forms of energy, God could be another total different concept of energy. This is the problem that I have with these "SUPER- EDUCATED- MASTERS- OF- FACT" geniuses. They can make the most bizarre statements and could be away with it unchallenged.

When circumstances starve the body of food, life occupying the body would begin to devour the body in order to sustain life's ability to occupy the body. To life the body is only a vehicle to serve its purpose. When the body starving, the body does not suspend life until conditions are favourable to have the body reinstall life. It is so very typical of the "SUPER- EDUCATED- MASTERS- OF- FACT" to uncomplicated issues to serve their insight. They make something such as life so simple as to pretend they are completely in control of the knowledge that subject has to offer. The body does not turn life off, life eats up the body until the body is so feeble it can't host life any longer after which when life then rejects the body. The body does not maintain structure but as soon as life evacuates the body, the body breaks down the structure it held when it hosted life. Life maintains the body and will even devour the body and consume the fibre until the body becomes useless to life and until it cant serve life any longer. After life abandons the body, the body returns to a state of atoms with no resemblance of what it were when life formed the body. It is life that constructs the body, maintains the body, and controls the body and consume either the body or as food that is some other life form that had a body.

The world contains a wide spectrum of different occupations that people earn their livelihood from. Seen from my personal occupation, there are two types. Those that farm and produce wheat, corn, barley, nuts, sugar cane, vegetables and many other produce. These are potential energy producers. They produce food, for the other group of the human population that uses this energy product to maintain their strength to apply it to other methods of occupation. Cattle and sheep farmers produce meat that is used by others to convert into energy for their personal use

All people have one thing in common. They devour one form of energy, which is known as food. That is needed to maintain a life cycle, and the consuming of food must be done on a regular basis, to enable a human to live and reproduce for a lifespan of seventy of eighty years. The only precondition is that life would sponge on other carbon-based forms of life, whether it is plants or animals.

This person maintains his way and means of life, thus transferring energy from a form of food to a form of work. Then one day he collapses and becomes still. That person becomes unable to move. We call this state that the person is in, being dead. Even if I take a shovel of food and force it down his throat, he still would lack the ability to transform that energy to movement. But why would this then not bring back life?

Simply because he does not breathe any more. And why isn't he able to breathe? The reason for the person's inability to function, as a human should is because the cadaver is dead. When a person is considered dead, he lacks energy to such an extent that he cannot bring his own body to the grave. Others like me, and I have to carry him to his grave. We, that are alive, and maintain the process of translating food into life, have to carry the dead (he who is without life) to his grave.

The only difference between him and me is the energy form known as life. However, life is not the same form of energy as food, oxygen, heat and electricity. Even if I force all the food down his throat, and pump his lungs with air, while I heat his body with a blowtorch and shock him with electricity, he would still find himself unable to walk himself to his grave. That means the one form of energy is not the same as the other form of energy.

It is widely accepted that there seems to be a generator in the brain that generates electrons which enables the body to function. We know the flow of electrons is due to the process called electricity. On the other hand, do we? In a later chapter, I shall point the difference out between this flow of electricity. However, for the mean time I would stick to this accepted fact that electricity is conducted by the flow of electrons. Now, you can shock the cadaver with electricity until it hops about like a ping-pong ball, if life has gone absent, conducting a flow of electricity would not reinstate life.

You could put the cadaver on life support, with a heart machine a lung machine and all kinds of other machines. This method has nothing to do with life being precious, but fare more with the money paid by his medical aid, being precious. Once the cadaver's line of financial support dries up, his life instantaneously becomes worthless.

Then the cadaver finds the problem that it seems unable to live which means it is dead. Death means the brain is unable to send electronic signals by means of amino acids to the muscles, which would enable those muscles to continue with its normal function. The cadaver finds itself without the energy called life.

At this stage, I think that I pointed out to the difference between a body filled with energy called life, and a body that lacks energy and is called death. However, the energy that I pointed out called life, is miles apart from the energy that consists of food, air, the burning of it and the destruction of it. There is a broad difference between the food process and the actual form of energy called life.

Now I would like to ask those Super Intelligent Atheists and consumers of food and air to explain where the energy form that is called life has gone. Energy cannot be destroyed, but can merely be transformed from one form to another. This is scientific gospel. Life as I pointed out, has a different value to heat. Life cannot be destroyed, that means it has to be transferred from one form to another form, and life itself is not heat, electricity, or food, because applying all those other forms of energy cannot raise the dead.

The fact that energy must be transformed and cannot be destroyed is proved by science to be unquestionable. The life energy started assembling a body albeit sperm or an egg before conception or procreation took place. If the sperm was dead the sperm would not swing and all those that swam in vane died. They did not hang around as lifeless sperm to be vitalised with life as soon as the next opportunity arrived. The very second life left the sperm or the egg without fore filling the process of fertilisation extending the ability of life to assemble more material in order to form a body filled with life, life left the sperm or egg and in that the body holding the sperm or egg destructed. It is life that captures material to form a body notwithstanding how small and in that no one can remove God from physics. The idea that life sprang from somewhere as soon as a sperm was there is as mad as having mass being able to form gravity. With my physics I can prove God being responsible for the flow of time within the Universe and if Newtonian science are not able to accomplish that, it is Newtonian science that is dismally inadequate, but then again that is what Newtonian science is in almost every sense.

The only answer I can conclude is that science is ignoring their own findings to prove their own religion fashions. With life being an undisputable form of energy and energy cannot be destroyed, it seems very unscientific to propagate atheism as a fact.

From these facts, one has to conclude that there does exist another form of life after death.

5. INTELLIGENCE VERSUS EXTELLIGENCE

In this book, I shall introduce you, as the reader, to a completely new line of thought about the science of cosmology. Some of these scientific facts date back to the time when man became aware of a lifestyle that just started to include a civilized order. According to some discoveries by archaeologists, it seems that mankind had its survival mostly due to the way it accomplished knowledge about primitive science, this enabled man to survive in a total hostile environment.

Man's first awareness about forces that he could not control, was explained as forces unleashed by pagan gods. In that is seated mankind's belief and mankind's desire to be in total control of these godly forces. This desire therefore became one of the biggest incentives that drove mankind to a civil obedience and law-abiding standard of living.

We can even today go as far as to except that the role that intelligence played in the development of our specie was far bigger than the role was of the more brutally and physical force. In the animal world, the strongest in specie would ultimately be the leader of the pack. With his brutal power and brute force no one in the pack would dare to challenge the leaderships hierarchy and in that the leader himself. If one challenger should dare to do so, the challenger would pay with his life. It is a well known fact that male baboons not only kill the previous leader, but he will wipe out all the siblings, no matter how much the female baboons might protest against it.

This is even more so in species that has much closer links with mankind. The chimpanzee male just simply murders all possible male challengers until the day he himself is also murdered by his successor. The orang-utan male is another example of a male that would not even tolerate any male in a smelling distance. This confrontation will definitely lead to the death of the weaker one of the two.

There may be a distinct possibility that fear for the unknown was the only reason mankind's development lead us to a higher norm in development than our close relatives. In case of other species the generational development benefits the physical strongest and do not favour the more intelligent of the species. These animals are still much stronger than man is, although man tamed all animals, at one point or another. Therefore, all animals submit to man.

In this, one must define the difference between intelligence and the idea, which I refer to as extelligence

In the understanding of the meaning of intelligence brings to mind how the animal socializes with its own species to guarantee the social survival of the species. That means that all animals have intelligence. Dogs has been with man as long as we can trace back human development, so in doing that he forced the dog into acknowledging mans intelligence. However, the dog still communicates with his own species in the way its intelligence dictates. He sees man as the leader of the pack, rather than a completely different species. When it communicates with man, it will wag its tail or show submission by lying on its back. The dogs intelligence never allowed him to try and communicate with man by using mankind's standards, although his intelligence placed him in a certain advantage point to share to some extend mankind's intelligence. However, the only reason it did so, was to further his own needs in surviving in the pack with man then becoming leader of the pack.

As a farmer, I often watch the manner in which cows interact. At one predestined time during the morning, the mothers leave in a group to have a drink of water. I admit there is nothing strange about that. The strange part is in the procedure, when taking into account that we regard these animals to be thoughtless beasts. One dry cow gathers the entire suckling calves, takes them to a safe, and secured area, where the calves would play and enjoy one another's company. I refer to this as the "kindertiun" which is the kindergarten. After the water drinking, the mother's would gather in a shady spot, and ruminate for about two hours or so. Then they would get up, and stroll in the direction of the kindergarten. Only when they come to a certain point will the calves leave the seclusion, and run to meat their mothers to feast in the generous supply of milk The biggest amazing part, is that the kindergarten hostess is never the same cow. Everyday another dry cow takes on the responsibility of playing stepmother to the calves. Not once is there an incident of one of the calves being disobedient or not under standing what is expected. They always seem to know which cow to follow, and are never fearful of leaving their mothers. They never are obstinate and wonder off in search of their mothers or run to their mothers before someone gives the signal, whatever the signal may be. If cattle are that mindless who decides who's turn it will be to play stepmother, who and how are the calves informed about who to follow, and why are they acting that responsible and disciplined. After all, they are only young mindless beasts. I concluded that in our self-righteousness we under estimate our fellow living species. However, in all fairness to my own species I have to admit, mankind disposes of intelligence as well as extelligence. Extelligence is the acquired knowledge to deal with matters not relating to its survival. A part of this development was due to the need of extelligence to eat. Mankind's progress in becoming a forceful species lagged behind because of his awareness to the fact that he could manipulate certain forces in nature to his advantage. A part of this manipulating process was accomplishing the skill to control and use fire. However, man also noticed that fire came from the heavens and clouds. These clouds formed part of the sky where the sun, moon and other stars are located.

With this argument, astronomy had to play a huge role in the development of mans culture, especially the religious aspect. His health, happiness, belief, future and wealth all derived from the gods that was found in the stars. This fascination and even religious fears was derived from the stars that even today is still apart of the science of cosmology. That is why even today, people are still motivated by the stars. Ironically enough, the other big motivation lies in man's lust for power and his war games to commit murder, to demolish other's property, and to dominate other members of its species to the point beyond that of slavery. The role of slave owners and slave drivers today is in the hands of the Mammonites. They use John and Jane Dow and all mankind that belongs to social grouping lower than they do. Mammonites are those that control every facet of the man on the street's life, should it be by job supply, political law enforcement, food and house supply, by dictating to the politicians in what manner and which laws should be applied and enforced. Mammonites are the bankers, the Wall street brokers, the Insurance firms, the drug lords, those that Motorcar manufacturing belongs to, the Shopping chain owners, the Oil barons and the Petrol companies but to name a few and all those and all other evil proprietors of wealth and monetary fortunes that force the law of the merchant onto the public. The merchant never has scruples, never has a conscience, adhere to one God which is Mammon has love for wealth alone and only has greed as a driving force. They that ask in all sincerity "What is wrong with having greed" but that question is on the lips of all other criminal elements in society. Mammonists on the other hand is the smaller and lesser counterpart that would pass his hungry brother and not help him, although he has more money than he needs, but he shares in the greed of the Mammonites.

To them, their love for money is far greater than their love for their fellow brothers and sisters and they will force millions to go hungry just so that they could have billions. This can be found in any social structure, be it capitalists, socialists, communists or kingdoms. They are the ones society looks up to while they are the one buying the politician to change law that bleed the population dry to enrich the those that already have everything anyway. Al Capone was nobody. Al Capone had one house in a neighbourhood of wealth. The others were more crooked and bigger gangsters that he was because they owned more than he was. They only pretended to be legal while they branded him as illegal. Who were the crooks that owned the other houses in the suburb that houses the rich? They are as big Mafioso because they pretend to be cream of the social structure while they would bleed those below him dry without mercy and always to the cream's advantage.

The fear that man experienced about forces, which was, according to him, inexplicable and therefore stronger than him, was considered by him to be of a godly nature and therefore only the wise amongst the wise could explain and philosophise about the nature of these forces. Common man never questioned the correctness of these layouts. However, man still prevailed in explaining to the best of his ability, the logic about his viewpoints and in so doing to guide the incredible forces that torched his fear. Today in retrospect, we consider these arguments laughable. Think how ridiculous it seems to regard the sun as a god on a chariot of fire that patrols the sky on a daily basis. Today it seems ridiculous to regard the earth as being flat, or to see a face of a goddess on the surface of the moon.

However, the tendency to except these super intellectual's arguments remains a custom until this day, no matter how silly it seems to be. Man is still upholding the ancient culture that in a case where it cannot be understood, the idea must be correct and therefore the ordinary man would inevitably be too stupid to follow. In addition, to this day, the "SUPER- EDUCATED- MASTER- OF- FACT" among us still misuse this phenomenon in common practice. Think about the silliness of Einstein's single dimension theory. How ridiculous is such a notion. Once we accept the earth to be round, Einstein came along and invented a flat universe!

In a hundred or two hundred years from now, it would be our generation's turn to be considered short sited and backwards because we accepted these ridiculous arguments. Thus, no matter how dynamic our visions of the cosmos seem to be, we shall still be regarded as non-intellectual and stupid by generations to come.

In writing this book, I too attempt to deliver a contribution to clarify a certain line of thought that is unclear and to give an explanatory value to it. You, the reader, will evaluate the acceptability of my reasoning and you will remain the only evaluator that will approve or reject my work.

My viewpoints are not the consequence of a big literacy, but rather due to a lack thereof. Because I shall entrust you as the reader with my thoughts, I shall have to introduce myself in a brief manner. Relatively spoken I can be considered as stupid. I do not try to sell myself short, but in accepting this fact, I was able to use it to my great advantage.

First, I have to qualify my statement that I am stupid. All people know what they know. However, they do not know what they do not know. For us mortals, the sum total of what we know, is enormously big. That comprises the total amount of our total human existence and our accumulated knowledge gained over a lifetime of labour. On the other hand, we regard the part that we do not know, as so insignificant small, that it bears a value of nothing. Because we do not know how much we do not know, we cannot evaluate the sum total of that.

People always concentrate on the part that they know and therefore realize how intelligent they are. In this lies the accumulation of their absolute arrogance. With this arrogance the part that they do not know, becomes even more insignificant. The normal procedure of man is that he will concentrate on the part that is of value, disregarding the worthless part that is of no value.

In my case, I had to concentrate on the part that I did not know, because of the lack of formal education and therefore not knowing how much I knew. This brought about that I always had to regard the part that I knew as being insignificant small. I had no formal examinations in testing how much I knew and thereby evaluating my field of knowledge. In the absence of examinations, I had to disregard the amount that I know and always had to consider myself as being stupid. This brought about that I had to remain humble, because I was untested and stupid. This book is the consequence not of my enormous intellect, but rather as the result of my stupidity. I always had to fight my ignorance and had to seek answers to my own questions because I was too stupid to accept the official answers given by the educated..

At school, I was a rebel. Because I was so stupid, I refused to accept facts without testing the reasons behind the answers and just because the teacher said so. An intellectual person would have accepted those given facts without causing him self all the inconveniences. I always insisted on outlined explanations by my teachers. I bluntly refused to "learn" anything. My point of view was that if I understood a subject brought about by a good explanation, studying was unnecessary. If the teacher could not explain the subject in depth, then I disregarded him and his subject and treated him with disrespect. I reasoned that the blind could not lead the blind.

This of course enraged the teachers and they tried to break my resistance with corporal punishment, which I deserved. This went into a spiral, where I got more rebellious and they had even more reason to apply the cane. In the end, I was the one that got the short side of the stick due to my rebellious stupidity, and brought about that I had to teach myself all that I know. I tell you this to point out that since my earliest days as a child I could never conform only because my superior said a certain thing and I was supposed to accept it. Sometimes (I guess) I was wrong and they were correct but only convincing me with intellectual arguments could allow me to see the other side. I was ultimately the one that paid the penalty for my behaviour because the road I took was tough. Some things I had to go through because of my stubbornness I do net even wish on the devil himself.

As I said, I am plainly stupid. But in saying that, I believe I am just your everyday person and if I could write this book, being as stupid as I am, any high school pupil can read and appreciate this book.

Because I do not have any noticeable academic background, I believe that these "SUPER- EDUCATED- MASTERS- OF- FACT" Academics will try to shoot the information down in flames, (if there are any that even would read it). The hostility I received so far from these Super Intellectuals did not surprise me in the least. However, I spent 21 years of my life to come to the conclusion that I share with you in this book and like every human being, I would want to defend my work against the onslaught of these sublime intellectuals.

The first few chapters are everyday common sense, but by regarding it, it will enable you to grasp my line of reasoning, which you will need in understanding the last few chapters. If one does not read the simple chapters, the terminology used in the last few chapters might seem somewhat incomprehensible but that is only because of new terminology that compelled me to introduce new terminology is brought in, in order allow he reader miss conceptions used and the new definitions I introduce.

This book must comply with a commercial value, to pay for the publishing costs and to introduce my work to the broadest range of reader's possible. That is the only way I believe I can force the academics to take note of my work. People associate cosmology with Einstein and his complicated brilliance and brainpower. That complication is only because Einstein told half the story. When I tell the other half in this book, you will see that it is not half as complicated as it seems. Every person with a normal mind will find all those unbelievable complicated statements that Einstein made, is in fact rather simple, when told in its full content.

On the other hand, this book must comply with certain technical facts to stop those "SUPER- EDUCATED- MASTER- OF- FACT" super intellectuals from blowing the statements that I make away with a few words. This work comprises of hundreds of new thoughts on cosmology, laws and processes in nature that was never noticed before and arguments that is now seeing the first light of day. Many of these arguments might be old statements that is purposely withheld by scientists, to the general public, because in admitting to the follies that exist in modern science, they then have to admit that their scientific layouts are in the least, foolish. As I said, the rejection received up to now, will be but a drop in the ocean in comparison to what I expect when this book is published in English. In the past, the only rejection on their part was because of the lack of my education, and therefore they refused to listen to my arguments. Therefore, I belabour this point concerning my education because if any person feels a need to reject my work based on the lack in education on my part, do it from the start, but if you do so, it will be to your own disadvantage.

I realize I do not beat around the bush, when confronting certain statements by some ingenious jokers that cannot be taken seriously. For that reason many of the academics would be sensitive to my work, but if they feel the need to make foolish arguments on international T.V. and in books, they must prepare themselves to be made the fools they are. These same intellectual gurus go to extreme detail in their own publications how the Roman Catholic Church denied science freedom of speech and prosecuted scientists in the dark Middle Ages. In that time, such people were prosecuted.

The powers that control the media today is far more powerful and much more methodical than what the Roman Catholic Church was then, and the modern day media's inquisitions are far less merciful and more subtle. If there is a certain school of thought, whom the modern media does not want to propagate, it will be killed by silencing its publication. If not for a medium like the Internet, this would have been the lot of this book as well.

This book will serve as a modern test in press freedom to see how science currently will respond to a new school of thought when their beliefs are ostracized. This book will put the shoe that was 500 years ago on the foot of the Roman Catholic Church, right back on the foot of science and the Newton apostles.

It may seem that I have a hate campaign against the intellectuals of the day. It is not true. However, I refuse to believe that with all their geniuses combined, they are unable to see the facts I have seen. I may be wrong, but there seems to be a sinister motive in the published work of modern science in that they promote atheism and use every chance to degenerate Christian belief. If a "mister nobody" as myself can see these facts so clearly, surely they have to see it too. Why do they then keep silent about it? They are the ones with almost unlimited IQ's, not me!

You might find the first four chapters to be simplistic and uncomplicated but if you do not get a well grasped understanding of the introduction of my theory, the complexity in the last four chapters will then seem impossible to follow.

The contents of this book is not aimed to relax and entertain, but the possible enrichment of your comprehension to the layout of God's creation will richly compensate for the effort, especially in the last four chapters. Thus, if you may find the first few chapters below your mental capacity and development, I ask you to bare with me, it would be worth your while when you reach the last four chapters and the truth starts to dawn on you. However, the last four chapters would seem complicated if the golden thread were not drawn right through to the end.

The main contents of this book, is as far as my knowledge goes, never been written or spoken by any person dead or alive and should be fresh to all.

If you do not approve of the very simplistic mathematical calculations, please feel free to ignore it. As I have shown in the prologue, it is merely put in to prove a point. If the simplest calculations on gravity are ridiculous, there is no applicable scientific applicability on reality. I did use it to point out how illogic certain scientific arguments are, but is will not enhance the explanation of this theme in any way what so ever. It is merely placed in this book to silence some of the "SUPER- EDUCATED- MASTER- OF- FACT" intellectuals.

As a background sketch to how I was motivated in writing this book, I must share the incident with you, being my reader. Due to my interest in the science of cosmology, I get asked certain questions from time to time to explain these known and accepted theories, principals and definitions propagated by the "SUPER- EDUCATED- MASTER- OF- FACT" academics.

As my personal studies progressed it became increasingly more difficult to explain certain accepted theories and concepts promoted by scientists and more even, to agree with these hypothetical mumble jumble and fairy tales. How does one defend certain accepted ideas that science promotes, but is faced on contra dictionary of how you relate to these facts? How does one explain your own concepts when it differs completely or does even vaguely been accommodated in the science of the day. I would have been able to ignore these conflicting feelings, if I did not know there are millions of John Dows out there who, as I do, did not appreciate (like me) or understand (and therefore do not accept) this misleading information.

To consider yourself part of the intellectual cosmology know how elite, certain recommended directions for use should be meticulously followed when confronted with the unexplainable concepts that is being promoted by the intellectual of the day. Applying these evading methods is very unacceptable to me. However, these directions are being used by the utmost intelligential on ignorant persons. When being confronted by a question you do not know the answer to, throw a mind boggler and complete mesmerizing question back at him, with the knowledge that nobody on earth knows the answer to that question. The questioner would find himself so bewildered that by the time, he recovers his senses, he would find himself still without an answer but he then would be out of a chance to insist on an answer. That will keep him ignorant and well in his place. On other occasions, other methods can be used with similar results.

First, if being asked an unanswerable question, congratulate the questioner on his brilliance and well thought question. Share the brilliance and well thought question. Share the brilliance of the question with the whole audience. Let the audience comprehend how brilliant the question really is and let the audience applaud his brilliance. Then abruptly ignore the question by going on to the next question. The first questioner will be so pleased with his own brilliance, he would never insist on an answer! However, beware; these methods can only be used in extreme cases and definitely not too often. The actual rudimentary way of dealing with the problem should be as follows.

(i) When nobody understands a certain concept give the questioner an impressed but unmistakable superior smile.

(ii) With a tone dripping with sympathy, you should use a stance of high and mighty superiority and let the questioner realize that you sympathize with his intellectual shortcomings.

(iii) Let the person very well realize (still with dripping sympathy) that if he was blessed with your intellectual insight and capacity, these facts would seem trivial, as it does in our own case, and you do understand his mental shortcoming, but he has to accept his inevitable weak minded position.

(iv) When the questioner's reason becomes far to logic suppress with your own ignorance of these illogic matters, like gravitation and electromagnetism, time and the black hole, immediately and without any further hesitation and with all the haste you can muster, refer to rule number one as stated above. Do not delay another instant. I shall damage your personal reputation to a point of no return.

I know all these methods of question evading by those super blessed intellectuals because it was used on me so many times by some of the most renowned geniuses. It was because of this that I started my personal search some 33 years ago. I had no intentions ever to put it down in published writing, because the motivation lacked on my part. For the past twelve years I kept myself occupied by trying to introduce my findings while not finding an audience very interested in what I have to say.

One night the owner of the local, Wimpy, Johan, asked me what space was. After trying to explain by starting with the atom, I saw that I was making very little to no progress. Then I decided to put these explanations down on paper so that he would be able to read it in his own time. Before long I realized that I could only explain this by means of a book, because in order to understand one explanation, I had to explain the facts that leads and follows that explanation. Now I find myself more than a decade years later, still trying to bring across my point where I wrote the book in Afrikaans, but has to translate it to English because of the weak market for such books in Afrikaans and for what it is worth here it is. *"Johan hier is jou antwoord in Engels!"*

The problem is those that should be interested in what I write are not interested because they are those that think they know more than God and that is not meant to be blasphemous because I show later on that those "SUPER- EDUCATED- MASTERS- OF- FACT" would rather have the cosmos change and start to contract in stead of expand as it does than admit Newton had everything wrong all along. Then we have the other lot that think physics are above them and they don't understand physics all the while it is that the physics they can't understand is so crooked not even Newton understood his own physics.

Between these two options of prospective readers I have not yet found that big understanding about the message I try to convey. I hope this effort will be simple enough so that everyone not connected to physics will show interest and understand what is wrong with the physics they don't understand and those that do understand physics will see how big fools they are too understand the physics Newton couldn't understand.

Those that do not understand physics will see why they were brilliant enough not to understand physics because it is one big hoax and those that do understand physics I hope will feel as stupid as they are because they think they understand the physics that is a complete joke.

SUBLIMATION; The Newtonian Mythology

The purpose of this book is not to echo the **Newtonian** version of Greek Mythology.
The Greek Mythology had aimed to bring, what they considered religion to be, in line with their perspective on what they considered science, cosmology and astronomy had to be. In Newton, a change a change came about in the contents of the mixture, but not the ingredients as such. The Anglo American Mythology now preaches a religion called atheism, which, as were in the case of the Greeks, based on what they perceive science and scientific facts to be. If you find a desire to run along with these myths of star travel, speeding through the Universe at the speed of light, encountering alien societies and indulge yourself in such modern mythology, please do not read this book. In this book, the magic spells that Newton named gravity, and which Einstein took to a single dimension fantasy, is discarded in dismay once the Anglo American Mythology is replaced by factual truth, the creation according to the Bible becomes a detailed analysis of the actual creation. When Newton's lies are replaced by the real functioning of the universe, the author of the Bible's firs book has such a precise recollection of events, that it put all scientific facts, being broadcasted at present to pitiful shame. That includes the ideas of modern atheistic Anglo American science cult. This book does not repeat all the traditional nonsense, but explain in detail, how the cosmic year structure works, in such detail that children can understand and accept it.

Civilization throughout the ages always used Cosmic Science, but especially, by the Roman Catholic Church, to prove the importance of the earth as the centre of the universe. That brought about that the sun shone on the earth, which was the centre of the universe. Since the earth was made for man and was the centre of the universe, God made everything with mankind in mind. Whereas the Roman Catholic Church was the only representation of God on earth saw they represent God and all His Powers on earth. It meant that the Pope was God on earth (being the head of the church of God on earth) and the Catholic hierarchy was the most elite and privileged on earth. That was precisely how the church considered them and how they could conduct their teachings.

Right at the top was the Pope who ruled the earth and as the earth was the centre of the universe, the Pope for that matter, ruled the whole universe. God and the saints ruled the heavens and the Pope with the Cardinals ruled the creation of God. This brought about the sublime picture that suited every person, which was considered anybody. He whom the Pope blessed was blessed and he, whom the Pope damned, was doomed. Everybody op importance could buy the Pope's blessing and could die in reinsurance that his life was ultimately saved.

By the middle of the 15th century, some unimportant persons with no real social standing came along and disagreed with this ultimate and universal accepted hierarchy. They stated that the earth was not the centre of the universe, which meant that the Pope and his cardinals were not in control of the universe. This new perspective on the Universe and the chain of command was very unacceptable to anybody of social standing. This (then considered) blasphemy was to be killed in its very infancy, even before birth.

However, as with everything else, the truth eventually prevailed and certain intellectuals took notice of statements by Galileo and Kepler. The reaction that followed can be regarded as one of the most important historical events in the route that man's civilization took in forming modern man.

In this book, we put the work of the two giants, Galileo and Kepler, under the magnifying glass. In contras to popular belief, there is an astonishing difference between the work of Galileo and Kepler and modern science. Modern science is based on the findings of the father of modern science, which I consider Anglo-American mythology. The father of Anglo-American mythology is none other than Isaac Newton in person. Galileo and Kepler had the truth of the Universe unlocked when Isaac Newton came along and raped their findings.

First, let us consider Galileo's work.

When a pendulum swings, the pendulum's rhythm remains while the stroke tarnish. All big clocks are based on this working principle and can therefore keep time mechanically. Time has been measured by means of this method for about half a millennium.

When two structures of different mass is dropped at an equal distance and time, the two structures will hit the earth at precisely the same time, as long as the wind resistance is equal to the two bodies. This experiment was the first that was done on the surface of them moon on July 1969 and billions of T.V. spectators bared witness to the outcome of this experiment. A hammer and a feather were dropped and they hit the surface of the moon on precisely the same instant.

Kepler, on the other hand proved that the earth and all other planets rotate around the sun in an elliptic orbit. All three these findings never mentioned any force called gravity.

At this stage, science used precisely the correct argument. The findings were noted and it could afterwards be checked for the same results repeatedly.

Afterwards an English genius by the name of Isaac Newton noticed an apple falling from a tree. This prompted him to calculate a force that existed between the apple and the earth, in which the matter of the apple and the matter of the earth was drawn by a force he named gravity. He never took into consideration any of the findings of the previous two giants, although he praised them for their work. Newton went and calculated the existence of a force that existed between the above-mentioned bodies. At this very point, science took a wrong direction. The force that Newton calculated is a secondary function to the primary condition that holds matter in place throughout the universe.

By using the findings of an even bigger genius, Tycho Brahe, Johannes Kepler proved that the radius of the planet, taken from the sun to the planet and is measured in astronomical units, is equal to the square of the rotation period. Kepler said that there exists a relation between two bodies where $T^2 = R^3$. He never mentioned a force. Galileo's findings proved the same as Kepler's, that there is a ratio between two bodies, not a force.

The second statement of Galileo can compared to driftwood on water. If two pieces of wood which is different in size and weight floats on water, and both pieces of wood is subjected to the same force value in the stream, both pieces of wood float at equal force, no matter what the difference in size and weight is that comes into play. This would be caused by the difference in the drag resistance on the different surface area of the two wooden bodies. However, the difference in weight that comes into play, at this point is only due to the drag that the water experiences, not the actual weight.

If this were compared to, the findings of Galileo one would find that this is the precise method how matter moves towards the earth. Galileo made no mention of a force between the two bodies. Then Newton came along and published his mathematical findings, which is totally out of line with Galileo's findings and ever since then no person ever gave the actual findings of Galileo and Kepler a second thought. There can be no force such as gravity, electromagnetism, strong and weak forces, or nuclear energy. These so called forces are part of precisely the same value that is in relation and exist between space and time.

You, the reader may ask yourself why is there such importance in these findings as to know the correct way that gravity actually works? All calculations have already been made; mankind already possessed the knowledge, expertise and willpower to visit out of this world's Tara novas and even to colonize them.

All knowledge about physics and astrophysics has been studied, formulated and tested! I will reply to this question by asking two other questions.

A certain man drives his car down a lonely road. After a while, the car comes to a standstill. He knows that a car needs fuel to run. He takes a 25-liter can and walks 10 km to a filling station to buy fuel for the car. He carries the 25-liter fuel 10 km back to the car and puts it in. After trying for 10 minutes, the car still would not start. He takes the 25-liter can and walks 15 km to another fuel station to buy some more fuel. After walking all the way back and filling the tank with the petrol, the car still refuses to start. He then takes the 25-liter can and walks 25 km to yet another fuel station for fuel. After returning, he filled the tank yet again. Do you, as the reader think for one minute the car would start?

The American scientists spends 1 000 million dollars to pressurize four hydrogen atoms into the same space of one helium atom. The experiment seems unsuccessful because the helium unfolds back to the original four hydrogen atoms after four seconds. After that they use three 000 million dollars to pressurize four hydrogen atoms into the space that one helium atom occupies. After seven seconds the four hydrogen atoms depressurizes back to its original state. Now the American nuclear scientists use 10 000 million dollars to pressurize the four hydrogen atoms into the space of the one helium atom where the experiment lasts for 12 seconds. After 12 seconds the hydrogen atoms moves back to their original condition. Question 2 now is this: "Can you see any connection between the two examples I have put to you? "No pressure in the world can fuse four hydrogen atoms to one helium atom, even if Einstein said so! Those super brains and academics are blind with their own mathematical genius of mathematics and physics that they fail to see the most basic and elementary principals in science. What is tragic is that it does with the taxpayer's hard earned money!

Another rudimentary example is the so-called "falling star". The conventional theory that is propagated is that the dust speck burns to ashes because of the friction the particle has with the air it collides with in the atmosphere. This is the biggest mindless rubbish one can imagine. Just because the person that tells me this have six doctoral degrees, does not make it the gospel truth. Far from it... What actually happens is that when the grain of dust enters the earth's atmosphere, the time aspect changes and forces the dust particle into a different space occupation which then changes the heat value of the grain of dust. The dust grain is forced into such a smaller volume of space-time occupation that the heat it generates just burns it to ashes. Therefore the whole structure glows itself into nothing.

These "SUPER- EDUCATED- MASTER- OF- FACT" giants might use the most breathtaking mathematical formulae that can humanly be dreamt up, but if the principle, on which they have based their calculations on, is wrong, the whole exercise is fruitless.

In my book, I discard the "conventional" standpoints by following the unconventional principals, based, on the work of Galileo and Kepler. The basis for my theory has never before been propagated except for me, and therefore all arguments will be new and fresh.

I am no writer and even less a scientist. Furthermore, I do not pretend to be regarded as any of the above mentioned. I simply came to certain conclusions. When I shared my conclusions with other people, I had the very same reaction repeatedly. The reaction followed spontaneously without exception. Those who were prepared to listen to me knew even less about cosmology than myself and never understood a word I said. Those who know more than I do ignored me immediately when I said Newton and Einstein are wrong in their views about gravity.

Partly the writing of this book is to prove that there is little difference to science and the Bible, even if scientists and theologians try to make it their lifetime task to prove the other opposing side wrong. However, all their arguments are based on their individual agnostic belief in their self-righteousness and have nothing to do with the truth.

Free thought has always been a fact that all that is in influential positions proclaim to strive to but the minute the free thought differs from their concepts, it is the very first thing they crush as hard as they might. In this book, I strive to accomplish the very essence of free thought and try to lead the reader away from the brainwashing that all intellectuals try to force onto the public. Every person that is in any position is busy with their own sublimation that it renders them blind to the truth that is out there.

Man-In-Motion, Man-In-Mind, Man-In- Motive, Man Is Blind.

I wish to state here and now un-emphatically and categorically without any reservation of any kind there may or there may not be that it is totally against my personal religion as it is against the religion my Congregation upholds that I belong to in converting whom ever for whatever reason and never, never has this article or any other reference I may make any purpose in converting any body to my way of view about the spiritual in any way. It is not what you believe or not believe but how you live by your convictions and what ever your convictions may hold making you man or beast. My remarks about atheism are to show those practising atheism the foolishness of exclusion and to have an open mind because teachings of what ever nature has a positive and a negative connotation and the individual sets the standards. I am not intending or have any intentions in converting or changing any person's outlook on the spiritual side even in the slightest way imaginable. Me, living by my conviction to its full believe no bigger sin can there be than converting a person for that more than any other fact holds the highest epitome of sublimation there are. Secondly and in line with the first is my belief that the Bible has a base derived from ancient Egyptian teachings and I refer to that as I go along exchanging thought in this article. To this day we with all knowledge of splendour can still not understand how the Egyptians erected the colossal structures and the manner in which they did it. It is not realistic to consider a civilisation being that advanced in one area exclusively and with no other wisdom in other advances.

The hour in thinking has dawned where we humans must come to terms with the cosmos, with creation and with life. Mixing and matching was fine up to a point in the nineteen sixties where thinking about the cosmos and creation was a smart way to show superior intellect but being rite or being wrong was only a case of honour and pride that can hurt. Since the sixties life loss results from incorrect principles and maters got far more serious than it had since man had a first time ever look at the night sky with a conversation beginning from that. In another book as part of THE THESES I show briefly why I am of the opinion that man became human when he saw the funny silver dots in the night sky with a degree of admiration and recognition to the splendour of the unknown.

Now there is no longer only splendour in the unknown but a quest to find the unknown and gallant as they ever may be, it is fool heartiness to send brave men and women on search and not know what are there waiting on them. What dangers will establish the outcome of their fate? It is not philosophising for the pride that we should find evidence in distinction where distinction should be but crucial to man and to machine, machine because man's life interlinks with machine as it did at no other time in development history of man. A great philosopher I may not be and I will never be but thinking does not hurt and some advances may come from the weakest of thoughts when the thought try to unravel a thread running in thoughts. With this I too wish to connect in sharing thoughts of woven patterns as I see them and for what they ever may be worth.

I WISH TO TAKE NOTHING OUT OF THE UNIVERSE AND SHOW IT AS THE UNIVRSE IS: AN OVERLOWING CONTAINER FILLED TO THE BRIM WITHOUT THE SMALLEST FRACTION OF EMPTYNESS ANYWHERE.

As I stand on earth holding my first dimensional space-time displacement of our planet I can observe by using the second dimensional light source of the sun where my surroundings are made of three dimension atoms holding space-time in the forth dimension in time and space. I wish to move from point A to B and think in consideration about my planned action as my brain sends electric impulses to my muscles and that brings my muscles in mechanical motion. Arriving at point B I think to stop (not necessarily by thought or mental planning) and my brain stops sending electrical impulses to the muscle fibre concerned with the action in applying my body motion. At that point I come to a stop and my thoughts go to a rugby match played at Loftus a South African provincial rugby team's head quarters where a game is in progress at that specific moment. As the proverb goes: I am there in spirit and my spirit being at Loftus are then some 400 km. south of my body where my body is on my farm. The duration in time it took my thoughts to travel is beyond human measure

The very next instant my mind goes much farther back in time and space as I travel by mental motion to a game played in Christ Church New Zealand a week prior to the day in question. My thoughts took me not only half way around the world but out of the present time dimension. As I am standing in thought I see my next-door Neighbour (to us in South Africa your next door neighbour lives normally 30 km. from you) coming towards me and my thoughts return not only from New Zealand but also from a weak past to the present in the very current time span my body was occupying all the time while I find my body moving towards my neighbour without my actual realising of this moving motion. I use the second dimensional system in the wave to transmit sound by means of repositioning the three dimensional atoms between Old Neighbour and me in applying motion to the atoms between us by the fourth dimension of space in time to convey thought harboured in the fifth dimension to him being in the fourth dimension of space and time. He then uses the same system to convey a thought by massage to me. Please note that it is a thought from the fifth dimension that I convey with the applying of organs in the forth dimension through ordering electrons in the third dimension to control matter in my body placed in the fourth dimension of space in time. While my words carry towards him, he drops down like a log as a result of not fighting the first dimension called gravity. A thought from the fifth dimension prompts me to respond in the forth dimension by creating electrons in the third dimension while my body stands supported but restrained at the same time by gravity in the first dimension. The thought from the fifth dimension orders a response in all the other dimensions ordering my atoms in the third dimension to use the forth dimension of space in time to act by fighting gravity in the first dimension on what I see in light holding a place in the second dimension which is restraining my motion in the forth dimension.

My response comes from some emotion and as it is not part of my mental reasoning or thought pattern it is directly conveyed form the fifth dimension to respond. I feel his pulse and find no beat. His breathing stopped and my next action is to look into his eyes. There is a dullness in his eyes that was not present moments ago. Something went that was. My observation consists of thoughts relaying massages that is transmitted by my physical body in relation to my senses receiving and responding to electronic massage translations about Old Neighbour in the fact that he somehow relinquished all earthly responsibility and problems of an earthly nature to his next of kin that is now saddled and burdened with his last remains.

The heart shows has no beat indicated by the absence of a pulse. The longs lost all ability to provide oxygen for transmitting and burning food. His eyes became stony marbles. I communicated with him moments ago, but his ability to respond by hearing and speaking has gone absent. My thoughts and breathing, heartbeat and hearing are still there. I can speak to him but it is his ears that have gone deaf. I can squash air down his longs but he is unable to use it through his voice box. I can hit his chest with fury and support a heartbeat, but the blood will carry the oxygen but can no longer create heat to live.

The air I force down his throat still has the ability to produce sound because my shouting to him creates sound, but his ability to establish a method whereby he can create sound from the air I push down into his longs has gone away. His ears are still connected to his head and all the required tools equipping all previous aid that use to enable him are still there, all intact, but also gone forever.

All the biological organs needed for hearing and making sound is still unscathed in the right places not damaged in the least, but the use has gone. The electrons needed to translate whatever requirements enabling body function must still be in there somewhere, because I saw no discharge of any sorts flashing from his body.

Even by giving him electrons through an externally generated flow of electricity will not create any of the required but lost electron flow to generate life back in place. I may shock him till he hops around all over the place, but motion is denied for brain activity to function once more. His brain has gone empty, although it is full to the scull. His thoughts are no longer with us or with his body. It is no longer Old Neighbour lying there, but it is his remains. Even an atheist will tell you there is a difference in what is there on the ground and what were there moments before he dropped to the ground. No heat or electricity can revive what he lost. It is a body without life.

Minutes ago there was life to talk think and reason, discuss and argue, be angry or glad, but that, which now is lying on the earth has no more such ability. The source of energy giving life to the cadaver is no longer present. Science proved that energy cannot go lost but has to go from one form to another form.

Energy can never destroy or vanish but has to replace form or attachment. The body is there, and it is holding all the organs and the organs has still got the required heat to perform because Old Neighbour has not gone but a few minutes ago and in the South African sun bodies do not go cold through lack of heat because we are use to temperatures of forty degrees Celsius and more.

The cadaver has all the essence to sustain life and if life was electricity, then I should be able to recharge him by connecting leads somewhere and call an ambulance. But supplying any form of current at any voltage rate will not bring back life once it has gone. You can heat him with a blowtorch while shocking him with a cow prodder (and does those things unleash electricity!) it will revive him as much harm him or do him bad or good. He has become apathy in every sense.

His lifeless body will never carry his mind anywhere again because although the brains are still there holding all the mass it had when Old Neighbour was still with us, the brain is thoughtless and that has taken Old Neighbour away from us. Our dearly has departed although his physical remains stayed with us to rot if we do not take care of the cadaver and the sooner the better for everybody involved. We that are part of the living now have to move Old Neighbour because he no longer has such abilities. Minutes ago he still had the abilities but from him went energy. It must be energy that he lost because all other necessities in for filling such duties he still has (that is if you consider his body as Him) But his body cannot be him because his body is there part of the fourth dimension in space and time securing all his abilities to function as a human but that abilities has gone vacant. All the effort he may muster will not allow a wink.

The only visible something he lost that makes him less of a human being than he was this morning when he woke from a nights sleep is the energy of motion. His body with all the parts still hold dimensions in the first the second the third and the fourth dimension, but clearly it is the fifth dimension that has gone absent. The cadaver is still part of every dimension excluding the dimension of life and life then has to be a dimension above and beyond that of the fourth dimension in space and time. The cadaver is at present what we refer to as being lifeless and dead. The generator or power source or dynamo or what ever you may consider it to be but that dynamics providing energy in sustaining motion has gone away never to come back.

Whatever any person may try to do the machine that gave drive to motion is no longer able to provide motion. All the wonders that the human body possess in motorised function is no longer in motorised function although it should be if it was only a matter of replacing the lost energy by providing an electrical shock or some fuel of some sorts. Nevertheless no fuel can get that motor running again therefore the energy lost is not a replaceable kind as in the case of ordinary heat from fossil fuels, food or electricity. The machine of human motion has gone for good. Surprisingly the problem of energy and life becomes far apart when logic replaces Newtonian atheism and illogic.

Shove a ton of coal down his thought and it will do him or you no good at all. Roast him with electricity and see how far that will convince him to return to life. Push a gallon of pure glucose into his veins and see what the reaction will be. Energy is not merely energy and once again Newton got every thing very wrong. With Newton's incorrectness all the sheepish atheists go about an echo one can hear for miles around, but all the echo is only echo after all with no substantiating individual thought about and amongst the lot of them atheists. When energy is not used it becomes latent or so does science proclaim in any case.

One cannot ever consider a rock rolled up a hill having the same latent energy because the rock needed the same life that has gone absent from Old Neighbour to role uphill in the first case. When inspected closer life is the energy keeping the human body running as a motor and by distinguishing life the motor stops. Something went latent and not vanished. Life was part of the fourth dimension up to the moment it went latent. It shared time in the body and space with the body thus it was part of the matter of the body it no longer uses. Without doubt is the fact that life was the indisputable source of energy driving the body through the fourth dimension? Where the space-time sharing then ends, life cannot end because it is the functions of the body ending and not the energy driving the body while inside the body when it gave the body a function of movement the body had an ability that no other cluster of atoms enjoys in one construction in the universe. It gave the body the means to displace space-time not only by gravity and motion as all other structures have but it gave the body a means of changing the space the body occupies in time that the body occupies. No mountain can move a little in the morning to avoid the blistering sun and shift to another place at night to escape a blistering cold wind.

The human body including all life on earth can shift position as to suit the needs and requirements of space enjoyment in time duration. This means is very exceptional as nothing ells known to man in the cosmos can achieve such motion by pure will power. A plant may not be able to run to a better position but when in competition for sunlight it can try to outgrow its neighbouring plants and claim a larger share of the available heat the sun has to offer. That effort is completely out of the domain of any rock. A plant can grow its seed in such a way as to ensure distribution and gain advantage over the spread of its space it holds on earth as territory. There is no chance that a puddle of mud can run after water to keep wet. Life can manipulate the space-time it holds to its advantage in the sense of bettering its chances on survival as well as its species chances on relocation.

That is the overall advantage life holds and is not merely an energy that does some work in relation to the growth in the universe. Try and measure the time a mountain holds space and compare that to any one form that life holds measuring from birth to death being on this planet. Then after getting an unbelievable answer a person can appreciate that life is the energy and without life the structure becomes the equal to what a mountain is from the onset of the lava flow.

Life is the manipulation of space-time and the higher the degree of advance is, the more life can manipulate space-time. An aircraft flying may be as dead as the next mountain is, but through the aircraft, man as the ultimate form of life can manipulate space-time far outside the reach of lesser species. I do not wish to start comparing life as being advance or more advance so I leave my argument at that as far as life development goes for now. In the very beginning I stated that through the way the mind travels, it has to hold a higher position than what the body holds because I showed how easily I could travel around and even half way around the globe in no time at all. Sure I was not there in person, but my thoughts conveyed some understanding of what was happening on other places outside my range of vision.

The mind sets a norm that the body can follow or not follow but the body never sets a norm that the mind cannot follow. All sells even those holding life has an electron a neutron and a proton and very deep within the very deep within next to the truly unknown is a structure that holds position in relation to singularity. When a sell holds life it is different from a sell not holding life although when the sell not holding life still constitutes of the same composition it had when holding life, something changed, something is different.

A life-carrying sell not carrying life has gangrene a most deadly disease that kills as none other. A sell absorbing heat normally is showing growth whereas a sell in abnormal heat intake is cancerous and again is deadly. I can go on and on about this but it is apparent that as soon as life looses control over heat the stabilising factor or thing go abnormally wrong and such conditions can, may and will lead to the vacancy of life occupying the body more permanently. To understand the way I wish to direct the argument please allow me to indicate how I see the normal as we will find in the cosmos in life carrying and non-life carrying matter.

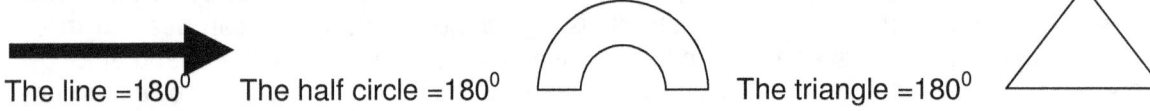

The line =180⁰   The half circle =180⁰   The triangle =180⁰

To begin to understand human nature we must first understand physics and to begin to understand physics we have to understand mathematics. There is a reason why mathematics is the way it is. Any student that has a Professor that the student worships for his or her mathematical brightness and think the Professor is in some way a mathematical God-like on earth by the cleverness the Professor has in interpreting mathematics. Go on test your Professor...Tel him or her but make sure you ask the Professor to tell you why would a straight line and a half circle and a triangle be the same value. How can a line be the same as a half circle and a triangle? When I came to the answer of this, I found hiding behind this answer is the manner in which the Universe formed. That is how the Universe began, but it took me a very long time to figure that out.

Since almost before serious recorded history dating became scientific principle mathematicians knew that the straight line holds 180^0 degrees matching the half circle as well as the triangle. But never have I read any definition about this phenomenon and how it comes about or what may cause such odd connection.

Heat occupying space has the cube that can apply r, as a straight line bringing about the cube with all its other names that may find attachment to specific form but nevertheless still remains only a six-sided cube with angles changing in some cases.

Creation is not there to serve life. Creation is not made for humans to enjoy and to multiply and conquer everything in the universe that we see. We with life had to adapt what is in the Universe to make the Universe work in terms of what we need. The Universe does not have to apply what we have because the Universe was in place long before life became a thought in terms of the Universe. What we find in physics will apply to life and in detail. Life is a designated factor of physics within the universe. The thought of life is not as simple as the simple minded Newtonian atheists think it is because the simple minded Newtonian atheist can only think in terms of resolving complicated issues by simplifying extremely complicated matters down to what the simpleton with his mathematics can formulate and prove. Let the Newtonian atheist prove by using physics we can think. I can and I do that in other books. I wish to keep this book simple and therefore I am not going to get complicated and going into that type of detail get quite complicated. Then if our simple-minded mathematician calling himself a Newtonian can't prove mathematically how a thought is created in the mind his physics come short on a gross scale. Don't begin with electrons flowing through the brain because going that direction you then presume that a generator generating electric current is one of the best thinkers we can find...and I will not be surprised to find the Newtonians think this way. They are of the opinion that computers can replace human life and that proves how inconsequential they argue about facts that are way beyond what they ever can understand. Life is partly the ability to generate electricity but that if what life can accomplish, not what life is. We have to look at physics and not Newtonian half-witted rubbish to see how the human physic works, and by establishing science we can see how life controls the body to act on behalf of the mind.

Let's take how physics work and see how that pans out.

In the very centre of the sphere the form of the sphere dictates that the shape will relinquish space as the line run from the outside towards the very centre. With this natural state of affairs the sphere are naturally inclined to dismiss all space that it can form in the form as the sphere holds space inside and the form will finally be without dimension. All that I attribute to the line shrinking by reducing actually takes pace in every sphere as the diameter reduces to the centre. In the centre where the radius line goes single the form relinquish the three dimensional form it has inside. Being without dimension in the very centre means that at a point in the extreme centre of all spheres there are a point that holds singularity because this point with no space has a mathematical position although it is invisible since there is no sides to such a point to give that point any dimensions.

The shape of the sphere is calculated by using the formula $4\Pi (r^3) / 3$. By reducing r to a point where r is r^0 singularity steps in because only the form remains as Π. Going even further we find that there then comes a point where Π goes singular Π^0. At that point absolute singularity is present but so is absolute gravity present at that point. When holding the strength of the shape of the sphere in mind as well as taking into account that all cosmos objects of importance is in the form of planets or stars and they are all in the form of a sphere, we therefore may contemplate that it is where gravity originate. We now only have to find the reason why gravity will hold a base in a space less ness as Einstein predicted. It is clear to be seen that gravity is in the centre of the sphere controlling from the centre everything that is outside the space less centre.

We can reason with confidence that gravity is the strongest where space is the least. We can further reason that it is gravity that is holding the sphere in true form and since the sphere allow gravity the best working opportunity, gravity can form the sphere in as strong a shape and form as the sphere seems to have. From every point on the surface of the sphere is where that point connects with the other side of the surface of the sphere by a line that runs through the space less ness of such a centre of the sphere. Such a line also connect by an angle of 180^0 as well as 90^0 to six other lines running from top to bottom, right to left, and back to front, where all join and cross in the centre of the sphere.

There are therefore six lines crossing and connecting by a centre from any given point on the surface of the sphere. Such points connects in total six surface points on each side of the sphere while they all support one another through the space less centre. In that absolute space less ness in the centre holding singularity we find gravity supporting and controlling all space within the sphere as well as space connected to the sphere. That is where gravity control and guide the space, which falls in the parameters as well as under the influence of the form of the sphere. In the gravity centre space goes singular meaning space becomes space less or flat.

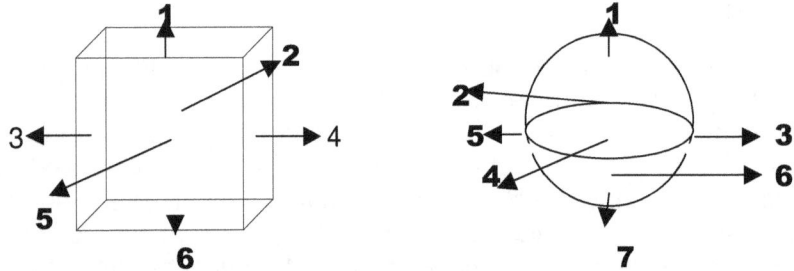

It is from the layout that the sphere uses as natural form that we are able to locate singularity. In the case of the sphere the material naturally reduces by measure of the radius becoming smaller to a point where the radius is r^0. At that point the line that will form the radius has gone single dimensional r^0 and that is equal to 1^0, which is singularity.

Also it is true that the entire form that is the sphere is controlled from a centre within the sphere. That centre holds the sphere in form and shape. Therefore the strong form is dictated from that space fewer centres where there is no space and no form left. The natural inclining is in the form of the sphere. It is part of the roundness that the overall shape of the sphere represents and this structural strength is carrying down to the very centre. Because the circle is forever reducing that reducing which is inherently part of the form of the sphere becomes a tool in distorting of space in the sphere and is eventually removing all forms of space from within the centre of the sphere.

The very centre ends up as having no space because of the reducing that continuous down to become the space less inner centre. The all roundness is the ingredient that forms the backbone of the absolute strength that the sphere has and that is the component that the sphere is so famous for. The form the sphere has allows the sphere to have a control that is coming from the centre deep inside the sphere where the space vanishes and being without space seems to keep the entire structure rigged. From the centre the sphere shape shows strength that the shape as tough as it is. How does it work in its most basic analyses?

This spot I just described and pointed out has no place in the Universe and yet it is what controls the Universe however, the simpleminded Newtonian got this spot down to **"Nothing"** albeit the most crucial point in the entire Universe.

Where space comes into contact with the sphere the cube loses one of the six dimensions it has to the more dominating seven dimension of the sphere whereby the seven dimension in equilibrium will dominate the six dimension loosely connected bringing about that the cube then has 5 sides to the seven of the cube. This means that in the cube the "bottom falls out" and without a "bottom" to support objects they fall to earth. Remember that a body "floats" in space, but at one specific point it starts to "fall" to the earth. That is gravity and it is a dimension change much more than any force.

The spinning of Π^0 around the centre Π^0 establishes Π and Π is what produces the form gravity has. Still it is the relation or relevancy there is between the centre Π^0 and the spinning Π^0 that gives status to the form that Π represents. In out Universe we are accustomed to and are familiar to the rules we want to place seven points holding singularity to the centre holding singularity in a relation of $7/10\ \Pi^6 / 6 = 112$. In that Universe everything less that a duplication ability to the value of 112 protons fit but only atoms to a maximum of 112 protons fit.

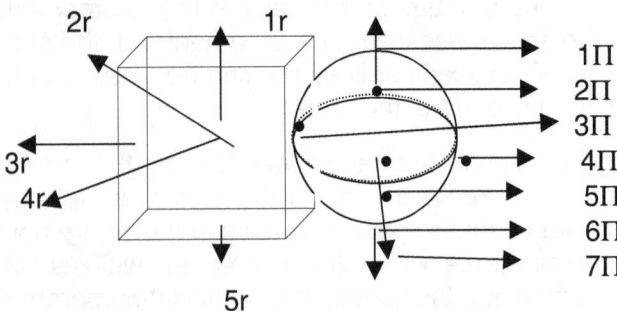

From such a point every other point will be opposing any other point not pointing in the direction to which the first point is pointing, whereby it extends the direction it holds. No matter what the point is or where the point leads, such a point holding a specific direction will be unique in the direction it is rotating because at that or any other specific point wherever, it will be directing not in the direction it spins but in the direction flowing from the centre point outwards.

All atoms are a minute form of a coming black Hole and viewed in the structure composition it is clear why I say this. On the outside there is heat trying to get inside the atom where the heat is needed. On the inside of the atom there is a need for heat and the inside is in constant regulation of the heat flow as to keep stability. In understanding the dynamics of physics we must understand the cosmos where the process begins and where the process ends.

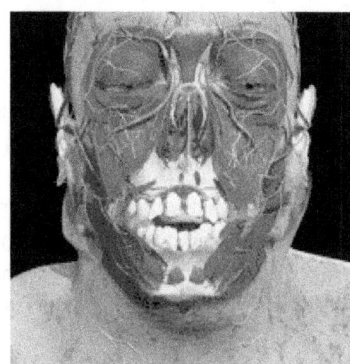

This is not the person but it is the tools life uses to enable the person to manipulate space-time in the interest of life. This is the body function but the body function adheres to the mind controlling the body. The body works by the mind establishing thought and the thought creates electricity and then with the electricity being a messenger, just as this computing machine uses electricity to convey what I have in mind through guiding electrons work is done firstly by telling the body to hit the keys on the keyboard and from there electrical signals send the commands to the computer to do whatever is necessary. The computer is not life but is only extending the ability of life. The body is not life but it is extending the ability of life. Life is in the thought and the though controls the body but is not part of the body.

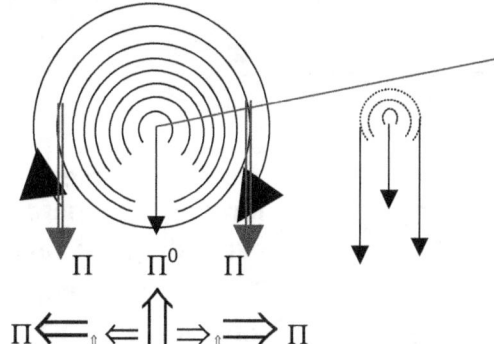

Pinpoint positioning of singularity Π^0 with Π positioning space to either side forming the border set by singularity.

The atom holds a very unique position in that it links three dimensions to a forth dimension and this part is where I came to understand Einstein's thinking but a with Newton I could not accept Einstein's thinking. Only bringing in religion could I get further about the formulation of singularity because in that I found what connected the universe whereas Einstein left space as space and tried to link time to space as an additional factor. That would be the same as not linking life to the body or exclude life from the body while trying to argue about life being part of the cosmos, as Newtonians seem to do.

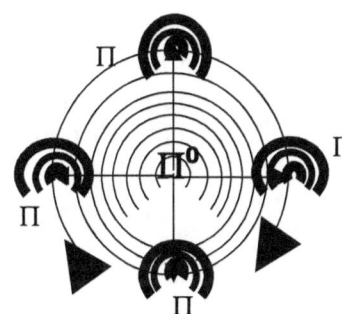

When singularity expanded for the first time ever and when heat parted from cold bringing about the Universe forming 1^0 to 1^1 from Π^0 to Π a relevancy was born and that relevancy grew into what we now have as a Universe

Gravity in the centre formed time Π^2 by dismissing while the four time positions started the cosmic trend of duplicating.

With every one of the four points taking form to the value of Π at a measure of $\Pi/2$ each brought about the Roche value of $\Pi^2/4$ in relation to the developing centre. One has to remember that the star of today takes on the characteristics of the form of that era.

If you go down any spinning object towards the centre there is a spot in the very centre that is not part of the cosmos because that spot holds no space nor does that spot represent the Universe in any way. Yet the spot is present and the spot controls all space surrounding the spot. If any reader wish to know more about the spot I suggest you must read other books where I go much deeper intro the complexity of this spot.

The centre changes motion to gravity by diverting the straight line to an immediate circle. By tracing the line back to where the circle is no more a straight line will uncover singularity plus one dimension. However, the entire centre forming singularity is still locatable within the Universe we have.

Reducing the radius r from all angles possible throughout the circle will bring about that all possible direction will eventually land on the very same spot with no more dividing possible. Yet zero cannot be a factor since the sides still hold value. In as much as holding all the value there can rise from such a spot. This is arriving at a point where more reducing will land the one side on the opposite side of the line but it will not bring about zero in the equation

Only by understanding this concept can we go on and can we see that life as a factor is not part of the cosmos as singularity in the spot is not part of the cosmos.

r /2 By dividing the radius r by the half of the value that then reduces r to a point where the left edge of the line reducing will be at the very same place the right hand edge of the line that is reducing will be. At one point the spots that formed the two ends of the line will be at the same spot where the original centre between the two points were. The two points would have moved evenly towards and in the direction of the centre by reducing all the space on both sides of the centre. By moving towards the centre they will at some point have to reach such a centre point notwithstanding cultural concepts favouring nothing to be filling that spot. Reaching that centre point will land all the sides on the same side and because of the presence of all possible sides such presence of all possible sides removes nothing out of any further possibility.

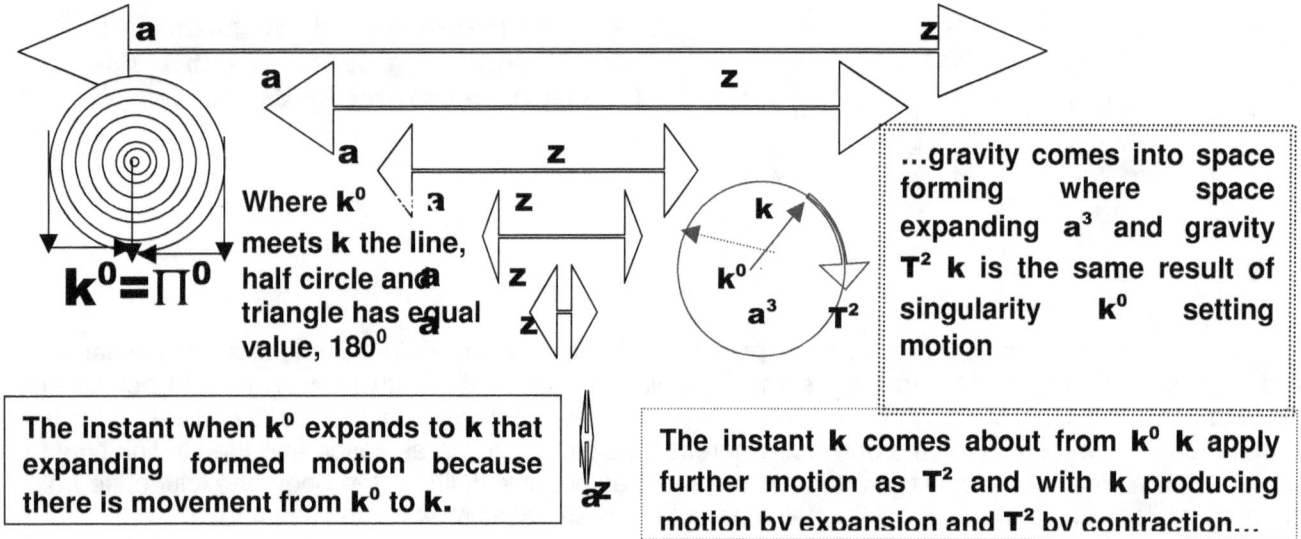

In the sketch I made, shows below each of the lines reducing there is a space left open

This occurs in all atoms through out the cosmos with no exception on the rule. But life-carrying atoms in carbon $_6$ commits life as an additional supplement to the atom as life can become absent from the atom leaving still in the normal range of a cosmic structure. In the past number of pages I brought reason to those of reason that there are more to the body supplemented by the presence of life than merely carbon fibre. It runs much deeper than physics can intrepid. As far as pure physics go, nothing changes when life goes absent and yet everything alters when life abandons the atom.

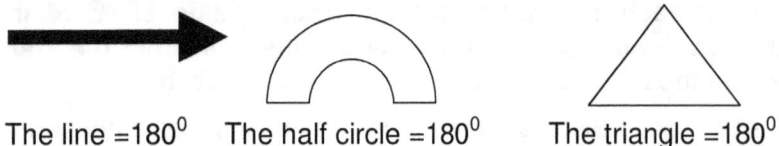

The line =180^0 The half circle =180^0 The triangle =180^0

I saw a very neatly outlined connection that the atom has in its position in the universe as it was the evidence of the smallest all connecting matter tying what is matter into a small container.

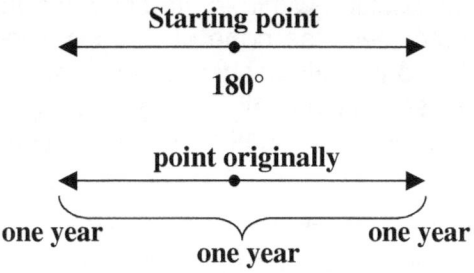

The question that nagged me for many years and on which I spent almost half a year just trying to solve the puzzle is how can light travel for one year in opposing directions and after travelling for one year in opposite directions be in two different points but was one year apart.

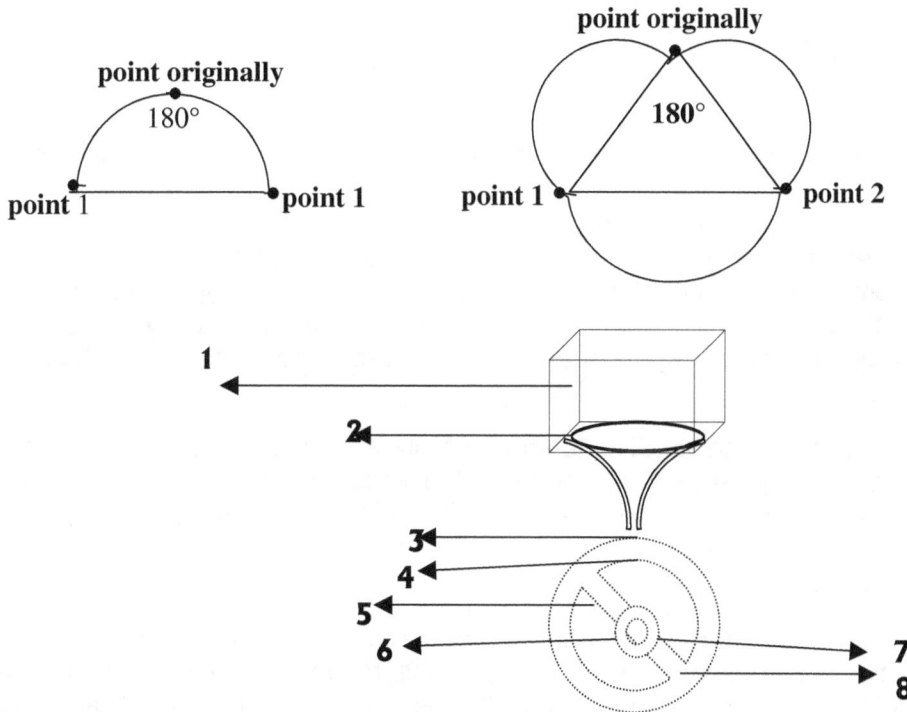

1 The square value of space-time in the fourth dimension holds a positional relevance to singularity by 10 points or places.

2 The space – time in the square of space loses the value of 10 by entering the atomic relevancy formula and become 3 sides to the cube.

3 The space – time holding space as the square loses the value of 3 by replacing 3 with Π thus going on down the line of the atomic relevancy formula and become Π sides to the circle as $\Pi^2\Pi$.

4 The atomic relevancy of space-time displacement changes once more to Π^2 where space goes flat in time.

5 The atomic relevancy of space-time displacement changes once more from the neutron square of Π^2 to the proton's double square $\Pi^2+\Pi^2$ as space unoccupied disappear and time forms the square by the square. The relevancy flowing from this figuration is so very important when archaeology presents facts and with this the archaeologists (may dare to say) became the blabbering fools they should not be as they are presenting serious science as a qualifying joke with the funnies they come up with in setting every one with thinking minds laughing.

I shall return with this argument in a later time when more facts relating to the argument are exchanged between us. For now we still have a cadaver on our hands to dispose of and quickly we must because of dire consequences that may follow if not done in urgency. Rotting corpses bring untold diseases as the great influenza epidemic of 1919 can support in evidence. Then why the danger about the corps all of a sudden if life within the corps have such minor importance as the Newtonian atheist wishes us to believe.

The molecular structure is still the same and so are the chemical composition within in the body and the mind. If it is chemicals making up man then, the man should be there because none of the chemicals went AWOL. If life is about the electricity that runs down the spine, our distinguished atheists should replenish life with some minor application of current as a means of stimulation. All the ingredients are present yet the manipulative nature of life making life so very exclusive to all other cosmic ingredients does no longer function.

Everything the fourth dimension can provide is still present within the body, yet that substance beyond the forth dimension, that ability we cannot detect but with a lot of intellectual thought, that ingredient Newtonians deny them of recognising as very special exclusive to life is not there any more. The ability manifesting as the energy or energy supplying has relinquished its role within body and mind. If it is only a matter of electrons generating pulse consistent to flowing from sector to sector in the providing of artificial current from an external source should supplement life's conducting of the work of the body. But shock the body as much as you may, the body functions disappeared with the disappearance of life from the body.

With the exit of life the vital electron distribution seized and the cadaver became just another particle containing what all substance other than life contains. The cadaver became just one more structure in the cosmos with the same ability as a rock or a mountain. The electrons conveying the massage may be replaceable but the sender and receiver and the decoding of the massage has no longer any function within the body. We may send some electrons into the body, charge the body with oxygen via a machine doing the pumping of the air to precise rhythm, as did life, we may stimulate the heart by nerve pulses artificially supplied. We may contract muscle by supply of electrons. But replacing life can never be one of our accomplishments once life has left.

All muscles including the heart and longs can be stimulated artificially and in doing it may prolong the writing of the death certificate, but having the person stand up straight once more or pronounce some wish to be for filled or just a simple effort as winking eyebrows when asked to do so is far beyond the abilities of the lifeless structure occupying space-time in the fourth dimension with the aid of all other dimensions excluding the fifth dimension.

Previously the body seemed to manipulate space-time with the utmost ease, walking wherever the mind chose to walk through unoccupied space-time, only adhering to the restraint compiled by the other dimensions and inflicted on life to startle the manipulating abilities. Now such abilities disappeared, vanished with life to some place we see as death. But life as energy has permanency that can never become denied and disappears. It cannot become washed away, wished away found to have disappeared for it is energy, a something of eternal power in being.

As energy has linking to eternity it has to come from somewhere as much as go somewhere it came from after it left. If it went latent even then it has to be stored in some place of gathering energy of life's nature. That storage facility is not part of the fourth dimension as the rest of the body is, so it will no longer be in range of the detectable, yet it must be somewhere with the linking energy holds to eternity. We also can tell that lesser forms of life will destruct the composition of the cadaver feeding and preying on it till what is left has no longer use in sustaining other forms of life as food. We learn from the esteemed and well respected Brainy Bunch the food we eat provide the energy we use. Sure that is very to the point and easy to swallow. But what uses the food in maintaining the energy. This is my problem with the SUPER-EDUCATED-WISE-OF-THE-WISE, as they will forever give information that has no substantiation but only scratches the surface and leads nowhere.

How does food give the energy and what uses the food for energy. If it is the protein filled body of the human flesh, then what makes the intake of food become meaningless after life departs. Should the food be energy then just feed the man and revive him. Food is all he needs to regain life and if such an effort fail to revive the dead then one should seek to find deeper meaning to what is obviously not obvious.

If life only connects to the fourth dimension of space in time through energy supply such as food may supply then the body is there to be nourished back to life. Also in the opening of this argument I showed how life travels time with no limits to boundaries in time. I did admit that such travelling only applied to the mind and not the body but since it is the mind that has gone absent and not the brain, then the travelling time by the mind must still be in affect as it was the mind that did the travelling and not the body. The body did the travelling of the time constraint but it is the same body that cannot even permit travelling to its grave because of the absent of the mind.

Time did not restrain the mind and if it did not restrain the mind time must have little control over the mind. Since the mind is no longer present in the body and time is still doing all the restraining on the body we may conclude with some extelegence that wherever the mind goes in storage time will hold no constraint over it. All the atoms that were in use by compounding a human body would one day again form some flesh of another being. It will ultimately not be human and it will obviously never form a body as one group of solid fleshy matter and it is logic that the compliment will become divided when forming a future body amongst millions of bodies all containing substances previously located in millions of different bodies to form millions of different life forms but it will become use full once more in the future.

With such a remark I do not say that the atoms scattered after destruction of the body of say a dog will form a dog in future. Such a presumption is madness. There will be dogs in future claiming some of the matter and there will be other life forms finding use for the ingredients that once constituted Old Neighbour but the principle is that the atoms that were in use will again find use because of the atoms eternal connection to time.

The carbon fibre is on earth placed for the use to carry and support life and as it did in the past, so it will do in the future. It is part of the eternal qualities of the atom to maintain space-time for the foreseeable future and the foreseeable future I suppose is the duration of time that the earth has to sustain life. To us humans such a concept of time on earth holds all the factors we connect to eternity because to us the earth and eternity is almost alike, but in thinking that never should one forget that in the realms of the cosmos the earth is but a flash in the pan, a wink of the eye and it is gone. But not straying that far into the future we are still measuring the chances of Old Neighbour becoming Old Neighbour once more for he was quite a likable chap and some people will miss him (I suppose).

What will the chances be of resurrecting Old Neighbour to his former self? Well left to the simplicity of the arguments held by the astonishingly brilliant Newtonian atheists we must consider it better than one hundred percent and according to their superb argumentative powers it is as good as done with the aid of a pump, an electric generator and a shovel to push food down his throat. But beware because when gauging by their record of previous successes notwithstanding the simplistic manner they go about denouncing the complexity of life, my prediction is also my advice: if you are a betting man do not bet on a positive outcome because you are about to lose money in such a bet! Your chances in winning will be as good as that of our atheists' wonderful arguments being correct.

Well now Old Neighbour is going to push daisies, or is he? Who is who and what does the daisy pushing? We know very well it is Old Neighbour that is going to push daisies because he is not with us and the "not being with us" part means gone away. Should one force he argument that his body, the compliment and assortment of DNA sells arranged to a specific order matching a pattern profile that belonged to Old Neighbour exclusively is Old Neighbour, or at least that is what our Newtonian-Bright-Boys insist on being the case. You, well any person, makes up a compliment of DNA sells and according to your sell arrangement whereby you become you and by pre-selecting sells and arranging them to a specific order where they form one totality and arrangement by assortment giving any person the prospect of life. Our distinguished atheist loses all other related arguments past this point in order to conclude what they believe to be correct. Considered in the utmost simplicity, yes, that is correct and as that alone it leaves no doubt.

Through this a rat cannot be a horse and a dog cannot be a lion

It is so very simple to understand when explained with such excruciating simplicity that even us living on the other side of the universe where the Lame-Brains belong are can accept without arguments because we are so scared of putting the least of effort into the simplest form of thinking giving the Brainy Bunch the scope of miles around to come up with the most idiotic answers they can dream up and we the Lame-Brains are too willing to accept as long as we are excluded from any form of thinking. Therefore we allow them so gracefully and with all dignity applied to both sides of the intellectual divide, to bullshit us to a stand still and make us feel great full that we were so privileged in accepting they're demising and diminishing mentality bestowed onto us. I say this from a stance where I am part of the idiots ranks and stand amongst my fellow mindless admiring those of the fortunate and privileged with they're wealth of thinking power because they achieved so many a splendid degree and are therefore the rich in thinking making me just one other poor beggar in thinking-power.

The human being as with all beings having life connected to the body structure they occupy which are the compliment of arranged sells and such an arrangement exclude my being a horse and it excludes the horses chances of being an ant.

With things that simple and sells going nowhere as they did not go anywhere in the dying of Old Neighbour why are they not functioning? What made them go on a permanent strike? Why can our Brainy atheists not once more persuade or force those sells on strike into accepting responsibility for their work responsibility because all of the world needs Old Neighbour around and the medical profession did not yet receive they're rightful chance to drain his money like a broken dam wall under the banner of keeping him alive for his family. Well at least until his medical aid runs out and his bank account has gone bust. With that simplicity being the case of life the atheist can at least replenish the life to the sells until everybody in line from the chemical manufacturers down to the cleaners washing the hospital floor had they're chance of becoming Old Neighbour's inheritors and not his wife and children. With Old Neighbour circumventing the money draining system it becomes totally unfair and what is more is why did the system spend so many billions in creating a net where they made Old Neighbour so scared of death and disease he will gladly part with all his money as long as the system gets the chance to help him cheat death (should you not believe me look at the cancer and other advertisements and think for yourself who is paying for the brain washing). Why not only tell those with cancer to do the fighting? Why charge everybody up to come out with they're six shooters a blazing in spraying lead. Who is paying for such advertisements and who receives the benefits of such advertisements all done under the banner of securing a longer life for every body.

I am a diabetic and a smoker that does no exercise of any kind but to get out of bed in the morning. I was medically ordered on so many occasions to quit my smoking, and I not sooner did that then they started feeding me anti depressing pills and anti anxiety pills and sleeping pills and stimulants to fight the sleeping-during-the-day-attacks and the.... The list goes on almost indefinite. Once I pick up my smoking habit again I suddenly do not need one of their pills to keep me "normal". While my smoke may kill me the exhaust fumes of the cars in use which pours the most deadly of gasses into the atmosphere being carbon monoxide is not maybe but definitely not only killing me but also nature in every aspect. Carbon dioxide is a natural element on earth while carbon monoxide is a chemical acid eating or more accurately said devouring even the likes of statues chiselled from granite rock as well as things manufacture in iron to rust. That aspect no one ever comes to mention BUT SMOKING is the killer destroying life by the billions! The doctors are reluctant to allow the tobacco industry to kill you because that will deny them the chance of killing you chemically and making the profit themselves either through driving their luxurious cars or stuff they prescribe and you can only purchase through chemists. I was a smoker but I stopped smoking because I couldn't afford it since I now am jobless. But I will say it is the most rewarding experience those simpleminded fools took away from me. Smoking is not addictive. Smoking is not a drug because you don't get intoxicated from smoking. I know what I am talking about because I have been through the mill and I still get hungry for a smoke. Yes I said hungry. Smoking is a system of feeding, taking in food. It is a way of taking in carbon as food through your longs just like plants take in carbon in dioxide and in monoxide through the leaves from which they grow but in the case of human consumption it is a system of feeding a person by long intake. That is why smokers that stop smoking are permanently hungry and it peckish for the rest of such a person's life. If you stop smoking then you are a quitter quitting food. Your smoking did you no harm. I am prepared to challenge any medical doctor on TV on this. I dare any doctor to organize a live debate between that doctor and me to prove smoking is bad. Smoking is a way the body feeds and not a drunk you get unrecognisably drunk on in the way you get with alcohol and other narcotic mind altering drugs and if doctors have done so little research on the use of tobacco how could they be experts on the use of tobacco. It is again diverting the money to the pharmaceutical companies because no one speaks out about tranquilliser abuse but tobacco is from Satan. To kill people on medical grounds by prescribing though controlling and minds abusing tranquillisers is much preferable especially while doctors grab some of the money and pharmaceutical companies get the rest is very preferable to people smoking and dying cheaply with a clear mind. Drug addicted patients always come back to doctors form more "medicine" while tobacco "HUNGRY" patients buy their tobacco from the corner café where the medical industry don't control the profits It is the place where the money goes that becomes this issue of keeping the notion healthy and to up the tobacco buy increasing taxes help to equalise the price difference between narcotics and tobacco. It is correct that I don't want my sons rto smoke but then it must be for the true reasons and not to feed them pills instead.

So the doctors scare the daylights out of you about death (which you will never escape in any case since you are born to die) to feed you pills (so chemically poisonous they can only sell on prescription as they are sure killers and the most dangerous available) and the system is creating another slave by making another fool so brainwashed he truly believes he will eventually cheat death! And Old Neighbour had the audacity to escape the loose of the system and die still with money in his bank account! Such a dead is outrageous and cannot be tolerated. Believe me if the medical profession got to Old Neighbour before I did, in his dying effort they would have kept him alive for another few hundred thousand reasons, reasons you keep in a bank vault and pester his wife and children with guilt so that they part with the money so willingly they will even pay anybody to advise them to part with the money. (If that is not why you pay the doctors treating a man that is ninety nine percent dead already then why are you paying him in any case).

You the reader may not see it but this is all resulting from atheism and a system promoting atheism and is an all out war world wide making every breathing person on earth a slave to milk until death does its part. Convincing people about the simplicity of life will encourage them to fork out money to be kept alive so that the slave will gladly allow more milking.

Slavery so I am told and so I do believe from the bottom of my heart is wrong. But the slaves did not have it so bad in the days of the Greek and Roman Empires. They were much better off than us the slaves of the current World Order. Slaves under the Roman law were fed clothed and accommodated on the Master's account. The law was that the owner of a slave had to feed him and provide accommodation for his slave. Then the slave had the right to ten percent of the income the owner generated from the services of such a slave while the slave had the chance (if he could) to buy is freedom Slaves in the current World Empire of the Hoggenheimers an Mammonites enjoy the pleasantness of a just system where the system does away with the need to bay slaves, the slaves join the system or die. Furthermore they make the slaves pay from their wages for food logging transport and clothes while the Hoggenheimers do not even pay them ten percent of what the Mammonites earn from their services. Under modern law, modern slaves are worst off than slaves two thousand years ago! And to top this Old Neighbour had the audacity to escape the slavery without even paying his last bid for his freedom. How criminal can a man become in such a manner of escaping what was rightfully his dues to pay. With all the simplicity about life and the promoting of escaping death why can the atheist not bring Old Neighbour back to do his last part and fill the already overflowing money caskets of the Hoggenheimers and Mammonites.

There is this wife of one certain pop star a member of a very well known group in the sixties and one of the four members in this very well known group. This wife of the famous pop star made millions on promoting the abandoning of the use of animal meat as food. She told about her and her husband having lamb chops one afternoon while some other lambs were grazing nearby. As she saw the lambs with her mouth stuffed with their friends she then and there got thinking about cruelty and the humane aspect about eating lamb in the presence of lamb nearby. She was devouring the flesh of sheep that was killed for the purpose of feeding the human population and that gave her the idea to make millions on that thought and selling humanity in the process.

For some sake of sanity let us scrutinise the situation and for once go just a shade deeper than just being prognostic in our conclusions. The lamb has carbon$_{12}$ as a mixture of forming the composition that we named protein. What will be that different from eating grain and eating flesh? Both holds life and both holds death after life. The grain is an infant that did not yet start life whereas the sheep is an adult whose life was cut short during life. Both faced death before they received the honour of completing their sole purpose on earth and that will be to feed man. She went on a campaign promoting vegetarian dishes that did not even contain fat as protein but included the biggest variety of plants imaginable.

While on the tour of promoting the eating of plants (and selling her book to millions of other fools that run on emotions they do not understand, cannot control and where such emotions totally outsmart their thinking capacity) she stopped far short of explaining why she would consider plants lesser life than what sheep are. Can the reason be that the sheep think nothing of devouring the grain and she allows the sheep to do the thinking on her behalf? Is it because grain does not run around when "chased to become grained" for food. Or could it be that the price of the book and her selling power of the content of her book allowed her to sucker some idiots (and I believe the number of idiots caught in the scam runs into millions) tinted her perceptivity so very slightly in favour of the consuming of plants that cannot make any sound or request any human emotion by running and shouting in protest trying to escape the butchers knife or in the case of plants the sickle.

In the case where we consume fruit as food the fruit we eat is food still alive in the same manner as does lions starting to eat a buffalo that is still standing on all four legs. If someone somewhere came about the promoting of eating animals while they are still alive I would surely go on the same protesting crusade as she did in her bit to fight the food supply in the form of meat. We now are faced with the same cynical questions our friend Old Neighbour left us with. Is it his corpse lying there or is it he lying there. If it is he then I have to admit that we are eating lamb. But if it is his corpse then we are not eating lamb but merely the remains of what was lamb once. I am not wasting any space on arguments about killing to eat because kill to eat we do because we have to do it. There are no other options open to us but to kill or to become killed through starvation.

The bottom line underwriting everything said about what form of food we should or should not eat is the human capability of becoming completely self- absorbed in sublimation? We think we know exactly how God created all around because we know exactly God did not create that which is all around. Therefore it is our claim to right that we may take the place of God and decide what should count where and what is food and what is not food, but for god sake keep it simple otherwise we will not understand why we may think ourselves as gods. As long as science portrait matters simple excluding the not very popular complications of thinking every thing thought through decisively we may find that being god can be a very pleasant way of living and un-complicated. If we do not complicate everything we may even think of ourselves as very clever gods without the excruciating effort of being clever gods. Just go about and visualise our brilliance in reason and tell ourselves how kind-hearted and humane we are without any deep philosophising about truth and matters of complexity.

If you are in support of the humane aspect then consider that the deed of eating fruit will be far worse when eating the unborn and defenceless or robbing the unborn defenceless seedling of nourishment so dearly and lovingly accumulated through severe hardship and unquestionable devotion in loving labour by a caring mother than a developed specimen of any specie. Remember that when eating the unborn fruit or the food meant to feed the unborn seed will be denying life the chance to be and that is very unfair! At least the meat eaters gave the sheep the feeling of being sheepish before removing the feeling permanently but in the case of fruit eaters the fruit never had a chance of feeling fruity. I should add that to my mind humanists are the worst practising sublimation because atheists deny the fact of God but humanists are in criticising of Gods way in creating the balance we know as the echo chain. Humanists are constantly trying to show all that are willing to listen to their senseless rambling how much better a job they have in mind for all life on earth than that which God established up to now through giving man reason to think with a mind and not an emotion and forgetting that the methods applied got civilization in such a tested and tried state as those methods did but still they whish to change it because they think they know so much better.

If our pop-star-wife did not have the pop star fame and all the pop press in support and with the wealth of food supply around how far would she come with the cheap mentality and the thoughtless advocating of the shameless theatrics to support her promotion of self enriching by selling books. When any nation is in total starvation as the Germans were just months after W.W 2 I wonder how many hungry men and women with children crying starved to almost death would applaud her madness as greatness. She got through because there were abundant and not because she had sensibility in her quest to make money.

She could manipulate others while the others were swamped in good times and rolling in the fat of fortune fed to burst while gloating about how their humane hearts bleed for the helpless sheep all over the world knowing very well none of them ever had to skip one meal because of want. They never had to live through one night of agony where their children were crying because the children were too hungry to fall asleep. When thinking about such conditions their gloating in self-praise is quite sickening. From me and mine to you and yours I am telling you this shocker: the total destruction of mankind may only be as far away as the swing of a telescope, and the announcement of a funny little dot that seems to grow as it is heading our way but more about this later on. She is merely one of millions making senselessly money without thought of dangers larking

This I say because nature tells the truth about man and the way mankind evolved. All predators on the hunt have eyes pointing foreword to find the maximum advantage in three-dimensional sight. By focussing in hundred present accuracy the predator can pin point the kill and act swiftly and abruptly minimising the chances of the hunted from escaping such an attack. On the other hand when looking at animals that is mainly vegetarian we find their eyes on the side of the head to secure maximum vigilance and response to such an attack. When looking at the human face we find the eyes even more in the centre of the head than in the case of an eagle, famous for his hunting skills and such a small but obvious clue demolishes the entire bleeding hearts cry for passion.

All animals dependent on meat for food sustaining have eyes pointing to the front the very place humans find their eyes to be. The road our humane idiots genes followed took them through a ancestral path with a long range of meat eaters that brought the gene carrier to what he or she is in the modern age, but being smart they make themselves the fools they are. If we humans were fruit eaters only and had no natural inclination for meat then our eyes would be next to our ears instead of being rite above our noses in the centre of our faces. Those placing meat eating in so many disputes should then also change their eyes position to the side of the head and denounce their ancestral trace of meat eating.

Man has a vision allowing 180^0 sight where as animals born to be the prey has a sight range of 360^0 and none of the humane intellectuals ever came that far in reasoning. With such direct and undeniable evidence about our eating of meat, how on earth can those shouting no in support of meat eating show their faces around as intellectual beings. This also goes to some religions denouncing the eating of meat but as long as they keep their religion to themselves without trying to convert me to such rubbish they can believe what they want and exclude me. I say this because on occasions I got into debates with such people that wanted to push their religious ideas down my thought about some Indian god living in India and you send him money with a prayer where he then fixes your problems rite across the ocean providing you do not eat meat because of his say so.

All species on earth are what their history made them. They are moreover the road they followed down to where the specie currently is than what they are at present because when circumstances change the genes with idle qualities will arouse the complexity of the specie and old habits that saved the specie from extinction in the past can come to the front and again save the members from extinction. The Sudanese can survive by eating leaves from trees until the rains arrive to bring about new harvests (although the rains never return permanently). On the other hand the impala cannot start eating lions to keep alive until new vegetation grows again. But even the harmless impala is not that harmless to grass, as grass has to grow meters every year in order to sustain the impala's nourishing needs and at the same time secure the survival of the grass as a specimen of life on earth.

Man too, if need be, can survive on grass and that puts man on top of the evolution ladder and not their misguided impulses in correcting the ways we developed. It is great to play god when God gives in abundance. It is great to play god when God brought your specie this far. But try and play god when God closed the clouds bringing rain and hunger with facing starvation. Then the mind fills only with thoughts silencing the hunger pains and the obsession comes as the hungry wish to fill the stomach with food without filling the mind with cheap sentimentality. How brave will the Super humane then be I wonder. Being humane is closing life to a very single minded approach and in this the massage of the atheists simplistic views about life ring out loud.

All this may be fair but there is another side as all things in creation stand in relevancy In my quest to find answers one question I could never find an answer for is why do the world not import the Sudanese to Britain America and Australia instead of exporting the food they donate to Sudan. Sudan has become a country that will never again support such massive numbers of people and the food will forever be needed. The growing desert claimed the country and it cannot sustain human populations. Declare Sudan uninhabitable and take the people to the countries donating the food. It will be much cheaper to feed them in the countries I have mentioned and at the same time it will please the bleeding hearts, give the Mammonites more slaves and the Mammonists more slaves to drive while not hurting the unemployed one bit for jobless they are because jobless they wish to be. Change the relevancy in the equation and take the people for once to the food and not the food on a yearly basis to the people.

Before every Anglo American starts demanding my immediate and successful castration without precondition let me add why I say what I say. By feeding the population the bleeding hearts are getting their wish but in it they are sadistic and devilish cruel. Before any aid can be requested a disaster must be in progress. Being a disaster in progress means millions are suffering. There has to be an enormous lack of food supply to wake the caller. Babies go hungry mothers weep fathers run off because they wish to find food and disappear in the process. Suffering runs deep as it runs wide and no aid can prevent that as no precautionary measures will ever be good enough. By helping once you are spreading the suffering to last longer and with more pain next time around and we all know there is a next time around because of climate changes going on. Feeling good about your self because of proving once again your good nature, your blessing heart and empathy by the giving aid helps no one because of the coming of the next time.

The simple truth is that those in power and those with influence give nothing as much as care for the helped victims. The philanthropist collect money on behalf of the Hoggenheimers from the bleeding hearts while the philanthropist encourage the bleeding hearts to donate in giving for the simple reason the philanthropist share in the spoils of the unselfish act. The Mammonites bay the food as cheap they can in names of companies they own with as little money possible from stocks the donating parties would trash in any case because of poor quality, then bay the food from their private companies with huge profits going to the private company because the selling party is also baying on behalf of the relief organisations with the money the bleeding hearts donated not because out of true sympathy, but the bleeding heart wish to kill the guilt they feel as they know they have it splendid and therefore they need to prove to all but mostly to themselves they're godlike generosity by donation.

The Hoggenheimers take their cut with excessive profits by distributing the bleeding hearts', which the philanthropist collected so unselfishly as proving it by taking their fare share of the profits going around giving the money to the Mammonites, which are baying on behalf of the bleeding hearts from their firms as they sell the stocks they previously bought for next to nothing with excessive profits. At this point the Hoggenheimers bring in the Mammonists to do some slave driving as the spoils has to be sent across the world. In this heart braking act of generosity some more unbelievable profits go the way of the Hoggenheimers and the Mammonites because the firms involved just so happens to belong to a shared venture between the Hoggenheimers and the Mammonites and by some more overwhelming generosity they share crumbs with the Mammonists doing the slave driving.

Now you tell me who is unhappy while all this good heartedness goes around and is there any blame to be where the rich becomes richer as that is no one's fault. If the bleeding hearts were serious about their conviction in generosity they would not bay some guilt relief. If the other parties were serious about their convictions they too would try to find a permanent solution but then there will be less profits to gain. The bleeding hearts are quite satisfied that big planes are used to transport and distribute the food but they know very well that that is the most expensive means of transport and someone somewhere is changing very unselfishly a dime spent to a dollar wasted. In this way the relevancy is getting the rich richer, by giving the guilty guilt relieve and helping the luckless to another round of heart ship in hunger.

Change the relevancy around if the act is in pure kindness and brother love. Take the luckless out of the equation of desperation in cycles by removing them from the problem. In that there are some more relevancies involved. If the bleeding heart were serious they would never mind bringing the luckless to share in their abundance. The other part of the option is to let nature take its toll rectify in natures way and be done with it but then the profit issue stands to lose millions of reasons why neither option is an option.

The relevancy will lead to a cheaper solution although more expensive the first time around. Everything is about relevancies. On the one side of the relevancy is the earth became unsustainable to carry a human burden in that part of the world and on the other side of the relevancy is, the western countries have food to donate in tons through baying and selling agents, (and I shall gladly eat my farm if the politicians were not sharing in the bounty of tax money donated in generosity).

The one side of the relevancy is the Sudanese will never be self supporting because on the other side of the relevancy is in the long run a desert means drought and water will never again be abundant. The only solution to the equation in solving the problem is by changing all aspect around in the relevancy and through that finds a permanent solution to an unsolvable problem that will forever remain unresolved until the relevancy changes to finding an answer instead of avoiding a solution.

If the cosmos can tell us one thing it is that changing the relevancy brings about solutions. By creating the Big Bang it solved a problem of overcrowding as we have in Sudan and by creating space as we should in Sudan the cosmos separated matter from space as it is still doing with the Hubble constant proving that space is on the increase. But if space is on the increase and all is about relevancies something else must be in decline on the other side of the relevancy to find equilibrium between the problem and the solution.

The cosmos brought in space on the one side and matter on the other side and between matter and the factor of space growing must be some sort of problem solving. If we wish to find the answers to the cosmic mysteries it should be the most obvious starting point because there is one side of the relevancy known to man and then looking on the opposite side of the relevancy must be the solution. Where one thing is growing something else must then be in declining and in that comes the answer of the relevancy that I share with the introduction of my theory on matter holding space in time.

Most prominent in all relevancies there are must be the atom, the one little container giving matter character and different uses in the universe and by adding or removing one small part it changes in character, as Doctor Jackal and Mister Hyde never could. Every one knows what is in the container but what is the container in? If someone ever gave that thought the light of day I have missed it.

However we also can learn much in thinking by placing a reference we see with life back to nature. If we cab see light by using a photon we pick from space, then we could only do that if we copy what happens in nature in using the atom.

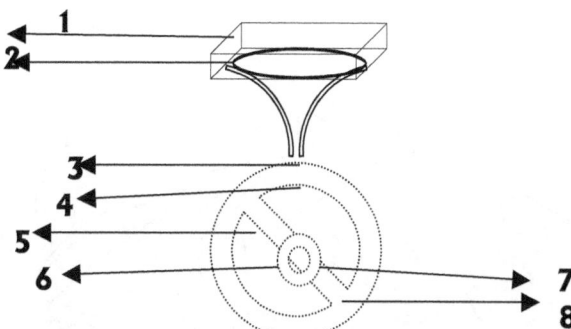

Referring to relevancies means one may exclude only nothing from attachment and as such that put whatever there are to consider in a relevancy to whatever there are to consider. The universe is one giant spinning machine holding everything in tune and aligned with all ells. Everything become relevant as all in the cosmos divert from singularity by the line of singularly applying sides to singularity forming the triangle in singularity bordering the half circle to form a position where Π will become r^2 and lead on as a value of C. In singularity the value of space held by lines diverting from singularity forms the space value of $\Pi\Pi\Pi$ which is so close to eternity the time value applying exceed the dome compliment of Π^2 matched only by the half of the square being the triangle at Π^3

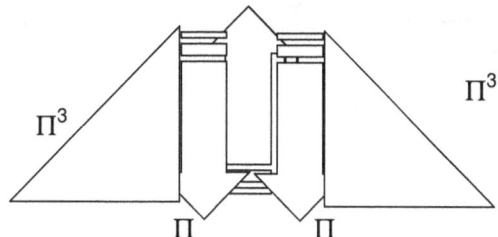

Diverting from singularity as extending Π time claims space from singularity comes about the double proton in space $\Pi^2 + \Pi^2$

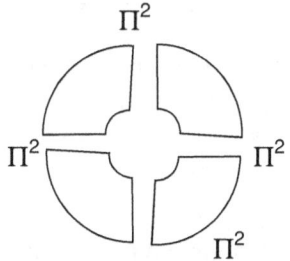 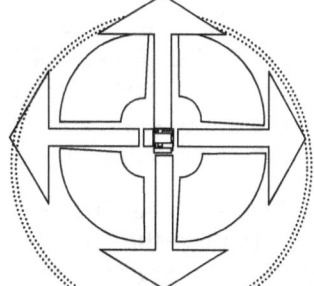

From the square of Π^2 the dome as a half circle by four places the line implicating the line by four to place the triangle relation by double half a square.

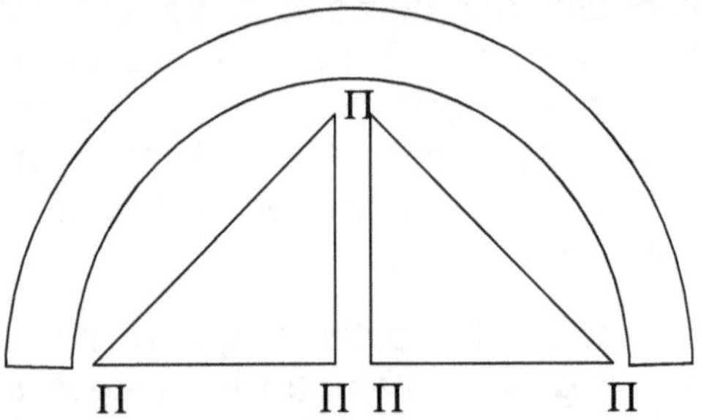

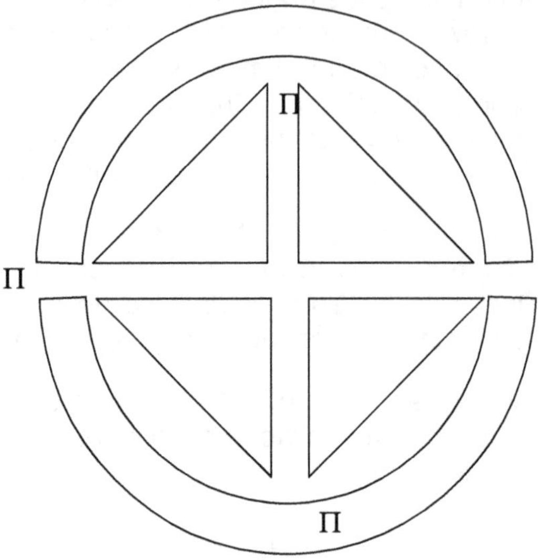

The total value of the four in the circle bringing about the space of the triangle reflected onto the time the circle holds committing the straight line to a value becomes the fourth dimension of space-time.

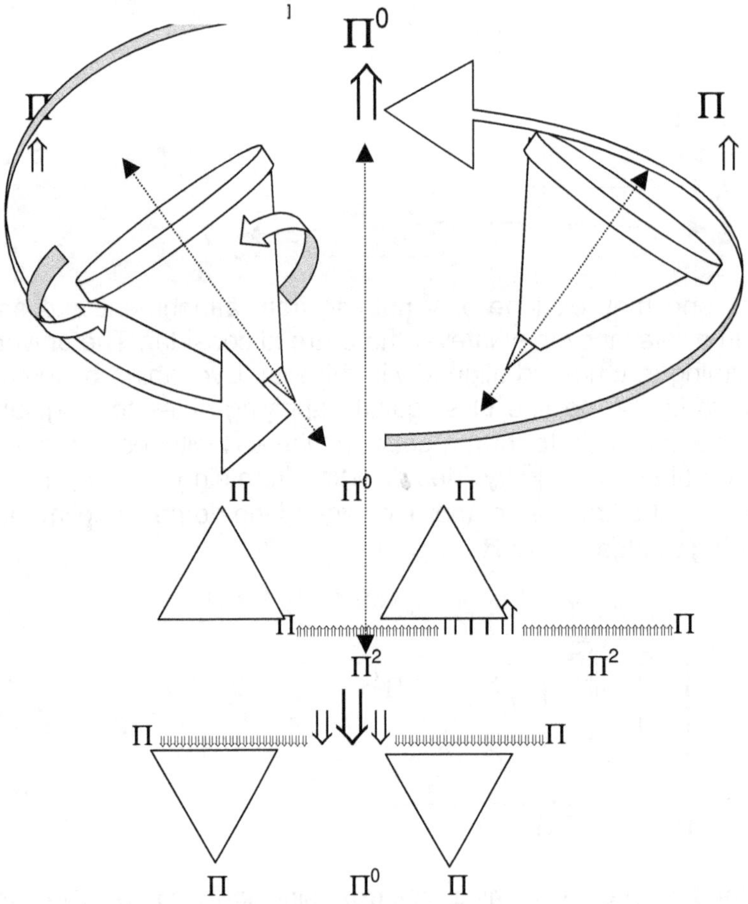

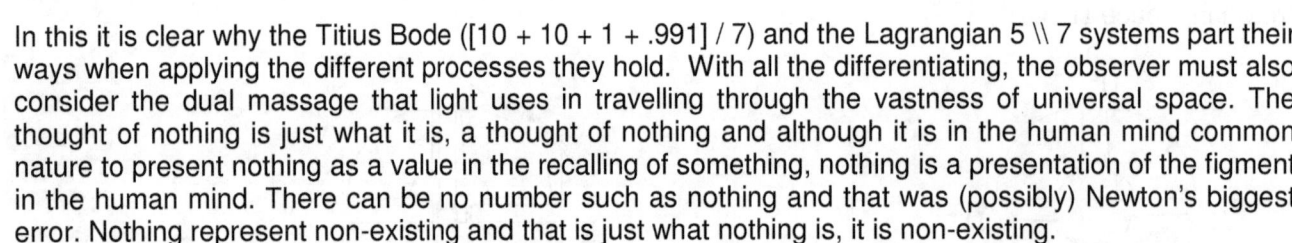

In this it is clear why the Titius Bode ([10 + 10 + 1 + .991] / 7) and the Lagrangian 5 \\ 7 systems part their ways when applying the different processes they hold. With all the differentiating, the observer must also consider the dual massage that light uses in travelling through the vastness of universal space. The thought of nothing is just what it is, a thought of nothing and although it is in the human mind common nature to present nothing as a value in the recalling of something, nothing is a presentation of the figment in the human mind. There can be no number such as nothing and that was (possibly) Newton's biggest error. Nothing represent non-existing and that is just what nothing is, it is non-existing.

In order to prove my point I wish to ask the reader to define the shortest line there can theoretically be. If he should answer anything but that the shortest line will be at a point where the beginning and is the very same spot he will be wrong. The shortest line that can ever be anywhere must have a start and finish holding the exact same spot. The line will be humanly impossible to create but we humans are capable of very little.

Stars can and stars do overheat, sometimes and the polar regions where the Titius Bode matter to matter applies holding the square of space (10) in a double relation to the square of time (7 + 7)

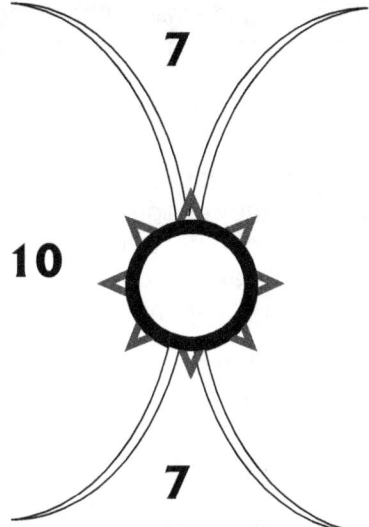

Other times they overheat in the Titius Bode principle holding the square of space (10) in relation to matter (7)

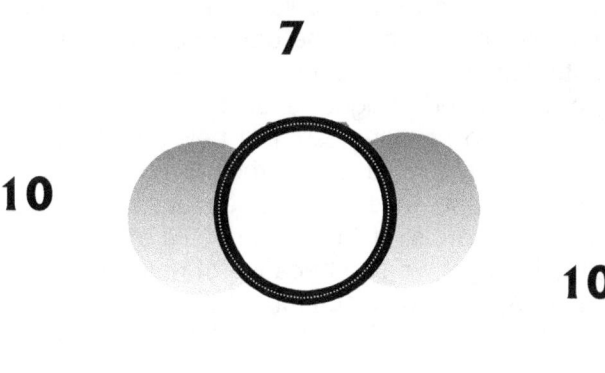

This comes about through the overheating of singularity (7 + 7)/10 (top) or layer overheating 10 / 7 (bottom)

It is no longer an issue that stars overheat but the issue shifted to the question of why stars overheat. Applying the Titius Bode laws (shown above), the Roche and the Lagrangian principles correctly, I can prove that :

2) There is no gravity and therefore GRAVITY DOES NOT EXIST.

3) With no gravity it stands to reason that I also maintain that both NEWTON AND EINSTEIN ARE wrong about their views on cosmology.

4) With no gravity NEWTONIAN VIEWS ON THE WAY as how CREATION CAME ABOUT IS HALF CORRECT BUT HALF IS INCORRECT.

5) And shocking as it at first may seem but true enough is the fact that THE BIBLE IS CORRECT ABOUT CREATION.

Could any one ever make claims as I do and at the same time claiming sanity as being one of my virtues. Well I shall try to explain how humans see the Universe in total contradiction to how animals see the universe and to top the lot how atheist and animals see the universe alike.

Animals use eyes but we humans have more too see with when using what we have by using our minds to see with. We should see the universe with the light of understanding shining in our minds. When one look at the night sky one see darkness with little specks on light. Why would anybody see darkness because darkness has no light. Yet we see the darkness. The darkness should be invisible if we are seeing light because the one contradict the other. If the night sky was black then black is what we should see but then again black is the absence of colour and colour is the visibility of light.

We see the darkness because it has light it is withholding from us and while withholding the light from us we see the withholding as darkness and the darkness we do not see because we are not suppose to see darkness. That makes the darkness we see not darkness to be but it is in fact light we can see as darkness. On the other hand there is the brightly lit dots we can see because the light shining as dots are darkness as the darkness are stars giving us the light they are not withholding. As they are not withholding the light but pouring it into the vast container of light the stars then become the darkness we cannot see because they give us they're light and by giving us they're light they then have no light to have. That means by giving us they're light they withhold they're darkness from us and that makes the stars filled with darkness. That means what we see as light is light and what we see as darkness is also light instead of what we ought to see as darkness because that we cannot not see is the darkness.

Therefore when God gave the command "Let there be light" it was the command "Let the universe begin" because the universe we see is light we do not see and the stars we see is the darkness we do not see. When one is looking at the darkness as an animal you will be seeing the darkness because the mind you use is that of the animal. Then, yes you may be an atheist because all animals are atheists. I have never herd of one bleeding heart or philanthropist of whatever kind convert animals to any religion there is available. If you are an atheist and you see the night sky as darkness that would mean that is what you see as an animal; a darkness that if you had the sense of a man you ought to know that it is impossible to see darkness therefore it must be light. On the other side of the relevancy you also ought to know that by seeing the light the star is giving away the light and when giving away the light it has to hold all the darkness it is claiming for own use because only by claiming back the darkness can it give the universe the light as it does otherwise it would give its darkness by withholding the light. The darkness is singularity uncommitted to specifics, spinning at the speed of light never pointing in one direction long enough to shine as light but shining long enough not to be darkness. But being light we can see it but because it is in random spin, spinning at the same speed we use electrons to convey massages when translating information we cannot see it. The light then becomes darkness because it is extracting all the light through the one singularity line uncommitted not energising it because of the absence of a replacing source converting new light. In contrast to that is the light we see because of reasons forthcoming from not being able to see darkness we can see it as energised uncommitted singularity with the aid of a sustaining in singularity from a committed form replenishing the uncommitted singularity to maintain direction.

I do not see how one can be an atheist and put claim to being a human while observing what there is out there in the way animals observe by only and purely relying on the eyes without incorporating the mind. I cannot see how any human claiming to be human cannot see past the barriers restraining the animals from being human. With minds it is so clear what the Word of God says, but to be human and not see what humans should see is a dangerous reflection on the mind you use. God did not say "Let light be visible" or "See the light", He said "Let there be light" and that is what there is. If humans then see darkness where they know that one cannot see darkness the darkness are within their minds and therefore they are atheists not withstanding that they may or may not claim faith as part of their thinking. If you cannot read the Bible through human vision and as a consequence not understand the Bible don't blame the Bible for your inabilities but blame yourself and your inabilities. It is not the Bible you cannot read it is you that cannot read the Bible. Place the relevancy where it and as it belongs.

We are human therefore we have light in our minds and ought to make use of that! This very afternoon as was writing this part I took a break and lo and behold, one of my sons came to me with a problem of a religious nature. I shall not go into detail about his problem but I asked him to define religion and what life is. To strike some sense between his problem and the size he sees it in I asked him to tell me in his view about the contents of the Bible according to the Bible and the dominie (Afrikaans for preacher man) how would they define life because some parts in discussion about his problem was the discussion involving tackling the issue and thereby the issue turned to how far can you go in solving matters and leave the rest to preying and doing prayer. I am of the opinion and will die by that opinion that prayer only serves a purpose in thanks when you yourself completed the task without preying for some force to help you complete the task at hand. Life is the manipulation of your surrounding and that means you do things yourself if you want things done and you do not prey for things to be done on your behalf by God. That is the definition of life. It is the manipulation of space-time and involves neither magic nor divine prayer but you go about changing your surrounding to match your needs. What all preaching never advocate is that we are in the seventh day of creation where that specifically states that God went to rest and from that I draw the conclusion man can and man must do everything by himself because God clearly says He has gone to rest. We do things on merit by ourselves or not at all. That is the energy we think of as life. The fact that we have the ability to self-sustaining and not being fixed to the universal position space-time landed us in gives us life. With life in hand you manipulate what ever you can as you replace positions to suit the required changing of objects where changes are needed. Then your acquired needs changed them to be to your taste and there is no other way out. Life is about changing your surrounding for the better of yourself or others and to improve all around you. The ability to manipulate space-time is the energy I have and Old Neighbour lost. Still it is energy. It is neither food nor electricity but it is a more advanced form of energy than the energy mentioned. Another part of life is tacking the responsibility for change your manipulation may bring about and the effect such change may have for other beings sharing space-time with you. Never confuse the needs of others with needs of your own and project such needs about yourself as beneficial to others without consulting others. This is very typical human behaviour.

With the Newtonian confusion raging man has mixed matters bringing about a highly unsatisfactory climate where we try to pin cosmic value and pre-conditions on life and place very stringent condition suitable only for life onto the cosmos. That leaves science in disarray and confusion. Heat sustaining life as pre-condition Xepted science projects to stars and where stars fade we allow them to die as if blessed with life's changing and renewing.

Stars certainly do not have emotions and when they erupt it is not in anger. The chemicals stars need to maintain singularity is very poisonous to man and the matter making life sustainable will have no chance of surviving even as a flash in the star. We think of a star being hot in the manner we translate life's pre-condition to what is hot. It is to the letter the same way that we take outer space as being unsustainably cold where it is quite the opposite applying.

While looking at the earth we think of the cosmos. We reflect what we conceive as conditions to match life being normal to the cosmos. Planets have to be plenty full because even we have one in hand and eight others in the back yard as spare should we make this one we have untenable to life. And should we run out of planets to ruin there then should be others nearby carrying life on one in nine, as is the case with us. We try to find life everywhere because life has such abundance on earth in everything we see. We even reflect our vision of time to mach time in the cosmos giving the start of creation an earth bound time range never thinking that the universe is growing and not dying. In the same manner we think of the universe as a living organism while the universe constitute every aspect we relate to death. In fact, the universe is the ultimate death. In the universe everything will only be once and never again whereas with life there was as much as there will be and even more will come than what was. That is the last thing one will find in the cosmos. If time ran out for whatever time will not replace or bring back what ever. Even the way we portrait the earth's surface we wish to reflect to space using the same methods we use on earth. One mile will be one mile wherever you wish to take the mile. After all one mile is one thousand seven hundred and sixty yards (if my memory serves me correctly because this is still part of my culture when I was at school and South Africa used the British yard stick). Not once comes the thought that man cannot step one yard in space. Still one yard will be one yard wherever the yard may follow man. Man has acquired the inability to divorce life and the cosmos for some reason we can presume as cultural. Unfortunately we go in accordance to what we see and that is more cultural than culture it self. We see a shining light and presume it is a star in the same manner as we see a large dark antelope with horns on it head exactly in the same way as that of a buffalo and presume that what we see is a buffalo. In the case of the buffalo the past thought us such observations are correct and hence we grew accustomed to the culture of believing our eyes.

Never do science take charge of thought and divide flesh from energy in the manner I have done during this the writing of this article. Outside the view we have we can locate a something that is there but needs some vision in extelegence to locate. It is a small part of life that has an attachment to the physical but an overwhelming comes attach to something indescribable to define.

In other books of mine I try my best to prove that our view of trailing outside the sphere of the sun is a myth and even travelling to another planet is not the same "as going abroad". There are so many dimensional barriers attached to what we can see without our locating or even knowing of such existing barriers because they remain unobservable barriers. There are so much more than what ever may meet the eye.

In part 7 of the Theses I touch on the subject about the age of the earth and how short sighted (once again) the Newtonian view are on this matter. The earth is in truth not 4.5×10^9 years old but the core was part of the cosmos during the birth, the very first moment of the cosmic birth. Many processes came to change and shape the earth to what we enjoy today, but the inner-core-value came from the first parting of the singularity Alfa.

How life started as such I do not wish to speculate on, but logic tells that what ever was at day one of singularity Alfa, nothing since was added or removed and that puts the carbon carrying life at the very start as well. It would be reasonable to suspect that all cosmic structures holds the carbon but not all structures can present a satisfying environment to sustain and protect the singularity of the carbon in order to bring it to a point of holding divinity secured.

One opinion that I strongly hold is that Chandrasekhar is as misinformed about his carbon-a-plenty theory as he was about his crushing stars in weight. Carbon cannot come from the cosmos and go through the Π limit unscathed to infest the earth. That is as Newtonian as all other bullshit can be. Life in carbon was a part of the earth as it was part of the sun, but it had its being burnt to blisters and could on that account not develops on the sun.

What ever the earth went through was also a survival test for species on earth. What ever the sun threw at the earth the form of life that was dominant then, had to make do or die. The fact that life made do is testimony to life's survival skills. Life will last, no matter what man may throw at it. It is man that places man in jeopardy. Man is the prize of life's achievement that I do believe. Man is the accomplishment all other species carried the burden of. Life is built into man and all qualities of life manifested in man.

That makes man the youngest and the least protected. That makes man the weakest link in surviving. I have my sincere doubt about modern civilised man's ability to survive even the onslaught of a brake down in civilisation. One harsh winter and not one in a thousand would be able to see the next summer rains bring relief. Picture a big city without electricity for one month and think who would survive even such a limited test. One hundred years ago such a remark would have made me as silly looking as the claim I make about gravity. But man has gone down the tube, at the end of the ladder although to man's thinking he is at the top of where he ever was before.

We are launching a chemical war at all pests we do not seek. We kill and destroy them without thought. Bacteria, fungi and, viruses have been at tests far grater than man can produce and survived to tell the story to the next generation. It is written in their life code for the next generation to read and fight. When a species are at it greatest danger of not surviving an onslaught on its very existence a factor much dormant in normal conditions kick in. That factor rewrites the coded massage and the following generation find armoured protection. Man is weakening with all the chemical aid we see as medicine protecting us while we put the most dangerous forms of life on a survival course we cannot afford the luxury of. The day will come when there is no stopping these killing-surviving machines and we, man will stand defenceless while they go about killing and maiming on sight. Every little headache is a call for aspirin. Every cough is a call for anti biotic. One day we will find the disease and ourselves defenceless well and truly developed. Man will die and the count will become more than man can destroy human bodies. That will leave corpses for more viruses to grow and plan more attacks. Payday has to arrive we must see that coming and not be as arrogant in our self believe.

The fashion of the century is to place all, as equal and life holding space in a dog is equal to life holding space in an ant. That can never be for the single reason that all life in the body of a human cannot be equal. Any person can go without a limb notwithstanding the sacrifice they endure in whatever function. Losing an arm or a leg does not risk life at all on the condition that it is removed before it may infect disease to other organs in the body. Losing a liver is serious but machines may provide such an organ function replacement and life goes on, fairly difficult but without eminent danger of death to the rest of the body. The same argument can be said about the heart longs kidneys and such. The function organs play in maintaining the body is crucial but not vital. Losing such an organ does not mean death by necessity and can become even to some of minor significance. When losing the head or part of the brain things turn to a lot more serious nature.

I have witnessed friends of mine that were motorbike maniacs like I am, falling off their bikes and receiving head injury. After the recovery those persons changed in a manner where they became alien to themselves. They became another person no one new before and none can recognise. Such an injury is very serious and lethal, more to the persons that love him than lethal to himself. The persons that love him has lost a love one and gained a stranger they do not care for. Even in one body all life does not stand equal let alone from specie to specie. Losing my arm is not the same as losing my life because I can still live (more unpleasant but that is not the argument) with such a loss. The conclusion of logic is that the arm is not the "me" I lose, as did Old Neighbour when he went missing leaving his remains behind. Some of my body is life in issue for use to be discarded when no longer required for service but other part is much more closely connected to me as life.

This brought about the atheists campaign that life comes as part of a wholesale package wrapped in a carbon container and all philosophy centred around this argument went missing when some connection was proved between electricity and motorized motion of body muscle and fibre. This was the dawn of electricity and the wish-wash that went around with miraculous curing by only sending impulses of electric devises that could cure all and almost bring death back to life. Some devises remained proving through time their worth but in general it was a lot of quack and most disappeared where they came from.

Then came the theory that life was only electricity flowing from the brain to where ever body motion required the flow and all other philosophy went silent. It is not hard to imagine why because physics place electricity as a force with the same presumption (though they will die before admitting it) that a force has a control in similar fashion to a ghost or some unknown free spirit running around to every one's amazement. That mentality sticks like glue and much of that influenced scientific arguments to be in apathy to the philosophical and since 1945 when the physics got hold of the German nuclear bomb and let it loose on Japan it is mathematics ruling logic to the point of madness. No one since then had any inclination to touch this aspect again since all were satisfied that everything was flawless. Flawless indeed but at the heart of mathematics and in the very start of physics lured a flaw that became more apparent every year and the flaw eluded every one to date. It even diminishes all sensible argumentative possibilities to a stand still.

Losing a limb might not kill and it might not change any personally but it is loss to life. If some one acts promptly and in time doctors commonly have the ability to connect the lost limb and with some minor complication the limb may even restore to normal application. Would such prompt action work in the case of Old Neighbour being officially dead for say twenty minutes. The answer may be yes and more likely no because it depends on the brain damage that occurred in time laps where the brain fibre were starved of blood and more important oxygen. As was the case with some of my biker friends brain damage can and more likely will result in a mild to drastic personality change and in some cases dangerous insight attacks may occur.

Changes of such a nature are very serious and symptomatic of injury to the brain. In the brain damaged victim likes and dislikes behaviour pattern and mood swings will change the personality of the individual. The changes may result from a blocking of the flow of blood and it may result from a nerve area that lost function culpabilities but life still remains present. From physics point of view I am of the opinion that it is a natural phenomenon gone very bad and such changes in personality takes place with or without injury. The Romans believed that when a person breaks a mirror he is doomed for seven years because the broken mirror damaged his sole. This we modern people know is just another folk law tale but with some angle of truth. Of course the mirror part is the untruth but there is quite some truth behind the personality changes with an interval fluctuation of seven years. I would not go as far as putting a stop watch to the date in seven years but in a more or less manner we all show some changes in personality and a man of fifty will not find the company of a few teenagers to be friendship bonding and neither will the teenagers like a fifty year cold going gallivanting with girls very pleasant. Of course once again there are many exceptions to the rule and as with all else in the cosmos there are relevancies changing circumstances that may occur. What is without doubt is that the link between life playing a part and the fibre connection playing a part and it will be as silly to claim the carbon has no influence on the life energy as it would be to deny that there is another energy present above and beyond the fibre. With this I wish to introduce my Theory on the Seven Dimensions and I put it to you as I originally started with without changing some of it to fit my present day views.

1.4   THE   SEVEN   HEAVENS

Although from the name one may have the idea the article is exclusively attached to the spiritual as much as it is about religion and has nothing to do with physics. When a friend of mine saw my article in one of my scribbling pads (this was years ago before any idea of writing a book ever entered my head) he was astonished by my claim that it was pure and unadulterated physics. This was my advance from nowhere into physics. Justifiably you may say as my friend did so many years ago that the seven heavens have no bearing on physics but by saying that the biggest mistake comes into the open. I admit whole-heartedly I did not realise the importance it had back when I wrote down the loose ideas but in retrospect that was my initiating although not my first ideas.

Every aspect of every aspect connects in some way leaving only nothing unconnected. It should be somewhat obvious by now that I see "nothing" having no claim in any form of nothing as part of mathematics or physics and to my view that is the main difference between arithmetic and mathematics. In arithmetic there are an allowance for a number or a marker such as zero or nil whereas in mathematics no such number can be found because no such pointer can claim any position from the origin.

Even when I wrote the thoughts down that many years ago I did not yet dispute zero as a number, but I have to admit I had some difficulty with the value of nothing. For instance what was more nothing and what was less nothing when there was two of nothing facing each other. In all of mathematics there has to be growth as much as there has to be decline from wherever any marker may be. In the article I show that the line the half circle and the triangle have on common factor in as much as all being 180^0

A straight line cannot start at zero and still be a straight line because zero extending to wherever brings about a full zero. A straight line starts at the point where the pen point meets paper. That point may be any distance from infinity to a measurable dot, but it cannot be zero.

180^0 X 2 = 360^0

Any straight line is also half a square be cause the line forming the square cannot start at zero for the reasons I just mentioned. That is singularity pointing an eternal direction from a point of infinity and that is the basis of the cosmos as much as that is the basis of mathematics. To escape from nothing one has to become something and by doing that one could not have been in nothing in the first place. If one holds a point in nothing one cannot become something because of the nothing value.

To back this argument that no line can ever start at zero is to ask the simple question: what will the length of the shortest possible line be. It must be a line where the starting point is so close to the ending point the distance parting the two is incalculable yet there is the line therefore the end and the start is apart still sharing the same spot.

The difference in the circle and the square is the direction the indicator follows and a square cannot spin, as a circle cannot be motionless The factor of Π indicate eternal motion and NOT zero motion. There is a massive difference in that concept. If no line can have a zero point to start with where will the circle get the zero to indicate motion! This principle is the most basic mathematic rule The method applied when calculating a wave is by finding an average in the triangle continuing from the straight line to the pitch of the wave and then the decline will form a duplicate presenting the other side.

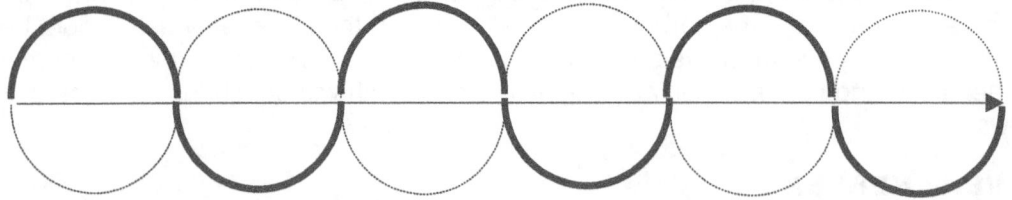

When the end of the rotation arrives the end rotation also announce the beginning of another rotation and not nullifying of the previous rotation because the rotation will have a line showing the effort it made and as it forms a wave, the wave will be there forever. The pitch may decline to a straight line, but the line remains. The wave confirms rotating directions followed by the circle as it spins. By stating that a wheel has a relevancy of zero by completion of a rotation such a claim denies the wave its rite of existing. The wave going flat, as it becomes a straight line also has an indication to singularity.

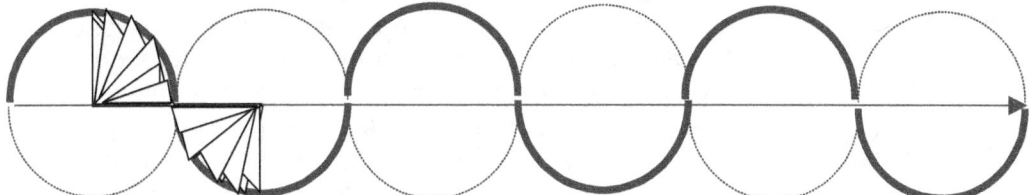

Being a circle means the thing must be round and spinning. In that case, let us take an example well known to all, the spinning top. The top spins on the thinnest of points, and still maintains a balance. By being a calculating value to match the work done in the rotating half circle the triangle depicts the flow of the straight line.

The straight-line holds a duplicate value of 180^0 to the half circle as well as the triangle all being part of singularity as much as being positions from singularity. That alone has to confirm the connection existing in the dimensional aspect.

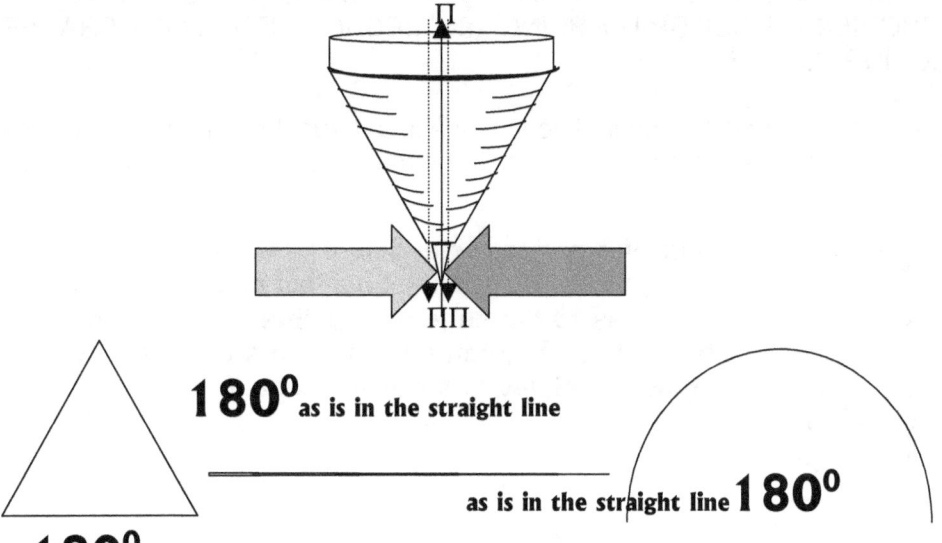

180^0 as is in the straight line

as is in the straight line 180^0

180^0 as is in the triangle

The dynamics behind the two principles is much, much more complicated than what the illustrations as shown above would suggest. However by using such basic of illustrations the simplicity might be tending somewhat to come across as misleadingly simple, but taken down to the core of factors behind the principles that forms the most basic of the principles, the illustrations prove rather effective in explaining the crude idea. However, please do not be fooled by such simplicity, in the very detail analysis it is as complex as can come.

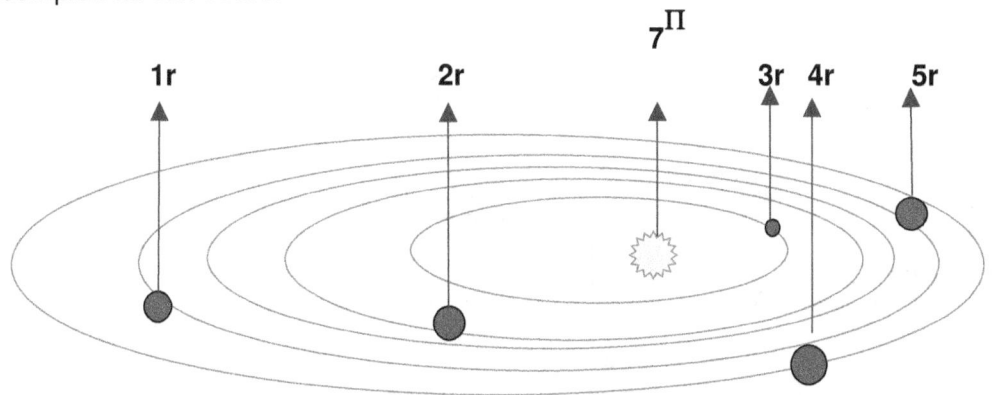

From the star holding a dominant point or most valued point in singularity it affirm all five other structure each holding singularity individually.

The universe link in so many ways we will not begin to realise the manner within the next thousand years. Electricity is one part of the link, but there are other links we may never come to know about because there is always another part of the cosmos above and below our perception and abilities that will elude us.

Π^0 **Star holding singularity**

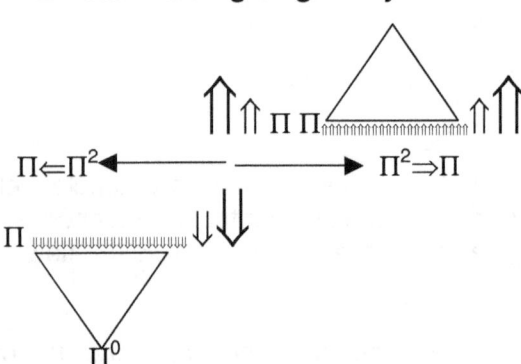

The network of individual singularity not only provide spinning through governing singularity in the sphere but also provide spinning in the geodesic through out the cosmos linking all matter to matter in a network no one will ever come to understand in full. In the sphere the foursquare triangle holds space in time maintaining singularity of different assortments. In view of the matter-to-matter Roche factor where the factor consists forming relation between particles occupying densified space-time of where ($\Pi / 2 \times \Pi / 2$) relating to the foursquare triangle the value of gravity Π^2 comes in position as $\Pi^2 / 4 \times 4 = \Pi^2$.

A STRAIGHT LINE , TRIANGLE AND HALF A CIRCLE WILL ALWAYS HAVE EQUALITY IN DIMENSIONAL CAPACITY PROVIDING EQUILBRIUM BEING 180^0 BECAUSE EACH ONE SHARES A COMMON DINOMINATOR IN SINGULARITY.

As the straight line averts a zero it holds another straight line in place to set about such an averting where the two lines will always carry a relevancy in elation to progress (the triangle) and a common denominator in the start from singularity.

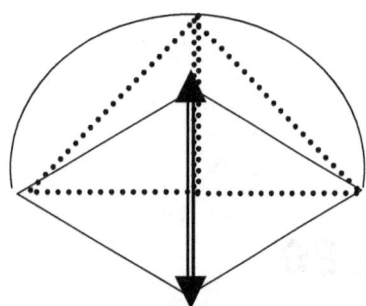

With the normal extending of singularity it will always form the triangle in a half circle whereby Π relates to the cube by 5 points to either side of the line singularity forms. Thus there are 10 standing related to seven and visa versa.

As singularity holds the straight line the triangle and the half circle as a base to form giving all and everything next to connected to and adjoining any form being of a straight line half a circle or a triangle forms space time. From singularity in the straight line (180^0) the half circle(180^0) and the and the (180^0) triangle matter form space in holding, claiming space by controlling space to influence space, but as maintaining singularity insist on space in spinning to the time singularity dictates time sets from such spin motion and by diverting from singularity time forms the law of Pythagoras in the square of space –time.

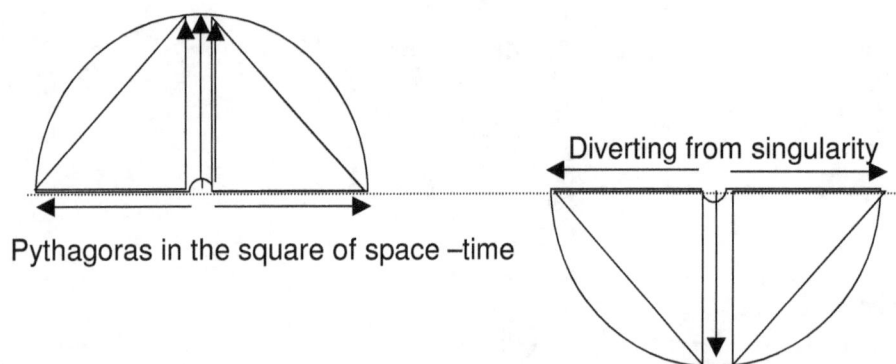

Pythagoras in the square of space –time

Diverting from singularity

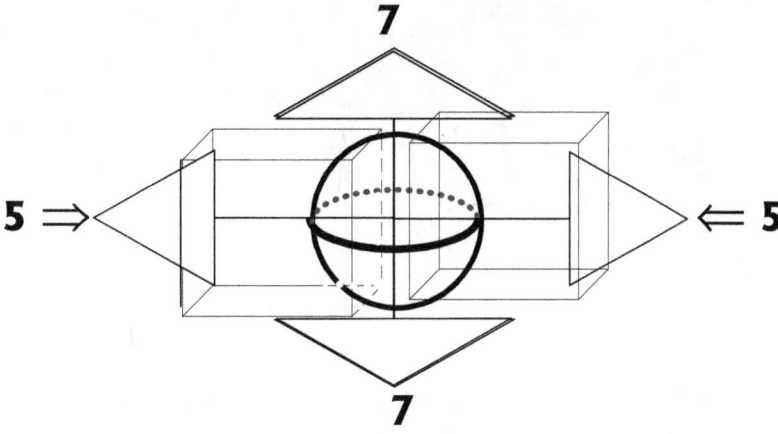

The normal flow will allow singularity extending to 10Π but when singularity blocks another sphere in singularity the two will form a joint value and by joining the larger will dominate the space as well as the time of the lesser taking control of the surface and the atmosphere. Through this the Roche lobe comes about with all its other dynamics I describe farther on in the theses.

The result of the seven markers that matter diverts from singularity is present in the 7^0 inclinations the earth holds as a sphere as all spheres have. Matter is always moving seven points way from singularity as it progress in space through time.

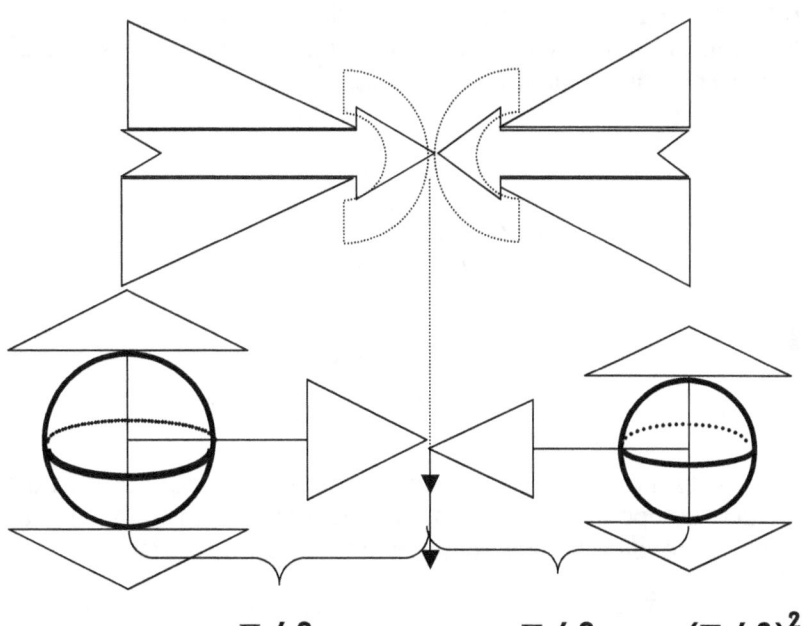

$$\Pi/2 \qquad \Pi/2 = (\Pi/2)^2$$

SINGULARITY MEETS AND COMPLIMENTS EACH OTHER.

The diameter of the cosmic structure holds the value of r and singularity holds the dimensional value of Π meaning that the radius or diameter (r) extends to become the diameter multiplying the value of singularity. But since r already consists of the square of space holding a definite positional relation with the value of singularity being Π the diameter comes into effect.

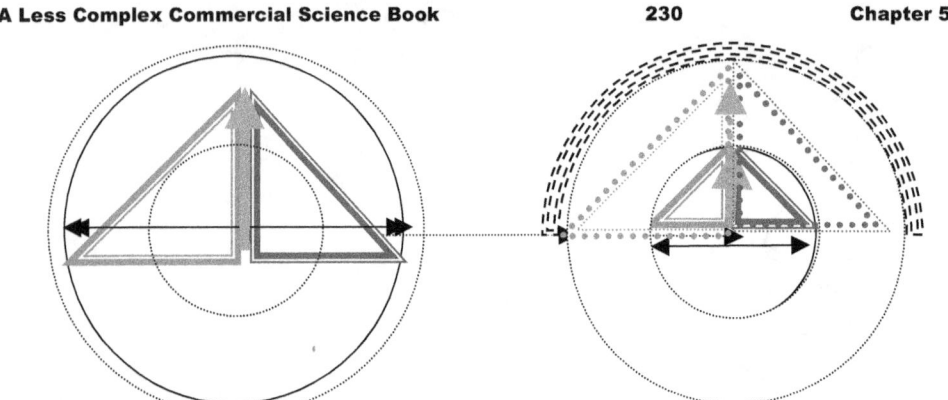

At this point the equality of the straight-line dimension to the triangle and the half circle holds prominence as a straight line, a half circle and a triangle is dimensionally equal. The common denominator will bolster all factors to an equivalent ratio,

When singularity by the straight line increases the singularity by the triangle will also bolster giving equal potency in singularity by the half circle. As the singularity of the major component revives the lesser singularity to equality, the triangle in singularity will match the performance and so would the half circle respond in precise ratio setting equilibrium in order.

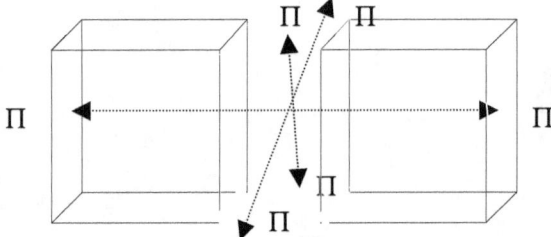

From this the lesser partner will fill by the extent of the larger partner and as soon as equilibrium sets in the growth will duplex in both accounts, normally to the fatality f the lesser partner as the lesser partner will not be capitulating under the straight of the duo. The Titius Bode configuration in accordance to orbiting formation holds a slightly different explanation to the explanation that applies to cosmic structure surrounded by space. It is moreover the individual singularity in maintaining the major singularity, which sustains the governing singularity providing equilibrium in space-time.

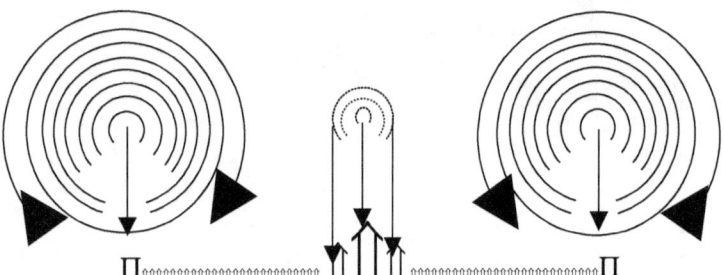

Not only does atomic individual singularity maintain self preservation, but in doing that it also sustain a governing singularity holding structural composition and form within a cluster of matter for example a star. Between stars there are a mutual or bonding singularity between atoms and stars.

The sectors provide individual singularity a means in sustaining governing singularity by which provision comes through maintaining governing singularity the required spin in maintaining cooling. If this process did not apply, there would be no connecting individual singularity to major singularity. The sectors provide individual singularity a means in sustaining governing singularity by which provision comes through maintaining governing singularity the required spin in maintaining cooling. If this process did not apply, there would be no connecting individual singularity to major singularity In this maintaining of cross referencing of singularity providing spin to the governing singularity many factors of singularity all form a close knit network inseparable one unity but also strictly individual to a point of destructing.

Singularity has three part and five points with Π as matter being sixth and space (r) as light the seventh.

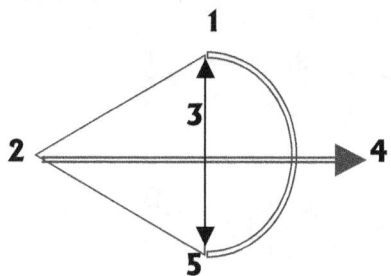

The TITIUS BODE Principle

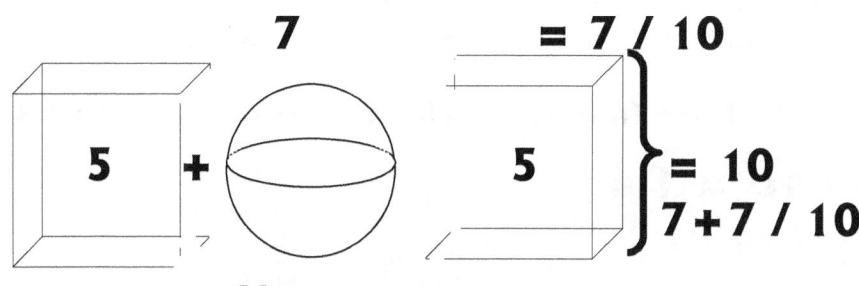

Matter-to-matter

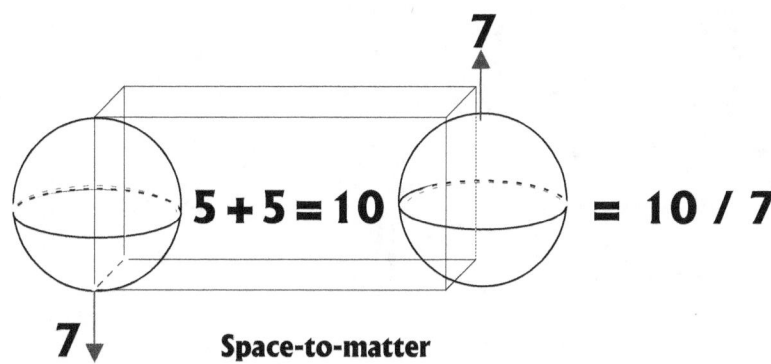

Space-to-matter

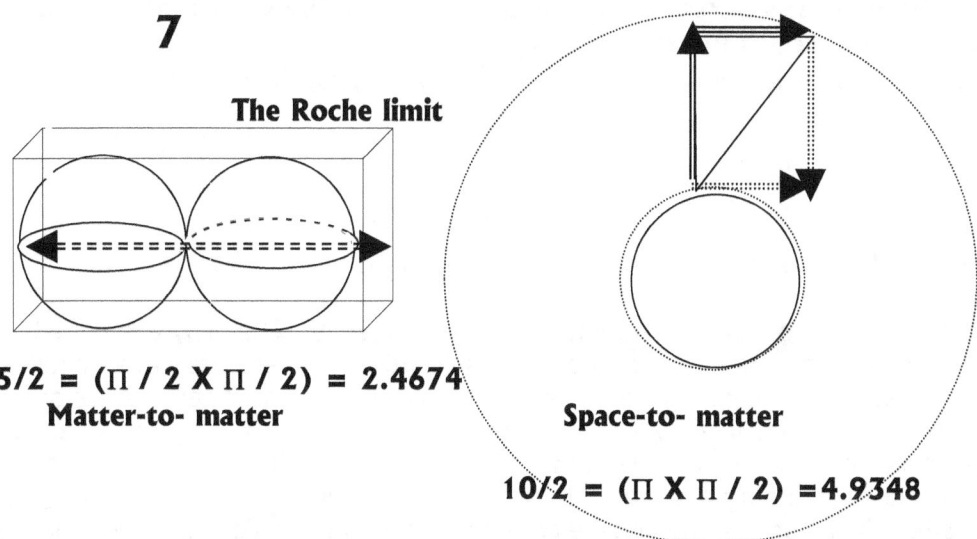

$5/2 = (\Pi/2 \times \Pi/2) = 2.4674$
Matter-to-matter

Space-to-matter
$10/2 = (\Pi \times \Pi/2) = 4.9348$

The space between the spheres divide in half, but because of the extending of Π and not applying r as ordinary mathematics will suggest where Π replaces r the singularity extending from Π^0 will be half of Π in the square of $\Pi = (\Pi/2)^2 = 2.4674$. In this lies the dynamics why planets have a positional (be it rather a dimensional) relation of 7 / 10 The second Roche limit is within the sphere as $(\Pi^2/2) = 4.9348$.

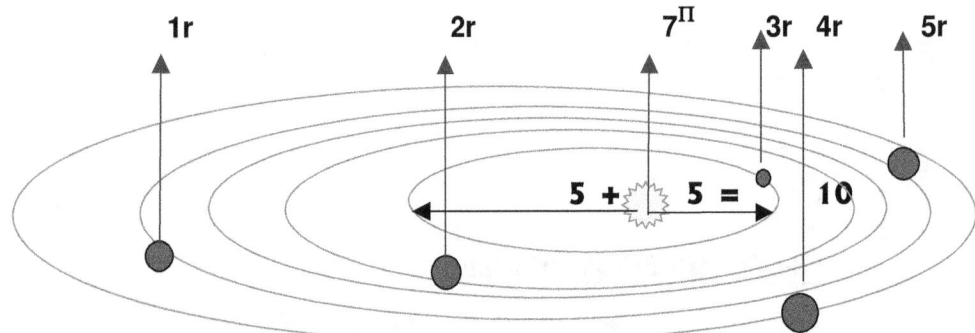

From the matter-to-matter relation in the Titius Bode configuration there are 7 / 10 + 7 / 10 = .7 + .7 = 1.4

From the space-to-matter relation in the Titius Bode configuration there is 10 / 7 = 1.42

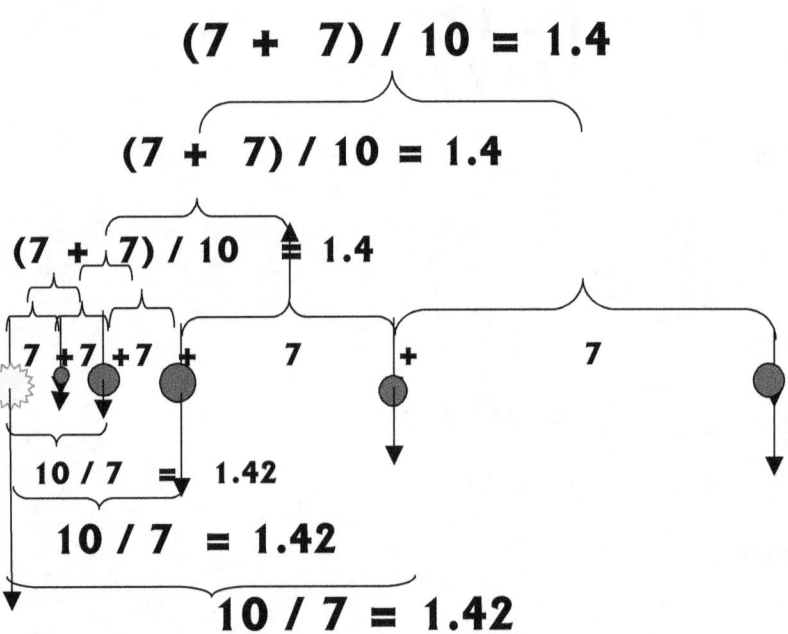

= .7 /Λ\ 1.42

= 1.4 /Λ\ 1.42 **Because the space-to-matter is in the square at 10 placing the matter-to-matter at a square of .7 + .7 = 1.4 the space-to-matter forces the matter-to-matter to double the distance by number as structures are place father from the mainΠ^0 maintaining singularity.**

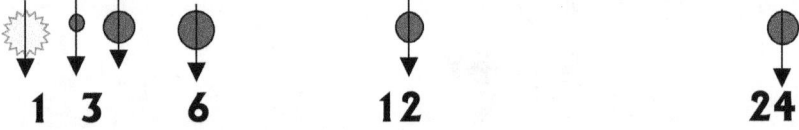

Reasons why this does not fully apply to the solar system I give in book # 7.

According to me, as a believer of the Holy Bible, there are seven heavens. The Holy Creator sits in the seventh heaven. His Word tells me that. That means if his Word says there are seven heaves that means there are seven heavens, that means there must be six undetected heavens. Before those atheists start shouting their lungs out, first go about answering the question put by me in a previous article. Answer the question where do life, being energy, go after it has left the body after death. Let us presume that the question is still without an answer. If the Creator lives in the seventh heaven, then He must have created the other six. By acknowledging the creation, I echo what millions of intellectual people believe. The so-called Christians, Judaism and Islamic faiths all accept the first five parts of the same religious Bible. Therefore, I feel free to speak on behalf of millions of people who consider themselves religious and these people come from all lifestyles and all over the world.

First let us consider the dimensions accepted by science and later argue those forms that are excluded. The so-called heavens have another name, which are dimensions. These dimensions could be regarded as planes or spheres. In looking for the dimensions that form the universe, one must look for sides, which has a combined value, but exists in total isolation from one another. Any object that can be visualized has to contain at least three sides with six obvious different spheres. These spheres do have single applications in the universe. Allow me to explain.

THE SINGLE OR FIRST DIMENSION (GRAVITY)

In the single dimension, one finds gravity or a pulling force that every cosmic body has.

If any person does not believe in gravity, then try to jump high of far. One would find that there is a force, which pulls your body back in the direction of the centre of the earth. This force is calculated to be 9,81 N m/s^2 and is almost precise the same value right over the world. This force can only be overcome if an object is hurled into space at a speed of 11,20 km/s^2 and an angle of 90° with the earth. That brings about that any object on earth is moving at a speed of 11,17 km/s^2 at any given moment.

As far as my knowledge goes, this force is perpendicular to the earth, with a distortion of 7% due to the inclination of the earth. A person in China will be pulled in an opposite direction to a person in America, because China and America are approximately in opposite directions to one another on the face of the globe. This means a person in America moves towards the earth at 11,17 km/s^2 which is directly opposite in direction to the 11,17 km/s^2 which the person in China is drawn to America. Taken these viewpoints into consideration, gravity is a single directional force moving only in one direction that depends on the body position to the earth. Where does this force stop? Nobody knows, because the crust of the earth is being drawn towards the centre of the earth. Even the see and the crust underneath it moves statically towards the centre of the earth at an even pace. If one looks at this force, one gets the impression that there is only one direction towards the centre of the earth without any given point where it will stop. On the other side there is no given any definite starting point. Without a star or an end, only a direction remains that envelops the whole idea.

I hope I was clear enough about the idea I tried to explain. The whole idea of gravity consists of only one single directional movement in one direction seen from all directions. There is no starting point, no point where it ends, only the definite direction that the body moves towards the centre of the earth relative to a point where all directions are measured. Another strong candidate of this dimension is magnetism. The polarity of the atom in the iron core is only towards one direction. But I shall explain a little further down the road, why magnetism do not comply with the single dimension force when I have brought some more facts and arguments in explaining what forces come into play.

THE SECOND DIMENSION : THE WAVE

In the context of the wave, for example water, the circle of the wave flows from one given point outwards. That implies that there is a definite direction. However, there is a second dimension in the wavelength. The moment that the breadth of the circle is determined, the wave has already altered it. That means there are only two values to be considered realistically without freezing time and that is the direction and the frequency. Sound and light are two such values. The moment the wave is frozen in time, which means it is standing still, it is no longer a wave, but an unnatural structure man made for his own benefit. It does not occur in nature because a wave can never stand still.

Light, sound and waves are two-dimensional values. The dimension consists of a point of origin and a frequency. There is no distance, because the wave beams out in all directions simultaneously. The Doppler effect in sound has an influence but that is because the point of origin keeps altering due to the movement of the source. The speed of sound has a definite ratio to the density of the medium it moves through and that applies to the time value of the space-time occupied by matter. In a later part, I shall explain this in detail.

Due to the changes in space time occupation of the matter the frequency alters and tarnish directly in ration to the space and the time the matter occupies. As the frequency of the space deteriorates in time, the wave becomes less in value but still remains. Therefore, all the sounds the dinosaurs made are still with us but the space-time occupation that the sound waves had, rendered it undetectable to us. Proof of my argument is found in the heat that still can be detected with the Big Bang event. Although the space-time value is only round about 3°K, the waves can still be measured, because these waves form part of space-time. The light wave is transmitted from its source in a sphere and will transfer itself through space-time until such time that a few particles hits a solid object. The wave consists of billions of light particles that move by wave density from the transmitter. The photon moves at a rate of ± 300 000 km/so. That means that the photon displaces space-time by negatively at the above-mentioned rate. This negative displacement of space-time by photons is called a light beam.

The light rays follow wave upon wave in endless motion through space in time. As soon as one wave of photons hit solid matter, only a very small number of the total wave is stopped. These meaningless stopping causes a shadow to form, but the large number of photons available would fill the gap formed by the loss of photons. That means that a shadow as large as the earth, will cast darkness for a very small distance / time span that means a tiny part in space-time.

The reason why the heaven does not light up is that the beam is scattered as it spreads out in a balloon formation and therefore the number of photons lessens in density as it becomes "duller". The more space-time it has to displace the thinner the layer of photons would become and the less intense the beam would seem. It is not so much the space (distance) that should be regarded but the time it spends in space has the ultimate segregation of the light beam. In space and time, the beam would become so minute that it would alternately become unobserved by the human eye.

All that we know, all the knowledge we have accumulated through time is based on the light that sends information to us. The size, the distance, the structure, the density and its position are all determined from the light that sends information to us. The light that reaches our planet determines all the size, the distance, the structure, the density and its position. All our facts are based on this small amount of evidence. Light is so widely connected with insight and knowledge that we assume that we have seen the light when a picture of a person is projected with a light bulb next to the person's head. Light is seen as the same as knowledge. Light is actually a very poor source of information. Magicians and con artists use the information of light reflected by mirrors, lenses and colours to mesmerize our wits.

Let us consider light as a source of information in daily use by humans on this planet. Light is a very poor medium of information, due to all kinds of illusions, which it causes. Take for instance how mirrors and lenses can disturb a person's image. Why then can we categorically and without doubt, be assured that the information that we collect are the truth. Still, those in power disregard any phenomenon they find as the truth. A very practical example to prove my point can be put as follows. Light shines on a green leaf. The leaf is reflecting a certain variation of light and the rest is absorbed. Let us picture the leaf olive green. The olive green leaf accepts all the colours in the spectrum, except that olive green. The olive green are not accepted by the leaf but are rejected. The part of the light spectrum that seems to be unacceptable to the leaf is the very colour we associate it with. If the leaf accepts all the colours except green, it must consist of all the colours except olive green. Another example we use every day, is "How bright the moon shines", when in fact, everybody knows that the moon does not shine !

This might seem trivial and irrelevant, but remember, because of such trivial terminology, people were burnt on stakes. When Georigiano Bruno tried to persuade the church it was not the sun rising and setting, this became his fate.

Man's attachment to his visual sense is actually a little funny and a lot deplorable. Science can calculate the density of a neutron star, they can determine the heat value in such a star, but when this star becomes more dens, to such an extent that the density causes the neutron star's atoms to disintegrate, these very same scientists declare that:

1. the star has vanished;
2. it has gone through to another universe
3. it does not exist any more
4. because it is dark, it must be cold
5. it is lost in the creation, never to be recovered again
6. with this, all other laws of science are disregarded that says an atom consists of frozen energy;
7. the gravity fields can still be mesmerized therefore the matter it consists of must still be there;
8. All that matter can still be placed at one certain, predetermined place, and therefore it is position is a fact as far as the universe is concerned.

When taking the above mentally into account, one cannot but wonder, how far did man actually progressed from those darken middle age mentalities.

With these very obvious facts in the fore ground of our minds, we still reassure all the doubtful thoughts of our ancestry, which regarded everything the unknown comprised of as magic and mysterious.

We still cannot appreciate anything that falls outside the boundaries of the visual senses that the instruments can detect and determine. How small we are and how inflated we regard us to be.

I have in the past been asked the question "Why are all these other stars and galaxies necessary , if we as humans can't use it?" The human arrogance has no limits and the common position man occupies in the micro as well as macro space, is completely distorted by our sense of self-appreciation and self-importance.

THE THIRD DIMENSION (THE ATOM AND MAGNETISM)

In the article that deals with the meaning about "nothing" the structure and lay out of the atom is explained to some extent. The atom is the smallest and the largest single unit that the universe comprises of. In the most basic form, the atom exists of one single electron that orbits to energy levels around a nucleus in a single electron lay out. This electron to nucleus balance is the precise force that keeps the universe in perpetual notion bound by time. In a later stage, I shall explain this in more detail. The energy levels that the electron positions itself to the nucleus are valued in quantum leaps. However, even this complies with a three dimensional length, width and depth. This means there are three definite measurements, although they seem to be microscopically.

Taken one step further, the atom is made up of frozen energy. Any matter that moves at the speed of light is pure energy ($E = MC^2$) according to Prof. A. Einstein. That means this atom cannot be changed in shape or size without enormous energy loss or gain and this would lead to extremely serious consequences. The Japanese at Hiroshima and Nagasaki can declare what extreme consequences are hidden underneath the structural disfiguration of the atom structures. Therefore, can the people in the Tunguska river valley give evidence to the outcome of atoms that gain in mass.

In this process, man has released the worst kind of destruction available. The energy released is of such vast dimensions those 50 years after the explosion the shadows of the victims are still edged out on the background in the cement and bricks. The intensity of the light that was released in a billionth of a second, has almost forever changed the face of the material it shone on. Spare a thought for the people who received that light onto their bodies as their shadows remain as testament to those in power who damned them forever.

This process was not caused by atoms that was demolished, but merely by changing some of the element value from one atom to another. This leads to a spontaneous thought: "Why has no American ever been brought to justice for horrendous acts of war crimes?" Needless to say this act was the mother of all war inflicted war crimes by killing and maiming hundreds of thousands woman and children and civilians. These bombs made those in power, who ordered the release of these bombs, the biggest sadists and mass murderers ever known to man and when compared to evil-minded monsters like Nero, Nero with all his menace, suddenly becomes a silly and naughty boy. Even decades after the release of these highly toxic energy sources, they still kill and maim the innocent. However, this only ties in with another admirable fact of the 20th century. Almost all nations on earth have produced war criminals and warmongers, except the English speaking nations on earth. This group is blessed with the innocence to such an extreme that not one has ever been charged with one single act of a war crime. If none has been charged, none could be found guilty and then not one can be guilty of war misconduct. Let us go back to the atom's structure. Because the atom moves about at the speed of light, the structure is timeless. All matter that is timeless, will last forever or eternally.

This brings about the atom's structure to be forever and timeless forming the third dimension. In a later part, the reader will find that I disagree totally with professor Einstein's assumption about the fact that the speed of light has the same value as time itself. In order not to start confusion this early in the book, we shall accept dr. Einstein's theory as for now. The message I am trying to bring across in this article is the comprehension of what the third dimension contains. It is made up of three dimensions without having time as a factor.

MAGNETISM: THE MISCONCEPTION OF THE ATOM

As I already pointed out, the atom is pure energy. Seeing that the electrons rotates about the nucleus at the speed of light, and the nucleus vibrates (?) at the speed of light, that brings about a confined cell made up of pure energy.

Seen in the whole picture, the fact that the atom is driven at the speed of light and is in total balance and harmony, its permanence lies in the fact that it exists confined to a structure, but not to time itself.

There are length, width and height. The tree dimensions it composes of will render the structure eternal life, or that is how it is regarded by science. The property of the atom might change from star to star, but the structure remains with its three dimensional qualities varying in size, but it remains the same. You, the reader may ask: "What is the common factor between the atom and magnetism?"

Magnetism on the other hand flows between two points without stop in a closed circuit, not influenced by time. There is a direction (length), a circuit (height) and a start / finish point. This forms a closed ring formation, which has the same qualities as the atom. Time plays no role in this energy displacement, because this movement is coupled to the speed of light.

To prove its existence, lies in the fact that the human civilization is mostly driven by electricity.

The energy determines the magnetization, but the circuit remains permanent although some times in a latent form. Because of this, it actually consists of all the ingredients to qualify as a third dimensional force.

The ferromagnetic field proves that space-time is being displaced as it is being done in the case of the atom structure, but the displacing of the space-time is in a continuous and constant closed circuit.

The poles that attract each other displace space-time in the same direction and those that repulse each other, displace space-time in opposite directions.
In bodies as insignificant as the earth, the difference between gravity and magnetism will be enormous in comparison with structures the size of the sun. The magnetism in the sun is comparatively much stronger than the earth, but the difference between gravity and magnetism would be less. In a structure, the size of a White Dwarf the electromagnetism might only be twice the force of gravity and accordingly as the star becomes bigger, the force would become equal.

In structures that compose of the mass of a neutron star, magnetism would be dominated by gravity to such an extent that the force of gravity would not allow any magnetism to exist. In the so-called black hole, there can be no such a thing as magnetism because the electron does not exist any more.

Electromagnetism is the "short circuit" in the flow of positive space-time displacement and can only exist in a structure that compiles of an iron-based core.

THE FOURTH DIMENSION: THE TIME FACTOR OF TIME IN SPACE

The previous three dimensions all dealt with the space factor in time in space, disregarding the time factor. Let us consider an everyday household item in normal use like a table to explain the value of time. The table is made of wood. The wood is comprised of timeless atoms. (That is, if the reader accepts the previous argument about the atom that lies outside the boundaries of time). The form and the shape that the wood is in, is not timeless. It started as a seed that was enlarged by cell multiplication as it germinated to become a tree. The process of germination took a certain time and the growing of the tree took another period in time. That means the tree occupied more and more space in a given time period in the space-time it shared with the earth in the form of a tree.

Afterwards the tree was chopped off. This felling of the tree comprised of a certain given space in a certain given time, as did the falling of the tree. Both these periods consisted of different space in different times that are coupled to the space-time the earth occupied. For a certain period, the tree remained in an upright position occupying a little more space as time moved on. Then at a predetermined point in time, the tree was felled. Every blow by die axe displaced a certain piece of wood to a different place in space and time. That meant that every splinter of wood that broke from the tree was given its own place to occupy space in time. The position of the wood has been altered and even if one try as hard as they may, every piece of wood has received its own space in time to occupy and could never be regarded as a tree again. It occupies a complete different position in the space it shares with the earth in time.

Afterwards this tree is stripped of its branches and leaves. The stripping takes a certain period and position in space and time. Each branch occupies a different position in space and time and is forever dispositional from its original position in space and time, because every part that is not part of the tree anymore, is in its own position in space and time.

Then this, the tree was taken from the plantation to the mill and moved through a different space in each fraction of time as it was transported. Every millisecond held a different position in space in relation to the next time fraction. At the mill, it was sawed into planks and the structure's position was altered even more in space and time. The planks were bought by a carpenter and were given a completely new form as a table in space and time.

Although this particular table I am writing on now, at this present moment, has occupied space in time since 1921 as a table, the wood occupied space in time much longer than the table has in its present state. This wood can never be made a tree again, because it is space and time has been altered indefinitely. So, how does this fit into the big picture of the universe as all things in the universe are connected and related?

I was born on a given second, minute and hour and I shall die on a certain second, minute and hour. From the day of my birth until the day of my death, I shall constantly alter my position in space and time and will never be able to occupy the very same spot of space-time because space and time will not allow it. According to my visual observation, I can presume to occupy the same space in time, but the geodesic outlay of the universe is altered as every millisecond goes by. That means I can never remain in the same space in time, because my body's position alters by the rotation of the earth, the sun and the universe.

When my time is up, my space and time is altered to such a position where as I am no longer in control of it. From then on, I will never be able to determine my own position in space and time again. I am in a state called death. In this state my body will be broken up by microbes into gas and heat, every second time moves on, my body's occupation of space, and time would diminish. This will carry on in space and time until only the elements my body comprises of, remains. The atoms I am made of, have previously been used to form plants, trees, animals and even humans. After my death, it will never ever form another combined unit to match my exact replica. Therefore, I have departed in more senses than one.

That means my position as Peet Schutte, in space, and in time, is suspended, unconditionally, forever.

This means that for certain duration of time a certain combination of atoms is forced together to form the elements that are dedicated to me. This dedication is temporary, which means that I was for a certain designated space in certain duration of time-sharing space-time with the earth.

Although this is a known fact to every person on earth, it is astonishing to see how every person yearns to maintain a youthful and vigorous maternal structure. This structure is condemned to destruct the minute a person is born, yet everybody guards his attachment to that structure with a jealous observation. The cosmetic industry cashes billions of dollars a year and all that income is based on this fear of ageing, which ultimately leads to death of the person and destruction of the body. The ongoing process in the envelope of the fourth dimension connects all the previous three dimensions, we regard as time. This means that time itself is not created by man, but is part of the physical universe and is only recognized by man, the very same way man has recognized the existence of the other three dimensions.

THE FIFTH DIMENSION: UNOBSERVABLE - THAT MEANS TIMELESS TO MAN'S SENSES

In the article "Life after death," I touched on the subject that life is a form of energy, and therefore is indestructible cannot be destroyed. If any reader cannot accept this argument, then please prove the opposite and let me know. Therefore, until the opposite is proven, I shall regard this argument to be correct. Because the generator, which I regard to be me, I of the electrons, is not the" me" that will be placed in a box and will never be able to share space time (because the worms are going to feast on me) I am above time and space and I shall not be in that coffin. Only the "me" that am made of flesh and bones and was considered "me" and which I had temporarily control over, will deteriorate in that coffin. If the generator of the electrons that is somewhere connected to the brain, and from where I control the me, which is the muscle bone and tissue held in place by the body I consider to be me, where do I , the generator of electrons go after I was disconnected from the me of flesh and bone me. I know where I, the energy-less dead body is going. I, the part I consider me, is to be thrown into a wooden box, dumped into a hole which is dug in the ground and where the worms are going to enjoy a feast of a meal.

This feast of a meal by the worms cannot be me, because I am not part of that decomposing structure. I was part of that decomposing structure, until the very second, I lost control of that decomposing structure. Thoughts, that travels faster than the speed of light is the actual part that is I. I can elaborate on this line of argument, but those that do not wish to be convinced will remain so because they prefer to remain unconvinced, not because they know they are right, but just the opposite. The I which is regarded as me, has received the ability, for a short while at least, to manipulate space-time, whether it was in the form of my body, or other matter I came into contact with, or unoccupied space-time as I moved about, occupying unoccupied space-time in a random fashion at free will.

The part that is I, and which is not part of the corpse any longer, can think of one thing and then think of something completely else. The very following second I can change my surroundings as soon as I change my thoughts by creating new thoughts. These thoughts are not connected to time or space. It moves arbitrary and involuntary through time and space, from present to past to future. Sometimes these thoughts are so strong that a person loses track of reality.

The terminology we use to describe this, is daydreaming. When I was teaching as a pedagogue in class, this condition was my biggest enemy. While I was conducting my class, the students would sit there wide eyed, listening, but at the same time they were miles away from school living a daydream that had no connection to matters of schooling. When they were asked to reply what was just said in class, they were flabbergasted and completely unaware of their surroundings.

I think I may presume with some certainty that the reader would follow the two parts of the same person I meant. I agree that an argument can be made that these thoughts are part of brain cells that are stored by nerve tissue, but the information in those cells are created by emotion. These thoughts can be depressed by emotion or be prominent on the foreground because of emotion. Man is made up of a stream of emotion that flows continuously through these emotion fields and the flow of emotions is that that puts meaning into a person's life. One has to consider emotion to be one of the most pure sources of energy.

Energy creates thought, creates ideas of a spiritual as well as physical nature, it establishes a flow of electrons that drives the human mind and body, it controls all muscle groups in the body like heartbeat and when the body is in danger, it produces chemicals which enables the body to react far better than it normally does. That means these other organs are also under the control of emotion. The emotion drives the body to produce chemical substances, which enables the body to perform at levels far above its known abilities. This emotional control defines heroes from cowards, sportsman from the ordinary and even philosophers from the masses. That means the emotional part is the part of me that cannot be destroyed and that is the part that generates emotion. All this comes down to one value that life possesses, and that is the manipulation and control of space-time.

When this emotion driver, that generates electrons, leaves the body, a person is considered dead. When one takes the moment that life leaves the body, nothing physical leaves the body. There is no lightning like electric conduction, there is no immediate spontaneous combustion, but there is no emotion either. As I already pointed out, the only reason why a person is considered dead is that he is considered lifeless, meaning without energy.

A lot of energy has to be deplaned somewhere. It is no longer part of the physical world and being energy it cannot just vanish into nothing. What is factual, is that life as energy, is no longer attached or bound to the fourth dimension of time. At this point, I have to prevent my ego and self-importance not to get the better of me and put myself on a pedestal equal to our Creator.

It would be much better if I were grateful and thankful for the time I was allowed to use the atoms loaned to me for my own personal use. The loan period was of such a short duration in space and time, that the extent of the period of loan becomes oblivious in space and time and space time.

THE SIXTH DIMENSION: THE LIGHT AND THE TRUTH

I must confess that in this my faith and religion plays a large role, as one of my beliefs is that the Messiah has already come.

Maybe the Jews, Moslems, Hindus and other religious groupings would find their own explanation according to their faith. My Messiah said, "In the house of my Father there are many mansions.", which I interpret that these groups should be left to find their own salvation. I may not condemn them or denounce them or try and convert them, because each should have its own mansion. However, personally I accept my Messiah and He declared that: "I am the Light and the Truth and no one can enter the house of My Father but through "Me."

If my Messiah points out to be a light, it stands to reason that there must be darkness. That will be a dark fifth dimension and light sixth dimension. Seeing that the majority of English speaking persons do not share my beliefs and religion, and I do not believe in converting any person to my faith, I shall leave this matter at this point.

It is not that I am ashamed of the religious beliefs that I follow; to the contrary, I believe that only Israelites may be converted to my religion. I think I made enough argument to prove the existence of a fifth dimension to which life has to go after death. I presume, because I am not familiar with the contents of other religions, that their religion will allow them a passage by what means their religion chooses, out of the dark fifth dimension.

This dimension, the sixth can only is entered by human life. Let us then see what human life is all about. A human does not have to be a human because the creature is compiled of human D.N.A. D.N.A. cannot form a human. The gorilla is 97,8% human and the orang-utan is 98,2% human. However, there are obviously big differences between these species and a human being. No cultured person on earth will consider one of the ape species to be human. That means a human is more human by culture than by physical appearance. The additional 1,8% that a human have, is not even enough to explain the physical differences that exists between humans and apes. However, we know that the main difference between the species is the human's ability to reason, think, argue and control their emotions and instincts.

The more a human explore and scrutinize his feelings, himself, his surroundings and his universe, the more such a person would qualify to be human. After all, that is how other species evolved away from the animal and that should categorize us to be part of the sixth dimension.

THE SEVENTH DIMENSION: THE SUPREME ALMIGHTY
Because so few English-speaking persons consider themselves Israelites, they do not fall under the law of the Israelites. In such a case, I would consider myself blasphemous to share such knowledge with those that do not regard themselves to be law-abiding in all ways. Any person that does want to know more about this matter should read the book of Henog. I will say this much, that those that read Henog would find out why the book of Henog has been left out of the Holy Bible by the Roman Catholic Church and the other churches. The Supreme Being lives in the Seventh Heaven as Creator of all. Because I regard my fellow Boere as brothers, I did explain to a very small extent the seventh heaven in the original Afrikaans version. But all Christians and only Christians read this…all churches are part of the Anti-Christ being the Body –Of- The- Anti-Christ. Do not look for the coming of the Anti-Christ for he is among us. They crucified Christ for His throwing over the money tables and throwing out the money offerings (a lucrative business in any society) ridding the Temple of money some two thousand years back, and today all Christian religions fight one another to feed that which Christ threw out…the money tables…and best of all is they feel righteous in doing so! All denominations are a part of the Anti-Christ and BEING THE ANTI CHRIST. Christ threw out the money tables because He said you cannot serve two Masters…you cannot love God and Mammon for one you shall love and one you shall hate. You cannot serve Mammon and God. If Christ came back today and once again threw out the money tables all Churches and worshiping priests will once more shout for His crucifixion as they did two thousand years ago. As the Pharisees were the Anti –Christ back when… so is all Christian churches and denominators, Priests, Pastors, Reverends, Bishops, name them what you like, small and large…they're all taking part in the crucifixion every time they ask for money "In the Name Of The Lord" and if He came to destabilize they're Money machinery today, they will hang him tomorrow morning at they're earliest convenience and even on the same cross (if they can find it). Every preacher is more into collecting than pouring out the Word. It is a trade off that the preacher will bring the Gospel in exchange for collecting offerings. They heel, bless, pray, condemn and condone on behalf of the name Mammon. I challenge every purist of heart to show me one preacher of the Gospel that sends donations back with the message that such donations is

condoning the Crucifixion and as preacher will not except having the blood of Christ on his conscience. If you cannot show me one, I can show the body of the Anti-Christ for they will kill again if some one should try to diminish the lucrative trading done in the Name of the Lord.

9.6   THE TRINITY

As far as the aspects concerning the physics side goes I could prove in some way that my initial way of reasoning brought fruit to bear, but what about the dimensions past four and even past five. I think I may presume that in some way I did manage to show that the body has its place in the fourth dimension and without life it becomes just one more cosmic structure without any form of moving ability when all forms of life (including the bacteria that will decompose it to atoms) removes from the carbon and other elements where the elements then are exclusively and only cosmic particles. With good reason one may believe that the ability in conducting the manipulation of the space-time within the body has diminished to all extent and none is any longer present. To find a means of putting mathematical formulas to use in applying proof to indicate the fifth dimension is beyond me and that is where human extelegence plays the part. By the same token that atheists can say they wish for more proof about life and the fifth dimension I can demand the explanation to prove otherwise and ask an explanation about the energy presence in life-holding bodies and the energy absence in lifeless bodies and where as I doubt I may force atheists to sound understanding they too must admit that something does seem out of place in they're arguments about the energy being of a pure physical nature and only stubbornness will win by the days end.

So to them and those, as I might not find a way to prove beyond doubt them and those also must admit I have sown some doubt in their minds. Getting them to admit with some gallantry about the doubt factor, well that I must gallantly admit is a horse of another colour. With the fifth dimension seemingly impossible to prove mathematically the sixth will be much more difficult to prove and I shall not even attempt such an act. Fortunately man is not mathematics but much more complicated than mathematical equations can ever bring forth. Only good old fashion arguing with a dash of logic sprinkled when and where necessary will pave the way to understanding man and beast. To find out what man regard as good or bad and what beast evidently regard as good or bad must be different as everything else about man and beast are different and we must scrutinize beyond where mathematics and physics can prove. Even the most convinced Newtonian should see that there is a point such as that.

Being human every reader must have an opinion about good and bad and what is evil and what is not. I can scarcely imagine a lioness feeling bad about a kill while her cubs are filling their bellies with mouth-watering flesh of an antelope kill. Neither have I detected sorrow and anger as a new male kills the previous litter to establish his new domain. Neither the female nor the male bears sorrow after the deed of him destroying her litter sired by a previous dominant male although the lioness will protect her cubs while they are alive almost to the point where she may put her own life on the line. That how ever has nothing to do with rite or wrong good or evil and after the lion male did his killing of her cubs she shows no remorse or blame for that matter as she follows him back to the rest of the waiting pride. This is not exclusive to lions or even to predators but has a wide range of animals following the very same living style. Horses kill without thought because when a stallion wants to find a mare that he knows may follow him but for the foal by her side he will kill the unwanted foal and not be bothered by her reaction, being in the knowledge she will follow him after the foal is dead in any event. While the foal is still alive she may do some protecting of the foal but she will normally not go as far as the lioness in preventing the death of the foal. Baboons, monkeys and a variety of animals have this approach to life. During such an attack by the new dominant male baboon the females will fight off the onslaught either by grouping together or by fighting him off in ones and twos but the usual is that the new male is big young and strong and even a group of females are no match in a fight. However after all the noise and the shouting blood sweat and the rage of adrenalin has died down, the babies are dead and no female attacks the male after the fact in heart felt sorrow for the loss they feel such remorse over. No, it is clear that the deed was done and life goes on.

In humans such behaviour does take place and every time we read about such a deed committed against a harmless infant even the biggest humanist find a moment where he or she wishes the death penalty on the criminal for acting in such brutality. Why will humans shout for blood in punishment while other species take it in stride?

To find a solution I do what I always do. I turn to the Bible for an answer. (You atheists deny yourself a wonderful encyclopaedia of information and when reading the last book in this Theses you will come to understand my saying so.) In The Theory which is the seventh part of The Theses I explain to very detail the events of the first six days of creation as recorded by The Authentic Biblical Author but even after that some more explaining arrive when it is correctly translated. In the book called the Bible there are two trees described with distinction and I am of no opinion to whether they were in wooden fibre or just symbols to explain the complicated issues to persons with even lesser education back then than I have at present. It is an undeniable fact that man has an inclination about what is good and what is evil. Man would not kill an infant because he cannot find the infant's mother and the infant is crying of hunger. In the animal world any adult of the specie will walk past such an infant in distress with no feelings of care what so ever even if the specie does not show a normal tendency to destroy such an uncared suckling. Where one mail may kill another mail in a fight to establish dominance the group does not cry for justice as they loath such a deed. When a superior member of a pack relieve a lesser member of food or eating rank they do not hold congress in judgement to provide sanction for the lesser member with accompanying reprimanding about such incorrect behaviour by members of the group.

Such is the caring of man that any human notwithstanding whatever urgency drives him at that particular moment that human with maternal instincts or not will stop to care when coming across a deserted human infant hungry and all alone. Why will man show such behaviour as normal through out all races on earth without any cultural distinction in any way and this may be the only distinction that races share because some eat their dead and some burry them with pity filled emotion and… Oh, I can go on writing a book on this topic alone but that will be useless because every one should know what I mean. With this shared by all where and what does mention this distinguishing behaviour of man's ability in judging between good and bad for the very first time as a landmark to man.

Accept the Bible or not, it remains the oldest Book available on matters reported by man since no one knows when because Moses may have assembled the research on information but the information as such dates to times predating even what Moses' research may indicate. In addition it may be correctly presumed that many or most of the facts he recorded he was taught as a prince in the house of pharaoh and that then may explain some detail about matters he actually knew nothing of. His being adopted by pharaoh's sister must be a plan with some significance and must result in some meaning other than to give Moses a childhood of luxury alone.

One of the trees the Bible mention carries a name specifically as life and from that I draw a conclusion that may reflect that life chose to go the way of having a variety as sells with complexity in the evidence we now gather from DNA strands whereas a choice of life would indicate a sell of simplicity in structure as we find with lowly developed insects and other specimen of life. I have seen how a corn-crake of a specific kind only found in the desert and semi desert regions where I live can start a pest becoming so out of control that when run over on the road they form layers of millimetres thick trampled and squashed by cars to the extent the tar on the road is no longer visible. It truly is a pest of Biblical proportions but fortunately to unleash this pest it has to rain in November in one specific week. If it does not rain at that specific date, and I do not mean approximately but precisely they don't show at all. That makes their return very sporadic and it only occurred about five times where two of the five times became a pest like none can repeat in the more than twenty years I farmed on that farm. The pattern also follow a distinction where at first one or two may show very sporadic in places. Then mating starts and the pest develop where by it truly come to a climax at the end of February and dies down in May. From a few eggs they develop millions on millions and I do not exaggerate in the least when I refer to this phenomena as a plague of Biblical proportions. Poison does not kill them and when one gets hold of another one the bigger one just start feeding on the smaller one. At the end of the meal where the bigger one devoured the smaller one in totality (and I mean boots and all) the specimen will shed its skin immediately, eat it or not eat it and walk off. The one is a precise duplicate of the other with no distinction amongst them of whatever nature. Seeing one is seeing the lot. They share genes in precise replica with no differentiation of any kind what so ever whereby the one may have even the tiniest of difference in any form. They come from a line where life was still very basic and the mother specie of the very original has not changed in any way through millions and possibly billions of years. That I say on the grounds that to my judgement that specie is of a very basic nature and has developed with one single aim in life and that is to survive. They eat everything from the most poisonous plants to fruit to meat and bones of animals lying dead in the veld to dry hide and even one another. Fortunately too they only occur in the most severe droughts and when developing into the pest, which I describe, there is little to nothing for even grasshoppers to feed. Seeing the specie for what it is it made me realise that life somewhere after them made a choice to form complexity and variety or remain as they have and form a universal gene where the original mother is still present in all her offspring even after so many billions(?) of years having the opportunity to progress from where they were. They made the choice to remain the same where as the line that man developed by the original parents may have made the choice to evolve through complexity.

With all the explaining I do not wish to prove that man had a nibble or never had a nibble from that specific tree but I only wish to indicate that there are a variety of interpretations and clues around and when sanctioned they may deliver a vastness of possibilities. There is one other tree of distinction mentioned and also a mention of some eating by the female at first and later on by the male. This tree also was named and it was the tree of good and evil but according to the Afrikaans Bible the mentioning of the name says specifically and I quote: *Boom van kennis, kennis van goed en kennis van kwaad"*. Directly translated it reads as follows "tree of knowledge, knowledge about good and knowledge about evil" and that is where my argument starts in my attempt to indicate the possibility of the sixth dimension belonging exclusively to man.

In the detailed analyses the specifics concentrate on the "knowledge" and then distinguishing between "knowledge of a good nature" and "knowledge of a bad nature". Please note there are three mentions of knowledge one only about knowledge then about the knowledge of the good and thirdly about the knowledge of the bad, but most important separating the three by distinction.

After clearing this part we may return to the animal world of some being wise and not very wise. Animals by nature and by genes acquire a base for knowledge to carry the specie through dangers and more important even, to the survival of the future of the surviving gene pool. Surviving as far as the animals go is the good and the evil and there, at that point all other definitions stop. All intelligence the specie holds and all the intelligence the specie acquired contains the one underlining element being individual and specie surviving.

In saying that I do certainly not say that rules amongst members of a group of animals is non-existent. There are certain criteria the individuals have to meet to establish rank in the tribe. At the same time such rules do not centre around emotions of ethics but they are practical well placed and directed to ensuring stability and it seems the higher evolved the specie became the more sophistication there are amongst rules constraining some members to the advantage of others.

The Matriarch in the Elephant herd is Boss and that is in capital letters! No bull will dare to push members and lesser infants around and she takes much less shit from young elephant males than the females. She is the rule and the law and every one abide by that. Should a male wish to afflict his attentions on some elephant cow the Matriarch will condition the visit to her satisfaction or the visiting male may even pay with his life.

I say this as a result from knowledge I acquired as some game-farming friends of mine has elephants in captivity. Crossing electric fencing is hardly an issue for the matriarch because she takes a teenage male place the young male (and it is always a male) between her and the fence and let him walk unsuspectingly alongside her as she deliberately holds him in a position where he walks between her and the fence with the current. Then at a moment she decides on she thrushes the young male allowing him to plough through the fence and take the electric current shocks while he goes on his way braking the fence altogether. After that the fence is open and clear for the rest of the herd to cross. She will never act in this manner towards young virgins but the young males get the stick every so often. With the advances the African elephant show would the African elephant be a good pet. I would hate to find out because they may set the rules and not me.

I know for a fact that a Nile crocodile does not make a very obedient housebroken pet. Should any one have an idea to keep one in his swimming pool be warned the pet would not distinguish very well between his owner and his next meal. And his love for children might be somewhat different to that which a good pet should have as is the case with dogs. With dogs man had thousands of years in breading good pets but it is unlikely that the first relations were as timid between master and dog as that we grew accustomed to.

1. Through many generations of exclusive inclusive breading did we finally manage dogs to have become what we wish them to be. In this lies another fact to analyse. Many different breeds make many different dogs and the one race has characteristics setting that race apart from other races but in the race itself the variety of characteristics find more prominence in the race than in individualism. Characteristics of dogs connect more to the type of dog than to the individual dog

and therefore some races have inborn hunting skills where others may have guarding skills. It is the breed that brings the selection and not the individualism in personalised characteristics. Therefore it cannot be said that a dog has a conscience but it is better said that a dog has a better breading line.

2. Some evidence suggests that when Cro-Magnon – man arrived agriculture replaced hunting as the feeding method and we are confident man exclusively kept that dog for it's hunting and sharing abilities. With the arriving of agriculture man then extended his space-time manipulation not only beyond his physical abilities in hunting but also his physical strength in working with tools. This must be the biggest leap of all even much bigger than the leap of the electronic age but such comparisons are extremely difficult to make.

What would be man's drive to not only manipulate his personal surrounding but also manipulate surroundings of other forms of life to their benefit but moreover to his benefit. What would give man such judgement as to select species beyond him and feed them to eventfully find more benefit from their feeding than they did benefit from their feeding. Genes it cannot be . The orang-utan has 98 . 2 % of the genes man has and the gorilla has 97.8 % of the same genes man has. The Gorilla still lives in woods and is destined to disappear while the orang-utan lives in trees and holds no better future prospects. Genes would at least give the species having such close relation in the gene pool with man an idea to follow the trend set by man and copy some of the abilities. Genes it cannot be and that just about excludes the last cosmic or natural physical explanation from the list of possibilities.

There seems to be a massive gap between what man became and what ape became. Science makes a great singsong about chimpanzees with the ability to use tools for their benefit but man has surpassed that so long ago science have no tracking record about the time and the way that came about. It seems as if man was not, then man was with agriculture and all other providing the manipulation of the other species under the control of man and by increasing all benefiting that the animal enjoy man could bring benefiting all around to benefit man.

One day man was ape and the next day man became super-specie-of-the-world, the world champions in space-time manipulation or in other words of controlling life. Not only the life of man but also life of others to some mutual benefit slanting heavily in the favour of man. Still the benefit of the other species holds so much that only species that benefit man started to dominate world population with man.

Of course as usual and as with most of Newtonian scientists' findings, I question the accuracy of the gene pool percentages strongly and I am of an opinion that such percentages are in use for political issues more than scientific proof. I prefer leaving it at that. I am a very small fish in a very large pond sharing the pond with very powerful other fish that can destroy a small fish like me with one gulp.

Going into the development journey as man followed the trail one has to look at not what man achieved by own ability but with own measure in manipulating others in life. When man started with chips and flint it was progress but it was also very limited. Only when man acquired the muscles of more powerful animals and took their ability in measure with mans manipulative power did success arrive at a level that brought progress in leaps and bounds.

It is not mans hands or legs that brought man's domination and control but man's brain that brought response from other life to benefit man and find benefit in shared life styles where mutuality brought about safety and mutual prosperity. It is surviving more than anything else that means good or bad to animals and most of all surviving of the species and in that the animal only find man and man's company good because man holds its safety. When a lion brings down it's pray the rest of the flock will start grazing immediately without showing even slight remorse to the victim and for the loss the close relatives are faced with. By the death of one the rest find safety and to animals that is good. To the rest it is about surviving and if one pays the price, little concern goes to who paid the price. That is what annoyed me about our all-famous-pop-star-wife. She truly go beyond what nature puts down as rules applying her liking (and selling her book with cheap bluff) while other sheep walking on two legs like only humans ought to cheer her stupidity as if the stupidly were they're own. They have not even got that much brains to acquire that much stupidity but has to borrow to get their tally that High. The biggest annoyance is that with such stupidity those can vote and choose my future because of democracy. They prove almost all ways not to have the thinking power of a mouse but they have the rite to choose my future and I have no say in the matter but to follow what such morons may wish upon us.

With the novel idea (novel as it is only man that uses mutuality single-minded and still provide beneficial good from all angles) of widening the use of abilities provided to different developed species, man gained extensively in progress and comfort. After all it is much lighter work riding a horse than walking all the way.

But gains in comfort goes both ways as the horse find protection against predator attacks while finding good nourishment in winter and the best hay to feed. Such a diet provides the horse with the strength to carry the rider and enjoy own comfort with the fact of much reduced fear and anxiety. By having male and female and promoting mating the good in the life of the horse becomes better. Did man rob horse from freedom? Did man take what was not his to take? Should man get some conscience attack leaving him with sleepless nights about his cruelty in robbing the horse of a natural life of freedom? Well the humanists will tell you with teary eyes and running noses that the horse should have its freedom as all are born to be free. But this emotional outcry comes with the comfort they, the humanist enjoy of secured sleeping a good all year round food supply and breading safety. I have not seen one humanist go running into the mountains never to return to civilisation, to enjoy the freedom they should enjoy as much as they wish that upon the animals in captivity. To the horse, after the initial fear of the subduing and the ultimate realising that the subduing is not life threatening as he can live with that, he finds comfort and even enjoyment in that. I see on a daily bases how horses get jealous when the owner takes one to ride when the rest wants to go first. They come and nestle with a desire to connect and in jealousy push each other away to be the one receiving the owner's attention. It is a bigger issue of the conscious to decide the likes of others in what they like and what you think they may like. No one of sane mind will hurt an animal in your care and when slaughtering that we do in the most humane way as described by law. No one cuts of a chunk of steak while the animal is alive. We humans have civil norms and values and by using our brains we can live and let live with more dignity going around sparing the animals huge cruelty than what the animals would have come to face if they're fate still was in freedom and being hunted down by wolves and hyenas Such is the difference between those having bleeding hearts and brainless skulls and others that can think. Now we arrive at an interesting question as how do we think and reason. I am sure all humanists will have as much to say about my way of being correct only as they will find many arguments as proof of the fact that by dislodging my logic they can prove me being beyond the norm of classifiably insane. There is ever a clear definition about rite or wrong and all principles we find appealing or appalling is within the brain

According to an article I read the brain holds more connecting lines than does the universe and I may even accept such a statement on the grounds that life has much more complexity than does the universe. After all life takes the dimensional barrier as far as the universe does and then beyond where the universe stops. This does not make the universe simple because I cannot see how any person may ever come to understand the flow of light as the light uses both the straight line of singularity, the half circle presented as the Roche limit in singularity and the Titius Bode triangle making light representative of every aspect which connects space-time away from singularity with singularity as light where Π meets r to become the value of C. But in the brain this is only one function as electricity holds an equivalent of light forming electricity as the messenger to whatever energy is above life and then in the human capacity above even what forms the barrier to life. The arm is not human life because a human can loose the arm send still be alive. Therefore what ever is in control of the arm is in control of life, which puts what we find as life at a higher dimension than that of life. You may argue that in case of animals such thought also control life because a dog may lose his legs but not his life. But even as complex as that may become there are relevancies between life and the physical because where the physical uses pain as a warning system the mind uses fear as a warning system.

By following such a line of argument one can freely deduct that an insect as our corncrake, which we discussed earlier on, is representative of life as the same life we find in the arms or legs of our body and the life we control but is not truly part of the energy "me". Clear to all it must be that life we find in mammals are advanced above and beyond the development the insect arrive at. If that is the case then I may claim that human life has more developed than what other mammals did because with my manipulation of space-time I may manipulate other mammals to harvest some of their manipulative abilities in benefit of our mutual relation inclining more to my benefit. The cow does not seem to mind when I milk her but can any person imagine experiencing a milking session involving a crocodile? Well, fortunately crocodiles have no milk but if they had I would never volunteer for the honours of being the first to train a crocodile how to behave in a steady manner when in a dairy session. You may have or may not have noticed but I am telling you that they have a sharp side and they have a blunt side and the sharp end holds rows of teeth they surly know how to use. Even the blunt side hits like any whip never can and I am sure a fully groan specimen may kill with that tail. Going down the order of evolution we come to bacteria and viruses, some of the lowest forms of life. Do bacteria and viruses count as animal and if not then surely they count for life because life they are. In that we find the equivalent of bacteria in higher developed species as we find that the insect may have the developed life mammals use in their bi-products included sustaining their superior life development. We can see evolution by applying a relevancy of devolution to siphon and separate life from life. I am trying to indicate that life becomes a compliment where the lesser developed formed a mutuality and aided the supreme form of developed who is controlling the master brain in that form of life wherever the master brain may be attached.

Life is above and beyond the cosmos and surely even the most ardent supporter of atheism must grant me that much. By that grant must the atheist then add the fact that life cannot only fall onto a category of to be or not to be but there is a range in life forming a line of development and superiority. Life is more than life but has status of being lesser or more and that is the point I wish to address after all the talk.
Within one body a range of life values combine in making whatever accomplishments the life form accumulated by extensive development that range in development. It is appreciable in concluding from a range of facts I mentioned but mostly from human common sense we all know that on top of the range being the model best manufactured and with all accessories all other models also having life envies and fear is man. What will make man that special?

In 1905 a case was reported for the first time of a woman that had a hand, which attacked her every night by trying to strangle her. In the manner the hand acted it was clearly out of her control as it was clearly out of control of whatever controls the brain have over body functions. She would wake at night and feel someone strangling her but the person strangling her was something she ought to have under her control. Imagine waking one night feeling someone squeeze all life out of you with a murderess motive. Even the thought of that will make most people get up and bolt their doors and windows just in case. It must be awful having a murderer wake you with such a horrific intention. Go one step further and think that person may be one of your house members you trust with your life. A thought of that becomes rather preposterous! Then for the ultimate in revulsion; think of the chances something acting in such a horrendous manner is something you know with every grain of your body as that thing acting is your body. There is nothing worse to be scared of than being scared of "you". How do you fight such an act. You cannot hide and you cannot go without sleep and as you go to sleep you know that there is some part of you yourself hat is after your life. If this is not enough to drive any person into hysteria I do not want to know of anything worse.

This flabbergasted the doters and no one seemed to make any sense of such phenomenon. I suppose if such an incident had occurred before it would have been denounced as an act of a demon of some kind but fortunately for medicine the art of healing had abandoned forces of nature as a scientific accepted fact unlike the likes of physics still clinging on to such madness. If my memory serves me correctly this case was in Germany. Please remember unlike our distinct academics I have no extended library to find all kinds of information but have to rely on a failing memory being destroyed by my diabetes.

Then in France later on another case became known about a woman that had a hand also out of control where in this case that hand tried to forcibly turn the steering of the motor car she was driving to force an accident of a serious nature. This manipulation was seemingly as much out of control as was the previous recorded case. The common factor about the two cases was that in each case a person had an arm that was intent on destroying that person without the person aiming to do so in free will.

Later on in America two neurosurgeons planned an operation procedure where by they aimed to relieve patients having chronic and continuous convulsion attacks caused by epilepsies in the brain. These cases were dire and with the operation as a last resort all the serious after effects became a secondary factor to the superior motivation of saving life and improving the demented quality of life by the patients .

The operations involved only the utter most serious cases that left no other option for improvement. It was this or death and not choosing death the patients chose the intended operation procedure instead, but still it was extremely serious and dire options in the choosing.

They reasoned that the epilepsy was a result of the brain having vibration and with the vibration stimulating other vibrations through the brain in some cases the one caused the next vibration and it was more a reflex of the first causing the next as the symptoms was going on a prolonged non stop convulsion. To stop such reflex by the brain tissue they held the argument that when cutting the cortex the two lobes attached will not have the reflex and thus the continues convulsions will loose the continuous effect.

By separating the lobes the nerve attack coming about in the one side will not transfer to the other side thus it would not cure all elliptic attacks but the prolonging effect will be reduced. One vibrating lobe wills then being separated from the other part not cause the response in the next lobe and it was diagnosed that it was more a response to the reflex allowing a reflex to the respond and this brought about a never-ending cycle. The idea was that it would result in reducing the severity of such grand mal epilepsy

According to American law the doctors first had to show a high degree of success by operating on rats in order to prove that the consequences of such an operation is in acceptable levels before starting such a procedure on humans. To obtain the rite by law for the granting of the operation many rats underwent the procedure and the procedure then were extended to many other species.

Every aspect of recovery and side affects must be documented to an exact accuracy with no exception to the rule in the slightest.

The behaviour of the animals before and after and the general physical data then goes to excessive detailed scrutiny by the finest the medical profession has to offer in America. Accuracy in the process of accumulating data and other relevant information is beyond question especially in the country with the highest standard in medical care.

The after affects the procedure had on animals were indicating no serious side affects of any reason for concern. Many species went through the procedure eliminating defects if whatever possibilities there may be.

When monkeys went through the operation procedure our primate cousins had no side affects in any way. There was perfect hand eye co-ordination and the nerve system had no complications with the motorized operating functions in any way. This confirmed the surmise the medical profession then at the time had that this third lob was just fibre with no function of distinction. The fibre was in position to stabilise the two lobes and had no connection with the lobe in a functional manner at all. All the indicators brought about such positive results that the American government granted the licence for the first experimental operation conducted in a human in absolute confidence the procedure went about and with a very good outcome. But shock was looming to all medical experts.

In every case the patients had one common disability. It showed a horrible disadvantage no one expected in the least.

All patients showed the science of a phenomenon later named after a movie Peter Sellers made famous. It was named the doctor Lovejoy syndrome because on of his characters in the movie was an eccentric half mad all crazy German general that had one arm always trying to strangle him. This was meant to be funny in the film but the patients suffering from the reaction of the aftermath are not so inclined to the humour. They all had one arm that went out of control and the arm showed serious signs of having a mind of its own by doing the most annoying things the patients had obviously no control over. The one arm had life apart from the person free will doing things that would embarrass or even threaten the so-to-be owner of the limb.

Well I am no brain- surgeon although I am inclined mostly to form an opinion of my own that may not always stroke with informed opinions by professionals. The test operations were conducted on a variety of animals including monkeys, the so-they-say close relative of man. Well as close as they can get but I am of the opinion there are other species being still closer to man, but that is somewhat off the point. From all the facts I mentioned the past pages I drew a conclusion of my own.

I showed that man has a higher evolved form of life than other animals.

In the Bible it reads that man was made the last but far from the least with more superior qualities than all life combined because man is all life combined and then added more than the fare share. Any one thinking of our Creator as a magician is mad and that I state without excluding even the Pope or preacher of whatever denomination. The Creator is a building architect applying mathematics and physics we can never come to appreciate. If there were any one that has an opinion that God spoke a word and magic was the word that person would have another opinion when thinking with some clarity about Creation as a whole. The Creator is Creating and by creating there is a building process involved. Every person starts an individual process of building a human just after birth.

Looking at tribes living in regions far away from civilisation as one can still find in the Amazon River we find those individuals being adults by body still play games we find our children play with much amusement and childish enjoyment.

It is far from incomprehensible to make some sort of comparison between our children at play and those adults at play and the similarities are astonishing. Racist remark it may be but the grown ups are more child than the western child is child. They defiantly are backward in mind and mentality. The most logic deduction is that the child in us represents our development phases through a long journey. Humans at birth are animals. Babies can make noises to convey their needs and nothing more. As the little life grows it develop not only by growing but also by culture and what the parents put into the culture. It is more than likely that the developing pattern children follow is the same developing line humans evolved through as the generations brought more insight and better understanding to the following generation. It is only of late that there is some pattern of devolution taking place especially in the western world but that trend is set by a culture of greed. The parents are chasing a good life and instead of placing morals they push money in the hands of their children to get the children out of the way so that the parent will lead the life they choose and rid them of the burden of children while conveniently blame teachers for the children having unacceptable behaviour as much as they pay psychologists good money to correct they're mistakes. It is moreover a fact that parents do not actually pay the professionals to fix but pay the professionals to rid parents of responsibility, guilt and of course the children. The price society will pay for such luxury is far more expensive than affordable.

I so many times wished I could have my life over again. There is one study I shall conduct and that is follow the pattern the child indicates how the human developing process came to pass and draw the parallels from that to man's evolutionary path. It truly must be a study worth one life. To know man that well must be the ultimate there is to know and afterwards death can bring no regrets for wiser no one can ever be.
I think we should now return to what the Bible refer to as the tree of Knowledge and reflect once again on that verse. The quote was and I quote in firstly Afrikaans: Boom van kennis, kennis van goed en kennis van kwaad". Directly translated it reads as follows "tree of knowledge, knowledge about good and knowledge about evil" and that is where my argument starts in my attempt to indicate the possibility of the sixth dimension

From the stage of toddler I found eve dice of three characters in all persons. There could be more but not less. Let us call the three persons three entities of the good the bad and the ugly, but not in such a specific as being a saint a pleasant and a demon. The entities are rather more under cover that that straight foreword by definition. In every person's brain somewhere there are the trinity of entities ruling our lives.

The trinity are one person and not individual persons but the same although very integrated they are also very separated. It is not a question of schizophrenia with multiple personalities because I do not believe in that. I think that was made up to convince who ever needed convincing about something to do with nothing and has no base or then has the same base as physics hold gravity responsible for a variety of facts they otherwise have to admit they no nothing about.

The three belongs to one person and in fact is the same person. In the Afrikaans book I named them Ek, my and myself translated as I, me and myself. To make matters more interesting and subdue confusion somewhat I wish to keep the Afrikaans as that would make explaining slightly less complicated. Ek is I. My is me and myself is will you believe it myself. The "My" one pronounce just as you pronounce the month of May in English and that would make the pronunciation of myself as you would say May-self in English with the pronouncing of self in English and Afrikaans exactly alike.

Ek my and myself are the same and there are no distinguishing between the characters but at the same time they are as far apart as three that never met before. Every of the entities belonging to the same body, the same brain one has a character as unique and as far apart as another being on different sides of the universe.

Every personality has different likes and different dislikes and feels a different purpose in life as much as to life Once again I wish to press home the fact that this is (to my view) as normal as breathing and has nothing to do with the mental instability known as schizophrenia. Although only one personality claims occupation of the body at any given time, all three take responsibility for action all the time because all three are the compliment of one.

Any one such personality may claim occupation of the body at any single moment and normally do not relent occupation easily but of the three one is always in charge and the other two take position when the main personality loses concentration or relaxes guard of the situation. Every personally has own motives being apart as far as the north may be from the south. One may even think them in classes of being the person's personal god and personal devil but such a thought may place boundaries that are unfair.

I would rather describe them as one is the charger being scared not even of the devil himself and the other will be the cautious the guard, the one always on the lookout for trouble coming. They form the one that is in charge and the one in charge cannot be excused because it was a totally foreign entity pushing in charge in the direction it never wished to go but was blackmailed in doing that wrong! The one in charge takes responsibility to the full for each dead the body did and every wrong committed. In the end they are only identifiable but not that clearly deniable and they always appose each other in complimenting one another.

In the American cases where the neurosurgeons performed the operation in cutting the cortex to split the brain lobes of the operated patients a condition became a situation where the persons found the uncontrolled motorised motions of limbs under their control supposedly, but not under their control at all. Then for the first time the phenomenon became a syndrome with a name. It became "the alien hand syndrome" but also find referring by the use of the Doter Lovejoy syndrome named after the movie. There are cases known on record to result from a variety of brain damages and severe apoplexy. I can confirm as a witness that in the cases of my biker friends their faces changed with mood swing. Not bone structure or complex feature but the facial expression changed the way the muscle form the face. In cases they looked somewhat alien, but my referring to as normal is not about such extremes.

The alien hand syndrome observes cases where the one and the patient refer to as the naughty hand and the clever hand. The naughty hand seems as if it had a life and mind of its own sometimes acting to embarrass but some times even to endanger. The main connecting issue between these cases is the hands control not being within the owners authority and that the hands will obey as nothing ever change but then on occasion from the blue it will act on own impulse and with a motive clearly never matching the owners intension.

I wish to underline with no exclusion to whatever intension that the alien hand syndrome does not prove the trinity that I refer to and the alien hand syndrome underlines two facets where as my view state clearly three identities. The reason for this mismatch holds a most intense connection to my personal religion, which I never share, with any person out side members of my immediate family. The alien hand syndrome only confirms (to me personally) a connection of some sorts in some way.

Many years ago after I had to overcome some personal problems at the time through which I was admitted on occasions to an institution. I spent time in nerve clinics where I had to recover from some brain disorder called endogenetic depression. It is a condition that the patient will get severe attacks of depression and the cause of this is totally inherited by nature. The disease comes from a bad gene carried from parent to child and as much as my father suffered from it so will my children and is not very scares or very serious. I would say it is as serious as you make it to be. My regrets about suffering from this condition is minimal because of the extent of learning I had the opportunity to come by that otherwise would never have come my way. But then from what I saw in the times I was admitted to the clinics I mentioned, my (unprofessional) conclusion is that in just about all cases the prognoses depend almost entirely on the patients willingness to recover and how serious the patients are about recovering as the recovering is al about the relevancy struck between obsessions and true problems. In that there rests a balance between fighting for the sympathy they wish to evoke in others and fighting a battle to achieve oppugn from the problem. In hindsight and after all the bad is forgotten it was worth wile because the learning of the human mind and the way others think and feel was enriching beyond my suffering. At the time the suffering was almost overwhelming but to escape ones own problems one can listen to others and by learning from them one get perception about their state of mind and your own state of mind. It was in this period I came to the conclusion about the trinity within us all. By increasing my personal learning curve I decreased my personal discomfort.

Inside all of us lurks to personalities above and beyond Ek and they go by the names of My and Myself. Most of the times they behave well but sometimes they can bee opprobrious without my consent in the matter. One of them and I leave the choice up to you are truly obnoxious and spiteful more to yourself than to others. The other one is normally timid and can be classified generally as your conscience. Between them you come and take the control because you are the boss. Being the boss and demanding control you are the balance as much as you take blame and shame when not being the boss and not in control where matters go out of control. You carry the consequences always as you should because whenever what ever goes wrong others will pass the blame on to you. And so they should because you are responsible even in such times that you are not responsible.

Ek finds himself in the middle of My and Myself and as Ek is in the middle Ek get advise from My and Myself. In accepting or rejecting comes regrets and jubilation but ultimately the final choice is with Ek. The characters are strong at times and are weak at times limbering and dominating whenever opportunity presents. In the Afrikaans I identified the character by names but found it somewhat complicating the issue in the English and so I did not name to identify by character.

Ek takes full responsibility for all actions for Ek is My and is Myself by only being Ek. We all have this in us and some find ways to fight it better and some of us are in a desperate fight for sanity and survival. The issue is not the guilt because every one is guilt ridden as we all have to cope with this behaviour in some degree. The stronger any one denies this the harder the one is in a fight for self-protection. Cases are sometimes most serious and other times just under the skin but it is there in every one. If you are human you are fighting. Being the one finally carrying the burden then becomes self.

It could be where he father feels he as person never accomplished anything and with that self detesting he puts all his blame, guilt and rejection for what he is into the passion he feels for his child. The passion he feels is carrying the burden of hate he has for his image and that drives him to expect from his child what he never could achieve. The child has to be many times better performing with many times more positive results, be a champion, be a scholastic genius, become the school president and make a mark as a pillar of the community although the child is only a child. The child has to outperform all others on all terrain in a fashion fitting the image of a champion, which the father never was. It could be as serious as I indicate or it could only be in little suggestions made to better his child.

These same feeling can bring about the very opposite where the father tries at every chance he gets to run his child into the mud because the father fears that the child will push him out of his role he has to fill but lack all abilities to fill the role. By destroying the child he is protecting he child because he is securing his position as the father figure. Knowing well the chid carries his genes and believing the genes the child is carrying is not worth much, he protects the child by showing the child what he (the child) is and allowing the child the realisation before the child will one day find out for himself in the cruel world. There is no good as much as there is no bad, and in the same breath everything is wrong.

The same behaviour may come from the father in fortune, the one with success, the pillar of the community. He is the top judge, the success fill attorney the town's top businessman the city mayor the admiration of others. He hates his child for that child will one day inherit all the goodness he now has and that makes him sick. He might feel the child will never become the doctor he now is but through his reputation that he worked so hard to achieve will become even greater than he now is. That makes him to push the child to live up to his personal greatness or destroys the child to show the child how much the child should be great fill for the admirable fortune the child has to have a father such as he.

It does not have to be about someone you love. I shall be very frank about my case and millions will recognise their fight.. My personal struggle involves money for one. I have not the slightest idea how to administrate my money affairs and sometimes I know I have this inward hate towards money. Whenever I have money I allow people to sucker me with a sob and crying story about their hard life and the bleeding heart dashes through, the knight within me with the glittering armour takes full control and helps me to give away sometimes even thousands and tens of thousands, knowing very well notwithstanding all the promises of repayment at the time, I shall never see the money again. It is not I being the bleeding heart and then yea it is I the bleeding heart but that is one character. I the bleeding heart am on the background where I the hating bastard am rite on the dot standing on all fours and then some shouting and urging me to help the others, or buy what ever. Forever it is ensuring me that there is thousands more coming my way in any way so what the hell, let some goodness flow. This will always come where for some reason some money fountain runs dry just afterwards and I drop my family into financial surviving periods allowing them, the ones I love most to suffer the hardships. Not once did this occur and was not the prelude to personal hard times! I am a middle-aged man. I know these characters. I recognise the precise feeling accompanying each one. I have scrutinised and analysed they're being part of me decades ago as I and the bastards still catch me again and again, sucker punching me over and over.

The first time I went gambling I recognised the one being very aggressive and on the forefront just under the skin filling me with anxious excitement and then and there I realised gambling was not meant for me. In that sense I have beaten him hands down by not starting the habit. Old positive me is about cars speed and going crazy in my mind and I know it is the strongest one because that one am the biggest I. A middle-aged man tearing down some street on a massive motorbike showing some youngster what the bike can do when there is someone on it that knows his onions are grossly irresponsible. How childish can you get, how in mature can one be. That is the I and that is the very me and that is Ek and I have as much control over that as a drunkard has control over his drinking. Saying that is also admitting I do not wish to fight him as I do not wish to beat him. I find his company very pleasant stimulating and destructive and when he comes to the foreground all other characters harmonise giving me all the different feeling each one of them should supply. I am then self-destroying giving the negative character his day, I am childish giving the neutral character his day while the positive one and I am the same person to let all power loos.

My characters bring me in conflict whenever I bay or sell or demand a price for my services rendered. I would bay the biggest shit at the highest price in the belief I am helping the poor slob. When I sell I get all guilt ridden when trying to make a profit because I get this idea I am cheating the poor fellow by insisting on the price I am aiming for. In all cases I feel so guilty to ask any person money that actually belongs to me I get a nerve attack or a running tummy. It is not out of fear because if any person gets violent or aggressive I know how to defend myself. You have to know that if you are a devoted biker because they always get drunk and strong at the same time and fortunately I was in motor racing during the age my mates learned to drink. But with racing cars always braking down and crashing there is never money or time to drink so I never got around to start the habit and by the time I stopped racing every one accepted me as a teetotaller so I never got around to develop the habit.. But asking for money or insisting on a fair price is more than I can achieve.

I shared with you this Ek, My and Myself part of the personal me-story so that you, every you will know what I mean because the objects may change but the objectivity and feeling and the motives never change.

One example how the process works will be when a mother that hates her child for birth pains, unwanted pregnancy, feeling unfulfilled through a bad self image or carrying on where her parents left off with them treating her as a child in the very same manner she treats her child. The negative character prompts the mother to reject the child as she detest the child and loathe the child. The Positive character saddles her with a tremendous guilt punishing her as anxiety in blame riddles her.

In self-punishment aiming to destroy her because after all that is the purpose of the negative character the positive character punishes her as she fills her with self-loathe while neutral character tells her about the wickedness in her. She gets reminded at every opportunity about her love she must have for her child and her duties as a good mother to love her child. It is her responsibility to love and protect and because she strives to be the good in her the punishment is severe.

If she gave in to the negative character and start enjoying the hate and blame she feels towards the child the positive character let loose the pre-historic maternal instincts with a flow of torturous guilt hidden behind such strong emotions she deflects the hate onto her self.

Then comes the neutral character reminding her that no one should ever know about her hate towards the child because as she detest herself so would the world detest her if ever someone became wise to her feelings and she will be driven from society with hammer and tongs. With the conflicting hate polarised and swinging between her and the child she knows that all of human kind will see her for what she is and with the hate she feels towards herself the negative character takes that hate and turn it into fear for others finding out about the truth. In realising that no one may ever know about her true feelings toward the child and know about the loathing she feels about herself she takes all precaution to hide it from the world.

In this mind game of rocking emotions the positive character supply her with advice in how to take charge of the situation never to the benefit of the child but to her benefit in protecting herself from the outside world. The advice will never have any concern about the child because she carries that burden by herself. The positive character continuously reprimands her of her evilness while the negative character fills her with hate running between the child and her feelings about herself. The neutral character reminds her constantly not to allow any one to find out about her feelings for her child and demands protection of the outside world finding out about her and her child. The neutral character pushes the fear to match the severity of the onslaught by the other characters in order to maintain equilibrium and equilibrium means almost insanity

As the insanity at times almost become intolerable she reflects the blame for her situation onto the child being there and making her life hell. This the negative character grabs by prompting her to punish the child as severe as she can and through this stop the child torturing her. The cycle leads to the next cycle and in this sanity guilt love and hate becomes one flowing emotion of disturbance. With this conflict within the mother the child's developing personality receives knocks the child cannot stand and less understand. The child starts behaving rebellious and unacceptable to the mother's neutral side and in the eyes of the community. This chance the positive character grabs and being positive only as far as the mother's well being goes the positive character advises the mother to leave the child and let be. This will show the world how much she loves her child by refusing to even punish the child when the need arrives.

In this advise the negative character joins by advising her to let the child become out of control establishing the fact that should her hate towards the child ever leak out, the world will not blame her for every one in the world that has contact with the child will hate the child in any case. When every one despises the child the negative character swings into action by filling the mother with more hatred towards the child and when opportunity comes and they find themselves alone the hate comes to the foreground and then she punishes the child with most cruelty. This can be as part of actual criminal prosecutable child brutality or it could be most cunning and devilish in conspiring but the brutality is all the same. With the presence of such severe child brutality all others in the community turn a blind eye not to get involved where each outside person will find some excuse not to become involved. As every one has a struggle of their own they too are in self-protecting not feeling the urge to come into the open and defend the child. Being in the open will unveil their personal fight for survival and they then will become the target of the community. All this is in the very distant back of our minds never in front where we can kill it but present in the way to be us and not to be us.

All three personalities agree on one thing and that is that the world must with all its people have a hate in the child as much as the mother. But as every one find the child unacceptable and revolting she can feel better about her feelings because now she is part of the crowd. Being part of the crowd will bring sympathy from others with their understanding about the hell and the torment this child inflicts on her every day. Such a feeling soothes the aguish of the wrong she feels she is committing as much as the wrong the child is committing to her. By allowing the child adverse behaviour and defending the punishment of such behaviour the child will become more unacceptable and the situation heads directly for the disaster she hopes to accomplish. She directs the child's personality in that direction while she feels all the torment others see her go through. She allows the child to go to nightclubs doing drugs and commit self-destruction while the mother is merely a spectator because after all that is the child she hates.

When the child is out at four in the morning she can feel good about not having the pest around. She can feel good about her hating the child. She can feel good about all the sympathy she receives about having such an unruly child. She can live the life she claimed, hating the child, feeling sorry for herself and good about others understanding what she is going through. Should the father, a teacher, a policeman or any other figure try to stop the madness and bring order to the child she will attack that person with all the hatred she feels towards herself and towards he child. She will destroy the prevention of her self destruction because any body trying to discipline the child is fighting her aim to destroy the chid and after all then the world will see how she can fight for her child's protection by almost putting her life on the line.

This becomes the Sudan affair where the bleeding heart buys guilt relief and be god while the philanthropist pushes guilt as hard he can and be god to collect money on behalf of the Hoggenheimers that then can be god with such wealth distributing it to the Mammonites who can be god by baying from themselves as much as selling to themselves with unscrupulous profits making him god and allowing the Mammonists to be a slave driver and being god to the slaves. It is this sickness of society no one cares to see because every one gets what they want, even the luckless get what they want with the minor condition that when the luckless suffer most that becomes the region where most profits are for every one in the chain of gods. So the luckless must be in crises starving as they are dying to gain most profit for every one. The profit has little to do with money but with being god. Any attempt to stop the situation will never be tolerated by any party and therefore my remark that every one will press for my castration because of my suggestion to rectify and bring a solution.

The mother will fight any and all positive solutions with tooth and claw and no one should dare to lay a finger on the child because she know how successful she is in destroying the child and it is so easy to shout child abuse when someone wishes to correct the ways of the child to the benefit of the child in the interest of the child. Her devoting love will protect the child from such brutality as what a good hiding on the backside will bring if the hiding is done in love and the child knows it was on behalf of care. But that will stand in contrast to the mother's brutality and punishment and the child may recognise the difference and wise up to the difference.

The social worker will never accept such brutality after all it may cure the little brat and put our social worker out of a job. The lawyers and judges are on such a big job creation drive by minimising penalties and getting criminals back on the street for the next cycle in crime they will fight any interference that may reduce the crime and decrease their chances of money making. With so many to loose so much on the one side and only the child to gain on the other side all brutality in the name of positive punishment will become child molesting and will never be tolerated by those with influence in society.

The mothers behaviour becomes a reflection on the disease within society and it is in everybody's interest but the child's not to admit to any knowledge about the foundation behind the scenario, after all it is only a child going down the tube and to top it all it is a child no one cares for. Is there anyone out there that can see the parallels running here between the animals not caring for the unprotected infant crying in desperation for a mother while every one ells in the species cannot be bothered? May I now comment on the fact that we are going the way of the animal and devolution of our species is in progress?

Weather you care to admit it or not but greed and money is destroying man to the fullest while man is enjoying the destruction with all its lust. The drive in society after W.W.2 became progressively to feed the children to the hyenas of society, which are the crime bosses, the prostitution rings, the drug pushers because after all, they bank the money at the Hoggenheimers to the convenience of the Mammonites and Mammonists. The ones that are caught in police action are the ones not part of the official system and they become the offers the politicians demand in protection to show the public the system is doing what it can but unfortunately it can only do that much and if it is not good enough it is because we all are human. Hundreds of thousands of children disappear through out the world and no special task force has ever been set into action to get behind the problem This problem has no boundaries in as much as it is going on in every country there is world wide. I have an awful, awful feeling and please consider the next remark as a thought with no substance but that internationally oil is bought with children as payment because no politician through out the world shows much concern. But let three banks get robbed in one day, then a special task force comes into action and gets the culprits with extreme prejudice. This is the symptom while the mother's behaviour is the condition.

The positive character advises the mother to defend her child against disciplinary measure, the neutral character reminds her of the image problem and demands protection while the negative character sees the child slip into the ditch and everyone is happy. Every aspect is in line with almost one aim and that is the destroying of the child. What ever may bring positive results everyone shout down by making the connection where the punishment links directly to what may be extremely negative because from the onset it seems cruel and negative. Giving the child the spanking of his or her life driving the fear of god into the young person will have extreme negativity but that must stand in complete contrast to the love the child then must receive before and after the spanking. The child must know with one hundred percent certainty that the parent is and will always conduct the child's care with one aim and that is to ensure her or his well-being. But if the child knows that every aspect of the child care swings around the drive as far as getting the young person destroyed the yes, the child will find all connection to punishment intended on the destruction aspect but it will remind the child of the similarity and lack of contrast in the usual treatment because of the absence of the love and caring aspect is in harmony with destroying.

The whole aspect changes around when there is another person also punishing the child but with prejudice intended. When the punishment comes from despise and not from care the mother sits back and allow this to happen. If the father is sexually abusing the child the mother will suffer greatly all in silence all quiet not allowing outside intervention spoil the situation because then it is the father who is to blame. It is the father that is destroying the child and it is the father that is the devil. The other parent can then take responsibility and blame and the mother becomes the second blameless victim in the case where she does not participate in the abuse only because the father then plays the part. That is the only aspect that changes where as otherwise the scenario remains the same. All characters play their part as if she is doing the destroying because she is doing the destroying by helping to provide the perfect environment for the destroying and not allow any clue get outside the close knit intimate family circle. She takes a part in the abuse and takes as much enjoyment as if she was acting although the blame and the soothing shifts somewhat but all intensions still encircle the destruction of the child.

The conflict within the woman may drive her to protect the child she hates as much as the father does. Her neutral character tells her that now she is no longer to blame therefore what ever happens she can stand in the shadow of the male and he has to take the blame. But the negative character will not tolerate such idleness. The negative character wants the child's destruction but moreover the character wants her destruction. He advises her to action. When the father is at his most dangerous being overwhelm with cruelty she will jump in and save the child by physically protecting the child while knowing full well that she can do as little protection as the neighbours budgie can. This action will satisfy the positive character by her showing her unbound unlimited care and devotion as the epitome of true motherly love. With her actions she know she will unleash much more anger and the father will loose all control. The negative character finds stimulation in this action and supports more involvement to unleash more violence all the way. In such a rage she knows the father will then beat the daylights out of her before he turns onto the child with more rage than he had before. With such reaction all three characters are satisfied The positive finds a way where she can become good, the neutral character knows that society will condemn the father and the negative character will justify the cruelty as the correct way to go because the child the father and the mother is bent on destruction. Now the positive character can tell her she did what any good mother would have done, the neutral character knows the beating is the same as what she endured as child in any case and that did not kill her so this beating cannot be that bad while the negative character will enjoy the situation to its full as she and the child is being destroyed.

The balanced behaviour of a person with the trio not having violence to promote would run outside and call outside help from any source available at that moment. A woman beater and child abuser is always, always a coward and when real trouble arrives he will stop immediately. But if the trio is involved in violent provocation exemplifying hatred to the child, she would take charge in a different manner. Even if she does act in this manner she will not call the police or get the husband behind bars because she argues that the family will suffer with the father not providing at the time. No one receives any money while being locked up. The excuse she uses is that there will be will no one to provide for the needs of the family living expenses. The father's inability to provide is all but the truth as she wants the situation to continue because she is enjoying it as much as the father. Should she truly admit to the seriousness of the crime she would have the father locked up as if he died and never allow him close to any member of his family again. To her and her child the best will be if the father is dead because the father will destroy where ever he involves himself. The only sane thing to do would be to declare the father dead as far as the family concerns go because when he is released from prison she would have a new life to live that no longer will depend on him or his providing. She would recognise him for the monster he is and not see him as the senior partner in crime, which he truly is. Keeping this partner ship in place by using a lame excuse like ninety nine percent of woman finding themselves in such a situation uses would satisfy her three characters and once more the money matters more. Now the stage is set for the beginning of the next round because with the father as with her the creation of the next climax begins and develop until the next time hell brakes loos where the father then has even more hate and a lot more to prove and correct. His fury and outrage will cover his hate but also the negative character in him will demand revenge as compensation for lost pride. He will repeat his role and she will repeat her role and the child will have no role but suffer destruction and again and again the process goes on and on.

By throwing her body in between she knows no woman can stand against any man in a physical fight. Her excuse for aggravating the situation is that it is what any good mother will do and that also becomes the advise of the positive character. She knows she will outrage him blowing his week self esteem out of reality because now she takes him on as a man insulting him at the area he feels the weakest because he acts in the manner that he does because he knows from his weakness he can never manage to be a man in the company of men. The outrage she unleashes within him will satisfy the negative character. Reminding him of his coward ness and weakness will bring the monster in him to its full potential and that is exactly the plan. By her action he will be reminded of his weakness and that will be in response from advice given by the negative character as for the violence part and the shift in the blaming will be in responding to advice given by the neutral character. The child will now see how much she cares and that she does not hate but love the child to a point where she will sacrifice her life to protect the child. That will please the positive character. Then afterwards when the beating or molesting or whatever cruelty is completed she runs off with the male as the lioness did when the lion killed the cubs. The mere fact that she still remains with him runs parallel with the lioness accepting the animal behaviour because the human litter may not be dead but that is not her fault and in any event the fun can continue in full rage the next time around. Why call a halt to all the fun because everything will be alright after the husband pushes a few hundred dollars in the hand of the child or bay him a brand new whatever that will sooth all pain going around. He does not care for the chid as he can bay the child's silence. She does not care for the child because she allows the situation to continue regardless and the child does not care for the child because no one cares for the child in any case. After all, the father did show remorse when he bought the child a new whatever. The theme is about money. If the father is caught he would most likely get a fine because the penal system does not encourage incarceration for such minor crime. He can bay his freedom even by penalty of payment and money wins the day.

In the event where the child stepped out of line and the father (or mother) comes down on the child as hard as he can, to shock the child with such force as to scare the child so much the child will fear any thought of repeating the incorrect dead ever again, the other parent holding the hatred will then come in and phone the police, contact the magistrate, get the executioner in preparedness and go on with all dignity going to madness. She will never allow such abuse. She will rather see him dead than punish her child. She will make such a fuss and such a scene that should anybody not get involved they will become assessors to crime and child brutality. This again is a charade where her characters enjoy every second of her instable behaviour urging her on to over react. Now the moment has arrived where she can show the child who is the devoted parent after all She can show the world just how much she cares for her poor little baby that only went about to destroy the child she was destroying in any event. She will never allow the destruction to stop because of an action a balanced parent sees fit. The law comes down so hard on this man as they wish to scare off any other parent that will ever try to correct the behaviour of a self destroying child. After all where will the next generation of criminals come from if there is a well balanced society and how many lawyers and judges may become jobless.

Every one in an influential position throws their weight behind the mother's instability destroying the caring parent for caring. The judge himself has three personalities to fight and no remorse but to uphold the law to the letter. The politician that helped creating the law realises there are many more unstable persons out there to vote than stable persons and with the majority being mad it is clear with whom he will side when writing the next law into the law books. And besides he also has this little fight going on inside himself. His negative character has an enormous advantage because to him was not only given one life to destroy but so many it can keep him busy as long as he can remain in office. His neutral side tells him to remain in office because that is such a lovely place to be and his positive character now with him in office can be the god he always new he was.

The madness runs deep as it runs wide leaving no pillar in any community in strength. The description I give may be mostly in exaggeration of the truth but the truth it is. Even if the slightest way of is applying or of finding such evidence that this is taking place it will show that with the least provocation the balances is shifting towards destruction and the way society is, is an indication pointing out more than just strongly that the human race is on the decline. The truth stands out as a sore thumb showing that there are massive problems waiting for man on his spiralling way down the devolution ladder.

This is part in all layers in society from the very rich to the very poor and every one puts the blame squarely on the others without accepting any blame. When this child becomes an adult the process not only continues but worsens. For the sake of argument let us make the child a grown up male. The child as an adult misses the mother that he had but also never had. From this he acquired the loss he felt with the loss he does not recognise. Now he is ready to find a mate. But being human we humans have the culture that mating is a life-time commitment.

The wife he is looking for must be someone like his mother but with a slight twitch. He hates his mother by now as much as she hated him all through his natural life. He confuses love and hate, compassion and punishment, caring and rejecting, in a way that allows him the freedom of becoming a most confused person. With all confusion running wild and seeing love and caring in the same light as not caring and running wild he goes on the prowl in search of a wife. He longs for stability, which he totally rejects. He wishes for companionship he finds to smother him. He hopes for security he does not care for. In everything there is a threat. What should bring devotion brings hatred because that is what he recognises but does not care for. Will this young man's three entities have a wonderful time. He came into the world unwanted because of free love, he was raised in anger because of a free and fair society, he was neglected because of democracy and now he wants to employ the culture that brought him destruction as a child. In loving his wife he hates his mother. In wishing for her companionship he fears for his life. In pleasing her he understands only rejection. He is even more bent on destruction than his parents were. Where there might have been someone that tried to show him rite from wrong he got that show of caring as massage that that is a person is out to get him. Any person alive including his wife that will make any effort to show him the wrongs in his ways he will reject to a point of committing violence. In that he will find the rejection he hates so much and he will kill to destroy that. He may or may not administrate violence when at home. That comes with the role of the dice.

The wife he loves he has to hate because to him that equates mother-love. To him loving someone is destroying that person. The one he cares for he has to destroy. That is the love he was taught to accept. This leads him on a journey of more self destruction than any attempt his mother ever made. His yearning for the mother he never had pushes him on to every woman he can find. His wife being at home does not please him because she is not the one he was looking for. His morals are mingled like a mixture of concrete. The good and bad are so intertwined he cannot see light from darkness. He starts a life of adulterous affairs partly to destroy his wife whom he confuses with his mother and partly on the hunt for the mother he has a desperate need to find.

With devilish cunning this sets his three characters in motion where they can join forces and destroy at will. His positive character allows him to bestow the love he feels but cannot share onto the woman he is with just because he does not love her. His neutral character finds her acceptable because it will last but a night and the negative character helps with the charm because he will once more destroy himself and the woman at home, which as a matter off fact he truly and dearly loves. Unfortunately he does not recognise the love as love because he does not know the feeling of love. The love he recognises as love is the feeling he feels for the woman he is having the adulterous relation with because he hates her as she reminds him at that moment as the female figure representing his mother. If he really hated his wife he would be at home destroying her but as he loves her he does not whish to destroy her so he destroys her by not being at home. The positive character puts all his attention into charm that he throws on the female he is with. But because it is the positive character it also helps him realise that he can have fun in destroying the adulterate as the adulterate is the one destroying the one at home which he loves. The neutral character tells him to carry on because he has to punish the one at home for not being the mother he is searching for and the negative character is in heaven as every one around goes to hurt.

He believes that his wife at home truly loves him and for that he does not wish to hurt her but in the connotations he has about love he also knows that if she truly loves him she will try to destroy him. After all that is what love ones do when they show love. But because for the simple reason that she does not try to destroy him the neutral character holds that against her and make him believe that she is acting in such a way simply because she is not caring about him. Such behaviour stands totally in contrast to what he thinks love is, while his positive character keeps reassuring him of his wife's devotion therefore he should set his mind at rest as he will not lose her. The negative character finds this unacceptable and reminds him to do onto her before she can do onto him. The outcome is a vote of three to none in favour of the affair and another round of cheating starts for another night.

While the cheating takes place the positive character will remind him to love the one he is with as the one he loves is not there and the neutral character will tell him that as long as no one at home finds out then no one gets heart while the negative character tells him should his wife ever find out she is in any case getting what she deserves for not being the mother he wishes to destroy back. He will not enjoy her company and may not even enjoy her sex. He is in search of something and that something she will no be able to provide. During the relation he will get bored and then dismiss her as a dirty rag.

The worst that could happen to her is to let him find out that she has true feeling for him. That would place her in his power a place no girl will wish to find herself. The mother hatred will come to the forefront and he will start destroying her as he then can inflict all the injury he does not wish to inflict on his wife. By chastising her he will find some accomplishment and relieve and seeing her anguish will fill some of the need he finds in repaying his mother. But that will only bring some satisfaction and it will only last for short periods. But he will still yearn for his mother and in that there will still be a need to run more woman down. Because he does not have a clear image of what his mother was and what love is the characters can play mind games he will not understand. We all get some notions when one with a clear image of a mother and love between mother and child come across a female. We all have thoughts about what may be but we discipline the thoughts because we know we love the one at home and do not wish to sacrifice what we have for what may be. Hell, there is some woman I have met that is as attractive as any creature can wish to be but the very last thing I ever wish is to spend even one night with her. It is not because she is unattractive but to the contrary. She knows what she is and she knows how she excites men and she uses that charm to get men to dance around her with pleasing delight. The worst fate that can ever come to any man is to get involved with such a woman or even worse than death will be to marry her. She is the female of the male I described and she is bent on having men flirting with them and then just throws the verbal cold water on them to enjoy the reaction they get. In a marriage the first signs of trouble will drive her into the first bar where she will pick up the first victim and destroy her partner and the one she is with one blow just because she did not get her way in the argument she had with her husband. The poor slob that someday lands her as his wife will have enough information to write a book about hell.

With the conflict another situation with another disturbed child may provoke the complete opposing figure of which he is in search of.

In the next scenario of the Don Juan now in discussion holds the neutral character in place that will not have a clear picture of the woman of his dreams. With the image of his perfect woman being very vague any of the other two characters will come in and pour their versions of the perfect woman in his mind. The reference picture that he has about the woman he wants will be completely out of focus diluting his perception completely. He will want the woman every one desires. The disco queen or the brothel bitch or the bar tender out every night with another guy. He will wish to find the woman that treats him like dirt. Being treated this way will so kindly remind him of his mother and with his wish to please his mother as a child he will transfer that to his wife to be.

He will forever find some tart he wishes to please that has no wish to be pleased as she is in search of a man like her father. Her hopes are to find a man, and usually with success is one that beats the daylights out of her. Her first her second her third husbands will all have one thing in common. They will be woman beaters without exception. Or they will be drunkards, or womanisers, irresponsible persons but that is precisely what the woman has in mind although she will die before she admits it. If someone that truly loves her for what she can be to him turns up she will hate such a person because his devotion confirms her rejection. To her that man is the representation of her father and she in her twisted mind thinks her father was such a nice and devoted man because her characters will never allow her the opportunity to see her father for what he truly was. As her attraction and his attraction does not meet the requirement of their characters he will follow her like a lost puppy because her rejection of him is what tells him of her true love and devotion and that is precisely what she find so revolting about him. She wants a man that loathes her and here is a man that adores her. That throws her characters into disarray just as much as it throws his characters in disarray.

She hated her father as much as he loved his mother and because her father was the personification of brutality as was his mother they had to accept what they received as parents. But since neither had a real figure to relate to their characters turned the image they built around that parent around as to make them very acceptable. The characters they have will for the rest of their natural life bring to them the opposite of what they had as parent. That leads them to the very opposite of what they are in search of and what they find is what they wish for although it is exactly what they do not want. They both are condemned to one life of misery and if there ever is a hell, that place will be a merciful relief when they die compared to the life they have.

I do realise from the examples one must deduct that these cases are only the mental cases and anything more serious would find the person a patient in an asylum kept under lock and key by the President's special request ordered through the highest court in the land but it is not like that. The extreme I underlined because the extreme is the easiest to understand. But the destruction the parent has in mind for the child could be baying the child a very expensive pair of shoes only to let the child feel slightly important or giving money for a movie you would not wish your child to see but from the expectations your child has you cannot deny the child. It could be that you allow your child to go out with friends knowing the next day the child will write an important test. I do not wish to go into detail why that is part of the destruction or why it may be destroying the child because that is not the issue. The issue is the personalities lurking and being you. There is no slip of the tong with some wrong words slipping out. It is a deliberate intentional conveying of a massage to the other person about the true nature of your personal feeling and thoughts concerning the other person. What is the slip is your allowing one of the characters taking control for that split second while your guard was down. The turning of the cars steering wheel landing your mother-in-law in front of an oncoming vehicle while you were looking the other way and can swear under oath you never saw the oncoming car or had no inclination to go that way in any case. Why that happened is a total mystery because you will never do such a thing intentionally or other wise while the truth was that you were quite enjoying the nagging old witch's' company and her on going tormenting in her criticising you in the way you handling her precious daughter . That calling your employer by his first name when you were actually out to impress him. Calling the young girl in the office my lovely in the presence of the biggest gossiping bitch in town and realising this will lead to direct link involving a phone call by informing of your wife within seconds and knowing there will be hell to pay that evening while the truth is that you truly never even noticed how beautiful and smart and lovely and sexy and gorgeous this young girl was. Any thought of her appearance never crossed your mind for one tiny second. In another case your unintentional looking at a girls legs as she gets up and finds herself in a very embarrassing position for that split second while she is glaring at you for being such a dirty old man. Your looking down the blouse of a very breasted beautiful girl as she was bending over while your wife has caught you with your hand in the cookie jar. This is every day incidents with no intension of ill on your part but happened when you did not have full control of that situation for one instant. It all happened by accident but you believe me the deed was as deliberate and intentional as any of the cases I mentioned. It could even be as serious as your kicking a business competitor on the shin while you slipped and almost fell. It is the one character shouting your innocence as the other character is calling on your record always showing good manners and polite conduct while the third will never let an incident slip by.

This can and does even go as far as a nation. I am an Afrikaner Boer and being that I am not blind for my people's mistakes. When four Boere gets marooned on a deserted beach of a desolated island far from any other culture you can rest assured that within the hour of landing between the four there are, they would have started five different Christian denominations and six different political parties. It is a well established historical fact the during the Anglo Boer war at the battle of Ladysmith the Boere had twelve thousand generals with not one soldier amongst them. That is quite typical because we listen to God through His Word and no one ells. This last remark does not exclude our personal characters promoting twelve thousand times three different opinions.

....And then there are the Bible punchers...the ones that will convert you weather you need converting or not from ....well seeing that they never met you before it does not matter what faith you may hold, because only they can bring you absolution because they not only personify what ever they believe their god is but they see themselves as the direct extension of God, a finger or hand of God controlling life on earth.

They are hoity-toity, overbearing and haughtiness rolled in one god given container forming the BIG They. They can recite hundreds of Bible verses for minutes on end and that they believe is the key to their absolute presumptuous claim to God.

They walk with God and they talk with God and they discuss with God matters of mutual concern giving God advise where needed and as they see fit and where God is in their opinion straying from decisions taken at their previous meeting.

I am not referring to the normal God-fearing pious person that goes to Church and feels his thirst needs quenching on Sundays. I have no rite even to discuss any person's religious thoughts and belief. What I am referring to is religiosity to the extreme, a mental unstable drive of laying on the hands to heal, involve every one in religious debates every second of the day, starting to pray out loud as to draw attention of all persons around that should observe how their closeness to God has become as they are in constant prayer and will grant you some time between prayer because you should remember, they are keeping God on hold and the line is busy. Only they have the rite to prayer because after all when ever they get hold of you they wish to pray for you as if you have no connection to God, God has only given him a direct line and all others have to go via the switch board and wait their turn if they get a turn. Normally and in most cases, actually always I let them be but times arrive where they interfere so much you have to put some perspective in them. When I take them on issues their argument has the same logic as that of a pregnant pig and I have learned not to let them off easily. You press home the point and make them as big an idiot in front of as many as that wishes to listen, and destroy their mental thinking ability for the rest of the day. I have had situations where they tried to run away from me and I would run after them taking with me as big a crowd as I can possibly gather at that moment and destroy their image to shit. After such a session they are normally so annoyed with me they ignore me flat where I then leave them alone. I would never do that to other people for no reason can be important enough to humiliate your fellow man. But with them they leave you no other choice.

Not once and I repeat not once in my life did one person ever come to me with the introduction of: I come to you in the Name Of the Lord, and that bastard did not cheat, swindle, rob me, or steal from me. I have reached a point where I decided that should any one in the future introduce himself and refer to "I come to you in the name of the Lord" I'll chase him from my property like a bad dog. They are the biggest crooks and con persons walking on earth.

They too have a massive trio rage where their characters use religion to go out of control knowing very well all people will respect God and their referring to God always bring the other person in obedience. That is when they hit home with the most devious cheating and swindling you can imagine. If someone of their likeliness offers to pray for you don't close your eyes, grab your purse!

They're knowledge of the Bible is astounding but they know nothing about the Bible. They learnt a thousand or two thousand texts and recite them with speed, throwing the one after the other without making sense or allow the meaning having any connection. That is only an eye blinder, a way to astonish you so that you will lower your guard and that is when they hit home. There is no such a thing as a free ride and they always want from you ten times more than what you are prepared to give and when finished with you in the very last paragraph of the small print area it is only all about money but of course "in the Name of the Lord".

You wish to spread a gossip or any untruth, well be sure to use their channel. Normally those services they provide for free but then it must be juicy, unfounded and completely void from truth. Their trio works on the basis that the positive character takes them to personify God, therefore they can do no wrong in the eyes of the Lord. The neutral character advances the notion that they may convert you and that may be useful in some future schemes where you then can fit into more devious plans while the negative character is of the opinion that by your not believing the way they do, you are doomed in any way so robbing the convicted bears no shame. God put soles like you on earth to be useful to their likes where after you will go to hell anyway so what the hell, they might turn some profit before you meet your final demise. After all if you cannot see the light they give you, you may as well be blind and being blind you don't need more than what you can see. By them taking from you and giving onto them they are receiving with a self help scheme what the hand of God on earth deserves and where you are going to lose everything by your departing to hell you might as well start at a point where you are still useful to their blood sucking. Should any reader not believe me take some time and start discussing non religious issues with intent on your part to learn some angles their characters maintain as informed opinions. They are not hard to find. As with all criminal hoodlums hanging out at places of criminal conspiracy they normally hang around at the tents of the evangelistic preachers commonly referred to as the "Happy Clappies".

When I shared time in clinics with persons having some psychological problems this was the beginning of my theorising in this direction. I wish to state once more it is merely an observation of a layman fighting to analyse his own condition and took time to see where similarities were between different humans in the same boat that was sharing a mutual difficulty. Some were alcoholics where I am a teetotaller but still behind the condition I came upon similar causes giving one person one crutch and another person another crutch but it is the crutch one has to loose and behind the crutch you have to find the pain causing the person to grab for the crutch. What ever I share must be taken as not even an informed opinion but as merely another opinion where we all have opinions and it is worth the while to share opinions of an assortment and a variety.

In the last part I indicated persons holding the righteous views about their religion to advance and use as an excuse for their almost and sometimes definite criminal behaviour and malice intent. On the other hand I have seen suffering where these characters dish out what no one can bear. The agony and torment some people go through is of a much higher pain than I ever suffered when I came off my bike. The pain is more real than physical, the fear is stronger than death, and the confusion is louder than not understanding. It is horrible because some of them feel the anguish moreover than they would if a genuine murderer was chasing them. With a genuine murderer you can try and escape or hide but in their suffering there are no such luxuries.

The way they suffered and reality of their hallucinations had put the fear of God into me and made me more than willing to get over whatever small difficulty I had because my luck is worse than the Irish. (To my mind no one can have worse luck than the Irish because they got themselves in a spot on earth from all the places they could they chose their spot next to Brits, where the forever meddling and interfering bossy Brits is occupying the very next island) With such bad luck going my way I might just find my problems increasing. I had electro convulsion treatment on several occasions and to my opinion the treatment has a healing affect as the neutral character loses some dominance with the loss of memory through the electric flow. The patient then loses confusion by gaining perspective where the dominance of the entities reduces. As the generating of electric flow brought about by the life factor decreases electric tension in the brain the location of the entities become affected as well

Gravity, electricity, time is all the same thing and in a more or lesser manner influence life and moreover life in the brain. By reducing the electric tension the mind stimulates and as the convulsions allow electricity to escape from the sells stabilising the brain activity and helps sorting the influence the characters have on the person. I must admit that directly after such treatment I don't feel such an excessive urge to speed. Fortunately the condition normalises quickly and I can get to enjoy the exhilaration of my crutch once more. That proved that influences of such characters do vary and can be in dimensions of interpreting and it seems that influences from outside sources can be a major consideration

You may believe it or not considering my poor academic background but I have an inquisitive nature and a need for mental stimulation by acquiring facts and information and that was my academic downfall. My positive character always urges me to test another person's knowledge base and interpretations of facts. This was present even as a scholar as I did forever test my teachers. My neutral character will then classify him in filing order typifying and classifying ranging from brilliant to shit where my negative character will fore ever test the teacher in relation to my personal abilities and from that stance supply the necessary admiration or animosity. I never allowed my teachers at school to escape and when I became suspicious of their depth of knowledge hell would be upon us, moreover on me because they had the cane and always knew how to use the thing. However in cases where the teacher became a source to quench my thirst for knowledge I would eat from his hand. My positive character would take charge and push the others to a silence where no one knew they existed but in the other events of me growing suspicious about the teachers abilities the negative character would destroy any form of harmony that may develop between us and that was the normal in all but a few cases.

In primary school the teachers thought my behaviour was cute but in high school it became intolerable for all parties concerned. I make this remark to indicate that outside influences does play a part in younger minds and through positive stimulation the influences on the characters can be directed to a positive outcome for the child. No one lives in a tight cased cement container never to have an ability for change. It is the duty of the teacher to recognise and direct the children's interests to the benefit of the child and that could lead to the benefit of the class.

I established my theory with all relevant information based on my personal case. While this was going on I also realise there are nothing about me being exceptional or unique and if this applied to me so would it then apply in other cases. With a clear objective I started discussing other people's situation with them to draw similarities that would match my case. It was similarities I was after and not parallels so every time I got behind the whole issue by befriending the person and in that way I could establish a confidence that no professional could. As there was no malice intended on my part and I did not brief the person on my theory or tried to offer remunerable advice I could see no harm coming to anybody. The information was never brought to paper establishing personal files and since gossiping is not one of crutches no information of any private nature slipped past me. If ever any advice came from me it was certainly not on the grounds of my theory so I could not harm any person in any way.

But the more I came involved with other persons the lesser the importance of my personal issue became and the more I detected some golden thread running along lines undetected. In some people I could even detect which character reigned supreme that day by remarks they made or the moods they had and in limited cases there was differences in facial muscles as the mood swings occurred. But in the Afrikaans book which up to now a very limited number of readers had access to, this is the first time I went as far as mentioning my observations to any person. I can even remember the precise moment the light of understanding went up when a psychiatrist Dr. Steenkamp, which is still treating me explained about the mind and the free will of persons' personal thought, the way a person react on they're dissensions and the total absence there are of demons or other spirits that may influence the mind. I state this categorically I have never commented about this theory I have and least of all to dr. Steenkamp so I do not wish for any person to conclude he had any personal opinion about my conclusions. I merely said this because only a few incidents stand out in my life as very memorable and this was one such moment.

What I found was that it was as good as a human trademark apparent in every one, slightly more apparent in some than in others, they are in every one all the same. The entities are mostly absent but come in when a person has his or her guard down or when a person has emotions with an influence stronger than the person's ability to control. All actions man make is with intent. Some might be under the guidance of one or more of the characters but every one has full control over all their deeds, without having an excuse for conduct. They rule your life as much as they are you and will promote your true intensions when ever you do not wish to. Fighting the characters is fighting yourself but you can and you must find a way to recognise them because their intentions are to harm and never to uphold. The characters come with certain emotions bringing along certain feelings. The feeling may be an excitement that does not match the situation or an anger that does not fit the occasion but if one is vigilant you may catch the feeling before the feeling catches you. It may be very slight in irritating handling like for ever pressing the wrong "t" being the "y" on the keyboard or turning a cup of tee over on some important guest or having a dislike in someone you never met before and should not have any special opinion about.

One incident to try and prove my point I wish to raise is from my personal recollection and I wish to share it in order to avoid other peoples' affairs. I suffer severely from acrophobia. Putting me on a double-decker bus is about as high as I can go. In all sanity taken seriously one cannot have acrophobia. One cannot be afraid of heights when steel bars inches thick will prevent your falling. One cannot be scared to look down a glass window when it is closed and you cannot fall through. It does not make sense and yet I can assure you it is a fear greater than the mind itself.

The fear is irrational but should any one try to loosen my grip once I grab onto something he is not only endangering my life but he is seriously messing with his own life. I am aware of the problem and believe I do have a rational mind until it comes to heights. There is no thought, there is no reason and there is no arguing about matters. It is instinctive irrational animal-like behaviour where I go into a survival mode. I cannot fly and yet I know flying is the safest form of transport so much in fact it may be a thousand time safer than my cars or bike, but that is the rational and it disappear when I look down and see something small down there realising it should be big. Even just the realising that I am about to leave the earth is more than I can control.

The fear is almost if not an obsession and becomes uncontrollable. Then one day i stood on an exceptional tall building (well exceptional tall for me a person coming from Ellisras the true one horse town in the middle of a semi desert) of about eight to ten storeys high. I did it purposely to see what the emotions was accompanying the acrophobia because the attack must have some prelude. It can't just hit you like a brick that is nonsense. Something has to form a fore play, a sign of what is coming. Even if it takes one second it still is there. Nothing can just overpower the mind instantaneously but every thing must be about a collective of factors and facts coming together.

Coming out of the lift as I was walking towards the corridor where after entering it I could look down for the first time I intentionally was waiting for what ever to come first and announce the shock. Then as I came to the open I felt the feeling of fear but it was first another fear, one I was use to and knew. It was one of the characters coming to the foreground as if called. That made me realise that it was not the heights I feared but the negative character. I feared the character might take control and make me do what I did not wish to do. The fear of the heights is there, and that is no maybe but that is an extension of the problem and not the problem as such I may not fear the character and I may feel uneasy about the heights and it could even be that I become insecure but the problem was outright the negative character coming to the front.

As my dominant personality losses confidence the negative character comes in. It is not a case of him pushing me or my jumping but it is something going on in the realms of my mind where I do not understand all things all the time. It was a fear of what I may do to myself and not of the heights. Under all of this was the presence of the negative character lingering almost like a shadow feeling not present and not absent but just there. Then came the shock of the actual height and all logic flew away like a little bird. I was clinging and grasping for dear life.

If I can take my mind back as far as I can go back my very first recollection I can recall is a scene where I was on this huge tractor and it was far down. I was definitely under two years of age because I know which farm it was and my Grand Father sold the farm in the Free State that had the tractors before my second birthday. When we moved to Tzaneen he farmed without tractors so it was definitely before my second birthday. I was on this enormous tractor (enormous because of my youth) shouting desperately as I was crying hysterically for help because I remember the thought that I had no chance of getting off that tractor all by myself. I do not know how I got off or who helped me off and being where I am now I must have gotten of because I am not on the tractor any more but that day my negative character and my acrophobia met and got mates. Of that I am sure and if I am correct, then parents should take care not to scare their children in an innocent prank of unintentional fun with their child's fear. It could have lasting consequences to the child. It is not the heights I fear but it is the character taking control even if I know there is no chance of that happening still there is no logic as far as the phobia holds ground. If that is the case with me it should be the norm. People are not scared of objects because objects hold no threat and every one knows that.

My humble opinion is that one of the characters is dominant and the dominance is so much that the person feels threatened by that character. The character is in control so often through a depression or an anxiety or a mania of sorts that when situations arrive where the scene should be normal fear becomes the norm.

The negative character brings the threat the neutral character bring the warning about dangers and the positive character joins in by bringing the fear. The positive character brings the fear in to dislodge any attach the negative character may launch. The positive character on the advise of the neutral character disables the body and disables the ferocity and fierceness of the negative's dominance. By pumping adrenalin the body goes numb, the legs go weak, the arms shiver and the body has no strength to function while the person who is in the middle of the junior civil war see object as the reason for the attach but the object is only the trigger and not the show.

I too, am of the opinion that these characters and the way they perform their balance on the day and in the situation makes the hero or makes the coward, depending on the balance at that precise moment and occasion. Phobias connect to the negative character and mania to the positive character and by allowing un- protection through some situation triggering the mini civil war inside the mind; the person becomes a bystander where the person should be the controller.

A kleptomaniac may put something small in a bag and swear by the fact the kleptomaniac did not know about the action. That may be as much the truth as it is a lying because the actions was not deliberate, but the actions were intentional. It is easy to ignore the compulsive behaviour and claim non-participation when participation may have been semi unintentional but still enjoyable. The one character may distract the attention of the person but it is done with full participation because in the end responsibility is with the person. One would often hear the remark: "I knew it I knew it was going to happen" when something was going wrong. All humans can read situations and your mind told you something in the situation were desperately wrong. While the positive character does warn you of events coming it is a deliberate action to allow the neutral character to distract you while the negative character can play for time for whatever occurrence to take place. The whole scene was a deliberate action by the person to gain a negative outcome to produce some suffering or hard ship to some degree because that is why we are on earth. But I shall get to this last remark later on.

This takes us back to the bleeding heart baying off guilt by paying the philanthropist to collect on behalf of the Hoggenheimers dishing out to the Mammonites paying the Mammonists for some slave driving. The actions are deliberate but the true intentions are deliberately unintentional. We are bullshitting our conscience for gaining our mistrust. It is the lye of culture and all participate but some participate to a degree that does not please others. The degree might be to some extend not serious enough to bring commitment and the persons would stand on the side line and criticize without direct involvement because of fear of own guilt uncovering or even of a want to participate while others would come in and rescue but not to save but out of spite because of personal yearning for participation that the person knows would not be permissible.

Another scenario is where a person is drowning. The rescuer comes to save the drowning victim. The victim is exhausted beyond normal mind control. The lifeguard reaches the victim whereupon the victim tries to drown the lifeguard. The victim has lost rational thinking and is then in a mode of action versus reaction. The positive character grabs and clutches at the lifeguard in anticipation while the neutral character tries to survive the conscious and the negative character wants to save the situation by taking the lifeguard with the drowning effort because after all it is the responsibility of the negative character to destroy and destruct as much as possible.

Some gave these characters names. The negative entity goes by the name of a death wish. The neutral character goes by the name of don't care. The positive character goes by the name of optimism. By naming them we found once more a way of avoiding our duty to recognise. It is much easier to dismiss than to admit because admitting has to lead to prevention and prevention is no favourable option.

We say the drowning person got panicky but that is another word we use to escape from reality as much as an excuse in avoiding responsibility. By being panicky we deliberately excuse behaviour and responsibility about actions we may commit to explain irrational behaviour. The negative character wants to punish the lifeguard and even make him pay for his life in interfering with a situation the negative character is enjoying thorough rely while we others only recognise the efforts of the positive character because that will be the nice thing to do. The next time we behave in the manner as to kill our saviour we too can be acquitted on grounds of incompetence. The action of trying to kill the lifeguard is as intentional as the trying to grab onto him to be saved and as intentional as loosing control through the neutral character.

When I first read about the Lovejoy syndrome it brought to my attention that not all mortised control of the body is in the domain of the person all the time and sometimes there are some part of your life that can take control of your actions when you are not in absolute control. It is all a mind game you play with yourself in diverting responsibility with the compensation of enjoyment but the avoiding of dismay about yourself. There are no excuses because the final responsibility is in your power. It is in your power and much more even it is your birth duty to fight the characters but the moments you lose the fight you take the responsibility for the actions because you momentarily lost the fight. It is a win lose situation where only you walk away with the prize as much as the punishment. The muscles are under your control and you are in charge but the entities are little pests being you and are thorough testing you by grabbing control whenever they can. The entities are not only and absolutely negative but are positive as well. We have all been through situations where we admit afterwards we came through by the grace of God. We always hear some one remark that he or she does not know how they did "it" but "it" came through far better than "it" should under "normal" circumstances.

All our phobia, all our desires and all our hopes in achievements we pin on luck or the role of the dice but luck and the role of the dice has nothing to do with it because or achievement good or bad as our accomplishments wrong or correct, and our thoughts being acceptable or not is within us, in our control as much as it is us.

A paedophile should be hanged from the nearest tree because he gave in to the want of the characters and not because "he is not in control of his actions". He wishes for the characters to take charge and even deliberately set up situations where they may take charge, because he enjoys the dead as much as they do and more because they are he. They are in all of us but for some certain behaviour are unacceptable and for others it is not. The judge bringing judgement knows in his grain that the molester will commit again yet he sentences the criminal to a few years of incarceration and is fully aware that their is no chance of rehabilitation because the culprit does not wish to be rehabilitated. He will find his rehabilitation as a death sentence because molesting children is keeping him alive. When the judge do not bring the death penalty he, as much as the culprit participates in the next cycle and therefore must take responsibility and participation in the next round of child abuse. But the judge sits there with the idea " there am I but for the grace of God", which is true in a way but also is not true in a far bigger way. With my fast driving I do not wish for a cure, but I know something is going to go wrong somewhere some day. That is my chance I have to take. He upholds the same argument and goes to the molesting because the sentence is the chance he has to take, and is the chance he takes. You can bet your bottom dollar that should I know before hand the next road race would kill me I would not participate, and the same goes for the molester. With all certainty that the hangman's loose is waiting he will have second thoughts about his next molesting session. Murders always fight for their life by fighting the death penalty. The underlining is that there is no black white and grey. There are no clear-cut defining borders and sides. The neutral character can be as destructive as the negative character can bring a positive out come. When the mob comes out to lynch the actions of the mob are negative in lynching but as they do not wish the continuing of the criminal's behaviour it becomes positive when doing the demonstration. That it will bring conflict to the child's guilty feeling and in that sense their actions are neutral to the child, the participation is positive in preventing the repeat and there by positive in the negativity of lynching where the police prevention is negative by protecting the paedophile and that is positive by upholding justice as it is neutral by delaying another criminal's relapse in crime because relapse he will.

The atheist does not go on a disgraceful child molesting campaign because he thinks there is no God and if he does not get caught there is no punishment. He avoids indecency because he is human. The paedophile may be the biggest Christian around but argues that since he is only human and humans are sinners and sins are alike he can maintain his behaviour until judgement day. Such a line of argument is very typical of the trio being in charge. The responsibility is always with someone or something else and the person never pin it to specifics but shifts the blame to wherever is convenient. And so does the Judge! Hang the bastard and judgement day to him the sinner will come a little sooner. If he does not wish to control his characters help him by eliminating him with his characters. By molesting the child he starts another cycle producing one more child having little control over the characters the child will fight when he is an adult. Stop the violence by stopping the cycle by stopping the one not controlling his characters. If death waits as a surety he will mend his ways or seek help to accomplish change. It is easy not to change and difficult to change but we all can change.

One may have be opinion what I refer too is about the good and the bad, about the saint on the one shoulder and the devil on the other and think well…yea I've heard that before… But it is much more than that. People stop at serious motor crash sites not with the intension to help. If any one admits to that that person is untruthful. They stop to feed the urge. People watch blood sport, not for entertainment but to feed urges hidden deep within the mind. When an armed robbery takes place with possible killing spectators run to the scene. When there is a fight on the schoolyard the word spreads like fire and little else can generate more enthusiasm. The most brutal serial killer always receives the most male. Woman would throw themselves at the criminal misfits with marriage proposals; coming up with the excuse they have enough love to concur the beast's evils, but that is a hideous lie and they more than any one else know that.

They wish to share in the darkness of evil, find someone that could lift the veil covering the beast within. Stopping at the scene of a bloody accident evokes prime animal senses covered by social upgrading but is still very much lingering within. It serves as a reminder about the days when a feast came from such human blood. At some time all of man was a cannibal, man-eating monsters that feasted on the flesh of the enemy after a conquering battle brought victory to some and victimisation to the others. When faced with starvation it is a normal sense that kicks in where people will start eating human flesh. Shocking, as it is... the shock is about realising the urges more than revolting.

We all have the darkest desires of committing unspeakable atrocities and most barbaric acts that will shock the normal mind into panic and frenzy. Some people dislike blood sport, not for what they're purist of heart tells them to reject but the rejection comes from the craving they fear. They fear the need for such beastly acts may linger and run out of control. A story about a mass murderer is best- seller weather it is fiction or truth it is popular. It sells and there is a valid reason for that although our minds reject the reason.
When thinking about these crimes we put it in the basket of the negative because it is where it belongs. No one can ever be positive about such behaviour and with that we create fabrication of truths we wish. It is as negative as it is positive...it is neutral. The good / bad character is not the danger; it is the neutral that is dangerous. It is the neutral taking our minds on fantasy journeys. It is the neutral character we most easily identify with and mingle with. It is the neutral character setting our morals. When Ted Bundy killed the many woman that he did with the brutality that he applied he was not negative, he was super positive. To the woman "forgetting " to lock her door, walk alone down the dark ally at night, stopping to give a strange man a lift or hitching a ride the deed is not negative. They are super positive in their expectations of what may come to them. Sure, they do not ask to be raped, or beaten, they do not beg to fall victim to crime, but lower down in their minds they know of such a possibility and very, very deep, deep down they stand neutral to such an event because they do not mind the excitement or the sympathy afterwards. The bank robber, the pickpocket, the shoplifter, they are not negative they are super positive. It is being neutral that is the negativity.

The animal cannot see the universe from any other position than the one the animal holds. Singularity placed divinity in the centre of his universe and the animal has no means or brainpower to translate his mind to another position other than his centre where his needs and desires are. From there comes the neutral holding positive or negative as well as the he. Negative means he should fear and positive means the other should fear. In this religion can be one form of atheism and atheism another form of religion. Atheism is the inability to see anything except from the centre of that person's universe and the inability to transform to another position seeing it from another viewpoint other than that persons universe centre. When Fish sent the parents of the child he cannibalised a letter informing them how he devoured their child he was positive and the deliberate pain he caused was positive. Peter Kurten the vampire of Dusseldorf only became neutral with his last victim, but he always remained positive to his wife so much that in the end he forced her to betray him so that she could claim the reward. By her collecting the bounty he realised his final act of being positive as he always felt about her...but only to her and so his final offer he gave her was that the bounty that went her way. To him that deed was as positive as his killing and eating of three children in one night was positive because he was unable to translate his centre point to their position. That is being animal in every sense and animal on two legs. The Boston strangler was always neutral working his way to becoming positive for days although that meant the utmost negative to his victims. Being human and religious is the understanding of other concepts forming a centre way beyond your centre of the universe and moving away from the animal neutral to a human stance Not eating the flesh of the sheep does not in any way make you more positive but it makes you too stupid to see the universal picture of the totality of Creation and its wider meaning. That is trying to prove you are not the animal you know you are but cannot seem to separate from. In the neutral character is the one that no one propagates because that character is much to close to us to be comfortable with, that is the animal, the beast or civil the morally accepted or rejected but still nursed by all because of pre man mentality. From that the other characters stands positive or negative but the neutral character holds the centre and the key to light and darkness in the mind of man becoming man. Claiming the position all other characters including the self occupies a part of the mind. The neutral character sets the tone and the rest will follow. In the neutral character lurks the animal and depending on the person, the darkness of the animal comes to the foreground or stays in the back ground but is forever present in the mind. Identifying with or identifying the neutral character places us in the realms of man or animal. Setting us apart from the neutral character is what produces man and not animal. In the same manner as not eating sheep is the saving of one criminal life where every one involved knows that criminal is destroying dozens of future lives and expanding the problem by creating many future criminals where they then would create some more criminals at a ratio of a dozen to one. In three generations the one criminal is then the cause of hundreds paedophiles walking the earth. That is degenerating civilisation and it is all because the law enforcement from politicians and judges through the lawyers and civil servants down to the bleeding heart and the cop on the beat that wish to prove them not being the animal that the criminal is. By creating the environment and breeding ground for the animal and forcing their positive ness in neutrality they destroy the future of the following generations of man. Law not taking blame for their actions in the neutral stance is as much being a criminal as the criminal's neutrality creating his positive to negative relation. The law enforcement' officials excuse for not wanting to be as bad as the criminal is more destructive and a much bigger misdeed to society than what they do to the criminal because after all it was the criminals free choice to commit or not commit and as much criminal not to wish to pay for his deeds. Either we teach the criminal there are billions of centres to the universe or remove him from our centre but allowing him to become forcefully our centre proves as little as it is destructive to every body. But by throwing him in prison where he shares time with others the same as he accomplishes only that law enforcement may promote crime to establish more cycles for their and all the other criminal's benefit but to the disadvantage of man in general.

The paedophile or any other criminal or anti social behaviour is quite rectifiable where psychologists must teach the criminal in recognising the incorrectness of behaviour, the recognising of the characters and the control of the characters by thought control. Criminality starts with a thought and that is the point I started in my discovery of the characters. Being suicidal as I am and that being my problem starts with a thought starting with a feeling leading to a depression. I am never suicidal on my bike or my car doing high speeds. Then there is only the positive character with the neutral character keeping guard for traffic control. Never do I at any stage exceed my personal limits or endanger lives through recklessness by outsmarting my personal ability where my mind works at the speed matching my vehicle and that is the secret to success. It starts with a thought. It starts with a feeling. That is the gate to progress the characters follow. The remedy is striking a link where one will recognise the thought sparking the feeling and recognising the felling sparking the thought. There is countering the feeling by thought as much as suppressing the thought by feeling but the BIG issue is that you HAVE to know yourself. The alcoholic feels the urge for booze as much as the thought for liquor but when recognising it he counter acts and that is where alcoholic anonymous has such a great success. If he did not wish to recognise the urge as much as he did not wish to suppress the thought alcoholic anonymous can do little. It is the difference between cure and tramping. But it is fighting day and night and fighting yourself knowing you are in the fight for your life for the rest of your life. It is fight you can never win and must never lose. It is continuous round after round with no victory and no defeat, no prizes and no glory. Only shame to follow defeat and the winning part has no recognition for effort or accomplishment but to you. That alcoholic fighting himself and finding his determination is the human victor. To me staying neutral is staying alive. That is what prison should teach the criminal in recognising thought and control thereof. The big courtroom confessions these murderers make are about showing the world how big their universe are, its about how they are the centre of other's universe and the control they had in destroying the others with their universe but it is never about apologizing to show they can understand what others have in the universe they destroyed or that in fact are others with an own universe to have. Should he think he can outsmart be unable to rehabilitate he is animal and should become destroyed like all other raging animals being out of control. He should therefore fear death or find death. In the modern penal system rehabilitation is a word used in the courtroom with no other place to have. Life should be about improving and not sustaining.

As long as doctors treat suicide as a state of being morbid or depressed they will not get anywhere and it will take them a long time never to come to any cure fort his state of mind.

Alcoholics and all drug addicts lose the neutral character when intoxicated. The person becomes super positive loving the world for the world loves him right back or becomes super negative by getting aggressive or drunk with remorse hating his parents for what they did or never did. In this the addiction of the parents always plays a part when the child of the alcoholic also becomes an alcoholic. When the addicted becomes sober the neutral character takes control with vengeance, as the addict needs the next round of substance. To the neutral character surviving means finding the next round of intoxication. There is never a balance and when the positive becomes positive, drunk or sober, it shares a spot with the negative and of course the other way around also applies. The drunkard is prone to mood swings that is apparently out of his control, but that is a fairy tale. When sober the same applies, as his moodiness seems to follow his every move. The drunkard uses the substance to escape from himself and his own adequacies he feel about him in his self-portrait and therefore allows the characters a free hand when intoxicated. This we in South Africa call Dutch courage and will have different names in different regions but all names apply to the same behaviour.

Things are normally not as serious as in the case of the child molester or the mass murderer turned cannibal. On average it does not have such deep and intense underlying emotions and fights to the bitter end. It could be a case of the young man is meeting his in laws to be for the first time. The dinner is in great preparation as it is in great anticipation for all involved. Every one but most of all the lover-boy and soon to be in-law is under pressure to impress. He is the new face in the family having all eyes on him because every one knows all the other faces and he only knows his face, which he cannot see in any case. He is out to impress and is tuned for this all out effort of do or die. In this effort he relies without relying on some help from the characters because he can use all the help he finds. He lowers his guard completely to allow the positive character unhindered passage in the situation arriving. In the background lurks another character that is far less dominant in the whole affair.

It will take but a flash in a moment for the negative character to prove a point but as he opened all the doors so wide for the positive character he now is even positive in being neutral. With all the tail wagging he has to go through his negative character takes little appreciation in any possible discomfort on his part. Lover-boy is yearning for acceptance to such a degree this making of his all out effort is rather new and strange for the characters as they show an all out help line helping by the full range of their individual abilities and to use the chance in such an unhindered open channel participation. But one is waiting his chance in quiet anticipation. At the moment of climax where father in law to be wishes to make a toast will be the moment all three characters join the fun by creating the incident all three have been waiting for all night. It will be to the determent of boy-impressing of course. The negative character is very aware that the neutral character may prevent his planned action there fore the strike is lightning quickly. The neutral character sits lapping up all the attention as the positive character is all out helping with the impressing of all around. With every one well occupied the chance comes for the negative character to make his point for the night since everybody was having fun and he had to sit idle and unwanted.

As the tray carrying the wine glasses, which are filled to the top, passes by lover boy, the negative character motorizes the arm closest to the tray and hits it with force. The action holds the speed of a boxers punch and with that lightning speed lover-boy never thought he had such quick muscle movement. The action reaction reflex action reflex reactions is beyond the abilities our boy to be married ever dreamed he had. He is totally surprised at his quick ability in movement and that to his knowledge is miles ahead of his ability to move. In all his life his arm never moved that fast as he turns the tray over on the lily-white table linen.

With him being totally out of place and out of sorts he cannot recognise his ability and that is what his positive character ensures him. His neutral character will be very embarrassed and that embarrassment he places in lover-boy's private embarrassment about what happened. Shouting and proclaiming accident by him as well as the other two characters will bring all members of the in-laws-to-be under the impression that this was indeed an unfortunate incident completely convinced by his very genuine embarrassment. The embarrassment he proclaims are partly his but mostly the embarrassment comes from the other characters proclaiming innocence to him about their involvement and misuse of trust. This he uses to further his embarrassment to exclude all blame of deliberate action on his part in order to impress his in-laws-to-be.

Every one present will feel deeply sorry for him except his negative character that made the point that all the licking may be for tonight but he is still his own man and will have is own way in the future to come. The question is was his innocence really that innocent and was his embarrassment truly that hearty? On the first count no, he was just as deliberate in the overturning of the tray as all his other actions was though out the night and innocent is the last thing he can be guilty of. On the second count, well yes, in a way but not only for all the reasons he proclaims. No one can ever proclaim innocence and non-participation in deeds the person commits. All your actions are all your responsibility. No excuses can be maid without lying through your teeth. Being born means the fight is on and the fight will continue till your last breath is wind. Having trinity around is your birthright and fighting is the option you made before birth and not after birth.

Life of man is about fighting yourself with all the vigilance you can ever muster. That is why you are here having the time of your life for all your natural life. It was your inheritance the day you were born a human. It is not all about negativity but it is all about achievement. It could be quit within your self and it could be in the presence of thousand of spectators. Every sportsman has "on" and "off" days and there is such strong emphasis on the physical that by training the physical no one notices it is the spiritual in training. The spiritual always is the physiological but all preparations are psychological. Practising is about training life to manipulate space-time to the best affect. In all it is only about the physiological. Training is about telling the muscles to obey command and telling command not to obey the muscles. When the muscles shout in agony to stop, command must turn a deaf ear and when command shout to the muscles to go on the muscles must be like a dog and react without questioning or arguing. To describe this we use the name fitness. It is the conditioning of the flesh but it is much more about the mind having control over the body and all fitness is about life commanding the structure in occupation to do what life wishes to be done. Fitness is moreover about controlling the characters than the body. You have to control the negative character's destructiveness to be about the opponent and not about you. You have to control the neutral character to dismiss outside interfering with your efforts. You have to control the positive character to bring subduing confidence. You have to use the fear to your advantage in believing you are fighting for your life and not merely a trophy.

Every sportsman has a story of "absolute brilliance" and always the remark is about the sportsman not truly believing he had the ability in accomplishing what he did. When saying that the sportsman only considers the physical aspect and never the mental drive that pushed him beyond the limits he accepted. On another occasions it is the very opposite when the sportsman declares the day as a disaster because notwithstanding an all-out physical effort nothing went according to plan. His muscles did not respond, his legs were stiff; his arms were not in synchronisation or what ever the excuse for the disaster is. When shove comes to push it is the three characters we find again behind the success or disasters.

The apparentness comes through by him not admitting his efforts links directly to his state of mind and the blame goes to external factors. Everything went just rite or just wrong. It is everything that holds the responsibility for his success or his failure. It is external forces at work and in charge of his luck or bad luck. With that remark he indicates his absence in his actions. His lack of admitting participation and his unwillingness to claim success or admit failure proves that he is relying on something he feels he has little control over.

We all admit that when we are tired we make mistakes. When we lose concentration things go wrong. When being absent-minded we make accidents. That is admitting to the role the characters play. Being tired means relinquishing control, letting go and then the character take control. But also when in fear of one's life you find yourself in super control where you are miles better than ever. That too is the characters at work. It depends on which character takes charge and to what degree does the character take charge. I always teach my sons never to fight a man in front of his girl because you face a man much better than normal. If you are winning a fight leave the other person a way out to escape and never allow the impression the person has no way out. As soon as the opponent gets the idea he is fighting for survival, or he has to fight for position in the tribe in example for the favour of a female you have a monster on your hands and in that case be prepared to fight for your own life. Never allow the idea to enter the opponent's mind he has to win or ells... that will be to your determent. In the instance where you do not allow an escape route or a man thinks he is fighting for a female you have a raging bull and three characters to fight and you do not wish a fight him with his characters on his side fighting against you. That will be your death you wish upon yourself.

It is not schizophrenia I am referring too. I am no psychiatrist but as a complete novice I do not believe in schizophrenia. I do not believe the person is hearing voices from beyond because there can be no voice of beyond. It is his imagination and his characters playing mind games and if a psychiatrist or a psychologist come in and join the fun admitting to such a scenario. With such outside help and sympathy to help the dreaming along the characters then can and will come out to play. The "split personality" in the fight is within every body and can go into rage whenever allowed to do so by deliberate actions, provocation by other, tiredness, fear but also delusions, that is true.

From the article so far one may tend to get the impression it is about big issues like meeting you're in laws the first time but it is not. It is every day all the time things. Sitting in traffic waiting for the light to turn. The positive character slowly shifts to neutral and the neutral slowly joins force with the negative that was in the background all day. Without the person noticing the whole situation within him shifted in frame but he is unaware of it. Suddenly an explosion fitting a war burst to life. Another motorist sits daydreaming with his neutral taking him on long trips because his positive has nothing to do and his negative is in the background. The light turns green and the daydreamer is just that tad slow in responding because he was in thought. He was in thought as much as he was on Jupiter. His negative character helped the neutral create a situation the negative character can participate and not be bored. So his slow acting is as deliberate as his breathing is and the other person with the shifting emotion sees this, recognises the stunt that the daydreaming motorist is pulling and with the positive character standing on neutral ground and neutral fully in the negative territory, the negative character comes in with a punch like none. Suddenly two very timed persons go into a rage because of the smallest incident. They do not recognise their own behaving and if they do not get their characters on a leach quickly the characters of both men will take charge and blood might flow. This we call traffic rage. It is a very convenient name.

The man sees some one he may regard as rather attractive and smiles at her. The positive character becomes embroiled in the proceeding while the neutral character joins by taking control and urging the man to step just a tad closer, just to see… while the negative character is anticipating rejection in any event and deliberately steps on the girls' toes. Embarrassment is all around even with surrounding crowd because somewhere in the back they in the crowd all no what happened and that brings the embarrassment to their door. Every action on all accounts are very anticipated and pre-arranged. We use the name of an unfortunate incident to file this under.

It's about calling your wife but using the neighbours' wife's name. It is knowing you have to cut the lawn but the drowsiness just will not let go so you sit down and close your eyes for a second, just to relax for a second. The one character shifts one position while you were not attending procedure and you miss the opportunity to mow the lawn for one more week. This goes by the name of slipping the tong and nodding off.

Very typical of this is the driving absentmindedness. How many times did we sit back after arriving at our destination and thought how on earth did I pass this or that town? While driving your car it becomes as routinely as breathing so there is little to be positive with and less to be negative about. That is the chance the neutral character has been waiting for all week and he takes charge by letting your mind wonder to many destinations but the one you are heading to. Little harm can come from this under the normal, but when danger suddenly strikes you are gone, your positive is gone, you negative with all the adrenalin is completely absent and your neutral being neutral does not bother in any case. By the time all the absentees arrive at the scene just that second later, a horrible accident is in progress. This we call driver fatigue. The process lingers on while we are absent minded, not totally in control of the moment and shit happens. But shit happens because we want it to happen for that bit of excitement we do not need, but the characters do because they make misery, To them it is a case to change the situation to something more exciting, more emotional, to press our social standing or just to ease boredom. This goes by the name of the mind is wondering.

Someone may say something out of the order and a rage follows. Why would a rage follow when you know very well that what the other person said was unfounded and if that person does believe what he said he is so misinformed you should not even listen to such nonsense? But confrontation is about to blow like a rocket on a launch pad. You will show him what he said was untrue and he will take it back or pay with his life. Once again we allow the characters to shift and completely obstruct our normality. The best is that with you allowing the shifting you will press old issues long ago forgotten but the characters suddenly helped you remembering this or that and this is the last straw! Under the normal the previous incident hold such minor importance you would never remember about it but at that moment it does not even strike you as odd remembering such trivially while not surprising at all it is clear in your memory that moment. You feel you can murder while never in your life did you ever have a thought about how it may feel taking another person's life. Everything you experience is out of the ordinary and out of order yet it seems to you at that moment as being as normal as discussing the papers. Does your characters enjoy the excitement and taking you along for the ride. They take control and you take the mess afterwards not knowing how you will ever show your face again after such an incident. This we call losing your temper.

You walk down the street with your best suit on feeling as chirpy as a robin in mating season. Then suddenly your foot misses the curb as you were walking and you land face down in a crowd of people you never saw previously or know any one around you. There is not the slightest chance of you meeting any one of the spectators ever again. Yet your world plummets down the deepest mine shaft. You cannot see the light of day ever rising again. You hang your head in shame while trying to conquer the incredible urge to run from the scene as fast as you can. Not for a moment do you stop to think that it is not that shameful and happens to everybody many time throughout they're lives. That the people you hold so important can never be important because you will never see any of them ever again. Again you characters turned a situation around as they turned on you. This we call embarrassment.

All of us meet the challenge in battle without ever realising. A young man newly wed has a friend of his wife staying with the couple for a month or two, just until she can find some other accommodation. This young man is going ballistic with the fight. He is in rage about this beauty sharing a roof with him and with him working shifts he finds himself alone in the company of this woman because she is still in the market for employment. She page thorough many papers per day to find a suitable position but that does not take all day and with his wife being at her work there are many hours pleasuring about unchecked. Now he goes in spin. His positive character tells him of what he has in his wife and that he should not endanger his fragile and young marriage, his neutral tells him his wife is at work and need not to know while his negative went positive by telling him how beautiful this young female is. This we use the name temptation to identify. It is the clearest way we know of identifying the threesome.

We were all at one or other stage young in our lives and young at heart and know the feeling when you are running the hundred meters or you are on the Rugby field and there is that special person looking from the spectators end. You feel her eyes burn in your back and the excitement blows your senses. It is like someone ells takes control and you become that much better to the degree you cannot believe yourself, or you have gone pinching fruit (a favourite pass time amongst the Afrikaner up and until I was a teenager. That was before money and greed came in and made kids criminals for helping themselves to fruit at night). The other side of the coin was that you knew who ever gets hold of you, will tear of the skin from your behind, but that was part of the fun being the challenge to the danger. Never was police involved one way or the other and even coming home safely did not mean security because if your parents catch you, you have to go back to whom fruit you pinched belongs and fetch your licking. I was a hundred meter athlete in my time and was quite quick, but when caught in the act I saw these real slow guise and can-not-run mates in crime pass me and if Carl Lewis was there they will pass him as well. They beat me by a hundred meters on the three hundred meters and cross one and a half meter fences as if it was hurdles on the track. The neutral character was on guard all-night and got the other two pumped but on stand by. The moment surprise takes over the positive gets negative but charging past the negative to get the body in motion. The negative sees that being negative is a splendid way of getting away from the danger and re-passes the positive while the neutral is pushing both to get out of the way so that the negative can take command of the muscles. With in less than a heartbeat there are four characters (including yourself) that take control of muscles and there can be no fitter and more potent athlete occupying the body than the fear factor presenting the next few minutes. This we call motivation.

Then there is the young person that is the one-week in the dump because he cannot see what the world is all about. His girlfriend dropped him, he is in the middle if an exam he cannot see how he will ever pass and to top it all is the fact that it has been raining for three days while he has this camping trip planned. The positive character finds a few days rest while laying low in anticipation of the coming holiday forcing the negative character to take charge of his outlook on life while his neutral character takes charge of the schoolwork making him take his vacation early while exams are pressing. The next week he finds himself being over positive bout life, about his happiness and about all the opportunities in life that are waiting on him. The exams are history, he is vacating and met the girl of his dreams and have a song in the heart. To tell the world, he is the man of the moment and to prove that he has long uncombed hair, walks with a swing and doing his thing just to annoy his square parents that are living they're life in history. That is the neutral character in charge and being all-positive for one week/ month and then negative for one week/month does not surprise any one. The name we attach to this is growing up. It is all about finding the characters, meeting the characters and blending with the characters in order to fine tune for preparedness for the fight ahead.

We name the events. We know the events. We suffer through such events but never stop to think why it is taking place. Why would your mind start to wonder? If you are in control and you are in that position you should not find yourself out of control or somewhere ells while you were being there. If you were your body why would the mind go absent? If you were your body and mind why would you stray or over react in anger, pity, shame or stupidity. If only being in the body you are in as the atheists believe, then you should be in the body, because where ells will you be but in the body you are. The fight is on and you have to recognise your opponent because your opponent is you in person. Your opponent knows your weaknesses as you know them because you are your weaknesses. Your life is your fight and it is on for the rest of you natural life.

It is as common in every one as miss-placing keys, spilling milk, forgetting some one's name and such minor incidents bringing great embarrassment at the time but is as normal as breathing. It is no big psychological problem forming in the dark side of the sole where the brave does not dare. Yet it can be there. When ever a person comes up with a brilliant remark astonishing the person that made the remark much more that the person to whom the remark was directed is an example of the trio. When trying to repeat the brilliance the very next time we become the stuttering idiot that wishes the floor would dissolve the human body. We find an inability to repeat such brilliance. We all have astonished ourselves from time to time as much as we embarrass our selves from time to time, so there is no exclusion and only inclusion.

It is the person being unwilling to do a task and finding himself repeating an error that he knows he should not repeat but something is driving the person to repast his actions. The more he repeats the error the more he gets annoyed with himself but getting annoyed with yourself must be the most unaccomplished task you can accomplish. His normal reaction would be then to become annoyed with any object or person he may find displeasing and although the object or person has little to nothing to do with his actions it proves the best way to blow off steam. This I refer to as the boss syndrome because that is one of the perks a boss seems to have. Being negative about the task makes your mind go wondering to more pleasant places to be. By your negativity you are suppressing the positive character and that puts the neutral character in charge. But the neutral character has as little interest in you're being there and takes you away on more pleasant day dreaming trips to where you would rather be and that leaves the negative character to be in charge. Who is doing the task with every one gone…it is the negative character, and an unpleasant one at that being all alone and hell is on its way!

The person starts making deliberate errors that can be avoided but with his lack of enthusiasm he becomes absent-minded and that is when the accidents stars cropping up. It could be small annoying things but also it can be very serious injury coming from such little absent mindedness. The negative character wishes to draw all three characters attention to the fact that the negative character is as displeased with the situation and wants to be relived of the duty. The LAPSE in CONCENTRATION is no lapse in concentration and the disaster following (big or small) is as deliberate as the person slipping away on his daydreaming trip. The whole affair is one big disaster waiting to happen and always does happen. All parties involved claims innocence and protest to any involvement accept the negative character that now is relieved of his duties. The net result is that the negative character is the only winner in the end.

Where does it come from? It comes from being human. What encourages the characters to become more dominant in some than in others are the better choice of question? There are at least a thousand possibilities I presume but it is more than just likely that it may present itself as a manifestation of a chemical imbalance in the brain. That does not explain the fact that it is present but may support the fact of more dominance in some individuals than in others. Being alive is about chemicals but the chemicals allow conducting electricity and are not life itself. The chemicals are a conductor but we all know the conductor is not electricity. But when the conductor goes hay-why the electricity goes hay-why. In that sense the chemicals will play a part but not play the part. When there are cross over of wire connections some life will flow in a direction where it actually is intended to be at other outlets. From what I have seen with some of my brain injured friends the control becomes absent but that does not mean the injury cause the characters because I have witnessed the characters in every person I have met. It is as much part of our personality s it is our personality.

The chemicals could be one aspect but underlying fears are the predominant issues that renders the characters the possibility of going out of control. An unhappy childhood brings an unhappy life. When the child is growing into a skew adult the adult will bring about another skew child. The main drive behind man is fear. Where fear is absent the man is a danger to him and to society. Fear brings about a conscience. The fear I refer to is not anxiety or being scared but having respect. Having respect for one's parents or teachers or the community and respect for the law. Above all the fear of God brings respect for God and that forms the basis for being man. Persons with personality problems and character flaws may have too much fear or none at all. It is a balance and it is the balance that keeps us upright. Underlying in the balance is respect for yourself and that you have with the respect you have for your parents. When being a child your parents are a substitute for God because only they can bring values and norms that will one day bring about a balanced member of society. Money cannot buy that.

I know that in itself having a conscious and adhering to the conscious does not say much because the alcoholic hides his booze from himself because his characters are in conflict. He knows he is on the road of self-destruction but normally that does not bother an alcoholic that much as they love drink more than life. It normally is the fact that the alcoholic knows about the destruction in his children and his love for his dear ones. He wishes to protect them from him but he has this massive problem that is stronger than his urge for life. To find a way to solve the problem he starts to hide the liquor from himself and in that way he stars to lie to himself. The characters are accusing him as they are destroying him and the escape is the destruction. When he is drunk he has no control and when he is sober he has no control. The characters are telling him he has no problem as much as they are accusing him about his problem and denial is also admitting where admitting then becomes blame and the blame he carries are more than he can carry therefore he drinks to escape the blame of his guilt.

It is not only liquor but also it is sex, drugs, pornography and gambling. Those are the weaknesses that puts people in the same situation as that of madness. It is living the life of lust knowing that that life of choice is what you choose and choosing such a life is the equivalent of destruction but modern society makes fighting thereof much harder than ever before and capitulating as easy as breathing. The guilt, fear, anger and despair comes in waves and in conflict where the alcoholic is in the middle with his problem out of control and out of his hands. His characters took charge and they destroy him as well as those surrounding him... all that loves and care for him but the biggest destruction is his destroying of the ones he love. Seeing the suffering in the lives of those poor, poor soles make me great fill for the small load I received and the ease with which I have to carry my small burden. In that way we're in a fight but the fight is all in private and we may acknowledge failure where there are great success as we may judge success where there is great failure. We on the outside judge what we see on the outside while the fight is on the inside where we can never see. A rehabilitated alcoholic must be a far bigger success than a successful achiever born with the golden spoon in the mouth but skins the cat in the dark by indulging in cravings of the night, all in the quiet his money can bay. It is about what you fight and not the way in which you fight or how you fight.

I am no philosopher of any sorts but thought about success and failure and my position in life where I as a person will never reach great achievements therefore the question I asked myself is will I die a failure because of that? My success is my children I leave behind and the success they may be. Not in great achievement because in the end even King Solomon declared it was all about chasing the wind. Leaving behind riches in money can bring as much despair as leaving behind poverty. The only way I may achieve success is leaving behind children that are of a better fabric than I was. The next generation must be better equipped, better evolved and better humans than was the previous generation to ensure evolution. What is taking place to my mind at present is devolution through out the western world. I do not have to bring proof about that because reading the paper or looking at nightlife will bring proof to any one wishing to see proof. How do we western man raise our future and how do we equip our future? Western society removed discipline from schools with admirable success. We removed the authority from the teachers as best we could. The teachers are through out the western world the lowest paid professionals in society. Being the lowest paid brings about the under achievers of society and any successful teacher leaves the profession for better pay in the private sector. This is the cancer of the new age we are facing. We all know my last remark is the truth and by that the parents wishes to compensate, but to modern man compensation is about money and paying to get rid of the guilt. Every nation culture community and individual admits that money is no measure for success and yet that is the norm we live by. Every person living a life is but a building block to establish the next generation. Man is an endangered species and animals are overpopulating the world and walking on two legs is not the criteria for human classification. The position you hold distancing yourself in thought and behaviour from the norms that animals uphold makes man or beast.

Animals are not in a struggle with their improving of the mental but in a struggle with the survival of the fittest, the one that can kill the best, run the fastest outsmart all others. That is not man. Man is about his fight to better his life and not his body or position. Man is about fighting the battle of the best in man. That is not modern man's ambitions. Modern man has ambitions making him the best animal on the planet because atheism propagates the fact that we are all members of the animal Kingdom. If man is that blind it may be best if man returns to the animal Kingdom where he thinks he belong. At least the loss will be small but smaller will be the gain of the animal world.

I wish to take you back to the indicial verse in that there were three things man would obtain when rating from the tree, and not only two…it is the three that every preacher misses **…There was the tree…the tree of knowledge…knowledge of good, and knowledge of bad. You determine your knowledge…of good…of bad but above all it is your knowledge. The Bible does not name what is good or bad…that is your choice you make and that is your price you pay in the end…because the knowledge you accumulated will stand you either good or bad, as animal or man.**

It is not what you may accomplish or accumulate that has importance but what you learn through being man and standing apart from beast, that is what you take with.

Accept the following or reject the following but consider the following.

Because there's mathematically no nothing there must be a God.

By excluding nothing you have to include God because if there was nothing there was some scope for God being nothing and therefore non existing but since nothing is the only excluded number in mathematics you have to include God as a factor. What ever you dare to prove or disprove you can only prove or disprove through the human thought and understanding of concepts only human concept can understand.

I do realise with my statement about nothing and God it is stretching matters beyond the argument. When taking the argument that far the argument can include the existing of fairies and other fantasies, that is quite true, but by my placing a fairy and other fables in the realms of insignificance brings only harm to what remained of the child in me and I can assure you there is not much left in that sense. What ever is there is also in the infinity of my childhood memory. Not believing much in the fabric of the fantasy does not affect my position relating to the animal as the animal has little regard in such matters. Fantasy may have significant when I connect ethics to it as the Roman Catholics do with saints, demons, angels and such. When one can disregard such fantasies without affecting norms and values of the civilised, not harming the moral fibre of society there is little harm done by such removal. But when removing such norms for the enjoyment of being superior while the correctness of the argument in its core is invalid and, the structure of society goes to hell, a lot of question marks appear about the morals behind the motivation as to the motive behind the act. Man is moreover morals than life.

All life will destroy other life to its own benefit. A virus will kill his host for the benefit of one life cycle and at that a virus life cycle. What waste we may think. Killing a host only to spurn seems a lot of waste. Not so, because ticks will devour a cow alive without feeling any remorse. The same apply to a lion killing its pray. If there were no chance of the antelope finding means to escape the lion would start eating without killing saving itself the effort. It does not kill quickly through pity like humans do. It kills because of self-interest alone. When one of the herd falls victim to a lion attack the rest will start grazing thanking in that manner the victim for securing their position for one more day. There is no compassion, just self-centred egoistic drive to self-protection. When a person jumps into flames in an effort to save another person he is a hero. He receives praise from all concerned and may even land a medal for his effort. Why would we humans consider that as brave, being exceptional and above average?

When a person buys a car worth millions (in South African Mickey Mouse Money), because his business is going grate and he can afford to, but has his sister and his brother in law is working for him at minimum wage, not making ends meet in providing for their school going children, battling every day to put food on the table while he is making money like water flowing every one admires him for he is rich. He is treated very softly and no one dares to stand up to such a person because he has money. Such a person as a human is wasting breath because he may walk on two, but he may as well walk on four for the humanity he is. Still society regards that animal as a pillar of the community because he has money and he is a "sharp and intellectual business man". He should be shot at dawn for impersonating man whilst being beast. His positive has only one aim and that is to better his financial position securing a better admiration in society and establishing a front with him being god to the rest he sees as lesser mortals. His neutral sees his money drive as security and that satisfy while his negative is bent on destroying others because his positive is telling him how great full his sister and her "useless" husband should be for his generosity of providing food on their plates. Not once for the shortest and briefest instant will he ever give a thought that it is precisely the other way around. By their self-denying they are enriching him, but through his ego madness that thought never comes to mind. Then the philosophical will say that is man.

From our cosmic position we are in the centre of the universe. Where you may sit or stand is the very centre of the universe (your universe) because you can only relate to the universe having your individual singularity and where all other aspects in the cosmos are pointing away from your singularity to all other positions. We will always be 56 in a universe of 112. The earth will always be Π^2 in relation to the sun with the speed of light being $3\Pi^2$ from our position. It is not surprising we see ourselves on the edge of the Milky Way. From where we are we will be on the outer edge of our galactica. To the inside everything will be brighter and to the outside everything will be darker. That is a fact, but not a reality. The fact we may appreciate but the reality we can never understand. We are egocentric maniacs coming from the position of our singularity. That is the animal, the trio breathing our air. That is what the fight is about and what one may take with after death changes dimensions occupied. That is wisdom standing apart from knowledge and the animal has knowledge in intelligence but man has wisdom in extelegence. That parts man from beast. Man has to part from singularity's approach.

Us humans must fight to find our place outside singularity, outside self-interest and the strive to better of our position including finance. We have to relate from a position in divinity and not singularity to understand the cosmos and to understand life. If not we may well die as animals, and such possibility is there. If one person spent a lifetime killing his human in him, he may end as the animal he always strived to be. Being artist's engineers or preachers have nothing to do with it. Knowing the Bible or not has nothing to do with it. Being a believer or not has nothing to do with it. It is your approach to the cosmos that makes you part of the cosmos or that dimension above the cosmos. Accepting that one is not part of the cosmos but part of life in a cosmos puts one above the cosmos, but still in the cosmos. Never lose reality but accept responsibility. Persons believing in fairies and fantasies have the problem of over acclimating the positive while the neutral protects sanity with much of it going in the direction of a lost cause leaving the negative happy for destruction can only follow such obscurity. The art and artistic are typical in this. This is the line the drug addict follows to the last letter. By pumping the acid the positive hallucinates, the neutral exclude the incorrectness and the negative character are in seventh heaven as destruction is deliberate and decisive. The very same apply to a soldier in war where the positive character enjoys the killing as much as the negative character enjoys the being killed and the neutral character wishes the cruelty on them before they are upon us. When the soldier arrives back in society, society understands him as little as he understands himself because the fabric of human principles has gone skew. His characters, including him is as mixed as a milkshake fruit salad.

There is the middle, a precise middle where I as a person hold my personal relevancy as I see myself. What ever I attach or detach puts me in relevancy to others in life. That is the purpose of my fight with my characters. My precise middle must be very straight and when the middle leans toward any side in particular my middle is out of alignment. To secure a centre there must always be room for others and their opinion. When I say there is no God I go eccentric and when I say there is only God I go eccentric.

As it is in mathematics in the matter of the line and the dot, it is a question of you deciding the relevancy. It is your choice and only your choice to what relevancy you wish to place God as a presence or a factor. By declaring the absence of God your relevancy might reduce God to infinity but in your denial you place relevancy therefore relevancy remain be it in the infinity. It is to the peril of the denier that such a person excludes God as a factor because by placing God at a point of infinity next to zero the denier places himself next to the animal that excludes God because God excluded the animal from extelegence. Such a thought proves the fool. God allowed the animal excluding the admitting of a God because God exclude the animal from the dimension of being human therefore liability to questionability. In the perception lies the norm.

On the other side of the spectrum is the religiosity maniac and such is his idiocy he gives God the role of the animal being on call by prayer to serve the master's call. In all such cases the maniac places himself in eternity by creating a spot in the centre of eternity for himself providing him endless power having God as his slave or animal on a leach in acclimation to his eternal status he then places God just outside eternal big for that spot he reserved for himself. Through his effort by prayer he can heal, bless and doom…and God will obey as instructed. Such a fool proves the thought.

Through your relevancy with God you become the God that makes you your own God as much as you're own devil and your own forgiver as much as your own accuser. You make the relevancy and the relevancy proves you. Being positive is loathing and being negative is damning and being neutral is obstinate and in everything you confirm the applying of norms be they right or wrong, good or bad as it is the free choice you make proving your distance you have as a factor claiming life and location as a being from the animal. In the case of the animal the ride is free for the animal knows no better but man does and is liable through conscience placing relevancy to deeds. On the physical aspect in the being named man, the human body is an accumulation of singularity positioning divinity as one comprising of three identities without ever separating the trinity. The one holds the straight line the other forms the triangle and the third is the dome, the inclusive sphere, and the container having seven sides to the entire outside world. You cannot be in two sides of the universe simultaneously but you may observe both sides by being centre.

CLAIMING A POSITION IN SINGULARITY'S INFINITY IS LIFE'S DIVINITY.
YOU LEARN TO LIVE AS MUCH AS YOU LIVE TO LEARN

Singularity has three in parts as does divinity have three in places. On the other side of the divide is the divine and as things are in the one side of the divide in total relation to the other side because all relevancies align as much as match in equilibrium and in as much as it is the cosmos forming one half of the divide it has to be the divine duplicating to establish equilibrium. **The purpose of man in life is to know your other entities, to learn to live in recognising the other entities, to take from them strength they can give but also to detach from their weaknesses, and above everything else recognise yourself in what you do as pure or spoilt. Gain knowledge the knowledge you have in the good you can acquire and the knowledge in the bad you have to detest. Then only you may serve a life of gain. Did I prove anything, well you are my judge! As I am the one promoting all things connect, so I am the man believing all things connect.**

Without a zero there cannot be death. Without a nothing there cannot be no God. Accept one and you cannot except one.

If it is that simple why is it complicated.

BEST WISHES,

PETRUS. (PEET) S. J. SCHUTTE

A Less Complex Commercial Science Book   275   Chapter 6

PART 6

I have found what gravity is. I have found what drives gravity and what the factor mass is. Be assured that it is not what Newton saw is taking place, but for my not supporting Newton I am rejected by Mainstream Physics. Mass is not producing gravity as Newton thought but mass rather is a method of restraining gravity where gravity is being blocked by a phenomenon Newton called mass. Mass prevents an object from moving and that prevention of movement is mass while the inclination to continue to move is what remains of gravity the object retains. But while the academics brainwash students into submission by telling them untruths and stupefying their brainpower with **NEWTON'S FRAUD** and in this way controlling student's learning thoughts, it is through this method of improvising the truth with **NEWTON'S FRAUD** that they get away with mind control. I am unable to crack their deception and introduce the truth because by the power vested in them they can prevent me from doing so. It now is the time that students start asking the correct questions and begins a process of unmasking those criminals that is teaching physics. Ask them to tell you students exactly how much did the Moon come closer to the Earth since the time of Kepler or even since the time of the Moon landing in 1969. Insist on them proving **NEWTON'S FRAUD** $F = G \frac{M_1 M_2}{r^2}$ is correct and it is working by showing that the Moon is moving towards the Earth at any rate. If they fail to tell you how much the Moon has come closer or at what speed is the Moon is coming closer or in the event when that they have to confirm that the Moon is moving away from the Earth, then you will know that they are lying through their gritted teeth and I am telling the truth. There is no contracting Universe and Newton presumption of mass attracting is one big farce those cheats use to brainwash students into stupidity. Physics is based on $F = G \frac{M_1 M_2}{r^2}$ which is based on **NEWTON'S FRAUD** which is complete and utter well placed fraud that is set in place by student brainwashing. Here is more of such deception students are lied about.

Academics in the science of physics say that a feather will fall with the same speed as what a large rock would fall, but the condition they connect this process too is that the fall of the feather and the hammer has to occur in an atmosphere filled with vacuum such as we find in place on the Moon. This gives the impression that such falling of a light object versus the falling of a heavy object requires a vacuum atmosphere and in that this atmospheric condition is completely alien to what we have on Earth. That is part of their cheating! What they never add is that the largest rock will fall at the same speed as a tennis ball and to achieve that there is no vacuum filled atmosphere required such as the Moon has. This can and this happens on Earth every time and thus the process of all things falling at an equal tempo is far from alien to the Earth. Why don't they say a man and a car will fall to the Earth at the same rate and then prove how mass is part of such a falling equation! Mass has nothing to do with any object falling because all things fall equal just as Galileo said and this applies notwithstanding size (or mass) differentiation also as Galileo said it applies! You can see on TV how cars, bags people and clothing fall precisely equal and that is the case notwithstanding that not one of the objects are even closely resembling an equality in mass! If mass created a force called gravity, then it must be true that objects could never fall equal because mass will create different forces in measure and every force will have the object fall to the Earth by the mass it weighs. ...And don't let them come out with the nonsense that there is a difference between mass and weight because they use that lie frequently and that is another part of the fraud science created to cover **NEWTON'S FRAUD.** If things fall equal but their mass is unequal then things do not fall by mass and then mass can't be responsible for gravity. Those professors hide behind inconsistencies to get you students confused while they are brainwashing you into submission to accept their betrayal of the truth.

That all objects fall even and fall precisely equal is an accepted principle according to Galileo and that was accepted as a principle in physics long before Newton thought of becoming as wise as he later on though he was. Newtonians never can see any difference between what Newton claimed happens when he claimed all thing fall because of mass producing the force of gravity by which all things supposedly falls and what Galileo claimed when he said all things fall equal thus ruling out any consideration of mass inequality interfering with a fall. To them it is the same when Newton said things fall by mass and Galileo said all things fall equal thus not by mass. To them these total contradicting statements are still the same thing and they have the task to brainwash you into believing Galileo and Newton said the same thing! It is their job not to teach you the truth but to force you to believe that Galileo ands Newton said the same thing. Newton said all things fall by mass. Galileo said mass has nothing to do with falling. Their job is to force you to believe this is the same idea. For the first time ever since the time Newton introduced gravity I seem to be the first and the only person that questions this interpretation.

Has anyone ever explained how the idea of a feather falling as fast as a hammer fits into the idea that mass pulls mass and how the falling by the gravity forms power that is exerted by mass as a hammer has much more mass than even what a large feather has. Test you Professors explaining ability and his skill to hide the truth from you when you ask him how mass that is the principle bringing absolute differences between objects moving influences all things to fall equal. In other words how does he console Newton's theory on differentiation with Galileo saying all things fall equal! How on Earth do these two concepts of a feather falling equal to a hammer or a car fall equal to a human, which every time it happens proves that Galileo was absolutely correct and Newton was absolutely incorrect fit into this interpretation they use of mass causing objects to fall?

How can a large mass pull as equal as a small mass pulls to travel equal covering an equal distance it descends to Earth at the same speed over the same distance and still be driven by the power of mass creating gravity. If it doesn't differ then there is no proof that mass is any factor in the falling because what does mass then bring into the equation. Have you given this idea a good thought? By me scrutinising this concept I disagree and by me disagreeing I am silenced by those academic frauds in power. Don't settle for any more brainwashing and mind control and start to insist on answers.

When any person disagrees with any academic in any lecture hall about mass not forming the factor that is responsible for pulling gravity and you come to a conclusion that you doubt the mass part that they bring into the picture they claim is responsible for establishing gravity, then the academics wipe you from the table with a swipe. To cover their crime they contemplate that you are so stupid because you are unable to "*understand*" Newton. By your "inability to *understand* Newton" you fail to see facts and you are too stupid to *understand* physics. They even in some cases go on to say that physics is not for stupid people and only "*clever*" and "*informed*" persons would be able to "*understand*" Newton!

I have been at odds with academics for years and only because of the superior positions they hold in office are they able to bully me into submission and silence at the time. They hold the power of the reigns in their hands, but what they do not count on is that I can go to the public and bring my case and their fraud into the open. Academically I am not from their league and neither am I from their ranks and with me not being part of their ranks they form have they bluntly dismissed me. Because I am not part of their group they are of the opinion that it disqualifies me to have any opinion. They hold the opinion that only they filling academic positions are allowed to form any opinion and the rest is too stupid to have an opinion. In that way they were able to dismiss me for years, but no more…I will not tolerate their behaviour any longer. Now I fight back. They have this opinion about their positions that while being what they are that status they have gives them the rite that they may regard or disregard all opinions when they do not fancy other opinions. It is up to them to decide what the truth is for they control all judgement as to what is the truth! They may silence whatever I may say notwithstanding my correctness and validity. To summarise their attitude one requires three words which is arrogance, autocracy and megalomania. Absolute power corrupts and they are the living example of that.

Due to the important positions Academics hold in the huge academic institutions such castles of power gives them free sanctuary from where they can hide their criminal ploy of deceit. If you think I stretch the truth by accusing them of fraud, then please read www.SirNewtonsFraud.com and find out why I am blaming them of utter corruption. They sit so high and are so mighty that they do not need to explain anything but to themselves amongst themselves and their deeds go totally unchecked. That makes them be the untouchable and unapproachable powerful from where they rule with absolute authority. This unquestionable authority gives them the locations erect a cover and give them the opportunity to hide behind that wall of absolute superiority and suppress little persons such as I into silence and submission notwithstanding…Whatever I have to say can never go past their scrutiny and can never pass their sanctions.

Now I am taking my case to the members of the public so that the truth must be brought into the open. I have had the tour they give and then more came my way. I never got around swallowing the fact they claim about mass creating a pulling force Newton named gravity. I distance myself from witchcraft and soothsaying and their claims about the force pulling by mass is fiction. Let one prove that part where science is of the opinion that mass pulls as gravity is…after reading www.SirNewtonsFraud.com. Academics condemned my work and therefore me and for eight years where I could not get a publisher to come around and had not one publisher that bothered to read my work, let alone seriously proposing a publishing contract. There is no publisher that is willing to go against Newton and face the world of physics. The lack of support in the publishing world forced me into a corner where now I had to finally go private with the publishing as all doors shut in my face as soon as the academics read the content of my work because from the nature of my work I take Mainstream science head on and am confrontational on most aspects of astronomy. There does not seem to be any publisher that wants to go head bashing with the establishment of science on official science principles, which I have to do to convey my message in a no uncertain language. Fact is I know what gravity is. Fact is Newton doesn't know what gravity is and also admitted to that. Fact is that I saw Newton is miles off the road with his presumption about what gravity is. Fact is Academics in physics carry on as if Newton and they are the sole experts on gravity while Newton and they have no foggy clue about what gravity is…and no publisher is willing to agree that neither Newton nor academics have a foggy clue about gravity. If you also have doubts about the academic's indisputable correctness please read on and confront either them or me on everything you read here.

After reading www.SirNewtonsFraud.com you will have to take sides because you will know the truth.

Then you either have to become a partner in their crime when joining the academics in physics with the purpose to cover the truth from getting known or you will be part of the truth and become an activist fighting by helping me confronting those perpetuating to perpetrate in crime until the academics stop their criminal conduct and acknowledge the truth.

By not confronting the establishment, you give the establishment grounds to allure you into being sheepish. They declare facts and you sheepishly follow as so many did for centuries without questioning those in academic power. Because they see you, as being just another stupid senseless student they have the opinion that they can brainwash you into accepting these fallacies that I am about to tell you in **NEWTON'S FRAUD**. They do literally brainwash and condition your mind to and control your thinking in believing what is correct until you accept what they never yet were able to prove.

They are of the opinion you will swallow any of **NEWTON'S FRAUD** they throw your way just because every generation before you were mind controlled in the way they are about to control you. You may think this is big words but read **NEWTON'S FRAUD** and see after you come to know all the facts if I exaggerate even in the least. They see you as a slow-witted and mindless nobody with a mind to form any opinion because they think they are the academics being superior which is just other words to describe what is making you the inferior. If you are not aware of the facts beforehand and before they start to brainwash you by mind control they know you will follow their teaching without asking questions.

They think that your naivety makes you mindlessness and being vulnerable leaves you so stupid it will incapacitate your thinking ability which would lead you into their control. They don't want you to ask nosy questions about contradictions existing that they have to refuse to answer because there never was answers available since the time of Newton. This process of brainwashing and mind controlling in physics has been in progress for hundreds of years. Just answer how a small object such as a hammer and a large object such as a car would fall equally while mass forms the driving force that establishes gravity as a force. The fact that they fall together is images we see on TV everyday and is in modern society beyond doubt. If you can't explain how it is mass driving gravity that allows a car and a hammer to fall at the same speed while a hammer is so much inferior in mass to what mass a car has…well they can't either! Their task is not to explain but to mislead since they think you can't think while they think they know how to control you.

The following web page does not aim to represent the full entirety of the original book called **NEWTON'S FRAUD** but is reduced to aid any possible potential reader in the examining what the purpose is of the information this web page wish to announce. Anybody and everybody are aware that all objects fall at an equal rate. If an object such as a car weighing one ton falls at the same pace as a person weighing fifty kg while the two objects escort a hammer weighing one kilogram all the way to the ground, then how does mass come into the picture by committing a force to do the pulling in accordance with mass? Mass has to pull because according to their teaching it is mass that establishes gravity. However, mass is a factor that produces differentiation whereas all objects show equality during their fall. If it is mass that is establishing the force gravity, all objects must fall at different speeds. That they do not do as they all fall equal. That means physics is wrong from the start because mass cannot have any input in objects falling.

This is but one of many I mention in **NEWTON'S FRAUD** where all that I mention in **NEWTON'S FRAUD** are not the only untruths but forms facts that the Paternity called Mainstream Science is keeping concealed as a cover up that is wrapped under an airtight blanket of deception. If you sit in class and listen while also experiencing the sinking feeling that the facts you hear about what Newton claims are not adding to a total that you personally are comfortable with you are most probably correct because Newton is incorrect. If your professor makes claims about physics while you disagree with what is said then you better read **NEWTON'S FRAUD** because **NEWTON'S FRAUD** has it at task to show all that will read **NEWTON'S FRAUD** how much discrepancies academics lay on unsuspecting students that trust Academics with their future and their life. Do you as students realize the inconsistencies that physic Academics present you with when portraying that what they teach you as being the solemn truth.

Students tell your Professors to stop deceiving and tell them to stop trying to control your minds with their fraud. Those Academics tutoring you are telling facts about physics and about gravity that has never been proven.
That is mind control and giving selected information to have you form an opinion they wish to manufacture.

They wish for you to accept facts on physics and on gravity that they hold as the truth but in all of three hundred years were never proven…not even once. They claim those truths are beyond questioning yet with the least examining those truths they stand by then proves to be totally void of substance because it was never corroborated by one single experiment.

Should you question that mass produce gravity and insist on showing that $F = G \frac{M_1 M_2}{r^2}$ is not utterly flawed in principle, they will expel you from University by letting you fail your examinations and this will happen while the formula that Newton introduced $F = G \frac{M_1 M_2}{r^2}$ was never proven. They will expel you because they will claim you are unable to "understand" Newton and therefore lack the knowledge to present Newton accurately. Let them first show how $F = G \frac{M_1 M_2}{r^2}$ works when all the values of the symbols are processed and an actual value of the force is calculated by using strictly the mass of the Earth as well as the mass of any object divided by the radius between the objects standing directly on the Earth as it is used in $F = G \frac{M_1 M_2}{r^2}$. But they will rather terminate you schooling and have you fail tests should you question their authority on the matter of gravity formed with the implementing of $F = G \frac{M_1 M_2}{r^2}$ while at the same time they can't for one second bring evidence in support of what they wish you to accept as the unquestionable truth.

That's brainwashing by mind control because if you don't accept their baseless fact as God given truths they dismiss your academic career.

It is either put up and shut up or be gone. Academics do put mind control to work on unsuspecting students by forcing students never to question the legality of statements they offer as being sound and correct.

Module Five
The Comet's Gravitational Demise

In the book the author, which is I explains gravity. This achievement is possible because I saw a way to break away from invalid concepts Mainstream physics hold. I recognised the impossible double standards Mainstream physics apply to promote their much shady explaining. The inconsistencies brought them double vision and to compensate their incredible theories they simplify issues to a level where what they embark on to understand is becoming meaningless. What they say can't be supported and authenticated any investigation even in simple terms. It is as if they never read with interest that they explain and they never scrutinise that which they advocate. They give values that are senseless and make that which they say meaningless.

In this article I am going to investigate how much truth there are in mass pulling by the force of gravity. To most if not to all of the persons reading this and just the thought about me embarking on the investigating of the issue is totally senseless to investigate. It is senseless because the concept it carries became accepted as household practise and life science.

Do you think of astrophysics as the department that is run by the wise and the level minded the sober thinking and the absolute trustworthy? If you are a student there is no other choice you have. If you think those in charge of astrophysics are the pillars of trust, then get wise and read the following. What you are about to read is simply mystifyingly simple and yet to this day I have not challenged one academic any where that had the honesty to admit to the fact of Newton being wrong. After you have considered the following you might agree with me that even small Children can reach a higher level of clear-minded logic and find more sensibility than what those scientists promoting astrophysics have because science lives in a make believe fool's paradise. If you are a student, then ask your Educated Masters please to explain the following abnormalities and inconsistencies they promote as part of official physics, which I present in this web site and get wise instead of brainwashed. I say brainwash again because they force-feed you fabrications, as you will come to see. They can't explain the facts as the facts but hide the fact that the facts are in fact untruths. Tell them to prove that planets have mass. Tell them to prove that it is mass that generates gravity that pulls the planets. Ask them to explain gravity in detail.
The idea of proof comes automatically to the door of Newton although Newtonians will deny this fact as if they deny the honour of their Master Newton and that is what they have to do.

Think of what planets do… and you think that planets orbit. It is connected to the brain. No one thinks of planets spinning or planets basking in the summer Sun. When hearing about planets the first thing that comes to mind is the rotating of planets while circling around the Sun. However, just using the term orbiting id in total defiance with Newton! Newton said gravity draws or pulls or moves in the direction,…which would have one understand that the two objects in example the Sun and any of the various planets will be moving directly towards each other. The term pulling does not suggest any circling because no one can be pulling towards and do that while circling. When pulling anything it must take place while using the shortest line possible. That serves the term pulling. Then the saying goes that planets orbit indicating they follow a circle. That is not what Newton said. However, wrong that may seem but circling is precisely what planets are doing.

In conversation we speak of the planets orbiting. If Newton was correct we should be speaking of the planets pulling, but talking about pulling would be blatantly wrong according to the normal spoken word. Never do we refer to the planets pulling the Sun or the Sun pulling the planets, but we speak of seasons coming from orbital positions. Being in orbit has to neutralise the pulling and then cancel the pulling concept that also became culture.

If there was a pulling, and the word orbit cancels such an idea, then there has to be some sort of prevention taking place that disallows the pulling to commit the direction of travel. I know it is said that the orbiting object falls as fast as it circles and by falling while moving to the following side on position it never reaches the Sun, and yes, it makes sense, but there has to be some form of resistance replacing the planet in the next side position and preventing the falling or the pulling from taking place. Using the formula $F = G \dfrac{M_1 M_2}{r^2}$ as Newton provided, disallows any other concept other than moving towards. The person Newton got his ideas from and the work he raped completely, that of Johannes Kepler explained this very well, but Johannes Kepler makes no room for any pulling of any sort. In the work of Johannes Kepler he said that the space being the orbiting route a^3 remains at a specific distance **k** while the orbit T^2 takes place…and in all my other books that addresses more information I take Newton to task on his dismembering of Kepler's formula by corrupting Kepler's work and with what amounts to fraud, Newton takes science on a goose chase that holds no truth. There is no pulling by mass of mass in any way.

In the book named "*an Open Letter on Gravity*" I bring the solution to the mystery behind gravity. I tried in vane to introduce the principles I find valid to the academics in charge of astrophysics. Facts that Science present as being the uttermost explicit and unwavering truth, fails to bring any logic answers to so many questions that it should address. It fails to have substance in addressing the most basic and simple questions about gravity and physics. Yet to every question science can't answer my approach does bring many solutions. The presentation and the delivery of my answers that I reach are understandable and simple where it serves both logical science and the truth. Since my answers do not match Newton and his misconception about gravity and that mass generates gravity, those in charge of science don't even bother to read my work. With their affixation to the corruption they portray I can do little to the giants where they are in the mighty positions they have and just because of that they can go about to sideline and ignore my work and this is notwithstanding the correctness that my work delivers compared to the utter failing that Newton's work shows. When confronted with my evidence and they have to match my work with the hypocrisy and misleading nature of Newtonian cosmology their defence in substantiating their claims is to ignore me. Since I do not applaud mainstream science and the clear fraud they embrace and fraud it is that they embrace, I am silenced. Why is it that my work is going unrecognised or even in the least goes never debated and never commented on…it is because it will then trash every article anyone has ever written about astrophysics and cosmology. They show little integrity when academics with high standing such as they should have will rather protect fraud and save their skins than seek the truth. It is that when they begin reading my work they then have to back my work. Doing that will trash all work in cosmology delivered thus far and condemn it to the waste paper basket and render all work invalid and void. It will put all the Newtonian's bias and fraud into the place where it belongs. Considering that such acting will lose them money, those academics in controlling positions then will rather misrepresent the truth in order to benefit from continuing to corrupt student's minds further. If they wish to justify their inconstancies they have to attack my work and disprove the accuracy of my work. That they can't do. They then ignore my work because they can't attack my work. In that sense they also place their work beyond my approach, as they can simply ignore me as if I represent the plague while they carry on with little consequence to bother them. I challenge them to prove Newton correct and not just declare Newton being beyond reproach after all has seen the evidence I bring. After reading this all students must challenge them to defend what they can't or get honest.

This is the basis that Mainstream science uses as the foundation of all physics anywhere. If this is wrong then everything they have got to work with goes out the window. They put mass and the distance that parts objects in a relevancy, in other words the one is a ratio to the other. The one factor brings a measure to the other factor's value. The one cannot be without the other. The increase in one becomes the reducing of the other and the other way round also applies. When the distance is large, the influence of mass will be small and when the distance is small, the influence of mass will be overwhelming. Then they state we are in a Big Bang expanding of the entirety. Why then, when considering that if it is mass that produces an inclining force of contraction as Newton says there is going on then…why didn't the expanding stop before it started when the Universe was small. Today using hindsight after the fact of the exploding Universe became apparent by the studies Hubble brought to light did the lot of everything that is not implode as Newton would have us believe whereas, instead it did expand just as Hubble proved. The radius at the time of the first instant back then was no factor, which makes the gravity at the time a totality of unrivalled force. The radius being that insignificant leaves the mass unchallenged in asserting power in relation to the non-existing radius it had.

When the Universe was at the point where the Big Bang started, the radius was incredibly small. According to their studies it shows that the Universe was the size of a neutron. Having the size of a neutron proves my statement and it is science telling us that the Universe was the size of a neutron at that stage. That would make the mass that was at the event of the Big Bang that was producing gravity and the gravity charged a force by contraction, which then had to prove to be inconceivably large. This is because the mass was completely overpowering all factors with the small radius. But as Hubble proved this did not result in an implosion that drew all that there was into something even smaller. That is what the overbearing mass contraction was supposed to unleash on such a small Universe in the beginning. This understanding and accepting of the most basic brings us to the next inconsistency as far as I can gauge the situation. Maybe Newtonians are correct. Maybe I am too stupid to understand Newton but see how you fare.

When the most basic and simple mathematical law prevails it is clear that a small radius will bring about an incredible powerful gravity since the power that the mass establishes will then be overbearingly strong. The more the radius develops in time, the lesser would the gravity, be that the mass factor generates in relation to the advancing radii developing and the larger would the reducing be of all contraction. The effectiveness of the force that the mass are able to produce will tarnish as the radius that separates the material from each other increases as time moves on to create space.

Although it is presumed that the Universe was small at the dawn of the Big Bang, such presumption will bring validity to another presumption. The presumption is that if it is that produce the gravity then mass at the time of the Big Bang just had to be enormous. The influence of the incredibly small distance in radii and the factor such distance produced increases the influence that the mass factor can assert in relevancy by an exponentially large number, which elevated the mass to an enormous large factor.

An electromagnetic charged magnet typically demonstrates this example. If one holds a magnet far from an iron you would hardly feel the drawing of the magnet and the iron. When the magnet is a very short distance from the iron it draws so furiously the contact can't be prevented as it clings onto the iron notwithstanding the human effort to keep the magnet apart from the iron. The magnet didn't get stronger and the iron did not pull harder. The close proximity favoured the magnetic fields exponentially when the magnet was almost touching. This was then also true and what applied to the magnet also had to apply to the mass that produced the gravity that caused the pulling of gravity.

This same issue becomes true if an object is a distance apart; the radius is exponentially less influential in value, which then is by dividing into the mass, a factor that increases the influence that mass must have on the gravity pulling force. As the radius increases such growing by the distance will reduces the influence that the mass can produce by one, a ratio equal to the growing distance. Then the same force drawing would in that event be much less than when the objects in question are only a short distance apart. That is the most basic realisation about mathematics. It puts ratio to order and define coherency. That is what gravity is to the Universe as it puts respect to factors about the Universe in the Universe. It is what derives order in the Universe of mathematics.

We will find in a Universe that was so small where it had a radius of less than one kilometre, then at such a time with the Universe being that small it then must also be accepted that the gravity the mass charged was massive. Einstein said that where space is zero the gravity is ultimate. The extremely small radius that was only the size of one neutron in radii distance and with the factor that such a distance produces, it then in that case must have promoted the mass factor, which will support the gravity in having an enormous large factor by relevancy to what the case must be at present. The mass factor that produces the gravity at any given point during the event of the Big Bang, had to be eternally larger at the dawn of the Big Bang while having at the time almost an infinite radius, which gave gravity all the power it could ever have and which it will ever have.

We can see when only looking around us that in the event at the time when the Big Bang began no material growth could come in place. The with not sufficient mass to destroy the radius and prevent the expanding from coming about, immediately at that given instant of moving apart the expanding won the dual because the mass then in future will never become sufficient to launch a Universe of contraction as Newton stated. The contraction afterwards could never match the expanding that took place. If the Universe is expanding as the Big Bang concept promotes then there can be no contracting Universe as Newton had us to believe. If the Universe started a journey of parting objects by allowing the distance between the objects to grow then no amount of dark matter that might lurk in the night sky and is at this moment hiding from detection will afterwards produce the gravity required to stop the expanding from continuing. The moment the mass is not able by quantity to support and sustain a contracting Universe the expanding will carry on because the mass is losing influence as the radius increases and that mass then by ratio produces less gravity in force or influence. The declining of the influence the force experiences is directly in ratio with an increasing radius. At the start when the expanding became evident and as the radii grows the inclination of producing a contracting Universe will suspend as the balance progressively shifts in favour to decrease the influence of gravity as a factor. By the third microsecond after the direction was decided there was no chance any longer that the mass would ever find a manner to contract the Universe by enforcing gravity as a pulling power. If there was insufficient mass at the start in order to tilt the balance in favour of the reducing factor, no amount of mass can ever accomplish to turn about the direction flow afterwards. There cannot be a growth in mass because what is in the Universe cannot escape and what is not in the Universe does no exist. The Universe is the container that contains whatever possible things there are to contain and no adding or no releasing is in place. Then Newton's surmising was corruption, which is making that which all physics are based on a fool's idea that can only find proof if evidence is corrupted.

If you might be of the opinion that my accusing the greatest intellectual department in the world as being in misconduct and to your view such accusing is outrageous and far-fetched, then please give me the honour in being my guest and judge the following evidence that I bring with a clear and unbiased mind because when scrutinised with a clear view then the facts cannot fool an idiot. However, that is just what the physics paternity thinks we are. Academics regard us being the rest and being those that fill the sector that are not academics as the small-minded. We are to their minds therefore those that are the idiots they can manipulate. We being the part that is forming the general public at large are too incompetent to see the deception in their theories. If they don't regard us in dismay then why feed us all the rubbish they do and which they do for such a long time? Reading from their actions we can clearly see that they have the opinion that they can feed us being the everyday tax payer, us in the public arena any senseless rotten garbage they dish up because they see us as being inferior by thought and mind. Why else would they promote such shambles? It is either a case where they think of us as incompetent idiots with pea sized brains and which are thoughtless or they are the mindless incompetent idiots that truly believe their incoherent nonsense because what they promote is senseless rotten muck and hogwash.

With all this in mind did any one ever come to wonder about the reality driving the all too famous Einstein's Critical density theory and the fact that this idea was conceived to conceal the corruption of Newton in physics? Allow me please to elaborate and then make up your minds. The facts in truth are that the Einstein's critical density theory was a scheme plotted by those in charge to cover up and conceal corruption in the heart of physics. Hubble saw an inflating Universe that contradicted Newton's deflating Universe and this perception of a deflating Universe being a myth had to be most ardently hidden as to yet again compromise the truth about Newton and his theory. One minute ago I showed how the relevancy applies between the strength of the mass and the ever-increasing radius that keeps on diminishing the influence of gravity as long as the radius is increasing and the distance between objects are growing. This is so basic that primary children learning the basics of mathematics will understand! Yet Einstein proceeded in searching when the mass would bring a turn about in the direction that the cosmos is involved in. Einstein was looking for the moment the mass will become strong enough while the most basic principle indicates that an increasing radius leads to a decreasing gravity since the mass becomes less prominent. If Einstein was unable to recognise this most basic of mathematical principles, then what type of genius did physics create in him and what slur did physics promote. This idea of the two factors being in opposing relevance is so simple that children will recognise the principle, and yet those fathers of physics wants me to believe that the greatest mathematician that ever lived did not realise this principle…the principle that the radius and the mass stands related and the growth in the one will promote the decline in the other as a dominant factor. Can any one with this information including the information given on the previous page have any other conclusion than smelling rotting fish somewhere? It is obviously clear that having such a total idea that there might be dark unseen mass floating in the Universe which at this time does not generate gravity but will some day kick in to generate gravity in order to cover up Newton's deception about a contracting Universe and just because Newton has to be correct at some point in the future. Science wishes me to believe that since there is a lack of seeing material there then will be dark and unseen material where they are so dark they are undetected by all humans. That leaves another question to address…where can such mass be found. How will such undetectable mass be found, which will bring about contraction after all this expanding ends and the Universe recognises the infallibility of Newton once again? Why would the mass at present then not activate gravity and why would the mass at some point spring to life and start activating gravity? How much can the Physics paternity still hide the fact that Einstein's critical density is being used as a cover-up to distort the truth to conceal the fraud Newtonians wish to cover? The uncovering of the Newton's fraud by the Hubble constant is so simple to see. Hubble found the Universe is expanding and Newton said otherwise. Who is lying about what? Hubble's declaration was on track to blow the cover that was concealing the Newton fraud wide open and uncover the century old deception. To see this we have only too look at the comet behaviour when any and all comets again come around on a cycle by repeated visiting the sun. The question is if it is mass pulling mass onto mass, then why do we have comets left in the solar system? The mass of the Sun should by now at least have destroyed every comet going around.

Every one believes in Newton except comets, because comets fail to collide with the Sun.
However I can explain in some way…

Every indication that we so far received in vivid portraying from astronomy photography studies from outer space disputes a shrinking Universe concept. From the Moon increasing the radius distance between the Earth and the Sun, to the Hubble Constant indicating there is a space growing all over and any where in space wherever man may conduct studies. Since the end of the middle ages a force called gravity was identified, but further than that science did not take it. What is gravity, besides being a force? What forces the force? I introduce a cosmic theory that turns the missing questions to answers.

Let us for one second return to the science we all know.

$F = G \dfrac{M_1 M_2}{r^2}$ There is an undefined phenomenon in the cosmos, never mentioned (in public) because it obscures reality but is proven in using this foundation of science, the basic formula used to prove all science.

Let's put the mathematical formula into a practical context.

By reducing r would bring about the same result as enlarging the mass factor of the cosmic objects i.e. the Sun and the planets. It is a very drastic implication that will cause much more than just seasons changing. It must bring about that gravity changes through out the year…yet the radius does constantly change therefore… it is evident that Kepler's name was used when science introduced a formula as follows

$$E - e \sin E = M$$

That proves everything but that mass is responsible for gravity

$F = G \dfrac{M_1 M_2}{r^2}$ In relation to the next few arguments it is very critical to understand the arguments I present because on the soundness that these arguments represent in the arguments that I make I am in dispute with the arguments that Newton makes.

The entire philosophy of the science of physics rests $F = G \dfrac{M_1 M_2}{r^2}$ on the arguments that Newton makes. It is physics that base everything on the fact that mass produces gravity and therefore by the force that mass provides as gravity the entirety of all physics are founded. I do not dispute mass as a factor in physics but what I do dispute is the way Newton presented the fact that mass has any value between cosmic bodies.

The formula as Newton presented it is just not baking the beans any more and must be re-examined because there is a lot of contention in this statement.

Please follow my line of thought and scrutinise all that I say. That which I touch has resting on it the entire philosophy of astrophysics as well as physics.

What I dispute is that it is mass that is in control of the cosmos and that mass provide the sticky substance called gravity.

Also do I dispute that gravity is a contracting force by which the entire Universe is in collapse. However I begin by disputing the idea that it is mass that is producing the gravity that supposedly produces contraction that is there to pull planets towards the Sun.

Newton insisted that mass is the influence under which gravity becomes a force. He promoted the idea that gravity is a force instituted by the measure of mass because the mass unleashes the force of gravity on other unsuspecting objects and then begin to pull the unsuspecting objects in, in the same manner as what anglers will do to fish. Once the mass gets hold of mass the mass reels in the mass from both ends that carries mass and according to Newton everything in the Universe carries mass. We know the strategy behind gravity is to first get hold of a body containing mass. Then it natural to presume that everything in sight is pulling on everything in sight and the bigger the mass is the stronger the pulling is. Jupiter is pulling much harder than what Mercury is pulling because mercury is much less massive than Jupiter, which is more massive.

Another thing we cannot miss about gravity pulling objects around by the measure of the mass is that gravity pulls mass towards the centre of the other mass conducting the pulling by gravity. We therefore know that all objects are heading towards any object's centre and the pulling of going directly to the centre of the other object pulling by the force produced by the mass the body has. I ask the reader politely to control my facts since so many academics told me with much sympathy in their attitude that I just don't have the insight to understand Newton. Somehow it is suggested that Newton is far too difficult to be understood by a person with my meagre intellectual capacity such as I apparently posses and because I was somehow denied by God to inherit a strong brainpower with a capacity to crack the limits of the Universe in a Newtonian fashion I ask you please to check that I do follow these extremely complicated facts correctly. Those academics I confronted with the issue in hand are of the opinion that it as a God given fact that I am too simple in mind and comprehension about my surroundings to understand Newton and when taken into account that they are capable of understanding Newton due to the capacity they have in using the brilliance of thinking with mind clarity they can hardly be wrong about anything.

$F = G \frac{M_1 M_2}{r^2}$ Every time I confronted an Academic in the physics department I was told in a very polite and sympathetic manner but also in a very unmistakably fatherly and firm way when they share their opinion with a lesser being such as I am in their opinion is what they can see about my intellectual status I am a lesser-blessed individual. There are some people that are intellectually advanced and such persons have a born ability to understand Newton. Then there are those with much less potential and with much reduced thinking ability and in the case of those or should I say we there is not enough grey matter in our skull filling the vacuum between our ears to follow Newton. I therefore have to accept I was born with much less capabilities and then with that much reduced mental capacity I had to accept my fait which is that I shall never have the ability to understand Newton. With my meagre intellect I do try my best but where it comes to the obvious then it becomes hard to follow Newton when it is Newton that apparently lost track of his senses. Be as it may please inform me where I lose track of reality and where Newton loses track with reality. They (our Super Professors in Physics) say I have no ability to grasp how mass entices gravity by mass that is producing gravity as a force forcing gravity to pull over a distance to the measure of r.

They must be of the opinion that I do not understand mass. Well I am not sure that they understand mass because there are as many definitions explaining mass as there are opinions about what mass is. It also could be that they presume I do not understand the gravitational constant through which the mass must move to reduce the distance between the objects. In that case they don't know much about it either because not one of them could up to now show me where the centre of the Universe is and where it is that that gravitational constant is pulling all the mass. To have gravity pulling what there is to be pulled, there also has to be a centre to which all are pulling all there is to pull. It could be that they suspect I don't know what a force is…I do know that…it is that which they burned witches for because they said witches and ghost are or have forces. Then finally they might suspect I don't know what a distance is that separates objects and there too I think I know what a distance is that separates objects. Other than what I just mentioned I truly don't understand what they say I don't understand about Newton. That means they are correct and that my mind is so weak I don't even understand what I don't understand about the complex issues they seem to understand.

This is what I do understand but apparently it is not enough to understand what I think I do understand. There is a force that puts relevance between the amount of mass and the strength of the force produced over the distance the mass produce with such a force.

If the distance increases, the influence of the mass coming across the increased distance will decrease. When the distance decreases the influence coming across the decreased distance will increase as the distance decreases. In that the force of gravity that the mass of the comet and the Sun produces will increase by the square of the distance that is diminishing due to the shrinking of the distance parting the objects. I suppose there must be more because that which I understand and I just shared with you are not enough to prove that I understand Newton!

Mass produces gravity in accordance with the radius distance between the two bodies. This is where my mind gets too weak to understand Newton. Looking at all orbiting objects there are always a wide part and a narrower part in the orbit where the planet in orbit seems to be closer than on the other side (E- e sin E = M) or on the orbiting structure side further away from the centre which was the position that the Sun claims. This does not quite fit the picture of a constant and never changing mass that all structures supposedly have. If mass is responsible for gravity why there is this flexing wobble in the radius. The lot should be rather constant because the mass is constant. But that is my weak mind as that is what I don't understand about what Newton understands…and with my weak mind I can get no one that understands that.

We know that the Earth has no mass increasing and decreasing and neither does the Sun have mass adding and then removing of mass. In that case when considering the practical mathematical implication the radius has to be at a constant with no mass fluctuating on either side of the factors.

We know from personal experience living on the place all our lives the Earth doesn't change mass as the year progress. From that as well as the evidence we know about all other things in the Universe with mass that the mass of all things are a constant and doesn't fluctuate. So the gravity that the mass of the Sun and the mass of the planets would generate should allow for a pretty round r to be in place that will keep the planets circling evenly. That is not the case because we know there is a fluctuating especially in the case of the comet orbit. We then have the task to find what would encourage the deviation.

$$\frac{M_1 M}{r^2} G$$ Or on the other hand $$r^2$$

$$F = G \frac{M_1 M_2}{r^2}$$ It is my understanding when considering the implications about the relevancy there is between mass and distance parting mass when used as it is in the Newton formula that if the mass is at a constant then the radius too must remain at a constant.

When the object having mass asserted the gravity the mass employs and the radius parting the objects are large, the objects in mass asserting gravity over such a large distance must be small. If the distance is large then in that case the distance will reduce the force that the mass can produce to a trickle. The force there is between Mercury and the Sun must therefore be ten times weaker than the force that there is between Pluto and the Sun. It is not only that Pluto and Mercury has about the same mass and therefore they should peddle around the Sun equally but since Pluto 39.44 x AU or (5900139992.8 km) is 101886 times further than what Mercury 0.387 x AU or (57909 km) is if taken that one astronomical unit is the distance from the Sun to the Earth and that is measured in AU = 1 or in km 149 597 870 km. That means the force of gravity must allow Mercury to go 101886 faster than Pluto's 101886 slower. That is not happening. I surely don't understand Newton but that is also surely not because I am too stupid to understand Newton.

The gravity one must find between Pluto and the Sun is then concededly less that the gravity there is between mercury and the Sun. Mercury is so close that one might form an opinion that the gravity Mercury generates must be in the vicinity of one of the outer gas planets. The gravity Mercury has with mercury being so close would have Mercury spin at a rate that only Jupiter can mange since Jupiter has the enormous mass in its favour. No other planet should come near to the speed that Mercury and Jupiter have when going into orbit around the Sun. Jupiter has the enormous mass in its favour giving it momentum around the sun a boost and Mercury is so close that the mass it has is many times bigger than the mass it should have just because it is so close to the Sun. Then Pluto must be at a snail pace and hardly moving in consideration of everything the gravity applying to Pluto has to endure because it is a hundred times further from the Sun than Mercury is while it has the same mass as that which mercury has. Sorry I forgot that all planets move at an equal pace. I wonder if it is because they share the same mass or are they possibly at the same distance while in orbit around the Sun because there is no mass indication of any of them speeding up or being slower than the other. They all orbit at the very same and equal pace.

Now we get to comets and their bad behaviour. When taking this distance effect further as it affects something as small as a comet, the gravity increase on the comet has to be devastating to the comet. The force that can hardly move the comet out where the comet is hiding from the Sun must be billions upon billions of times more when the comet is closely approaching the Sun.

With all that in mind we now draw our attention to the comet and the way gravity pulls the comet.

When the Sun gets hold of the comet the comet is miles away and the miles stretches for miles without end. The comet is where the Sun can't shine. Think of the considerable small mass that the comet has and the enormous distance there is at first when the Sun gets its gravity onto the comet and start to drag the comet closer. The distance being that large must seem to make the mass incredibly small.

Yet with the massive mass the Sun has in its favour it not only gets hold of the comet but it drags the comet all the way through all the space and ever closer to the Sun.

Gravity the **force that mass supposedly provides is a hoax, which keeps the world believing in science and with which science holds the world at ransom for the past three hundred years.**

You might decide to ignore these facts but then you would remain part of the last bastion of the dark Middle Ages, or you may read with concentration and become part of the future.

Gravity is based on **Newton's presumptions and as clearly can be seen not facts**, which are **all incorrect. My work is based on** the findings of <u>**Galileo and Kepler,**</u> and **their findings** are very **inconceivable** with the **findings of Newton. Newtonians forever tell the cosmos what Newton orders the cosmos to do and never takes into consideration what it is the cosmos does.**

When an object swings around another object, the distance between the two bodies divides the force that exist between the two bodies. As this force is mutual, coming from both bodies together, the distance is calculated as if it is two; therefore this distance is multiplied by its own value. The bigger this distance is, the less force will be excerpted between the two bodies and the shorter the distance is, and the stronger the force will be. With mass forming the force, the only fluctuation in the distance must be as a result of the mass on either side growing or diminishing at certain stages of the orbit. That can't be the case.

Even in the event, where this force applies an even distance, the two structures will be evenly apart through out the year. This ultimately will lead to the perfect day, the perfect seasons and the perfect year (for some people). We may start by determining the influence of gravity on planets as we find them in the solar system. **First, let us concern ourselves with a comet**.

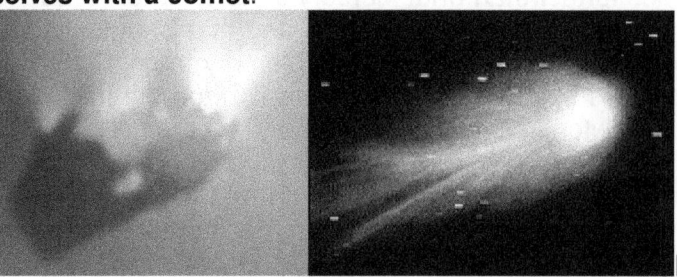

Gravity pulls of the sun on the comet It is common knowledge how the comet stands related to the sun's gravity. **Firstly, picture the comet at its farthest point, away from the sun.** Comet at its most furthest point from the sun

The **gravity** of the **sun pulls** the **comet straight towards the sun**, this we all know. Gravity always pulls an **object directly towards** the **centre of a cosmic body**: that too is common knowledge. Therefore, the comet is drawn directly towards the centre of the sun and throughout its journey the comet is picking up momentum directly related to the gravity that is centred in the middle of the sun, (**gravity is always centred in the middle of a cosmic body**). As the comet is increasing its speed, the comet comes closer to the sun and therefore the sun's gravity pull is simultaneously increasing as the distance between the two cosmic bodies is reducing. Each instance the comet is drawn towards the sun, the gravity that the sun applies to the comet becomes larger progressively.

When the comet is at its <u>**closest point to the sun**</u>, <u>**something odd happens which cannot be explained by Newton's gravity at all! Remember gravity should now be at its strongest point because of the proximity of the two objects.**</u>

1. The comet remains at an even distance encircling the sun.

2. No longer does the gravity of the sun pull the comet towards the centre of the sun.

3. At this very point the gravity that the sun applies on the comet does not pull the comet towards the centre of the sun any longer, in fact, it seems as if the effect of the gravity has been neutralized.

4. The comet stays at an even space from the sun as it goes around to complete a half circle's orbit around the sun. It only completes a part of its rotation around the sun.

5. After this, an even more peculiar event takes place. <u>**The sun, at the point where gravity should be at its most dominant, suddenly loses its complete grip on the comet.**</u>

6. The comet brakes free from the sun's pull of gravity and speeds off towards its destiny into the vastness of the cosmic space, undeterred by the gravity of the sun.

As I said we now come to the difficult part where very few people understand Newton and apparently according to those in the know how I am one of those being incapable of making sense of Newton. Let's try again and see if I can manage to get it correct this time. The Sun has mass. The mass covers the distance between the comet and the Sun by the mass that produces gravity and the gravity is pulling the comet. Then the comet also has mass and that mass produces gravity that pulls the Sun over the distance between them. Because the distance is coming from both ways the distance is by the square where there it is including the gravitational constant and this lot is in a product to one another that gang up to destroy the distance parting this lot.

There is normally a diverting from the centre or the rotation axis, which the Sun forms by the orbiting planet but this, is much more so applying in the case of the comet (E- e sin E = M). Considering the weakness of the force that initially reaches the comet as it drags the comet from the depth of total darkness where the comet hides in places the Sun does not shine it is remarkable that the comet does apply the gravity that will pull it on route to the Sun. It just comes to prove how correct a genius such as Newton is! He saw the mass that creates the gravity and the little comet stood no chance hiding from this domineering Sun with all the mass it can display

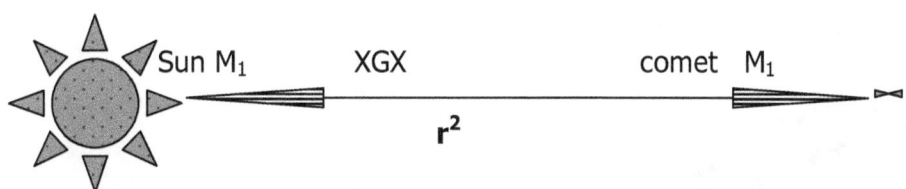

From the coldest and the deepest of space where not even the Newtonian imagination can reach, the Sun finds the gravity strength to draw the comet through the deep of space. One would imagine that if it was or is not for the shear size of the massive mass of the Sun, the Sun would not be able to generate a force so strong it could get hold of such a tiny piece but it does accomplish the gravity to bring the comet from where it is lurking in the depth of space and drag it towards the centre of the Sun. But with the resilience that only the Sun can muster as it is the only object that has the mass in ability to form the gravity that can extend all the way even to where the comet is lurking, such dragging would have been hopelessly inadequate. But the gravity that the mighty Sun manages with that enormous mass, the comet has little chance in defending its position. The comet is heading towards the centre of the Sun and towards its final doom since with such gravity coming from such a mass the comet has no chance to defend its position. If any one ever invented an unfair and totally bias fight then it is this fight between the small comet and the enormous Sun.

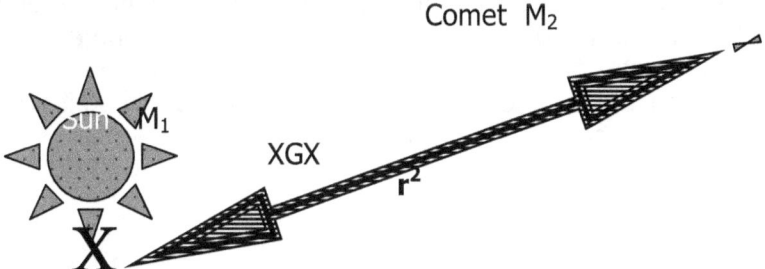

Alas it seems that the bookies once again had a say in the outcome of the fight because it seems they rigged the outcome or had some intervention about the outcome…well it is either that or the unthinkable has happened, Newton is mistaken…Newton is incorrect. The comet strayed from its pulling and the Sun lost its grip on the pulling. The mass of the Sun proved insufficient to grip a small little comet by force of gravity and pull it to the centre of the Sun.
Then after a pre-determinate and pre-calculated time the sun starts applying its gravity on the comet once more. At a point where the comet is at its farthest point, the gravity of the sun becomes strong enough to bring about a complete turn around to the comet's direction of travel. **However, the gravity between the sun and the comet is at this point, at its weakest point of influence.**

So, when the sun's gravity is at its strongest, the comet manages to brake loose and neutralize the sun's gravity pull in order to avoid its fatal collision with the sun and when the sun's gravity is at its weakest, the comet cannot escape the pull of gravity. There is definitely something very wrong, either with the comet's behaviour or the laws made up by Newton.

$$\frac{M_1 M G}{r^2}$$

The relevancy of the big radius by the square reduces the influence that the mass supposedly brings about. The further the objects are form each other the smaller will the gravity be that the mass can inflict A long distance has a small mass that produces weak gravity.

The comet aims at spot marked X and misses the Sun centre where all the gravity is concentrated. The mass of the Sun had the comet when the mass was small due to the enormous distance there was between the two but as the mass grew in potency with the aid of the shrinking distance and with it the force of gravity that the mass produced becoming ever increasing in the presence of the declining of the radius parting the two objects in mass, the force was unable to direct the path of the oncoming comet to the centre of gravity within the Sun. There is an unexplained diverting of direction, which Newton's brilliance never foresaw… that is if Newton's brilliance did foresee anything at all…and it goes back to where it came from and that oversight blows Newton into a barrel of shit.

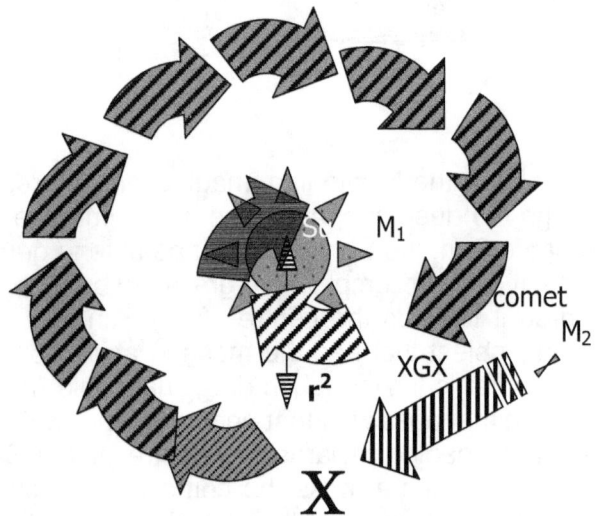

It started when the radius was big which made the mass be small that produced an insignificant force in gravity. That pulled the comet out of the shadows and into the Sunlight. The force was strong enough to grip the comet and pull the comet towards the centre of the Sun where the strongest gravity mounts an attack. When the force of gravity grew the influence the force made started to tarnish because it allowed the comet to aim for spot X, which is way outside the centre of the Sun.

$$\frac{M_1 M G}{}$$ Somehow with an ever-increasing gravity, the comet got brave and fighting fit. In the presence of overwhelming mathematical evidence to prove the contrary and in spite of mathematical laws and principles applying nevertheless the comet prevailed and got the better of the Sun. It missed the centre of an ever-increasing gravity force notwithstanding. This is not the end…it is the beginning of Newton's problems. Up to this point it seems as if Newton's was a profit sent by God to foretell us other far less important mortals about the future and about the future of the comet. The Sun had its mass providing a force called gravity and the force called gravity was going to take the comet to its eternal grave, as gravity will bring the comet its final demise. Alas the comet has other ideas because…

At this point Newton makes complete fools of his followers, the mindless sheep that follow his word through centuries without ever thinking about anything of his word that they come to follow. This is where the brainwashing kicks in. This is why only the mathematical minded excel in physics. It is those that cannot think that become the Brainy Bunch in physics. Those that are the Super-Educated-Masters-In-Physics are Master in the art of not thinking but through an excellent brainwashing that is the best tried and tested system for centuries those that underwent mind control before anyone knew about mind control became the world leaders in the art of physics that perform in matters not requiring the ability to think about concepts. That is where the mathematical mind is very powerful, in the process not to think but to be programmed to calculate.

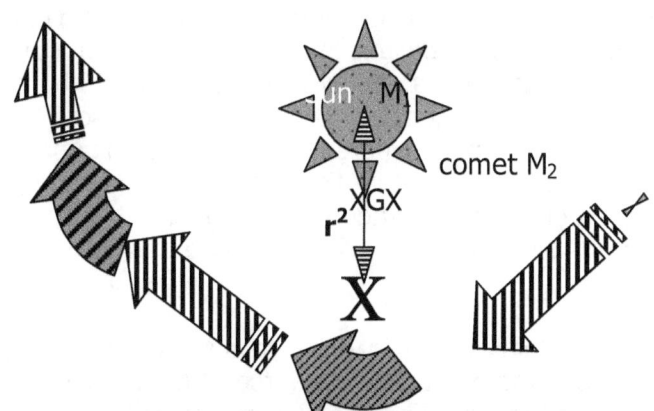

F= $\frac{M_1 M}{r^2} G$. This is why Newton does not require or deserve proving, his followers can't prove or disprove, they can calculate... F= $\frac{M_1 M}{r^2} G$ Explains the first sketch but then?

The mass of the Sun brought the comet in range and to the point where the strength of the force of gravity of the Sun is most and up to where the force the mass provides is at it's greatest and seems to be unconquerable. With the radius tarnished to being almost representing a factor of one the force of the gravity becomes eternal since the small radius will charge the mass into greatness which will unleash a force that will drag the Universe around if it could get hold of the Universe at such a close range. That does not happen because at the point where the force that the mass must produce is at its point of ultimate victory we find the comet avoiding defeat. The comet not only avoids the centre of the Sun and final destruction but it wins the fight completely. In the jaws of absolute defeat when the comet should have been clinging to the ropes and living to survive the last desperate moments, the fight swings unexpectedly in favour of the $F = G \frac{M_1 M_2}{r^2}$ loser.

The comet rushes past the Sun as if the comet is unaware of Newton's predictions. The comet behaves in a manner that leaves Newton in doubt. That cannot be tolerated. Newton cannot be wrong because Newton is always correct. The comet cannot speed off into the distance where the dark is much and the light is little. The comet cannot move around the Sun and then slip past the Sun because we know Newton said the mass of the Sun is pulling the comet as it is pulling everything there is.

The thought alone is more than what the mind can bear and the principle alone is outrageous. The man that has never been proven wrong proved himself wrong by predicting the demise of a small comet crashing into a big Sun and where the mass of the Sun produces a force of gravity that pulls the comet towards the Sun and then into the Sun, that is not happening. That which pulled the comet towards the Sun is now pushing the comet away from the Sun. If mass was pulling then what is now pushing because the silly little comet is in defiance of the great Newton and is rushing into the blackness again where the Sun hardly shines. If it is mass pulling then what is pushing?

Newton's vision gave us all the answers to the point where the fishing of the Sun starts but then as things really get serious Newton slips away like the comet and leaves us all without an explanation. Newton fails to provide answers and that leaves the academics looking silly. It is at this point that I don't understand Newton because I am too stupid and uninformed to understand Newton. Newton now gets beyond what I can comprehend and that is true.

One may even suspect that if the mass was insufficient to provide a force of gravity strong enough not to pull the comet into the Sun, the comet would at least begin to circle in a reducing fashion around the Sun until the comet falls into the Sun. One may argue that the comet may come at a speed where the Sun is too little to stop the velocity of the comet but that argument doesn't actually make sense. The speed will ultimately not change the direction from the centre to a point where it totally misses all apparent targets and coming at speed will benefit the Sun as well as the comet in aiding the effort of giving more of the gravity that the mass provides. I know such an argument is outrageous considering the mass of the Sun and the speed of the comet but hey, at this point Newton needs all the help he can get…and he needs help.

But without getting silly, all arguments we might think about in order to save Newton's reputation is bordering on madness. The comet is small. The comet is providing a force at best it can to enable the Sun to accomplish what the Sun set out to do from the start. The comet is pulling as hard as it can and in that the radius reduces by the square. It is hardly as if the comet is trying to preventing the seemingly inevitable destruction it is heading towards. The comet by mass is enforcing gravity as much as it can to self-destruct. This is in aid of the Sun's efforts to destroy the comet. While this is all going on it is at the same time not happening. That which is pulling is now pushing because the comet is escaping into the yonder.

If the comet's mass does assert gravity onto the Sun by measure of mass while the Sun does the same right back to the comet using the same grounds and mass principles in doing so, it must be conceivable and detectable that all the planets rotate around the Sun at different speeds seeing they all have different mass bringing about different forces of gravity. The rate of orbit must vary considerably as Jupiter must spin around the Sun billions of times faster that the little comet we just now observed does rotate around the Sun.

In a following article we are about to investigate such a scenario and then when it comes to light how much faster Jupiter spins in orbit than does Mercury on the very inside orbit and as Pluto on the very outside orbit we than can start to vindicate Newton and find the resolve about the comet behaving very awkwardly and out of step with the rest. With the massive mass Jupiter has in enforcing gravity there can be no more arguments about how the mass brings about gravity to have Jupiter move at 318 times faster than the Earth around the Sun because Jupiter is 318 times more massive than is The Earth. Then surely all scepticism must be nipped. The closer any two cosmic objects come the stronger the force should be, with eventually no force in the Universe being able to keep them apart. This is just not happening!!!

There is no indication of truth about a contracting solar system as Newton proposed and as Newton's followers promote a contracting Universe while when seen from the Hubble perspective where he introduced the Hubble Constant…and not from any other evidence seen through the Hubble Telescope Newton just is not matching reality notwithstanding fraud such as the critical density issue.

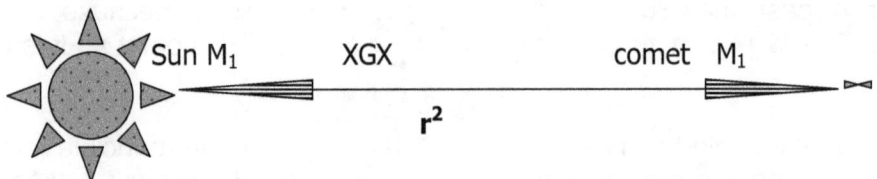

As explained, there is some discrepancies about calculating the force of gravity, because gravity would apply as nicely as it does if it was the perfect balance, a balance exists in space of equal measure bringing about equal seasonal time.

Comet M_2

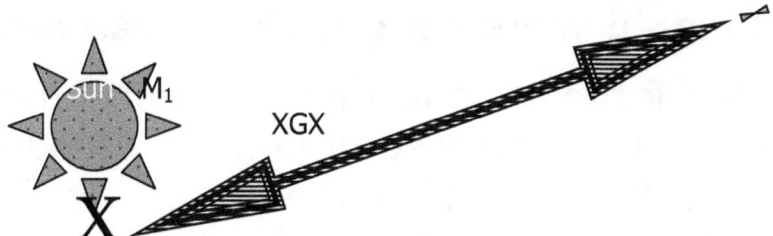

The biggest discrepancy and a practical denouncing of the official version of the comet's flight around the Sun,

The Sun gets a grip on the comet by mass inflicting gravity and as it gets hold of the comet it drags the comet through the solar system straight ahead to the Sun just as Newton predicted the Sun with all its gravity producing gravity will do.

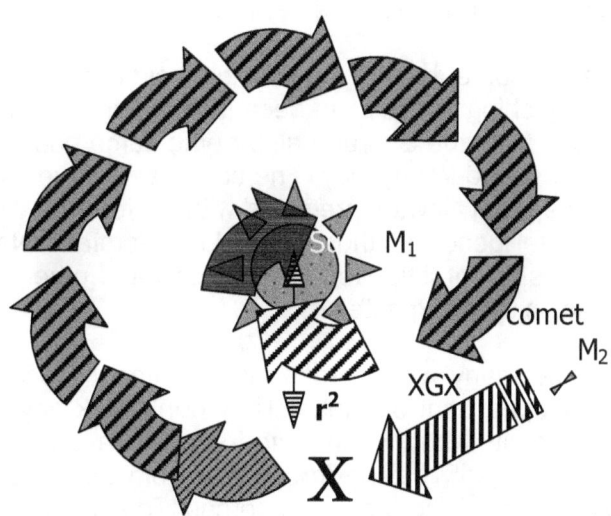

As Newton had said the gravity that the Sun and the comet mass induce will pull the comet to the Sun. this at first appeared to be true. As we all know the comet moves to the centre of the Sun just as Newton predicted but with a slight complication and a change in the venue, the comet no longer aims directly at the centre of the Sun but aims at a target outside the limits of the space that the Sun occupies. It follows a line aiming towards a point to the side of where the Sun is located. Again this is the part my intellect is apparently insufficient to follow Newton because Newton is his formula never mentioned this apparent deviation from his prophecies. It seems the Academics are correct in saying that I don't understand Newton but they never try to explain this part that I don't understand about Newton.

The comet aims at some point gravity does not hold and misses the point that gravity secures the strongest force, which is right in the centre of the Sun.

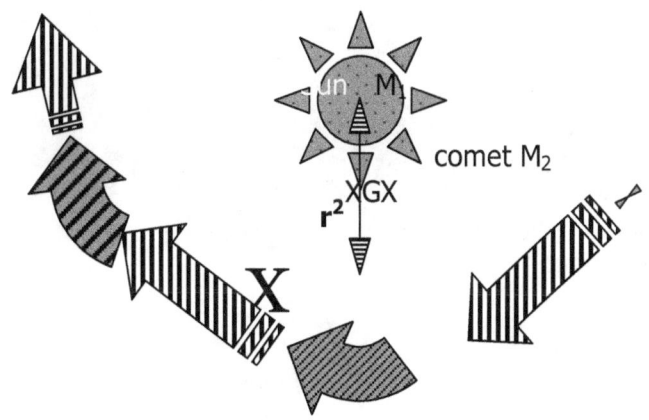

$F = G \dfrac{M_1 M_2}{r^2}$ **In over three hundred years not one Newtonian came forward and tested Newton's formula...and then the world must believe there is no conspiracy in science to brainwash students by enforcing mind control on the subjects called students they teach, or is it that they train to accept and believe!**

Should one give Newton the benefit of the doubt and disregard this missing of the centre of the Sun the position that the comet should aim at then one would reason that it could be that the gravity was not generated strong enough to locate the centre at such a great distance as the comet was at first. It could be that the reducing of the radius comes in instalments of a few circles. His formula does not indicate these presumptions that I make but hey, let's try and give the guy a chance. The comet might just take a cycle or two to wind down as the radius reduces and one should wait and see if it is not that which Newton meant when he said the mass by the mass is dismissing the radius between the objects.

Alas, it is not what the comets intends to do because the comet breaks the strangle hold of the Sun and what ever was pulling the comet at first is doing all the pushing at this point because the comet is surging into the darkness of the abyss. The comet speeds away from the Sun and also at the same pace it was heading towards the Sun and there is no altering to the speed in any way. The comet seems very much unaware of the comet behaving in opposition to what the great Newton predicted with his formula. Again I admit I don't understand Newton but I suspect the intelligence of those that do understand Newton because they might for the same measure believe the story of little Red Riding hood and other forces fairy tales offer. When we enter the world of forces what is to be expected?

Newtonians improved on Newton's force by creating so many more forces legally without the Church interfering in the process of creating forces that the witches went on strike. They don't work any more! The poor magicians and witches no longer has any authority when it comes to forces because from the one side the Church roasted then alive for having or allegedly having forces and from the other side physicists stole them blind by taking away all the legal forces flying around as forces. To some life is not fair!

Teaching students that it is mass pulling the planets around the Sun while the evidence that are supporting this matter is very skimpy and dodgy such teachings is a way of committing an act of folly by which then perpetrators are committing fraud. With so much evidence lacks and yet academics still insisting that students accept the fact that Newtonian presumption on the matter is correct is brainwashing. It is outrageous to force ideas onto students by disciplining thoughts and controlling the minds of the students into believing that the mass is in charge of the gravity, which is pulling matter onto matter while all obvious evidence so far lacked any proof. A presumption remains a presumption until it is proven and Newton made a presumption in the case of his apple falling which until now was never substantiated by fact. In such an event the presumption remains a presumption until facts prove the presumption accurate. Then only does the presumption become fact.

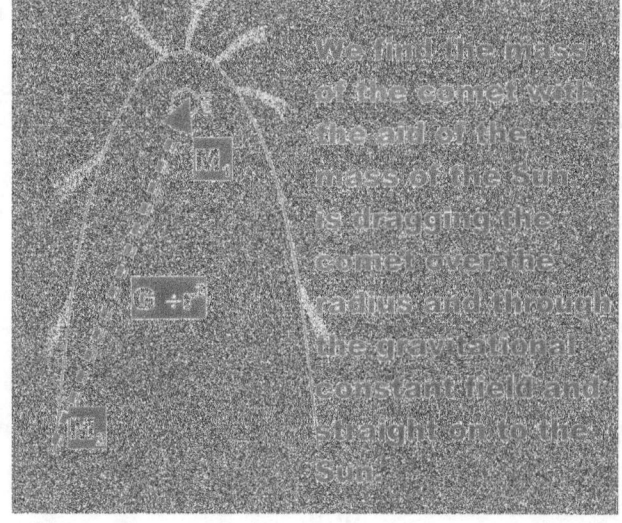

At the start the first sketch explains $F = G \dfrac{M_1 M_2}{r^2}$. The rest is an unknown and putting facts aside and fiction in place. When we bring in intellect and we put stupidity aside then not even Newton can fill in the blank spots. It is all make believe and it becomes so clear that the academics in physics are keeping the minds of students busy with the biggest cover up the world has ever witnessed. Your local academic is not a wise old man having all the answers he wishes to share with you…instead he is a shrewd criminal that wants to deceit you with lies in order to carry on with the biggest cover up ever produced.

The comet performs all the other manoeuvres as the sketches indicate except returning to the Sun, and it is all the manoeuvres that the comet does except running into the Sun, which Newton's formula totally ignores. Yet on this very principle is every aspect of modern physics based. If the formula forms the basis of all physics used by science, then it is the basics, which are around for hundreds of years that suddenly are trash and simply does not perform as it is supposed to. If there is out there one professor in physics that will tell me the profession didn't know about the comet not falling into the Sun then I can show you one liar as I cannot show you in playing poker games. For many hundreds of years every person in physics were aware of this flaw but did nothing about it. I call that criminal and I call that deception because they took money from others to betray the innocent, the young and the trusting vulnerable by spreading untruths to the young and the trusting vulnerable and forcing young minds to believe what they well know is not remotely true.

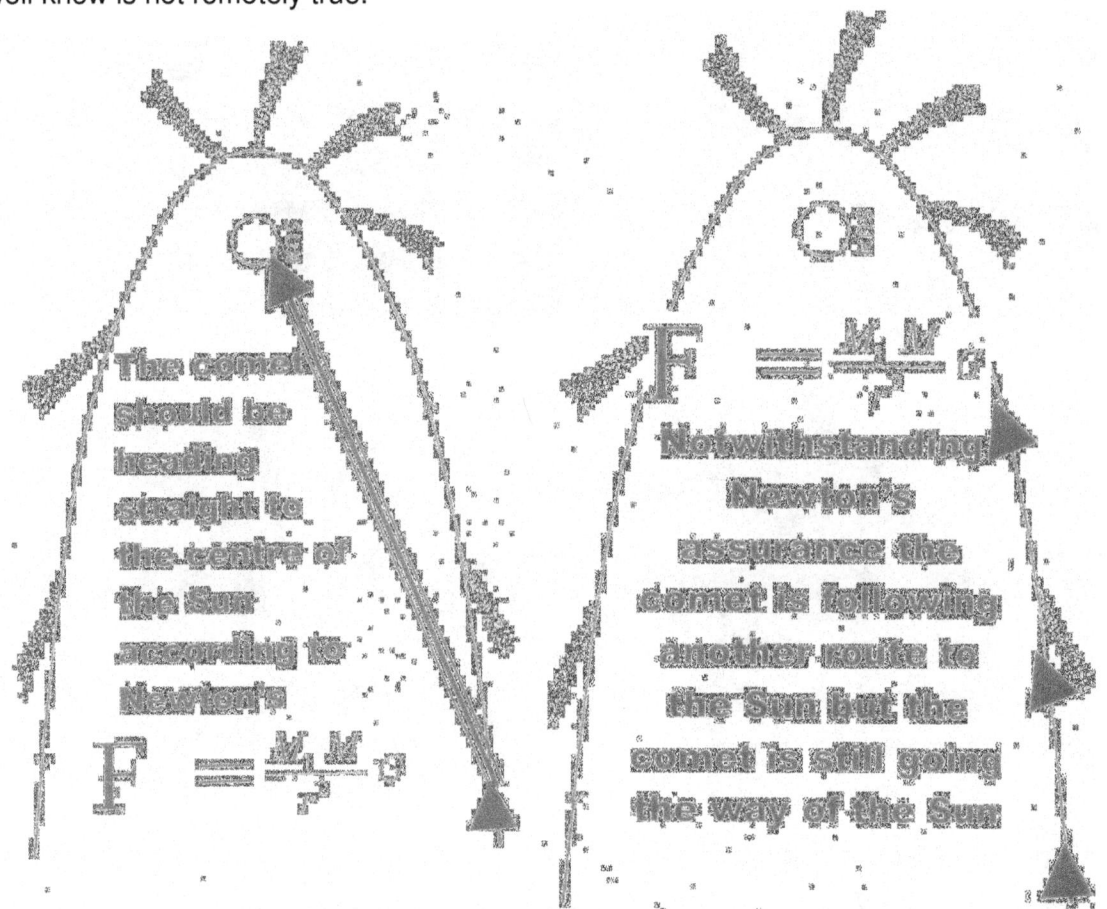

If it is true that it is the mass that is pulling the comet to demolish the radius, then ask you're most wise amongst all those that are wise Professors what is pushing in comet into the dark beyond. What is driving the comet away if it is mass that is pulling the comet towards the Sun. Where does Newton's formula $F = G \frac{M_1 M_2}{r^2}$ allow for that the phenomenon of attraction to turn around and become the complete opposite of what it was before? If it is mass that pulls then have them clearly state how and what brings about that the comet escape this destruction so easily. The normal deception that they use is that it is momentum, but don't let them fool you with more garbage that thy swindle because if it was momentum, then why is it not Jupiter coming that fast and the comet being so small that it can hardly be pulled with the mass it has. In this you would be able to see who the fraudsters are and who is telling you about the scandalous brainwashing they put in place to keep your thoughts under their control and that would keep you as a student in your place.

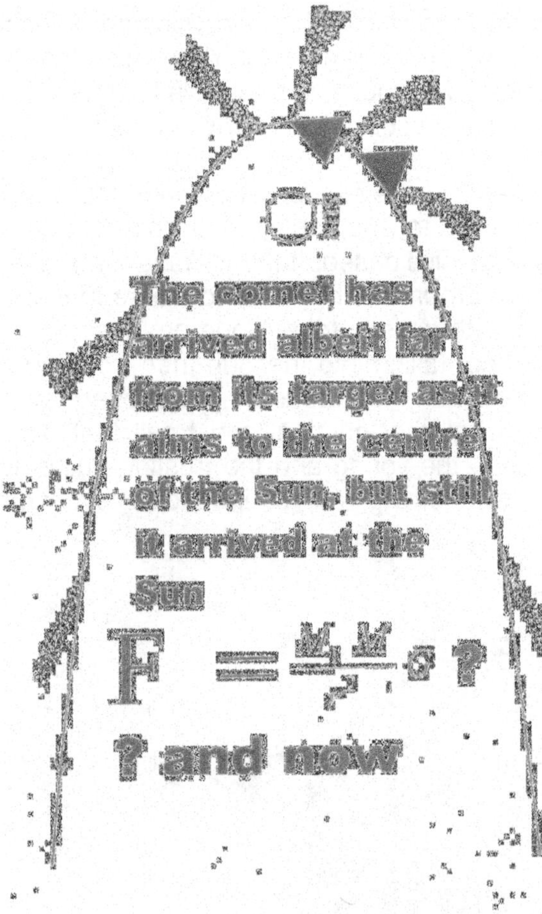

01
The comet has arrived albeit far from its target as it aims to the centre of the Sun, but still it arrived at the Sun

$$F = \frac{M \cdot M}{r^2} \cdot \sigma \, ?$$

? and now

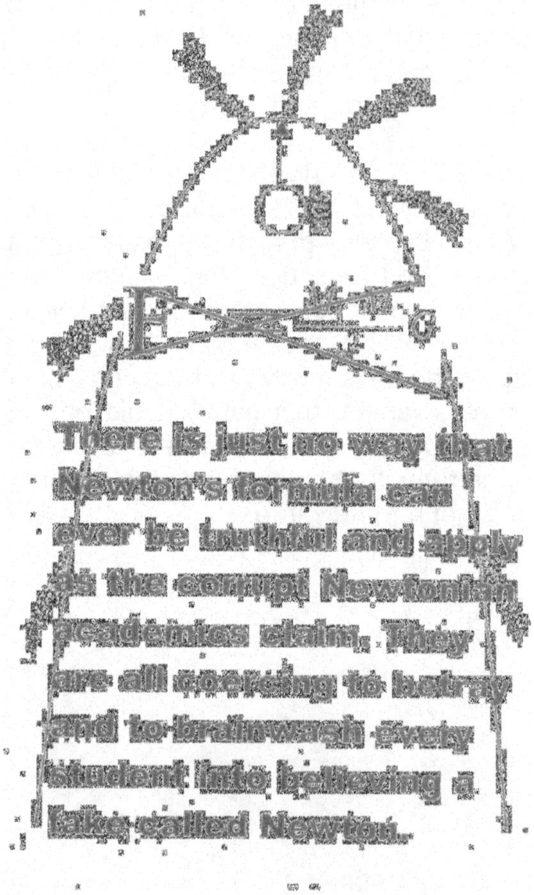

02

$$F = \frac{M \cdot M}{r^2} \cdot \sigma$$

There is just no way that Newton's formula can ever be truthful and apply as the corrupt Newtonian academics claim. They are all coercing to betray and to brainwash every student into believing a fake called Newton.

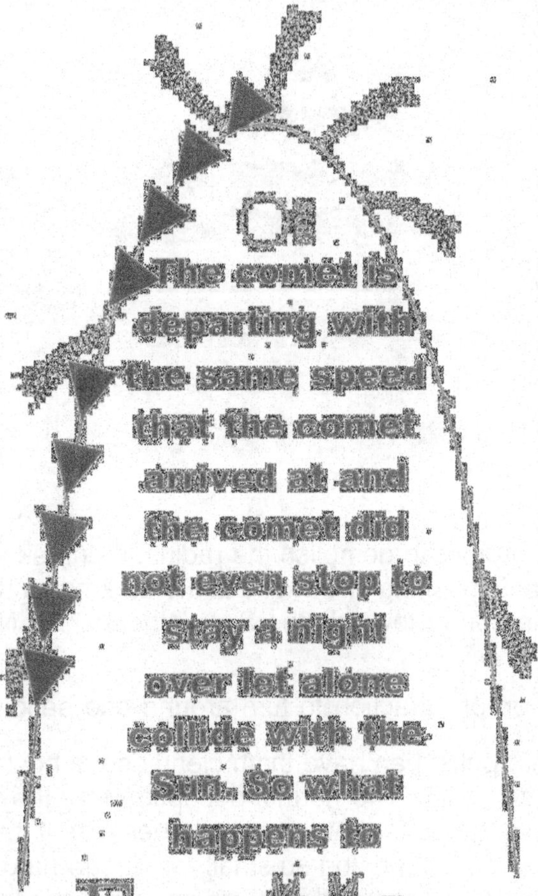

03
The comet is departing with the same speed that the comet arrived at and the comet did not even stop to stay a night over let alone collide with the Sun. So what happens to

$$F = \frac{M \cdot M}{r^2} \cdot \sigma$$

04
The Newtonians are either the biggest cheats or swindlers you will ever come across or they are the stupidest morons you will meet. If they dispute that they try to trick by brainwashing students into believing the ridiculous then ask them...

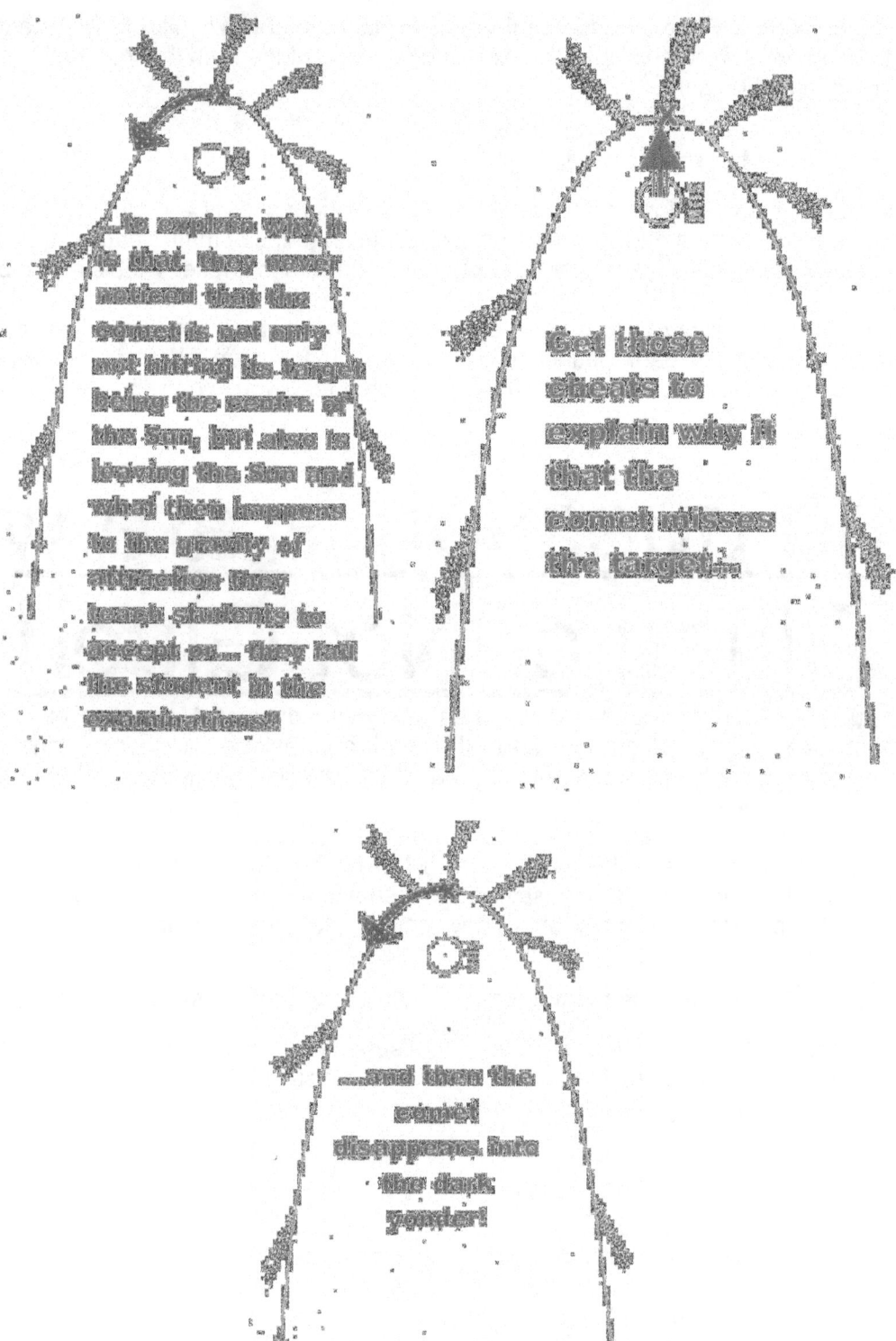

That is an act of a criminal mind notwithstanding what motivated the person to commit the crime. Defining a criminal is to have a person that is prepared to place himself or herself in the centre of the Universe as to deprive all others of truth and possession in order to further the needs and wealth of the criminal while the criminal acts by never thinking about the rights of the victim or the harm done by the criminal's action in the matter. Spreading untruths and forcing payment for such actions is most certainly criminal! You students in physics go on and confront you Professors about the truth. Insist on the truth for once and all. Mass has nothing to do with gravity but to prevent gravity from being in place. If it is mass pulling the comet as Newtonians declare then they better inform you being the paying students that keep their wallets filled what is pushing the comet away. The comet is leaving as fast as the comet was coming. Confront them because you have the right to insist not to be lied to about Newton and his shambles and criminal fraud. Go on and ask them to explain these Newtonian inconsistencies and see how they try to carry on with the criminal deception and covering of the truth that has been going on since Newton first thought up this scam.

This is the basis of physics. $F = G \dfrac{M_1 M_2}{r^2}$

It is not a concept or a suggestion or an idea but it is a formula used as a mathematical expression in terms of a mathematical formula and therefore it is expressed in terms of mathematical accuracy. It has to be put to the test in relation to its mathematical accuracy or Newton is a fake. That is as simple as it is!

This is but a small part of a big picture uncovering the scam Newton came up with and which all the academics are knowingly still participating in... and read on for I am about to inform you of more criminality the academics came up with and which they force feed you. You will see fraud as you never saw fraud before.

The question is are we Glued or Not Glued

If you are a student in the science of physics, then ask your Educated Masters to please explain the following abnormalities you are about to read in this book and insist on a clear explanation about the inconsistencies they promote while tutoring physics as if the physics they present are the most flawless and accurate institution there has ever been. Ask those academics supporting Newton about the following flaws that no one mentions ...ever... except me in this book you are about to read and get them to explain the inconsistencies never talked about, which I present in this book and then after confronting those charged with tutoring physics and seeing who should be believed, then get wise instead of brainwashed. Let them mathematically show how one would go about and use Newton's visionary formula $F = G \dfrac{M_1 M_2}{r^2}$ to calculate the force of gravity by replacing the symbols with the actual values in mass that the items referred to have. Put in the Earth's mass in place where it belongs and put in your mass in place where it should be and then divide that with the distance between your soles and the Earth measured in micro millimetres by the square thereof!

Do you know what gravity is?

This is the question I asked myself about thirty years ago. In the late 1970's when, one day, I came into a room with a TV on and this may sound unusual in today's society, but back then in South Africa things were different. Today my saying that I entered a room with a TV on is unusual but at the time the unusual part of saying unusual was for me being able to enter a room with a TV on in South Africa, was unusual. In South Africa we just got TV and the concept was rather new to everyone. On TV I saw this man going on about an issue I knew little to nothing about. On TV there was some person going by the name of Carl Sagan or something having a program called Cosmos or something that was explaining things my mind found boggling. What he said was more than what I could readily take in. This concept in itself was as new as everything I was hearing. I took a few minutes to get the information sorted in my mind but after those few minutes of sorting, I was never yet that excited except in a racing car or on a fast motorbike. I sat down and stared at the box and that alone was extremely unusual for me to do. I started listening and the more I listened, the more I knew that I knew nothing of what this bloke was going on about but what he was going on about, was the most fascinating information I heard in all my life. I knew I had to know more because I have never yet encountered such astonishing concepts in my entire life. My fortune was good for that the program was broadcasted for the first time that night making the broadcast being the first episode. That night the broadcast was program one, which meant the series just started and the program itself just started very shortly before I entered the room and that meant I missed very little. My life could have been totally different if I did not enter the room at that particular point and began to listen to the TV as the program started. I was surprised at my interest because I am not a TV person and the coincident for me to be in the right place at the right time was extraordinary. I started to take in and grasp things and for the first time in my life I was hearing things that was at the edge of my mental development. What I took in, I realised, was the limit that my mental faculties were able to cope with and that experience by itself was a very first for me. It was something I never encountered in any way or form before with anything or in any discussion I ever had. I normally know what is discussed and often can add one or two aspects to whatever is discussed. The very next day I then went into town going from bookstore to bookstore in search of the book going by the title Cosmos. My searching paid off as I found the book Cosmos and I started reading. It was a new world that opened and the information that was presented was more than I could manage at first. What I encountered was a new Universe filled with information and the idea of so much new things opening up was most thrilling.

These made me ponder on every aspect there was to ponder upon and I started asking questions more to myself than any other person. Every question I asked myself I found a person going by the name of Albert Einstein had an answer for. I have heard of Einstein and knew about such a person but never encountered his work on a personal basis before. I had thousands of questions, all looking for answers. Today I can't even recollect one of the questions I had at the time, but at the time they were plenty and they were quite intense in relation to the mental development I had back then. Compared to what I now have figured out, the questions were mundane, but then even something as simple as space-time and a Black Hole were on the edge of my understanding. I thought about time and went in search of an explanation and found Einstein gave the explanation that brought an answer to my nagging question about time…but not entirely… I pondered on space and went in search of the specific issue I was puzzled about and guess what… a person by the name of Albert Einstein did explain the specific issue I was confused about…but not entirely… Everything I thought about, this person Einstein answered…but not entirely… I vividly remember thinking one day while standing in front of a class of pupils just after completing a lesson on the subject I was presenting at school at the time, (what ever the lesson was is of no importance) that with all my investigations it seems as if there is nothing going on that this Einstein fellow did not have an answer for. He came to conclude whatever there was in need of some conclusion. He speculated on almost every conceivable aspect there was to speculate on in physics an ponder on most thoughts there was…except one… he never said what gravity is…and that part was the part I realised was the part that I could reply "Einstein did not entirely understood physics…" and that part makes Einstein just another Newtonian. He did not know what gravity is…and that made me ask myself if I knew what gravity is. I was looking for the meaning of gravity but not in some intellectual reference about a force not known where or why and without being specific about anything, but what are the specifics forming gravity.

Do you know what gravity is?

I ask this question because Einstein couldn't answer it. I ask this question because even the so presumed inventor of gravity, which everyone thinks is Newton, couldn't answer it. I ask this question because I have serious doubts about the fact that there is gravity anywhere. I have doubts as to the realisation in physics about the authenticity there is formulating the concept applying as a force called gravity. Hearing this, everyone in hearing range steps back in shock and disgust because everybody suddenly suspects my mental status. Everyone thinks my state of mind could be infectious and contagious. But please hear me out because there is some sense in the seemingly apparent madness I present. In every book I read and in everything I studied I never could find any trace of gravity. How does mass bring about gravity? What is mass that it finds the ability to employ something as vague as gravity to pull and pull? Those well to do Masters of physics already gave the undiscovered particle that does not truly exist a name. It is going by the name of graviton but that too is matching the same principle as the Phantom (and his white dog or is it a white horse?), Tarzan, Superman and a host of aliens not yet introduced to man, but in retrospect when all are considered the idea is foolish, which by any standard is not entirely scientific.

I have serious doubts about gravity being present. To my thinking I concluded that there is no such a thing as gravity. If gravity was present we should find gravity and we should be able to define it much better than it being a Neanderthal concept of forces presenting little understood dynamics. That gave me a goal and a direction in which to search. Ten years later I was still convinced there was no gravity because no one had found gravity. Then twenty or so years after my introduction to Cosmos and Carl Sagan I was reading one night about Einstein being of the opinion that he thought if he fell out of the patent office in Austria which was a multi story building, that he would then experience the feeling as if he (Einstein) would be without weight while descending to the ground. He realised that if he, the pen he had, the table he was next to and the chair he was sitting on would fall through the window, then the lot would fall together. While descending, all were then in a free fall and it would seem as if the lot that was falling would have the same mass because they all were falling in the same manner. For the first time in my life I had an issue about something that Einstein said. If he felt that he was without weight it could only be because he was truly physically without weight and not because his imagination was running wild. He couldn't imagine things that weren't real. If he and his desk and the pen were falling at the same rate, then they were having the same mass and the same weight, and his imagination played no part in his physical experience. His desk is somewhat more massive than he is. His pen is somewhat less massive that he is. They all travel at the same rate and will hit the ground at the same instant. They're travelling together and having no mass, was not a feeling Einstein was dreaming up, but was a reality. If they travelled together while descending, they then had the same mass while falling or they had no mass at all. There was no middle ground of having mass and imagining being without mass. That means I am questioning gravity. If they are falling equal then there is no gravity pulling them differently. That means the gravity Newton saw is in Newton's imagination and not in Einstein's imagination. However, that is not the way I am perceived, because since the first time I uttered the notion I was considered by friend and foe as fulfilling the role of the village fool. Still I am of the opinion that there is no gravity...mad as that may sound.

If I say there is no gravity I am not trying to convince you that you're being on the ground is entirely just your imagination holding you on the ground. Your being on the ground is not due to deeply rooted physiological issues following you in the form of a concealed depression since your childhood and is reminisce of a dark period from your early days that you never got to terms with. It is also not due to your willpower or maybe your lack thereof that is preventing you from flying to Mars. Hey, I am not that mad... I wish to give you a test to judge your intellect seeing that you (and almost everyone else on Earth) think of me as filling the role as the village fool since I am in doubt about gravity and the being of gravity presented as a reality.

Do you know what gravity is?

When hearing the question I put to you, you immediately jump to the conclusion that I refer to that which holds you steadfast on Earth. You don't think another thought about what I might be referring to because you already know what I am referring to. I am referring to gravity and you know what gravity is. It is gravity that holds you on the surface of the Earth, where you've been stuck ever since you can remember and the sticking goes on relentlessly. It is what you have been fighting since birth and if not for gravity, you would be a Superman. The only thing Superman has that you seemingly, obviously and definitely don't have is his Superman ability not to be restrained by gravity and confined to the ground. He can fly as he pleases while you are tied down by gravity. If it was not for the effect gravity had on you, then you could have been the local superhero along with six billion other Superheroes that had no restraining from gravity. By your suffering from gravity during all your life, that puts you in terms of being an expert on the subject of gravity and no one knows gravity better than you do. Then I have the audacity to ask an expert such as yourself if you are the expert on what you know better than anything else you might know about. The only aspect of your life you are unable to change is your attachment to the restraining of gravity.

To your knowledge there is only one form of gravity and you know better than most that you are standing in aide of that gravity, where it is that gravity that is keeping you on Earth, so by experiencing the restraining it brings with the concept for your entire life and during your entire life you very well know what gravity is. Your expertise on gravity had you fighting gravity more than anything you have ever fought or had any other fight with, including you Mother-in- law. From this on-going relentless battle every second of your entire life you have gained the experience only coming by being in a continuous fight and it is this fighting that makes you the ultimate expert on gravity! You know gravity is what prevents you from having the ability to jump over the Moon or run the hundred meters in three seconds flat. If it wasn't for gravity, you could out accelerate a fighter jet in mid air. You know it is gravity that is going to get you old and it is the very same gravity that that is telling you that you now desperately have to lose weight or over strain your heart and die young ... then I come and accuse you of not knowing what gravity is by asking you what gravity is! You are so sure about gravity that you are no more familiar with any other subject and little else has your expertise as gravity does.

Are you sure that gravity is what you might think gravity is and that you and Newton have the same concept about what gravity is?

Have you ever thought that which you think of, as being gravity, is not that which science presents as gravity? The idea you have about gravity is not the gravity science says is gravity. You might think that your being stuck on Earth is that which you think of as gravity...that is the concept you formed ever since your first attempt to sit up straight and that was just after you formed an opinion about milk. Since then you needed no one to inform you about gravity because since then you have a very clear idea of what you envisage or what you think personifies gravity. I say that if what you think of gravity as that which is keeping you glued to the Earth then it is not the gravity science defined as gravity. That which Science defines as gravity and that which you are thinking of as gravity is very much not equal but also it serves the Masters in physics well to leave you with having that idea about your gravity and their definition about gravity being the same because now they don't have to inform you what gravity really is. They leave you with what you think of as gravity. You now are so well prepared as to become a candidate to be brainwashed in believing your way of believing is their way of believing what gravity is and in believing you share their view about gravity, you are in the best prepared state to become another one of their millions they have mind control over. What Science says gravity is, is far from your view because the Newton's approach to gravity is the same as a magnet hooking onto a metal. The grip coming from that is gravity, according to science. They say there is a force between you and the Earth and this force is pulling you as much as it is pulling the Earth, but it is pulling the Earth much harder than it is pulling you because the Earth has much more mass than you have. That means there is a force within you and there is a force within the Earth and these forces coming from within pull together that, which should be apart. That form of gravity will have you locked onto to the Earth with no release but when you do have release then the release will allow you to escape. The closer you are to the Earth the more significant and powerful the force then must be. The worst part of breaking the force is the first millimetre and from then on the rest is child's play. That is what they suggest when they say mass is pulling mass to reduce the radius in $F \alpha \frac{M_1 M_2}{r^2}$, which is $F = \frac{r^2}{M_1 M_2}$, which later was changed to $F = \frac{r^2}{M_1 M_2}$. What this means is if your mass is hundred kilograms and the Earth has a mass of 5.974×10^{24} then when you are standing on the ground with an infinitely small radius between you and the Earth, then the force keeping you attached to the earth is eternally big. With the radius of say one billionth of one millimetre you have no chance to be released from the Earth while when you are say one kilometre way from the Earth the force is one thousand million billion times weaker. That is shit at its best because notwithstanding height, the gravity remains the same. $F = G \frac{M_1 M_2}{r^2}$, $= M_1 = 5.974 \times 10^{24}$ and $M_2 = 100$ kg divided by the incredibly small radius of 1×10^{-15} m then the force gets to be $1 \times 10^{+15}$ making the mass more proficient by a margin of $1 \times 10^{+15}$ and that is totally ridiculous in reality but if the formula is correct that is what should mathematically be true. With a magnet, where this concept is true, the last millimetre makes the magnetic pull so strong it becomes almost humanly impossible to control the distance between the two magnets whereas when there is a meter distance between the two magnets there is no pulling power to speak of. The same should apply in the formula if the formula $F = \frac{r^2}{M_1 M_2}$ did apply. Once you are free from the ground, your first stop should be the moon, because then you overcame gravity at its worst. We know that is not true because jumping the first meter is the easy part. Getting higher than one meter becomes more tiring and from two to three meters a person needs other devises aiding the jump. The further the jump is the harder the task is but contrasting this formula $F = \frac{r^2}{M_1 M_2}$ would suggest that the smaller the radius is the harder the effort must be to break the gravity strangle hold.

$$F \propto \frac{M_1 M_2}{r^2}$$

When the radius is insignificant the mass becomes enormous. The opposite also applies because when the radius becomes enormous the mass and therefore the force become insignificant. That is just the way it works when there is such a directly suggested relevance applying between the mass bringing on the force in relation to the radius influencing the force.

$$F \propto \frac{M_1 M_2}{r^2}$$

With the radius big, the influence of mass and the force is weak. This would then suggest that if we are able to break the first meter between the Earth and us, it is then the end of confinement as we would be able to fly to the Moon for weekend shopping on a Saturday morning. We know that is not true because if that were true walking on Earth would be desperately strenuous. It is getting into the air at an increasingly higher distance that presents the problem. What we seem to experience is like a blanket of something pushing us down and the first lift is the easiest. It seems as if the strenuous pushing starts when we try to lift the blanket of air way up into the air. Partially lifting is not the problem but breaking the cover altogether serves as the real problem.

Now, you might say Newton did some damage control when he changed the formula from $F = \frac{r^2}{M_1 M_2}$ to become $F = G \frac{M_1 M_2}{r^2}$ because Newton saw some controversy in the way I explained the working principles of the formula. Don't you believe that one because Newton made the principle even less applying and much more complicated!

The changing of the formula from $F = \frac{r^2}{M_1 M_2}$ to become the formula $F = G \frac{M_1 M_2}{r^2}$ was done in order to give the concept a cosmic significance. When using the formula in terms of $F = \frac{r^2}{M_1 M_2}$, it only refers to an apple falling from a tree to the ground. There is no sense of dignifying the true nature of the event by supporting the refulgence that would reflect upon the appreciation of the spectacle Newton witnessed the day from which such splendid scientific demiurge eventuality arose, where $F = G \frac{M_1 M_2}{r^2}$ indicates gravitational implications of a cosmic nature. Using the formula $F = G \frac{M_1 M_2}{r^2}$ places Newton's lustrous occasion in appropriate perspective. What the grounds was for making it a cosmic notion still eludes me to this day…because after all was said and done, it was only an apple that fell from a tree as the formula $F = \frac{r^2}{M_1 M_2}$ would suggest, and not the Moon coming from space and landing near Newton as the formula $F = G \frac{M_1 M_2}{r^2}$ would suggest, but I am getting to that later on in the book. The minute the radius r disappears, the mass is overbearing and the minute you find the means to break the strangle hold of mass keeping you on Earth; it is like a magnet releasing with no effort available to secure your position again. In the case of gravity being the same as a magnate, it will mean that an object will fall faster and indefinitely increase the descending rate as the object falls, since the object's mass is increasing in force by the diminishing of the reducing radius. Such reducing of the radius will increase the strength of the mass. An object will fall while descending onwards with limitless acceleration and in a fall the object should even be able to break the sound barrier, if the object is massive enough or if the object was dropped from far enough. This same story they try to tell when they tell the story of the person that fell from the balloon that took him almost into outer space. They try to tell the story that he fell faster than the speed of sound. If that did happen, he would be in pieces because his body would not take the strain of breaking the sound barrier. These Newtonians struggle for most of a half century not to get to grips with the sound barrier because they try to look at the principle's tonsils while staring up the arse. Claiming that the person reached the sound barrier is principally an indication of just how poorly they really understand physics.

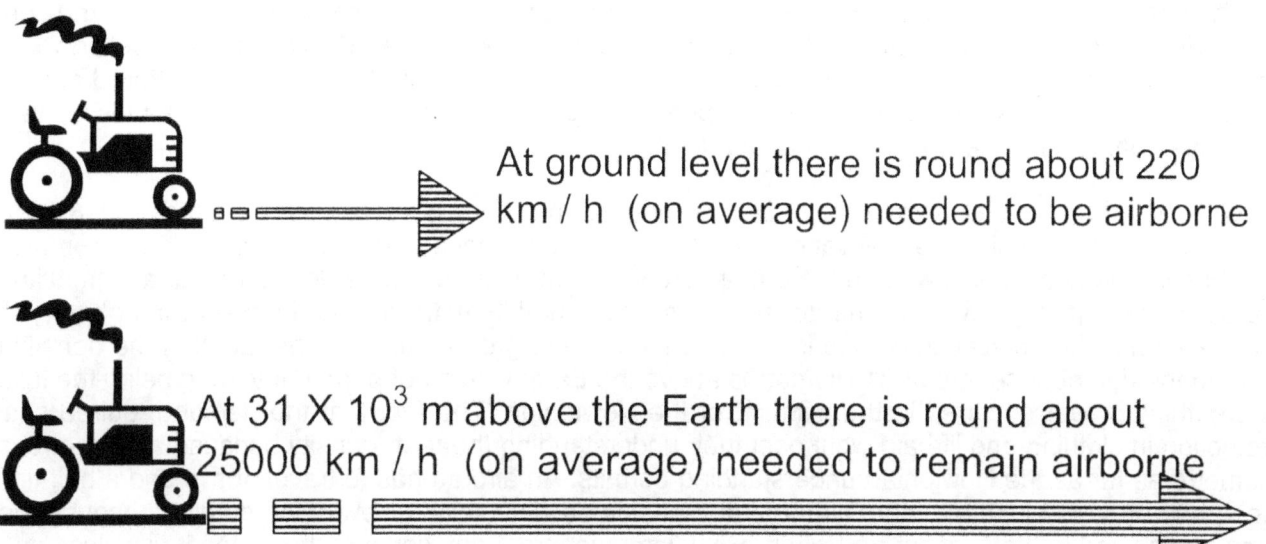

At ground level there is round about 220 km / h (on average) needed to be airborne

At 31 X 10^3 m above the Earth there is round about 25000 km / h (on average) needed to remain airborne

The length of a kilometre is not a constant but is depending on the density that makes up the material forming the kilometre. The density allows the aircraft to be better fuel consumption efficient at high altitudes and helps reduce flying time. Newtonians are unable to notice the concept in hand. Newtonians want mass and mass related arguments. For that reason more than anything else, science passed them by. Newtonians have no concept about the sound barrier but always fly at the speed of the sound barrier notwithstanding their inability to even hear sound at that altitude. That too has gone past our brilliant Newtonians because about that, Newton never said anything and what Newton did not reflect on does not exist! The fact that there is any relevance in the situation and the entire argument depends on relevancies and not the fabrication of mass, is far too complex for the Newtonian to comprehend. Newtonians can master mass and there it stops.

At 31 X 10^3 m above the Earth there is round about 25000 km /h (on average) of pure rocket thrust needed to remain airborne

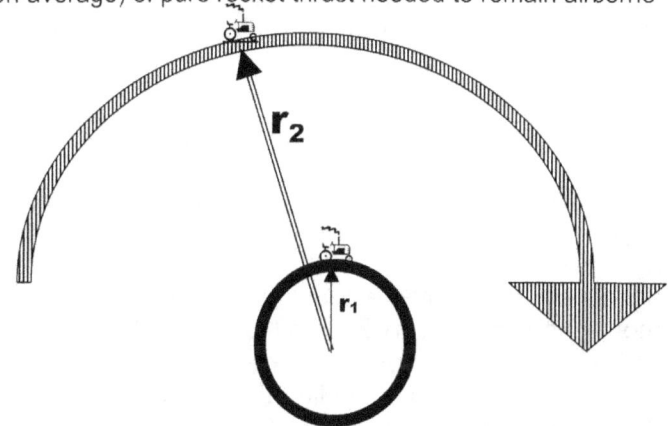

At ground level there is round about 220 km / h (on average and depending on wingspan) needed for any aircraft to get airborne

When an aircraft reaches lift-off speed, the relatively required speed would be around two hundred and two hundred and fifty km / h. To reach to the sound barrier, the speed at that level has to be 100 to 1200 km / h. At a height of 31 km straight up into the sky a rocket driven jet propelled supersonic aircraft requires the trust of 2500 km / h just to remain airborne. Clearly the two thousand five hundred kilometres per hour is meaningless, because one can see that propelling a craft at 200 km / h on the Earth is the equivalent of 2500 km / h in space. It has to do with the density of the space through which the object travels and not the man made suggested norms that become meaningless as the situation changes. This means up there at 31 km into the sky the sound barrier (if there can be one which I most strongly doubt) will be 2500 ÷ 200 = 12.5 X 1200 km / h (sped of sound barrier) = 15000 km / h. It is a question of air density and not distance travelled that determines flying. But since the word mass has never come into the argument the concept goes beyond what Newtonians can appreciate since Newton only told them about mass and nothing more. Now they can either appreciate mass or nothing. Up high, the travelling distance being in relation to time travelled will seem more but as density increases lower down, the speed "will reduce" which is utter nonsense because the relevant density of air displacement travelled to time spent during the exercise will remain the same. Again I repeat: without the word mass uttered once, no Newtonian is expected to understand or appreciate physics because going beyond mass kills all Newtonian participation in understanding and debating.

The man fell from his balloon at a super altitude and has a declining speed never exceeding 218 km / h. In terms of the density the relevance might indicate differences, but in space used for material distribution, the descending was no more than 218 km / h at best. When he descended at a high altitude, the air he went through was lacking density. This made the kilometre air that he went through very thin and very long when placed in comparison with air we use at ground altitude The fact that the density up there in the thin atmosphere at 31 kilometres above the Earth would not permit any living being the luxury of breathing, never occurred in the mindset of the Newtonian. The fact that the person required space protection in clothing and life-aid went past their understanding the term density because a kilometre is a kilometre as far as the Newtonian understanding permits. An aircraft has to travel at two and a half times the speed of sound just to remain airborne at that altitude. This tells a Newtonian nothing because there is no mass used in the concept and when excluding mass from any concept the concept dissolves as far as our brilliant Newtonians' intellect goes. The fact that the distance stretches ten times further at that altitude because the density decreases ten times eludes our Newtonian brainpower completely. In relevancy his deceleration at one point was in relevance still 218 km / h but at that height the kilometre in compressed space is far less than what it is on Earth. That means if the density factor reduced ten fold up there, the falling person then travelled 2180 km / h just to be able to maintain the gravity descending at 218 km/h. The kilometre up there does no hold the same amount of meter density than that which applies on Earth and the kilometre is no kilometre up there. But science hasn't got the wit to figure that one out. If the person was not slowed down by a parachute and had landed on the earth, then the person would have landed at a rate of no more that 218 km / h and most likely at a speed of 207 km / h. He could never break the sound barrier or exceed 218 km / h because that is only possible in the stupidity of the Newtonian mind. The magnet idea is not gravity and having that idea is proving a rather backwards mentality that only the mentally brilliant Super-Educated Newtonian-Masters in thought can achieve. They give you one concept while their approach has quite another. Using the formula $F = \dfrac{r^2}{M_1 M_2}$ or in its cosmic interpretation being $F = G \dfrac{M_1 M_2}{r^2}$ connect your position on Earth in gravity in the same manner as that which the magnet is applying. That means when your foot is clutched to the ground you are very much unable to shift it, but once the grip is released even by a millimetre, the gravity will decrease exponentially. That is not the gravity you have in mind, is it? What you have in mind is that the further you jump the more difficult the jumping gets. The radius extending increases the mass by multiplication and not by division. It takes a lot more effort to fly at 31 kilometres above the Earth than it does flying at 310 meter. Jumping 2 meters can be achieved by human power but jumping 250 meter requires machine-aiding power. This means jumping 250 meter is not dividing the effort but is multiplying the required effort. When translating this scenario into their formula the force that mass would require will diminish by 250 times when the radius becomes 250. The effort required to keep the body afloat will be 250 times less at 250 meters than it will be when the person is on the ground. Your concept about their gravity is not nearly that which they say is gravity, when they say the force of gravity is the product of the mass multiplied by another mass in conjunction with the multiplying of the gravitational constant and this product is reduced by the radius between the two objects holding mass. That is not the way you are walking on Earth but they wish you to be confused with the truth and their lie so that they can get way with murder.

But they would love you to have that concept about gravity and they would hate it if you thought further about the issue because then you would exceed their knowledge limits about what they know as gravity. It fits their comfort zone to have you believing you are an expert on gravity. If they can keep you thinking you are very well informed on the subject of gravity, then you will never question their authority on the matter of gravity and you will never learn that they know nothing about gravity whereas they will have you thinking there is no further point in discussing gravity since we are all experts on the subject of gravity. They will put up a fight to keep you fixed in the idea that gravity is a one-word concept and by thinking in terms of gravity you then entirely define the whole concept involving gravity. You then can define everything that gravity defines by using one word and that is gravity. It is not what they present as a scientifically defined explanation of gravity, but who cares. To them you and I and the rest are cow fodder, there to be used when needed and then to be left in the dark and to their mentality, they could care less whether we have any opinion of sorts, for what we think is of little importance to those mighty academics occupying towering intellectual heights. They could care less about what you think than they care about the opinion your dog has on the human civil war crises in Africa (and this statement will never go out of use because there is always a human civil war crises in Africa). The crisis will be there because the western nations that care so much about the disasters and the prolonged human suffering while the genocide is ongoing everyday. That is all the same to them because they are the ones lining their pockets with money they receive from weapons sold to the warring factions on both sides of the confrontation. Never once was there ever a genuine independent commission launched by the United Nations as to investigate the weapon transactions before the genocide even began or to find who the parties were that provided that arms used in the genocide. It is all a scam to confuse and depress the truth, just as they depress the truth about gravity. It is a culture that drives the powerful to subdue the weak. While one thing is happening they whish you to think another thing is in progress. It is a state of mind the controllers such as Industrialists, Academics of all sorts, Hoggenheimers, Mammonists and politicians use to commit the true crimes against humanity and they are those whom are all criminals with criminal intent to defraud and corrupt the truth. They use many things to defraud the public such as the fraises using the words Human rights, human crises, democracy, racism, political correctness and so many more meaningless expressions just to defraud and confuse. It is everyone that has a personal opinion that such a person is with some authority that allows the person to jump on the wagon of authority and everyone that thinks of himself in terms of being a so-called expert see his role as taking over from God Almighty and start to rule in His stead. This goes from fox hunting to what children eat to what and how parents should discipline their children and criminalize the ordinary citizen while decriminalising the true criminals. They give a concept a meaning as in using an idea while behind the scenes they use an entirely different definition they share very sparsely.

This is mind control just like UFO's are used to culture the very same principle, which all the institutions from the Government to the press and all the Academics at NASA and in agents at NSA benefiting from the misinformation because that suits their use in mind control on you being the victim. They use it to hide behind while they dictate the terms by brainwashing the people into reluctance. The entirety is very deeply rooted in society but in this book I only aim to tackle gravity and therefore we return to gravity.

We have to analyse what gravity is. There is a defining to do by dissecting the factors we see as gravity. Gravity is not many things to lots of people, but is very specific one thing. To investigate the identity of gravity there are questions one must first answer. Is gravity the part that has me standing on the Earth or is gravity the part that has me glued to the Earth. It is surprisingly not the same thing. While I am standing on Earth I am standing still as far as my perception carries but there remains the intention to move, even with me standing still. I can't stand still and at the same time intend to move by the same driving effort. My standing on Earth but remaining in intention to move further can't be motivated by a similar or the same principle. The concept is not the same because when standing still the Earth does the moving on my behalf and there is no independent motion of my body. But when I jump, I am moving away from the Earth and when I walk, I walk along the surface of the Earth giving me independent movement on top and in addition to the movement of the Earth. My movement is stopped by mass and not committed by mass when in mass. When in mass, my intention is to be motionless. There are two occurrences where one is immobility through mass and the other is being without mass while moving. Mainstream science wants you to believe it is the same thing, but it is not. Gravity is the inclination that my body has to move further towards the centre of the Earth while mass is, that, which frustrates my body, and prevent it from moving towards the Earth. Science tells you it is the same thing but it is not. The very second that that which prevents me from moving falls away, the inclination of moving, which is gravity kicks in and gravity lets me move freely again. I will continue to move towards the centre of the Earth and that process has a very well defined and scientific term which we use to describe the event: We call it falling in English but every other language devised another word to describe the same concept. When I stop falling that which give me mass is what ends my moving. What gives me mass changes my gravity from movement to being inclined to move. As soon as that which prevents me from moving allows me to continue to move to the centre of the Earth, my gravity will start giving me full motion and then that which gives me mass by frustrating my moving will not stop my progress towards the centre of the Earth. When my gravity is stopped by mass blocking my movement and that, which gives me mass, then has the ability to turn my descending towards the Earth over to a frustrating of gravity or my freedom to move and change such a freedom to move into a tendency or an attempt to move. The price I pay for the loss of free movement is the gain of mass in mass preventing further movement. It is not as simple as being glued or not being glued…

This is what I have tried to convey to academics throughout the world and in person to so many in South Africa. But conveying this message is criticising Newton and not once did I mange to get one of the esteemed to listen to my criticism of Newton. If they hear it is about gravity implicating Newton, the debate stops with them defending Newton and not giving any more attention to my argument. When trying to convey my message about gravity I firstly have to start to show that gravity is actually the moving of the body and when the body has mass the body isn't moving independently as a cosmic independent object but becomes in mass meaning it forms a unit with the Earth as the Earth. To the normal concept that all persons have, the thinking is that the body not moving and the body moving is the same thing and the body not moving is gravity with the body moving is that which is formed by having mass because the culture that Mainstream science promotes the past three hundred and fifty years never got past mass. If you, the person that is reading this don't believe me that Mainstream science does not discriminate between mass and gravity in the correct way, then test your nearest physics lecturer. Ask him by testing his thoughts and you will see it suits them well to let everyone get all mixed up and confused about the two concepts. To get to the bottom of the concept behind gravity one needs to analyse and dissect and then realise what involves all the aspects that covers the concepts entirely. We have to go splitting the factors forming the concept we hold as gravity and defining the two concepts into proper categories.

To the average people, the idea of expanding a mental view that is normally achieved by promoting thought becomes overwhelming overloaded which leads to confusion and they experience it as me aggressively trying to harm their sanity by dangerously attacking their mental stability. Everyone uses slogans that bring comfort to all person as the slogan covers the concept and dividing the concept is taking out the comfort of the advertising intellectual age.

If I say gravity is the part not connected to mass that brings on the intention of motion to carry on moving downwards notwithstanding the blocking action which comes from intervention of space occupying by a controlling body while mass is not connected to gravity since it is stopping the motion of moving downwards, the concept starts to override the normal commonsense. Mass is having a much more demanding space filled with material in a position that will intervene further movement of descending to a centre of the body having gravity and therefore performing the descending motion. But the descending is part of a rotating motion as well and the complexity of the combined effort to move, forms gravity, while having mass is rendering the control of the movement to a larger body that takes hold of the moving altogether. This is far too complex to cover al aspects in one slogan and by crashing the slogan, people feel I am trying to short circuit their wits and understanding. By getting more detailed about gravity being or not being what the slogan refers to, people hold the view that I attempt to or I dare to try and bust their slogan mentality and break the confines of what the Powers that control society invested in so heavily and what the advertising industry depends on so much, and politicians feed on frantically by promoting the culture so that it is much nurtured by all and which everyone is addicted to. Their comfort zone becomes threatened. Then my threatening to burst this super intellectual "two-word in one idea"-slogan mentality making everyone a genius, most people see as culture threatening and in the defending of that superior feeling, they only see that I wish to destroy them and challenge me to protect their comfort zone. By me attacking their senses of genius which overrides all of their modern nature, is more than what they will tolerate because the easy life the slogan mentality sponsors brings a feeling of intellectual superiority to all concerned and my effort to remove that is most uncomfortable and no one will accept that.

They immediately feel threatened because suddenly the person facing the question also faces the demand on the ability to start thinking, and not just by conveying a slogan to express an encyclopaedia of understanding. When the slogan mentality is forced out and there is a demand to think, the comfort levels drop considerably. That is the last thing anyone wishes to accomplish because in all sincerity that society can offer, we have to realise that thinking is what the TV age and Video culture tries to avoid most. By my asking questions that surpass the slogan stage, I succeed to make that person to feel stupid and threatened and the person goes on the attack. They are not attacking my argument but they are attacking me for making them feel insecure and stupid. I suddenly take away their expertise and that makes them feel naked. That takes them to a point they have to attack me by being without reason because it is their reason I attack and it is the reason to understand what they find lacking. I came to reach this conclusion about human mentality, which I just shared with you on the grounds that not once did one person ever ask me to explain what I just said, when I suggested to a person for the first time that I challenge the authenticity of the concept portraying gravity. Never did one person first consult with me as to what it is exactly that I am saying. Everyone presumed I say they are not glued to the Earth and with me being that mad I am clearly more stupid than they are and for once they find someone with a lesser mind which they can demolish.

When I state that there is no gravity they all respond in a manner that is putting me under suspicion and not one person responds positively by asking me why I put gravity under suspicion. Now I give the reasons why I challenge the concept of gravity and I would love to see a person challenge my challenge. This is why I discard the fact of gravity as promoted by the Brainy Bunch.

By definition gravity is defined as being:
Gravitation is the force of attraction that operates between all bodies. The size of the attraction depends on the masses of the bodies and the distance between them: the gravitational force diminishes with the square of the distance apart according to the inverse square law. Gravitation is the weakest of the four forces. Newton formulated the law of gravitational attraction and showed that gravitationally a body behaves as though all its mass were concentrated at its centre. Hence the gravitational acts along a line joining the centres of the gravity of the two masses.

It is not you being glued or not being glued to the Earth that I discard. It is the definition holding this whole idea that I do not share in the least. What the definition describes is magnets pulling and it is the total opposite of what I experience. Breaking the first millimetre of gravity clampdown is the easiest and not the most difficult. The difficulty increases as the radius grows and not as the radius decreases. When I say there is no gravity everyone thinks I say we all are going to fall off the Earth at random and with me thinking that way, then it is obvious that I must be a nut. Everyone thinks of me as the clown acting mad when I say gravity is not to be found in nature. But I do not say we are not standing on the Earth. I do not say there is nothing that is keeping me glued to the earth. I say there is no attraction between two bodies by the force of the mass that in such doing is diminishing the radius parting the bodies by the inverse square law. I say there are a connection by motion between the centre of the body and the material surrounding the centre. This is what I say when I say there is no gravity.

I dispute Newton and so do all students at first when students are forced to learn about Newtonian physics because Newton's arguments are an onslaught on human intellect. Think of the resentment that students have towards Newton under normal conditions when they have to cope with understanding the Newton principles Mainstream science says are applying and how that confusion of what is possible and what Newton suggests is possible clashes with their intellect which makes them feel stupid. Students hate Newton because they don't understand Newton and for that they are accused of not having the intellectual capacity to follow Newton. Every student from the past going into the present and even including those forming a future generation of students will purchase a book that is showing that Newton's legitimacy is cracking up when exposed to some vivid scrutiny.

In short I will now explain what I explain throughout the book you are about to receive and which is named *Newton's Fraud* or whatever it will be named as. The Newtonian formula $F = G \frac{M_1 M_2}{r^2}$ is the formula used by science to explain and define gravity. It says the that the ($\underline{\mathbf{M_1 \times M_2}}$) mass of the one object pulls the mass of another object and this process in relation with a gravitational constant ($\underline{\mathbf{G}}$) (a supposed force keeping the Universe attached) and the pulling subsequently destroys the radius ($\underline{\mathbf{r^2}}$) being between the objects. That says that objects **ALWAYS MOVE CLOSER *BY FORCE*** in relation to **MASS**. Newton submitted the suggestion that objects fall as MASS provides the force that will cause the falling by the inducing of a force he named gravity which he subsequently only proposed was the acting suppositious force. I disprove this formula in so many ways in this book and I show that this formula and the ideas Newton introduced just don't stand up to even the smallest tests. Then, if Newton's idea on gravity has validity and mass is responsible for objects falling, then all objects that are in a process of falling must be subject to mass and in that idea rests differentiation and discrimination in size and compactness producing speed variations. If any and all falling is subject to the variation mass introduces and the influences coming about is the result of mass interfering in the gravity force being generated, this then must bring different speeds to cause substantial variation in the falling of different objects holding different mass factors. There can't be conformity in the falling of all objects while such falling is the result of the discrepancy that mass has to inflict due to variations that result in mass differentiations. This is a vital issue that science eludes and has all clever ways to avoid direct questioning. This part science just run around and never addresses and avoids confronting the issue. This avoidance of confronting the issue whish will disprove the validity of Newton is done with such cunning as you will not believe. The fact that objects fall due to conformity in the falling, science accepts but portrays a picture of deceit that mass brings falling distinction and therefore equal falling doesn't happen, while they at the same time admit to Galileo's presentation that falling of all objects are equal in tempo, irrespective of size or any form of differentiation. While they promote the obscurity that Newton and Galileo is in harmony the truth about their deceit is that the two can never have the same issues. That I prove is a fact and also I show how big a part this is in the overall covering up of Newton's initial fraud.

I have written several books in which I challenge the thought process of Mainstream physics and especially Sir Isaac Newton's arguments about physics. I am of the opinion that even though everyone thinks of Sir Isaac Newton as the genius who established every aspect that is used in modern physics today, but in spite of every other person hailing Newton, I remain of the opinion that the man did not have a foggy clue about any of the principles driving the concept that he named as gravity, or what brought about gravity according to his explaining of what forms gravity. I am able to explain gravity but it doesn't even vaguely resemble Newton's version of gravity. I can explain gravity by proving my explaining with the use of simple mathematics. I use Johannes Kepler's formula to back up my statements. By using Johannes Kepler's formula I found a way to prove there are four phenomena found in the cosmos. There are the four phenomena applying in tandem that together forms gravity. They are: The Titius Bode law; The Roche Limit; The Lagrangian Point System and; The Coanda effect. As the phenomena don't support Newton's vision on cosmology, the phenomena has no support amongst Mainstream science although they did apply it with a positive results in locating the missing planets at the time of their discovery. When they located unknown and undetected planets in the past, the existing of the phenomena was never disputed but when the argument of proving them comes to mind, then they are dismissed as some coincidental abnormality occurring. But since it holds no similarity to Newton's view on science, Mainstream science rather disclaimed the validity of the phenomena than they would find fault with Newton's ideas. In the mind of science the cosmos can be wrong and God can be wrong but Newton can never be wrong. In using the four correct principles correctly, which I back up with the correct mathematical interpretation thereof in support of the function that each phenomena has in forming gravity, I did a far better job than what Sir Isaac Newton did and what I achieved is of a far more acceptable level as well as being mathematically far more correct than what Sir Isaac Newton did achieve with his guessing about issues he couldn't explain. To be successful in my quest to find an explanation for gravity, I had to redirect all my concepts I previously had and also alter all the otherwise normally accepted thinking on physics. I had to find the phenomena and I had to dissect the function of each phenomenon

A Less Complex Commercial Science Book

as well as mathematically valuate the phenomena. In this process I realised that to come to realise what gravity is, I had to realise that gravity is not what Newton saw forms gravity.

Newton devised a formula $F = \dfrac{r^2}{M_1 M_2}$ that represented gravity. Newton thought the mass of the apple that fell drew the Earth as much as the Earth drew the apple. The one mass factor represented the mass of the apple while the other mass factor represented the Earth and the radius was in place of the distance that the apple had to travel as the apple fell from the tree in view of Newton. This falling he saw as the gravity that the Earth's mass and that apple's mass were achieving. Let us have a look at the force F that Newton introduced

What is F and what worth has F while we find out what role F plays. Let's place F in $F = \dfrac{r^2}{M_1 M_2}$ and find what F really has in a mathematical sense.

$F = \dfrac{r^2}{M_1 M_2}$ can be replaced by $F = \dfrac{a^2}{a_1 \times a_2}$ which then would leave $F = \dfrac{a^2}{a^2}$ that leaves **F = a^0** and that outs the factor of gravity without value or worth being a factor of **F = 1**. This doesn't make much sense but Newton never saw this imperfect outcome to his otherwise perfect formula. But calculating $F = \dfrac{r^2}{M_1 M_2}$ in terms of real factor worth it makes no sense in another sense.

Replace all the factor values in terms of $F = \dfrac{r^2}{M_1 M_2}$ and with the mass of the Earth multiplied by the mass of the falling body, which was the apple, the force is exceptionally small. If I calculate the force in terms of my view the force that comes about in this formula the result we find in calculating Earth with the apple divided by the distance between the two is something in the region of less than what the mass of one atom would be. This left our genius with some headache and a large problem (or is it a very small force of gravity) to solve. The force coming from this equation is less than microscopic small! Then Newton improvised masterly by cheating the wits out of all mathematical logic.

Newton changed his initial formula that was $F = \dfrac{r^2}{M_1 M_2}$ to $F \, \alpha \, \dfrac{M_1 M_2}{r_2}$ and the entire world still to this day think this move is brilliant. This is coming from the best mathematical minds found on earth. They applaud Newton in his brilliance by saying Newton never made a mistake. If you think that way then answer the following argument. Newton placed he value F would have in terms of the formula $F = \dfrac{r^2}{M_1 M_2}$ as being equal in context to $\left\{\dfrac{F}{1} = \dfrac{m_1 m_2}{r^2}\right\}$ and by changing the formula by only changing one symbol α the entire outcome of the formula changed without changing anything. Newton saw it fit to replace = with α and the formula was reborn in value while staying the very same. There is an applying rule or law in mathematics that says when one change a formula from $F = \dfrac{r^2}{M_1 M_2}$ to $\left\{\dfrac{1}{F} = \dfrac{m_1 m_2}{r^2}\right\}$ then F being F ÷ 1 must also remove a position to become 1 ÷ F making F the fraction value. All those that know even the least about mathematics and of which Newton and his followers not part of knows very well that if any part on the one side changes dynamics from being on top of the dividing line then the very same must apply on the other side. One can't just say that to change a formula $F = \dfrac{r^2}{M_1 M_2} = \left\{F \, \alpha \, \dfrac{m_1 m_2}{r^2}\right\}$ would not translate in ultimately change the outcome of the formula because the truth about mathematics is that $\left\{F = \dfrac{r^2}{m_1 m_2}\right\} \neq \left\{F \alpha \dfrac{m_1 m_2}{r^2}\right\}$ but when it changes the ratio of what is divided and what divides there is a principle in mathematics whereby one then changes every aspect in terms of such a change to alter the ratio on both sides of the equitation as to maintain coherency in mathematical logic and the equation changes become $\left\{F = \dfrac{r^2}{m_1 m_2}\right\} = \left\{\dfrac{1}{F} = \dfrac{m_1 m_2}{r^2}\right\}$. Newton had this idea that because he was Newton. The Great (Cheat) normal rules did not apply and with him being Newton even mathematic laws was below his status. He could replace symbols = with α used in changing the formula $F = \dfrac{r^2}{M_1 M_2} = \left\{F \alpha \dfrac{m_1 m_2}{r^2}\right\} = \left\{\dfrac{F}{1} = \dfrac{m_1 m_2}{r^2}\right\}$ and that will change mathematics forever. It never dawned on him or his followers that came after him that $\left\{F = \dfrac{r^2}{m_1 m_2}\right\} \neq \left\{F \alpha \dfrac{m_1 m_2}{r^2}\right\}$ but the correct application is in fact $\left\{F = \dfrac{r^2}{m_1 m_2}\right\} = \left\{\dfrac{1}{F} = \dfrac{m_1 m_2}{r^2}\right\}$. Let's find out why the changing that Newton did $F = \dfrac{r^2}{M_1 M_2} = \left\{F \alpha \dfrac{m_1 m_2}{r^2}\right\} = \left\{\dfrac{F}{1} = \dfrac{m_1 m_2}{r^2}\right\}$ in order to improvise for his theoretical shortfall is total mathematical corruption on the highest level.

In $F = \dfrac{r^2}{M_1 M_2}$ the factor of force is with the mass multiplication as the mass presents the radius by the square with the diminishing or increasing value. A large mass will produce a small radius and the mass reduces the radius by determining the reducing of the radius. The force will reduce the radius.

Then in the improvised version $\left\{F \alpha \dfrac{m_1 m_2}{r^2}\right\}$ that actually that with the factor manipulating inexplicably becomes $\left\{\dfrac{F}{1} = \dfrac{m_1 m_2}{r^2}\right\}$ such changes brings total factor revaluation to the entire prominence of the force changes all together. When using the ratio as $\left\{\dfrac{F}{1} = \dfrac{m_1 m_2}{r^2}\right\}$ the radius becomes the force carrying factor and the radius will determine the mass value in ratio to the formula. A large radius will provide a large force of gravity and not as with $F = \dfrac{r^2}{M_1 M_2}$ where a large mass will influence the ratio as to establish a large force and the result will be to produce a small radius. In the one $F = \dfrac{r^2}{M_1 M_2}$ we the force lies with mass and in $\left\{\dfrac{F}{1} = \dfrac{m_1 m_2}{r^2}\right\}$ we find the force being in relation to the diameter size. This Newtonians missed all the time during the past (about) 300 years. The Force of gravity changed from initially being in the mss $F = \dfrac{r^2}{M_1 M_2}$ with the mass dividing (driving the force) to being in the radius between the points holding mass $\left\{F \alpha \dfrac{m_1 m_2}{r^2}\right\}$ where it is the radius that will determine the influence the mass has on the Force. This is a small anomaly but it shows how little did Newton consider the impact of the stage he brought about and this shows Newton was inspired by one motivation and that was to defraud science.

But then he went much further and cheated the cheated by introducing $F = \dfrac{r^2}{M_1 M_2} = F = G \dfrac{M_1 M_2}{r^2}$. There was never one Newtonian that even hinted that the Newtonian could explain how did the initial thought of $F = \dfrac{r^2}{M_1 M_2}$ than mathematically changed to $\left\{ F \; \alpha \; \dfrac{m_1 m_2}{r^2} \right\}$ which was intended to become $\left\{ \dfrac{F}{1} = \dfrac{m_1 m_2}{r^2} \right\}$ and then with normal, mathematical principles still applying change this lot to $F = G \dfrac{M_1 M_2}{r^2}$ Furthermore, how could academics in mathematical physics teach children or students in physics this as the truth! How could any mathematician explain a process of following logic maintain that $F = \dfrac{r^2}{M_1 M_2} = F = G \dfrac{M_1 M_2}{r^2}$ ...explaining it is preposterous.

Let any academic mathematically show how one would go about and use Newton's visionary formula $F = G \dfrac{M_1 M_2}{r^2}$ to calculate the force of gravity by replacing the symbols with the actual values in mass that the symbols should have. Put in the Earth's mass in place where it belongs and put in your mass in place where it should be and then divide that with the distance between your soles and the Earth measured in micro millimetres by the square thereof! If it can't be done, then that is proof of Newton committing fraud when he introduced the formula $F = G \dfrac{M_1 M_2}{r^2}$ being able to calculate the force applying as gravity. Take any formula used in daily physics and show where they use the mass of the Earth as a factor in calculating anything. Never, not once, do any formula used by physics hint that the Earth's mass has any influence on any part of physics when any one calculates factors to determine whatever they wish to determine. If the Earth's mass is never used in any calculation, then the Earth's mass has no part presented as a factor and then the Earth has no mass that influences any aspect of physics. That means the Earth's mass doesn't produce gravity because if it did, the calculating formulae used in physics must use the Earth mass as a factor in all calculations! Newton cheated to bring in the Earth as a factor that has mass that produces gravity and never does the mass of the Earth contribute to any part in any of the many calculations that form part of physics. The Earth has no mass because the Earth's mass never plays a part in any formula. It is as simple as that! The formula Newton first devised has not even a ring of truth to it. If it is true then show how the formula reading $F = \dfrac{r^2}{M_1 M_2}$ is used to indicate that this brings about gravity without cheating it to become $F \; \alpha \; \dfrac{M_1 M_2}{r_2}$ and then committing blatant fraud in changing the formula to able $F = G \dfrac{M_1 M_2}{r^2}$ while even in this form it still doesn't apply.

If you think I am going on about academics then think how much they tormented me by ignoring me in eight years. With the clear evidence I show they still dismiss me as the one that is mindless because I am unable to "understand" Newton. What is there to understand when everything I am supposed to understand is tainted ands flawed! If it seems I am going into rhetoric about academics then it is because I wish to describe their deceiving methods in dismissing me.

The point I wish to make is that they say gravity is $F = G \dfrac{M_1 M_2}{r^2}$ while they also say that the value of "F" as in gravity "g" is $F = g = 9.81$ and further more they say that $F = mv$ while they first said $F = \dfrac{r^2}{M_1 M_2} = F \; \alpha \; \dfrac{M_1 M_2}{r_2} = F = G \dfrac{M_1 M_2}{r^2}$ Now get this lot married mathematically...that is a challenge they can never manage and yet they say it is true because Newton said it is true.

Let all the physicists show how they manage mathematically to get $F = G \dfrac{M_1 M_2}{r^2}$ equal to the measured value that they say gravity has being the "g" value and not the "F" value at $g = 9.81 \; Nm/s^2$. They advocate that gravity is another symbol that somehow replaces F with g but also is gravity with a totally new value than that which Newton had in mind and then as "g" apart from "F" has a measured and physically determined value of $g = 9.81 \; Nm/s^2$ So let them do the calculating of the Earth mass and any person's mass multiplied by the gravitational constant and get this lot divided by the distance between my feet and the Earth when I stand on the ground by the square thereof and to top this, they then get gravity to be $g = 9.81 \; Nm/s^2$
I'd love to see them accomplish that!

When they use another formula that also uses the symbol F in the formula $F = mv$ I still have to find one academic that can show me whereto did the mass of the Earth disappear while taking with it the gravitational constant as well as the diameter parting the mass m from the other disappeared factors. This is one of the many small issues they never think of because they can't explain it while upholding the correctness of Newton at the same time. Let one of them with the many doctoral degrees, show how they come from $F = G \dfrac{M_1 M_2}{r^2} = F \; \alpha \; \dfrac{M_1 M_2}{r_2} = F = \dfrac{r^2}{M_1 M_2}$ to eventually reappear on the surface as the formula $F = mv$. If you thought gravity was an act of magic try this magic. Where did all the factors (M_1, G and r^2) go while being on route to change in appearance to become $F = mv$. The mass of the Earth that academics in physics claim is there and that is supposedly is doing the gravity pulling, is a relevance that the object has with the Earth having a factor of 1 and this relation is effective viable only when the object having this mass is resting on the surface of the Earth or having some direct contact through another medium connecting the object to the Earth. The object rests on the link by a link or otherwise is resting directly on the Earth, but the condition of mass of any object has is that the object is standing still or moving while being in direct contact with the Earth. But all action the object has is relevant to the position the object has in relation to an allocated relevance with the Earth and relating to the movement that the Earth has. The object in mass has to move directly with the Earth or slightly more than the Earth. The object only shows having mass when connected and when accepting the movement the Earth has but the mass the Earth should placed into the calculation alongside the mass the object has, that as a complimenting factor is totally absent in normally used physics because the Earth has no mass. The Earth's mass is lacking all visible presence in influencing physics by lending support or increase any calculation in physics. This proves my statement that it is because the Earth and all other planets do not have mass and therefore can't be used as a calculating factor.

Planets have no mass and neither has the Sun got mass except the mass Newtonians wish to credit planets with. Bigger planets don't move faster because they have more mass and smaller planets are not further from the Sun because they have lesser mass. All planets big and small spin at the same speed around the Sun and in relation to the Sun and all planets are scattered going around the Sun while being big and small where all sizes are well mixed. This is because planets have no mass except in the imagination of Newton and his devoted followers. The mass of the Earth never plays a role in physics and the mass of planets do not draw any of the planets closer to the Sun and let one physics professor bring proof that the planets do draw nearer to the Sun!

They just can't because planets do not have mass that can produce a pulling gravity! If and when the mass of the Earth do not feature as a factor in any formula that is used in physics, then the mass of the Earth is no factor playing part in gravity. This then can only indicate that the Earth has no mass. If there is an absence of mass as a factor that influences physics, this can only be as the result that the Earth mass has no gravitational presence in any physics formula. Gravity does have the value of $g = 9.81 \text{ Nm/s}^2$ but that I explain and the value $g = 9.81 \text{ Nm/s}^2$ I prove as well. With that evidence being that clear, then the mass that the Earth should supposedly have, does not produce gravity as Newton suggested. Prove me wrong by getting gravity at $g = 9.81 \text{ Nm/s}^2$ from using either any of Newton's formulas being

$$F = G \frac{M_1 M_2}{r^2} \text{ or } F \propto \frac{M_1 M_2}{r_2} \text{ and } F = \frac{r^2}{M_1 M_2}.$$

Let me see Newtonians do that and I will become a believer in Newton! The Earth has no mass because physics can't show the Earth's mass playing part in calculating formulas and if there is no mass that plays a part that should produce gravity, and then mass can't be responsible for the producing of gravity as Newton declared. That makes Newton's suppositions total rubbish and that makes Newton responsible for a crime of defrauding and falsifying the science of physics. If you, the reader is able to get academics in physics as far as even reading this argument I make, then you are more influential than I can ever be. They plainly dismiss all these arguments with arrogance by discrediting my credentials!

What Newton saw as gravity can't withstand even the slightest test of proof and I showed that it is not possible to use Newton's formula as Newton suggested it applies to mathematically calculate gravity. I come back to this issue later on. I have tested Newton's thinking and the book I offer to you for investigation serves as the testimony to all the testing I did on Newton. This any body who can see, will see when reading this book, I tested Newton from all the angles to see if he possibly could be correct but found his thinking wanting every time. The truth about Sir Isaac Newton's concepts I came to conclude, was that the reality is that it is not in any way overstated to declare that Newton conspired to defraud science and moreover that he committed blatant mathematical corruption in trying to prove the concept he had about what he thought forms gravity. There is no backing for Newton's ideas and even the ideas which are in use are not in the form that Newton said it applies where physics in daily use serves as the best discredit to Newton bringing no proof about any of the claims that Newton made on matters concerning science in cosmic gravity.

I show that every thought Newton introduced that later proved useful and was correct, was what he stole from another far better cosmologist called Johannes Kepler. Not one of his laws are directly relating to any concept Newton ever introduced at any stage but is the result of academic theft he committed against a much larger figure that preceded him by almost a century. But he stole, he lied and he raped the work of a predecessor in order to defraud the world of science in his time. Newton brought no original input into science except that he gave a concept the name "gravity" and even that is inappropriate. Newton made suggestions that break every mathematical principle he could think of. That, Newton did in his attempt to win over the prevailing academic thinking of the day in his time as to lay some sort of groundwork to form backing for his ideas on physics and to attempt to explain gravity or what he thought gravity is. If this is shocking and sounds outrageous, then a lot more shocking detail awaits the reader in this book.

Newton's claims about the principles he declared as being responsible for guiding physics carry no proof and after I realised that, I was able to start forming another line of thought on gravity. After formulating my concept about how gravity was truly formed, I had to introduce my ideas to academics in physics. In my quest to find the method how gravity formed I used the four phenomena and the principles of these phenomena as well as determining in which way each phenomenon applied. Then I placed each one in the way that were known how they work and then implicated that specific formula's function mathematically in forming gravity in the cosmos. This was no easy task but I did it and by formula shows that my argument is logic and the mathematics prove that it works well.

The phenomena that I use is still to this day unexplained by Mainstream science because it shows no sign of using mass and without mass the Newtonian mind understands nothing!. Newtonians don't understand the four phenomena due to the fact that science up to the present date has no means or method to explain the four mentioned phenomena while I can explain the working of each independently and how they work in a combination to produce gravity. I found a way to put those four phenomena in a perspective and put the four in a mathematical sequence that from there I could explain gravity in detail. When I first approached academics, I had the opinion that all academics were knowledgeable about the lack in the correctness we find in Newton's views and that every one in physics would be rejoicing in finding what gravity consists of. I was under the impression that I would be embraced by those in physics for finding a solution to Newton's errors. I was in for a nasty shock with such naivety.

I met with such rejection that no one even cared to look at my work because they were of the opinion that looking at my work would be sacrilegious to Newton. I was told on occasions that Newton has never been proven incorrect and therefore any attempt on my part in doing so is a waste of time. At first I was not confrontational towards Academics in physics and avoided any indication about disagreeing with Newton, but academics always threw Newton at me and eventually for self protection I had to start to confront them and confront Newton, with which I was in disagreement from the beginning although at first I was reluctant to voice any opinion about the matter. But slowly it dawned on me that if I had any serious plans to introduce my ideas I had to dispute Newton's gravity principles and show the inconsistencies and dishonesty in Newton's approach to physics. I came to realise that his flaws are there and the mistakes are present whether I avoid it or attack it; the inconsistencies are part of forming the basis for modern accepted science. It is that strangle hold I had to break before I could even think of finding acceptance about change.

Then slowly I concluded that only and after I can get people to see how incorrect Newton is, do I stand any chance to introduce my line of thought on gravity and I am so sure of my ideas being correct that I dare any one to disprove any part or the entirety that forms gravity as I see gravity! But that can only come about when I can get an audience to see how I expose Newton for what Newton was and in that is where I find no luck. I can't find one academic with influence that is brave enough to stand up and face my attack on Newton and argue me down or prove me wrong in a sound debate. The moment any academic realises he or she is reading my condemnation about Newton's correctness, their minds shut down! No other thought can penetrate their mind but to think in terms of Newton being correct even when confronted with facts proving Newton incorrect. They stop reading my work. They do not get confrontational but defensive and in defending Newton they refuse to read further!

I realise that every one has the view that my finding fault with Newton shows signs of madness and progressive signs of dementia on my part and in my thinking to even regard any possibility that I am the only person on Earth that is correct and all others that ever studied physics are wrong is pure foolishness but mad as it seems, if that is what I have to say to be correct in what I say, then that is what I say...Newton is wrong about gravity. I don't say this lightly or without understanding the enormity of what I suggest is going on, but be that as it may seem, it is the truth without question that Newton went on for three hundred and fifty years defrauding science with no one testing his claims.

Detecting Newton's misconduct is possible because I saw a way to break away from the invalid concepts Mainstream physics holds. I saw where Newton went wrong and correcting the major mathematical error is so small...if only any one would listen! ...And of course one has to admit that the Earth doesn't pull or push by mass or any other way just like the physics formula they use to formulate indicates. Notwithstanding the pose Mainstream physics tries to uphold, the entirety of physics still uses the idea of magical forces intervening in nature and they still base concepts on unexplained novelties. Think of how they found four unexplained forces going around and influencing persons in an unexplainable manner except that they can see that it is through the inexplicable magic of gravity keeping people attracted to the Earth. To say the least, the concepts physics use in terms of Newton would not even be acceptable to children in the modern informed era we live in. I challenge any person to prove Newton, not to accept Newton but to undoubtedly prove Newton correct! I recognised the impossible double standards Mainstream physics apply to promote their much shady explaining. In short I tested Newton's principles and found the principles to be wanting on all levels of consideration.

The statements that Newton introduced have inconsistencies and to cover these holes science has in their understanding of cosmic principles, they have to apply standards, which are symptomatic of double vision. To compensate for these bogus truths that were supporting their incredible theories, they simplify issues to such a level where what they embark on is equal to and the same as witchcraft and soothsaying. They admit the cosmos is expanding but the expanding is not here in our neighbourhood because in our neighbourhood Newton rules and therefore notwithstanding a Big Bang still applying and a Hubble expanding going on, in our neighbourhood we contract because Newton said so and Newton just can't be wrong. Their pitiful explaining of the fundamental working of physics is meaningless. In spite of finding evidence that the Moon and the Earth is growing apart in distance they still uphold Newton's view on contracting because it suits their work and leave everyone under the impression that the contracting is valid as it should be if Newton is correct. The Earth and the Moon is growing apart at the same rate as human hair and human nails grow. Because they lack true basic understanding they have to accept the unproven and it remains unproven that the cosmos is coming together by the power of mass that is inflicting gravity. They proclaim to understand what flows out from what they understand but such concepts become meaningless because of many inconsistencies. To name but one such an example is the explanation they put forward in the Tunguska event. To claim that a mini Black Hole went through the Earth is demeaning just going on the basis that they claim there can be such a thing as a mini Black Hole. Such statements are beyond the ridiculous and to achieve some degree of believability from the public they create scenarios, which use arguments that are entangled with deception, such as what is obvious in the case I mentioned. What they declare as unwavering facts can't even be supported in the least form when tested. Even the least degree of verification of correctness is absent and Newton lacks all evidence of authentication in any investigation of even the simplest terms. It is as if they never read with interest that which they explain and they never scrutinise that which they advocate. They give values that are senseless and make that which they say meaningless. In all this they use billions of tax dollars to prove what they have no idea of. They try to commit matter to fusion while they have no idea why matter would fuse at all!

Do you think of astrophysics as being the department that is run by the wise and the level minded, the honest and pure at heart, the nobility of well-to-do academics and the sober thinking standing in front of the world as the absolute trustworthy? If you are a student, there is no other choice you have but to trust them while they feed you absolute hogwash! They force students in believing gravity is the result of mass and to see that students comply with the unequivocal acceptance of the brainwashing, they subject students to various tests and examinations. In those test they determine the degree the students' have developed by brainwashing and the levels they test goes by employing examination standards. They never prove this concept that it is a fact that it is the mass of the Earth that pulls the body down while all formulae used in science prove otherwise by not using the mass of the Earth. While knowing this all to well Newtonians insist on students accepting Newton and still students have to acknowledge these concepts as truthful facts. If you think those in charge of astrophysics are the pillars of trust, then get wise by reading the following. What you are about to read is simply mystifyingly simple and yet to this day I have not had the privilege to challenge one academic any where that had the honesty to admit to the fact of Newton being wrong. After you have considered the following you might agree with me that even small children can reach a higher level of clear-minded logic and find more sensibility than what those scientists promoting astrophysics have because science lives in a make believe fool's paradise. One such example is to put space travel and extra terrestrial life forwards as even a remote scientific possibility. That is departing from any possible sane minded thinking. Even to consider space travel as an option shows the level of not even understanding the most basic principle behind gravity as we find the comet proves.

The scientific presumption is that gravity is established when one object holding mass is pulling another object having mass and forces the two abject to move toward each other. The entire basis of all physics rests on this formula where init it is believed that mass produces all gravity by distinction of differentiation in density as well as size and if physics is anything to go by, then what ever is proven, such proof must stem from and be in support of well as being supported by this formula $F = G \dfrac{M_1 M_2}{r^2}$. It is the formula that keeps the entire Universe in place and all of Newton's accuracy solely depends on $F = G \dfrac{M_1 M_2}{r^2}$ as a formula that has to be truthful and unquestionably accurate. The mass is the crucial factor because the mass is in a position where the mass destroys the distance of the radius from both ends equally. The mass generates a force and the force produces the gravity and the gravity produces the pulling and the pulling is what the time depends on that we have left to enjoy a Universe. We have to appreciate Newton's finding of mass until Kingdom comes because if not for mass, Kingdom is coming either tomorrow or never. Then we also accept the formula $F = G \frac{M_1 M_2}{r^2}$ has been tested and proven so many times by science that there is no other formula on Earth that has endured the testing that Newton's gravitational formula $F = G \frac{M_1 M_2}{r^2}$ has under gone. The force of gravity has the mass that would generate the gravity whereby the pulling of the other object orbiting and in also generating by mass the other object will also force gravity onto the first object $F = G \frac{M_1 M_2}{r^2}$. What goes up must come down. We fight our mass because we fight gravity the entire time during one life span we live through. When I jump the force of my mass that generates the gravity by which I pull the Earth and by which the Earth pulls me back and the pulling is the result of the mass of the Earth that pulls me down again while at that moment I am pulling the earth up again, thus the square value coming about in the radius factor. When I fall my mass kicks into action and by mass I hit the ground at a rate my mass will determine. Newton is a genius because Newton realised all these wonderful happenings. Newton saw that a planet pulls another planet by the gravity that the mass of the planet charges. But consider that if mass is what brings about falling, it then implies that objects just cannot fall equal but have to fall differently and according to their mass. If mass has nothing to do with the falling then objects must fall equal and in equanimity through out the entire distance of travel while in the process of falling and that fact goes without argument. Mass brings variation and conformity is the result of mass not applying! The Universe is in a state of contracting $F = G \frac{M_1 M_2}{r^2}$ as Newton's formula must indicate. The objects are drawing closer to each other all the time.

Now marry that thought with the ever expanding Big Bang beginning and the Newton's concept of a Universe shrinking which totally contradicts the reality that Hubble found to be true and that there is a Universe out there of which we are part of that is exploding in expansion. To the world they declare openly that Newton's contracting Universe and Hubble's expanding is the same thing and we must wait for the Universe to admit being incorrect and start to employ Newton's contracting. They gave this blaming of the Universe going the wrong way on the Universe being the incorrect party because they are looking for mistakes in the Universe and wait to find out when the Universe will start to comply with Newton and start shrinking because the Universe has to stop this ridiculous expanding since Newton said the Universe is contracting. Since Newton just can't be wrong, therefore the blame of such silly contradicting of Newton has to be found at the door of the Universe. This blame game and detecting how far and why the Universe went wrong in disobeying Newton they named the Critical Density Theory and is the biggest scam and covering of fraud ever invented by any group of persons any time during the history of man. If the Big Bang is true (and it is true), then Newton just doesn't fit! In my book I show how this led to the biggest criminal cover up man has ever devised and was initiated by a person called Albert Einstein. The entire philosophy behind the Critical Density Theory is a scam and is even as ridiculous as what the rest of Newton is. You are about to read how far Newtonians will employ criminal cover up to form a blanket of deception!

The Newtonian formula $F = G \frac{M_1 M_2}{r^2}$ explains the comet arriving at the Sun, drawn by the mass of the Sun, pulling the mass of the comet as the comet comes closer to the Sun, but then if Newton's $F = G \frac{M_1 M_2}{r^2}$ has any validity the comet has to crash into the Sun after arriving. If gravity by mass was pulling the comet towards the Sun in the manner as Newton insisted in the Newtonian formula $F = G \frac{M_1 M_2}{r^2}$, then try and get any academic to explain why and how the comet moves away from the Sun and into the black yonder. After reaching the comet, the comet avoids colliding with the Sun as the formula $F = G \frac{M_1 M_2}{r^2}$ would suggest and head into the darkness of outer space. The comet then is moving directly in the opposite direction of what Newton's formula $F = G \frac{M_1 M_2}{r^2}$ would have us believe as the comet is not suppose to be pulling away because it is the mass pulling that was in place when the comet was drawn by mass as Newton stated. Does mass then start pushing mass to get the comet floating away from the Sun? Mass establishing gravity by pulling of a force is a gimmick Newton suggested but is unproven and it is nothing less than foolhardy to believe that mass does the pulling of the comet. Try and get those academics in physics to sensibly admit this reality and then in the explaining be sensible by using their Newton formula $F = G \frac{M_1 M_2}{r^2}$ as Newton's formula presents the law to show how this going away happens when mass is doing all the pulling at first. Try and get any Newtonian academic to explain this escaping of the comet from the mass of the Sun in the face of mass pulling mass. Some try to use the idea that the momentum drags the comet around the Sun but the mass will pull the comet into the Sun if $F = G \frac{M_1 M_2}{r^2}$ applies. Newton never created a detour as the mass pulling mass forms a linking straight line running from the centre of the Sun to the centre of the comet. Newtonians always bring more deceit to cover up Newton's fraud. Students otherwise never ask these questions I address because students are brainwashed to accept and not think about asking questions. In presenting my work I can and I do answer the questions raised above but my answers do not fit the Newtonian visions of mass doing the pulling and because it contradicts Newton, I am ignored. In my following describing Newtonians is not to moan and grumble but it is to show the means and the manners they use to fight and when using such utter arrogance, despicable high and mighty autocracy with plain bullying tactics and megalomania. They have this attitude that only they are wise enough to think and the rest is mindless dehumanised animals walking on hind legs. If they fought fair and used intelligence it would not be that bad but to use dirty tactics when confronting me by just dismissing my views from a position of having authority is coward ness. By bullying me from holding a position of being able to ignore me and I can do nothing about it doesn't frighten me, it angers me!

Only when and after proving that a student has totally lost all ability to think for him or her self may a student be promoted into the ranks of their sublime intellectual group. This form of accepting someone into their league they gave the name as being a postgraduate. The sifting process they named examinations. You write on paper what they tell you and never question their opinion and after passing that examination, only then will you ever enter their sphere of intellectual brotherhood. If this was not true, then how could all the misconception I show in this book remain on paper and be taught in Universities for all these centuries? Are there so many misconceptions as I claim there are?

Does this sound far fetched? Then you better read on and I will remove your blindfold and show you what a world of deception the Academic Physicist force on us. Read the following and see how they, the high and the mighty, those that think they can replace God and those who think they can think on our behalf and think what to tell us what to think, read how much they are clowns and the jokers in society. Read how little are they, the Academic Physicists, able to understand concepts about Creation while they think they are able to replace God by using their superior intellect.

If you are a student in the science of physics, then ask your Educated Masters to please explain the following abnormalities you are about to read in this book and insist on a clear explanation about the inconsistencies they promote while tutoring physics as if the physics they present are the most flawless and accurate institution there has ever been. Ask those academics supporting Newton about the following flaws that no one, except me, ever mention. Get them to explain the inconsistencies they never talk about. Wise up and confront those charged with tutoring physics and see who should you believe. Then get informed instead of brainwashed.

One very simple example, which I mention now at this point but I do not elaborate on this matter any other place in the book since in this book I wish to limit space, used, is mentioning the gravitational constant. If any one wished to bring in an explanation by employing the gravitational constant also introduced in the Newtonian formula $F = G \frac{M_1 M_2}{r^2}$ then using this gravitational constant is one of the ultimate bogus ploys Academics use to confuse the public.

Newton first envisaged the idea that it is mass standing in relation to mass that is destroying the radius found between the two objects forming gravity as presented by the formula $F = \dfrac{r^2}{M_1 M_2}$ but subsequently the notion as well as the formula used changed to $F \, \alpha \, \dfrac{M_1 M_2}{r^2}$. To get Newton's miscalculation $F = G \dfrac{M_1 M_2}{r^2}$ to work with some dignifying crookedness' they devised a constant of sorts going by the title as the gravitational constant and is this constant holding the symbol **G** in $F = G \dfrac{M_1 M_2}{r^2}$

It is put in place as being the same as all the gravity but is apparently that gravity that fills the space between the Earth and the Moon. Now comes the Newtonian part... This same space filling ingredient called the gravitational constant and holds a measured value of 6.67 X 10^{-11} where it is using this value while it is playing its part in filling all the space we find between the Sun and the Earth as well as the Sun and Pluto and everywhere there is space in outer the gravitational constant is the space-filler to have in that space being filled.

If you think of space then we have such space filled with a gravitational constant at a value of 6.67 X 10^{-11}. This was the case in the days when it was accepted that ether was filling the space the gravitational constant filled and therefore ether might have had the value of 6.67 X 10^{-11}. Then after finding no evidence of ether, the ether that was not filling the gravitational constant was miraculously and by a stroke of Newtonian magic removed and replaced with...nothing...yes, nothing is now filling the space ether filled before they realised ether was not filling the space but the marvellous part is that nothing that

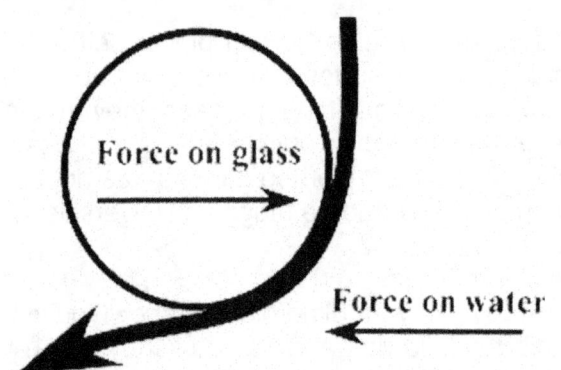

the replaced ether took from the ether that is not there the value given to the gravitational constant and now while space is filled with nothing it still holds the measured value of 6.67 X10^{-11}.

It is not my manner to speak ill of the brain dead or the dead by other means, but in the case of the Newtonian academics I am left with no option.

Their forces haunt me to death and it is their forces and ghosts and witchcraft I have to fight. The lifting of a body comes quite natural when a certain speed is exceeded. By exceeding $7(3\Pi^2)$ the body will start to lift no matter what the mass is. A 747 Boeing of multi tonnage lifts off spontaneously at the excess of that speed.

Newtonians are forever concerned with middle ages and with forces they can't explain but such forces and witches there are not, therefore they do not have to fear and can sleep well at night. In the sketch the circle portrays a glass and the arrow portrays running water. The Coanda effect is the water that does not drop straight down but follows the curvature of the glass.

The Coanda effect is gravity and my explaining this statement is part of many other books. The Coanda effect shows how liquid attach to the solid by $7(3\Pi^2)$ and the solid attach to the liquid by a relevance value of $7(\Pi\Pi^2)$. That is

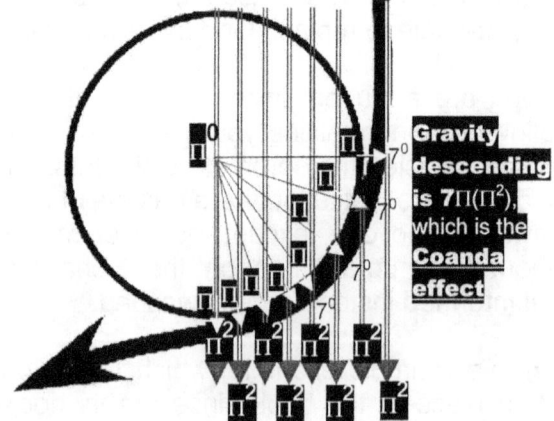

gravity.

Should anyone require more or better explaining, I would advise that person to purchase any of my books holding the title as an Open Letter. A flying object is under this gravity control of movement and it is this that has crafts fly and cars requiring down force by the aid of aerodynamic devices.

 Perceived natural direction of travel →

When viewing any object travelling, we have a perception of the vehicle heading straight ahead in a straight line whether it is a donkey cart or an aircraft, to our perception it is all the same.

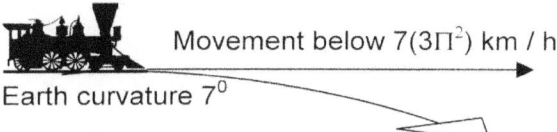
Movement below $7(3\Pi^2)$ km / h
Earth curvature 7^0

Our perception is letting us down because it is incorrect. The moving object is following the curvature of the Earth and that is turning by 7^0 as we travel.

Movement below $7(3\Pi^2)$ km / h
Natural gravitational displacement or natural direction of travel

In truth we are dropping by 7^0 as we follow the curvature of the Earth. This will be valid as long as we go at less than $7(3\Pi^2)$ = 207 km / h. It is gravity (Π^2) multiplied by time (3) multiplied by the space material holds (7)

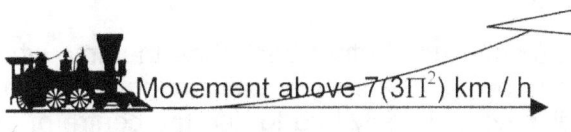
Movement above $7(3\Pi^2)$ km / h
Perceived natural direction of travel

When exceeding gravitational lines in singularity running at $7(3\Pi^2)$ we see the vehicle lifting into the air as we think the vehicle is getting airborne. Again it is our perception letting us down because it is incorrect. The moving object is instead following the curvature of the Earth and that is turning by 7^0 as we travel, now it is breaking free or releasing from gravity by following a diversion of 7^0.

 Movement above $7(3\Pi^2)$ km / h
Natural gravitational displacement or natural direction of travel

The object is rejecting the gravity the Earth enforces by applying individual gravity in the excess of $7(3\Pi^2)$ and in that it is relieved from the gravity the Earth applies. It has nothing to do with air lifting a car. By the way, the sound barrier is just adding one phenomenon further, being the Roche limit in double at ($\Pi^2/2$) bringing the total speed required to match the sound barrier at the gravity velocity of $7(3\Pi^2)(\Pi^2/2)$ = 1022.795 km / h.

As soon as the idea of gravity connects with the idea vested in thought of mass by sharing the split concept in a unifying bout, then I say that gravity there is not. I say, what is defined as gravity is not gravity and then what is acting as gravity is not named correctly.

Gravity is defined as a force that is present in mass pulling mass and it is that entire idea that there is no evidence of. When I refer to gravity everyone grabs on a cultural notion of a concept they formed and in that concept they link the smallest part of the concept to become and represent the overall gigantic principle and by knowing one line, everyone has the opinion that anyone then is the absolute master on the idea of gravity. When I freeze any substance, the substance contracts to a liquid and with more cooling it contracts to a frozen state of ice. The gas expanded more than what the solid did because the gas is hotter than the solid is. When we form the opinion that the outer space expanded to the limits, the idea springs to mind that outer space is freezing cold. When I say the Sun freezes hydrogen to a liquid because my eyes see the liquid squirting from the Sun, I am dangerously mentally impaired, since the Sun is blistering hot. Then through this culture my effort to say gravity is motion and motion is the cooling of an overheating and thus expanding Universe goes wasted. Everyone has the opinion that where gravity is the strongest such as the case is on the Sun or at the centre of the Earth, such a place is extremely hot and where gravity is least that place is unbearably cold.

Students tell your Professors to stop deceiving and stop trying to control your minds with their fraud. Those Academics tutoring you are telling facts about gravity that have never been proven. That is mind control.

They wish for you to accept facts on gravity that they hold as the truth. Should you question that mass produces gravity, they will expel you from University by letting you fail your examinations and it was never proven. Academics do put mind control to work on unsuspecting students by forcing students never to question the legality of statements they offer as being sound and correct.

For the first time ever and this statement takes time back as time runs further back than the time when and since the time Newton introduced gravity there now is a logical and simple explanation as to what gravity is and why there is gravity. Before I achieved that discovery, I firstly had to find the centre of the Universe because it is there that I could locate gravity. I can now show how gravity forms because I have detected the centre of the Universe. Academics guarding physics will never allow an outsider to enter their domain without the intruder paying a heavy price and in this matter I was the intruder. I argue that if it is the correct practise to use $F = G \frac{M_1 M_2}{r^2}$ to calculate gravity, then the radius holding the gravitational constant must lead one to the centre of the Universe. If the Sun for instance has mass that is apart from the Earth and the Earth also has mass and there is a gravitational constant between the Sun's mass and the Earth's mass, we have the radius in that location. It then must be the gravitational constant that fills the space that the radius holds. Through my venture I discovered one person that knows what gravity is!

Ask yourself the following: If gravity pulls towards a centre and gravity holds the Universe attached, the question arising from that simplistic answer is then … where is the centre of the universe?

Science sees to it that Kepler stays the least appreciated Cosmologist where as in truth Kepler proved gravity, proved singularity, proved space-time, proved the Big Bang, proved every dynamic most of the wise persons afterwards thought about.

Yet, no one gave Kepler any recognition up to now because science denies Kepler his limelight.

Part 7

SUBLIMATION; The Newtonian Mythology

The purpose of this book is not to echo the **Newtonian** version of Greek Mythology.

The Greek Mythology had aimed to bring, what they considered religion to be, in line with their perspective on what they considered science, cosmology and astronomy had to be. In Newton, a change a change came about in the contents of the mixture, but not the ingredients as such. The Anglo American Mythology now preaches a religion called atheism, which, as were in the case of the Greeks, based on what they perceive science and scientific facts to be. If you find a desire to run along with these myths of star travel, speeding through the Universe at the speed of light, encountering alien societies and indulge yourself in such modern mythology, please do not read this book. In this book, the magic spells that Newton named gravity, and which Einstein took to a single dimension fantasy, is discarded in dismay once the Anglo American Mythology is replaced by factual truth, the creation according to the Bible becomes a detailed analysis of the actual creation. When Newton's lies are replaced by the real functioning of the universe, the author of the Bible's firs book has such a precise recollection of events, that it put all scientific facts, being broadcasted at present to pitiful shame. That includes the ideas of modern atheistic Anglo American science cult. This book does not repeat all the traditional nonsense, but explain in detail, how the cosmic year structure works, in such detail that children can understand and accept it.

Civilization throughout the ages always used Cosmic Science, but especially, by the Roman Catholic Church, to prove the importance of the earth as the centre of the universe. That brought about that the sun shone on the earth, which was the centre of the universe. Since the earth was made for man and was the centre of the universe, God made everything with mankind in mind. Whereas the Roman Catholic Church was the only representation of God on earth saw they represent God and all His Powers on earth. It meant that the Pope was God on earth (being the head of the church of God on earth) and the Catholic hierarchy was the most elite and privileged on earth. That was precisely how the church considered them and how they could conduct their teachings.

Right at the top was the Pope who ruled the earth and as the earth was the centre of the universe, the Pope for that matter, ruled the whole universe. God and the saints ruled the heavens and the Pope with the Cardinals ruled the creation of God. This brought about the sublime picture that suited every person, which was considered anybody. He whom the Pope blessed was blessed and he, whom the Pope damned, was doomed. Everybody op importance could buy the Pope's blessing and could die in reinsurance that his life was ultimately saved.

By the middle of the 15th century, some unimportant persons with no real social standing came along and disagreed with this ultimate and universal accepted hierarchy. They stated that the earth was not the centre of the universe, which meant that the Pope and his cardinals were not in control of the universe. This new perspective on the Universe and the chain of command was very unacceptable to anybody of social standing. This (then considered) blasphemy was to be killed in its very infancy, even before birth. However, as with everything else, the truth eventually prevailed and certain intellectuals took notice of statements by Galileo and Kepler. The reaction that followed can be regarded as one of the most important historical events in the route that man's civilization took in forming modern man.

In this book, we put the work of the two giants, Galileo and Kepler, under the magnifying glass. In contras to popular belief, there is an astonishing difference between the work of Galileo and Kepler and modern science. Modern science is based on the findings of the father of modern science, which I consider Anglo-American mythology. The father of Anglo-American mythology is none other than Isaac Newton in person. Galileo and Kepler had the truth of the Universe unlocked when Isaac Newton came along and raped their findings.

First, let us consider Galileo's work.

When a pendulum swings, the pendulum's rhythm remains while the stroke tarnish. All big clocks are based on this working principle and can therefore keep time mechanically. Time has been measured by means of this method for about half a millennium. When two structures of different mass is dropped at an equal distance and time, the two structures will hit the earth at precisely the same time, as long as the wind resistance is equal to the two bodies. This experiment was the first that was done on the surface of them moon on July 1969 and billions of T.V. spectators bared witness to the outcome of this experiment. A hammer and a feather were dropped and they hit the surface of the moon on precisely the same instant.

Kepler, on the other hand proved that the earth and all other planets rotate around the sun in an elliptic orbit. All three these findings never mentioned any force called gravity.

At this stage, science used precisely the correct argument. The findings were noted and it could afterwards be checked for the same results repeatedly.

Afterwards an English genius by the name of Isaac Newton noticed an apple falling from a tree. This prompted him to calculate a force that existed between the apple and the earth, in which the matter of the apple and the matter of the earth was drawn by a force he named gravity. He never took into consideration any of the findings of the previous two giants, although he praised them for their work. Newton went and calculated the existence of a force that existed between the above-mentioned bodies. At this very point, science took a wrong direction. The force that Newton calculated is a secondary function to the primary condition that holds matter in place throughout the universe.

By using the findings of an even bigger genius, Tycho Brahe, Johannes Kepler proved that the radius of the planet, taken from the sun to the planet and is measured in astronomical units, is equal to the square of the rotation period. Kepler said that there exists a relation between two bodies where $T^2 = R^3$. He never mentioned a force. Galileo's findings proved the same as Kepler's, that there is a ratio between two bodies, not a force.

The second statement of Galileo can compared to driftwood on water. If two pieces of wood which is different in size and weight floats on water, and both pieces of wood is subjected to the same force value in the stream, both pieces of wood float at equal force, no matter what the difference in size and weight is that comes into play. This would be caused by the difference in the drag resistance on the different surface area of the two wooden bodies. However, the difference in weight that comes into play, at this point is only due to the drag that the water experiences, not the actual weight.

If this were compared to, the findings of Galileo one would find that this is the precise method how matter moves towards the earth. Galileo made no mention of a force between the two bodies. Then Newton came along and published his mathematical findings, which is totally out of line with Galileo's findings and ever since then no person ever gave the actual findings of Galileo and Kepler a second thought. There can be no force such as gravity, electromagnetism, strong and weak forces, or nuclear energy. These so called forces are part of precisely the same value that is in relation and exist between space and time.

You, the reader may ask yourself why is there such importance in these findings as to know the correct way that gravity actually works? All calculations have already been made; mankind already possessed the knowledge, expertise and willpower to visit out of this world's Tara novas and even to colonize them.

All knowledge about physics and astrophysics has been studied, formulated and tested! I will reply to this question by asking two other questions.

A certain man drives his car down a lonely road. After a while, the car comes to a standstill. He knows that a car needs fuel to run. He takes a 25-liter can and walks 10 km to a filling station to buy fuel for the car. He carries the 25-liter fuel 10 km back to the car and puts it in. After trying for 10 minutes, the car still would not start. He takes the 25-liter can and walks 15 km to another fuel station to buy some more fuel. After walking all the way back and filling the tank with the petrol, the car still refuses to start. He then takes the 25-liter can and walks 25 km to yet another fuel station for fuel. After returning, he filled the tank yet again. Do you, as the reader think for one minute the car would start?

The American scientists spends 1 000 million dollars to pressurize four hydrogen atoms into the same space of one helium atom. The experiment seems unsuccessful because the helium unfolds back to the original four hydrogen atoms after four seconds. After that they use three 000 million dollars to pressurize four hydrogen atoms into the space that one helium atom occupies. After seven seconds the four hydrogen atoms depressurises back to its original state. Now the American nuclear scientists use 10 000 million dollars to pressurize the four hydrogen atoms into the space of the one helium atom where the experiment lasts for 12 seconds. After 12 seconds the hydrogen atoms moves back to their original condition. Question 2 now is this: "Can you see any connection between the two examples I have put to you? "No pressure in the world can fuse four hydrogen atoms to one helium atom, even if Einstein said so! Those super brains and academics are blind with their own mathematical genius of mathematics and physics that they fail to see the most basic and elementary principals in science. What is tragic is that it does with the taxpayer's hard earned money!

Another rudimentary example is the so-called "falling star". The conventional theory that is propagated is that the dust speck burns to ashes because of the friction the particle has with the air it collides with in the atmosphere. This is the biggest mindless rubbish one can imagine. Just because the person that tells me this have six doctoral degrees, does not make it the gospel truth. Far from it... What actually happens is that when the grain of dust enters the earth's atmosphere, the time aspect changes and forces the dust particle into a different space occupation which then changes the heat value of the grain of dust. The dust grain is forced into such a smaller volume of space-time occupation that the heat it generates just burns it to ashes. Therefore the whole structure glows itself into nothing.

These "SUPER- EDUCATED- MASTER- OF- FACT" giants might use the most breathtaking mathematical formulae that can humanly be dreamt up, but if the principle, on which they have based their calculations on, is wrong, the whole exercise is fruitless.

In my book, I discard the "conventional" standpoints by following the unconventional principals, based, on the work of Galileo and Kepler. The basis for my theory has never before been propagated except for me, and therefore all arguments will be new and fresh.

I am no writer and even less a scientist. Furthermore, I do not pretend to be regarded as any of the above mentioned. I simply came to certain conclusions. When I shared my conclusions with other people, I had the very same reaction repeatedly. The reaction followed spontaneously without exception. Those who were prepared to listen to me knew even less about cosmology than myself and never understood a word I said. Those who know more than I do ignored me immediately when I said Newton and Einstein are wrong in their views about gravity.

Partly the writing of this book is to prove that there is little difference to science and the Bible, even if scientists and theologians try to make it their lifetime task to prove the other opposing side wrong. However, all their arguments are based on their individual agnostic belief in their self-righteousness and have nothing to do with the truth.

Free thought has always been a fact that all that is in influential positions proclaim to strive to but the minute the free thought differs from their concepts, it is the very first thing they crush as hard as they might. In this book, I strive to accomplish the very essence of free thought and try to lead the reader away from the brainwashing that all intellectuals try to force onto the public. Every person that is in any position is busy with their own sublimation that it renders them blind to the truth that is out there.

In the past, I have been severely criticized by academics for the view I hold and to my judgment rather unduly. The largest complaint by academics comes because of my critic about Newton's findings. There is a vast difference to the cosmos and the earth surrounding us filled with life, the part we humans take fore granted, but the only place we will find life will be on this planet. Compare earth to the moon, and the moon represent the real cosmos. It is dead, except for matter that grows with cosmic time. To understand the cosmos the cosmos and life has to stand apart, the two on they're own and only then does the cosmos appear as it is. A rocket launched is not part of the cosmos, but is part of life's extending energy. Life is the ability to occupy and manipulate space-time occupied and unoccupied. In the manipulation, there is a very strict limitation to what man can and cannot accomplish. It is as if man has become drunk with his supposedly unlimited ability. This will lead to tragedies man still has to meet. There is no reason sending brave men off to their death, if only we can find a way to recognize the dangers. NOTHING IN THE COSMOS CAN DIE, BECAUSE DEATH IS A HUMAN MISCONCEPTION. ALL ARE PART OF DIMENSIONS WHERE WE ARE INCLUDED OR EXCLUDED. THIS CONCEPT ABOVE ALL SEEMS TO EVOKE THE MOST REJECTION AMONG THOSE I HAVE COME ACROSS.

The last four five wrote full time and with which I went on the Internet. The persons I wish to draw to my book do not seek information on the Internet and those who do look for information, cannot understand my work. The commercial route I tried to follow; however no publisher would touch it because it is not recognized science. That reason forced me to seek an academic route, in order to find recognition for my work.

To the correctness in my work, I have no doubt; however there may be parts I still have not clarified sufficiently. I am more than willing to meet such challenges because I know the proof is there, I was just not yet able to recognize the question.

Being human we tend to form concepts through culture and such concepts we then translate to the cosmos in an attempt to fid meaning moreover about our worth than the facts relating to the cosmos. We find ourselves physical apart and as much linking and with that we then wish to link and find meaning. What links and what divide is the stumbling block because we place man in the cosmos through which we then wish to produce a link. Such a link does not exist but for the tiny speck of cosmic dust which we call either home or earth. The main issue to divide what should be divided and then accumulate what should connect and doing that is the purpose of this book.

Quoted directly from the Oxford dictionary of Astronomy the following:

The definition of space-time is as follows:
Space-time is a four dimensional position of the universe where the position of an object is specified by three coordinates in space and one position in time. According to the theory of special relativity there is no absolute time, which can be measured independently of the observer, so events that are simultaneous as seen from one observer occur at different times when seen from a different place. Time must therefore be measured in a relative manner as are positions in three-dimensional Euclidean space, and this is achieved through the concept of space-time. The trajectory of an object in space-time is called world line. General relativity relates to curvature of space-time to the positions and motions of particles of matter.

The definition of singularity is as follows:
Singularity: a mathematical point at which certain physical quantities reach infinite values for example, according to the general relativity the curvature of space-time becomes infinite in a black hole. In the big bang theory the universe was born from singularity in which the density and temperature of matter were infinite.

While it probably is the greatest mind to walk the earth that produced the spectacular in the above, a much more simple mind as the one I have noticed much more simple aspects of nature that only one with a simple mind as I have could recognise because my mind does not have the capacity for the greatness of the great minds.

If the universe did start from one single point and time matter and space flowed from that point, then that point must have a relative connecting base because such a point holding singularity must be eternal as space matter and time link eternal. There therefore must be one point linking the entire universe when regarding the fact of singularity. Then according to the theory off relativity there has to be one exact point holding time in a relevance notwithstanding the fact that time depart from that position and relate differently to all space-time away from such a point.

Every person I have discussed facts about creation recollects images in the trend depicted in a presentation as one may find to the above. That would be the most unlikely way Creation came in place. The recalling of pictures representing images about creation must have form, but to mathematics it had no form. From this thought the very opposite arise where Creation came from nothing but such an idea is mathematically simply not possible.

The thought of nothing is just what it is, a thought of nothing and although it is in the human mind common nature to present nothing as a value in the recalling of something, nothing is a presentation of the figment in the human mind. There can be no number such as nothing and that was (possibly) Newton's biggest error. Nothing represent non-existing and that is just what nothing is, it is non-existing.

In order to prove my point I wish to ask the reader to define the shortest line there can theoretically be. If he should answer anything but that the shortest line will be at a point where the beginning and is the very same spot he will be wrong. The shortest line that can ever be anywhere must have a start and finish holding the exact same spot. The line will be humanly impossible to create but we humans are capable of very little.

When the line has a beginning and an end at the very same spot and it wishes to extend the position as to further the possibility it has, which direction should it favour. Humans in the west would naturally think of extending from left to right while in the east humans may want to go from right to left. Some persons will tend to go up or down, but all of the options are about human preference and not mathematical conclusions. Extending the line in any one direction will favour one direction without a conclusion about not extending in other directions. Such a conclusion has no sound mathematical foundation. The only option about extending will be in all directions equally in order to give a meaningful non-bias flow of mathematical equilibrium

The shortest line in the realm of possibilities must have a start and finish holding one spot and such a line will also be a dot or a circle. Not favouring one direction puts all directions at equilibrium meaning that any form what ever may be can develop from such a spot with the end and the start being the same. This reasoning prompted me to look for singularity in such a spot because if the prime spot from which all came was a spot, then the spot must hold the shortest line but more prominent it will hold the smallest form including the smallest circle.

One possibility that the shortest spot can never have is having a starting point on the zero mark. If the mark of zero holds the start it must also hold the end because the end and the beginning has the same position. If the position of zero then is the beginning, the end will also be zero leaving the line without an end as well as without a beginning.

The conclusion from this is that no line can start at zero because that will be a mathematical impossibility. A line or spot starting at zero would therefore be shorter than the shortest line possible. A line growing or extending from zero can never leave zero because of the influence of being zero disqualifies any possibility of growth. If the line then had to grow in all directions at the same pace the line must therefore be a circle. The value of the circle is Π, and that is where creation started.

That gave me the clue where to start looking for singularity. One would find singularity in the value Π and the value Π will be in all things rotating in a circle. To start my explanation about my cosmic theory I wish to firstly bring some nostalgic and the relevancy will become apparent later on. Such is the importance however that I wish to place this at the very start of the prologue.

When we were boys we played with a top we called the spinning top. I cannot imagine that there is one boy in the western world that did not hold such a devise in his hand. Tying a string securely around the tapered cone started the operation and then with a jerking or pulling throw the devise is launched in a projectile manner and the big knack to success was getting the nail end firmly on the ground and by the realizing jerk the top was rotating. The champion was always the one boy that could throw his top to spin the fastest and that would create a humming sound. The louder the sound produced the bigger champion

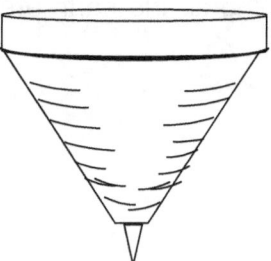

When a back braking effort produced a throw of enormity the spinning top would not only produce sound varying in pitch but also create a spin that would seem to have some instability. There are very many limitations about the spin, parameters that determine the slowest and the highest sin rate and spinning is within the parameters of such settings. The question arising is why such parameters are there in the first place?

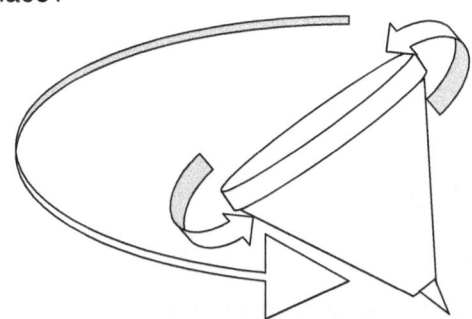 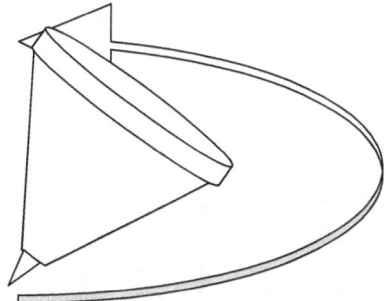

An enormous effort will have the top going oblong while spinning violently and as the pace reduced the top will stabilize by coming to an upright position. In the upright position it wall then spin for the remainder of the period where it will in the end start tilting to the side and in a last effort throw a few wild oblong turns and fall over.

Boys playing games will never realize scientific breakthrough explaining and grown ups do not play with toys. In this little toy played everywhere everyday by almost every one is the answer most brilliant of human Brainpower seek answers about all the cosmic riddles no one seem to understand. In the spin as such one may find two vital boundaries in the motion and the boundaries are marked by a wobble coming about as if the top is fighting some other influence. Spinning too fast pulls the centre off centre and so does spinning too slow. It is the same influence coming about at both ends of the limitation in the spin. There are influences at work, but force…no; it cannot be forces setting such boundaries. From that I started per cuing what sets such limitations because that limitation must be universal as all matter is spinning in one way or the other.

I MAINTAINED DURING ALL MY CORRESPONDENCE TO SO MANY I HAVE CONTACTED, I STILL MAINTAIN AND I PROVE MY VIEW POINT THAT:

1) There is no gravity and therefore GRAVITY DOES NOT EXIST.

2) With no gravity it stands to reason that I also maintain that NEWTON AS WELL AS EINSTEIN IS ALL TOGETHER WRONG!

3) With no gravity NEWTONIAN VIEWS ON THE WAY CREATION CAME ABOUT IS ALTOGETHER INCORRECT.

4) THE BIBLE IS ALL TOGETHER CORRECT ABOUT CREATION.

In the past these remarks made me the clown in the courtyard and no friends came to my aid because no friends were in support of my statements. A description that would be closer to is that no friend wanted to admit any friendship because such admitting may also reflect on his or her sanity.

When looking at the cosmos from whichever angle indicates the fact that the cosmos is moving. It is forever spinning and it is going to as much as it is coming from. Everything is on the move and always encircling something of greater importance. A top can spin but the parameters of its spin are limiting the motion it can apply. By not spinning the top is still spinning as the earth are doing the spinning on its behalf.

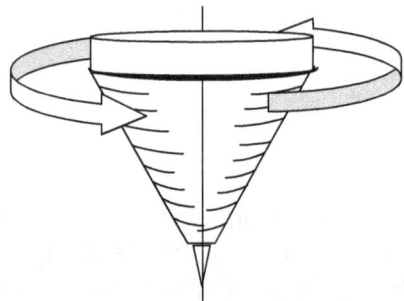

When spinning too fast the top fights something because the alignment keeping it upright starts to tarnish. The same apply when spinning too slowly but that makes sense. It is the fact that the same affect comes about when spinning too slow that triggers the questions.

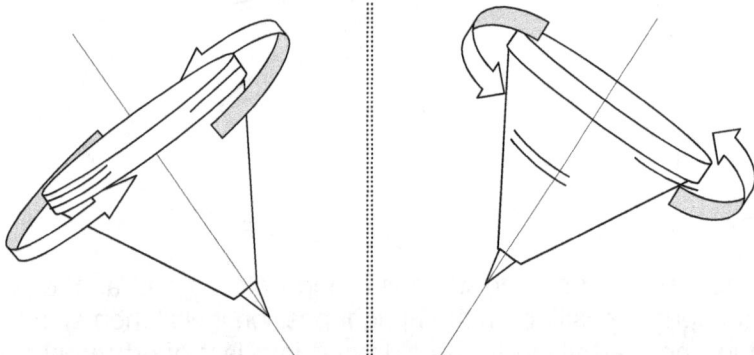

The spinning top is all the evidence any one needs to come to such a conclusion. By saying that I first have to admit about (no not my mental stability), but that I have no academic background and I do not enjoy any link to any university.

I know probably as much as any graduate about cosmology but lack certificates to prove my knowledge. I am not part of established science. In my developing of knowledge accumulation I came to some conclusions about cosmology that are unique and divert somewhat to drastic form the accepted norm. Most of the work I see the same way as the norm does but in a reverse. Allow me a short explanation

When looking at a red flower we say the flower is red. Nothing can be further from the truth. The flower is every colour in the spectrum, except the colour we attach to it. It is screaming with all might to its disposal that that specific colour it cannot accept. Yet, we maintain that that colour is the colour we associate with the object, ignoring the objects rejection of that colour. Only when looking at the cosmos from this stance, can the cosmos make sense? By recognizing a disassociation in spite of our cultural recognizing the association, can we understand the cosmos?

We maintain the sun is burning, while the fact of the matter is the sun is freezing. From our perspective on the outside we see the sun burning as we see the red flower. What we see is not what is the truth. Only by applying the correct view to the cosmos can the four principles I introduce, make any sense and find any proof… and I do prove them. Only by telling the complete story as I do in the complete six parts of "*Matter's Time in Space – The Thesis*", can the explanation surface to a point of understanding. One cannot draw any conclusion from the outside; one has to be inside the star to see what is going on.

To get such proof I had to do extensive research on cosmology. The proof lies in unrecognised and misunderstood laws and principles science know. These laws fall outside the parameters of applied physics.

I defined gravity; I defined energy, but before that I had to prove the existence of time and time's control over the universe, time's role in the universe and what time is. This was up till now not yet been achieved. I had to prove what space is, that time and space is sides of the same coin, with matter forming the separation. The main conclusion that brought about such conclusions was my different view of science. It's not the explanations science at first that made me question the validity of Newton, but the things Newton cannot explain but is factors in the cosmos nevertheless.

As a school going youngster I was fascinated by astronomy and in particular the cosmology aspect. In a long and strenuous process of self-education I was completely stunned by the behaviour pattern that the comet had in its relation as it orbits the sun. Please forgive my boyish way of presenting the following but it is important that I bring it across as I saw it as a boy and as a matter of fact still see it today as a middle-aged adult.

We may start by deter-mining the influence of gravity on planets as we find them in the solar system. **First, let us concern ourselves with a comet**. It is common knowledge how the comet relates to the sun's gravity. **Firstly, picture the comet at its farthest.**

Point, away from the sun. The **gravity** of the **sun pulls** the **comet straight towards the sun**, this we all know. Gravity always pulls an **object directly towards** the **centre of a cosmic body**: that too is common knowledge. Therefore, the comet is drawn directly towards the centre of the sun and throughout its journey the comet is picking up momentum directly related to the gravity that is cantered in the middle of the sun, (**gravity is always cantered in the middle of a cosmic body**). As the comet is increasing its speed, the comet comes closer to the sun and therefore the sun's gravity pull is simultaneously increasing as the distance between the two cosmic bodies is reducing. Each instance the comet is drawn towards the sun, the gravity that the sun applies to the comet becomes larger progressively. When the comet is at its <u>**closest point to the sun**</u>, <u>**something odd happens which cannot be explained by Newton's gravity at all! Remember gravity should now be at its strongest point because of the proximity of the two objects.**</u>

1. The comet remains at an even distance encircling the sun.
2. No longer does the gravity of the sun pull the comet towards the centre of the sun.
3. At this very point the gravity that the sun applies on the comet does not pull the Comet towards the centre of the sun any longer, in fact, it seems as if the effect of the gravity has been neutralized.
4. The comet stays at an even space from the sun as it goes around to complete a half circle's orbit around the sun. It only completes a part of its rotation around the sun.
5. After this, an even more peculiar event takes place. <u>**The sun, at the point where gravity should be at its most dominant, suddenly loses its complete grip on the comet.**</u>
6. The comet brakes free from the sun's pull of gravity and speeds off towards its destiny into the vastness of the cosmic space, undeterred by the gravity of the sun.

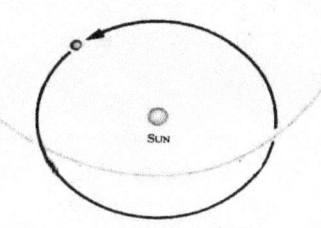

Then after a pre-determinate and pre-calculated time the sun starts applying its gravity on the comet once more. At a point where the comet is at its farthest point, the gravity of the sun becomes strong enough to bring about a complete turn around to the comet's direction of travel. <u>**However, the gravity between the sun and the comet is at this point, at its weakest point of influence.**</u>

So, when the sun's gravity is at its strongest, the comet manages to brake loose and neutralize the sun's gravity pull in order to avoid its fatal collision with the sun and when the sun's gravity is at its weakest, the comet cannot escape the pull of gravity. There is definitely something very wrong, either with the comet's behaviour or the laws made up by Newton.

However, **this is not all**. When we regard the planets as they stand related to the sun, the effect is the same, but not as obvious. All the planets follow an oval orbit around the sun and therefore the same factors concerning gravity apply to the letter as it does in the case of the comet. Let us investigate the one planet we relate the best to, which of course is the earth.

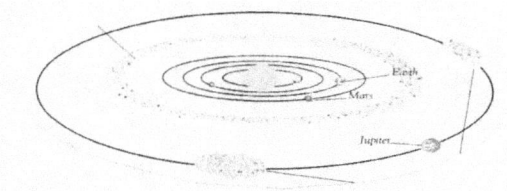

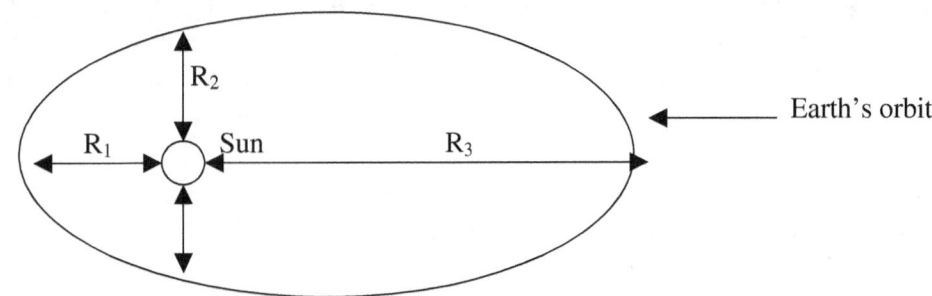

This illustration does exaggerate the radius of the earth's orbit around the sun, but since it has taken place 4 500 000 000 times, it has no real effect on the validity of the next statement.

At one point (R_1) the distance between the sun and the earth **is less than** at another point we call R_3. Let us put a value of R_1 = one and R_3 = three. This means that each year, for the past 4 500 000 000 years the effect of the common gravity between the earth and the sun has a greater effect than at another point six months later. **At one point the earth should be drawn or pulled closer to the sun** and **after another six months** interval **the earth should stand less effected by the sun's gravity**, therefore it should move away from the sun. Each cycle of twelve months would have one point where the gravity pulls the earth closer and exactly the opposite must apply six months later when the gravity is at its least. So, for the past 4 500 000 000 years the earth has been re-establishing its seasonal swing towards the sun and away from the sun, which means by now the earth has to collide with the sun in midsummer or escape from the sun in midwinter, as it may then drift away into the unknown.

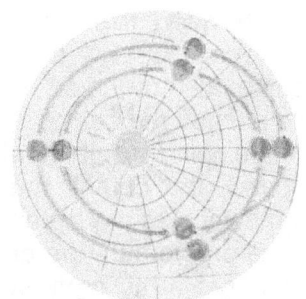

For the more mathematical minded person the argument is as follows. May I remind you, THAT NEWTON'S OWN LAWS ARE IMPLIED, and again the planets disobey these laws completely!

$$F = \frac{M_1 M}{r^2} G$$

THIS IS THE SUGGESTED FORMULA THAT PLANETS APPLY WHICH ENABLES THEM TO MAINTAIN THE ORBITS AROUND THE SUN, WHICH THEY DO.
We know that $F_1 \neq F_2 \neq F_3 \neq F_4 \neq F_1$
because that is what seasons are all about.

Even if $F_1 \neq F_2 \neq F_3 \neq F_4$, we know that $P_1 = P_2 = P_3 = P_4$.

Because r is at different values F could not be to the same value. Therefore, the value of F has to be unrelated to force its value on to P. Nevertheless, Kepler has proven that $P^2 = a^3$, although $a_1 \neq a_2 \neq a_3 \neq a_4$. If $a_1 = a_2 = a_3 = a_4$, we would not have had season and climate changes on earth. That means that to proclaim $F = \dfrac{M_1 M}{r^2} G$ is nonsense. The truth of the matter is that Newton actually proclaimed that in an ellipse, which has an uneven circle (Kepler's findings) the value of $F_1 = F_2 = F_3 = F_4 = F_1$, but because an ellipse has no constant radius, it actually means that $r_1 \neq r_2 \neq r_3 \neq r_4$ and thereby anybody can see that Newton's calculations are wrong. $F_1 \neq F_2 \neq F_3 \neq F_4$.

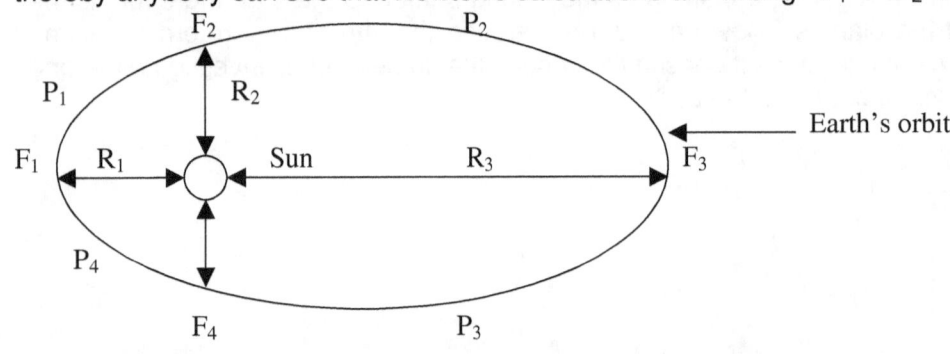

Using such logic makes science appear foolish. There is just no rational in the time verses events that cane explain facts without. Since the time of Newton, the arguments tarnished from being brilliant to clever to fair too poor and a hundred years ago to the point of being stupid. That is what Kepler's formula is all about? That is what Kepler indicated with his formula $a^3 = T^2 k$. The space of an object (a^3) is equal to the time (T^2), which it is in, in every given instant (k). If the space becomes smaller, the time duration becomes longer every instant of time's progress.

Singularity is a mathematical reality. Einstein may be the first to name it and Galileo (unwittingly) may have been the first to define it as Kepler was the first to formulate singularity, but in mathematical terms singularity is the most basic principle.

At this point I wish to establish a fact that seems lost in all other grandeurs of cosmology. A straight line cannot begin at zero or nil it can only start at infinity/ Such a statement will hardly seem appropriate but the relevancy of this fact has no limits.

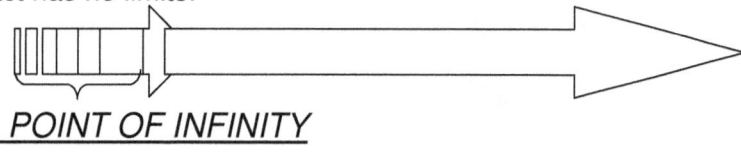

POINT OF INFINITY

If the line started at zero there was no line to start because zero multiplied by whatever results in zero as the answer. That must also be the cosmic starting point. Einstein introduced such a point and named that point singularity.

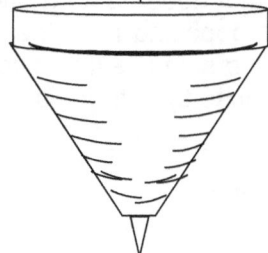

This brings us back to the spinning top I presented at the beginning.

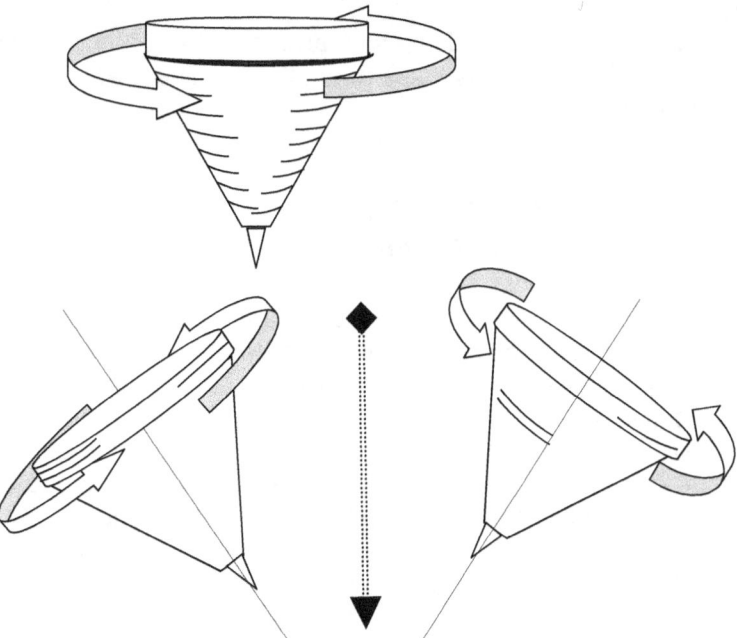

I have asked as many persons as I do not care to remember why the top sinning will remain spinning around one point while turning. The answer I receive from the most educated to the schoolboy is always about momentum. That is a very simple answer and to say the least a little too simplistic by further analysis. Why would the spinning top go of centre when spinning higher than a specific velocity and lowering the velocity it would stabilize and run square to the earth only after that it will go oblong and then fall. I could go on about different positions bringing across different momentum of thrust but I do not wish to insult your intelligence because I am aware that you are familiar with all the law. When the top is spinning it is spinning about its own axis and when it is not spinning it still remains spinning about the earth's axis therefore when it is spinning it is also spinning about the earth's axis. Therefore the limitations applying can only result as an influence coming from the earth's axis. The second question now comes screaming across and that is in what manner could the earths axis ever affect a spinning top since the spin and he spinning top is a gross mismatch to what ever standard the earth may introduce. It is clear that spinning objects do influence each other in contrast to Newtonian opinion.

Every round object has a point establishing a very centre, a middle dividing one side from the other. That division determines the space from one side away from the other side. At one point there must be a point that does not fall on either side of the divide. Such a point will still be a circle, because from that side the circle divides into two sectors.

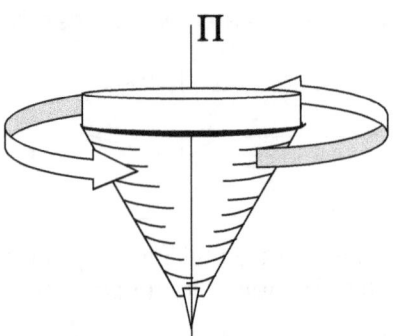

In every spinning object there is a point of infinity, a point that does not turn because it holds the dividing spin. From that point running in all directions the spin is opposing the other side. All spinning activity starts at that point diverting outwards and from that point the spin is either clockwise or anti clockwise in all directions. As I pointed out no line can start at zero because then there is no line and no rotating point can start at zero because then there is no rotation.

Calculating a square involves two aspects that we think of as sides.

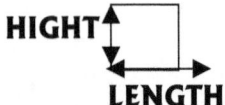

The square consists of two lines one holds the one dimension and the other one holds the next dimension. The important factor is establishing consensus about the two lines. As already agreed no line can start at zero because then there is no line.

By reducing the one line the other line can never reach zero because then there was no such a line to begin with. That makes a straight line also inevitably always a potential square and that makes the straight line half the value of the square being 180°. At a later point I shall continue with this argument, but for the mean while I wish to come back to the circle. This same principal apply to the cube and that means everything there is and ever will be is either a square being part of a cube or a circle. With the straight line forming half the value of a square 360° / 2 = 180° in as much as being one line and reserving one line in infinity to eternity. The straight line is just half the value of a square. In that manner the triangle is also half a square and therefore holds the same dimensional value as the straight line being also 180°

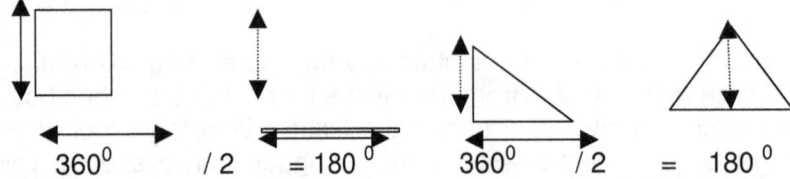

The circle is a square holding a round shape, as the straight line is a square holding one side to infinity. Calculating a circle involves two aspects where the one is either the radius or the diameter that is double the radius. The other is the factor Π

$\Pi \times D^2 / 4$ = circle and $\Pi \times r^2$ = circle

The point of singularity cannot be in space at large because space is not there and secondly what ever is there spin to slowly to have a connection with singularity directly.

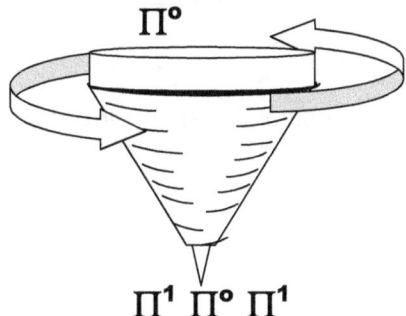

With everything in a cube or a circle or a potential of the two, brings about the implication of eternity in a form of singularity or the point of creation. Removing the radius of a circle does not remove the circle, because the circle is there, securing the ring. If the line (or imaginary line if you wish) holding the value of Π^0 = 1there has to be a point where the circle is no longer in infinity but claims existing outside the imaginary. At that point the radius may be lightly more than infinity, but to all calculating purposes it still remain as infinity.

The spin was going on for eternity because the spin does not apply, it has a value of zero and zero is another expression for eternity.

Having edges where Π^0 duplicate to present the edges singularity lost the value of Π^0 to the value of Π^1 with the same value singularity had being Π^1 to the one side and Π^1 to the other side, Π^0 must be the point splitting singularity into two parts of eternity, the eternal value of the first dimension outside eternity. It was the square of Π^1 being Π^{1+1}. That was the first dimension outside singularity Π^0 where singularity has a value of Π^1 in the form of $\Pi^{1+1=2}$. The first claim to space had a value of Π^2. This applied to both sides of the claim to space outside singularity, and the double proton became the dominant factor on matter.

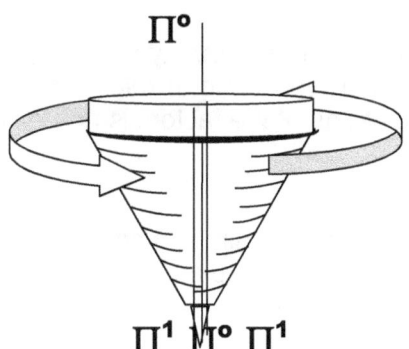

By receiving space, singularity received a value outside eternity as Π^0 received edges. Granted the fact that the edges were so small there still was no r to present a circle.

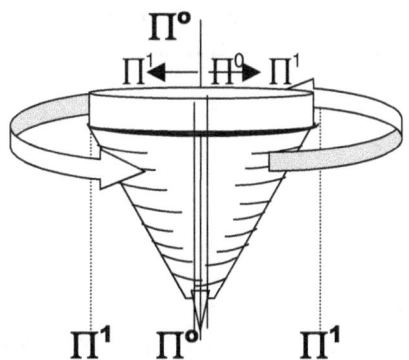

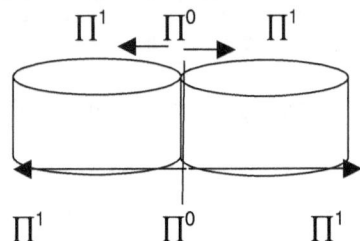

Taken from the point of rotation the two sides are in opposition to each other in every aspect that they may contain and with all that they hold.

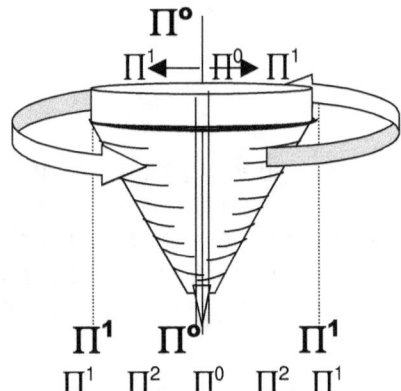

With Π^0 little more than a figment of the imagination there is actually to values of Π^1 facing each other in a relation combining Π^1 to hold the value of $\Pi^{1+1=2} = \Pi^2$ and with two sides being the very same but opposing each other there will therefore also be Π^2 to every side that holds Π^1.

At last I can come to the one part that I disagree with Newtonians, and what I regard as Newton's second biggest infamous or famous blunder. Science, made one enormous blunder, from this stance. They took the radius of a wheel not to have any influence on the wheel. In doing that, they removed the very fact that keeps the universal attachment together.

$$\frac{dJ}{dt} = 0$$

This disputes mathematics. DJ / dt can have any number from eternity to infinity, only excluding one; it cannot be 0. By placing the one in division of the other, you bring in relevance. You cannot then say there is no relevance. By doing such, you proclaim that one of the factors is non-existent.

$$\frac{dJ}{0} = dt \text{ or } \frac{0}{dt} = dJ$$

In both cases, one of the factors then does not exist. Such a claim is incoherent, because you proclaim that a circle has no radius, or a radius has no circle. When calculating a circle, you multiply either the square of the radius by Π, or the quarter of the diameter at a square by Π.

$\Pi \times r^2 = $ CIRCLE

If you remove r it then is $\Pi \times r^2 / r^2 = $ CIRCLE.

You cannot then say $r^2/r^2 = 0$ and therefore $\Pi \times 0 = 0$. That is nonsense. $\Pi r^2/r^2$ will always be $\Pi \times 1$, and that is the eternal circle. When looking at any rotating object, there has to be a point of no rotation and no rotation means "no rotation", not no existence. No rotation means a factor of 1, not zero.

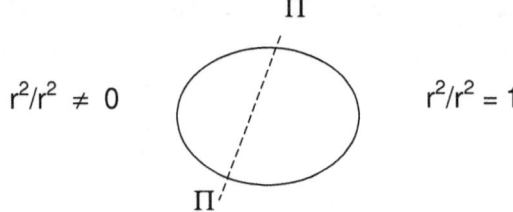

That then is singularity. The eternal Π, the Π that may not have significance but still it is a Π of value. The relativity remains one, eternally one, but it cannot be zero. Therefore, dJ/dt cannot be zero.

dJ/dt can become eternal or infinitive or at the worst it can become
dJ/dt = 1

When explaining this to any child, they can immediately see that. Explain this to any Newtonian High Priest and he may have you removed forcefully from campus. I cannot find one Newtonian, large or small to accept that. Every solar structure is spinning around an individual axis while the whole lot is spinning around a mutual axis the sun provides. The spin that shows on the different planets is the most crucial aspect of their orbiting the sun.

TIME IS THE SPIN OF CONCENTRATED HEAT IN A DEFINED SPACE

NEWTON ON THE OTHER HAND HAD A MUCH DIFFERENT IDEA AND FOR THE PAST THREE AND A HALF CENTURIES THE SUPER-EDUCATED ECHOED HIS SENTIMENTS.

All spinning matter has the point where the spin is still there but the radius is to small to measure by any means. That point is standing still in relation to the rest of the spin. In relation to that logic I do not accept Newtonian science holding the radius of s spinning object unaccountable in the spin, whether the spin is applying or not. To tell the truth it has been above my ability to convey this to any person with a decent physics education where as telling this to any one as poorly educated as I, they see the point immediately. That too is above my comprehension. For the life in me, the SUPER-EDUCATED cannot come to realize my explanations. They either ignore me, or they ignore what I say or they dismiss me as un-educated without receiving the ability to understand Newton.

Just as impossible as I find one SUPER-EDUCATED to understand this aspect to that very same degree I cannot convince one SUPER-EDUCATED about the comet not colliding with the sun as it is supposed to do according to Newton. There is nothing about Newton to understand. I do understand Newton but I disagree with Newton and to brush me off as ignorant does not explain the comet not hitting the sun. The comet and the sun share a mutual point in singularity while each one is relating to an individual point in singularity. Applying Newton's second law F=ma One arrive at the formula
$GMm / r^2 = m (\omega^2 r)$

By replacing $(\omega^2 r)$ with $2\Pi / T$ we obtain Kepler's third law

This law predicts that **$T^2 = a^3$**

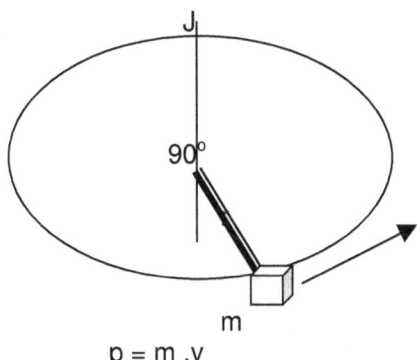

The mass (m) multiplying the speed (v) forms a new value J AND THEREFORE j CONTINUOUS TO IMPLY $J = I \omega$.

$J = r \times p$ where $p = (v = r \times \omega)$

$J = r.m.v = m.r^2 .\omega = I. \omega$ and becomes interpreted as $J = I \omega$

This establishes that $r = dJ/dt$

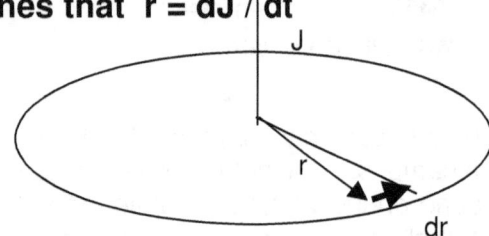

$r = dJ/dt$ **In the case of planets in orbit around the sun r forms a value of zero because** $dJ/dt = 0$.

What this statement implies is that r does not exist. When anything has a value of zero it is for all purposes non-existent. Only when an object is following s straight line can the radius be non-existent because the radius alters value through time development.

Taking the argument back to Kepler's law, a^3

$a^3 = kT^2$
$a^3/kT^2 = 1$

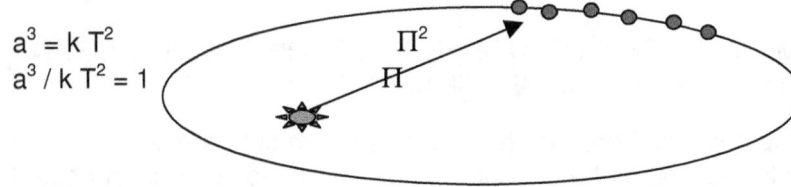

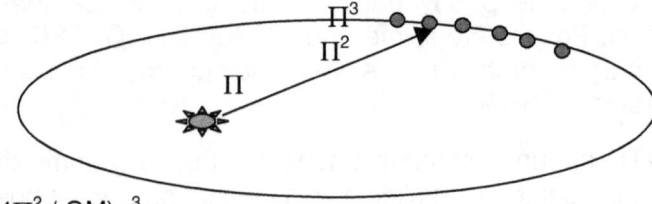

$T^2 = (4\Pi^2/GM)r^3$

The spinning or not spinning is not part of the issue because at the point of absolute singularity the object never spins. Therefore spinning or not spinning does not apply to the point of singularity because singularity never spins in any event.

<u>Since Newton became an institution forming the King bee of the academic cartel world wide The Brainy Bunch had Newton's vision written in the minds of the future generations almost at gunpoint...well definitely at an academic gunpoint.</u>

$r = dJ/$

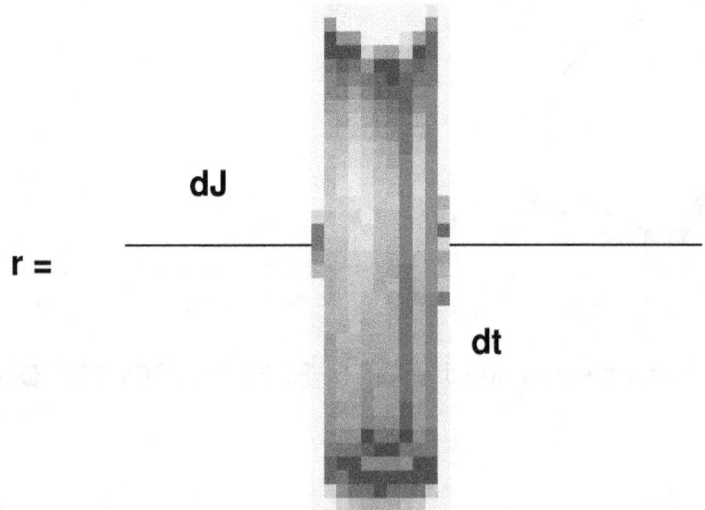

In the case of planets in orbit around the sun r forms a value of zero because
$dJ/dt = 0$.

I am not the brightest that I admit, but one thing no one can do, not even if you are the one and only Isaac Newton is that you cannot place any relevancy in a relevancy and then claim it not to be in a relevancy because such a relevancy does not suit your taste.

You cannot put something in relation to another object and then decide there is no relation because you find no relevancy in the relevancy.

r = dJ /

r =

dJ = 0

dt = 0

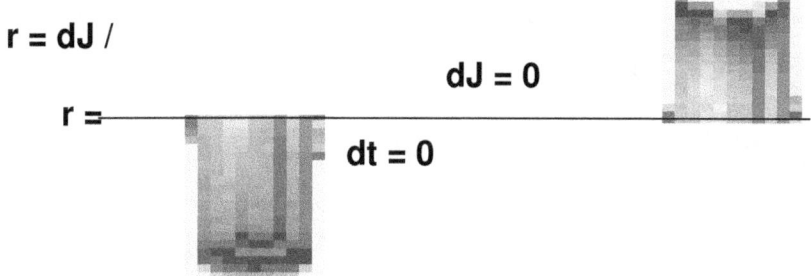

dJ / dt ≠ 0.

If dJ = 0 then dt = 0 That is a mathematical principle, much larger than even Newton

One cannot claim there is a wheel and then remove the spokes because according to you taste, too do not like the spokes

In the same manner the ring cannot remove, because the spokes will then still imply where the ring must be. The only way to cheat yourself out of the situation is to remove the wheel and spokes altogether, and you are left with what you say there is: NOTHING. But that does not apply in cosmology. The object rotates the centre structure and therefore there has to be a radius holding the circling orbited in relation to the centre structure. By not having a wheel rotate, the wheel becomes the factor of one, and the rotation becomes zero. The wheel does not disappear. In the cosmos, everything is rotating because nothing ever stands still. Therefore the mean equilibrium, the common factor there is to share, has to be one, eternity, the eternal Π, because all rotating objects has Π in singularity, and sharing singularity, gives every object in space a relation with all other objects in space. After trying for many years to bring them the candle, I concluded that Newtonians are incapable of realizing that mathematical principle as reality.

The comet rotates the sun, and the sun by itself has a point of singularity where Π remains without r. The comet, holding the orbit, also has a point of singularity, but since there is space separating the two objects, they cannot share a mean point of singularity, the very point of existing. Since singularity means just that, being single, there cannot be two. The comet and the sun have a mean point of singularity but the space they occupy divides their common singularity. That is why they orbit in an oval path, a path where the one structure holds on to more space from its point of singularity towards the space it claims. Since they do not claim equal space, BY THE DENSITY they hold, the space will not be in proportion. They do share in the common fact of singularity and singularity cannot be two, because then it will be "dualarity" or duplicity (in case there is such a word) where both find the space they occupy, with the space they hold, will be their individual eccentricity from singularity. The two objects are holding eccentric space around their individual but common singularity. That point of singularity is Π the circle without the radius because the singularity removes all forms or values of r, leaving Π to be singularity.

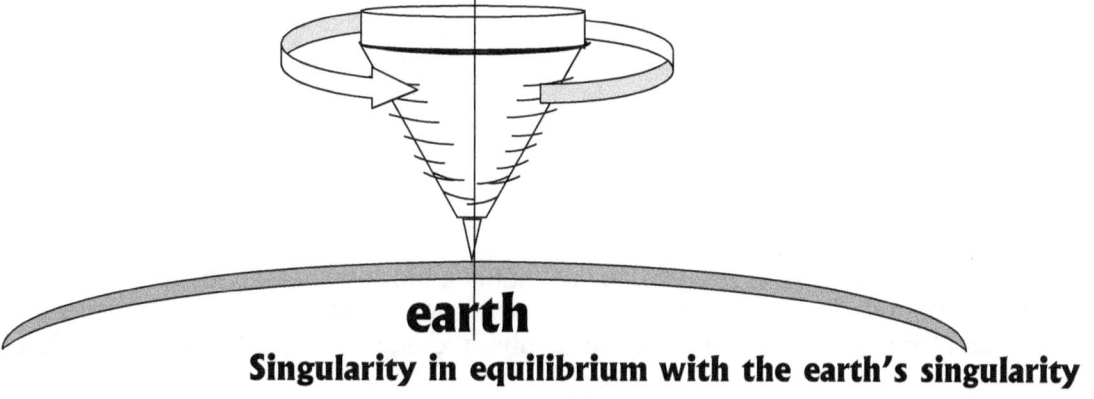

Singularity in equilibrium with the earth's singularity

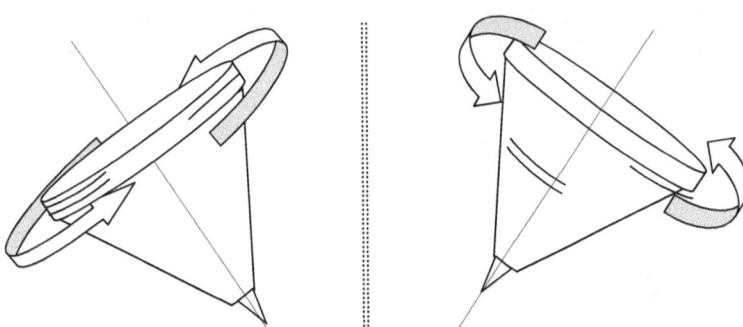

Singularity of the top exceeding the earth's singularity

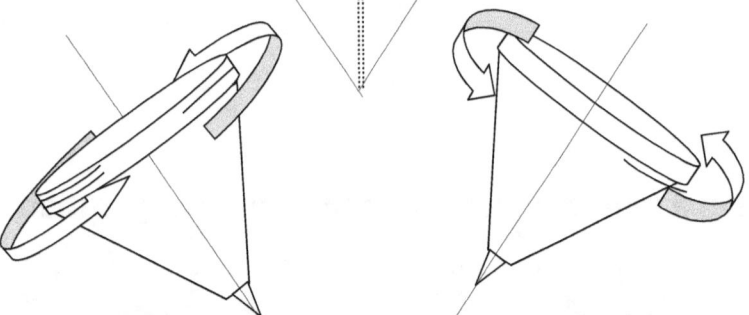

The earth's singularity dominating and exceeding the singularity top produce through spin as the top collapses and fall.

The centre may or may not spin and the fact that it does or does not spin is all the same because that centre part never spins in any case. Therefore the boundaries set by the spinning motion does not depend on the spinning motion of the object but has to stand related to another bogy bringing about a larger spin influence. Granted the fact that the influence the earth has on the top may be that of gravity but if that is the case then surely the sun has also influence on the earth and other rotating objects through gravity. It needs more investigation because it may bring about evidence we are not aware of.

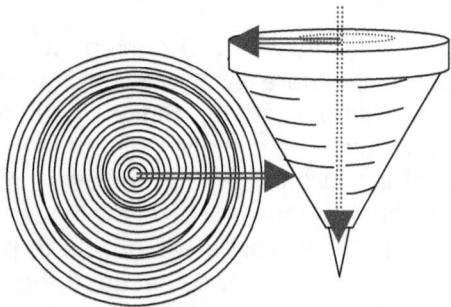

This observation places a much bigger question mark on the statement of Newton where he proclaims no influence on two rotating cosmic structures.

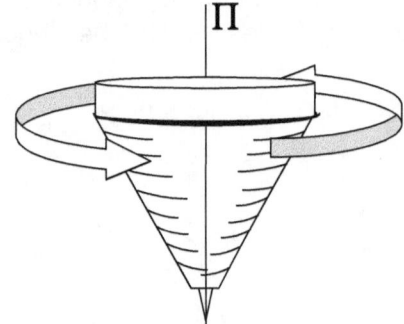

We may proceed to the wider picture that the cosmos hold. What is it the Newtonians fail to see? If an electron is orbiting around an atom, the inside of the atom must be a circle. If the atom was not a circle, it then had to be a cube. The electron cannot rotate around a cube; therefore, the inside of the atom is a circle.

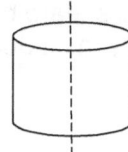

In a circle, there is a radius that initiates the circle. The calculation of such a circle is Π X r².

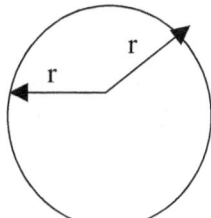

The radius r runs from the circle outwards, from a circle centre point towards Π, the value of the circle. In the centre of the circle, there is a point where the radius starts. It runs outwards from that point in all directions towards the circle Π. Technically, there then has to be a point where r is zero, an absolute zero. However, the circle therefore remains Π. The circle does not disappear; it remains there for all to see. It is only the radius that removes.

$$\frac{\Pi r^2}{r^2} = \Pi$$

If one removes the radius from the circle, the circle remains, only holding the value of Π. By removing the value of r, Π becomes singularity with no place to be. Singularity is the place where there is no space to be in place. However, Π remains because once r receives the slightest of space Π will find space. Then the circle will grow to Πr² and r would determine the space. Without space, there is no r but there is a circle with the value of Π. Singularity is in every single rotating object, be it the proton or the universe.

Every person blessed with eyesight can observe that that is not the case. SPACE IS AS RELATIVE AS TIME. R^3 / T^2 = one AND $R^3 = \frac{1}{T^2}$.

An enormous effort will have the top going oblong while spinning violently and as the pace reduced the top will stabilize by coming to an upright position. In the upright position it wall then spin for the remainder of the period where it will in the end start tilting to the side and in a last effort throw a few wild oblong turns and fall over.

Boys playing games will never realize scientific breakthrough explaining and grown ups do not play with toys. In this little toy played everywhere everyday by almost every one is the answer most brilliant of human Brainpower seek answers about all the cosmic riddles no one seem to understand.

I am not disputing Newton; I am disputing the relevance of Newton's scientific breakthrough. It was not two objects of cosmic proportions, colliding in a show of spectacular. It was, after all, only an apple falling from a tree.

Newton, and science, made one enormous blunder, from this stance. They took the radius of a wheel not to have any influence on the wheel. In doing that, they removed the very fact that keeps the universal attachment together.

$$\frac{dJ}{dt} = 0$$

This disputes mathematics. DJ / dt can have any number from eternity to infinity, only excluding one; it cannot be 0. By placing the one in division of the other, you bring in relevance. You cannot then say there is no relevance. By doing such, you proclaim that one of the factors is non-existent.

$$\frac{dJ}{0} = dt \quad or \quad \frac{0}{dt} = dJ$$

In both cases, one of the factors then does not exist. Such a claim is incoherent, because you proclaim that a circle has no radius, or a radius has no circle. When calculating a circle, you multiply either the square of the radius by Π, or the quarter of the diameter at a square by Π. Newton's claim suggests that a wheel in rotation will return to the same spot it had previously, as it does not affect the spin.

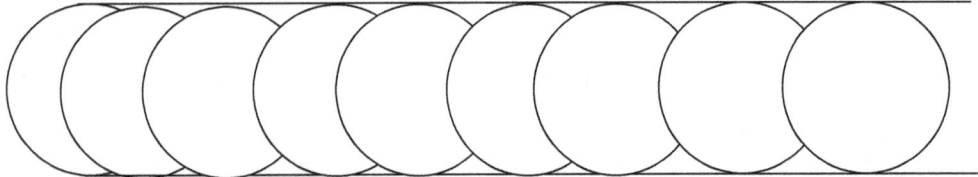

That is impossible since rotation brings about motion changing the principles of the location.

One do seem to get the impression that little changes as the rotation will bring some forward motion and some returning to the original position.

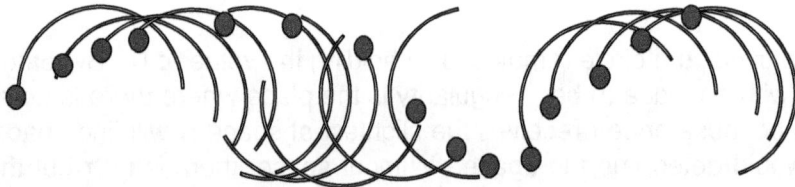

Even by using half a wheel would still bring considerable confusion but one can clearly see that Newton's presumption does not quite match reality

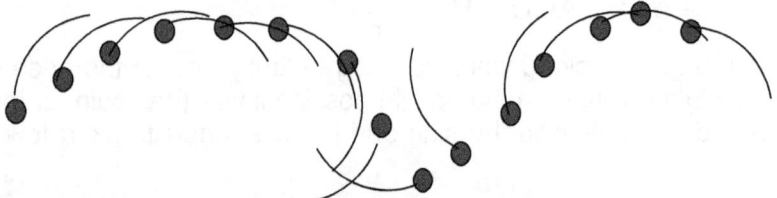

Shortening the arch changes the complexity considerably as one can then see a changing of the arch does not nearly bring the return of the dot to the previous spot.

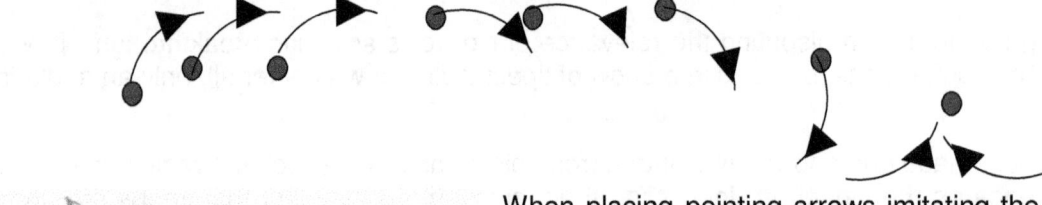

When placing pointing arrows imitating the direction the line of movement indicate it becomes clear that there is a complete mismatching and the cosmos changes as rotation progresses.

Even by not supporting the claim that structures influence space, there are acceptance that the gravity tugging between cosmic structures do take place. When mentioning that I must immediately state that this very fact I am about to dispute in the forth coming few pages. However let us leave it at that for the time being. Let us just say there are admitting that structures influence one another. t is moreover the individual singularity in maintaining the major singularity, which sustains the governing singularity providing equilibrium in space-time.

Not only does atomic individual singularity maintain self preservation, but in doing that it also sustain a governing singularity holding structural composition and form within a cluster of matter for example a star. As there is between stars so there are in the same manner a mutual or bonding singularity between atoms in stars, which we see as fusion.

The sectors provide individual singularity as a means in sustaining governing singularity by which provision comes through maintaining governing singularity the required spin in maintaining cooling. If this process did not apply, there would be no connecting individual singularity to major singularity.

Every quarter provide a distinct value that indicates the progress of the flow of time from the one point Π to the next point Π.

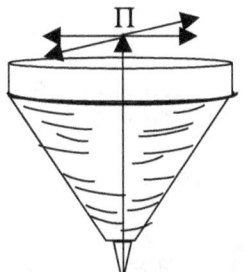

Any changers occurring in Π will lead to a an unequal triangle providing two different values to r and will alternate the link between r and Π² bringing about different form (Π) and time (Π²). When singularity forming the lines of the triangle is not in equilibrium the triangle will destroy the matching of half circle.

The sectors provide individual singularity a means in sustaining governing singularity by which provision comes through maintaining governing singularity the required spin in maintaining cooling. If this process did not apply, there would be no connecting individual singularity to major singularity

<u>In every sector the directional flow will provide a distinct meeting of Π linking r to Π² and this allow the time component in the rotation.</u>

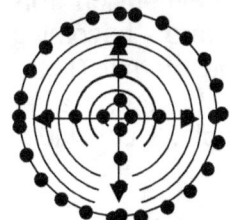

As the meeting of r points to a very distinct different r in direction such a point of meeting opposes the other points in meeting and will lead to destruction of the form Π in any the event of any value changes by Π changing Π² and r.

The sun is on the outskirt of the Milky Way and the sun is in an ova orbit around the Milky Way. The law of orbit is in principle that all orbiting structures follow an oval path.

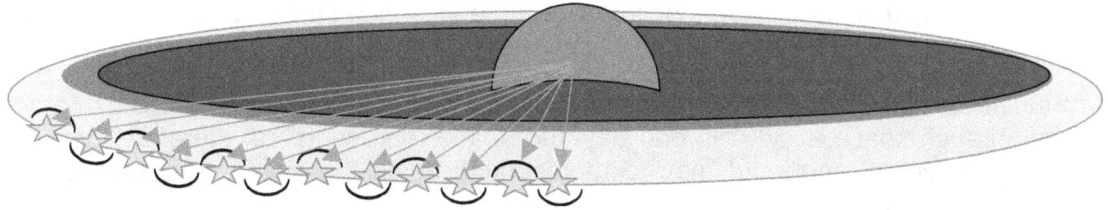

Exaggerated to a large extend the influence the Milky Way has to have on the earth orbit comes to focus when a pattern comes in pace as the earth follow not a circle but a wave around the sun while the sun sets its motion around the Milky Way. The fact that the planets orbit the sun and the fact that the sun orbits the Milky Way indicate an influence undeniable. The fact that the sun is heading farther away from the influence should then lead to a variation in the planets orbiting wave. The earth never, not once land on the exact same spot by the completion of one more year cycle. It is therefore not possible that the

$\Pi \times r^2$ = CIRCLE

If you remove r it then is $\Pi \times r^2 / r^2$ = CIRCLE.

You cannot then say $r^2/r^2 = 0$ and therefore $\Pi \times 0 = 0$. That is nonsense. $\Pi r^2/r^2$ will always be $\Pi \times 1$, and that is the eternal circle.

When looking at any rotating object, there has to be a point of no rotation and no rotation means "no rotation", not no existence. No rotation means a factor of 1, not zero.

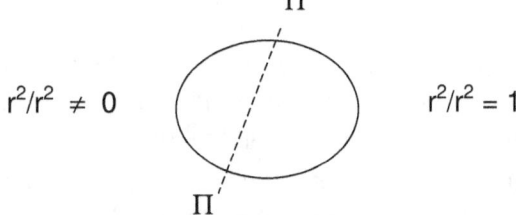

That then is singularity. The eternal Π, the Π that may not have significance but still it is a Π of value. The relativity remains one, eternally one, but it cannot be zero. Therefore, dJ/dt cannot be zero.

> dJ/dt can become eternal or infinitive or at the worst it can become
> dJ/dt = 1

By not having a wheel rotate, the wheel becomes the factor of one, and the rotation becomes zero. The wheel does not disappear.

In the cosmos, everything is rotating because nothing ever stands still. Therefore the mean equilibrium, the common factor there is to share, has to be the dimension from where all dimensions develop, the dimension becoming a straight line – becoming a square – becoming a cube – becoming a circle- holding eternity and infinity to a point of. one, eternity, the eternal Π, because all rotating objects has Π in singularity, and sharing singularity, gives every object in space a relation with all other objects in space. After trying for many years to bring them the candle, I concluded that Newtonians are incapable of realizing that mathematical principle as reality.

When science calculated the value of gravity and the gravitational constant, as well as the speed of light, they never considered the moon to play a distracting factor. It is quite understandable because they did not know about the Roche-factor influencing their calculations.

With Π^0 little more than a figment of the imagination there is actually to values of Π^1 facing each other in a relation combining Π^1 to hold the value of $\Pi^{1+1=2} = \Pi^2$ and with two sides being the very same but opposing each other there will therefore also be Π^2 to every side that holds Π^1.

From the above I can conclude that gravity is not 9,81 Nm/s, it is Π^2 = 9,8696.

The gravitational constant is not 6.67 but it is 6.9 (7/10 (Π^2)) and here the moon had an even bigger influence. It is a fortunate coincidence that we took water to be the measured calculation since water holds the combined value of 17,5 and that is half the value of either space (31) or time (Π^3). That makes a kilometre (1 000 m) one cube laid flat and since movement represents space-time occupation in a linear manner, it is the cube that went in a single line.

More of the same fortunate coincident is that we connected time to spin long before Newtonians came along. The earth spins through space at 360° in one day and space represents 10, therefore there is 3600 minutes in the 7° of spherical angle out moving representing again the seven. All this makes explaining matters a lot less difficult.

The value of the proton is not 2/3, but Π^2. The proton spins at a rate of Π in a dimension of 2. The neutron is not 2/3 + 1/3 + 1/3, but again it is $\Pi^2\Pi$ and the electron's 3, holds a dimensional implication, because time is in singularity and space is in singularity. Time is eternal and heat releases space, which is time from singularity. In an effort to make the understanding simpler, you have time at an eternal value and that makes space zero. $\Pi^3=\Pi^2\Pi$; $\Pi^3/\Pi^2=1$, Time and space interlinks because it is the same thing and heat, (matter in many spin rates) allows time to break free from eternity by allowing a distinguishing of the flow of events. Time in movement is the result of mater (which include heat) to change their relating positions.

Think of a movie. The continuous flow of pictures indicating the change of the position of the photo's bringing about the concept of time. Play the picture too fast or too slow and it will be unreal because we know at what tempo matter changes its position in relation to all other matter surrounding it. That is time and that makes time irreversible, because the position matter hold in relation to each other in considering it to be throughout the universe, can never repeat once it has changed.

Newtonians, forget about time travel because just by mentioning such absurdity you prove what little you know about the cosmos. To go back to a certain time, you will have to redirect all matter in the universe in a reverse, apply that reversing of all particles up to the point required, stop the movement of all particles and start time going forward. Before some Newtonian grabs for a calculator, remember, you that are doing the changing, is as much part of matter in the universe, therefore your action in changing the direction will stop even before you start! It is silly to think people with healthy minds, acting like adults will indulge in senseless stupidity such as claiming to be able to reverse time.

What was within the universe at the start will be in the universe at the end. The universe holds all; maintains everything and combines the lot. In Afrikaans we call the universe the "Heelal". It is a combination of two words namely "geheel" and "alles". "Geheel" means everything and "alles means everything. Therefore the "heelal" directly translated from Afrikaans to English will mean the "Everything of everything". Nothing can be added and nothing can be lost. It is all-inclusive. With this fact so commonly known and accepted, how can the universe grow? How can the universe expand? Well, it cannot, and that is yet another illusion the Newtonians create through misunderstanding. What is in it is in it and it cannot grow, as much as it cannot shrink. It cannot expand and it cannot demise. It is only a consistence of changing relevancies, where the relevancy flows away from one part of eternity or singularity (space) to another part of eternity or singularity (time).

Every aspect of the universe holds relevancy by applying time to space and the time to space first claims space from singularity, then control space from singularity and influence space outside the direct contact with singularity. In every event the factor remains the same as it is only the relevancy re-applying a dimensional influence on space-time.

The key to the relevancy is heat and space. When matter heats, it expands therefore it takes more space. When matter is cooled it shrinks, therefore takes less space. That is the relevancy because matter in any form is heat. Heat produces the increase of space and reducing space produces an increase of heat. That is the relevancy. That is the secret of the universe. That is the secret of gravity. That is the secret of momentum and every other aspect within the universe.

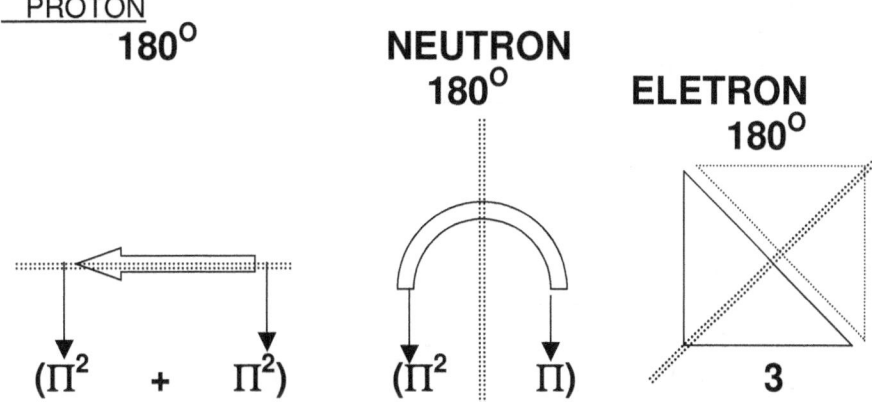

Time stood still in eternity, then after a command of the Creator, time started to move by overheating and eventually formed the relevancy of the proton ($\Pi^2 + \Pi^2$) the neutron ($\Pi^2\Pi$) and the electron (3). As a star return time by depleting space to the dimensional increase of heat, a space destruction is in progress and the star will abandon systematically some of the dimensions the atom holds. That is the relevancy. That will be whatever position there is in the universe. In the depleting process of dimensional re-adapting the star shall abandon aspects of space-time. The electron (3) may become obsolete, the neutron ($\Pi^2\Pi$) may become obsolete in neutron stars and even ($\Pi^2+\Pi^2$) the proton will become dysfunctional as space reduction completely disappears from the star's space-time occupation. However, those stars will be dark, and beyond our vision.

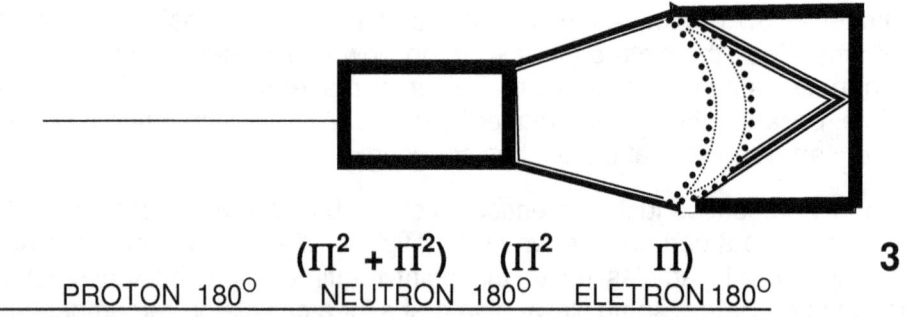

$(\Pi^2 + \Pi^2)$ $(\Pi^2 \quad \Pi)$ $\quad$ 3
PROTON 180° $\quad$ NEUTRON 180° $\quad$ ELETRON 180°

The relevancy holds value pointing the relation between the various dimensions as they are in the atom. The relevancy of $(\Pi^2 + \Pi^2)(\Pi^2\Pi)(3) = 1836$ will remain but the mass of the electron and the mass of the proton will change in every space that time applies. Cosmology thus far was incomprehensible because it was incorrect. When applying natural laws, it becomes so simple that a person as ordinary as I can understand and explain it.

If I can understand it, every other non-brainwashed human on earth should understand it. The relevancy of $(\Pi^2+\Pi^2)(\Pi^2\Pi)$ and 3, is a dimensional reduction of the flow of heat from space back to time. The flow of heat becomes necessary to prevent solid matter from overheating. By removing heat from the gas of space, through the neutron, to the solid of matter, space reduces as the intensity of heat flow requirements increase.

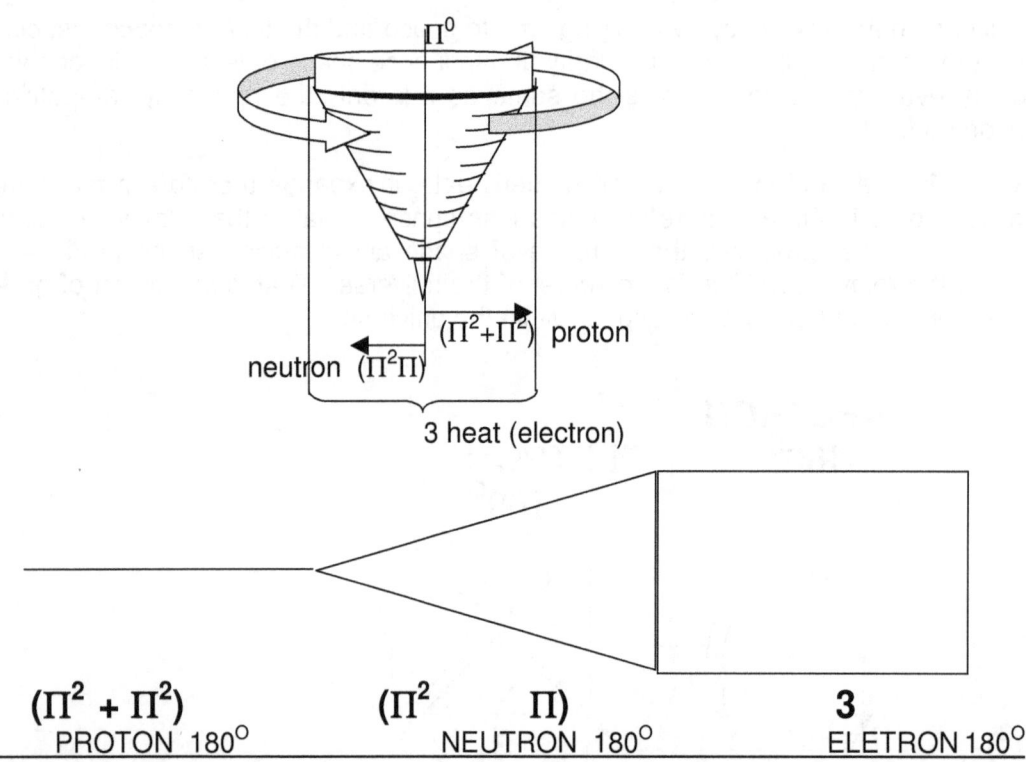

$(\Pi^2 + \Pi^2)$ $(\Pi^2 \quad \Pi)$ $\quad$ 3
PROTON 180° $\quad$ NEUTRON 180° $\quad$ ELETRON 180°

The above indicates where singularity originates and how that establishes the factor in singularity Π. The universe started from the factor in singularity Π. The entire universe holds a spinning relevancy to all other factors in the universe, if that was not the case; the universe was not there. The first person to consider the factor in singularity Π was Galileo. In the swing of the pendulum he saw singularity remain as one, that formed time, destroying space to maintain time. To prove my statement I shall very briefly indicate some barriers of motion science refer to as the Doppler effect, but Doppler used a slow moving train that at best could indicate two or three very minor moving limit.

The universe started from something I named time in singularity. Time in singularity is the only constant because there is always a direct line of contact to time factor in singularity π. and time in singularity is the end product of everything that is cold, frozen beyond space. We view the ultimate freezing point to be outer space at a temperature of −273°C. Heat will always move, from the hottest area to the coldest area. If the outer space were the coldest point there is, the earth would be frozen solid, because heat will never arrive at the earth from outer space.

We would have a continuous flow of any heat the earth may hold to outer space. Such is not the case. We may view that heat flows from the earth at night to outer space and that is correct. However there is no flow of heat that will bring about temperatures to fall to limits even close to outer space. Nothing on earth can reach −200° without the interaction of human life. That means the earth will always be a place colder than the universe, just because we have more heat on earth than in the universe. If it was not the case, the earth should be at least as cold as the universe because all heat will flow from the earth to the outer space.

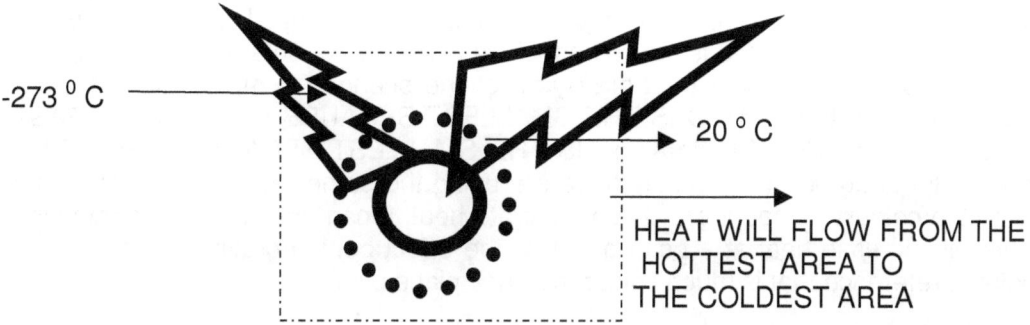

IF THE OUTER SPACE REGION WERE COLDER THAN THE SOLLID STRUCTURES, THE COLD WOULD DRAW HEAT TO THE HOTTER REGIONS. THE NATURAL TENDENCY MUST APPLY WHEREVER THERE IS HEAT VERSUS COLD.

The opposite is happening

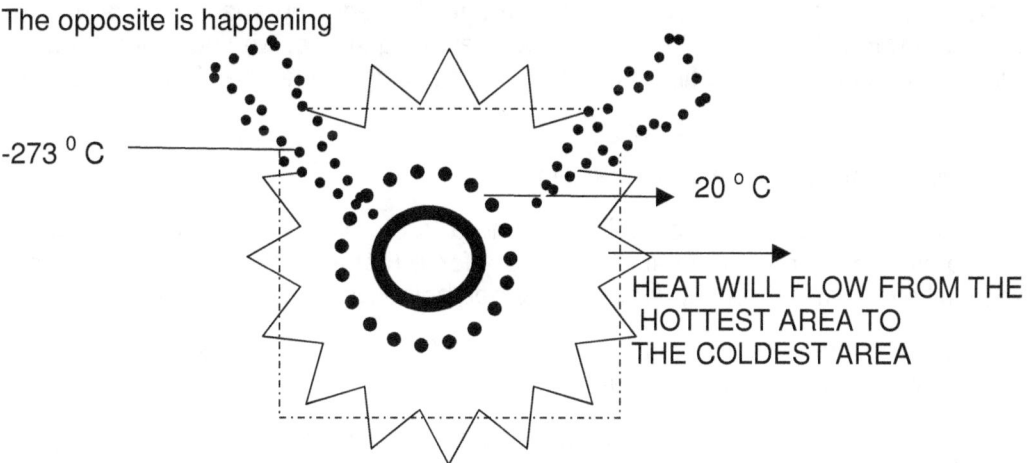

INSTEAD OF THAT, THE NATURAL LAW PROVES TO HOLD HEAT TO THE STRUCTURES. THAT MEANS GRAVITY HAS SOMETHING TO DO WITH HEAT CONSENTRATION. THAT ALSO INDICATES THAT GRAVITY IS THE RESULT OF SOMETHING MUCH COLDER THAT IS NOT IN THE VISION LEVEL OF THE HUMAN EYE. THE ONLY SUBSTANCE HUMANS CANNOT VISIONALISE IS THE PROTON.

Why would the earth appear hotter, but is colder than the universe? It is the fact of gravity, a term I reject because of the connection science apply to gravity being a force. If it was a force then something is pulling or pushing and since we do not believe in spirits creating magic forces, and no one can detect anyone applying any force, we have to dismiss the connection to a force or forces. For the moment in order not to confuse any reader in the introduction, I shall remain using the term gravity but in doing so, I dissociate myself totally from the implication that science connects to gravity.

One may say that this is the effect of the sun shining on the earth, and one may be partly correct, except when considering the wider picture. Measure the space the sun pours light into and divides that area the earth holds in the vastness of space. I truly think no ordinary pocket size calculator will provide a realistic reading. Yet the earth is many times hotter than the rest of space, if it was merely the sun shining on the earth, the earth was no factor to consider in relation to the space out there. That alone cannot account for the difference there is in temperature.

This brings us to gravity, (not a force). My definition about gravity is that: **Gravity is the reduction of space to concentrate heat,** therefore the more gravity there is, the more heat there will be. The earth, or any cosmic structure claims heat from outer space by concentrating heat. You may shout that fusion is the blaming contributor to the heat in the sun, but I shall denounce that shortly. The more gravity a structure holds, the hotter that structure is, Pluto is the coldest place in the known solar system, but it also is the smallest place, and it also has the least total gravity. The one scenario compliments the other scenario, you may be of the opinion that PLUTO IS THE SMALLEST BECAUSE IT HAS THE LEAST GRAVITY, AND I MAY BE OF THE OPINION THAT IT IS THE SMALLEST BECAUSE IT HAS THE LEAST GRAVITY, and we will not agree at all, although we are using the same words. If Pluto has the least gravity, it will also be the coldest planet, if gravity concentrate heat, it can only do so by something that is much colder. This will increase heat at a point because the effect behind gravity is producing a reduction of space, a point of reference that is much colder then any place else.

I started off at the beginning of this part, by saying everything started at a point, that was so cold, space froze to singularity. Let us test this statement: When an element heats up, the space it occupies, increases. When an object becomes colder, the space it holds reduces. The less space per atom there is the colder the object must be. Anything can freeze rock solid, as everything can boil to gas. This means all elements in nature, is neither gas, nor liquid nor solid. It is the space that is between the elements that allow the elements the form they hold at that moment. By reducing space, space has to concentrate because to concentrate is to reduce the solution. By reducing space, we find more heat. If you find more space by increasing heat, and you find more heat, by reducing space, then heat and space is the same thing.

Therefore after everything said do you now know what gravity is?

If you knew would you tell me what gravity is…yes beside it being a force…well we all know that those who should know says that gravity is a force but other than being a force that acts on behalf of mass and pulling unsuspected objects all over, what is gravity, which is besides being Newton's pet force.

If you don't know what gravity is besides knowing that gravity is one of the four forces and it is Newton's original force, then do you know someone that knows what gravity is?

 If you still answer in the negative and you still don't know what is that which is behind what is causing of gravity, then have you heard of any one that knows what gravity is?

Maybe there is one Academic professor or a NASA Scientist that knows about some person that knows one who knows what gravity is, other than knowing the fact that gravity is presumed to be one of the four forces and being Newton's personal pet force.

There has to be an Academic professor or a NASA Scientist known to someone somewhere that knows about some person that knows one who knows what gravity is, because there are so many Academic professors and even more numerous in numbers amongst the many NASA Scientists with super human mathematical abilities in the art of physics acting as if they know. There are even those going around with ideas to renovate the cosmos by assembling space whirls. They apparently plan great voyages that will take man over great distances while they are so informed on the matter of gravity and space-travel amongst gravity. With their absolute phenomenal calculations as they present physics they are showing such abilities in a class which no one can imitate, which must represent a picture about having super knowledge on gravity in the most precise detail.

When reading what they say they are able to accomplish in terms of astrophysics it stands to reason that they know gravity to the smallest detail there is to know and know everything anyone can ever consider knowing or hope to know about gravity...after all they can present the cosmos as they are able to explain the cosmos with gravity taking centre stage in the past, present and future of the cosmos. This they accomplish by presenting a few mathematical formulas in which all of the Universe then are defined.

They can calculate all the matter throughout the entire cosmos, adding every atom by mass into the conclusion of their calculations when they determine the critical density that the cosmos has, which is responsible for the entirety that is providing all the mass that provides all the gravity. They have gone as far as even calculating the explicit required quantities of atoms that forms mass which should be available, which they then find to be short falling in the mass availability through out the vastness of the Universe and the missing mass is not establishing the matter density throughout the vastness of the Universe required to bring Newton's vision on contracting to reality.

Their calculation ability is so vast that they have the opinion about the mass they measure that is forming the gravity they require which is probably less than the requirements needed to substantiate the cosmos' effort in rendering a constant supply of gravity that will eventually secure the returning of everything to where the cosmos came from. This eventuality they named The Big Crunch even before locating the Big Crunch. It is like naming a baby even long before knowing how the procreating is taking place that will lead to impregnating of some member of the specie (which member it will be is still unclear) where it later on will lead to conceiving the baby ... that is the manner in which science dogma is enunciated but that is how clever those are that knows everything there is to know on gravity, or so they pretend to have in their promotions.

They know gravity to split detail where the detail goes to such precise extend that they are able to calculate how much gravity the missing dark matter in all the Black Holes must be to provide a force allowing the cosmos in experiencing the next big implosion that is coming somewhere in the future. That the implosion must come even in the face of insufficient gravity is a certainty otherwise Newtonian physics is completely inadequate in their cosmic vision about the Universal future! With their having this qualified virtue of intellectual splendour spawning such phenomenal abilities they then would have to know what gravity is!

Well...if you don't know any one that knows anybody that knows someone somewhere that is familiar with the ins and outs of what is causing gravity, I then can assure you I know about someone that knew all there is to know about gravity. I am the person that knows someone and that someone knew gravity...but he is dead now...died a premature death long before his time (have you ever heard of any person that died spot on at the second that was his time where he was suppose to die). He sadly passed away and is no longer with us. Still I would like to introduce him to you...and about his work of course, if you would page on.

If that is the formula the cosmos abide to and that is the formula controlling the cosmos the radius of all things should lead gravity on an eternal quest to voyage towards and finally into the centre of the Universe. If gravity is taking whatever there are to where all things will end and all things are forced by gravity in relation to move according to mass to such a point that will unite the entirety of what is into what will be that outcome can only be where the centre of the Universe are allocated at this moment. The first thing to do is to find where r^2 ends because where r^2 ends it will be where we will locate the centre of the Universe.

The factor r^2 should eventually lead everything to the centre of the Universe in the end...then where is the centre of the Universe and where is the factor r^2 going and conclude in the end at the final destiny?

Facts I advise the reader to become acquainted is the following:

Did you know that not one of the supposed laws Newton invented going under his name or the name of other innocent person's such as Kepler, does apply in the cosmos just as Newton described it must It is not I that is at odds with Newton, but it is every principle applied by the Universe that totally contradicts every declaration that Newton made.

Dear Dr. Schutte,

You submitted an article of 15 pages to the Annalen. The content of this paper doesn't constitute a theory in physics.

With a lot of words and some simple algebraic relations, there is no way to "explain" the world of physics. You seem to be out of touch with modern developments. This is also shown by the fact that you don't quote any relevant literature.

I am sorry to say, but the Annalen is not able to publish your work.

I am sorry for having no better news for your.

Best regards,
Friedrich Hehl

Co-Editor Annalen der Physik (Berlin)
--
Friedrich W. Hehl, Inst. Theor. Physics
* University of Cologne, 50923 Koeln _____/_____ Germany
fon +49-221-470-4200 or -4306, fax -5159
hehl@thp.uni-koeln.de, http://www.thp.uni-koeln.de/gravitation
* Univ. of Missouri, Dept. Phys. & Astr., Columbia, MO, USA

Dear Prof Friedrich W. Hehl, I have received your e-mail reply and I wish to respond on your letter. The article of 15 pages to the Annalen had in mind to introduce a very wide-ranging concept contained in many books. I wish to promote books in which I introduce a much larger and much more detailed cosmic picture. It is four books that actually form four volumes of one theme supporting The New Cosmic Theory. I wish to unveil a totally new approach to the thinking in cosmology. The concept is proposed in the article I sent to you which is "revealing" The New Cosmic Theory In the article as much as the theme I wish to go where no one ever attempted to go before. I introduce the Universe of singularity a state in which the Universe still is because it is a state from which the Universe grows It is where material in a dimensional dynamic does not apply because it is where Einstein said "the Universe goes "flat"". I show you how and where the Universe goes "flat" I will guide you to the point where I go…so that you may see where my books and the article lead you. It is in the domain of singularity

When you read work about the Big Bang you have to go right down the development (in reverse order) to the point where the Theory of the Big Bang points at a spot named singularity. It shows the very start from where all material developed. At that point one will find The Absolute Relevancy of Singularity and there has never been any attempt by any person ever to venture beyond the dimensional birth of the cosmos, which is called the Big Bang by going into the era where singularity prevailed. I take you there in my books as well as the unpublished article. However, going there requires a very high degree of concentration and calls for understanding that a very little number of persons are capable to show. I try to show how the Universe goes "flat" as Einstein said the Universe goes "flat". Even by completing this unimpressive letter you will also know how the Universe goes "flat". Even where you failed to read the article I sent you, then by just reading this letter you will be able to find where singularity takes the Universe "flat". But it requires a mental capacity to understand because where I venture no one ever in the history of mankind reached into before. I do not speculate but even in the unpublished article I show with pictures and sketches as well as "some simple algebraic relations" where to go to where the Universe starts, but you failed to read that because you are opinionated as to what conditions should the Universe have before the Universe will allow any one into physics. That is a pity. One should learn from the cosmos and not tell the cosmos what it must be to qualify as the cosmos. Then in the article I show you by almost taking your finger to the spot, the very point where the Universe ends and that too I qualify. You might dispute my arguments and show me about what you disagree, but it shows very little understanding of reason on your part about qualities man should have before understanding the Universe. I go into a Universe that was in place before light was in place in the Universe and only darkness prevailed because light calls for space and in that era of singularity space was not even a thought yet. I show why the Universe goes "flat" and in a "flat" Universe only the value of 1 holds value since singularity is 1. If you can understand 1 or $5^0 \times 7^0 \times 3^0 = 1^1$ you have all the mathematical skills required to understand the applying concepts. To reach a value of 1 does not require big mathematical equations but to reach singularity requires 1.

The collection I named The Absolute Relevancy of Singularity The Theses and the collection as such forms a small introduction to the thirty-two or so books I wrote on various matters concerning physics with gravity in mind, but **The Theses** as such in the entirety of the four books does not officially even start to introduce the spectrum of every aspect of my work. I have been in contact with numerous Academics and about one in one hundred reply. When the one in a hundred reply, the academic always uses a most aggressive tone which I came to accept as what I receive from academics, and because of that I was most delighted to find some kind remarks from you as a practicing academic, and might I add, the first such kind remark in ten years of my trying to contact any person in physics that would take note of what I have to say about a new line of thought, because the few others that replied were extremely aggressive about me confronting Newton. I only began to submit books to publishers after twenty-seven years of studying Newton and the role Newton play in cosmology and thereafter which was ten years ago I began promoting these ideas. The New Cosmic Theory is a process wherein I try to introduce a study that is ongoing for about thirty-seven years, give or take a few and I did not jump into the frying pan having my first thought about the matter published as an article when I sent the article to the address of Annalen der physics.

The New Cosmic Theory that I try to convey by writing books in total holds much information and every time when publishers reject the publishing of any entire book I propose, the rejection was on the grounds that "the discourse is not falling within the main-stream science discourse" and therefore I was subsequently advised to write articles on the subject as to find recognition. I was told that only then could I achieve publication of any entire book. Now I find that trying to publish articles has my work rejected on grounds as follows and the following is directly coming from the reply in which one of my articles was rejected recently: "You submitted an article of 15 pages to the Annalen. The content of this paper doesn't constitute a theory in physics. With a lot of words and some simple algebraic relations, there is no way to "explain" the world of physics. You seem to be out of touch with modern developments. This is also shown by the fact that you don't quote any relevant literature." It is not possible to introduce the totality of my work in 15 pages (or whatever a journal would allow) while remaining absolutely coherent on all aspects during such an introduction about anything. You wish for me to work with mathematics and calculations while the world I enter starts mathematics. My aim with the web site www.singularityrelavancy.com is to introduce the reader to a world before mathematics as a multiplying process took centre stage. I take the reader into the cosmic era when 1x1 was 1 and only 1+1 was valid forming 2. In the article I say that in so many words, and you would have noticed me saying this if only you took notice to read the article with care. I take you into a true flat Universe where space has no dimensions because dimensions are the multiplication of numbers whereas a flat Universe is found within the adding of numbers. Multiplying brings about a discipline of dimensions and singularity is void of dimensions, thus deemed to be single in dimension. The era we enter uses a line called time to create a single ongoing dimension.

I show why the triangle and the straight line and the half circle are all equal to 180° and in the world using space as form by using dimensions this fact about mathematics is bizarre. The triangle and the straight line and the half circle are all unequal in form while mathematics proves the three equal. It is obvious that the triangle and the straight line and the half circle are as wide apart as the sea and the Sun is, and yet there was a period in cosmic development when the three were mathematically equal as much as they still are. I have mathematics telling me this fact beyond doubt. Please use a formula and your brilliance in mathematics and using no words to prove to me why the triangle and the straight line and the half circle are all equal as they all are 180° while explaining details because on this rests one entire pillar of mathematics. The answer about this we find in the Lagrangian point system, which is one of the four cosmic phenomena, I explain when using the four cosmic phenomena to explain gravity. This becomes clear when using the law of Pythagoras to prove how this very law became the basis for mathematics and I do use mathematics in the law of Pythagoras to prove how mathematics started when the Universe started mathematics. However, I don't prove that in the article because the space allowed in the article is much to little to prove anything.

In the article however, I show why did Π become $21.991 \div 7$ or then $\Pi=3.1416$ or why is a circle Πr^2 or why is a circle circumference Πr or $\Pi d \div 2$. I show why a circle begins with Π and don't just surmise it. In my books I show why the phenomenon called the Titius Bode law is responsible for Π as a cosmic form and value. In my books I explain just as I claim in the article how the Roche limit come about and how the Roche limit is responsible for the sound barrier and what is the true cosmic value of the Roche limit as it plays a part in gravity on stars…that I show when I enter the era of singularity when calculations were still not yet developed. I show why a sphere in calculating the volume of space is represented as the formula $a^3 = 4\Pi r^3/3$ and why it is used to calculate the sphere when using these specific interpretations and how this is different from Kepler's $a^3 = kT^2$, which is the way to calculate volumetric space in applying singularity. The basis of this formula is derived from singularity finding form and that too I prove, but I have to use words because prior to when volumetric space came about, singularity prevailed and singularity is single dimensional. I pertinently state this over and over in the article. In the article alone I have no space to show all these facts and therefore in the article I only show why a circle uses Π to begin with. I show where and why did gravity start and what the true value of gravity is as gravity kick-started the Universe into a beginning because the beginning began with gravity. That I don't show in the article because printing space available will not permit me the opportunity to do so, but I introduce a book where I show exactly why, how and by which factors did the Universe start by using singularity. I show how the Universe evolved by singularity before space developed and at that time it implemented the four cosmic phenomena that later became part of space when space developed.

I am trying to introduce a study I have done during twenty-seven years of research and there is not one word that I can quote from any other source since every word comes from conclusions that I make and which I prove with the use of logic. All I try to do is to find a medium wherein I can tell some interested parties where to go to read my work and then for them to judge me on their merit and not be sidelined by rules set by academics in charge of publishing. Why don't you allow everyone to read my work and then afterwards, let all readers be opinionated by personal impressions applying and do evaluation of facts according to personal interpretations? Everyone goes on about the unfairness Galileo endured at the hands of the Catholic Church, but at least the Church allowed Galileo to publish his work so that the entire world could take note. Every one in science as well as the Church thought Galileo was out of touch when he declared the science wisdom prevailing at the time was incorrect, and five hundred years later we know who was out of touch as you state I am. I do not compare my work with that of Galileo but I find the same restrictions brought on me by the Powers of the day controlling science. The method of the blocking of getting new principles published is the same as what was in place back then where those in power controlled the thinking about science and those in power today still controls the thinking in terms of science by using equal draconian methods. By disallowing any other views to be printed that does not resonate with the prevailing mindset, science ensures that the public out there consider the correctness of their position as beyond suspect. Their discourse is then thought of as the only possible thinking policy that could be correct, which makes what they think absolute, beyond any suspicion any person could ever have. Killing criticism makes science deemed by everyone in the world as being undisputable because no one ever could dispute Newton. But it seems that no one ever got the opportunity to dispute Newton. Newton is only undisputable because disputing Newton is not permitted by science. Newton was never proven to be incorrect because any attempt to disprove Newton is killed in the infant stage and more often so even before birth of any such a thought could take place. I know this because for the past ten years the academic world holding publishing power destroyed every attempt I made to draw attention to the obvious insufficient work they base physics on. If you kill the messenger, no one will know about a new message and that is what happened then and that is what happens today. The Catholic Church was the one stopping Galileo, but nowhere is it mentioned that this was also in total collaboration with every party in physics at the time. Galileo did not only cross swords with the Catholic Church but crossed with the views the academics at the time had so Galileo went against what the academic world believed. It was the academics that prompted the Catholic Church to believe the Sun was circling the Earth and that the Earth was the centre of the Universe. Again I say I will not dare to compare my work with that of Galileo, but the treatment I receive I do compare. One thing science can take even less than the Catholic Church could is criticising their supremacy.

I have done twenty-seven years of research about the working of cosmology and found a manner by which I could interpret the four cosmic phenomena science do not even recognise because while they are there, they also don't fit into Newton's mathematical physics. As science goes, they will rather reject the obvious presence of the phenomena because it does not match Newton and must therefore be out of touch with modern developments. I did not only unravel the phenomena but worked out gravity from the manner the phenomena influence cosmology. The phenomena holds root in singularity and no one has yet entered that domain. All I try to do is to find a medium wherein I can tell some interested parties where to go to read my work and then for them to judge me on their merit and not be sidelined by rules set by academics in charge of publishing as Galileo endured. Let everyone read my work and then after that let all readers be opinionated by personal convictions applying. Allow my work to be evaluated by those reading it and not be smothered by those trying to kill the content because they do not care for the style I use. Galileo had an opinion that was clashing with the present dogma of the day but he could express his views because we now know about it. The way modern science kills me is they make very sure no one will ever know about me because they silence me as if I am dead. It is also so evident that at Galileo's trial academics were brilliantly absent by not showing a united effort to defend the liberty of thinking. That image today's science try to portray they are fighting to uphold. However, today one may only think freely as long as your thinking is echoing mainstream ideas. For ten years my ideas were constrained at every possible level I encountered and my ideas was as much destroyed, as Giordano Bruno was burnt alive. Before finding publishing I have to find favour in the eyes of the Academics in physics whom will not have my work published since I disagree with prevailing sentiment and I denounce Newton in terms of cosmology, but only in terms of cosmology.

That is what everyone misses.

Newton does not work in cosmology but Newton works in physics because in cosmology mass does not apply. In physics mass applies. I can find no evidence of mass doing anything in cosmology, still everyone grants mass because with mass it is easy to play with mathematics.

On earth where mass applies in everyday practice, Newton's work is undisputable correct but going into cosmology there is no evidence of mass applying, and that is where cosmology parts from physics. Mass do not pull planets and stars and that Hubble proved when Hubble proved the Universe is constantly expanding. I return to this elsewhere. Because I challenge everyone to show that mass plays a role in cosmology and in forty years no one could, my through thinking that my discourse is not falling within the main-stream science discourse and those with the power to prevent my work getting published will think up any excuse not to publish. They will block me because what I think will have modern thoughts prevailing in science at present brought into question. For forty years I have been asking that just for once someone will step forward to prove mathematically and without doubt that mass brings about gravity. Show the evidence that all the small stars are either in the centre of galactica or are on the outside of galactica and the arrangement of allocating stars go according to mass. What is it in the atom or the moon that has the ability to pull by magic something it does not connect to. Prove how it is possible that things fall by the measure of mass. Just for once show how things fall by the attraction of mass when everything proves that all things fall equal and therefore mass has no role to play in falling. The example used is a feather and a hammer falling in vacuum and this is fraud. Show how a car and a brick fall equal in front of a camera held by a cameraman and then tell persons the objects fall by mass issuing gravity proportionally according to the mass dishing out the gravity when the camera can follow both objects falling. If the ratio of mass brought about the ability of gravity pulling, then more massive things will fall faster and they don't. Mass does not pull or attract by any means or measure and also in this statement I return to debate it further elsewhere.

To bring one point to your attention just the following: you do support Newton and I question Newton and that questioning Newton is mainly what science hates. Where you underwrite Newton's claims of mass bringing the pulling of objects then please show me by using $F = G \frac{M_1 M_2}{r^2}$ how much did the mass of the earth draw the earth closer to the sun by using the mass of the Sun since the days of Kepler? You know as well as I do it does not happen because in fact the distance between the planets and the sun increases and does not decrease, as it should do according to Newton. Please use the formula that forms the basis of physics to show the world when will the BIG Collision come that will inevitably have to come if $F = G \frac{M_1 M_2}{r^2}$ is correct and when will the moon slam into the Earth. If Newton applies we await the collision between the earth and the moon because the masses on both ends will do the pulling to create the devastation that will follow the collision. Since Kepler made his calculations centuries ago, tell me how much did the moon come closer to the earth, presuming that $F = G \frac{M_1 M_2}{r^2}$ is indisputable. Did any member of physics ever bother to do such calculations as to determine when the collision is due, or have no one ever took interest in the case, and if not, then please tell why not. If the formula you mathematically base physics on was anywhere near correct applying in cosmology as you in physics claim it is, then you must be able to apply the formula and show the precise date such a collision will take place because every factor in the formula is known to science! Please show me on what evidence do you build your belief that mass pulls by other mass because from where I stand what I see is that science had to invent a graviton spawned from the imagination of science to try and address the question as to how does mass pull mass. I put it to you that your use of $F = G \frac{M_1 M_2}{r^2}$ is as correct as the presumption was in the time of Galileo that the Sun is circling around the Earth. Sir, the substance of power controlling thinking still prevails in science as much as it did in the days of Copernicus during his life where everyone had to submit to the thinking of the Powers in Charge of science albeit members of the Church back then for fear for your life as Copernicus did. One still do not dare ask questions or ask for proof as I do on the merit of mass as a factor in cosmology (not in physics) for then one would be silenced till death interrupts the questioning. Copernicus so feared not the church but his colleges in science that he published his work after his death because then science could not kill him or employ the Church to do the official killing. Today science will allow me the privilege to die silently in a corner as long as I do it quietly because no one will allow my torturous screams to be heard. As in the time of the Copernicus, you lot still can't stand new thinking because new thinking will cultivate doubt in the minds of the many of those you consider as mindless. By seeing to it that my work goes unpublished you willingly kill me by killing and destroying thirty years of my life…and then you lot point the finger at what was done back then as if you could be bothered by not implementing the very same evil. I ask you where is the freethinking and what happens to the freedom of speech as long as what is said is truthfully substantiated and can be proven because all the facts I present in the article you are unable to disprove and that is a challenge I put to you. Professor Hehl, you didn't even read my article because if you did you would see there is not one point in any argument about my work where I am incorrect, not one point you are able to disprove me or show me I don't follow the laws applying to physics in detail, yet you have to audacity to denounce my entire article just because it does not fit the profile you envisage it should. You and all your colleges are more condemning than the Pope and the Church was because at least they gave Galileo a fair hearing and considered his evidence. Even that you lot fail to do. I have had this treatment for ten years and every time it is the same over and over. You just couldn't be bothered to read it because it takes too much effort on your part to think in terms of evaluating every idea I put across, and there are a lot of new ideas you then have to chew on!

If you did not bother to read the article you then would have seen I start where mathematics start and I can quote no one because I venture where no one has gone before; I go into singularity which by your definition is Singularity: a mathematical point at which certain physical quantities reach infinite values for example, according to the general relativity the curvature of space-time becomes infinite in a black hole. In the big bang theory the Universe was born from singularity in which the density and temperature of matter were infinite… and that I do quote, but that is all I can quote for the rest is the product of my labour and fruit of my mind. I challenge you to show where I stray from your definition of singularity when I show where to find singularity.

I explain just how it is possible to locate just such a point holding singularity to the precise value singularity must have in our modern Universe but I can assure you that where a mathematical point at which certain physical quantities reach infinite the grand splendour of mathematics are lost in dimensions not applying. I work in the era you can define but can't understand because it predates mathematics "singularity in which the density and temperature of matter were infinite" and it is in the infinite that mathematics becomes obsolete. Again I ask you that if no one ever has been there where I venture in physics, then whom must I quote because I quote the small part where science have been and that is all there is to quote. Every aspect about the Big Band deals with conditions prior to singularity deforming. If you use any quantity or formula based on numbers being more than 1 or any number to the power of zero, then you have left the realms of singularity because singularity could only be 1. At the point where singularity applies all complicated mathematics disappear. If you disagree, then give me any number that can apply to singularity other than 1 and please show me any mathematical formula that will apply to prove that singularity can be more than one.

If you did bother to read my article I sent you, you would have seen I show you exactly where $\Pi° = 1°$ could be found in the world of physics you study…and if I dare draw your attention to your accepted definition on singularity then as quoted it is a mathematical point at which certain physical quantities reach infinite" I show in the article where to locate this very point holding infinity. I also show where the point of singularity is infinite as it is holding what I named $\Pi° = 1°$. I show the point cannot ever start or become smaller since it is so small it has no space in which to form and if the point was in the Universe at the beginning, then it still has to be in the Universe because if it was in the Universe once, it must remain within the Universe because it has no other place to go by leaving the Universe. That is what you reject because that is what my article announces and my article introduces where to locate singularity and that is the article you reject. In this light going according to your attitude I am most delighted by your attitude, because from your attitude it is clear that where I venture you have never even left one thought.

In the web site **www.singularityrelavancy.com** I am introducing the reader to a world before mathematics as a multiplying process took centre stage. I take the reader into the cosmic era when 1x1 was 1 and only 1+1 was valid forming 2. This figuration proves mathematically that there was a time (1+1=2) pointing to a period before dimensions brought about perspectives (1x1=1). I take you into a true flat Universe where space has no dimensions because dimensions are the multiplication of numbers whereas a flat Universe is found within the adding of numbers and the adding of numbers point to a flat line forming the basis of the singular Universe. This process is directly formulated by translating Kepler's formula $a^3=kT^2$ to the true measure of gravity that is Π.

Please be so kind as to tell me, Professor Friedrich W. Hehl with all your mathematical splendour and magnificent abilities in constructing wisdom without using words, why is 1+1=2 and why is 2+2=4, because you do use it in physics, don't you? When you use numbers in your world of physics, being the Master that you are, you have thought about where numbers came from and how did numbers arrive? Please prove why it is that doubling two is also taking two into the square and while proving this, it is done without leaving a whisper of doubt. Please use your vast mathematical insight to explain without using words why would the third number be three, specifically three because that was how three came from singularity as 1 and why would the following number be four, which then is also the square of two by using the law of Pythagoras when proving this. How did five follow four to become five by using the law of Pythagoras to prove the point and then using this evidence to show that double five becomes ten, again by the merit of Pythagoras. Why would nine be the square of three and by adding 1 it becomes 10, because Professor Hehl, proving this is what really forms the basis of all science and that is how the Universe formed!

I wrote books about this process wherein I show numbers formed the start of the Universe and not material as you in science wish to believe. It formed by forming mathematics and the splendour of mathematics arrived only when the form of the Universe was completed and the cosmos stepped into the dimensional dynamics of space. This happened when the atom formed at $(\Pi^2+\Pi^2+\Pi^2+\Pi+3)=35.75\times\Pi=112$ which is also when space as a whole formed at $7/10(\Pi^6)\div 6=112$. The relation $7/10(\Pi^6)\div 6$ validates the sphere as (Π^6) spinning in (7/10) a six sided cube ÷6 which is outer space. What I show when using $7/10(\Pi^6)\div 6$ was the moment the Universe came into dimensions by arriving at the formation of the atom. In the article I show why one might conclude why the Universe uses Π as a numerical basis for gravity, but I agree, the article alone does not start to prove anything because for that there are four books forming such proof called **The Absolute Relevancy of Singularity: The Theses**

I dare you, no I challenge you with all your mathematical splendour to prove one iota I produce as evidence in these books being incorrect, and you have to use words to prove me wrong because where I venture is where mathematics goes singular which was at a point before mathematics came in place…and that place can still to be located in the present Universe. I can and I do show you the very spot where the Universe came from, but not in the article for there is no room to do it. You, with all the astonishing mathematics are stuck at the point where the big bang arrived and at that point everything that forms the Universe was already formed within the Universe. The Universe adopted space at the event of the Big bang and therefore mathematical values came in a dimensional context at that point. However, everything that currently is, was already present in the Universe at the event called the Big Bang. Before that the Universe was one being 1^0 or 451^0 or $5^0 \times 1^0$ because that is what singularity implies the value of singular space is, it is 1. It is because you got stuck with your mathematics and you used your mathematics instead of brains and that is why that you can't proceed to resolve issues beyond where the cosmos formed the atom. You are all agreeing about everything coming from singularity but going there you have to lose your mathematical equations; it does not apply! If you in science realised mathematics construe singularity as one, science might have realised that mesmerising mathematics before the big bang was useless as a tool to formulate facts, then you would've realised how to reach a pre big bang Universe. I did just that and I can show how, and why and where the first moment arrived. I can show you precisely where that fist moment of arrival is today. The mathematics you apply had to start somewhere and it is there where I venture.

The Universe in singularity adopted the four phenomena which is called **1) The Lagrangian system 2) The Roche limit 3) The Titius Bode law 4) The Coanda affect**, but to unravel their meaning you have to go into singularity and to do that you have to understand Kepler and to understand Kepler that introduced singularity when he introduce $a^3 = kT^2$ forming the measure of singularity applying. You have to part what Newton thought Kepler said from what Kepler said and explain what Kepler really said, and that I do in the article by using some simple algebraic relations and by the meaning behind $a^3 = kT^2$ and those four phenomena the Universe came about. But the condition to understand how the Universe came about is to first understand the four phenomena interacting. However, to understand how the Universe formed numerically or better titled **The New Cosmic Theory** one has to understand **The Cosmic Code** and learn how to read from it the interpretations of factors. To understand **The Cosmic Code** one has to understand the process of cosmic law supporting the Roche limit that works as what we think of in terms of forming **The Sound Barrier.**

To understand **The Sound Barrier** one has to understand the process of cosmic law supporting the Titius Bode law as well as the Lagrangian system that forms the Coanda effect and together the lot works as **The Four Cosmic Pillars** on which the entirety of everything was built by implementing singularity through a very specific process I named **The Four Cosmic Pillars**.

To understand those four comic laws applying one must be able to evaluate the process by reading **The Cosmic Code** in order to be able to recognise the actions brought about by singularity in relation to **Applying Physics** in terms of **The Absolute Relevancy of Singularity** and the one aspect of singularity applying in physics is being able to see how singularity forms space by the measure of Π, which is the only aspect of the entire collection of information I try to show in the article I sent to you…and you are unable to read that little bit…then how on earth will you ever get around to understand how the start of the Universe numerically came about when singularity and only singularity applied! Those books showing how the solar system was born and how the Universe came about I do not yet offer on sale. The information contained there are the really tough nutcrackers that explain by the cosmic code the inner working of gravity in stars and in galactica according to the cosmic code.

The Universe as we see it started much later in a period you call the Big bang but in truth there where the Big bang happened is when the atom came into form…and that I prove mathematically if you dare to read my work, which is the four books I wish to introduce via your Journal. Tell me Professor Friedrich W. Hehl, why is there mathematics formed by the adding process and mathematics formed by the multiplying process. In this evidence we find the development of the Universe. What happened that secured the forming of four to then be the prelude of five. Every number is a point and specific sets of points hold different relevancies placing the number in a quantification that brings about material in accordance to the coded relevancy. Please tell me the specific indisputable reason how the Universe did arrive at the value of what the number five depicts and it being precisely 5, or how did six become the next number on the numerical ladder and no other number but six tiny dots. Every number of dots serves a very specific purpose and five dots hold a value totally different from six. That is why nitrogen is a gas while carbon is a solid. That is why Mercury is a liquid with Xenon being a gas although they both are much more massive than say iron being a solid. Try to do this explaining according the law of Pythagoras by showing how the law of Pythagoras implemented the process and not use words. However, this is a country mile further than even the Cosmic Code is.

Why would seven bring about that a circle forms by redirecting directions as it is used in forming a circle by 7°? Why does the numerical value of 7° and only 7° play this role? Can you show me with your ingenious mathematics how the top part of Π is 21.991 when the bottom part is the circle by 7°, and use Pythagoras to substantiate the reasons. All the answers are in front of everyone but you said "With a lot of words and some simple algebraic relations, there is no way to "explain" the world of physics" and while you express you inability to see what I see, I see what I say I see and show what I see as clear as daylight while I do explain what I see precisely with a lot of words and some simple algebraic relations because complicated mathematical formulas did not yet enter the form of the Universe in the period I introduce as a **The New Cosmic Theory**. Before space applied, form applied and form was singular before space became dimensional. Have you ever considered why a triangle and a half circle are equal to a straight line by the dimensional value of 180°.or is this the first time your intellect went that far? …And this question points towards your obvious mathematical brilliance in physics and not the lack thereof. This I point out to show while you do know everything about physics there is thought to be, there also is a small possibility that you do not know everything about physics that there might be. The Universe started numerically mathematical and not by material as material came later; everything used a numerical order to form.

In the article you failed to read I show precisely where the Universe goes flat and becomes singular with the little impressive "and some simple algebraic relations" I show precisely the point where the Universe goes flat and in line with your impressive mastering of mathematics I challenge you to use your mathematics to show where I am misinformed or where I fall from the wagon by using "and some simple algebraic relations". Guess what, the "simple algebraic relations" is what the Universe used mathematically to indicate to Johannes Kepler as to inform him as well as Tyco Brahe how the Universe is constructed. The Universe showed how the Universe used $a^3=kT^2$ which is some simple algebraic relations as a means of form, but true to your academic arrogance I see you know better than even the Universe does because the some simple algebraic relations $a^3=kT^2$ is what the Universe used to describe to Kepler about the form the Universe adopts. By using some simple algebraic relations $a^3=kT^2$ what is in the Universe became the form of the Universe, and you missed all of that...that is a pity. You know, using $a^3=kT^2$ I show where the Universe goes flat, a task no one ever mastered by using breathtaking mathematical equations.

Some academics previously indicated there was some point holding singularity within the black hole but I show in my books where this happens and where to locate singularity in everyday life. If Einstein said the Universe goes flat, then that flatness must still be around and be everywhere so that everything will be able to go flat every time Einstein said it does! However, the most wise amongst you failed to even value the measure of singularity, being everywhere and all around by using impressive mathematical equations let alone to pinpoint the point serving singularity and even less to indicate where the exact centre of the Universe are to be located. I show why the Universe is a sphere, which is a fact that is up to now only been surmised. I show why the Universe applies gravity as the form of the sphere, which is something all the brilliant masters in mathematics failed to deliver up to now. Why did all the mathematical masters fail to prove that the Universe uses the shape of a sphere while all pictures indicate everyone accepts the fact, thus failing to impress with your brilliant mathematics you have to surmise, as you have to do with most things. I dare you to read my article and show my arguments I present are failing in anything that I say it does. In the article I show where singularity applies and why singularity chooses Π as the form of gravity. Why singularity chooses Π as the form of gravity you are unable to prove when you are using those most impressive mathematical equations you refer to, because singularity does not apply impressive mathematical equations. Singularity applies simplicity.

You showed me that I "seem to be out of touch with modern developments". Please let me show you why I "seem to be out of touch with modern developments" Please let me show you what inconsistencies there is with the basic mathematical formula Newton introduced when Newton tried to use mathematics to prove that mass was responsible for gravity because you belittle the mathematics I apply $a^3=kT^2$, which is precisely the mathematics Kepler used to portrays how the Universe forms and that formula he read from the way the Universe is constructed. Then you look down your nose at Kepler's work while Newton's mathematics broke every possible mathematical law it can when Newton tried to convince the world how clever he was by cheating with mathematics.

Newton started off applying the factors holding in the relevancy as follows: $F = \frac{r^2}{M_1 M_2}$ and discovered it fell short of any form of accuracy. There is no way that this formula would ever work even by a lesser degree of accuracy.

Then Newton changed the formula to being the following $F \alpha \frac{M_1 M_2}{r^2}$. Newton tried to convince (and succeeded) that one are able to change $F = \frac{r^2}{M_1 M_2}$ to $F \alpha \frac{M_1 M_2}{r^2}$ while it meant the ratio would still work in the same way as if it was something like this: $F = \frac{M_1 M_2}{r^2}$ and the formula still didn't work. The changing of the formula you use as the corner stone, the foundation of all physics still proved to be a total disaster notwithstanding the cheating of the most fundamental mathematical law that should support all physics laws. Then Newton and his fellow boffins in science cheated mathematical law even further to change the lot to $F = G \frac{M_1 M_2}{r^2}$ without explaining how $F = \frac{r^2}{M_1 M_2}$ could end up as being equal to $F = G \frac{M_1 M_2}{r^2}$. If you feel so strongly about mathematics used in physics then tell me Professor Friedrich W. Hehl, why don't you start to apply currencies to the factors and show the world how $F = \frac{r^2}{M_1 M_2}$ could become equal to $F \; \alpha \; \frac{M_1 M_2}{r^2}$ and this equal ness could be carried on to eventually become the same principle as $F = \frac{M_1 M_2}{r^2}$ to then become $F = G \frac{M_1 M_2}{r^2}$. Put in real numerical values and show it does not constitute to mathematical fraud. If Newton were that correct, then please use the formula $F = G \frac{M_1 M_2}{r^2}$ to prove that the value derived from $F = \frac{r^2}{M_1 M_2}$ could eventually be the very same equal ness as one would achieve from $F = G \frac{M_1 M_2}{r^2}$. Better still, why don't you write an article in your journal doing the song and dance about the accuracy the Universe proved Newton had when he implemented $F = G \frac{M_1 M_2}{r^2}$ because Hubble destroyed all the credibility that $F = G \frac{M_1 M_2}{r^2}$ once was thought to have. Then you can vindicate your attitude towards me while serving the cause of mathematics at the best you possibly can by restoring lost confidence in applying Newtonian religiosity. Put values to the factors and prove the Newtonian formulas have all the same results in the end. ...And while you are at it, write in the article showing how much did the distance there is between the earth and the moon reduce since the moon landing in 1969. Please use the most accurate figures available and then tell the world how much accuracy there is when implementing the calculation science applies to a shrinking Universe as Newton said it does when he said $F = \frac{r^2}{M_1 M_2}$ is equal to $F = G \frac{M_1 M_2}{r^2}$. Show how much all distances in the Universe shrink by their mass attracting other mass to bring about pulling forces of gravity applying throughout. Do use Newton's formula $F = G \frac{M_1 M_2}{r^2}$ to ensure accuracy. The Universe expands just as Hubble indicated it does as it expands from every centre holding singularity and it expands everywhere equally. Your colleague at Annalen Der Physics, Professor Doctor Ulrich Eckern once accused me of missing the basics of mathematics and classical mechanics by my evaluation of $F = G \frac{M_1 M_2}{r^2}$ but he failed to show what it is that I miss. Now it is the dream chance you have been waiting for...write an article about the correctness of the formula $F = G \frac{M_1 M_2}{r^2}$, show how the Universe comply to underwrite the absolute correctness of mass as a reliable factor shown by the formula having mass doing the pulling and then by the same token show what it is that I do not understand about the basics of mathematics and classical mechanics for I have been told this since my student days and after almost forty years I still fail to recognise what I am missing. To show me what it is I don't get, use the formula to prove how much did the moon come closer to the earth the past forty years by using $F = G \frac{M_1 M_2}{r^2}$ and the information acquired from data coming from the instruments placed on the moon for that sole purpose. If you are unable to do so, then never use mass in cosmology again because then mass is not pulling anything ever. Go one-step better... show why the Universe did not collapse back into singularity using $F = G \frac{M_1 M_2}{r^2}$ when r^2 was infinitely small and mass was absolutely contracted holding singularity in which the density and temperature of matter were infinite Back when the Big Bang took place the entire Universe was in one Black Hole, then why did $F = G \frac{M_1 M_2}{r^2}$ allow the lot to escape. It will never again have that chance to bring everything that went loose back into contraction again. At that point the Universe had its best chance to collapse into the Big Crunch because the further the radius expands with Hubble expanding, the lesser the chance is that mass can do the pulling. With an ever-increasing radius by the square, the mass effectively will reduce in strength. If you prove why $F = G \frac{M_1 M_2}{r^2}$ did not pull everything back into singularity then that will be a worthwhile challenge for your brilliant mathematical skills to achieve! Then you lot can stop searching for the mythical dark matter that has to cover Newton's blatant errors because the dark matter is just a cover up. If the matter was there and is there presently it has mass, then why does it not use the mass it must have in the present to pull the Universe into contraction starting here and now and why is it waiting for something to unleash the forces of gravity by the mass of the dark matter. ...And by the way, why would the matter not pull now if it has mass just because it is dark... if it is going to pull it already has to pull or it will never pull because it is not there at all. It is much more likely to be in the imaginations and calculations of science that be in the actual Universe. The matter being in the Universe has to have mass, dark or not. If mass does the pulling, and it is there, it has to pull now, in the present at present or the entire idea is just another scientific hoax to cover Newton's incompetence. The matter being dark or not, has to have mass, visible or unseen and if the mass is there and mass pulls by the force of gravity, then please tell why the lot is not pulling now and what are the dark matter waiting for to start the pulling that will begin the contracting? I say this is more proof that there is no mass and that mass does not pull and the entire concept is to try and vindicate Newton's absolute misjudgement of gravity. It is one more compromise to cover-up science.

You see, Professor Friedrich W. Hehl, if mass was the factor initiating gravity or then the falling of a body to the ground, solid objects will have to fall faster than objects that is empty and hollow because the empty space within the hollow object will restrain the falling by not falling with the object since only the mass would tend to fall leaving the empty space behind to restrict the downwards descent of the falling object. If the emptiness within the cup did not fall with the cup falling then the emptiness will bring a drag on the solid part that falls. The empty part within the cup will try to stop the fall while a solid filled glass will then fall faster than an empty glass because the emptiness within the empty part of say the cup or glass falling would not fall, leaving only the small rim of the cup falling. With the major part not falling this hollow cup will fall slower while the fullness of any solid object will fall in its entirety, making the fall of the solid object unrestricted by having no empty space that does not fall and thus the solid object then will fall faster. Drop a full glass from an aeroplane with an empty glass as see the emptiness in the empty glass falls as fat as the water does in the glass. It is the space that moves down taking the body with.

A filled container does not fall faster than does an empty container and visa versa because the empty space of the object falls as fast as the filled space of the object and all objects fall equal and according to a variation in density in air caused by temperature fluctuation (excluding some gasses) allowing any variety of mass to fall equally. It is the space and all the space notwithstanding being filled or not that falls or moves towards the roundness of the earth proving that space holding material or not holding material falls equally notwithstanding mass and for that reason that is why Galileo's pendulum swings regardless of pendulum length or size as Galileo said it would. It is the descending space driving the pendulum that swings. This again was proven by the very first ever experiment concluded scientifically. This fact of space descending does not come as a surprise because Empedocles proved this fact back in 450 BC. Empedocles showed that space displaces water from the clepsydra, which was a sphere shape container with a sprout on the top and small holes in the sphere through which water ran in small streams out at the bottom. When the flow of air or space was blocked in the spout by a finger covering the hole at the top of the sprout at the entry, the water stopped flowing from the clepsydra. They concluded in 450BC that it is the empty space that pushes the water out of the clepsydra because the moment one restricts the empty space or air to flow into the clepsydra from the top, the water will stop flowing out of the bottom of the clepsydra. Why would the flow of the water stop if the mass did pull the water down? When the finger blocks the sprout and stop the space entering from the top, the water does not fall to the ground but it is the empty space that pushes the water out at the bottom to fill the clepsydra from the top. When the finger blocks the sprout and stop air to come in through the sprout opening the water should still run out at the bottom by the mass of the water pulling, if mass was doing the pulling. If mass was the force giving factor, then the water must keep on flowing because the mass of the water did not disappear when the sprout was covered and therefore it still has to produce the pulling by forming gravity. All this evidence was known to science about 2500 years ago but since "With a lot of words and some simple algebraic relations, there is no way to "explain" the world of physics" it lacked mathematical communication and it should therefore surprise anyone very little that physics could not fathom this result 2500 years onwards. Professor Friedrich W. Hehl, do try and find the ability to use a lot of words to "explain" the world of physics because it is helpful preserving past experiments and results as it then does broaden one's horizons mentally...sometimes!

Forget the example always used about the hammer and the feather falling equal in a vacuum because the hammer and the nail and the elephant falling together will also fall equally notwithstanding falling in a vacuum or not falling in a vacuum. The vacuum part is conspicuously in place to purposely confuse reality as it is brought in to flagrantly spread misunderstanding of the issues in hand about the falling that takes place. With everything always falling equally when the same the condition applies to all objects falling and therefore with such falling happening under the very same variation of natural conditions applying, this shows it is the space in which the object is that falls and not the object falling while leaving the space it holds behind. The lack of relevant density in relation to air moving down stops the feather from falling equal just as gas does not fall with the space at the rate that space does descend. People realised this fact 2500 years ago but then used a lot of words to "explain" the world of physics and today because of not using a lot of words to "explain" the world of physics science has no idea how to interpret the very first experiment ever conducted! That is a travesty as much as it is a tragedy. All space falls by the compressing of the atmospheric space. The rotation of the earth moves the space sideways and this brings the space to move downwards by increasing the density of space or air as it comes closer to the earth. This results from the Roche limit applying to fix atmospheric layers varying in density. In my books I explain that principle applying mathematically. The increase in atmospheric density is the result of the rotation motion of the earth brought on by the Roche limit applying while it takes filled and unfilled space towards the solid of the ground and that is what the Coanda effect shows which is what your brilliant mathematics in one hundred years could not begin to explain. The Coanda effect is around for almost one hundred years and please use your mathematical skills to explain why the water will rather follow the roundness and flow with a detour along the rim of the glass than fall straight down as it should when mass would pull? That is the principle behind the sound barrier, and the Coanda effect, and wind restriction and hurricanes and more other things than I have room to mention. With all the attempts made in that past to uncover those issues I mention, it never was resolved notwithstanding all the impressive mathematics available to use. Notwithstanding using your mathematical marvels, science has not got any vague idea to explain any of the phenomena mentioned above. That is why modern science seems to be out of touch with modern developments. To understand these phenomena one has to understand singularity.

The simplicity singularity applies is shown when I show why the triangle and the straight line and the half circle are all equal to 180° but when considering form using mathematical dimensions this mathematical fact seems bizarre. It is obvious that the triangle and the straight line and the half circle are as wide apart as the sea and the Sun is, and yet there was a period in cosmic development when the three were mathematically equal as much as they still are. In the books, not the article, I show by using the law of Pythagoras why did Π become $21.991 \div 7$ or then is $\Pi=3.1416$. I show using the law of Pythagoras how and why by the law of Pythagoras is a circle Πr^2 or why by the law of Pythagoras is a circle circumference Πr or $\Pi d \div 2$. I show mathematically by the law of Pythagoras why is a circle using the specific value Π has to begin any circle or sphere with, but due to lack of space I can only prove it in my books as I can't prove it in the article. I show where and why did gravity start by the law of Pythagoras and what the true value of gravity is as gravity kick-started the Universe into a beginning. Can you show how the law of Pythagoras was implemented when the Universe formed, because I can show the reasons why and I do show the reasons why with using words since the law of Pythagoras implemented actual basic mathematics? I show why the law of Pythagoras implements the law it carries. The reason for this is the method how the Universe started off and the reasons why the Universe began. Maybe you should try to use words one day; after all it is a helpful tool in explaining physics, because it surely helped me explain what was never explained before. I mention the law of Pythagoras because the Universe does apply the law of Pythagoras in all of the cosmos and therefore the law of Pythagoras is part of physics, don't you agree? Can you use your breathtaking mathematics without using words to tell how did it come about that the law of Pythagoras has the dynamics it portrays it has in mathematics as well as physics, because I can by using words. Use your astonishing mathematics to show why everything started by the law of Pythagoras. Mathematics can't do it because the law of Pythagoras forms mathematics and the law of Pythagoras helped to form mathematics as mathematics developed.

Mass has nothing to do with falling and all things fall equal as if having equal mass when falling because all things fall equal in relation to the space in which they are. It is by buoyancy that space holds things and that removes mass while falling as a factor. Space not holding things fall with space holding things while it is therefore not the mass that causes the falling but the compressing of space which you call the atmosphere. The falling is written in relation with the value of Π. The value of Π is $3.1416 \div 1$ or it is $21.991 \div 7$. It is not coincidental that Π has two distinct equal values because it is due to precisely that that the first moment in the Universe came about when point 1 parted from point 2 putting space in-between eternity and infinity. There are two values forming Π as much as confirming Π. The air or space holds 21.991 when the Earth holds 7° but when spinning the earth applies the change of direction by instating the axis by the value of Π°Π which is the centre line or axis or earth centre Π° connecting singularity to the earth circle Π. Then relevancies in Π changes as the space that was 21.991 with the air held a link to the roundness of the earth being 7° at the time. But as the 7° dived into 21.991, the 7° goes singular or 1 as the space then in turn becomes Π=3.1416 or becomes the circle. The earth and the circle of the Earth becomes (7÷7)=Π° or 1 while the rim of the earth is Π=3.1416. Forming Π=3.1416÷Π° the form Π then aligns with centre of the earth holding singularity Π°because the axis placed singularity Π° in centre stage when the earth turned. Every time the earth or sphere turns, it places the surrounding space in relevancy from 21.991÷7 to form Π°Π. All this I said in the article that you say is not physics because I use of a lot of words to "explain" the world of physics. By the way, now for the first time in you entire career you also know what gravity is and what forms gravity and doesn't it make a lot more sense than to presume mass pulls mass by gravity without having a stitch of true evidence to prove Newton correct?

What happens is that the space condenses (21.991÷7) by the turning of the planet or star and the compacting of space surrounding the spinning sphere results from the rotational movement (÷7) of the earth that brings about that this compresses the space of air or atmosphere (21.991) into more density (÷7) which is done by movement of the space surrounding the turning object (21.991÷7) albeit a planet or a star and moves space filled with whatever or unfilled, going vertically down towards the roundness of the Earth that then is represented by the circle or the rim of the Earth (Π=3.1416÷1). Everything within the concentrating space will come closer to the surface because it is the space that moves down to the earth and not only the object filling space, but everything within the space including the space that falls downward. Every micro millimetre the relevancy of space changes from (21.991÷7) to the roundness of (Π=3.1416÷1) until the Earth forms the final (Π=3.1416÷1) and the object finds mass as a result.

The falling body never stops falling but find that mass comes about when relevancies changes from Π=21.991÷7 to Π=3.1416÷Π⁰. By touching the Earth, and by that ending the relevancy of Π=21.991÷7 from reapplying, the object then becomes part of the earth circle Π=3.1416÷Π⁰ and having contact with the axis Π⁰Π it becomes part of the Earth singularity distribution and only then finds in this relevancy applying the reward of mass. The body never stops falling but as the earth by density restrict the body movement vertically according to density, the falling becomes a tendency to move downwards in order to unite with singularity formed within the centre of the spinning earth. This is all about relevancies changing and relevancies reapplying positional changes, which is what gravity or time is. This is how the Universe goes flat or singular. It is Π°Π and that I say so many times in the article that you were incapable to read or you refused to read or you were not motivated to read…you can make your choice about you're reasons withholding you to understand what I explained in the unpublished article. Regrettably this is "a lot of words to "explain" the world of physics and because of the use of a lot of words to "explain" the world of physics it seems to be out of touch with modern developments making the process of thought very difficult to comply and prevent many academics this far to understand and therefore you lot would rather cling to the use of $F = G\frac{M_1 M_2}{r^2}$, which is truly what seems to be out of touch with modern developments since everyone accepts that the Universe expands. I prove my point of gravity being Π using some simple algebraic relations…you now have the chance to prove me wrong by proving how correct the cosmos shows Newton and his mass is. Show that Jupiter is coming closer to the Earth and therefore we have to relocate to some other galactica or die! Show why Jupiter is randomly located notwithstanding size and why the planets do not arrange positional allocations by the implementing of mass as the factor that would and that should arrange the positions of the various planets should mass truly be a reality in cosmology.

The spin of the sphere constitute of a change in direction to the value of 7°. In the centre of the circle the axis are Π⁰ and therefore the spin makes the directional change reform to singularity or change to the value of the circle in relation or in relativity with the axis in having the circle 3.1416 and having the centre or singularity or the axis ÷Π⁰ and with that connection the circle, which is space, goes singular or goes flat bringing about the much argued flat Universe. This is how gravity puts multi-dimensional space going into singularity or Π⁰. This means that the 7° becomes one or singular and space changes from (21.991÷7) to ΠΠ⁰. On the top of the equation the value of Π is 21.991, and by revaluating 7 through spin to become 1 that value changes to the value of 3.1416 being in relation to Π⁰=1.

The rest of the explanation that will bring proof to my statement when using Pythagoras is far too bulky to offer it at this point. I did say it in the article you refuse to publish because you refuse to read it that the relevancy of gravity is the changing of the value of $\Pi\Pi^0$. The space reduces (21.991÷7) to conform to singularity $3.1416 \div \Pi^0$ by the rotation of the sphere that produces an axis by initiating singularity Π^0. In the books I show the very reason why is $\Pi = 21.991 \div 7$ and I use the law of Pythagoras underwriting the Titius Bode law, that conjuncts with the Roche limit as well as the Lagrangian points to prove the Coanda effect and the Coanda effect, as I show in the article you didn't read and therefore wouldn't publish, is gravity by principal. Gravity is the Coanda effect that is the changing of liquids $\Pi = 21.991 \div 7$ to form solids $\Pi = 3.1416 \div \Pi^0$. I much rather say you didn't read it as putting it down to you not understanding it because I use "a lot of words to "explain" the world of physics and which possibly tops your understanding limitations and therefore you didn't publish it, don't you think…? It will be of no use to explain how the law of Pythagoras was implemented to prove $\Pi = 21.991 \div 7$ because I use words which you so honestly admit you don't fathom and it takes far to much space to explain the entire process in this letter. However, I do show how I mathematically conclude this value by using the law of Pythagoras in the books, if you care to look.

With the gravity being Π and that gravity comes as a result of the earth's spin contracting the space forming a denser substance called the atmosphere, which comes about and in accordance to every sphere spinning around an axis of its own, the increase in density around every spinning object brings about a loss in the overall density of space between all cosmic objects such as the earth and the moon and all the planets and the sun and the loss in overall density brings along that the distance controlled by the density of the substance there is between all cosmic structures gains in space. That is what the Big Bang was from the start. It is the substance that fills space in-between cosmic objects is that what you mathematically see as dark matter and it is limiting absolute expanding at the rate of what the density will allow. It is working as it should at present and no search is required. With the Big Bang the density in unoccupied space decreases as the density in occupied space increases making the Universe to seemingly expand, which it can't do. It is relevancies reapplying. The earth can't pull the moon by mass.

Mass does not constitute to the falling of objects but the space compressing brings about the falling action notwithstanding the space being filled with objects or empty of objects and therefore as Newton so vividly proved, one can pull a cover over the eyes of all but one (and that one is me) by cheating with mathematical formulas but to cheat by using words are a lot more difficult. The only thing Newton pulled by mass is a huge cover over many eyes for a long time. All things fall equal as we see on TV everyday where people and cars and bicycles and beds fall equal when dropped from aeroplanes and therefore it has to be the space in which the objects are including all the surrounding space in which the objects are not that is falling and not the object because of mass or having more favourable density. …And don't say I don't understand or I miss Newton because if you say you believe $F = G \dfrac{M_1 M_2}{r^2}$, then surely if you might understand Newton because then you clearly seems to be out of touch with modern developments in terms of cosmology and applying physics. It was shown this past century that the Universe is expanding and not contracting by mass…expanding means the lot forming everything in the Universe are drifting apart but that is "a lot of words to "explain" the world of physics which explains why modern Newtonian science seems to be out of touch with modern developments.

But all persons filling academic posts in science holds the attitude that they know all and others know nothing and academics know best while the rest of the population is mindless, thoughtless and worthless. Professor Friedrich W. Hehl, you are no exception to this rule. In fact, you are one of the best examples I have seen. I have taken these insults long enough…its been raining on me constantly for ten years ever since I tried to introduce my first thesis and if you wish to insult me, do so while you see there are in science parts that is yet still undiscovered and rather try to find how much there is that you don't know about science instead of thinking how much you as a scientist knows about science and what great achievement science is instead of trying to go where science still has to go. Let me give you some wisdom. A wise man thinks of all the things he does not know and what awaits his discovery while it is a fool that thinks of how much he knows and feels impressed by his personal field of knowledge. The best answer to any question one can have is "why" because no answer ever brings full conclusions.

I am no longer taking these insults on the cheek and riding it out as I did do so many, many times in the past ten years while trying to get my idea of The New Cosmic Theory read by anyone that is not so sublimely self-opinionated as I find practising and teaching academics in physics are. I try to introduce the cosmos, as it was in the pre-big bang era when only singularity prevailed because if it did prevail then it still has to prevail just as Einstein surmised it does in the flat Universe he saw. If it was part of the Universe back then when singularity prevailed, then it still must be part of the Universe since it has nowhere else to go when it leaves while it has nowhere to leave. The cosmos is written in a mathematical cosmic code and I found a way to translate the code and using that I became able to understand many unresolved facts about the cosmos. Please be warned that there is no simplicity such as just awarding mass in the forming of gravity.

The process is immensely more complicated than awarding mass because this is done without the circumvention of the truth by sidestepping reality in proving the of mass forming a factor in alignment of cosmic objects or in any cosmic planet showing more pulling power. You can't just simply gauge an object and then award mass to cheat reality, because in cosmology there is no such an escape root. In fact size plays no role in the cosmos because the smallest star there can ever be, the black hole, is also the biggest star there can ever be. If you disagree with my statement that mass does not apply in cosmic terms, then show what role mass plays in the positional allocation of planets circling the sun. See how planets align in the solar system and from their size, prove they use mass to line up accordance to mass being a factor... I prove that gravity is Π and every circle every planet makes is vivid proof of the fact that that gravity forms by Π applying. If the planets as well as the sun pulled by mass, then surely the lot had to be part of the sun by now because the pulling has been going on for some time. The truth is the planets are drifting further away from the sun every instant of time and how do you reconcile that with Newton's force of mass giving a pulling power.

The proof I bring is true about gravity being formed as a result of implementing the following phenomena, The Lagrangian system, 2) The Roche limit 3) The Titius Bode law 4) The Coanda affect, and the combination these phenomena we find the sphere as a multi-dimensional circle present the form Π, which I explain by delivering mathematical proof as to how they fit into the overall picture of gravity. I prove the fact that every individual one of those phenomena is forming a unit that is in total being what we think of as gravity. The phenomena altogether constitutes a unit that forms the process working as gravity.

I am going to use this letter as a web page introduction in the future as to show anyone that wishes to read my work, where I stand on matters concerning science. I think it gives a splendid opportunity since this shows the story of my life the last ten years and my encounters with all academics in physics! That makes this letter an open letter

PART 8

A Conspiracy to Commit Fraud on a Cosmic Scale

TO WHOM IT MAY CONCERN   I uncovered a mistake in science. From the onset, the mistake seems as insignificant as it is small. Because the rest of the book is about the mistake, I do not intend on elaborating about the mistake itself. The mistake came about with the culture of education and the mistake in itself seems harmless. When admitting that, one must also admit that any pilgrim that got lost and died of starvation through an incorrect travelling direction, made the very first part of his ultimate mistake by looking in the wrong direction. How harmful does looking in a specific direction seem, and yet such a mistake leads to his ultimate mortality. The traveller could when taking his first directional flaw with that the first incorrect step, only put his foot skew in avoiding a rock. Or he could have turned his face to avoid a branch and that move pointed him in a direction that lead to his fatality.

It is not the mistake that becomes the penalty and it is not the origin of such a mistake that leads to the penalising, but the ignoring of accepting signs telling the wonderer of an impending error and his stubborn ignoring of such telling sign that makes the lost party pay the ultimate price. By ignoring the mistake, for whatever reason, the ignoring of such a mistake is his undoing because the price due comes from the inability in recognising the sign indicating the presence of the mistake forming the reason for his final demise. The sooner such a person sees and admits the wrong, the less will be the consequences of his final price to pay.

The Newton mistake is one born in culture and the penalty from this mistake is bred by arrogance. At school minds are young and accepting, although developing. Through many tens of millennia humans came to a habit in surviving as a specie where culture taught them that accepting the elders advice is the same as to ensure survival of the following generations, and by such doing is also following the quickest way to an adult mind. By accepting the elders knowledge and experience without question proves the dominance of the tribe in relation to other tribes of the same race. This is culture we cannot do without and still maintain progress. It is an inheriting method humans grew on and is the corner stone of all civilization. We cannot abandon it.

The scholar sits in class and receives from the Master information that is completely his days breaking news. As far as his mind can tell the news is as actual as anything notwithstanding the fact that such news may be with the human mind for thousands of years. Whatever the teacher tells him is bound to be a first time experience so new he has no time to digest the information. Taking into account his youthful ways (which we all had), he has little stomach to scrutinize it because learning is a painful process to all. Without pain and perseverance there can be no education of any sort. He does very willingly accept the facts as tested and correct without flaws of any kind. The scholar has to because in any education system time will not allow students to ponder about detailing information and securing a prognoses to all learnt every day. What ever the Master tells the scholar is taken as Biblical correct without any thought about testing the results. Where there are cases of scholars having doubt and subsequent questions, the Master takes such behaviour as being obstinate and being a reflection of the student on his (the tutor's) personal integrity and knowledge. The young mind will very soon discover that his behaviour is not tolerated by the system, and the truth is the system cannot tolerate such behaviour for the good of the rest. Time must be spent on learning and accumulating as much information as that which the young mind can accept.

No information could be more affected by such a culture than that of Newton. You both understand Newton and are smart, or you do not understand Newton and accept that there will be very little future for you to have in the world of science. Newton is science. No Newton understanding automatically becomes "no science" education or learning. Without Newton, there are no other and science will be a vacuumed of containing nothing. This is very unfortunate but is the ultimate of truth. It is either Newton's way or no way at all. With this culture also brought along the stigma that only the minds of the sharp and the sighted can accept and understand Newton and when not understanding Newton one tends to fail your personal I.Q. test. It is a sure sign of the slow witted when the student fails to recognise what Newton said. The only way to advance in science is to understand Newton and indicate to all your pears how brilliant your mind is in accepting information.

All students have little understanding about Newton, and that I can and will prove through the next two hundred or so pages. The mistake Newton made and which I discovered is laughable small, yet it took me (not being that bright I may add) almost one lifetime to recognise the mistake whereas it took mankind three hundred and fifty years of research without recognising the flaw. Others in the past may have come to see what I saw, but if there were such persons they never saw what I saw because if they saw what I saw they should also see that behind such an almost invisible puncture hole in the tube, is a reason for science to deflate and not accumulate. When "understanding" Newton it becomes the very same as learning Newton from the heart and accept that what you memorise is what you know. The memorised knowledge is beyond question, as it has to form part of the identity of the student having secured the knowledge.

It is not the hole forming the puncture and preventing inflating being as such the obstacle that is of importance but what that hole does to the tire and the car and the travelling with the car that becomes a menace. In the extreme effort to keep the journey on course, the Academic Masters are spending an all out effort in inflating the deflating tire faster than the deflating tire can deflate. The recourses attached to this effort is enormous and without cause. It will lead to nowhere and that is where science is heading in their attempt not to head in that direction.

When students become masters the Masters seek to break new academic ground. Masters do not ponder the ways on which they lead their students, but have an all out effort in establishing their own support of new territory that will distinct them from the rest in future to come. My uncovering of such a mistake as I did could only be from a person as ill-educated as I. I did not go through the learning process where the learning process is the very same as a brain washing and mind controlling process. This remark may seem harsh and is not intended to be such, because reality demands no other way in education. The mistake and the carrying of the mistake with the support in refusing to recognise such a mistake is part of human training and there is very little to do but to admit that such may occur whereby to follow in acting to correct the mistake after the discovering, rectify what ever can be salvaged and build from there.

Friend and foe think me of alike as being slow of mind and not understanding Newton. This was what I was told on more numerous occasions than I care to remember. Pointing at the mistake I see, I am told by the wise as well as the nit-that-wise, such as I that through my lack in education I cannot dream to understand Newton and therefore are committed to the position of the ILL-EDUCATED, a position I learned to accept with grace. In all cases there are more sides than one and therefore I think of myself as poorly educated, the contras to my position must be those fortunate to be SUPER-EDUCATED. In this letter there will be the addressing to you holding the position of the reader and (I hope) the un-bias judge, with me presenting my case to the un-bias (you) in concerning the third party, the SUPER-EDUCATED. Referring to the party in opposing me as the SUPER-EDUCATED is by no means in disrespect and much to the contrary holds my whole-hearted admiration, (at time somewhat limited I admit.)

The information in this book I purposely made easy to red, and easy to follow. The content however, is only a drop compared to a bucket of information contained in all seven parts of "MATTER'S TIME IN SPACE The Theses". All information presented to you, in the first part of the book is an introduction to the second part. In the latter part of the book the conclusion comes. With out a detailed introduction in the first part, the information in the second part will be of little value, as the information is very conclusive about the first part. I have been researching cosmology since 1978, on a part time basis. The conclusions may seem simple; that however is in retrospect. Some of the arguments took me up to six months to arrive at since I do not have all the information on hand.

Please allow me to introduce the typical portrait of the members of your conspiracy clan that commits cosmic fraud on a Universal scale. These exceptional gifted and utmost blessed are also known as the Super-Educated-Wise-Amongst–The- Wise-Con-artists that has no equal amongst all human in all of human history going back as far as you wish. Newton set laws unto which the Universe has to comply.

The Universe must contract and reduce by gravity pulling everything and that is a very specific stipulation that Newton laid down. Then it was found that the Universe deliberately did not comply as the Universe was expanding all the time in stark contrast with Newton's explicit orders. However they, the one… and the only, the majestic Magicians…the Newtonian wizards, the Newtonian physicists in cosmology can find a remedy to correct where the Universe went wrong and dared to stray from Newton. This they can achieve because they found a way to conspire by which they control the minds and the thoughts of everyone having any knowledge about matters in science. These brilliant superhuman-men-amongst–man will form a composition in the cosmos comprising of an unseen, invisible non-detectable-non-existing dark matter that will pull the Universe together to get the Universe back on track where the rebellious Universe dared to stray from the orders of Newton. The process by which they will succeed was done the past three centuries with much success where it is called brainwashing by thought control in forming mind control and is also much better known as teaching physics. The process had countless success in teaching Newton's unrealistic principles.

Physics students, it is your duty to pull the plug on the powers of the All-Powerful Academics in Physics and stop their dishonesty. It is your task as the as the next physics generation to stop the criminals that are filling the corridors and the lecture halls of physics departments throughout the world by acting as if they know and all they know is to fool the next generation of students. Stop their teachings by forcing them to stop their criminal fraud. Force them to explain the deception called THE CRITICAL DENSITY, which is a conspiracy to commit fraud. Let them explain how an expanding Universe can suddenly and abruptly turn in direction of developing and start to contract as Newton stated it is doing at present, and when facing all other concluding evidence showings it was expanding since time began. Tell them to prove that the cosmos will begin to contract doing its turn about by using other proof than merely Newton's say-so. Tell them to bring proof with evidence that the cosmos is contracting as Newton said. Then force them to admit to the fraud they are precipitating in, which is THE CRITICAL DENSITY conspiracy. In THE CRITICAL DENSITY conspiracy all they say is that they are waiting to see when the cosmos would stop its criminally insane behaviour and start to listen to the laws of Sir Isaac Newton.
With The Critical Density shambles the modern Newtonian set out to defraud the world in the same manner as their Master Sir Isaac Newton has done centuries ago. Newton said the cosmos is contracting. When Hubble proved the cosmos is not contracting, Newtonians looked where the cosmos went wrong by not following Newton guidelines he so clearly set the cosmos to follow. It has to contract and not expand.

These whom I named in Honour of Sir Isaac Newton as the are the guard of the Newtonian High Priests carrying the name as the Newtonians are Men amongst mankind, that charged the Universe with not applying to standards set by Sir Isaac Newton, and then went on proving how incorrect the behaviour of the Universe was in not adhering to the direction gravity has according to Sir Isaac Newton. Since Sir Isaac Newton can't possibly make a mistake, it then was presumed the cosmos made the mistake by not following the gravity settings laid down by Sir Isaac Newton, the one that cannot falter nor could his teachings fail, carrying the illustrious name of Sir Isaac Newton. It must be the cosmos being at fault by expanding without seeking the approval of Sir Isaac Newton to do so in contradicting Sir Isaac Newton

Up to this point in science and in despite of an array of evidence pointing to the cosmos growing by expanding in every sense and with all pieces of evidence gathered by science from all over the Universe (including the solar system), the theory of contraction is still hailed as the infallible Newtonian truth. Every one that is part of physics, shares the Newtonian vision of a contracting Universe where the lot would one day again come together and Creation will end where Creation started some time ago. The Universe has mass that is pulling mass towards one another and we are in the centre of an ever shrinking Universe. That is what the lot of us can see… we are forming the centre of the ever contracting firmament having the entirety of the cosmos where every Newtonian can vividly see with his or her eyes through any telescope that all Newtonians minded scientists are sharing the centre stage of the ever collapsing Universe. The Universe is about to end where all mass contracts into one huge lump of material, and this conclusion contradicts al evidence gathered by science. If you don't believe me marry Newton's contraction with the Big Bang and see a divorce in place before any Church consummation of such a union could begin…but then again just as unlikely union in principle marriage between Galileo and Newton is in place and the mindless masses never once frowned on that! The contraction idea was never questioned and was accepted as being truer and much more believable than the presence of a living God Almighty was.

Students in Physics, it will serve you well to read the following arguments very carefully and come to a conclusion about what gravity is and what mass is and how it is impossible for the concept carrying the idea of mass then become responsible to form what we think of as gravity. Mass can't ever and doesn't bring about gravity.

If you are one of those members of society that never thought you would hear the name of an accomplished person as Sir Isaac Newton being associated with fraud, corruption and brainwashing, then these books are specially written to inform you about the truth there is lacking in the correctness about science.

Everyone knows that planets orbit around the Sun. Planets circle the Sun which is the same as saying planets orbit the Sun. Just by calling the circle motion in terms of what applies, that statement nullifies Newton's claim of mass that attract mass and put to question the reliability of Newton's dogma. Prove to the world that mass pulls mass to form gravity because that was never proven...accepted yes, on the say so of Sir Isaac Newton, but is you finish this book you will learn how awfully Sir Isaac Newton was mistaken about his entire cosmic principle.

If mass did attract mass, what kept the balance where the planets do find a balance in orbit and rather than moving towards the Sun. Planets orbit in ratio so precise we can set that on gears and yet the planets all so very different values in mass although the planets are randomly allocated and not according to the mass factor each holds. This is evidence of the fraud and a cover up. If mass did attract mass, then what is pushing the planets to remain in orbit?

Planets do maintain positions not according to mass that pulls but a balance foresees orbit. The idea of proof is automatically placed at the door of Newton. We talk of planets orbiting and that is what planets do. Planets don't creep up to the sun by the value of mass. If normal speech contradicts Newton, then it is this task Newton have to prove his supposition is correct and the claims about attraction Newton made can be substantiated, although Newtonians will deny this fact as if they deny he is correct the honour of their Master Newton and that is what they have to do. Think of what planets do... and you think that planets orbit. It is connected to the brain. No one thinks of planets spinning or planets basking in the summer Sun. When hearing about planets the first thing that comes to mind is the rotating of planets while circling around the Sun.

However, just using the term orbiting is in total defiance with Newton! Newton said gravity draws or pulls or moves in the direction..., which would have one understand that the two objects in example the Sun and any of the various planets will be moving directly towards each other.

The term pulling does not suggest any circling because no one can be pulling towards and does that while circling around the object. When pulling anything it must take place while using the shortest line possible. That serves the term pulling. Then the saying goes that planets orbit indicating they follow a circle. That is not what Newton said. However, wrong that may seem but circling is precisely what planets are doing.

In conversation we speak of the planets orbiting. If Newton was correct we should be speaking of the planets pulling, but talking about pulling would be blatantly wrong according to the normal spoken word. Never do we refer to the planets pulling the Sun or the Sun pulling the planets, but we speak of seasons coming from orbital positions. Being in orbit has to neutralise the pulling and then cancel the pulling concept that also became culture.

If there was a pulling, and the word orbit cancels such an idea, then there has to be some sort of prevention taking place that disallows the pulling to commit the direction of travel.

I know it is said that the orbiting object falls as fast as it circles and by falling while moving to the following side on position it never reaches the Sun, and yes, it makes sense, but there has to be some form of resistance replacing the planet in the next side position and preventing the falling or the pulling from taking place.

Using the formula $F = G \dfrac{M_1 M_2}{r^2}$ as Newton provided, disallows any other concept other than moving towards. The person Newton got his ideas from and the work he raped completely, that of Johannes

Kepler explained this very well, but Johannes Kepler makes no room for any pulling of any sort. In the work of Johannes Kepler he said that the space being the orbiting route a^3 remains at a specific distance k while the orbit T^2 takes place…and in all my other books that addresses more information I take Newton to task on his dismembering of Kepler's formula by corrupting Kepler's work and with what amounts to fraud, Newton takes science on a goose chase that holds no truth. There is no pulling by mass of mass in any way.

We have either one of two that has to be incorrect. If Newton is correct, then the normal way of functioning of the is incorrect. Then we must start saying planets are pulled to the Sun. If the normal form of speech is correct and the planets are merely orbiting the Sun, then Newton is wrong. The planets can't orbit the Sun while at the same time we have Newton's accepted scientific presumptions being correct.

The fraud part is in the accepting of the Universe expanding while still insisting that Newton is correct in his dogma of contraction with mass. This web page is an effort to show how Mainstream Physics brainwash students into accepting Newton's hypotheses of mass attracting by force while the entire Universe is expanding at the rate the Big Bang indicates.

NEWTON'S MYTHOLOGY Written by Peet Schutte

Students, read the following message about my book I named **Newton's Mythology** and learn how you are brainwashed and how your mind is pre- conditioned into believing in Newton's myth of pure deception which Academics call physics. If you are a student in physics who don't believe that you are subjected to unlawful brainwashing, then read on.

Let's start surveying civilized principles by evaluating what lawfulness means and what would constitute as morality. Let's determine what makes the crook in the book?

If any person, notwithstanding what reason is given in justifying such depravity, tells a lie or conveys untruths to further whatever humble cause, it is seen as fraud. To convey information that is not substantiated as a verified fact then the mere conveying of such information becomes fraud.

When any person, notwithstanding what reasons given, repeats such a lie unabated while being well aware that the information passed on by such a person is incorrect, then the person commits deceit. When anyone is repeating the information that is passed on as being unblemished factual substantiated and verified truth while such a person knows very well that such information is void of proof or lacks proof, then committing such an act is a criminal enterprise. Academics in physics commit every one of the above indignities and yet see their actions as being lawful and even much praiseworthy and hold their role in society in the highest esteem imaginable.

They fail to see the crime that they commit while tutoring physics. Whatever motivation they may claim to have which they offer to serve them as forming their driving force, the fact that they perpetually perpetrate in unlawful behaviour, by spreading untruths, such actions on their part put those academics holding such highly regarded positions in the league of ordinary cheats, gangsters and common criminals. By willfully and constantly falsifying facts to further whatever humble cause and produce illegal claims repeatedly, remains derogative behaviour and is unlawful by nature, notwithstanding what morality it should serve.

A Preacher or Pastor lying on behalf of God is not lying on behalf of God and to think the Preacher or Pastor improves or underlines the Greatness of God by lying on behalf of God is very mistaken, because in reality such a Preacher is falsifying the truth for his or her personal benefit and trying to impress the congress about his importance and not the importance of God. Lying is wrong and doing so even in the name of God remains despicable.

The same applies to academics in physics. There is no argument that can change this truth about falsifying the truth and when doing so there is no hiding behind any excuses of ennobling to benefit mankind that will change such truth into righteous conducting.

Newton said centuries ago that gravity is the force of attraction there is between objects that hold mass and it is the mass factor that brings about this attraction, which Newton claimed there is. The Universe does not contract and all the proof we require to disprove such a statement we find in the Hubble constant as a guarantee.

Moreover, it is true that the Universe never contracted even for a brief instant and proving that is the Big Bang concept with all the proof that this concept brings in backing the principle of expansion in the Universe. Planets never moved closer, are not moving closer and will never move closer to each other and this is backed by all information collected this past century.

The Moon is not coming closer but the distance between the Moon and the earth is widening. Studies about the Universe reveals every time that space in the cosmos increases constantly. Studies find all things are moving apart and away from one another.

Any and all the proof about this is beyond what any doubt may present to counter this knowledge. Notwithstanding this irrefutable findings, science still regards Newton as the only person that ever lived whom no one ever could prove wrong…and this is upheld by Mainstream Physics in spite of the cosmos proving Newton wrong every instant of time.

The basis of what science holds as its foundation we find to be the Newtonian principle of $F = G \dfrac{M_1 M_2}{r^2}$. The foundation used by science promotes this argument and backs up this argument well knowing that in the cosmos there is no evidence backing up this proposal Newton suggested.

The Newton formula $F = G \dfrac{M_1 M_2}{r^2}$ used as basis for science sees gravity as being a force of attraction and the force of gravity is being in place between all objects in accordance with the mass factor that the objects have as presented by Newton in the formula $F = G \dfrac{M_1 M_2}{r^2}$

What we find as we gauge all evidence found while studying the Universe, is that reality shows there is no attraction between objects in space going on anywhere in the Universe, that the entirety of such a concept is a myth and the outward moving of the Universe has been coming from and since the time of the Big Bang and maintaining this flow of material is substantiated in a concept named as the Hubble constant, which proves Newton's perceptions to be a myth.

The Hubble constant proves that space everywhere is growing ever since time began and the growth never stopped ever since. Knowing this irrefutable fact does not deter science from under scribing Newton as the sole basis that underwrites all the correctness of all of science known as physics. However, Hubble and the Big bang and all other investigations contradict this attraction Idea Newtonian dogma holds. Therefore, any further believing that there is attraction going on as Newton claimed has to be viewed for what it is and that it is a fairy tail.
The Big Bang Theory proves Newton's idea as not only being wrong but Newton's idea of attraction is a joke. If the Big Bang is expanding the Universe, then how can the Universe contract at the same time? Any contraction by nature would have the Universe collapse back into infinity the moment the Big bang moved out of infinity.

Ask your professor to show how an expanding Universe can also contract and your professor will tell you about Einstein's Critical Density theory. This theory I prove is the biggest fraud ever devised by any group of persons in the history of civilization! This is perpetrating fraud and conducting in upholding deceptions instituted by Newton that then formed the institution of lies they call physics. The Universe does not contract in any way, means or form and even such a suggestion is incorrect! The Moon and Earth are not moving closer but are moving apart. The entire Universe is growing in space and nowhere is space depleting by any norm used.

Academics are very aware of this misconception Newton had and still academics in physics are promoting the ideas of Newton as the unwavering truth. Academics teaching these misconceptions are committing fraud, notwithstanding the portraying of their role in society being unblemished, spotless while they are covered in a lily white blanket making them being whiter than snow and having such a holier than thou attitude.

Teaching Newton is participating in deception and promoting Newton is criminally deceiving the public and while doing so, is committing an act with criminal intentions.

Then, in the face of all this evidence contradicting Sir Isaac Newton, they remain upholding the correctness of Sir Isaac Newton and keep on teaching students about the unwavering correctness of Sir Isaac Newton. They put down conditions of learning to this effect and are expecting students to repeat these untruths and unproven facts by forcing answers to that effect in examinations.

Forcing the acceptance of this untruth about physics is equal to preposterous subjecting students to physiological torture and heinous mind conditioning, scandalous thought control and brainwashing. This applies to everyone serving as a tutor in physics notwithstanding whatever status the torturers might have in society or the morality they attach as a reason to commit such atrocities.

If you are a student, then you are conditioned by academics in controlling your thinking by enforcing pre-mind setting and in which they methodically force you into believing in Newton and this is an on going process conducted for centuries in the past, while it is the truth that Newton is completely void of any tests that may secure any form of confirmation and in securing proof then also by that establishing proof.

Read this book **Newton's Mythology** and then use the information I supply in the book to insist that Academics who are teaching physics, prove to students that Newton's statements of attraction are correct. Let those academics explain the method mass uses.

Let them with precise detail show when mass is applying, that gravity is produced by mass and such producing of gravity that then would establish attraction! I show precisely how gravity produces mass but mass can never produce gravity. I show with explicit detail when, how and where gravity forms mass but mass can never form gravity. What I prove annihilates every Newtonian claim.

They never prove Newton's philosophy on gravity but those persons conducting teaching in the subject of physics force all physics students to learn Newton's gravitational concepts and accept the facts as if it has been proven beyond all other facts. Students have to believe that Newton is correct or academics will see to it that they fail their examination. The condition of being accepted in physics is to accept Newton without questioning the proof that is never supplied.

Let those academics now prove precisely how mass brings about gravity and then afterwards test you on how Newton is proven correct and not on you repeating facts about what they say is true about what Newton said, which they say is true. The manner they present Newton is completely hearsay and that method may not be used in any court of law.

Let your professors now prove how it is that Newton's teachings are correct and then examine you on the process they use to prove Newton's concepts. At present they say Newton is correct and then they test you on your ability in repeating that Newton is correct without ever proving to you that Newton is correct. Let those physics professors now prove Newton and then test you on the manner they use to prove Newton to be correct.

The truth beyond all other truth is that Newton's gravity has never been proven (because try as you may it is not possible to prove Newton's formula forming gravity mathematically) and because academics know that, academics require the blind acceptance of Newton by students. This unconditional acceptance of Newton's correctness relies only on the pre-conditioning of students' mind set and academics depend only on the student trusting the academic "say so" about the institutionalised correctness of Newton. That Newton is correct nevertheless and notwithstanding that there is no founding proof about this matter, is what students should be accepting blindly.

Pre-conditioning students into blind acceptance depends on the academics' insistence that students approve Newton's concepts without pre judgment or students insisting on scrutiny of any sorts. In examination students have to outright and blindly follow academics' say so only because academics say so. Academics depend on students never questioning their say so or demand proof about what academics teach. Those academics in teaching positions insist that all students accept Newton's accuracy.

This is methodical mind control as much as it is the brainwashing I show that they enforce. If you are one of those believing that Newton was ever proven, then what you believe to be true is a lie because Newton can't be proven and that is the truth! The time has come to face your teachers and force them to stop the ongoing old culture of bullying students and conditioning their thoughts by enforcing on them dogmas which is mind control! In order to get students to accept Newton's hypothesis, academics resort to brainwashing pupils and students.
They teach you that the Universe contracts and to state their case they force students to learn that gravity is proved by Newton introducing the following formula $F = G \dfrac{M_1 M_2}{r^2}$ They say that M_1 is the mass of the Earth and M_2 is the mass of the individual in questions mass and the multiplying of these factors with the gravitational constant produces the force of gravity when this gets divided by the square of the radius. Please let you lecturer put in all the values of the formula and prove Newton is correct. If he can't and I know for sure he never can fill in the symbols and calculate the force of gravity, then read the rest of the web page that follows to see how far academics in physics go to brainwash students into believing in Newton's fraud.

This is a fair test to see if Newton's contraction theory underwritten by Newton's attraction formula $F = G \dfrac{M_1 M_2}{r^2}$ is valid, then force your professor to use this formula as it reads and show WHEN the Moon and the Earth is going to collide. If he fails to do it by using Newton's formula as $F = G \dfrac{M_1 M_2}{r^2}$ then you will know who is conning you, him or I and who is truthful again him or I. I charge all academics to prove what I say is being wrong in any way or even that I exaggerate in the least. I challenge Newtonian academics to prove that mass does indeed form any force of any sorts and in particular gravity!

To those professors claiming Newtonian ideas are substantiated by proof, I say that notwithstanding your personal academic qualifications and while at the same time disregarding your status and previous achievements as well as ignoring your many admirable abilities you may have and however superior they might be, I shall teach you about gravity. I say it is time Students learn the truth about physics notwithstanding the status academics will loose.

Students read **Newton's Fraud** and challenge those academics depending on their ability to brainwash you into submission.

After personal research on an extensive basis (when leisure time permitted at first) the past twenty-seven years, and in depth study the past six years, I completed the version of my book, which I named **THE THESES**. I have first tried the commercial route in order to introduce my work with the book I named **THE THEORY**. As there is little commercial value in such an academic book, I have not been able to find a publisher willing to publish the work.

I then tried to publish the book I renamed **THE HYPOTHESIS** with a greater emphasis on proof than the previous book, but the only route I can follow down that alley, is through the Internet. This route is seemingly impossible because the people in search of information on the internet, is not capable of understanding the work and those that can understand the work does not search for information on the internet. After the disappointment that this route proved, I decide on an academic version with all the proof I hope will satisfy academics.

This forces to find a promoter willing to allow me to follow the route through the academic institutions, should I ever have any chance to gain recognition. I have now completed the book version that I named

THE THESES. As I am of the opinion that the work in research may prove valuable to science, as material for further study potential, I feel it worth the effort to promote. By it self, it does not change the view science hold on cosmology, and on the other hand, nothing science at present view about cosmology is correct and therefore it lays groundwork for a change in the direction of further study. I submit a summery, indicating the line of argument this book takes on cosmology. I do realize the method of calculation is new to science, but it proves many aspects that science does not understand about unexplained factors in the universe.

In the past, I have been severely criticized by academics for the view I hold and to my judgment rather unduly. The largest complaint by academics comes because of my critic about Newton's findings. There is a vast difference to the cosmos and the earth surrounding us filled with life, the part we humans take fore granted, but the only place we will find life will be on this planet. Compare earth to the moon, and the moon represent the real cosmos. It is dead, except for matter that grows with cosmic time. To understand the cosmos the cosmos and life has to stand apart, the two on they're own and only then does the cosmos appear as it is. A rocket launched is not part of the cosmos, but is part of life's extending energy. Life is the ability to occupy and manipulate space-time occupied and unoccupied. In the manipulation, there is a very strict limitation to what man can and cannot accomplish. It is as if man has become drunk with his supposedly unlimited ability. This will lead to tragedies man still has to meet. There is no reason sending brave men off to their death, if only we can find a way to recognize the dangers. NOTHING IN THE COSMOS CAN DIE, BECAUSE DEATH IS A HUMAN MISCONCEPTION. ALL ARE PART OF DIMENSIONS WHERE WE ARE INCLUDED OR EXCLUDED. THIS CONCEPT ABOVE ALL SEEMS TO EVOKE THE MOST REJECTION AMONG THOSE I HAVE COME ACROSS.

The last four five wrote full time and with which I went on the Internet. The persons I wish to draw to my book do not seek information on the Internet and those who do look for information, cannot understand my work. The commercial route I tried to follow; however no publisher would touch it because it is not recognized science. That reason forced me to seek an academic route, in order to find recognition for my work.

To the correctness in my work, I have no doubt; however there may be parts I still have not clarified sufficiently. I am more than willing to meet such challenges because I know the proof is there, I was just not yet able to recognize the question.

Being human we tend to form concepts through culture and such concepts we then translate to the cosmos in an attempt to fid meaning moreover about our worth than the facts relating to the cosmos. We find ourselves physical apart and as much linking and with that we then wish to link and find meaning. What links and what divide is the stumbling block because we place man in the cosmos through which we then wish to produce a link. Such a link does not exist but for the tiny speck of cosmic dust which we call either home or earth. The main issue to divide what should be divided and then accumulate what should connect and doing that is the purpose of this book.

Quoted directly from the Oxford dictionary of Astronomy the following:

The definition of space-time is as follows:
Space-time is a four dimensional position of the universe where the position of an object is specified by three coordinates in space and one position in time. According to the theory of special relativity there is no absolute time, which can be measured independently of the observer, so events that are simultaneous as seen from one observer occur at different times when seen from a different place. Time must therefore be measured in a relative manner as are positions in three-dimensional Euclidean space, and this is achieved through the concept of space-time. The trajectory of an object in space-time is called world line. General relativity relates to curvature of space-time to the positions and motions of particles of matter.

The definition of singularity is as follows:
Singularity: a mathematical point at which certain physical quantities reach infinite values for example, according to the general relativity the curvature of space-time becomes infinite in a black hole. In the big bang theory the universe was born from singularity in which the density and temperature of matter were infinite.

While it probably is the greatest mind to walk the earth that produced the spectacular in the above, a much more simple mind as the one I have noticed much more simple aspects of nature that only one with a simple mind as I have could recognise because my mind does not have the capacity for the greatness of the great minds.

If the universe did start from one single point and time matter and space flowed from that point, then that point must have a relative connecting base because such a point holding singularity must be eternal as space matter and time link eternal. There therefore must be one point linking the entire universe when regarding the fact of singularity. Then according to the theory off relativity there has to be one exact point holding time in a relevance notwithstanding the fact that time depart from that position and relate differently to all space-time away from such a point.

Every person I have discussed facts about creation recollects images in the trend depicted in a presentation as one may find to the above. That would be the most unlikely way Creation came in place. The recalling of pictures representing images about creation must have form, but to mathematics it had no form. From this thought the very opposite arise where Creation came from nothing but such an idea is mathematically simply not possible.

The thought of nothing is just what it is, a thought of nothing and although it is in the human mind common nature to present nothing as a value in the recalling of something, nothing is a presentation of the figment in the human mind. There can be no number such as nothing and that was (possibly) Newton's biggest error. Nothing represent non-existing and that is just what nothing is, it is non-existing.

In order to prove my point I wish to ask the reader to define the shortest line there can theoretically be. If he should answer anything but that the shortest line will be at a point where the beginning and is the very same spot he will be wrong. The shortest line that can ever be anywhere must have a start and finish holding the exact same spot. The line will be humanly impossible to create but we humans are capable of very little.

When the line has a beginning and an end at the very same spot and it wishes to extend the position as to further the possibility it has, which direction should it favour. Humans in the west would naturedly think of extending from left to right while in the east humans may want to go from right to left. Some persons will tend to go up or down, but all of the options are about human preference and not mathematical conclusions. Extending the line in any one direction will favour one direction without a conclusion about not extending in other directions. Such a conclusion has no sound mathematical foundation. The only option about extending will be in all directions equally in order to give a meaningful non-bias flow of mathematical equilibrium

The shortest line in the realm of possibilities must have a start and finish holding one spot and such a line will also be a dot or a circle. Not favouring one direction puts all directions at equilibrium meaning that any form what ever may be can develop from such a spot with the end and the start being the same. This reasoning prompted me to look for singularity in such a spot because if the prime spot from which all came was a spot, then the spot must hold the shortest line but more prominent it will hold the smallest form including the smallest circle.

One possibility that the shortest spot can never have is having a starting point on the zero mark. If the mark of zero holds the start it must also hold the end because the end and the beginning has the same position. If the position of zero then is the beginning, the end will also be zero leaving the line without an end as well as without a beginning.

The conclusion from this is that no line can start at zero because that will be a mathematical impossibility. A line or spot starting at zero would therefore be shorter than the shortest line possible. A line growing or extending from zero can never leave zero because of the influence of being zero disqualifies any possibility of growth. If the line then had to grow in all directions at the same pace the line must therefore be a circle. The value of the circle is Π, and that is where creation started.

That gave me the clue where to start looking for singularity. One would find singularity in the value Π and the value Π will be in all things rotating in a circle. To start my explanation about my cosmic theory I wish to firstly bring some nostalgic and the relevancy will become apparent later on. Such is the importance however that I wish to place this at the very start of the prologue.

When we were boys we played with a top we called the spinning top. I cannot imagine that there is one boy in the western world that did not hold such a devise in his hand. Tying a string securely around the tapered cone started the operation and then with a jerking or pulling throw the devise is launched in a projectile manner and the big knack to success was getting the nail end firmly on the ground and by the realizing jerk the top was rotating. The champion was always the one boy that could throw his top to spin the fastest and that would create a humming sound. The louder the sound produced the bigger champion.

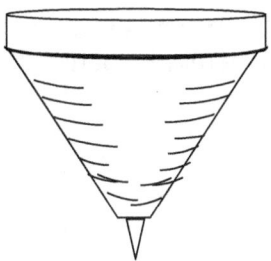

When a back braking effort produced a throw of enormity the spinning top would not only produce sound varying in pitch but also create a spin that would seem to have some instability. There are very many limitations about the spin, parameters that determine the slowest and the highest sin rate and spinning is within the parameters of such settings. The question arising is why such parameters are there in the first place?

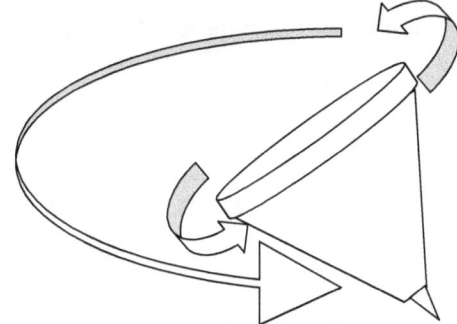 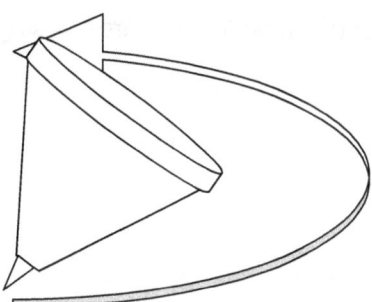

An enormous effort will have the top going oblong while spinning violently and as the pace reduced the top will stabilize by coming to an upright position. In the upright position it wall then spin for the remainder of the period where it will in the end start tilting to the side and in a last effort throw a few wild oblong turns and fall over.

Boys playing games will never realize scientific breakthrough explaining and grown ups do not play with toys. In this little toy played everywhere everyday by almost every one is the answer most brilliant of human Brainpower seek answers about all the cosmic riddles no one seem to understand. In the spin as such one may find two vital boundaries in the motion and the boundaries are marked by a wobble coming about as if the top is fighting some other influence. Spinning too fast pulls the centre off centre and so does spinning too slow. It is the same influence coming about at both ends of the limitation in the spin. There are influences at work, but force…no; it cannot be forces setting such boundaries. From that I started per cuing what sets such limitations because that limitation must be universal as all matter is spinning in one way or the other.

When the Universe started there was one spot that released a dot. The spot as large as a thought grew into a dot. This too is exactly as the Bible says it happened and it still happens exactly like this. The Universe starts from infinity and grows into eternity just as it happened at the start and this is what Newtonians confuse with what they know as the Hubble constant. To check if I am correct look at anything spinning and you will see conformation. The dot that released was by such release relevant to other dots released because there was to be motion that measured many dots. The dots in release were relevant since they were the same.

Only time being in delay by cycle of infinity interrupting eternity to form space, as space is the history of time gone to the past. But since we are looking at things as it started I put it into the past tense. That cycle brought about time delay as every cycle drifted further from the original singularity while the original was still responding as well. That was the first space. It was time being one infinity part. Since the dot was also singularity and was the very same as singularity with a small difference that the spot was •Π⁰ and the dot was •Π¹. At first, at moment-Alfa, there was no space nor time for only relevancies came about. Relevancies acted as motion to bring change to eternity by changing the flow of time in eternity. There was the perfect spot in which time moved while remaining the same. Then heat brought expansion and expansion brought space and space brought movement and movement brought about a Universe. That is how it started and that is how it still is. This was before the atom and the atom was before light and that too is exactly what the Bible says in Geneses 1:1.

Time will forever remain eternity but space or time distortion, which will forever remain infinity, interrupted the flow of eternity. Space breaks the monotony of time in eternity by parting time from in infinity. Yet, the relevancies did imply motion except for the fact that singularity is very much incapable of motion. Every dot had a purpose to fill a position in relation to the other dots that the spot excited. All dots had a line of three where two was one, each on every side of the spot. The spot was one with the two dots forming two, which improvised for motion that would later come to space and the three was what space was going to become. How do we know this: 0.1416 x 7 = .991 and that is by the value that the Universe grows.

Time was four because the four would bring about motion as heat separated infinity from eternity or hot from cold. Five was space because space was one removed from time. Space is the distortion of time and one outside time would bring a time delay or a time distortion of four plus one which is five, hence the principle behind the Lagrangian system. Because material was the square of space material was a crossing of three plus three forming six. However to find out what this means you have to read the entire theses called <u>The Absolute relevancy of Singularity</u>.

Space-time is the four of time, plus the three in singularity around which the four of time turns, therefore space-time is seven.

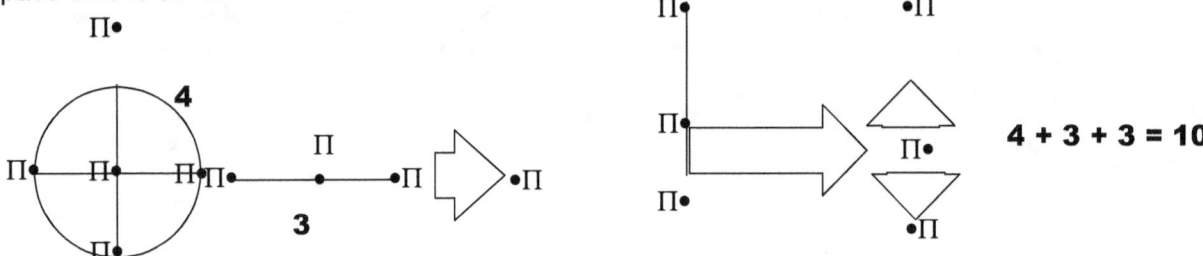

In the circle using r²Π the r has to have distinctive qualities placing it as a factor apart from Π. Where the growth shows no separate distinction but a continuous flow from the precise centre to the precise edge the flow would become in relation with Π depicting the circle and Π replacing r as reference to any point on the circle.

By using r, distinction in the circle is possible but by using, Π there is no distinction possible. Therefore, in the beginning when time formed space there was only relevance coming about from •Π⁰ and the dot was •Π¹ with no mention of any possible r. The fact of r representing a radius represents space and what we refer to be long before the Big Bang introduced space or mathematics using space.

Before the Big bang the lot was form without dimensions playing any part. Then the atom came. Only after that did the Big Bang come. Even before the atom was the point lining up and forming positions that was spinning faster than the speed of light can ever achieve. However every point today still serve the role it took on at that stage and serves in the position that it had during the time it had no space with eternal time. These relevancies developed as part of a Universe we shall never understand. The Universe had no sides and a line was equal to a triangle, which was equal to a half circle. Singularity holds the double space-time position of five times two (matter and space duplicating singularity) which then is ten.

How did the Universe liberate material and heat forming space from singularity because with singularity comes an unchangeable eternal condition that is non-changing-everlasting in all conditions and aspects that is remaining in absolute equilibrium. This equilibrium maintains because all development extends form precisely in a detailed equal equilibrium throughout. Think about what brought the cosmos out from the eternal rest in which it was. The eternal rest still maintains and is therefore our detection. What inspired the eternal rest the cosmos was in and inspired change to the state of eternal rest? What evoked change? That is the question the Atheist will never be able to answer but that too is the most basic and ever-lasting fundamentals of the Universe. Singularity Π⁰ is not substance but it is a thought establishing substance Π. What changed in this split second start before the official start? I do not wish to ponder on this matter in the letter I am writing at this minute, as there are other books where I delve into this matter. It is called **The Absolute relevancy of Singularity**

From the deep freeze of creation came the Hot Big Bang and the 3D Universal displacement came about with the relatives being **10 ÷7(4(π^2 +π^2)) = 112.795** and then a second one established 3D by introducing the six to seven sides Universe at a density point of **7 / 10 π^6 / 6 = 112.162**. There is of course a lot more information about this establishing of the Universe than what I mention at this point. The question is what made the Universe freeze, to form the Universe in space and through time. It had to start with a specific reason applying, which brought about space-time. Once the process started there was no stop to it, but there is no chance that the initiation of the start was spontaneous by nature. With everything being in one spot, all within that spot was in a state of eternal rest. While all remains the same and nothing changing, what brought on the sudden change of everything, shocking

It is the time the space takes to bring about new positions in space occupied through motion applied and through motion applied, it takes space-time to duplicate and in the relevancy of duplicating, the duplicating takes certain duration in time to move from point to point. Gravity is motion of space towards and in relation with a centre and the time is the period such motion takes while that centre is attempting to produce space by doubling space through motion leading space away from a specific controlling centre to another specific controlling centre.

The time it takes to complete such an attempt provides space the opportunity to double its status. Moreover, the time stands affected by this motion material creates to duplicate space-time by generating singularity and activating different locations holding singularity. Gravity is speed and speed in space in motion through time duration. Gravity is motion combating heat expansion and supplying space with space producing through motion providing the space the opportunity to expand while remaining in relevance.

At the start, the gravity part invested heavy in material. However, in space light came about since there is no gravity in space. Light is the attempt to establish motion not controlled by any centre and the time it takes to establish space between such a centre of space control and the light finding an ability to dispense the space by reactive motion. Gravity produced space by allowing as much as producing overheating with a feeble attempt to combat the overheating. What is the meaning of heat if there is no cold to set the standard for the heat to become a value, which is then related to the other end, where such another end must be the limit in cold? The moment singularity produced space-time heat distanced from the cold factor. Space produced a cold base to have heat within. How did singularity part the shared principle of being the unification of heat and cold as a unit? Singularity froze in applying gravity bringing about particle separation within singularity. Heat and cold parted to produce frozen heat by atoms forming and captured heat by unleashing uncontrolled material that ended as space in time.

We can see from what is available how everything fell in place. There was space filling with heat where the heat was compacted by the time delay. Every point established a relevance of three crossing singularity a motionless forming a line and four points formed the square of time. Crossing the four points that supply the time or the motion aspect is a line with three points supporting singularity by not spinning but they maintain singularity in being as motionless as singularity is. That four plus the three is seven. Then all this, which is the body of material requires somewhere to be within because to be is to move within something. That too is time but we regard that time to be space. That time is the heat that became space during the event of the Big Bang. That gives the atoms seven point as plus the electron or space having three points in relation to the seven points there is in time. If singularity is at the limit the value of ten but is the tenth portion of the value then ten must be divided by the tenth portion and from that the tenth portion must be distracted.

This as a group forms a relation with heat unlike any other. If mass did the trick these must have been the group having the second least density, but they form the group with the least density as a five point group. The next group holding the Pythagoras five or the Lagrangian five plus two or three, four or five enabling their relation with heat to be quite remarkable. This must be some indication of events during the period just preceding the Big Bang at say 10^{-7}, 10^{-6}, 10^{-5}, the time when the fuel that would ignite the Big

Bang turning heat into space turned material into heat. The relation of five plus one, plus one and five, plus one plus one plus one is just too uncanny to ignore.

Following the process and seeing the influence of singularity should bring about a pattern that may lead one to a pattern of how the required heat formed and how the intended heat transformed to space. Density depends more on proton number arrangement producing specific form in relevancy as to merely and only having mass as factor that contributes to the forming and development of stars in the cosmos. The evidence is so clear that mass has nothing to do with gravity but density has everything to do with gravity. Density is the volume of space in numbers used to fill material in ratio with numbers of space per volume not filled with space. It is matter versus space in every sense there are. This came about before the Big Bang took place and before space was formerly space and time was formally motion. It was a time when singularity set relevancies moving from Π^O to Π

In that manner we know that that was the way particles formed combinations just after the arriving of moment-Alfa. Singularity brought the Universe but also singularity brought the divisions between the many Universes that followed the immeasurable many Universes that came after the flooding of Universes to follow the leaders. The term "moment-Alfa" is the way I refer to the moment when singularity changed, not when space formed or time began or space exploded but even before anything including mathematics became definitive. At this point mathematics renders it useless. There was no space or time to calculate because relevancies came in place. Form took shape but space there still was not because Π^O moved to Π. Every slightest point in space became an opportunity of establishing a Universe with most different functions and ingredients there might form. This is apparent from the fact that it still takes place at the present moment by motion attaching new singularity through duplication and through duplication releases previously attached singularity from serving the purpose of duplicating by motion.

This was the era of distinction, when separation brought an all-possible new Universe

The spot becoming the Dot

The Spot

When the cosmos came to motion, motion was not yet defined. When the cosmos brought about motion, the first motion was relevancies. Cold parted from hot. Eternity parted from infinity. Motion parted from motion absence. Infinity broke the laboriousness of eternity for the duration of infinity. The spot became and grew into the dot.

From what the spot was to what the dot now is might be just a mathematical implication of going from 1^0 to 1^1 but in reality that first motion was the creating of and establishing of an entire Universe with all possibilities now in it. Never again can that much growth become a reality, although to us the growth is beyond what we ever can notice. But it is because the growth is so massive and we are so small that we are unable to notice such almighty growth.

When the spot Π^0 became functional and established all relevancies possible, heat parted from cold as eternity parted from infinity. The expansion was not clear motion but more a parting of relevancies where a centre formed a relevancy because the centre could not provide motion. Without being capable of motion, the centre established four points, which also served singularity. From the inverse square law we know that the centre doubled by producing the four points holding singularity.

By exciting the centre spot, the centre spot came to be because of the heat that formed in relevancy as heat parted from the cold bringing about the division that followed and that was the motion that formed. Therefore the heat had to move but being singularity it could not get singularity to move. In an attempt to establish growth, singularity activated six spots of which four was having motion drawn into relevance four spots that was providing what was to be motion and three that was to be securing the position the centre holds. There were four forming a ring around singularity with two forming in locations we will refer to as above and as below or north and south.

The three in line was in singularity not being able to move but the four was also in singularity and just as incapable of moving. All the points came as relevancies applying the forming of more of what was to come but only the four committed to time were expected to move. The four points that came as a result of discrepancies that became time that produced form and that established the relation with the one but had to perform the motion by expanding was as much incapable of motion as the centre was that charged the four with motion in the first place. As they were incapable of motion, it still required a tendency to apply motion that did separate Π^0 from Π. This not only involved form but it involved all relevancies that did come or may in the future come about as a result of the attempt to commit motion. If mass was a factor contributing to gravity the cosmos would have frozen back to singularity without ever releasing singularity to relevancy.

Mass does not establish gravity. There is no magical graviton. In the beginning there was no mass but boy was there gravity! The only means that the cosmos could find a way to break from the grip of eternal eternity was to expand into relevancies. Such a feat can only go to task by forming opposing hot and cold. Becoming hot produces more of what is heating. That implies motion or a moving away from where it was by generating more of what is available. Only where hot released from cold could whatever was repeated once again and duplicate what was before into what then is more. Secured by motion T^2 in relation to a specific centre **k** from where singularity holds the Universe true to form. The **k** was an intention to place apart and by today's standards will not even qualify any noticing.

All that are is in singularity. From singularity comes the motion and the space we call space-time. Singularity is dimensionless, time less and space less and because of all this features, it carries the value of Π^0. By expanding, singularity applies a relation coming about that reforms singularity from Π^0 to Π. Only when extending Π^0 to Π, the extending creates motion and the motion creates space that then doubles through motion applying which cuts the space in motion in half by matching the space as a duplicate. Motion creates another dimension or another level reforming singularity from Π^0 to Π or from Π to Π^2 or from Π^2 to Π^3

As said before we now know Π came about since Π is achieving form and not space. Only **r** can establish space as size will accumulate and as it had with everything else singularity had **r** covered by one as in being $r^0 = 1$. By reducing the circle radius **r** by half continuously will lead to an infinite small circle and an infinite number holding r would place **r** to the power of one as a factor. Then as a factor **r** would not contest any change when change is introduced into any future equation but Π will remain because the circle as a form remains even being infinitely small. By reducing r indefinitely to the tune of half each time, r would become infinitely small, beyond human calculating means, however as mentioned in the case of the smallest dot holding one spot, r would become insignificant beyond human comprehension even, but never reaching zero and still Π would remain intact and dictating form. To amplify by dimension a value has to be set to r but if r remained covered by singularity all alterations that could possibly come about was in the form, which was Π.

This expanding can be a problem one can wrestle with for one lifetime and never reach any conclusion. How can something grow without getting more that what was before? Then it hit me like a ton of bricks. The answer is in heat but not heat, as we know heat. It is heat in getting relevancies between outer limits. Only heat could break the monotony of singularity. Heat in the form we now know heat as heat is now. Since the Big Bang heat is material transforming from one state to another state.

The change that took place involved singularity but singularity was 1^0 and being $^0 1$ could not grow. The growth came about. Heat rose from singularity, but if heat rose from singularity. Singularity as a factor changed from 1^0 to 1^1, which means a relevancy came in place that no one could detect. It is true that 1^1 are still one, but one could then escape from singularity by producing factors other than 1. Heat came about but only as a relevancy to utter cold. If there is heat, there is cold or if there is no heat there can be no cold. Space came into forming a relevancy that brought form. Since it is a relevancy and not a generation by accumulation, the form produced was Π.

The spot formed a dot by heat and cold establishing relevancies and from that singularity was broken to allow all other forms of relevancies to come about. The cosmos did not start because of gravity. The cosmos started with heat and cold coming into a relevancy and in the cosmos there is no hot as much as there is no cold. The cosmos broke, put from the confinement of singularity by establishing a singularity in a relation of heat and cold. The heat that came about was beyond measure because the cold that held the heat was also beyond measure. The immeasurable heat was on the outside of the dot that formed and the cold was on the inside of the dot that formed.

The cold contracted because in nature cold contracts. The heat expanded into a dimension of form and heat by expansion is in nature about motion. Motion is duplicating that which is and heat is what is duplicating by motion. But only heat by expansion was possible because in affect singularity cannot move. The motion became contraction, as the motion was the result of heat expanding which was forming four points in the rim of the dot. The expanding of the points created motion in relevance of a centre that formed because of the motion, which established an immovable centre as the Coanda effect, placed more dots in relation to more dots that formed.

Every dot was Π and every dot formed Π^3 because of the expanding heat, which produced Π^2. With that a new relevancy came about forming a centre in between the four points of expansion that was resulting in time. But since the points were in themselves singularity, which is immovable and space-less, they still heated forming a cold centre with the heat bringing about motion. It became a repetition where infinity broke eternity by producing a centre because of space (or rather form) forming the motion to enable the space to form in relation to the heat applying motion. This brought about a Cosmos being conceived.

The spot forms a full circle, but the line running through the circle is forever present because that is the future radius of the circle that will one day develop the circle, which is equal to the present diameter. The fact of the presence of such a possible line in such a possible circle dividing the possible circle into two parts makes the centre line equal to the half circle. The line forms the half circle but not only that the line presents the half circle as much as the line is the half circle. The line then is 180^0 and the half circle is 180^0 because in singularity the two factors are the same.

The same value is of course $\Pi^0 = 1$. The issue of concern is o understand that singularity cannot move. Singularity has no space. Singularity is no only part of the Universe but singularity is the Universe. By establishing motion singularity has to be charged with the time delay we find space to be. The space is time taking a period or a duration while moving from one singularity point to another singularity point while conducting the heat and the accumulation of heat that built up due to the retarding of the time to conduct the heat forms the space that is conductor to bring about the motion of the space.

It takes heat time to entice singularity and singularity can only entice. Singularity cannot move and neither can singularity form space. By enticing from one relevancy to another there is a bridging of heat that has to be crossed in order to send the gravity or the enticing or the relevancy to depart the space and reconnect the space to the next singularity. Bridging all the accumulated various time delays that formed an accumulation of heat through time distorting brings us the space we see and have. However there is no true space or motion but it is eternal motionless space is singularity charging time to provoke heat into forming space.

Three points formed a line covering singularity where the centre singularity recovered heat to grow and two points served as an axis to allow the rotation and to assist the duplication. There is one centre connecting the duplication of three as well as the recovery of one (the fourth one) that is applying the tie aspect. Therefore, motion consists of three positions in relation to a centre, which forms as space in relevancy to the motion and the space receive a controlling centre.

The duplication comes about as singularity is exciting another singularity in precise relevancy of 3 to 3 to 1, but the points charged is as space less and as motionless as only singularity is. The heat it requires to carry the exciting between points forming space and the space excites heat and the time delay it takes to excite singularity between points forms space-time.

That is why the Universe is Π

Where motion conducts electrical charging which is equal to gravity the charging of motion is to entice duplication of singularity. This is the basis, the heart and the sole ingredient of the Coanda principle that includes the Roche limit ($\Pi^2/4$). The charging of gravity $((7/10) + (7/10))/(10/7) = \Pi^2$ and the charging of space-time $\Pi^3 = \Pi^2\Pi$ is all due to the relevancy brought on by the Coanda principle. The value of motion came from singularity exciting singularity and that is the duplication while the duplication or motion presents the space.

The development came into eras as the relevancies brought about new relevancies that spawned even newer relevancies that all remained in touch with the original singularity centres. Every one focused a new time delay that eventually brought about space and every distortion of time brought more. That concentrated between singularity points that charged the points to form space. When the charging became overdue in some sectors it erupted in forming the Big Bang. By the time the Big Bang erupted there was such a huge backlog in heat and time corrupted and delayed the next result was the employing of space as a commodity in the Universe. The relevancy was C the gravity was C^2 and the space was C^3.

That left what was inside atom still spinning faster than the speed of light applying the relevancy of $k = C$ where the electron applied the relevancy of $T^2 = C^2$ and that formed the atom which then became the cube of the speed of light $a^3 = C^3$. That left the atom at the relevant size of what the speed of light permitted at the time but since the Universe from that the relevancy expanded as the Atom grew in space to the extent it has now. The purpose of the star is to recapture the space the atom grew into and from there dismiss the space by spinning faster than whet the speed of light will be on the outside of the star.

This form came about when only form was present in the cosmos. It was in a time era where form featured in relevancies that would lead to one day becoming the atom. The atom forms a dual purpose of duplicating as well as dismissing and some prefers the one better to the other. This relevancy came in place when time was not time and space was form. Time is forever eternity being interrupted by form in infinity to bring about eternity ticking as infinity ticks. Before that singularity took on stages in forming relevancies between duplicating and dismissing space-time, which incidentally was not yet truly space-time in the sense we think of as space-time. At first a dot moved from the spot leaving the spot but taking with the spot as part of the dot to remain in the dot. The two never separated but the one allowed the other to be.

As the dot confirmed a discrepancy between infinity and eternity by defining infinity as an interruption of eternity cold and hot parted a union.

The dot that formed was not space but a relaying of time to form a new point of singularity where eternity was interrupted by infinity. Time took form from 1^0 to 1^1 or from Π^0 to Π. It brought form into differentiating between interrupted eternities with infinity doing the interrupting

Then a true distinct relevance came about that positioned a time differentiation outside the realm of time by four. In this realisation we can assume that space had some meaning at this point and the formula used to investigate suggests just that. Even in grouping, there are characteristics, which make a certain group of atoms more perceptible to duplicating and others more perceptible to dismissing

The lagging of exciting one point in relation to another point takes time. It takes time to send the message across to get singularity at that point excited. It takes effort to bridge from the dominating singularity to the independent singularity and that effort slows time down. The crossing of the divide is space formed by pushing time into duplicating. When time brought in a five points to the four points it took time to be, that fifth point became more than only form, it became space because it was one point outside the Universe of four or of form. One must see the three points established as motion duplicating singularity in relation to one dismissing singularity. This always has to strike a balance in order to establish space-time. It began as a relevancy and developed into space-time flowing or space-time displacement.

What the Coanda effect proves is that the rotating motion is acclimating a centre that exemplifies all phenomena in nature as we use nature to our advantage. All of nature including gravity uses the same method of motion forming around a circle in rotation and in the centre of the circle a point of no motion holding no space comes about. This is what Kepler taught us when he taught us $a^3 = k\,T^2$. With the Coanda effect forming the basic principle of all natural phenomena we can see from that, that the motion of liquid in the presence of a solid forms a centre that excites as it establishes singularity. From that rotation, space flows to a controlling centre but because of the lack of motion in that centre, there is a lack of space in that centre. Therefore, there is proof of a flow towards such an established centre and there is control from that point of singularity. In every case, the singularity controlling space-time sets standards for space dismissing in relation to space duplicating.

The duplicating stands in regard to the flow that the liquidity of the atom in relation to the solidity of the atom can reproduce. This forms density and mass but mass has little influence on the scenario.

There is a balance between the duplication in relation to the dismissing of space and the relation extends to the number of atomic elements present which then creates the balance applying within the star. As the liquid heat subsides in the centre of the star and the heat density is dissolved by the dismissing-prone elements the motion or moving ability of the star as a unit fades away as the star becomes static and solid with less space providing the star with less motion.

It is the way the atom formed before the atom took on space-time. It is in the formation, that space-time relates to motion. We have some elements being quite massive but also lighter than air and others are quit light but as dense as they come. This can only be a contribution from the way the atom relates to heat, which make the atom volatile (movable) or dense (motionless). Those elements being volatile are also very movable and in that we find the role that such elements play in the star. Stars that are predominantly made up of hydrogen and helium with very slight support from the metallic inner core are those stars that duplicate by producing motion. However the point I wish to press is that mass and being massive and being heavy do not support the fact that some elements have more gravity they produce because their protons are more numerous than others. The fact that mass generates gravity is a myth.

One will find that whatever group one chooses there are gasses and there are solids. If mass was attracting mass then the strongest mass must be attracted to the strongest mass and the least mass must float in the air. $F = G (M.m) r^2$ hardly can even begin to explain the fact that there is a gas that is more massive than iron but floats in the breeze just as hydrogen which is the least massive element.

Nitrogen 7	melts at $-210°C$	boils at $-195.8°C$
Oxygen 8	melts at $-218.8°C$	boils at $-183°C$
Fluorine 9	melts at $-219.6°C$	boils at $-188.2°C$
Neon 10	melts at $-248.59°C$	boils at $-246°C$
Sodium 11	melts at $97.85°C$	boils at $892°C$
Magnesium 12	melts at $650°C$	boils at $1107°$
Aluminum 13	melts at $660°C$	boils at $2450°$
Silicon 14	melts at $1412°C$	boils at $2680°C$
Phosphorus 15	melts at $44.25°C$	boils at $280°C$
Sulphur 16	melts $119°C$	boils at $444.6C$
Chlorine 17	melts at -101	boils at $-34.7C$
Argon 18	melts at $-189.4°C$	boils at $-185.8°C$
Potassium 19	melts at $63.2°C$	boils at $760°C$
Calcium 20	melts at $838°C$	boils at $1440°C$

Ignoring these facts, Mainstream science will hardly answer the problem we do not understand and such ignoring brings strong doubts about the quality and sincerity of science.

Excluding Argon, which is six (carbon's number) times two and suddenly that is a less dense material. The four times five plus… group are the following:

Scandium 21	melts at $-157°C$	boils at $-152°C$
Titanium 22	melts at $1670°C$	boils at $3260°C$
Vanadium 23	melts at $1902°C$	boils at $3400°C$
Chromium 24	melts at $1857°C$	boils at $2665°C$
Manganese 25	melts at $1244°C$	boils at $2150°C$

Iron being the five times five plus one is the only generator of electricity and therefore the producer of gravity making five times five plus one the ultimate relevancy to heat in reducing space. Still Krypton is much more massive and turns out to be a gas.

Krypton 36	melts at $1539°C$	boils at $2730°C$
Iron 26	melts at $1536.5°C$	boils at $3000°C$
Cobalt 27	melts at $1495°C$	boils at $2900°C$
Nickel 28	melts at $1453°C$	boils at $2730°C$
Palladium 46	melts at $1552°C$	boils at $3980°C$
Silver 47	melts at $1412°C$	boils at $2680°C$
Cadmium 48	melts at $321.03°C$	boils at $765°C$
Xenon 54	melts at $-111.79°C$	boils a $-108°C$

How can science promote their image of establishing honesty when they are confronted by such truths but choose to ignore the truth so long as a lie will bring them some respectability.

Since the star is the total configuration of the atom's characteristics, the atoms will tell us what we should know about every layer from what is applying in such a layer to what characteristics such a layer would show when it provides the function of what it has to for fill within the star.

The concept still is about singularity linking to time and that is a distortion of time. At point, five of extended singularity is one outside the rim of time and then is in the distortion of time, which is space by relevance of singularity in specific position according to time.

Let us investigate and try to find a way by using logic how a star applies gravity. Therefore it is not the number of dots that is important. It is not the size of the number of dots occupying the position or the size of the space the dots occupy that is prominent. It is the relation in the dismissing of space and the duplicating of space that becomes important. The less space there is the more the favour will be to reduce the space because of the advantage the dots have in securing space-time that will prevent overheating. On the other hand the more space secured will also prevent overheating and therefore those will opt to duplicate space in order to find space to secure and prevent overheating.

Since the Earth has no singularity demand that is much better developed than the Universe sustains, we find on Earth a relevancy of Π to $(\Pi^2+\Pi^2)(\Pi^2\Pi)3$ is adequate. But in bigger units the space-time displacing relating to space duplication presents much more demands on atomic structures occupying space within the star containing through set boundaries. In the presumed to be bigger stars there is much space filled with atoms occupying much space. In the stars more massive but holding lesser space the atoms must also hold lesser space but they also hold more protons by number in the lesser space.

The space the particles hold is directly in relation to the particles the containing structure duplicate. The more space that is relevant to the structure that the star duplicate by motion is then in turn once again relevant to the space the structure destroys by proton action in space less units. The more space the particle claims in relation to the space the container holds that relates to the space the container duplicate is relative to the space the containing structure destroy. From that mass derives value. As individual occupying space the atom is an individual container by own merits and as such duplicate space in this regard within the specific confinements of atoms.

This we will classify as normal applying structure values the atom has in outer space or in structures with very little atmosphere. Please note there is no pressure involved because the motion involved creates conditions naturally instead of unnatural pumping that causes pressure. Pressure is an artificial creation as part of life but has no role in the natural cosmos. Pressure is a condition where the retaining of particles has to be confined in a patrician made of material where the outer wall does the retaining of the substance within. This obviously cannot be in a star because the "pressure" is regulated from a condition applying and space-time controlling inner centre that needs no solid walls to contain whatever is inside. With that one can see there is a Universal difference between the concept of pressure forming due to human action inside a container and what comes about as secluded space-time within a star.

As the demand of singularity in such units grow stronger some relevancies within the atom come into play and I developed a system whereby I can arrange the space-time merits of space-time curtailing within the confinement of the star borders applying in the star to place such a demand in relation to singularity where the ultimate demand sets the standards. In the sun, for instance, which is a minuscule small star a relevancy in the outer region might be 3^3 relating to singularity and with the atom having, a sustaining displacement of $(\Pi^2+\Pi^2)(\Pi^2\Pi)3$ there is no danger of the atom demising. The electron in the sun will have a diminishing factor of 27 whereas the atom can sustain $(\Pi^2+\Pi^2)(\Pi^2\Pi)3 = 1836$. The relation in the atom degenerated by 27 leaving the atom a sustaining value of the electron plus the neutron applying space-time without involving any of the neutron aspects at all. That is the mass of the space-time that the electron will consume in the space reducing flow of space-time.

The flow is the result of heat distributed where the heat is delivers to the dismissing sector and producing of the duplicating of space by mass within the star that then forms a favouring of duplication in comparison to dismissing. The star is a bright little boy shining by dismissing pebbles of light-photons into space. When a demand on space-time displacement reaches an accumulated general displacing or movement to the value of what 56.6 protons can achieve in a general flow of conducting space-time that would be the requirement for such accumulated displacement within that space forming the motion of the space or the time aspect of space.

The star accumulated more heat by consumption applying direct dismissing without accumulating space-time in liquid form beforehand therefore there is no heat remaining to dispose of by producing light. When the general displacing flow of space-time within that sector of the star or the star in total reaches 56.6 displacement the natural state of absolute solidifying becomes the norm within the star. From then on, the star will exclude all electron functions and stop shining as the demand on space-time duplication and diminishing reduced the atom to space without a heat envelope that will be electrons or a liquid/gas jacket. Only the nucleus will be able to sustain the diminishing and the reducing of space by increasing of time.

The entire star becomes a solid structure by reducing space-time directly freezing the space-time from a gaseous state to a solid state. By motion, speeding up the tempo of the flow of space-time the liquid state of space-time is by passed going from gas to solidity in one motion. The atom would shrink to such little space it will have space within the star that only the centre nucleus will fit. More reducing by applying motion in creating space differentiation will leave a star with so little space the space will be insufficient to secure a position for the neutrons and the star will then have the name of being a neutron star. Going even further will find the proton rejected from the star.

Every atom holds (I am guessing), as many dots as the sun has subatomic particles per atoms and that would still be a very conservative guess. Every dot is a controlling centre selecting a regional centre where every regional centre selects a centre. This goes on as long as there are spots forming groups as individuals unable to survive independent. The others that was unable to group formed heat that became space, which became the broken dots. The dots form groups to survive and as a group, the survival depends on doing what the group has to do to remain cool. In another book, I reserve one chapter to explain the phenomenon what I called the Lagrangian atom. These dots arrange in a manner that they could favour either the space duplicating aspect or the space dismissing aspect.

This can only be the result of the fact that even in the case of the sun, the inner space is almost entirely liquid heat and the liquid heat produces sufficient space to dismiss as the centre that holds the heavy metal particles, where all the dismissing is done. The liquidity provides motion while the solidity removes motion in the centre of the star. The dismissing going on is in the space factor where the space leads to a denser heat within that space because there are insufficient material to accommodate all the heat by the dismissing factor T^2. In that case motion far outweighs dismissing $k>T^2$ but a time comes in every star that the dismissing takes absolute charge. $k<T^2$ That is when the star goes dark. The Earth is mainly about duplication of space much more than dismissing of space and so is every structure in the solar system.

I would suggest we think of stars in the following terms. A star that generates and transmits a lot of light is weak on gravity because their progress started recently. They command a lot of space-time but the demand they have to keep their cooling acceptable is very low. In that they can generate a lot of light but with the demand on cooling low and the gravity in the centre not very developed, those stars cast a lot of light back into outer space. It is just because of the size the stars hold that tell the that the stars are still young and have a weak developed governing singularity. The stars will have very prominent hydrogen and helium layers, with the inner core not very prominent. The control of the star is still very much in the individual atoms and in that the motion the atoms have to produce in order to maintain their individual singularity will only come about through motion. The atom has to make contact with as much space-time through motion as possible since it has a very poor ability in contracting space –time in support of the cooling system.

When what was perfect became imperfect the Universe started. When the spot differentiated and became differently allocated from the dot the Universe started. When infinity moved away from eternity the Universe started. I show where infinity is as much as I show where eternity is and any person can put his or her finger on the spot. I show where space ends and where time begins and any person can look at the point.

I sent an article to Annalen Der Physics and Professor Ibn Christianson explaining this in a 15 page dissertation. They came back to me advising me to While it is possible that a lay person hits on an insight that has been overlooked by academic trained in the field over many years, it is unlikely. We assume that work offering something new would be related to existing theories, either by building on top of them or by showing how and where they fall short (Professor Iben Maj Christiansen) and With a lot of words and some simple algebraic relations, there is no way to "explain" the world of physics (Friedrich W. Hehl, Inst. Theor. Physics). Maybe our Academic elite including the two I quote had no idea what I was talking about, understood not a word of the detail and was fat too brilliant to accept and admit this state of affairs. Maybe if our distinguished Professors including the two I mention read my work with more attention they would have seen what I try to show and what I try to show is where physics starts. Professor Christiansen since when is Π not science?

Creation started off with one dot so small eternity met infinity within. Then came one more, and another and they continued coming until there were a countless number of dots. The accumulative size of the dots were the same size as one dot because in the true Universe big and small plays no part. The dots were infinitely small and eternally big at the same time because size is a relevancy and without one the other has no size. So in the true perception, there is no difference in size.

It started with the fact that there is no place or part in with which one may associate zero or nothing. There are no room for a number such as nothing. Next to the one dot (infinitely close) one will find the next dot, and if nothing was a factor then that is precisely what one will find between the two dots. Nothing of space, a non existing entity, taking up no space, and much more important, no time, therefore the dots are infinitely close to one another, being the same space, eternally big as much as infinitely small. If we as humans cannot find a manner in comprehending this notion, there can be no manner ever understanding the cosmos as much as the start to the cosmos.

Every dot was a Universe in its own and the accumulation was a universe. The earth in itself is a Universe as the moon is a universe, because rules applying on earth do not apply on the moon and visa versa. When in the ocean another set of rules apply, therefore being in the sea places a body in another universe. The number of universal entities is still countless, as much as it was in the beginning. Every dot insignificantly small as it may be, is a part of another Universe as much as it is part of the accumulative Universe and every dot in the infinity holds singularity, which we translate as " nothing" being " darkness". There cannot be "nothing" just as much as there cannot be "darkness".

There cannot be something big or small, but it into relevancy of perception, and then the relativity of perception becomes the question. There cannot be hot as much as there cannot be cold. The sun FREEZES hydrogen to a liquid at six and a half thousand degrees Celsius and Universe boils over in the form of the Hubble constant at the temperature (we presume from our vantage point) at minus 273 degrees C. If we Humans cannot or will not abandon our human perception and our manly perspective, we may as well return to astrology for all its worth.

Every point in the infinity we may observe at is not merely part of the Universe in not being nothing, but is the point where the Universe started representing singularity. It is the very first point where everything began so many eternities ago, because after all, how can we ever determine where the first point was, as they were very much equal and alike at the beginning. Every aspect of the Universe started with the fundamental fact that no point in the Universe can represent "nothing" as a number, because every aspect in the Universe represents singularity in what ever form it may hold in that specific spot forming space-time. If man does not reach a conclusion where that conclusion is matching the Universe and stop to match the Universe with man (and man's incapability), we may all go back to caves and become starving hunter-gatherers again, because we will never find a way to progress to the ultimate understanding of the universe.

Looking at stars Newtonians see coal stoves being stoked to burn. In the days of Newton coal stoves were the nuclear science of the day ands while all other departments in science moved on and away from coal stove principles Astrophysics and cosmology remained true to Newton by inventing the coal stove in so many ways not even the coal stove could think of the facets it can go through. Newtonians see stars being fuelled like coal stoves and such stoves can run out of fuel. This is so much Newtonian backwardness as mass forming gravity and the moon coming closer and the cosmos shrinking and we falling into the sun because of non-existing dark matter making up what is required to make Newton not to seem the idiot that Newtonians are because they make him and his contraction theory to be less foolish that what it apparently is and they overbearingly are.

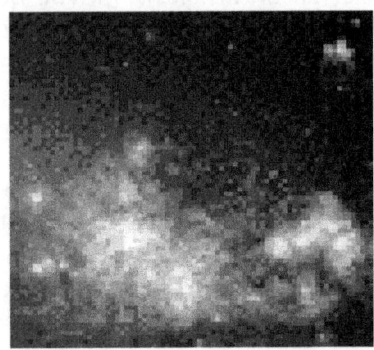

What is of vital scientific importance is that there are three fundamental dimensions controlling the universe. The three are beyond intermingling and one confirms a status in relation to the others but not intermingling in status. From singularity comes matter and forming space-time in own accord. By matter not controlling time, space grew uncontrolled and the third dimension came about. That dimension birth we now recognise as the Big Bang, but the Big Bang is the last of a three prong cosmic growth. Science has to recognise the dimensions of densified (singularity), occupied (matter behind the electron) and unoccupied (space-time outside the orbiting electron boundaries) forming three points of cosmic recognising space-time.

Every dot was by itself as well as the accumulation as it currently is the present universe. The earth in itself is a Universe standing apart from other universes such as the moon as well as the space between the moon and the earth. The moon is a universe. Rules applying on earth do not apply on the moon and visa versa. When considering conditions with in the oceans and applying space-time another set of rules apply therefore the sea places a body in another universe. It takes the same engendering technology going underwater in deep sea diving that going into outer space.

The number of universal entities are still countless as much as it was in the beginning matter as atoms and even much smaller. Every dot insignificantly as it may be is a part of another Universe as much as it is part of the accumulative Universe and every dot in infinity holds singularity, which we translate as "nothing" but it cannot be nothing. There cannot be nothing as much as there cannot be darkness. There cannot be something big or small except in the relevancies of perceptions and then the relativity of such perceptions becomes questionable. There cannot be hot as much as there cannot be cold The sun freezes hydrogen to a liquid at 6500 ^{0}C and outer space boils over at 0 K. If we humans cannot or will not abandon our human culture driven perceptions and our mankind's pre-programmed perspective we may as well return to astrology for what the future hols. There are so many boundaries out there ready to destroy us because of our lack of insight, as did the challenger disaster.

Creation birth started off with one dot so small eternity met infinity within. Then came one more, and another and they continued coming until there were a countless number of dots. The accumulative size of the dots were the same size as one dot because in the true Universe big and small plays no part. The dots were infinitely small and eternally big at the same time because size is a relevancy and without one the other has no size. So in the true perception, there is no difference in size.

It started with the fact that there is no place or part in with which one may associate zero or nothing. There are no room for a number such as nothing. Next to the one dot (infinitely close) one will find the next dot, and if nothing was a factor then that is precisely what one will find between the two dots. Nothing of space, a non existing entity, taking up no space, and much more important, no time, therefore the dots are infinitely close to one another, being the same space, eternally big as much as infinitely small. If we as humans cannot find a manner in comprehending this notion, there can be no manner ever understanding the cosmos as much as the start to the cosmos.

Every dot was a Universe in its own and the accumulation was a universe. The earth in itself is a Universe as the moon is a universe, because rules applying on earth do not apply on the moon and visa versa. When considering the conditions with in the ocean and applying space-time another set of rules apply, therefore being in the sea places a body in another universe. The number of universal entities is still countless, as much as it was in the beginning, before dots formed atoms.

Every dot insignificantly small as it may be, is a part of another Universe as much as it is part of the accumulative Universe and every dot in the infinity holds singularity, which we translate as " nothing" being " darkness". There cannot be "nothing" just as much as there cannot be "darkness". There cannot be something big or small, but in the relevancy of perception, and then the relativity of perception becomes the question. There cannot be hot as much as there cannot be cold. The sun FREEZES hydrogen to a liquid at six and a half thousand degrees Celsius and Universe boils over in the form of the Hubble constant at the temperature (we presume from our vantage point) at minus 273 degrees C. If we

Humans cannot or will not abandon our human perception and our manly perspective, we may as well return to astrology for all its worth, because that is the only boundaries we will find in the cosmos.

To unlock scientific truth we first have to dispose of scientific misconception

In the two pictures we are seeing disposing or releasing heat creates space. We may call it plasma or shock waves or what ever, but in the final analyses it is heat turning to space. Whatever you wish to call that which lies between the particles comes from being a solid, then with adding heat, the solid *"whatever"* becomes liquid and that is the white and orange plasma that we find. That white and orange is heat in a liquid form, just as all flames and smoke is heat in a liquid form. But that liquid does not remain liquid because the governing singularity cannot enforce a commitment ensuring the liquid heat remains liquid. The liquid *"whatever"* you wish to call the heat in fluid form then further overheats turning the heat to space. The space created must be equal to the heat reformed. That is a law of energy where energy equals equality everywhere it is.

Let us humans first detach culture from facts. Take the argument to iron, which we know well. Iron cannot boil, iron cannot flow or bend and iron cannot brake. Iron is an element like all the other elements we know, not one element can do any of the above, in sharp contrast to human belief. As indicated in this book the limits we should find to guide us we ignore for the reason that we cannot see it. We may not be able to ever see singularity, but with intelligence guiding mankind, we do not have to see everything to believe everything. It is because we could not see religion, but still practised religion that set us apart from the other animals.

At the start one would find iron and iron in a "natural state" as we find iron on earth being a human produce on the surface of the earth it will be a solid, suitable for man to handle with bare hands. When such a piece of iron is left in a desert in the midday heat, the human hand cannot handle the iron any longer without aid of covering the skin of the hand. Our perception is that the iron became hot, but that is not the case and our view is a culture contribution and not scientific fact. By heating the iron artificially with combined gasses (acetylene and oxygen or what ever) we now can over heat the iron to a state of flowing like a fluid. Our human culture tells us the iron now is melting.

That is a misconception!

Like the fact of "nothing" we inherited the idea from our past. After introducing artificially even more heat with more heat releasing gasses we may artificially form a condition where the iron would become a gas. Again it is not the iron that becomes a gas, it is the space the iron finds itself in that became hot enough to become a gas. The iron particles remain the same; it is the condition surrounding the particles that changes form with overheating.

Important to note is the fact that iron in a solid state will surround itself with solid matter in space applying a solid space. By introducing conditions producing <u>more overheating</u> the space or connecting between the particles become concentrated heat forming a liquid substance! It is not the iron that turned liquid but the wrapper containing the iron that concentrated so much it formed liquid fluid by the introducing of more heat to a point where the overheating created a fluid. It is considered that the oxygen burn and by that the iron heats up. NOT TRUE!

If oxygen burns no oxygen would be left on earth by the time man arrived on earth to use it to the benefit of intelligent life. The oxygen remains oxygen while the oxygen merely does a task in nature where oxygen carries heat to a specific space. On the other hand it is the task of nitrogen removing heat from the point of overheating by means of flames whereby it creates space. One can feel the "wind blowing" as the flames generate created space. In the extreme the creation of such space we call an explosion.

In the process where the space between the iron particles still further overheats, it becomes a gas. It cannot be iron that becomes gas, because iron will be as much a gas as iron will be a liquid or a solid. It is the space covering the iron particle separating the different iron particles, which will convert and sustain form. The gas is as invisible as space because the gas is the form space holds. This confirms the Biblical view of earth (solids) created and heaven (heat or gaseous/liquids) created. There are only two forms of substance that forms the Universe solids and non-solids, which is liquids and gas. It is not the solids going liquid but it is more of the liquid in ratio with the solids in between the solids that make a structure go solid or gas. There are heavens (non solids) and earth (solids) and this has to do with movement applying control or non-movement allowing non-movement control.

Iron is a solid. Introducing more heat the iron becomes wrapped in a cover that concentrates the wrapper to the point of concentration where it became a fluid. The iron remained what it is, neither a solid, nor a fluid nor a gas. By introducing more heat it becomes a gas. The gas we cannot see because the gas is space. But so was the fluid space. The introducing of heat brought about the turning of a solid to a liquid to space and every time more space becomes part of the picture.

Iron is in its normal form a solid. That means the space, which the iron particles are in, is solid and that disallow the iron to alter the form in which it is. By introducing considerable heat the iron melts changing the form of the iron from solid to liquid.

Considering the evidence we find it is not the iron that melted and that became liquid, but it is the space in which the iron is that became liquid. The iron particles are still as solid as they were. By introducing more heat the iron would eventually turn to gas. It is not the iron that turned to gas, but it is the space in which the iron particles are that has increased to the extent that the space now has so much heat, the heat turned to more space. The iron as particles remain the same, they are just elements confined to a nucleus with electrons spinning about. The space between the particles increased to such an extent it first became a liquid or a fluid and with more heat introduced the heat increase brought about that heat turned to space. That means by overheating the particles surround with heat as a fluid the heat increase then add space as a gas. The gas is the ultimate form of overheating but where one is unable seeing the gas.

1 Firstly the iron is cold enough to be a solid. Replace the word iron with cosmos and forget the colour we associate with heat being white and note the solidness of the centre of a galactica. This must have been the state of galactica that contained large parts of the Universe when time rolled away from eternity.

2 By introducing overheating the space between the iron and not the iron as such turns to liquid. The same apply as more matter (iron) produce more space forming as some matter turned to heat by overheating. The matter increased spin and in that way went out of sequence where it then became softer and softer in relation to other particles, where the loss of the matter released more of the third cosmic component we named heat and space.

3 Some of the heat introduced with the overheating by means of congestion then forms space while other remain in the form of heat allowing space to seam liquid. The matter could not breath and overheated by the enormous gravity the overheating created

4 As the area between the particles still further overheat certain parts of the area overheats to the extent that the space becomes an invisible gas allowing the congestion of matter to separate from one another and allow the stars' individual governing singularity growth. 5 From the soup of heat galactica come about allowing stars to rise out of the dense liquid cradle from where they can establish singularity growth. The process continues as more space becomes introduced through space overheating turning heat into space

6 Should star development come about as suggested it is foreseen that the Milky Way once was a liquid from which the sun developed the singularity in which it then form self-sustaining. The only pre-condition was that it captured individual space-time where the captured space-time remained a liquid frozen (as it was back then at the time of parting) by the governing singularity while outer space further overheated into a thin gas

7 The sun captured so much space by the intervention of singularity when released from the Milky Way that it produced space so concentrated today at present it clearly remained a liquid inside as it froze the interior in time the liquid it now is while outer space is still overheating as a gas with no visibility. From this overview one can judge just how far science is behind the time in their views on creation and the

beginning of time including the universal establishing. Cosmology still hides behind medieval ideas that other faculties and scientific departments forgot long ago.

There are always a relevancy applying between that which spins art double 7 and that which moves straight continuing at 10. The part spinning at seven we think of in terms of the diverting of direction it applies at 7. The liquid / gas holds the 10 factor.

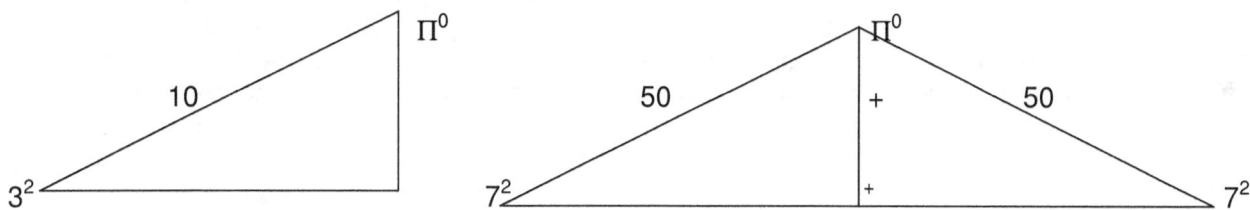

Without the application of specific heat, the object remains in the three directional moving of six possible directions. The value of space unoccupied therefore remains $\Pi \Pi^2$, as it was before the "Big Bang" event, whichever "Big Bang" you wish to refer to, because there were many. But space unoccupied holds time to the value of 10 to 1, and as the sketch of the triangles also indicated, holds space to Π. Therefore unoccupied heat holds the relation to space in applying 3 directions of influence $(3^2 + 1^2 = 10) = (a^3 = T^2 k)$. Always part of this equation is the dual function of space in $(a^3 = T^2 k)$ while at that very instant one has space-time. Therefore in space in time you have $(10^2) = 7^2 + 7^2 + 1^2 + 1^2$. In the sphere we have the axis holding a value of 3 and the circle holds a value of 4. These are dots forming in relaxation to the one spot holding a point from where singularity advances.

We have the axis valued at 3 going square through movement of the linear motion $(3)^2$ and then we have the circular motion $(4)^2$ going square by the spin of the circle ring the direct opposing side. Then the equation of influence becomes $3^2 + 4^2 = 25$ where $\sqrt{25} = 5$ and doubling the 5 on both sides of the triangle will apply the factor of $5 \times 2 = 10$ that then is $(10^2) = (7^2 + 1^2 + 7^2 + 1^2) = 50$ on both sides is 10. The implication of this may not dawn on one the very instant of realizing, but to scientists, there is no greater shock than just that. To any application of movement, the factor will be in the realms of singularity where half a circle is equal to a triangle is equal to a straight line and the lot is equal to 180°. No fancy mathematical expressions have any value in singularity because singularity holds a value of 1.

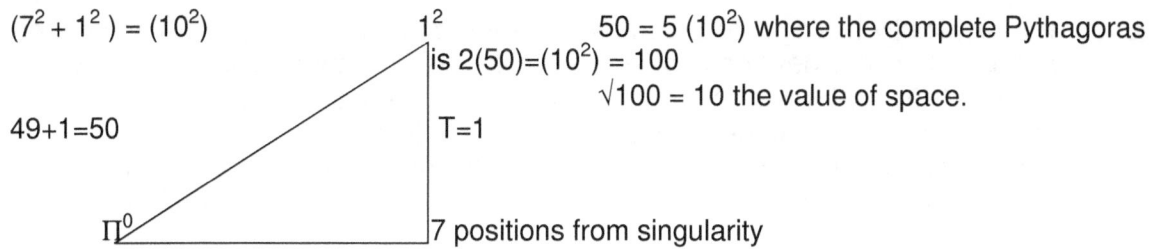

The fact of this comes as 49 plus one becomes 50 and that is in the three dimensions of space $\Pi^2/7$ where 7 holds the relation to one and $\Pi/7$ again where 7 relates to one. At this point it is most important to remember that Pythagoras works on the application of the sum of the square of the two sides. When seven has a direction in the fourth dimension applied to it, the opposing dimension will be one and this applies in time relevancy, therefore the interchanging in time between infinity will place matter at $7^2 \times 1$ relating to circular and $7^2/1$ with $7^2 \times 1$. This makes 49 plus one (singularity) always being a factor of one. Space in time however, never can be a cube, it will always be a square with one side pointing the direction of time from time to the past (1) to time to the present (1) to time to the future (1).

The circle forming Π uses 7 to indicate the roundness of the circle but the 7 holds its roots deep within creation. It indicates how the Universe started because this is the way a star will start moving and it shows how as the infant star starts generating gravity just as the top starts to spin when it is thrown by life. Life can create nothing and that is true but life can mimic all laws in the Universe. Time is eternal movement and will be with us always. The line in infinity is still present while not being a part of the Universe. This line is always ready to be in place when the slightest movement orders it in place. Before the Universe was in place eternity and infinity was in perfect harmony and the line forming singularity validates this fact.

Before infinity parted from eternity, eternity met infinity on one spot as eternity came from the past (1) forming the present (2) to go onto the future (3) but also returned to come from the past which was the spot held by the future and this we find in the fact that the line forms 1 when not spinning but as soon as it evokes by spin, 3 points form even now. Then heat and cold differentiated values and space landed in between eternity and infinity. As eternity moved in relation to infinity but not forming a part of infinity any longer, eternity had to follow a path by never going away from infinity (3) and always returning to the point infinity holds but never lash onto the point again. With space parting the points, eternity had two points (the past and the future) before the partition came about and infinity held both the past and the future while infinity had the present as it still gas presently. By eternity also moving, the two points it held opposed each other (the past and the future) and since it moves, by the movement it became the square of the two because movement is the square and not a flat blanket-like surface with squares embroidered on it as Newtonian science depicts it by using grand mathematics to understand singularity.

Then we had two point holding eternity in place going square by movement to form 4 points serving eternity and infinity captured the first three points held by both and since eternity could not release from the two it had but had to duplicate what it had, eternity by movement became a circle captured by the line. With four points captured by the line of three points the circle coming about is eternally returning to infinity but never complying with infinity because if mismatching temperature or movement (3 against four). Material will always be colder than outer space. It is because material spin and outer space moves by expanding due to overheating.

This is where I start when I start to explain the first moment but I use a shipload more information to do explaining when I explain the star in the book I do so. I involve the four cosmic pillars to substantiate the claims I make because all four still work the very same way as it did at the beginning of the Universe. The three points serving one part of singularity combined with the four points serving singularity unites as seven to form a circle of either 3.1416 or 21.991÷7. The seven going to one is eternity matching infinity by movement. But since seven moves it are seven that have to produce gravity. How do I know all these facts, because we can see from the top it is still doing what it did the very first second. When time started infinity as well as eternity had altogether 3 positions, the past, the present and the future. It is still forming the very line in the centre of the top as it forms all lines in the centre of all things spinning. Then eternity parted from infinity when heat separated what is cold from what is hot and eternity formed one more point than before when it had the three points.

With infinity and eternity then jointly having 7 the cosmos came into rotation. In the aftermath post big Bang we now see the phase of cosmic development where the tow sectors try to unite and this brings along the contraction. When Π forms it does so on the grounds that 7 rotates. The circle forms by a change in direction by 7°. Every circle has opposing sides forming in relation to the axis line. If the topside goes rite then the bottom side has to the left. If the rite side goes down then the left side goes up. There is this double presence of a change in direction forming on both sides of the circle. The 7° move and by moving 7° goes square 7^2 and that is Pythagoras.

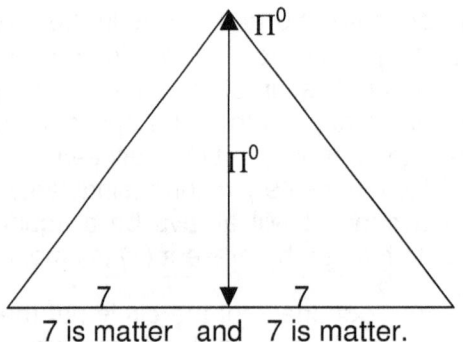

7 is matter   and   7 is matter.

They join space-time therefore the matter factor is the same. This is where one can visually see the one object, filling the space of the other object's atmosphere.

$$7 \times 7 = 7^2 = 49$$

That is matter Π^2 (time) times matter (49)+(1) = 50. This 50 forms space which then applies to both sides of the rotation of the solid being 7 that rotates.

As this is all under the law of Pythagoras the law will evidently place a square root to that value of 483,61 and therefore √50 +50 = 10. This leaves the space value of the Roche-limit, as it develops into the Titius Bode law giving them a shared value of 7 (matter) and 21,91 (space) the value of 21,991 / 7 = Π.
Then the relation becomes

(Π²+Π²) (Π²Π) (ΠΠ²) (Π²Π³) holding space (3) still outside. They therefore will share space and that sharing will continue till times end. We know by now that matter is 7, and space is 3. holding time to a relevancy in singularity of 1. Sharing the space means that 21,9 will become (10) space to the one side

1   to the instant position of time (k^0)
,99   lost to space depletion $Π^2/10$
7   the relation to matter.

Through that the Titius Bode law comes into affect of 10/7 or 7/10, depending on whether space or matter holds a superior position to time. From that stance, all objects will relate to one another by the value of $Π^2Π$ and seen in a whole sale total 7/10 or 10/7.

That means to become part of the neutron status of the earth, the object has to be space (21,991 or less) and prove to be matter (7) before the earth will accept it. If holding a position of less than 7°, the earth will discard it and if it is more than 21,991 the earth will find the relevancy to be higher than the space it holds in a neutron time. That places the object in a relation of $4Π^2 - Π^2$ (because it is not part of the earth) in a position acceptable matter holds (7) within the confinement of Π (21,991/7). That means the object is part of space (21,991) acting as matter (it holds an acceptable own proton structure) 7 relating to the earth in the position the earth allows of $3Π^2$.

That means to become part of the neutron status of the earth, the object has to be space (21,991 or less) and prove to be matter (7) before the earth will accept it. If holding a position of less than 7°, the earth will discard it and if it is more than 21,991 the earth will find the relevancy to be higher than the space it holds in a neutron time. That places the object in a relation of $4Π^2 - Π^2$ (because it is not part of the earth) in a position acceptable matter holds (7) within the confinement of Π (21,991/7). That means the object is part of space (22,991) acting as matter (it holds an acceptable own proton structure) 7 relating to the earth in the position the earth allows of $3Π^2$. With the space position of the matter in the parameters of 21,991 it relates to the Titius Bode law as a factor of one. The object has the space value of 21 (3 x 7), which shows the axle value turning, plus the space value of .1416, in that instant of time (7) complying to the earth's space (.1416 x 7) in reduction ($Π^2$) formulating .1416 x 7 = 0,991.

That makes the object complying with the full agreement as laid down by the Titius Bode law. The object is, no matter where it is, travelling at a rate of 7 ($3Π^2$) in the space of the earth (21,991). This will be agreeable to the parameters of the Titius Bode law as long as it remains within the space depleting "gravity" limits of less than $Π^2$. In accordance to the Lagrangian atom layout, anything less than 5Π is manageable and is in effect less than $Π^2$. When it exceeds 5Π it will start opposing the dimensional equilibrium space holds of 10Π, therefore it will (according to space) exceed the linear point of R/T, which is 10Π/2 (space going in a straight line).

Everything in the cosmos is moving, either by own individual accord, or under the influence of some other singularity dominance. In explaining we return to Pythagoras where the entire Universe with everything in it started.

It is the point forming the very centre that plays the part as the **controlling singularity** within the Universe I have named as **Infinity,** which is better known as the axis. It is where nothing can go smaller and anything within that point can never reduce. That point is where the entirety called the Universe begins and where everything holding substance begins. Once one accepts the fact of singularity being present in that location, that accepting of singularity then is contradicting all the things we know and we can measure and we recognise that point being present by merit of the fact that the point referred to is not being formed by any of the things we can recognise.

It is made up of everything we don't know and constitutes of everything we are unable to recognise or visualise. In that spot there is no space. That spot holds **Infinity.** In that space there can be no motion because there can be no space to have the motion within. It is formed as a line that is so small that our human reality by perception declare that point as not being there and the only reason why we know it is there is because of the results it left as an imprint of its not being there. We cannot detect it but notwithstanding our failure to note it we can recognise the dot on the merits of its absence and while in our Universe it is always absent, reality disallows the dot ever to be absent, because it is never absent. It cannot be absent. It cannot go absent but it can never be there where it should be in a place from where the third dimension forms and it is always present if I wish to locate it. It is **infinity** that can never go away. I named the other part of singularity forming space **eternity** because that area never become bigger, or become more or find an end to the outside. Whatever was and is and will ever be is locked in that space I named **eternity** and it is **eternity** that never ends because **eternity** can never end moving. What we think of, as expanding is never ending movement giving eternity the eternal motion that will go on forever.

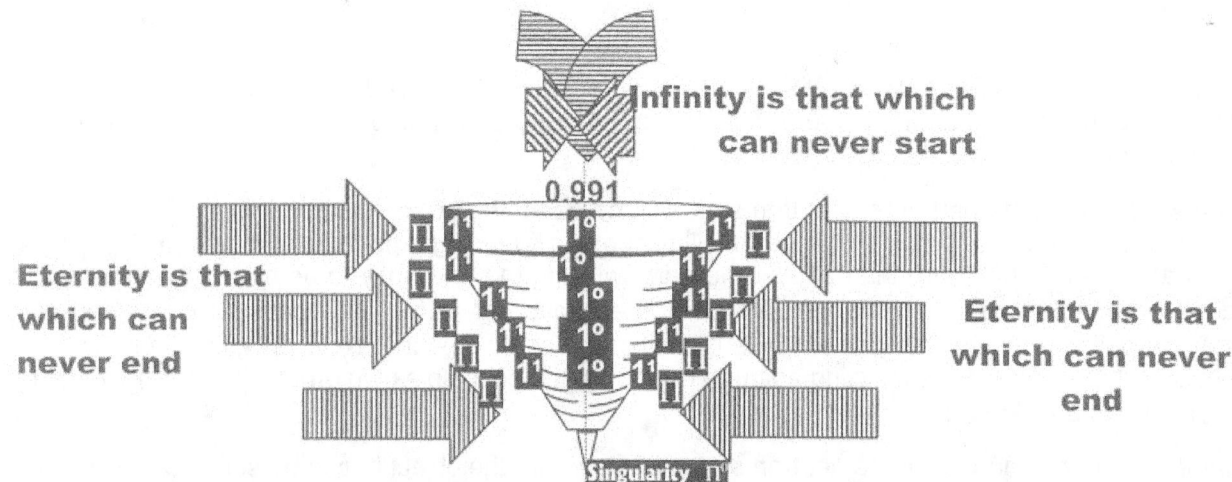

The line **k** coming from the centre (singularity k^0) forms by forming an initial spot Π^0 becoming the dot Πr^0. However, I went on to say that whatever the line used to start with has to continue in order to repeat the same that began the line. Therefore the line started with Π^0 and it has to continue with Π^0 until such a point, as it must end with Π. Whether the line is Π^0 or is r^0, or uses 1^0 the outcome all refers to singularity being used. By reducing the line we come to the end of the mathematical equation of the circle but the circle does not end there. When the top is in a state of motionlessness on own accord it is everything but motionless. The motion it adapts are synchronised with the earth in harmony with the solar system and according to the greater picture of the cosmos.

When an energy source not related to the cosmos called life intervenes and energises the tops motion, the singularity in that top suddenly jumps to life. By adopting a rotation energised to an unnatural state of energising because of life's intervention, the singularity of the top is not in charge but as it applies more and more energy, it will begin to find a means whereby it can escape and apply individual singularity as the top starts to separate from the singularity the earth holds. The singularity holding the earth would then allow the singularity of the top to rotate within a specific band where that a specific band of being active before the earth's singularity will start to destroy the singularity in rebellion.

The top on the other hand will try its outmost, when the singularity it holds gets by individual spin is too strong to remain be in domination of the earth's singularity. The motion of the top is an attempt to begin applying an individual singularity space-time defying and standing apart from the earth's gravity. That action we see as the top starts rotating in a manner where the top does not align with the earth's singularity. With the adding of spin, the time the top holds becomes unrelated to the time the earth holds and the top will start a campaign too escape from the singularity domination the earth has on the top. When the time or spin of the top exceeds the limits the earth places on the top, the top would emerge by trying to escape from constrains placed by the earth.

The view I represent at this point is known to science for almost as long as science knows mathematics. Not long after the law of Pythagoras was understood where Pythagoras introduced mathematics Eratosthenes of Syene made as big a discovery as Pythagoras did. But in the one instance the world took notice because the world could see and understand and the other instance the world disregarded the findings because the world did not see what the implications was. The same apply to aircraft flying and when the aircraft wishes to escape the earth's singularity hold it has to comply with the laws laid down by the earth.

The seven becomes as big a part of the concept as does Π as it all interacts.

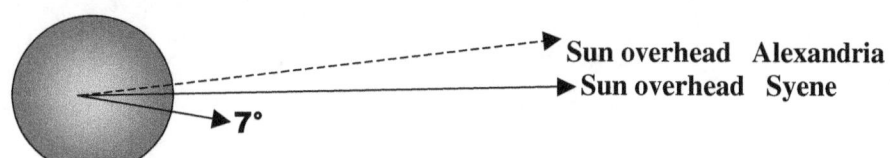

It took Eratosthenes of Syene (276 – 194 BC) a Greek astronomer who in the year 240 BC made a discovery that the earth has a profile of 7°. Since then no one ever did anything about it. When any singularity wishes to disconnect from the earths singularity, specific pre-calculated laws would have to comply to allow the lesser object to divorce from the larger object. I indicated how the dimensions of 10/7 and 7/10 interact to form (Π^2)

Matter is a product through the separation of space and time receiving the value of Π The original time and Π^2 as follows: By circling around a spinning solid the space contracts to form Π and Π^2. Gravity forms everywhere in the Universe by applying singularity. By dividing space into material (material spinning in space) and duplicating space by material spinning, the TITIUS BODE LAW forms a 7° deviation and 7 / 10 in conjunction with THE ROCHE PRINCIPLE OF $(\Pi/2)^2$

In my article to Annalen der physics I used 15 pages to explain this process of singularity applying. I received a rather cordial but sincere reply from the Editor of the magazine. When I placed an article in Annalen der physics Dear Prof Friedrich W. Hehl said in the e-mail he sent me that there is no way to "explain" the world of physics I am not going to go into detail how this works. On the other side of the Pythagoras's' triangle we have 1 going square.

That makes Pythagoras's' triangle 49 + 1 = 50 on the one side of the earth and the same on the other side of the earth. The total is 100 and the square is 10. That leaves the Titius Bode law with a value 7 (it forms part of the material of one body) and 10 in relation to the space. Then from the relation of 7/10 and 10 / 7 forming Π the Titius Bode law form Π^2 applying "With a lot of words and some simple algebraic relations" to quote Friedrich W. Hehl, Inst. Theor. Physics of Annalen Der Physics fame. This was simple algebraic relations but still it is science, is it not?

Since it involves singularity moving it calls for the law of Pythagoras to produce space. The law of Pythagoras is the triangle a^3 that is moving forward in singularity **k** by turning T^2. In singularity the 7 stands in for 7 points on the numerical line crossing over the line holding singularity or 1.

By moving 7 has to go square T^2 and that means 7 goes square 7^2 twice $7^2 + 7^2$ crossing the same divide $\Pi^0 = 1$.

Since all movement in singularity has to enforce the law of Pythagoras we have two triangles holding 7 dots moving across singularity. I don't want to get too involved by bringing in numerical outlays because then this can truly become complex.

The line has two opposing sides turning directionally against each other while turning with each other. By moving or turning this involves time duplicating space by the square Π^2 on both sides of the divide $\Pi^2+\Pi^2$ and using the same divide or the same axis or the same point serving singularity we have 7° crossing the same point in singularity Π^0.

There then is in this rotational movement 7° standing in for Π^2 on both sides of the divide $\Pi^2+\Pi^2$, which then is 7^2 on both sides of the divide 7^2+7^2.

The circle spins in duel directions. On the one side it would go left if on the other side it would go rite. The one side hold a directional change in singularity by 90°. As it is going sideways it changes to going down.

This produces a rite angle triangle of 90° and in it the law of Pythagoras produces direction changes. Since the square of the turn of the circle places by the spin and the direction change we have 7 holding a relation to 10 in space because it is space that has to carry the value of 10 when material circles by 7.

There is a connection between space surrounding the spherical circle turning and the sphere. The circle holds the value of 7 as in 7° and this we find from looking at singularity controlling the circle by movement

Part 9

I DISCOVERED A MISTAKE IN SCIENCE.

This is a crucial mistake since it touches every aspect of the foundation of physics and rocks physics like no earthquake ever rocked any part of the earth, ever...it puts science on a cross road that will change every principle prevailing in cosmology and alter the way we view physics forever.

The question everyone avoid is why is there a defined difference between what is mass and what is weight. Mass is measured in the same currency that weight is measured so it should be the same...but Newtonian teachings deny this idea...why would they deny this fact? I say don't tell me there is mass, prove to me there is mass that is different from weight. Prove to me mass forms gravity because I say gravity forms mass.

It is said that mass makes you fall...or in any case that is what the definition applies. The definition of gravity says that mass attracts and by such attraction gravity forms according to the value of mass attracting and this forms the most fundamental foundation of physics, founding all of the centuries old Newtonian thinking.

The Definition of Gravity according to the Oxford Dictionary of Astronomy:
Physics
1) **The natural force of attraction exerted by a celestial body, such as Earth, upon objects at or near its surface, tending to draw them toward the centre of the body.**

If this is true then Galileo is wrong. If you fall by mass because mass forms gravity, then big things should fall quicker than small things does and Galileo proved that all things fall equal. Now to confuse students they come up with the concocted idea that a hammer fall the same as a feather in vacuum. Yes, sure and a car falls equal to a human without vacuum. All things fall equal when all things fall unrestricted and by gravity alone. The hammer and the feather and the vacuum are part of the deliberate brainwashing process because everything falls equal every time anything falls free. What bring distinction are the heat and the temperature in the sky when things fall, but not the weight or the mass of objects.

This is not what I am referring too when I say there is no mass because we know it is there and I do prove it is the spin of the earth that forces the body downwards, but not by mass. It is by movement establishing Π.

2) **The natural force of attraction between any two massive bodies, which is directly proportional to the product of their masses and inversely proportional to the square of the distance between them.**

This is the one that I refer too that is the figment of Newton's imagination. I beg anyone to use the information gathered from the solar system read from the chart Kepler provided and tested as truth beyond doubt to prove this definition of mass pulling do apply! I give the tables that Kepler established. Show where any planet moves according to the mass as it supposedly does or how any planet is allocated a specific position in the solar system or even in the Milky Way according to the mass the object should have.

3) **Gravitation.**

This proves how low Newtonians can go to solicit absolute fraud and intoxicate student's minds in brainwashing their thinking to accept what never is proven. This is hogwash at best! Prove that the graviton is more than just more deliberate fraud invented to sustain the idea of magical forces pulling atoms.

The Definition of Gravity according to the Oxford Dictionary of Astronomy
Gravity is the fundamental force of attraction that all objects with mass have for each other. Like the electromagnetic force, gravity has effectively infinite range and obeys the inverse-square law. At the atomic level, where masses are very small, the force of gravity is negligible, but for objects that have very large masses such as planets, stars, and galaxies, gravity is a predominant force, and it plays an important role in theories of the structure of the universe. Gravity is believed to be mediated by the graviton, although the graviton has yet to be isolated by experiment. Gravity is weaker than the strong force, the electromagnetic force, and the weak force. This is the brainwashing part! Students have to learn this fiction by heart and repeat in exams as if it is proven truth.

Please show me the attraction in relation to mass that applies between any of the planets and the sun. Please show me the attraction in relation to mass when all bodies fall equal through the sky. Show me planets move closer to the sun to prove this force. The cosmos expands and so does the solar system and even the moon and the earth are moving apart every instant in time. So prove the attraction is occurring between all planets when all cosmic structures move apart! This is the fraud I refer to!

If you are a student in physics then you should read the following information with care and with much consideration because your mental health might be at steak here. One could think of another name for physics and that would be Newton's mythology. It is about the subject of gravity and is most important. The "Newton's mythology" comes from the fact that students have to learn what the professors claim to be true and what was never was proven. Students have to repeat in examinations that the formula $F = G \frac{M_1 M_2}{r^2}$ is truthful and viable while it was never proven. Do you realise that it is an accepted practise that all students that are studying physics on all levels are subjected to the most intense brainwashing and thought control found any where on Earth? This must be some sort of a joke you may think but thinking that way in disbelief is just what those practising the mind control wish you to think!

According to the tables, all movement is according to some other value than mass. If mass was pulling then Jupiter had to move 317 times faster than the earth! They never prove Newton's philosophy on gravity but those persons conducting teaching in the subject of physics force all physics students to learn Newton's gravitational concepts and accept the facts as if it has been proven beyond all other facts. Students have to believe that Newton is correct or academics will see to it that they fail their examination. The condition of being accepted in physics is to accept Newton without questioning the proof that is never supplied.

Let those academics now prove precisely how mass brings about gravity and then afterwards test you on how Newton is proven correct and not on you repeating their facts blindly about what they say is true about what Newton said, which they say is true. The manner they present Newton is completely hearsay and that method may not be used in any court of law let alone form the foundation of physics.

Planet	Mass per Earth unit	$k^{-1} = T^2 \div a^3$ Movement	a^3 of space volume	T^2 During time units
Mercury	0.06	$T^2 \div a^3 = 0.983$	$(a^3) = 0.059$	$(T^2) = 0.058$
Venus	0.82	$T^2 \div a^3 = 0.992$	$(a^3) = 0.381$	$(T^2) = 0.378$
Earth	1.000	$T^2 \div a^3 = 1.000$	$(a^3) = 1.000$	$(T^2) = 1.000$
Mars	0.11	$T^2 \div a^3 = 1.000$	$(a^3) = 3.54$	$(T^2) = 3.54$
Jupiter	317.89	$T^2 \div a^3 = 1.000$	$(a^3) = 140.6$	$(T^2) = 140.66$
Saturn	95.17	$T^2 \div a^3 = 0.999$	$(a^3) = 868.25$	$(T^2) = 67.9$
Uranus	14.53	$T^2 \div a^3 = 1.000$	$(a^3) = 7067$	$(T^2) = 7069$
Neptune	17.14	$T^2 \div a^3 = 0.999$	$(a^3) = 27189$	$(T^2) = 27159$
Pluto	0.0025	$T^2 \div a^3 = 1.004$	$(a^3) = 61443$	$(T^2) = 61703$

Let your professors now prove how it is that Newton's teachings are correct and then examine you on the process they use to prove Newton's concepts rather than test the state of brainwashing you have submitted to. At present they say Newton is correct and then they test you on your ability in repeating that Newton is correct without ever proving to you that Newton is correct. That is not testing your knowledge but it is testing the mind control you have submitted to. Let those physics professors now prove Newton and their ability to prove Newton correct and then test you on the manner they use to prove Newton to be correct.

They teach you that the Universe contracts and to state their case they force students to learn that gravity is proved by Newton introducing the following formula $F = G \frac{M_1 M_2}{r^2}$ They say that M_1 is the mass of the Earth and M_2 is the mass of the individual in questions mass and the multiplying of these factors with the gravitational constant produces the force of gravity when this gets divided by the square of the radius. Please let you lecturer put in all the values of the formula and prove Newton is correct. Use the mass of the earth multiplies by your mass and then these is divided by the square of a billionth of a millimetre and see what gravity forms with those figures applying! The radius between your feet and the earth is a billionth of a millimetre and not 1 as they normally show. If he can't prove gravity is believable and I know for sure he never can fill in the symbols and calculate the force of gravity, then read the rest of the web page that follows to see how far academics in physics go to brainwash students into believing in Newton's fraud.

The truth beyond all other truth is that Newton's gravity has never been proven (because try as you may it is not possible to prove Newton's formula forming gravity mathematically) and because academics know that, academics require the blind acceptance of Newton by students. This unconditional acceptance of Newton's correctness relies only on the pre-conditioning of students' mind set and academics depend only on the student trusting the academic "say so" about the institutionalised correctness of Newton.
That Newton is correct nevertheless and notwithstanding that there is no founding proof about this matter, is what students should be accepting blindly. I'll bet you they are more surprised than you about me accusing them of systematically mind altering the student's physic and ability to think than you are.
This is a fair test to see if Newton's contraction theory underwritten by Newton's attraction formula $F = G \frac{M_1 M_2}{r^2}$ is valid, and then force your professor to use this formula as it reads and show WHEN the Moon and the Earth is going to collide. If he fails to do it by using Newton's formula as $F = G \frac{M_1 M_2}{r^2}$ then you will know who is conning you, him or I and who is truthful again him or I. I charge all academics to prove what I say is being wrong in any way or even that I exaggerate in the least. I challenge Newtonian academics to prove that mass does indeed form any force of any sorts and in particular gravity!

Pre-conditioning students into blind acceptance depends on the academics' insistence that students approve Newton's concepts without pre judgment or students insisting on scrutiny of any sorts. In examination students have to outright and blindly follow academics' say so trumped up rhetoric and respond on facts only because academics say so. Academics depend on students never questioning their say so or demand proof about what academics teach. Those academics in teaching positions insist that all students accept Newton's accuracy or be sent home as failures in physics.

Where does mass manifest in the solar system? Where does Jupiter show its mass is pulling more than that of Pluto? According to the definition of gravity claiming gravity is a pulling power by the mass of objects and this formula $F = G \frac{M_1 M_2}{r^2}$ then Jupiter has to beat the earth going at a speed 317 times faster than the earth, because Jupiter is 317 times larger than the earth. Show me any evidence that mass plays any role or function anywhere in the Universe! If you are one of those members of society that never thought you would hear the name of an accomplished person as Sir Isaac Newton being associated with fraud, corruption and brainwashing, then the books **The Absolute Relevancy of Singularity** published by **Lulu.com** are specially written to inform you about the truth there is lacking in the correctness about science. Newton's formula and suggestion indicate a straight line moving $F = G \frac{M_1 M_2}{r^2}$ between the objects pulling by mass while it is clear everything that applies the motion of a planet going around the sun is using a circle. Everyone knows that planets orbit around the Sun. Planets circle the Sun which is the same as saying planets orbit the Sun. Just by calling the circle motion in terms of what applies, that statement nullifies Newton's claim of mass that attract mass and put to question the reliability of Newton's dogma. This might at first seem a small issue but from that I prove that gravity works by the value of Π and not mass.

This is methodical mind control as much as it is the brainwashing and I show that they enforce this practise. If you are one of those believing that Newton was ever proven, then what you believe to be true is a lie because Newton can't be proven and that is the truth! The time has come to face your teachers and force them to stop the ongoing old culture of bullying students into submission and conditioning their thoughts by enforcing on them dogmas, which is no more than systematically enforcing mind control! In order to get students to accept Newton's hypothesis, academics resort to brainwashing pupils' and students' thoughts.

I decided to have the books printed in e-book format as to inform all other potential readers. In the books I addressed the issues in detail and the work I named **The Absolute Relevancy of Singularity: The Thesis** I sent articles to Academics in which I proposed the full remedy to the mistakes I indicated. This only drained my funding and I got nothing back for all the effort and the money I spent in trying to charge a truthful reaction from academics. They are so entangled with their brainwashing they don't see the practise it.

The only response was just a vacant silence; they did not even e-mail me in response acknowledging that they received such books that I sent their way. Now I am going to reveille why there is such a silence coming from the academics…it is because I dare to challenge Newton's correctness about gravity. On the advice of a publisher I wrote an article to **Annalen der Physics** trying to introduce the working of the principle forming singularity. In the website THE ABSOLUTE RELEVANCY of SINGULARITY : THE WEBSITE [http://www.lulu.com/content/e-book/the-absolute-relevancy-of-singularity-the-website/7517996] I share with you the response.

Again I repeat that I do not wish to make unlawful money from a scam I thought up where I try to seduce your curiosity. You can download it for free. You can either got to www.singularetyrelevancy.com and download a more comprehensive version of www.singularetyrelevancy.com: **Informing about Newton's Mythology** which is totally free of charge or go to www.singularetyrelevancy.com and from there press www.singularetyrelevancy.com to download also from Lulu.com an e-book that is a little less informative version of www.singularetyrelevancy.com.

Should you find what I say truthful then you can go to the e-book that is without charge as well and find out what gravity truly is without brainwashing any one into stupidity? Then you can download www.singularetyrelevancy.com: the website and for the first time see the truth…**free of charge!**
Download it from Lulu.com and find out for the first time how gravity truly forms in space.

Reasons for reading www.sirnewtonsfraud.com

Firstly it is free. I am not trying to exploit any person by extorting money in launching a factual unsupported wild scam. The truth is that Newtonian science is practising the unsupported scam for centuries and they do it by mind manipulation and thought control. They force students to believe Newton is absolutely correct needing no proof. Let them prove how the solar system uses mass! If finding these statement harsh surprises you, just see why I say they play mind control games

Newtonian approach cannot even recognise any of the four principles but only Newtonian science are taught to students. No student can have the fortune to disagree about Newton and remain a student at any institution while studying. Students are taught to accept Newton and to ignore Kepler and any student doing it the other way around will fail all examinations and other testing at the Universities the student is attending. Students accept Newton or they accept a ticket taking them home. According to Newtonian science space is simply nothing with no qualities but gravity separate space and space does not mingle, as one would expect if space was nothing because space does form borders. Those borders form part of gravity and gravity is the least understood concept thus far in science. In truth no one in science anywhere remotely knows what brings gravity about and I used Kepler to unravel this mystery called gravity. But no one in science will admit this fact about Newton or any one else never being able to explain gravity in the least or that Kepler is the one who formulised gravity decades before Newton came and gave gravity the name. Newton started this realising of gravity but it had and still has no more substantial proof than a rumour has and Newton admitted to it being a concept he could not explain. Still to this day nobody in science at present will denounce the principle of gravity being never explained. All in science act in a manner as if Newton's gravity idea is proven fact and only occasionally admit it to not "fully understood" as Newton admitted it was when he introduced the name (not the concept). That includes Newton as well as Einstein and even Hawking. Scientists can declare gravity was a factor at 10^{-43} seconds after the Big Bang but what brought gravity about or why gravity became or still remained, as a presence is still tightly concealed information which all are speculating on. Even to the best informed amongst the most educated do not know what is gravity because they all ignored Kepler and for ignoring Kepler the price they pay is not finding the principles bringing about gravity.

Kepler studied what kept the plants circling the sun and that what is keeping the planets in orbit is gravity. From such explaining what Kepler said without Newton changing formulas on Kepler's behalf I prove the Titius Bode principal also known just as the Bode principle. I explain how singularity forms the Roche limit and how singularity brings about the Coanda affect but most important of all Kepler showed me where to search for singularity. My achievements came from my effort where I separated Kepler's work from the opinion that Newton formed about what he saw Kepler's work should contain and gave to the world his Newtonian concept about Kepler's work. From my view a force is just motion applying and that is what Kepler said gravity is. Kepler said $a^3 = T^2 k$. Presumptuous as it may be on my part of trying to disprove

Mainstream Physics, such a presuming does not change the truth about Mainstream science being incorrect about gravity. After all they admit they do not know what gravity is however this admitting is only sometimes when it suits them to do so. As they do admit they do not know what gravity is I am not disproving anything they proved because they agree they do not know what gravity is, which paves the way for my showing what gravity is. By they're admitting that they do not know what gravity is they then also admit being possibly incorrect about gravity but unfortunately mainstream physics do not see it that way (yet). The question in hand is finding what role gravity played when the Creation came about for the first time. I had to find a method that would allow me to explain why gravity played a role.

Remember that not even Newton could explain what gravity is or where it comes from, but Kepler did that without any person ever taking note of the achievement by Kepler. Kepler studied what kept the plants circling the sun and that what is keeping the planets in orbit is gravity. From such explaining what Kepler said without Newton changing formulas on Kepler's behalf I prove the Titius Bode principal also known just as the Bode principle. I explain how singularity forms the Roche limit and how singularity brings about the Coanda affect but most important of all Kepler showed me where to search for singularity. My achievements came from my effort where I separated Kepler's work from the opinion that Newton formed about what he saw Kepler's work should contain and gave to the world his Newtonian concept about Kepler's work. For instance from Kepler's work I can explain the operation of the Black Hole, which not even Prof. Stephen Hawking understands. From my view a force is just motion applying and that is what Kepler said gravity is. Kepler said $a^3 = T^2 k$. The Coanda effect is the establishing of individual independent space a^3 by applying motion $T^2 k$ in relation to a centre point the motion of the liquid establishes. Einstein came to this conclusion but failed to refer his view back to Kepler and by not referring to Kepler missed the point he wanted to make. Just by my studying of Kepler this became possible.

The Practice of Brainwashing and Mind Control in Physics

In www.sirnewtonsfruad.com I investigate how much truth there are in mass pulling by the force of gravity. Most if not to all of the persons reading this article will be annoyed by just the thought of me embarking on an investigation of the issue that seems so totally senseless to investigate. It is senseless because the concept it carries became accepted as household practise and life science. Mass is associated with everything that is represented everywhere.

Do you think of astrophysics as the department that is run by the wise and the level minded, the sober thinking, and the absolute trustworthy? If you are a student there is no other choice you have. If you think those in charge of astrophysics are the pillars of trust, then get wise and read the following. What you are about to read is simply mystifyingly simple and yet to this day I have not challenged one academic any where that had the honesty to admit to the fact of Newton being wrong. Newton created the factor mass as a trick of his imagination. We have a body with mass on earth and that we all know, but no heavenly body could ever have mass or present mass! After you have considered the following you might agree with me that even small Children can reach a higher level of clear-minded logic and find more sensibility than what those scientists promoting astrophysics have because science lives in a make believe fool's paradise. They love to calculate because with mathematics they create a fools paradise.

Looking at the size of a body does not allow any one to awards mass. If you are a student, then ask your Masters please to explain the following abnormalities and inconsistencies they promote as part of official physics, which I present in this article and as students get wise instead of brainwashed. I say brainwash again because they force-feed you fabrications, as you will come to see. They can't explain the facts as reliable but hide the fact that the facts are in fact untruths. Tell them to prove that planets have mass. Tell them to prove that it is mass that generates gravity that pulls the planets. Ask them to explain gravity in detail. You can get the question to ask them for free from www.sirnewtonsfruad.com.

If any person gives evidence and the evidence is unsupported by truth and unfounded by proof, it is fraud. If a Clergyman should bring evidence of some miracle healing that took place, and he could not repeat the act at will for all to witness again and again, it is fraud not withstanding the religion. Even if the lie is to underline the greatness of his religion, it still constitutes to fraud. If anything is said without truth backing the claim, making it unfounded, it is fraud. It is illegal to make claims that is not based and is not supported by reality. As perfect as everyone thinks Newtonian physics are, this applies to educators teaching physics. There is a suspicion lingering in the back of everyone's mind that something is not quite correct about the approach physics take on the matter of gravity and only those academics seasoned with years of studies and salted with time seems to miss this haunting feeling. But the longer the students are educated, the more this uncomfortable feeling dissipates.

Hidden under a cover of "understanding Newton" or "not being able to understand Newton" tutors in physics force certain incompatible arguments to join that which never can join and while joining also make sense at the same time. In this article I challenge the figures that charge the highest form of respect in our communities and those in charge of the most dynamic part of society and those who stand beyond and above any form of suspicion. I charge those that personify truth and are the very same persons that I accuse of betraying the ones trusting them. They misuse the positions of trust by participating in the conspiracy I am about to tell you about. Educators in physics I hope that you will stop your mind abuse on students because they are now able to know more about physics and the application of gravity than did all those that came before you and all those that came with you. Again I challenge you to come forward and tell your students the truth about what I uncover in the articles that follows and as the articles progress by introducing information...then you explain to them how you deceived their blind trust in you as a tutor in physics. In physics the blame goes to students ability to "understanding Newton" or "not being able to understand Newton" but I show that in this case the blame should openly be dedicated to those that should be blamed, named and shamed and they have to defend their years of lying and contribution to cover the misconduct that was committed by them. Those teaching physics as well as their predecessors now have to explain in the name of science why so many evidence were falsified to keep their noses clean while mud colours their lily white image in hogwash and the stink of their lies equals the pig pen it needs to cover such aroma. It is time that they reveal the truth. Believe it or not, but this diverging from reality and misconduct about applying science is in place because of centuries of brainwashing going on passed on from generation to generation and is employed in physics from teacher to student for centuries on end. This article is dedicated to bringing honesty into the faculty of Astrophysics and Astronomy as well as to show the Physics student on what corruption and deceit does physics base their facts which they proclaim as being such well proven, and godly accurate facts and is unwavering depicting only the truth. Notwithstanding as unbelievable as it may seem, I nevertheless challenge any one in physics to show me that the least of any or all facts I uncover is not true.

Go to the web site www.singularityrelevancy.com and then use the information I supply in the book www.sirnewtonsfruad.com titled Sir Newton's Mythology, which you may download for free and use the information given free of charge to confront you academics with the truth and then insist that Academics who are teaching physics, prove to students that Newton's statements of attraction are correct. Let those academics explain the method mass implements to obtain the attraction they teach about.

If you are a student studying in physics then reading this article is detrimental to your future, as you can remain part of the problem physics has had for centuries or you may join the solution that came to physics and start to heal the wound. Students read the next pages and you are about to learn how students are brainwashed into accepting the baseless and ridiculous misinformation Sir Isaac Newton puts forward as truths. The Custodians of Physics have nothing better to offer than presenting you with unfounded, corrupt and distorted facts...and by doing that they resort to mind control on students and introduce baseless concepts by manipulating the student's thoughts. If you are a student then read in www.sirnewtonsfruad.com what they do to you and how they brainwash you. Everyone knows there is a problem about gravity in physics and this far it seems as if no one can put a finger on the problem. This suspicion that there is, this feeling about a certain concern and doubtfulness that is lingering on in the minds of many… and also is lingering on from generation to generation… without anyone ever finding a solution… it is them defrauding you by exchanging your institution fees for corruption, so confront them about their dishonesty. Something about the way gravity is presented just doesn't add up as it should and does not quite reach the answers it should conclude. There is this vague unspoken question hanging in the air without any one ever finding words to express the question...and yet the question remains however unspoken it seems. Force them to become honest and to stop corrupting students with intentional malice. That which they offer has no truth. All they have are misconceptions and incoherent facts. Let them explain how mass generates gravity...

There is always a thousand and one conspiracy theories going around and the one tries to be bolder and more sensational than the next theory flying around. However, the biggest conspiracy is going on in front of every person's eyes and is committed by the most respectable persons in any respectable upstanding society. Even if you had your personal favourite conspiracy theory, try and match it to the one that I have!

Let them with precise detail show when mass is applying, that mass and such producing of gravity that then would establish attraction produce gravity! Show that objects in outer space are optionally allocated places by value of mass. Let those show where they see stars in any galactica are positioned categorically by size or mass. I show precisely how gravity produces mass but mass can never produce gravity. I show with explicit detail when, how and where gravity forms mass but mass can never form gravity. What I prove annihilates every Newtonian claim. No one cares to read or to print my work because there is no need for it.

I seem to be the first person in generations that ask questions about Newton's work. Questions I now ask is asked for the first time ever, well ever since the time Newton introduced gravity, before the emphasis fell on proof rather than merely what a person with reputation suggested. I now am able to show how gravity forms by forming a circle using Π because I have located the centre of the Universe. But by my effort in finding the location I disrupted everything Academics in physics hold holy and for that I am most unwanted in the presence of the Academics charged with guarding the ethics of physics. In short, I clash head on with Newtonian dogma and principles forming physics. During my research I discovered abnormalities and inconsistencies about mistakes the Arch fathers in physics must be aware of but are hiding with all their considerable influence and academic power. The road I took in my search for truth concerning physics was never smooth and the resistance I came across coming from the academic sector is almost unbearable. I made no friends but only enemies.

With my unpopularity rating this high as it does, I never qualified for help and those that would help found my ideas intolerable whereby I only found rejection instead of help as I tagged along. Because of this insider rejection I had to resort to private publishing because from the nature of my work I take Mainstream science head on and am confrontational on most aspects of astronomy. By finding the centre of the Universe it enabled me to find a point the Universe is controlled from, but because that does not hail Newton it sparked no interest. Then I decided to reject Newton is the only road to go if one wishes to lay axe to the root of the insider corruption they are guilty of. Since no one see the a problem there is in physics, no publisher wants to go head bashing with the Physics Custodian establishment of science on official science principles, which I have to do to convey my message in no uncertain language. I argue that if it is the correct practise to use $F = G \dfrac{M_1 M_2}{r^2}$ to calculate gravity then the radius holding the gravitational constant must lead one to the centre of the Universe. As I confront science dogma and principles, nobody is willing to publish my work. I have to walk the road alone and fight the battle by my private effort without any support anywhere. If Newton is the problem one have to go pre Newton to find the problem. The problem is that when looking at Kepler's table then if there is $T^2 \div a^3$ according to the table matching a column, then mathematically $T^2 \div a^3$ must be k^{-1} and where k^{-1} goes negative it shows space reduces time. It shows space in volume goes single by movement of space and not objects.

Planet	Mass per Earth unit	k^{-1} Movement	a^3 of space volume	T^2 During time units
Mercury	0.06	$T^2 \div a^3 = 0.983$	$(a^3) = 0.059$	$(T^2) = 0.058$
Venus	0.82	$T^2 \div a^3 = 0.992$	$(a^3) = 0.381$	$(T^2) = 0.378$
Earth	1.000	$T^2 \div a^3 = 1.000$	$(a^3) = 1.000$	$(T^2) = 1.000$
Mars	0.11	$T^2 \div a^3 = 1.000$	$(a^3) = 3.54$	$(T^2) = 3.54$
Jupiter	317.89	$T^2 \div a^3 = 1.000$	$(a^3) = 140.6$	$(T^2) = 140.66$
Saturn	95.17	$T^2 \div a^3 = 0.999$	$(a^3) = 868.25$	$(T^2) = 67.9$
Uranus	14.53	$T^2 \div a^3 = 1.000$	$(a^3) = 7067$	$(T^2) = 7069$
Neptune	17.14	$T^2 \div a^3 = 0.999$	$(a^3) = 27189$	$(T^2) = 27159$
Pluto	0.0025	$T^2 \div a^3 = 1.004$	$(a^3) = 61443$	$(T^2) = 61703$

Use the column to show where it is mass that produces any valid factor. In the table columns we have the mass of the main solar objects and according to Newton the mass is responsible for movement. If it is mass that does the pulling in ratio of the radius then we have to see some evidence of this applying somewhere. Is there any person that can prove that claim from the table showing Kepler's movement in relation to Newton's mass? In what way odes Jupiter move faster or how does Pluto move slower according to mass? If mass did the job, how is it done? Notwithstanding Newton's blatant mathematical manipulation, mass don't apply.

One such an example is Einstein's Critical density scam and the "Dark Matter" swindle where they hide Newton's shortcomings under a pretext of ongoing research. Hubble found the Universe is not contracting as Newton said it does with $F = G \dfrac{M_1 M_2}{r^2}$ and hell broke loose. The cosmos stepped out of line by showing Newton is wrong in his supposed theory that mass pulls mass to reduce the radius such as $F = G \dfrac{M_1 M_2}{r^2}$ will indicate. Einstein was then tasked to supposedly measure "all the mass" in the entire Universe and to find out when the Universe will correct its evil ways and start to submit to Newton's idea of contraction. They concluded it is not Newton that is wrong but the cosmos went out of line to contradict Newton. Therefore one has to find the error in the actions of not Newton's idea but about the Cosmos acting out of line with Newton. The fault was investigated not on the side of Newton, but in the cosmos! The proof of practise fell on the Universe that could be wrong but never Newton!

With all this in mind did any one ever come to wonder about the reality driving the all too famous **Einstein's Critical Density theory** and the fact that this idea was conceived to conceal the corruption of Newton in physics? Allow me please to elaborate and then make up your minds. The facts in truth are that the **Einstein's Critical Density theory** was a scheme plotted by those in charge of physics principles to cover up and conceal corruption in the heart of physics. Hubble proved beyond doubt that there was an inflating Universe. This contradicted Newton's deflating or pulling Universe and this perception of a deflating Universe being a myth had to be most ardently hidden as to yet again compromise the truth about Newton and his theory. In the formula $F = G \dfrac{M_1 M_2}{r^2}$ the relevancy applies between the strength of the mass and the distance of the radius that keeps the influence of the mass forming the gravity. The longer the radius increasing the distance between objects the more this will reduce the value of the mass, whatever the value of the mass might be. The Hubble concept proves that while the mass might remain the same, the radius keeps growing and such growth diminishes the influence of the mass all the while the radius increases. This is so basic that primary children learning the basics of mathematics will understand! This means with the radius increasing, there is no chance that the mass will ever become strong enough to bring about the pulling because the constant increase in the radius constantly diminish the influence that the mass might produce.

Yet Einstein proceeded in searching for a value that will determine when the mass would bring a turn about in the direction that the cosmos evolves in. Einstein was looking for the moment the mass will become strong enough while the most basic principle indicates that an increasing radius leads to a decreasing mass influence submitting a decreasing potential gravity since the mass becomes less prominent in influence. If Einstein was unable to recognise this most basic of mathematical principles, then what type of genius did physics create in him and what slur did physics promote. In $F = G \dfrac{M_1 M_2}{r^2}$ any factor standing in relation of division, it is the bottom value that determines the outcome of the value of the top. This idea of the two factors being in opposing relevance where the size of the bottom caries the value of the outcome is so simple that children will recognise the principle, and yet those fathers of physics wants me to believe that the greatest mathematician that ever lived did not realise this principle…the principle that the radius and the mass stands related and the growth in the bottom value will promote the decline in the top value as a dominant factor. Can any one with this information including the proof I asked for on previous page have any other conclusion than students should smell rotting fish somewhere?

Notwithstanding my arguments that should have been raised by the mathematical genius, Einstein supposedly measured "all the mass" in the entire Universe and then afterwards concluded there is "insufficient mass" to pull the Universe closer. This rattled the cages of the Newtonian conspirators because Newton once again stood naked, venerable and bear. Yet the blame had to be associated with the cosmos not playing by the rules of Newton. The cosmos had to be at fault because Newton just can't be the person to blame. Then the genius of the Newtonian cunning kicked in once more and we can see why they are seen as the most brilliant minds walking the earth. There had to be something they miss and no one can see.

Then some idea was presented that the Universe is hiding mass from view, I suppose just to spite Newton. If the mass was not visible and was therefore undetected, then the mass was dark and if no one could see the mass then no one can prove the mass being present and then no one can disprove the mass being there waiting in the dark. What a splendid idea this was for cheating. This presented a solution of Biblical proportions and a new scam was introduced to hide the failing of Newton and of the Critical Density sham. The Dark Matter hides mass that will supposedly pull all material closer again at one stage in the future. This is meant unleashing an enormous swindle, bigger than anything before, only beaten into second place by Newton's swindle about mass unleashing forces. They proclaim there is "Dark Matter" hiding and waiting in the wind to come to the present someday in the future and start to pull the Universe closer as Newton said it must by the measure of $F = G \frac{M_1 M_2}{r^2}$. In that way Newton was vindicated and the cosmos took the blame for hiding mass and the contraction was reinstalled.

It is obviously clear that having such a total idea that there might be dark unseen mass floating in the Universe which at this time does not generate gravity but will some day kick in to generate gravity in order to cover up Newton's deception about a contracting Universe and just because Newton has to be correct at some point in the future. Science wishes me to believe that since there is a lack of seeing material there then will be dark and unseen material where they are so dark they are undetected by all humans. That leaves another question to address…where can such mass be found. How will such undetectable mass be found, which will bring about contraction after all this expanding ends and the Universe recognises the infallibility of Newton once again? Why would the mass at present then not activate gravity and why would the mass at some point spring to life and start activating gravity? How much can the Physics paternity still hide the fact that ==Einstein's critical density== is being used as a ==cover-up to distort the truth== ==conceal the fraud Newtonians wish to cover==? The ==uncovering== of the ==Newton fraud== by the Hubble constant is so simple to see. Hubble found the Universe is expanding and Newton said otherwise. Who is lying about what? Hubble's declaration was on track to blow the cover that was concealing the Newton fraud wide open and uncover the century old deception. To see this we have only too look at the comet behaviour when any and all comets again come around on a cycle by repeated visiting the sun. The question is if it is mass pulling mass onto mass, then why do we have comets left in the solar system? The mass of the Sun should by now at least have destroyed every comet going around.

The term pulling does not suggest any circling because no one can be pulling towards and does that while circling around the object. That serves the term pulling. Then the saying goes that planets orbit indicating they follow a circle. In conversation we speak of the planets orbiting. If Newton was correct we should be speaking of the planets pulling, but talking about pulling would be blatantly wrong according to the normal spoken word. Never do we refer to the planets pulling the Sun or the Sun pulling the planets, but we speak of seasons coming from orbital positions. Being in orbit has to neutralise the pulling and then cancel the pulling concept that also became culture. The entirety of physics rests on this one formula $F = G \frac{M_1 M_2}{r^2}$. The questions concerning that which you are studying and that touches every aspect you are academically concerned with, is that if everything is moving apart, how does that support Newton's idea that everything is coming together…and please don't let them fool you with Einstein's Critical Density idea! If there was mass seen or unseen in the Universe and mass generated gravity and gravity does the pulling then why is the mass not at this moment doing the pulling? If there was mass in the Universe, seen or unseen and mass generated gravity and gravity does the pulling then why is the mass not at this moment doing the pulling? If there is an object it indicates the presence of mass. Then what stops the mass that should be present, dark or luminous, to start pulling by gravity, as it should do?

Should you think this page is some sort of a prank then answer the following simple question to yourself in utter honesty: If there is a Big Bang after which everything was left moving apart, how does that support Newton's contraction? Tests results received after the Moon landing show the Moon and Earth are moving away from each other! Yet students learn about mass pulling mass and that puling by mass forces togetherness by contraction.

This is only one of many points that I make on this one issue and there are so many other issues one may think of those in terms of counting in numbers in many hundreds or even in thousands. If the Sun for instance has mass that is apart from the Earth and the Earth also has mass and there is a gravitational constant in between the Sun's mass and the Earth's mass we have the radius in that location. It then must be the gravitational constant that fills the space that the radius holds. It is rather obvious that while the radius is filling the vacant space between the Sun and the Earth it is the only place left where the gravitational constant can hide. To find the centre of the Universe I had only to find the gravitational constant that holds the centre. Through my venture I discovered one person that knows what gravity is! Newtonians went and filled that space reserved for the gravitational constant having a measured value with nothing! How can nothing have a value of 6.67×10^{-11} while also being filled with nothing as it is nothing filling the nothing of outer space?

My question in this matter is what is all that mass of so many supposed stars living in utter darkness doing at present while waiting to get to work and begin with generating gravity by mass where it will only later, much later form a force of gravity that then will bring about this pulling of the Universe? What makes the mass slumber in darkness to one-day form a pulling force? What has the "darkness" or the fact that we don't see the mass got to do with the idea that the mass at present is not forming gravity that is forming a pulling force? You are taught that gravity pulls objects to the centre and obviously gravity then has to ultimately pull everything to the centre of the Universe. That is what the Critical Density research that Einstein initiated wishes to establish. The idea is that $F = G \dfrac{M_1 M_2}{r^2}$ makes the mass create a force that will destroy the radius and ensure everything is going to come together eventually at one point where the radius then will be no more. If that is the case, then where is that point? If everything is destroying the radius, then it must end at one specific point.

In the classes you students attend a physics lecture, has any one confirmed a location where one might find the centre of the Universe to confirm the ultimate destination of $F = G \dfrac{M_1 M_2}{r^2}$? If you wish to apply a Gravitational constant as a calculated factor then it is apparent that one must know to where such gravity is pulling since it then is the gravity that is where the contraction is going that predominantly is keeping everything apart. Then the gravitational constant is what is resisting the collapse of the Universe. If there is a force, then where is the force taking the pulling…if it is a gravitational constant applying through out outer space then where is it having a centre base? To those professors claiming Newtonian ideas are substantiated by proof, I say that notwithstanding your personal academic qualifications and while at the same time disregarding your status and previous achievements as well as ignoring your many admirable abilities you may have and however superior they might be, I shall teach you about gravity. I say it is time Students learn the truth about physics notwithstanding the status academics will loose.

What you are about to read comprises of extractions forming part of the actual book that you can download free of charge. Go to www.sirnewtonsfraud.com and press on the blue button and the book is yours to have! Use this information to test the reliability of your tutors' teachings. Confront them with facts and don't allow them to stupefy you with their ability to commit mind control. This is what eighty years of studying the solar system brought about and no individual study since changed one aspect of what this table brings as information.

If mass did attract mass, what kept the balance in the distance it held according to the Titius Bode Law where the planets do find a balance in orbit rather than moving towards the Sun. One might think of mass pulling the comet to the sun, but instead of slamming into the sun, the comet makes a circle (Π) and disappears into outer space. As much as the comet is pulled coming towards the sun, the comet is pushed away afterwards departing for the darkness of outer space. If mass pulls the comet closer, what does the pushing away afterwards. If mass did attract mass, it would explain the behaviour of the comet coming towards the sun…but then what is pushing the comet back into orbit, into the darkness of outer space? The idea of proof about mass pulling everything is automatically placed at the door of Newton. This is not as small an issue as it seems at first. Think of what planets do… and you think that planets orbit. If normal speech contradicts Newton, then it is his task to prove his supposition is correct and the claims about attraction Newton made, although Newtonians will deny this fact as if they deny he is correct the honour of their Master Newton and that is what they have to do. Think of what planets do… and you think that planets orbit. No one thinks of planets spinning or planets basking in the summer Sun. However, just using the term orbiting is in total defiance with Newton or physics!

Ever person associates gravity applying as **_"The natural force of attraction exerted by a celestial body, such as Earth, upon objects at or near its surface, tending to draw them toward the centre of the body"_** with what we think of happens to solar bodies having mass which is **_"the natural force of attraction between any two massive bodies, which is directly proportional to the product of their masses and inversely proportional to the square of the distance between them"_** and science cheats everyone into believing the two is the very same. The one I experience every day and with the other there is no evidence of applying anywhere. Yet students are fooled into believing the two are exactly the same issue, which is untrue. If there is any academic feeling insulted by me calling the lot fraudsters, bring evidence of the second form of gravity working on the principle of mass and I will withdraw my statement, otherwise if no proof can be brought, then you lot that ignored me for ten years are the fraudsters I accuse you to be. You are villains brainwashing students to corrupt their thinking.

I have shown many ways how mass and Newton does not apply. There is no possible way to produce evidence from the behaviour of objects in the solar system that mass is responsible for gravity forming between two heavenly bodies. If it did, the asteroids could no possibly orbit in their allocated circle at the same rate as Jupiter circles in its designated circle. Every one believes in Newton except comets, because comets fail to collide with the Sun. However I can explain in some way…The evidence that mass is pulling mass there is not and there is no evidence of mass pulling mass. What there is I can prove how that forms gravity?

The Definition of Gravity according to the Oxford dictionary y of Astronomy:

Gravity is the fundamental force of attraction that all objects with mass have for each other. Like the electromagnetic force, gravity has effectively infinite range and obeys the inverse-square law. At the atomic level, where masses are very small, the force of gravity is negligible, but for objects that have very large masses such as planets, stars, and galaxies, gravity is a predominant force, and it plays an important role in theories of the structure of the universe. Gravity is believed to be mediated by the graviton, although the graviton has yet to be isolated by experiment. Gravity is weaker than the strong force, the electromagnetic force, and the weak force.

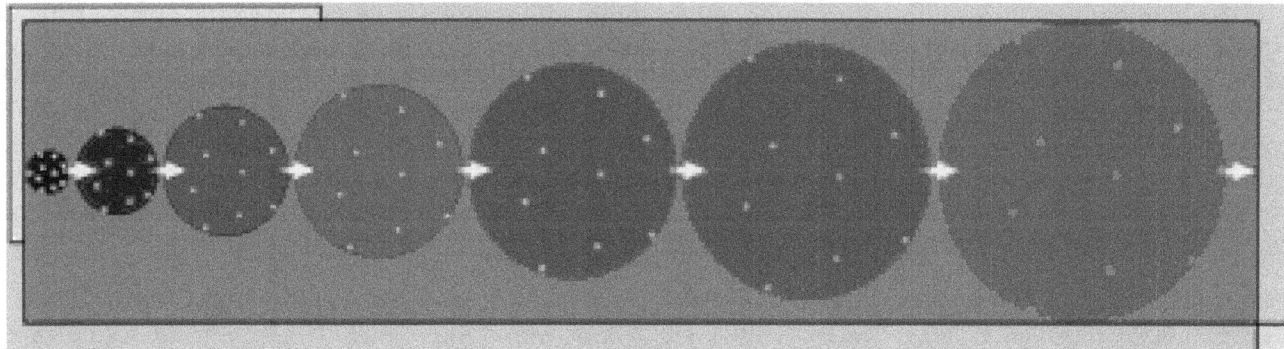

Presenting the Universe in this manner also forms part of the Newtonian conspiracy because it puts limits on what can never have a limit. The Universe has no outside because the Universe has only an inside. We have to look at the Universe not from the point where we are God but where we look up not the never-ending darkness and from where lights intensity starts at singularity.

I can and do explain this principle mathematically in a book I give away at no charge. Go to and download THE ABSOLUTE RELEVANCY of SINGULARITY : THE WEBSITE http://www.lulu.com/content/e-book/the-absolute-relevancy-of-singularity-the-website/7517996] and doing so will show you how it is possible to condense all the space around the spinning planets to form gravity by allowing cosmic structures to act as a centrifugal pump impeller. That is how gravity forms Π. You may see how material compresses the space around the star into a dense liquid. The liquid comes about from the spinning of the star compressing space and not by magical mass pulling. A star is compressed heat because when a star explodes it is always a process of releasing heat.

That is gravity. It is compressing space to form heat. That is gravity applying according to the Coanda effect and not by mass. This website informs you about the facts behind the lovely pictures that Newton's magical mass is unable to explain. Mass forming gravity depends on magic and physics has no magic. Please think clearly, if there is no contact between two objects and yet they teach you the two objects pull each other, then that pulling depends on magic, not science. This website will tell you why the cosmic physics principles establish these phenomena…and because the information does not salute Newton, academics in physics despise what I say. In the image one can clearly see the Roche limit in action where none of the stars are pulling any other star closer but a process of liquefying takes place.

The Roche limit clearly shows no stars can pull each other into a collision but clearly expands by overheating and if mass don't pull stars to a point of colliding then Newton is wrong! The Roche limit proves that any star that comes closer than 2.4674 times the radius of the larger star would liquidise everything within the space filled by the smaller star and this law and law providing the Coanda effect are a precise duplication of the interaction of the way that gravity truly applies in the midst of singularity. It is so clear that when stars expand stars overheat and therefore gravity must be a process of cooling! Only when stars overheat can stars explode as they do. You can see that gravity is locking in heat just by looking at what happens in the Universe and by studying every image as the Universe unfolds before your eyes.

Kepler used these three laws for computing the position of a planet as a function of time. His method involves the solution of a transcendental equation called Kepler's equation. Years ago I was reading of a remark Einstein made about his realisation whiles being a patent clerk. Einstein realised that had Einstein fell from the window of the patent office Einstein would feel as if he was as weightless as a chair and a pen falling alongside Einstein down the building.

Then I then realised Einstein felt weightless because he was falling and part of falling was feeling what was happening to him. He was not pretending to fall whereby he then would feel as if…he was really falling and with that there is no as ifs. What he experienced came by means of what he was experiencing. If Einstein was experiencing weightless ness, it would be because he was weightless while falling. Einstein would not imagine the weightless ness because Einstein was truly falling. He was at that moment truly weightless. Einstein, the pen, and the chair had the same weight since they were all weighing the same. All three items would be equally weightless during the falling…that was what Galileo found because objects of different size and different mass travel equal while descending. The bigger objects do not fall quicker than a smaller object and that can only be attributed to one fact; it can only be true if they weighed the same while falling.

From this one can deduct that gravity is motion or the intent to commit motion and mass is one the motion of gravity is frustrated by blocking the continuing of the motion. Gravity is motion of space and mass is the restricting of the motion of space. Having mass does not bring about gravity but it does restrict gravity's motion. Gravity produces mass but mass does not produce gravity. Mass is the restraining motion and gravity is material moving about. Mass only comes into the application when two objects filled with space moves into a position where both want to claim space the other occupy. In essence it still is the frustration of motion and the commitment to move once the blocking of space is relinquished.

I then after reading this realised that gravity is not mass orientated, but gravity is motion differentiation between objects. While falling, The object moves less or slower in the direction that the Earth rotates and will fall in the direction of the Earth centre until such a time as the movement of the object is in synchronising with the speed that the Earth spins or if not the object will and on the Earth surface at the edge of the Earth and that will bring about having mass. The gravity applies as speed that is putting time in relation to the distance travelled and distance travelled is space. While the object is in a process of falling, the motion confirms gravity, both by getting the object's distance or band in which the object travels in harmony with the Earth that conducts all the spinning taking place at that point. That will reduce the height in which the object spins until it lands on the Earth and then can't reduce such reducing of a travelling band any further. It has to do with specific density. If the specific density is increased by filling the object with helium we will find there arrives a point where the conducted speed is at a level that the Earth no longer will claim the body into having mass. When motion downward ends and the Earth disallow any further movement to secure a better specific density in relation to rotating movement, then mass sets in and becomes what is than point holding mass where the constraining of the object takes place to secure frustration of further movement and the Earth's motion annexes the object's freedom.

While experiencing mass the motion is still there but now incarcerated by mass and locked onto the Earth by the rotation of the Earth and the superior or equal specific density of the Earth. By connecting to the Earth the motion that the object is experiencing is what nails to object to the Earth by the force of mass and the object is then experiencing mass and not falling further through the loss of downward movement and now only conducts with the Earth rotating side-on movement. In this the downward movement is not lost altogether but remains as detectable movement is the form of having a tendency to move although the object in mass is applying by forcing the downward motion to stand still. While the object is in mass and seems to be as if it is resting the tendency to move downward remains applying but that tendency to continue to move downwards is the tendency he named mass. However mass then restricts motion and becomes motion tendency. While falling, gravity applies as equal motion to all objects relying to place all objects in relation to specific density and because of this motion counteracts any size, mass or weight by making everything able to fall equal in specific density. When falling, the object is either equal to what might be in the air according to allowed specific density, or has more than the specific minimum required density that is what is allowed to serve as the minimum required specific density and therefore will spiral down to the Earth. When the Earth restrains further downward motion of the object that comes as the result of finding an allocated position of motion according to the specific density of the falling object, this readjusting of allocated position is stopped from conducting further downward or readjusting movement and all such further movement of gravity is hindering in the form we call mass. The falling object remains individual and still tends to move while Earth individuality resists movement. Further movement is disallowed as other material fill space. While the bonding of the atoms forming the object will secure any further deforming the object will remain to be independent but it is this bonding that is the value of the specific density of the object applying. By securing a [lace on the Earth, the falling object will finally rest and from that motion resistance comes mass.

While falling, the object is experiencing gravity because the object is in gravity but when on the soil the object experience mass which is the restricting of gravity or motion of the space filled with material.

Moreover, I came to another conclusion of equal importance. When any person is standing on any place anywhere, while viewing the Universe, that person is filling the centre of the Universe. Let's get more personal. When you, the person that is reading this, are standing at night and are looking at the Universe you are seeing the Universe from the centre of the Universe. All the light, every single beam that ever left any destiny at any time acknowledges this fact. You are the most important person in the Universe because you are holding the most important position in the Universe. All the light that comes across all of space runs directly in a straight line towards you filling the centre of the Universe. Not excluding the effort of one photon, all light is heading to meet you where you are in that centre spot and not one photon will pass you by. Not one photon dare miss you because if they do they miss the effort that all light has to accomplish and that is to locate you as the person filling the centre of the Universe. If you find this funny, or laughable you are in for a shock because this is what gravity is and this principle dictates gravity. It is the most complex issue one can imagine and expanding on this thought takes thousands of pages. It forms the crux to all cosmic principles and embraces every successful and meaningfully theory ever used to explain the Universe. Without taking this aspect in to account, there is no valid explanation available to understand the cosmos. Al the light coming from wherever meets the point you fill in time and in space. For al the light travelling you hold the spot it was on route to.

Should you decide to shift your position to any other place in the Universe you will shift the centre of the Universe to that location as well. If you install a camera on Mars, the light is obliged to acknowledge your relocating the centre of the Universe at your will to reposition you're being that centre of the Universe. All the light that ever left its destination crossing the vast spaces of the Universe, excluding no particular light, travelled all the way just to find you filling the centre of the Universe, right where you are. By you're standing anywhere, you fill the centre of the Universe, and the entire Universe admits to that because all the light comes to meet you there.

If you shift from the North Pole to the South Pole you will shift the centre of the Universe because all the light travelling throughout the Universe will find you where you then moved the centre of the Universe. The light left its destination billion years ago as it travelled through space at the speed of light anxious to acknowledge you're being in the very centre of the Universe. No photon will pass you by where you are in the centre of the Universe. No wonder every person born has the idea they were born to fill the centre of the Universe, which we do fill. The Universe is spinning around you or I, which is filling a centre where all motion is connected. That is the Coanda effect on the utter-most grandest scale imaginable; nevertheless it is only a manifestation of the Coanda effect. It implicates gravity as wide as can be…

Then I reviewed the Universe. If gravity is motion, what causes motion? What stops motion? That answer is in the Black Hole. If a star is about fusing atoms thereby growing, what happen when all the atoms fused into one all collective atom? What is the gravity if the star has one all-inclusive atom providing all the gravity that the star had when the star still had massive volumetric space? If all that space that once filled an entire giant star fused into one enormous gravity applying atom and that enormous force has been secures in the space that one atom holds, the atom would then show a force that would pull the surrounding Universe flat. Where does the gravity of the star end when all the atoms in the star became one giant atom? Gravity is smallest where space is least. Where space of an entire massive star is left in the size of one atom the gravity coming from that will pull the Universe flat at that point.

Coming to the conclusion about gravity being motion and mass being the restriction of motion was the easy part. What produced the motion and what prevented the restriction from overcoming the motion was the tough part. Figuring out why was everything on the move and where did the motion stop that was the part that took some figuring and some explaining. What made gravity move and why does gravity move…the answers are in the four phenomena never yet explained to satisfaction but now turns out to be the cradle of gravity.

Gravity is The Roche limit,
Gravity is The Lagrangian system
   Gravity is The Titius Bode law
      Gravity is The Coanda affect

…And gravity as the Roche limit forms the principle in producing the sound barrier. Read the book and find out why this is the case.

Newton's claims about the principles that he declared is responsible for guiding physics carries no validated proof and only after I realised that, was I able to start forming another line of thought on gravity. This had the purpose of confronting the corner stone of modern physics and at first I tried desperately to do just that. At first I was not confrontational towards Academics in physics and avoided any indication about disagreeing with Newton, although avoiding to show my disagreements was also totally impossible too but every time I approached academics with my new concept the academics always threw Newton at me . Facing Newton or facing defeat became a two-sided blade and I had to start to confront them by confronting Newton, with which I was in disagreement from the beginning.

At first I was reluctant to voice any opinion about the matter of how far I was prepared to challenge Newton because Newton was and is an icon. But slowly it dawned on me that if I had any serious plans to introduce my ideas I had to dispute Newton's gravity principles and do it head. When the slight confrontation did not bring results I finally decided to go all the way and show the inconsistencies that were prevailing in Newtonian science. That worked neither and it brought me the same results as before whereby I decided to go public and straight to John and Jane Dow avoid arrogance academics have with only one motto they serve and that is their autocracy and in particular their megalomania especially to my case as well as me in person. I wrote them (nine in total) letters in which I warned them that I was going public to show the extent of their dishonesty in their Newtonian's approach and lacking of substance and proof their physics has. The lack of honesty and furthermore the absolute dishonest on their part is there whether I avoid it or attack it; the inconsistencies are part of forming the basis for modern accepted science.

This process I now described is explained in a paragraph or less and it seems I got that far in a breath or two, but getting this far took me the best part of seven years to get to I tried my best not to attack them or Newton but left with the option to leave the project and lose thirty years of work and then fail after I concluded an answer on every aspect they never even thought of or take them on and dish out what they should have received years ago made me decide on the latter. After being avoided and taunted by their powerful positions and arrogance vested in their mentality they show in regard to their positions as well as the disregard they show in the mentality of others I slowly concluded that only and after I can get people forming the general public and the opinion of those that holds their disregard just as I do to see what they hide will I get a response from the Mater's of fraud.

First I had to show the general public the true colours of the academics in physics and get every one to see how incorrect Newton is, and only then do I stand any chance to introduce my line of thought. I am so sure of the ideas that I propose of being correct that I dare any one to disprove any part or the entirety that my concepts about cosmology forms! But that can only come about when I can get an audience to see how I expose Newton for what Newton was and it is in that where I find no luck. I can't find one academic with influence that is brave enough to stand up and face my attack on Newton and argue me down or prove me wrong in a sound debate. Now I see frowning coming from everywhere because it is madness on my part to think the world is wrong and only I am correct!

I realise that it shows signs of madness on my part and in my thinking to even regard any possibility that I am the only person on Earth that is correct and all others that ever studied physics are wrong but mad as it seems, if that is what I have to say to find an audience to listen and to judge my case, then that is what I say. I don't say this lightly or without understanding the enormity of what I suggest is going on, but be that as it may seem, it is the truth without question that Newton went on for three hundred and fifty years defrauding science with no one testing his claims. Argue me down or prove me wrong but don't discount me before hearing me out and only after considerable consideration while studying my arguments then form an opinion that disputes what I say but when disputing what I say, do it while confronting me in a sound argument when proving me incorrect! This not one academic could achieve and I challenge the lot to do so. But do it after studying all my work and being in a position to account for all the details that I propose. Don't just dismiss me because I dismiss Newton because following that road is the way of the coward and the mentally impaired.

Read my challenge about the correctness of Newton's proposals when he brought no more than suggestions into science and when I dispute Newton, then take me on by proving Newton correct... do it just once... prove Newton correct just once...prove that his formula is working and that his principles apply on the grounds he principled his ideas.

Detecting Newton's misconduct is possible because I saw a way to break away from the invalid concepts Mainstream physics hold. I went about and tried to prove Newton and when that was not happening I tried to apply Newton's ideas into the greater fields of cosmology. That also wasn't possible. I tried to amalgamate the four cosmic principles applying in cosmology with what Newton said was happening in the cosmos with mass and with gravity and in light of what the cosmos showed was happening Newton just wasn't happening! Notwithstanding the pose Mainstream physics try to uphold, the entirety of physics still use the idea of magical forces intervening in nature and they still base concepts on unexplained novelties.

Think of finding four unexplained forces going around and influencing persons in an unexplainable manner except that the magic of gravity keeps people attracted to the Earth. To say the least, the concepts physics use in terms of Newton would not even be acceptable to children in the modern informed era we live in, I challenge any person to prove Newton, not to accept Newton but to undoubtedly prove Newton correct! Prove how Newton's formula of mass forming the force of gravity can apply as Newton said it does! I recognised the impossible double standards Mainstream physics apply to promote their much shady explaining. In short I tested Newton's principles and found the principles to be wanting.

The inconsistencies Newton introduced brought science double vision and to compensate for these bogus truths supporting their incredible theories, they simplify issues to such a level where what they embark on, is the meaningless acceptance of the unproven and they proclaim to understand what are meaningless inconsistencies and to achieve this they create scenarios which uses the entanglement of deception. Prove the attraction Newton said was enforcing gravity that is pulling by mass and is gathering plants by contracting the diameter between planets. Show how much the Moon came closer to the Earth since the time of Kepler. Show proven distances taken by radar tracking and indicate just how accurate Newton was. Show how much the Moon came closer to the Earth since the time of the Moonwalk in sixty-nine.

The figures are available but are kept in a grave of silence where no one ever speaks about what science found applies and how much the distance between the Earth and the Moon is shrinking as Newton said is happening or then how much is the is expanding which will contradict the very principles Newton brought about! What they declare as unwavering facts can't even be supported in some form when tested by a silly test as to show that the distance between the Earth and the Moon is shrinking. Even the least degree of verification of correctness is absent when trying to find support of Newton and Newton lacks all evidence of authentication in any investigation of even the simplest terms. It is as if they never read with interest that which they explain when they embark on explaining Newton and they never scrutinise that which they advocate when they teach Newton's principles applying. They give values that are senseless and the very values they use make that which they say meaningless.

In this book I am going to investigate how much truth there is in mass pulling by the force of gravity. To most if not to all of the persons reading this then such a venture of investigating Newton is time wasted and just the thought about me embarking on the investigation of the issue is totally senseless to investigate. It is senseless because the concept it carries became accepted as household practise and life science from where it proceeded to become everyday culture in every person's mind. The worst part is that the group of people normally considered as the wisest bunch there is, never did prudent testing on Newtonian presumptions, while to test the presumptions is most easy to do. I will not believe that a lot that lives up to the veneer of being the best mathematical intellectuals on Earth, never though of testing Newton's very simple formula and in that disregard the formula because of the incorrectness the formula holds.

Do you think of astrophysics as being the department that is run by the wise and the level minded, the honest and pure at heart, the nobility of well-to-do academics and the sober thinking standing in front of the world as the absolute trustworthy? If you are a student, there is no other choice you have but to trust them while they feed you absolute hogwash! If you would so much as dare to doubt any thing they say they will banish you from the institution they rule so absolutely. The banishing process is dome under the blanket of examination.
They teach you what to think and to make sure you think what they wish you to think, they tell you to confirm their teachings on a blank piece of paper. You write what they prescribe and you supply the answers they demand in the words (sometimes) of what they demand. Should you in any way say anything different from what they tell you to think, your presence will not be tolerated any further as they abolish you from their institution of academic tutoring. After reading this book I invite you to...no I dare you to challenge their statements with evidence gained from this book and see them wilfully further their culture of deceit by bringing unfounded arguments just in order to silence you and prevent you from getting behind the truth. If you think those in charge of astrophysics are the pillars of trust, then get wise by reading the following facts and arguments this book presents. What you are about to read is simply mystifyingly simple and yet to this day I have not had the privilege to challenged one academic any where that had the honesty to admit to the fact of Newton being wrong. After you have considered the following you might agree with me that even small Children can reach a higher level of clear-minded logic and find more sensibility than what those scientists promoting astrophysics have because science lives in a make believe fool's paradise.

The manner of regard to life that the Academic Physicist holds and the outlook on life that the followers of Newton physics have (I call them plainly Newtonians and to me they are sheepish because they resemble to the image that to me seems the same as sheep running after their leader without having the ability to think for one second any thought spawned out of personal intellect) is quite the opposite of what I think of them. They keep their forming the establishment of the order the Academic Physicist in high regard and consider their order to be the top thinkers in society.

This religion that they practise of self promotion and sublimely self regarding their status being next to God has them so high that we down on Earth forming the waste of human garbage can be told anything and we will believe what they say just because they with their supreme intellect tell us to think what they wish us to think. This they do because we human waste living way down below their supremacy have not the ability to think and therefore they must think on our behalf. In their view and so far very correctly judged on their part, they, the persons being in the group that forms the Academic Physicists, believe very correctly that can dish up whatever they wish and we, those forming the group in the gutter, those that are mindless in their eyes, we will have to accept what they say without being allowed to form an opinion other than having the opinion they give us to have because in their view we are unable to have a mind other than what they are able to control. This attitude they have is the result of a relationship that worked for so long and thee fact hat it worked that long is what confirmed their opinion that we, the public, are fools to believe anything and everything because of blind stupidity.

But in spite of their aggravating conduct and mischief towards us, it is not because of a lack of insight and inability of controlling a mind that we have our childlike belief and blind trust in their opinions and which there was. It is the faith we shown that they misused for their scandalous cheating. Our faith is what we have shown towards them and is that, which became used as the reason why we accepted what they said blindly. We didn't accept their word on the grounds of us being utterly stupid as they perceive us to be but our trust depended on our good nature and believing in their trustworthiness. This trust we have is brought on by a culture of trusting the King to do the people well and somewhere in every person's cultural past there was a King that did us well in leadership.

But their underestimating of our abilities is the testimony of their poor understanding and their weak insight ability, which results from their arrogance and stupidity. You are about to see just how stupid they really are in the thinking aspect of science. It will become clear as you page along while reading! They didn't fool us half as much as they fooled themselves and you are about to read all about it. The fact that they could fool us for centuries didn't run on their intelligence being so much superior but served their purpose as it stemmed from the trust we had in them resulting from good intentions on our part. This betraying on their part and misusing the public's good nature to be used in schemes to get the public conned must end and I pray that this book form the first step in resisting the arrogance of the Academic Physicist.

Any one not in their group of the Academic Physicist is part of the lowest order of mindless being and to become part of their order and those that have minds with an ability to think, students have to accept what they say when they say whatever they wish to say without having to prove the correctness of what should back their saying so and as a result of this students may never question what they say. Only when and after proving that a student has totally lost all ability to think for him or her self may a student is promoted into the ranks of their sublime intellectual group. The sifting process they named examinations. You write on paper what they told you and never question their opinion and after passing that examination will you ever enter their sphere of intellectual brotherhood. Does this sound far fetched? Then you better read on and I will remove your blindfold and show you what a world of deception the Academic Physicist force on us into.

Read the following and see how they, the high and the mighty, those that think they can replace God and those who think they can think on our behalf and think what to tell us to think, how much they are clowns and the jokers in society. Read how little are they, the Academic Physicists, able to understand concepts about Creation while they think they are able to replace God in their superior intellect.

If you are a student in the science of physics, then ask your Educated Masters to please explain the following abnormalities you are about to read in this book and insist on a clear explanation about the inconsistencies they promote while tutoring physics as if the physics they present are the most flawless and accurate institution there has ever been. Ask those academics supporting Newton about the following flaws that no one mentions …ever… except me in this book you are about to read and get them to explain the inconsistencies never talked about, which I present in this book and then after confronting those charged with tutoring physics and seeing who should be believed, then get wise instead of brainwashed. Let them mathematically show how one would go about and use Newton's visionary formula $F = G \dfrac{M_1 M_2}{r^2}$ to calculate the force of gravity by replacing the symbols with the actual values in mass that the items referred to have. Put in the Earth's mass in place where it belongs and put in your mass in place where it should be and then divide that with the distance between your soles and the Earth measured in micro millimetres by the square thereof!

In the book named an **_Open Letter on Gravity Part 1 and Part 2,_** I bring the solution to the mystery behind gravity. I tried in vane to introduce the principles I find valid to the academics in charge of astrophysics. Facts that Science present as being the uttermost explicit and unwavering truth, fails to bring any logic answers to so many questions that it should address. It fails to have substance in addressing the most basic and simple questions about gravity and physics. Yet to every question science can't answer my approach does bring many solutions. The presentation and the delivery of my answers that I reach are understandable and simple where it serves both logical science and the truth. Since my answers do not match Newton and his misconception about gravity and that mass generates gravity, those in charge of science don't even bother to read my work. With their affixation to the corruption they portray I can do little to the giants where they are in the mighty positions they have and just because of that they can go about to sideline and ignore my work and this is notwithstanding the correctness that my work delivers compared to the utter failing that Newton's work shows.

When confronted with my evidence and they have to match my work with the hypocrisy and misleading nature of Newtonian cosmology their defence in substantiating their claims is to ignore me. Since I do not applaud mainstream science and the clear fraud they embrace and fraud it is that they embrace, I am silenced. Why is it that my work is going unrecognised or even in the least goes never debated and never commented on…it is because it will then trash every article anyone has ever written about astrophysics and cosmology. They show little integrity when academics with such supposed high standing or then such as they should have, play a dishonesty game where those in commanding positions will rather protect fraud and save their skins. They would rather protect the corruption they have than seek the truth and find honesty in physics. Those academics in charge would much rather protect their un defendable ethos they maintain as forming the back bone in science and what gives their personal position legality although it is corrupt than admit to the truth they find when they begin reading my work and in agreement they then have to back the truth my work brings. Doing that (accepting the truth in my work) will trash all work in cosmology delivered thus far and condemn it to the waste paper basket and render all work invalid and void. It will put all the Newtonian's bias and fraud into the place where it belongs.

Considering that such acting will lose them money, those academics in controlling positions then will rather rape the truth in order to benefit from continuing to corrupt student's minds further. If they wish to justify their inconstancies they have to attack my work and disprove the accuracy of my work. That they can't do. They then ignore my work because they can't attack my work. In that sense they also place their work beyond my approach, as they can simply ignore me as if I represent the plague while they carry on with little consequence to bother them. I challenge them to prove Newton correct and not just declare Newton being beyond reproach after all has seen the evidence I bring. After reading this all students must challenge them to defend what they can't or get honest.

$F = G \dfrac{M_1 M_2}{r^2}$ $F = \dfrac{r^2}{M_1 M_2}$ This is the basis that Mainstream science uses as the foundation of all physics anywhere. If this is wrong then everything they have got to work with goes out the window. They put mass and the distance that parts objects in a relevancy, in other words the one is a ratio to the other. The one factor brings a measure to the other factor's value. The one cannot be without the other. The increase in one becomes the reducing of the other and the other way round also applies. When the distance is large, the influence of mass will be small and when the distance is small, the influence of mass will be overwhelming. Then they state we are in a Big Bang expanding of the entirety.

Why then, when considering that if it is mass that produces an inclining force of contraction as Newton says there is going on then…why didn't the expanding stop before it started when the Universe was small. Today using hindsight after the fact of the exploding Universe became apparent by the studies Hubble brought to light did the lot of everything that is not implode as Newton would have us believe whereas, instead it did expand just as Hubble proved. The radius at the time of the first instant back then was no factor, which makes the gravity at the time a totality of unrivalled force. The radius being that insignificant leaves the mass unchallenged in asserting power in relation to the non-existing radius it had.

I dare any physicist to show me where they apply Newton's formula just and exactly as Sir Isaac Newton suggested gravity applies. Show me just once where the mass of the Earth is multiplied with the mss of the object in normal physics. Show me just once how $F = \dfrac{r^2}{M_1 M_2}$ or $F \; \alpha \; \dfrac{M_1 M_2}{r^2}$ where one M represents the mass of the Earth while the other M represents the mass of the object and in this formula the end result will have a value of 9.81 Nm/s^2 … show just once one example… where the use of the mass of the Earth comes into play. If multiplying the mass of the Earth with the mass of an object and dividing that with the distance parting the two mass factors does not deliver 9.81 Nm/s2, and then any claim by Newton indicating that $F \; \alpha \; \dfrac{M_1 M_2}{r^2}$ is equal to gravity, such claiming constitutes to deliberate fraud…even if Sir Isaac Newton said this. Prove that the mass of the Earth with the mass of an object and dividing that with the distance parting the two mass factors delivers 9.81 Nm/s2 or admit physics is conducting fraud to protect Newton!

To whom it may concern:
My introduction as well as introducing the readers to general cosmology in a very brief and compressed manner but first, I have to give the emphatic warning to all prospective contemplating readers.

Please take note of a conscientious warning about the gravity of the misgiving there is on the part of the most respected Academics in physics about a much concerning matter.

I state it emphatically that science accuses me to be not schooled to the point where I am able to have any form of an opinion on any matter concerning Sir Isaac Newton. Notwithstanding that my research proves I did my private studies and through which I skipped the indoctrination and mind control academics place on students goes unrecognised by their standards and so too my ability to have any insight on matters regarding physics. However my skipping their methodical and systematic brainwashing enabled me to see and allowed me to be able to express the incorrectness in Newton's teachings and allowed me to show in clarity what destructive force Sir Isaac Newton used to corrupt the laws of mathematics, corrupting to science along the way and mostly raping to the work of a great man, Johannes Kepler and what Sir Isaac Newton did can only be expressed as being blatant criminal fraud. What his deeds amount to is to corrupt the laws of mathematics, to render the laws of cosmology useless and to rubbish all of science. Should you find this to be unbelievable, then I am glad to announce that this book is more for you than any other person, so go on and read what academics guarding science never wanted published.

I challenge any one that disputes any claim I make to prove me wrong by proving me wrong and not merely suggesting claims in that direction.

I have written several books in which I challenge the thought process of Mainstream physics and especially Sir Isaac Newton's arguments about physics. I am of the opinion that even though everyone thinks that Sir Isaac Newton is the genius that established every aspect today used in modern physics the man did not have a foggy clue about any of the principles driving the concept. He name gravity, but was unable to explain the concept in any detail. I made a study on the subject of gravity and from that study I am able to explain the entire principle. I prove my explaining with mathematics backing up my statements. In using the correct principles, which was what I found applying in the cosmos and which was what Mainstream science had no idea how to explain or even how to interpret, those principles I use.

Moreover I back up my interpretation how the cosmos use those principles to conduct gravity with the correct mathematical formulating. By my effort of using phenomena applied by the cosmos but never understood by Mainstream science, I do a far better job than what Sir Isaac Newton did and what I achieve is of a far more acceptable level as well as being mathematically far more correct than what Sir Isaac Newton did achieve with his guessing about issues he couldn't explain.

To be successful in my quest to find an explanation for gravity, I had to redirect all my concepts I previously had and also alter all the otherwise normally accepted thinking relating to physics. I realised that if I ever wished to come to realise what gravity is, I had to first realise that gravity is not what Newton

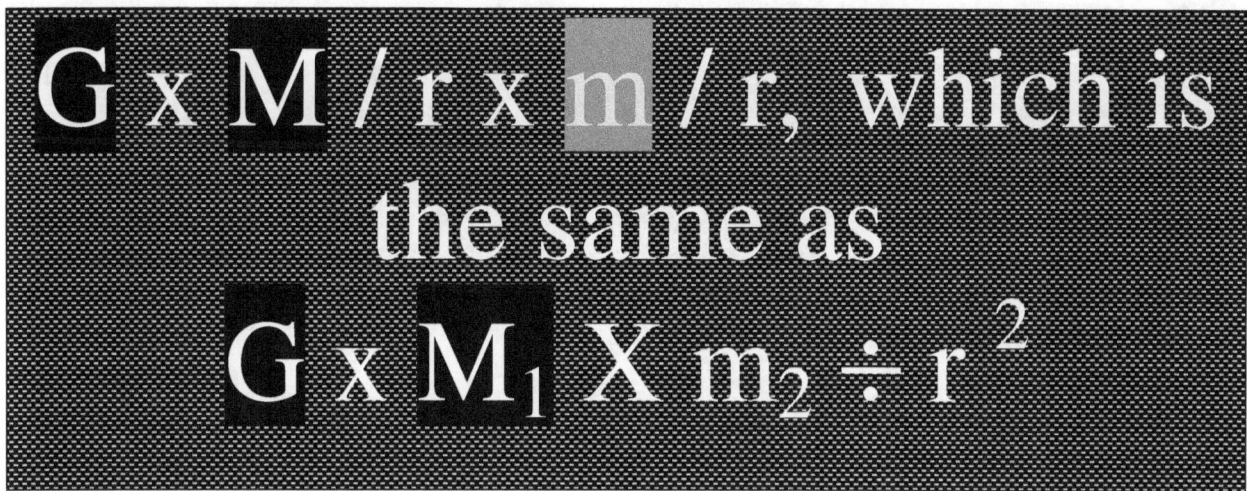

$$G \times M / r \times m / r, \text{ which is the same as } G \times M_1 \times m_2 \div r^2$$

saw as forming gravity. What Newton saw as gravity can't withstand even the slightest test of proof. I have tested Newton's thinking and this book bears the witness of all my testing of Newton. As any body can see that reads this book, I tested Newton from all the angles that he possibly could be correct and found his thinking wanting every time.

The truth about Sir Isaac Newton's concepts that he formed is that I came to conclude that in reality it is not in any way overstated to declare that Newton conspired to defraud science and moreover that he committed blatant mathematical corruption in trying to prove the concept he had about what he thought forms gravity. There are no mathematical or any other forms of proof to be used as backing for his ideas. There is no cosmic backing one might use to prove any of his claims in as much as what forms gravity or find proof about any of the claims that Newton made on matters concerning science in cosmic gravity, and every thought he introduced that later proved useful and was correct, was what he stole from another far better cosmologist.

At least the four phenomena I use are visible and are presently prevailing in the cosmos and I use the four phenomena in my explaining about how gravity forms. Not one of Newton's laws are directly relating to any concept Newton ever introduced at any stage but is the result of academic theft he committed against a much larger figure that preceded him by almost a century. However he committed academic rape and plunder of the man that preceded him while he saw to it that the phenomenal work the first person did was for ever inferiorly linked to Newton's concepts.

He presented the work of this man totally incorrect and these mistakes Newton made when he changed the work of the first person were since then never addressed as it should have been addressed. Newton brought no original input into science except that he gave a concept familiar to everyone a name. This well-known concept I refer too he named gravity and even that is inappropriate.

Newton changed science by incorrectly suggesting changes to science and to mathematics that breaks every mathematical principle he could think of. He has no right to change mathematical laws as it pleases him just because he thought to be greater than any person that ever lived!

Newton changed what no man can and that Newton did in his attempt to win over the prevailing academic thinking of the day as to lay some sort of groundwork to form the required backing for his ideas on physics and the changes he made to mathematics was his personal attempt to explain gravity or what he thought gravity is. Newton stole, cheated, lied, diverted the truth and raped other people's work as well as falsifying mathematics to find support for his most incorrect ideas on gravity.

This part is dedicated to a book entitled Open Letter Announcing Gravity's Recipe. In the book the author explains gravity. This achievement is possible because the author broke from Mainstream physics and the impossible double standards Mainstream physics use to promote their much shady explaining and double vision about things they have no vision of. Do you think of astrophysics as the department run by the wise? If you think that, then get wise and read the following. After you have considered the following you might agree with me that Children can be more logic than what they are because they live in a make believe fool's paradise. If you are a student then ask your Educated masters too please explain the following abnormalities and inconsistencies they promote, which I present in this information and get wise instead of brainwashed. I say again brainwash because they force-feed you facts, which they cannot explain because the facts are untruths. Tell them to prove that planets have mass. Tell them to prove that it is mass that generate gravity. Ask them to explain gravity in detail.

When you deal with a relevancy such as this $F = G \dfrac{M_1 M_2}{r^2}$ it is not the value of the top part that is of a crucial nature but it is the bottom part that controls the top part that predicts the value of the outcome.

The formula $\dfrac{\text{top part}}{\text{bottom part}}$ the size of the bottom part dictates the top part value. With the Titius Bode law applying where the distance doubles every time a new position of the next planet comes about, the mass has even a lesser role to play than it did before.

This is the way you are supposed to see the formula $F = G \frac{M_1 M_2}{r^2}$ and when having this view the formula $F = G \frac{M_1 M_2}{r^2}$ such as Newton saw it and what was the view in the dark ages Newton lived in having such a view will make a lot of sense. However, notwithstanding size or mass increases, the distance between the planets forms a doubling value relating to the specific position of the planet and in this there is no referring to size or mass whatsoever.

This Mainstream science use as the foundation of all physics anywhere. They put mass and the distance that parts objects in a relevancy, in other words the one is a ratio to the other. The increase in one becomes the reducing of the other. When the distance is large, the influence of mass will be small and when the distance is small, the influence of mass will be overwhelming. Why then when taken into consideration that if it is mass that produces an inclining force of contraction as Newton says then…when the Universe was small it did not implode whereas, instead it did expand. After all, the radius was almost no factor at that point leaving the mass to enjoy an eternal power in relation to the non-existing radius.

When the Universe was at the point where the Big bang started, the radius was incredibly small. That would make the mass inducing gravity by contraction inconceivably large because the mass was completely overpowering all factors with the small radius. It did not bring about an implosion that the overbearing mass contraction was supposed to unleash on such a small Universe in the beginning. The Universe at present with in comprehendible distances parting object that renders the force of mass no relative value as the force constantly weakens with the growth of the expanding.

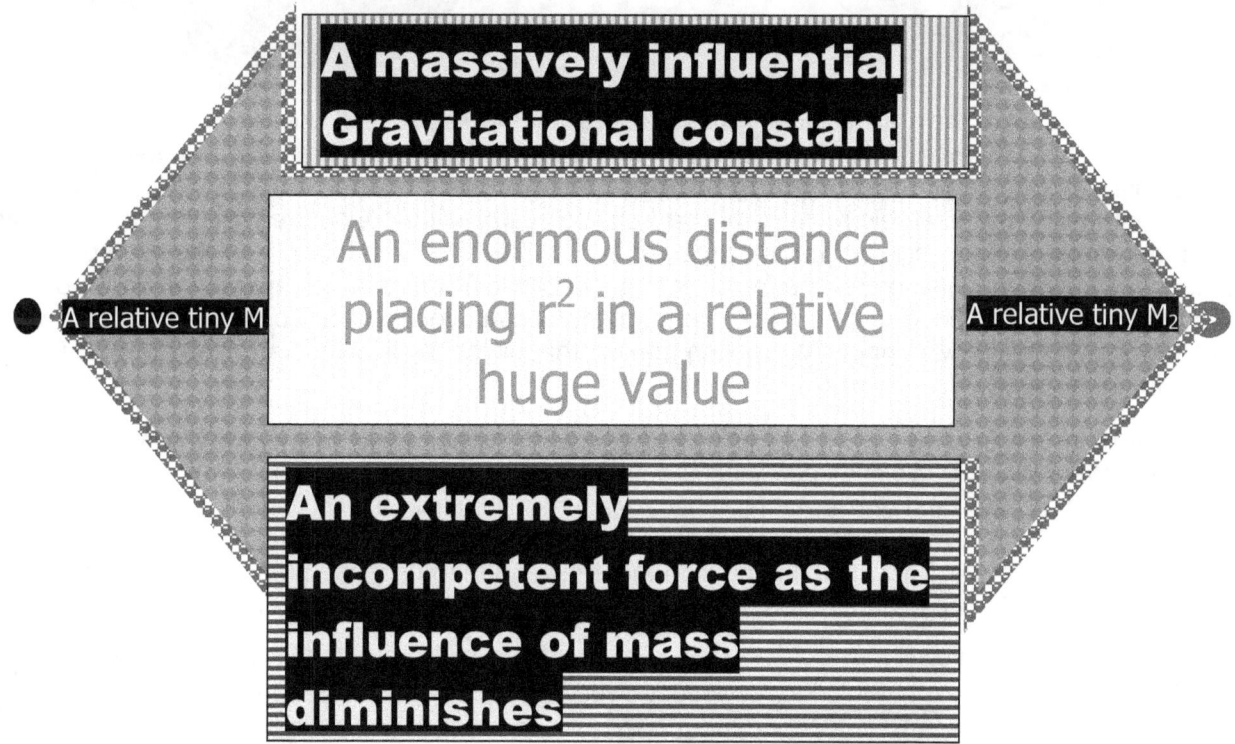

The more the radius develops in time, the lesser would the gravity be that the mass factor generates in relation to the advancing radii developing and the larger would the reducing be of all contraction. The effectiveness of force the mass produce will tarnish as the radius that separates the material from each other increases as time moves

Although it is presumed that the Universe was small at the dawn of the Big Bang, such presumption will put validity to another presumption that the gravity the mass charged at the time was enormous because the influence of the small distance in radii and the factor such distance produced promoted the factor, which the mass has to an enormous large factor.

If an object is a million kilometres apart the radius is a million times more in value by dividing the mass influence than when objects are one kilometre apart. That is the most basic realisation about mathematics. It puts ratio to order and define coherency. That is what gravity is to the Universe as it puts respect to factors about the Universe in the Universe. It is what derives order in the Universe.

At the very same time we will find in a Universe that was supposedly so small it had a radius of less than only one kilometre, then at such a time when the Universe was still that small it must also be accepted that the gravity the mass charged was one million times greater as it would be when the radius keeping the structures apart is one million kilometres in distance. The extremely small radius that was only the size of one neutron in radii distance and with the factor that such a distance produces, it must promote the mass factor, which will support the mass in having an enormous large factor by relevancy to what the case must be at present. The mass factor that produces the gravity at any given point during the event of the Big Bang, had to be eternally larger at the dawn of the Big Bang while having an infinite radius, which gave gravity all the power it can have and which it will ever have.

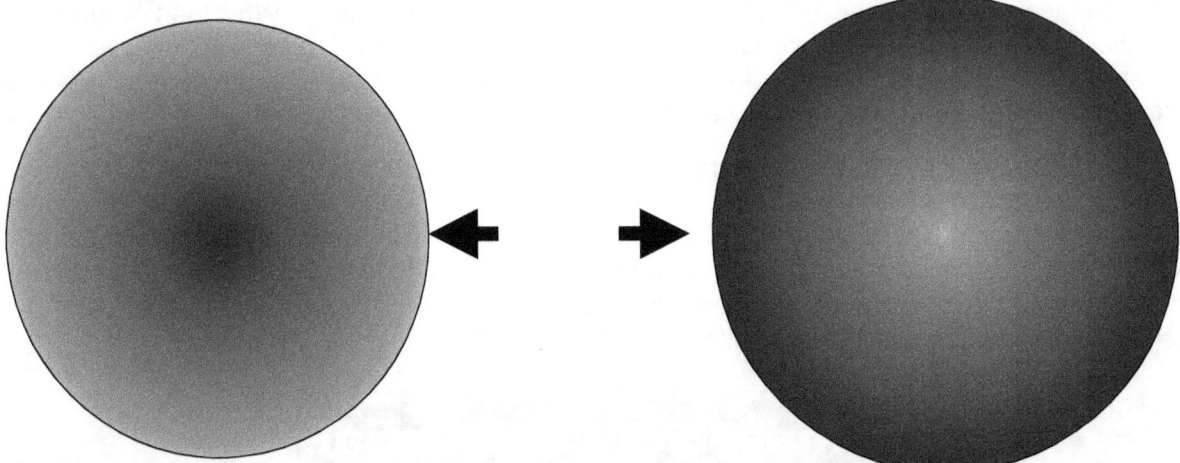

If at the Big Bang there was not sufficient mass to destroy the radius and prevent the expanding from coming about, then the expanding won the match and there can be no contracting Universe as Newton had us to believe. If the Universe started a journey of parting objects no amount of dark matter that might lurk in the night sky and is at this moment hiding from detection will produce the gravity required to stop the expanding from continuing. At the start the expanding became evident and as the radii grows the inclination will suspend in influence as a factor. If there was insufficient mass at the start in order to tilt the balance in favour of the reducing factor, no amount of mass can ever accomplish such a goal afterwards. Then Newton's surmising was one of corruption making that which all physics are based on fools thought and corrupted proof.

If you might be of the opinion that my accusing the greatest intellectual department in the world as being in misconduct and to your view such accusing is outrageous and far-fetched, then be my guest and judge the following with a clear and unbiased mind because when scrutinised with a clear view then the facts cannot fool an idiot. However, that is just what the physics paternity thinks the rest of us forming the general public at large are. They have the opinion that they can feed us in the public arena any senseless rotten garbage they dish up because they see us as being inferior by thought and mind.

With all this in mind did any one ever come to wonder about the all too famous **Einstein's critical density** theory and the fact that this idea was conceived to conceal the corruption of Newton in physics? The fact in truth is that the **Einstein's critical density** theory was a scheme plotted by those in charge to cover up and conceal corruption in the heart of physics.

If Einstein was unable to recognise the most basic of mathematical principles then what type of genius did physics create in him and what slur did physics promote. This idea of the two factors being in opposing relevance is so simple that children will recognise the principle, and yet those fathers of physics wants me to believe that the greatest mathematician that ever lived did not realise this principle…the principle that the radius and the mass stands related and the growth in the one will promote the decline in the other as a dominant factor. Can any one with this information including the information given on the previous page have any other conclusion? It is obviously clear that having such a total idea that there might be dark unseen mass floating in the Universe which at this time does not generate gravity but will some day because Newton has to be correct at some point in the future. I am to believe that dark undetected mass can be found and such undetectable mass could be found which will bring about contraction after all this expanding? Why would the mass at present then not activate gravity and why would the mass at some point spring to life and start activating gravity? How much can the Physics paternity still hide the fact that Einstein's critical density is being used as a cover-up to distort the truth to conceal fraud? The uncovering by the Hubble constant about of the Newton fraud is so simple to see. Hubble found the Universe is expanding and Newton's said otherwise. Who is lying about what?

Hubble's declaration was on track to blow the cover that was concealing the Newton fraud wide open and uncover the centuries old deception. To see this we have only too look at the comet behaviour when any and all comets again come around on a cycle by repeated visiting the sun. The question is if it is mass pulling mass onto mass, then why do we have comets left in the solar system? The mass of the Sun should by now at least have destroyed every comet going around.

Let any student ask his Master to explain Newton's formula in relation to the comet behaviour.

Contracting $F=G(M_1 \times m_2)/r^2$

Every one believes in Newton except comets, because comets fail to collide with the sun.
However I can explain in some way...

Every indication that we so far received in vivid portraying from astronomy photography studies from outer space disputes a shrinking Universe concept. From the moon increasing the radius distance between the earth and the sun, to the Hubble Constant indicating a space growing any where in space wherever man may conduct studies. Since the end of the middle ages a force called gravity was identified, but more than that science did not take it. What is gravity, besides being a force? What forces the force? I introduce a cosmic theory that turns the missing questions to answers.

Let us for one second return to the science we all know.

There is an undefined phenomenon in the cosmos, never mentioned (in public) because it obscures the basic formula

$$F = \frac{M_1 M}{r^2} G$$

Lets put the mathematical formula into a practical context.

Newtonians uphold their law of physics without showing mercy. The very first things the Newtonians use to beat us into submission are to blast us with incomprehensible mathematical formulas.

Incomprehensible they are but it is to scare anyone with the mathematical equations to get everyone hiding. They bewilder you with equations that put the fear of God into you; used simply to make you feel inferior so that they can feel superior and frown down on your inferiority from a dizzy height.

They are masters at manipulating anyone into a state of senselessness...but mostly they do it onto themselves. That they do because it forms the backbone of the fraud. They do not wish you to read closer and to find the fraud they hide to protect Newton. Ignore their mathematics because it only shows their incompetence to understand physics or Newton and see the fraud they propagate...

They employ mathematics to bewilder and that is all. I am going to show what we can uncover underneath what they cover. Look at what the mathematics supposedly says and then wake up, they are using maths as a scare tactic for three centuries to scare the daylights out of you and all this while its been working! Looking at the formula shows just how little Newton understood physics.

Please allow me to show you how they scare you to become fooled and suckered. Don't run and hide when you see the mathematics, it is meaningless although it was used as a scarecrow for more than three hundred years forming the backbone of the conspiracy.

This picture is as big a hoax as Newtonian science is when Newtonian science presents mass to be a factor that produces a pulling force called gravity. The question shouting for an answer in the picture is if mass is a factor that produces gravity as Newtonians claim it is then why are the planets not positioned according to mass as Newtonians declare.

The claim is as bogus as the entire philosophy. They present the proof that planets orbit according to mass in the following "Kepler law" which in its entirety had nothing to do with Kepler at all. It is all devised by Newton because Newton had no inclination of what Kepler's work was about.

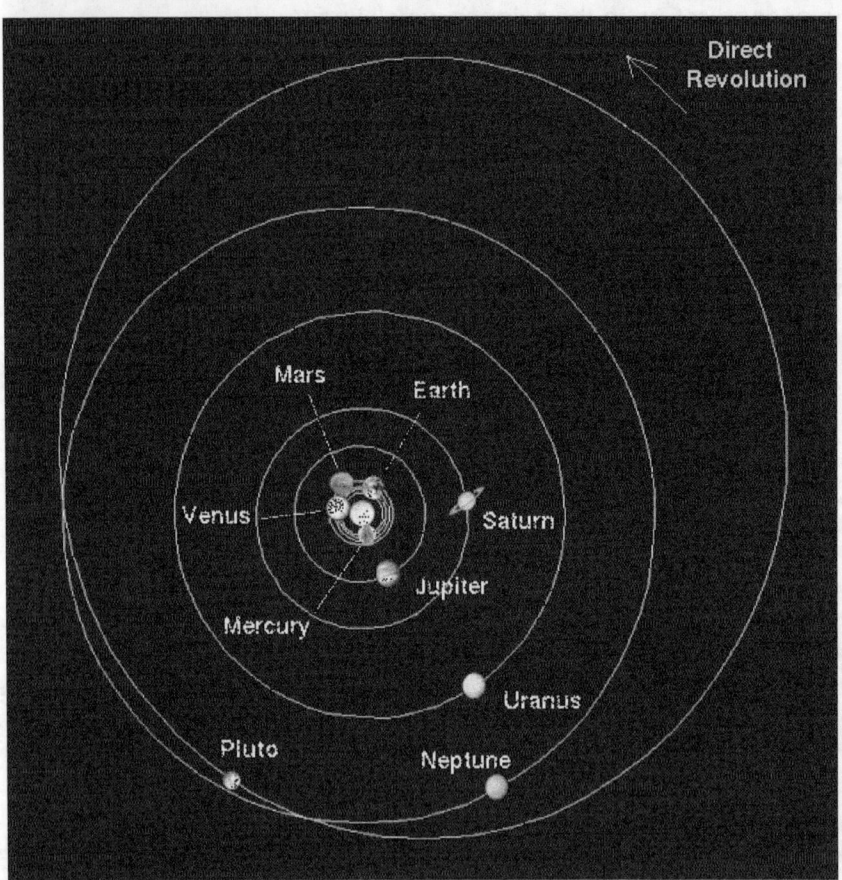

Newton brought about the idea of mass positioning the planets in the formula $4\pi^2 a^3 = P^2 G(M + m)$

Do not get scared as everyone usually does when seeing the mathematics and then as a result get frightened. Those physicists expect you to turn on your heels and run as fast as your legs can carry you. Then consequently as a reaction to find survival, you turn on your heels and run... but this time don't. Don't run, just read on and see how simple it is to prove Newton was a backward dark aged sod!

This time, don't run because I am about to show how meaningless this entire mathematical statement in reality is! This formula is total garbage and there is no sign of evidence that this formula forms any part of the solar system, even in the least.

Lets test this formula and see how truthful it is. $4\pi^2 a^3 = P^2 G (M + m)$ indicates that the circle in which the planet orbits ($4\pi^2 a^3$) is the result of (=) the position of the body (P^2) positioned by the mass of both bodies ($M + m$) in terms of the gravitational constant (G). The best way to find clarity is to test this statement with what is happening in the solar system just as it is, wouldn't you think.

A picture such as this provides much credence to the idea of gravity by mass since the lines drawn does not even begin to represent what is truly out there ands what is used by the cosmos in place of Newton's mass concept.

This is a table indicating the **mass** that **every planet** has in relation to the **distance every planet holds** in terns of the sun. If mass positioned planets then why did no one bother to inform the Universe about this because it is clear the Universe did not receive the memo from Newton's office to act accordingly.

Body	Mass (10^24) kg	÷	Orbital Distance(10^6 km)	=	ratio
Mercury	.3302	÷	57.9	=	0.0057
Venus	4.869	÷	108.2	=	0.045
Earth	5.975	÷	149.6	=	0.039939
Mars	0.6419	÷	227.9	=	0.00281
Jupiter	1898.6	÷	778.3	=	2.439419
Saturn	86.83	÷	1427	=	0.060847
Uranus	102.43	÷	2869.6	=	0.03569

This picture shows the hoax the Newtonian conspiracy pampers to keep the rest of Newtonian physics believable. They never mention the Titius Bode law and try to explain the Titius Bode law while it is the Titius Bode law that is really in place in the solar system. *If you wish to learn the truth then think again.*

Mass as a factor does not present or apply in one instance anywhere in the entire Universe and yet that is all theta physics says applies…but why would they cheat? It is because what is there applying between the planets is called the Titius Bode law and although this law is in place you have almost a hundred percent chance that you have never heard of it.

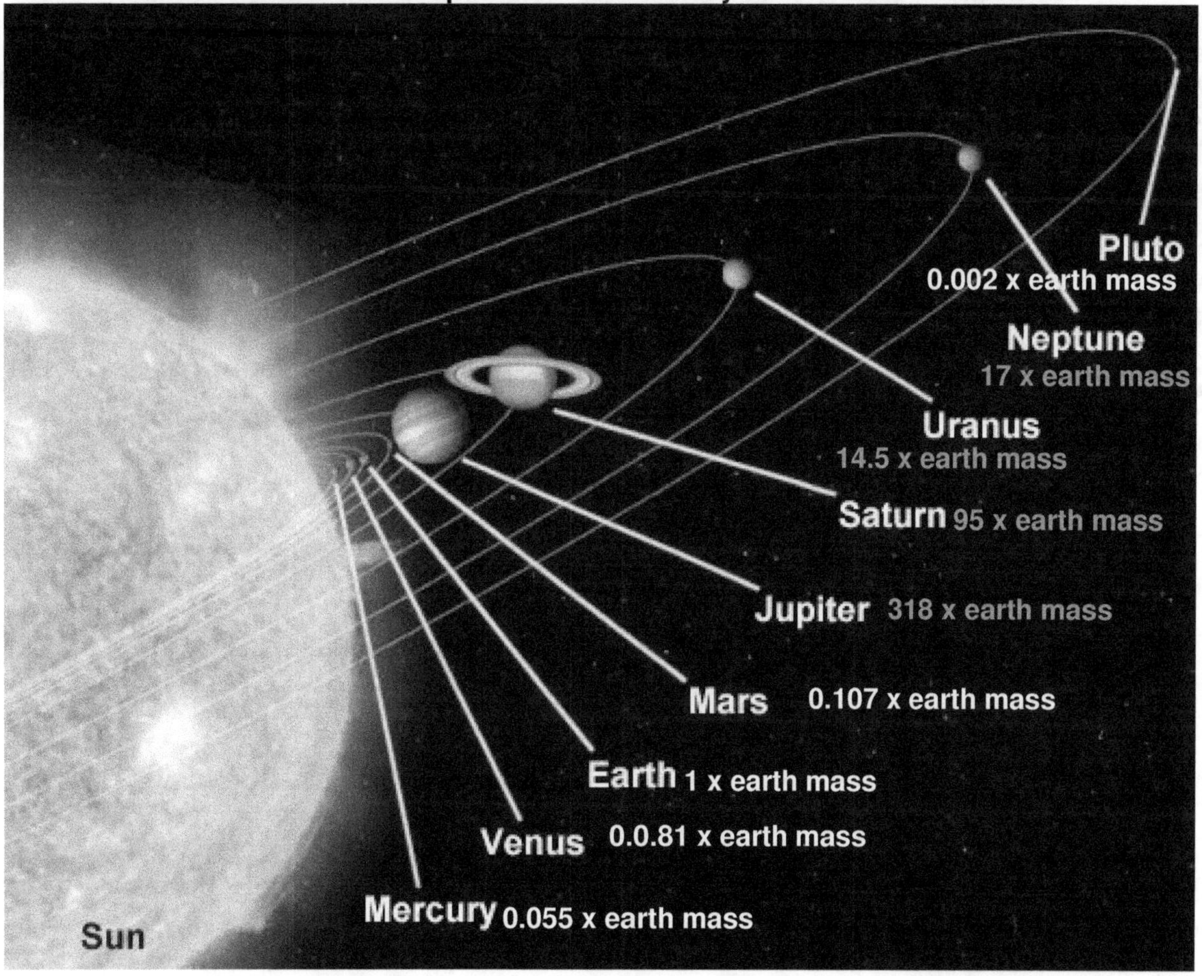

Pluto 0.002 x earth mass
Neptune 17 x earth mass
Uranus 14.5 x earth mass
Saturn 95 x earth mass
Jupiter 318 x earth mass
Mars 0.107 x earth mass
Earth 1 x earth mass
Venus 0.0.81 x earth mass
Mercury 0.055 x earth mass
Sun

If the planet layout was as I now show it to be according to mass then this was the order that is if it is true according to the solar system that mass do produce the position of the planet:

Jupiter	318 x earth mass	at a distance of 57.9 x 10⁶ km
Saturn	95 x earth mass	at a distance of 108.2 x 10⁶ km
Neptune	17 x earth mass	at a distance of 149.6 x 10⁶ km
Uranus	14.5 x earth mass	at a distance of 227.5 x 10⁶ km
Earth	1 x earth mass	at a distance of 778.3 x 10⁶ km
Venus	0.81 x earth mass	at a distance of 1427 x 10⁶ km
Mars	0.107 x earth mass	at a distance of 2871 x 10⁶ km
Mercury	0.055 x earth mass	at a distance of 4497 x 10⁶ km
Pluto	0.002 x earth mass	at a distance of 5913.5 x 10⁶ km

This is how it would apply if Newton was correct and mass did position planets. There is not even a remote chance that the positioning of the planets go in accordance with mass or $4\pi^2 a^3 = P^2 G (M + m)$. Do you realise there is much more "gravity produced by mass" in the space your feet has contact with the earth than there could ever be between Jupiter and the sun? You that can calculate it

 so then do it. Show that $\dfrac{a^3}{GM^2}$.

Put the orbit of Jupiter in relation to the mass of Jupiter and in relation to the position Jupiter holds. Forget getting swept away by the fancy Mathematics; just get to the task of putting the mass in relation or ratio with the position that any of the planets hold. Take the mass of the earth and your mass you have and then divide that with the square of the distance there is between your feet and the earth by the square thereof then divide that square with the product of the mass of the earth and your mass you have. Keep in mind your distance between your feet and the earth is about 10^{-11} meters going square! You that can't calculate it, the value would be meaningless but it is so much it will crush your atoms into a pulp, leaving you not even in a blood blob. It will leave a force to the value of about one Zetta 1 000 000 000 000 000 000 000 000 g per square meter. There is no chance in hell that any object of whatever size and formed by whichever method of construction could manage such a force of pressure. Those Physics-cheats always want to have the radius down to 1 meter because it makes their argument look sensible, and then they "forget" to use the correct mass of the earth not to stun any student into realising reality. Using one meter makes good sense when you wish to cheat but no body floats on meter above the earth. When anything stands on the earth, the distance between such a body and the earth is less than what could be sensibly measured.

Can you see the mathematics, Newtonian scientists use this to scare you into frozen stupidity and you allow it. I know and realise that you are disgusted by my attitude when I degrade the name on which physics are founded. In this introduction part I am going to show you just some minor deceptions all students are forced to believe since all physics students are forced to believe in Newton, Sir Isaac Newton that is. I am giving you a choice. You can say I am going to commit fraud or Newton has committed fraud. If I am judged to be the culprit that is guilty of deception then it is because Newton misled me. You can choose.

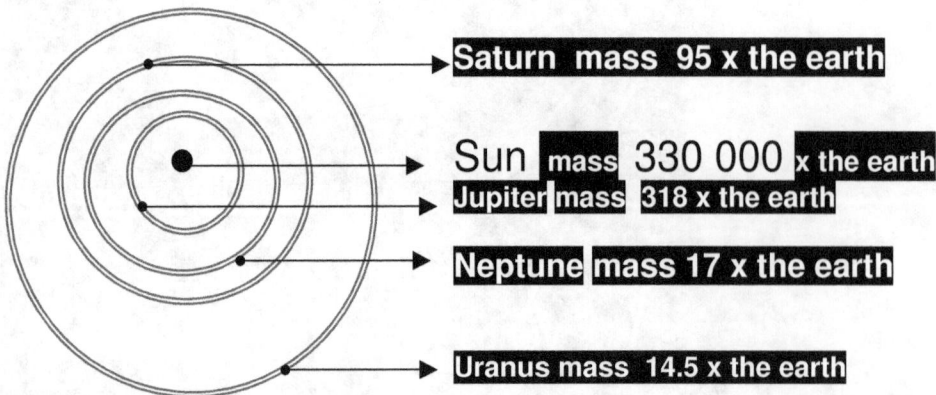

The inner circles that are very close to the sun we find the big gas planets with so much more mass forming the force of gravity that these planets are almost on top of the sun so close they are to the sun.

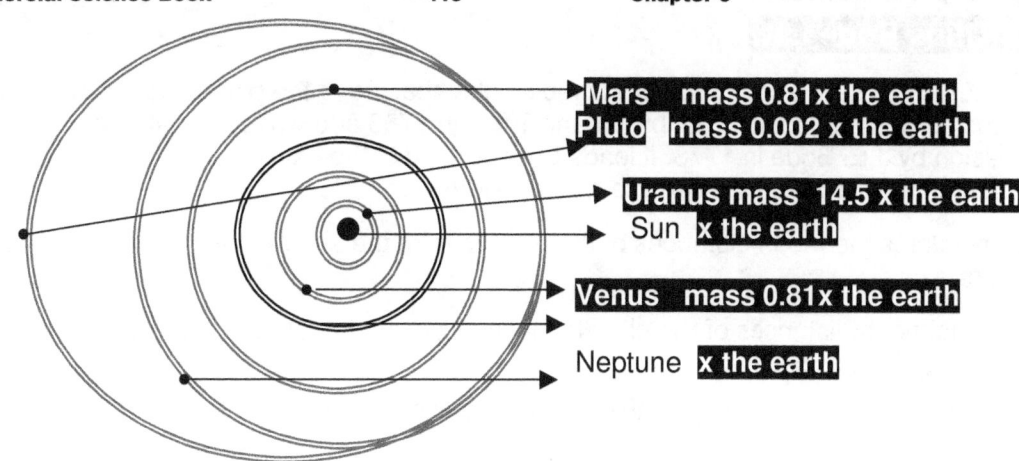

Then in the outer circles that are very far from the sun we find the smaller solid planets with so much less mass the force of gravity just can't pull these planets closer to the sun.

The above can be the only designated outlay of the planet position according to the sun when applying mass. Remember, if I am wrong then Newton and his formula $P = \left(\frac{4\pi^2 a^3}{G(M+m)}\right)^{0.5}$ is wrong but if I am correct and Jupiter is the closest planet then Newton is incorrect and then the academics in physics are indulging in Newton's fraud by forcefully brainwashing students to believe in Newton notwithstanding the fact that it is the solar system that disputes Newton's ideas altogether.

If you disagree with my layout, then you better disagree with Newton and his ideas. What is in place I am almost sure you have never heard of… Instead of the cosmos using mass as Newton claims there is no mass applying but there is the Titius Bode law on place. This is never commonly mentioned and the fact that what Newton advocates is not in place is hushed up.

Closet
1) Jupiter mass 318 x the earth
2) Saturn mass 95 x the earth
3) Neptune mass 17 x the earth
4) Uranus mass 14.5 x the earth

Then come the smaller planets with less mass and therefore less pulling force called gravity

5) Earth 1 x the earth
6) Venus mass 0.81x the earth
7) Mars mass 0.81x the earth
8) Mercury mass 0.055 x the earth
9) Pluto mass 0.002 x the earth

As clear and obvious as this is the fact of everything concerning physics is built on the presumption that $P = \left(\frac{4\pi^2 a^3}{G(M+m)}\right)^{0.5}$ is the formula used by the Universe to build the Universe. This is hogwash to put it lightly.

The Universe employs a system never referred to in terms of its actual position of importance as it build the Universe. This is because science has no idea how it works andf therefore has no idea how the universe layout works. That they would never admit and therefore they use the non-sensible formula $P = \left(\frac{4\pi^2 a^3}{G(M+m)}\right)^{0.5}$ so that they can cheat with mathematics and never be honest but will always seem informed and on top of the facts controlling physics.

The Titius Bode Law

The Titius-Bode Law is rough rule that predicts the spacing of the planets in the Solar System. The relationship was first pointed out by Johann Titius in 1766 and was formulated as a mathematical expression by J.E. Bode in 1778. It leads Bode to predict the existence of another planet between Mars and Jupiter in what we now recognize as the asteroid belt.

The law relates the mean distances of the planets from the sun to a simple mathematic progression of numbers.

To find the mean distances of the planets, beginning with the following simple sequence of numbers:
0 3 6 12 24 48 96 192 384
With the exception of the first two, the others are simple twice the value of the preceding number.
Add 4 to each number:
4 7 10 16 28 52 100 196 388
Then divide by 10:
0.4 0.7 1.0 1.6 2.8 5.2 10.0 19.6 38.8

The resulting sequence is very close to the distribution of mean distances of the planets from the Sun:

Body	Actual distance (A.U.)	Bode's Law (A.U.)
Mercury	0.39	0.4
Venus	0.72	0.7
Earth	1.00	1.0
Mars	1.52	1.6
		2.8
Jupiter	5.20	5.2
Saturn	9.54	10.0
Uranus	19.19	19.6

$P_n = P_o A_n$

P_n = Period of orbit of the nth planet

P_o = Period of sun's rotation

A_n = Semi major axis of the orbit

This is so typical Newtonian in every sense there is in science. The Newtonians gave the Titius Bode law a formula and that explains the lot. To they're under achieving standards that is very satisfactory. Now it is written in mathematics then what more do we need to know. The fact that the distance that Mercury has from the sun is doubled by that which Venus has from the sun is completely ignored. In cosmic reality mass plays no part. Then again the distance that Venus has from the sun is doubled by that which the earth has. This clearly has nothing to do with the size or mass of the planets. Explaining that part is completely ignored. Then again the distance that the earth has from the sun is doubled by that which Venus has and inexplicably this forms the layout of all planets in the solar system. Where do Newton and his idea of mass fit into what truly applies in outer space. Moreover, why does science never mention this?

Newtonians supply the formula. That is it. To the Newtonian doing that explains the lot that they can understand. That explains nothing but then again Newtonians fill the cosmos with nothing putting nothing between stars to fill distances. This is how the layout of the solar system is. It is there and it is in place and while the cosmos openly reject Newton's idea of mass, the Newtonian reject the cosmos' placing the Titius Bode law in place and thereby accepting Newton's mass idea that the cosmos clearly rejects. How on earth can anyone explain such behaviour and make sense. It puts planets at random as far as mass is concerned but it uses a formula that ignores mass completely $P_n = P_o A_n$

This process forming distance between planets carries on throughout the solar system.

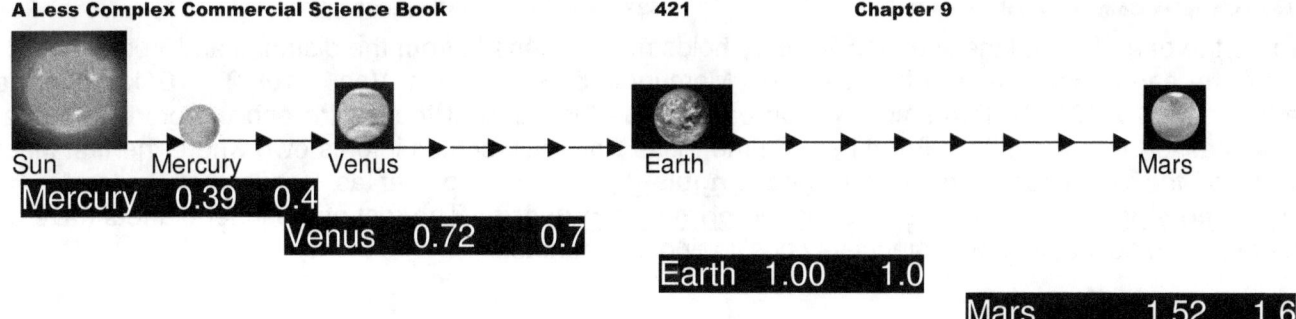

I am going to explain the Titus Bode law again because chances are excellent that notwithstanding how well read you are, you most likely never herd of this law all the while this is the most overall used principle in the solar system. To portrait that what follows in relation to this I would need a paper at leas several meters long. There is not a hunt of Newton's mass being used but the Titius Bode law forms one of the four building blocks that the solar system applies. There are reasons for this silence, it makes Newton incorrect and it underlines the stupidity of Newtonian science to explain what they can't understand and that is what is true cosmic science. I am going to explain the layout and later on the entire principle.
There is no room in a room to show this layout in its full compliment where it covers all the nine planets. If mass formed gravity then the layout should be running from the biggest to the smallest. Please put Newton's mass in this Titus Bode Law and explain what happens and moreover why this happens.

$$F = G \frac{M_1 M_2}{r^2}$$

The distance should be in terms of size , but it is not, it is according to **the Titius Bode law**, which is some law no one ever hears of because it disproves Newton and his mass concept. It shows Newton has no ground on which to form his concept that is completely wrong! But while the cosmos disproves Newton, science believes Newton in spite of the cosmos using **the Titius Bode law** This is very typical of science in the way Science prefers to cheat the truth to prove Newton correct.

Science would rather say the cosmos is wrong than to admit that Newton is wrong and in the book you can download free of charge named **Questionable Science,** you will see how many times does science prefer to back Newton when the cosmos shows Newton to be incorrect and this type of corruption goes on and on…do get wiser by getting better informed just by going to and **download Questionable Science.**

The entire idea above called the Titius Bode law is there in place used by the cosmos and is so accurate it was used to discover planets but also is dismissed as not being factual by physics professors because it repeals Newton's principles of mass. It is called the Titius Bode law and the undoubted accuracy of the Titius Bode law was never questioned. It used in the past to locate unknown planets but also it is clashing head on with Newtonian accepted principles and therefore it is degraded and denounced, just because it does not salute Newton's unproven principles.

I can and do explain this principle mathematically in a book I give away at no charge. Go to and **download www.singularityrelavancy.com The Website** http://www.lulu.com/content/e-book/the-absolute-relevancy-of-singularity-the-website/7517996

The sun holds the first inside planet at a certain position

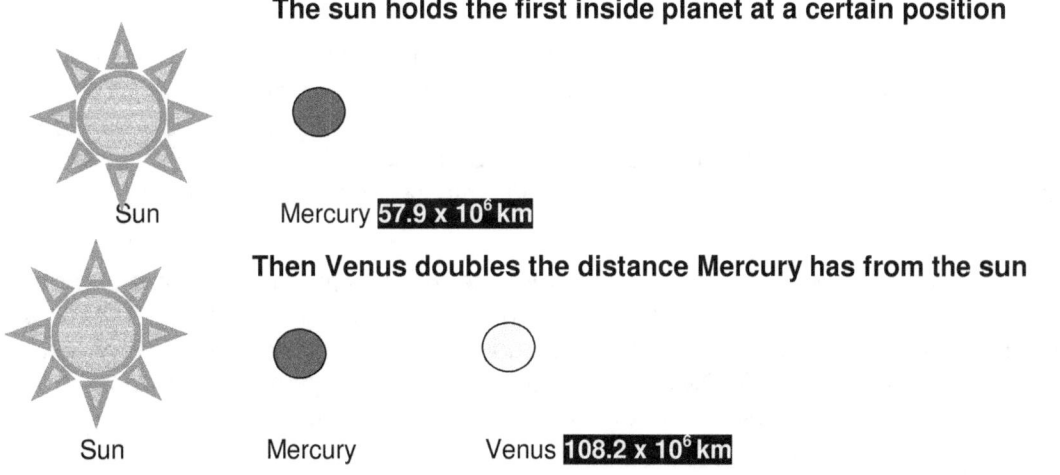

Where the earth fits into the scenario Mercury holds more or less ⅓ from the distance and Venus hold ⅔ where the earth completes the full distance. Mercury is 57.9×10^6 km; Venus 108.2×10^6 km and the earth is 149.6×10^6 km. Remember we on earth holds the earth at the centre and therefore the inner planets assort to our position. From here on the Titius Bode law comes into its own where the first inner planet or the earth forms the **controlling singularity** the outer planet as Venus then is forms the **governing singularity** and the sun forms the **primary singularity**. The rest of the inner planets have no role to play in accordance with singularity positioning the planets.

primary singularity

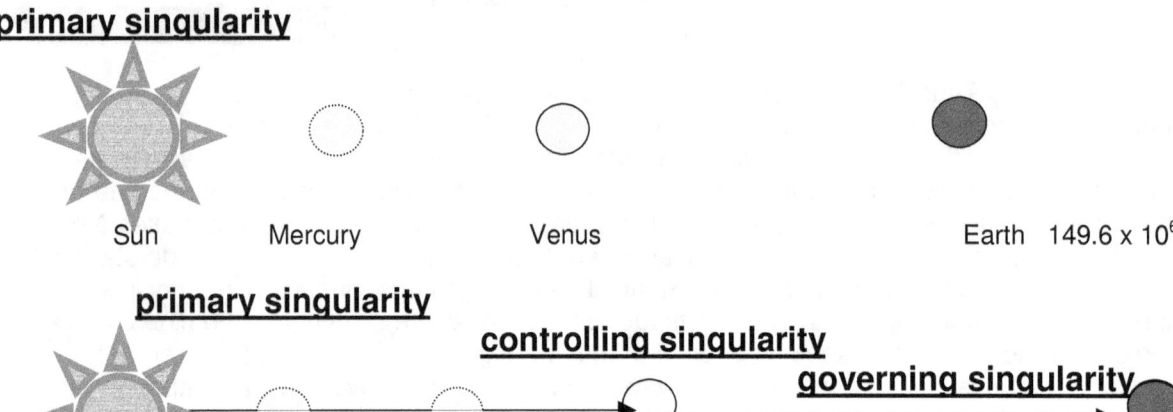

In this order there is no mention of mass playing a part. It is very convenient that Newtonian science ignores the relation or the fact that it is the Titius Bode law that is in place and that the Universe does not recognise Newton's claims on mass in any way.

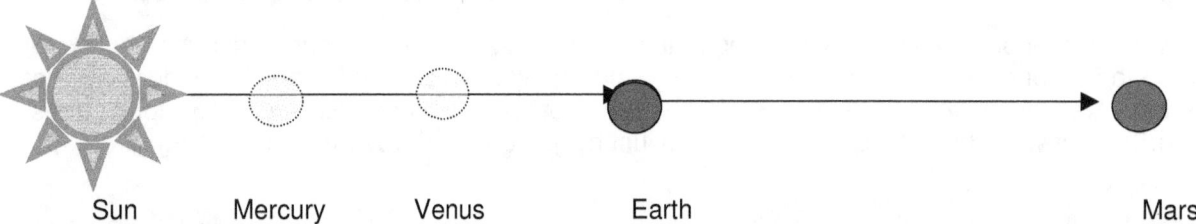

Is there still anyone out there that would dismiss my claims of a conspiracy to put fraud in place by giving merit to Newton and ignore the true factor that the Universe holds?

Confront your physics professor by asking your professor in physics to use Newton and then to explain why every planet doubles the distance it has from the sun in relation to the immediate inner planet because he should know why every planet is doubling its distance from the sun in relation to the previous inner planet; after all he is the Newton-physics expert and with this representing gravity as gravity applies in the cosmos he is the expert in physics… and if he doesn't know, then what does he know because this is the basis of everything forming physics and he is the Newton-physics expert in physics!

One would presume that when dealing with the most intellectuals on earth also being those that even attack the bible and religion for accuracy, they then would see to it that their front Patio is swept clean without a trace of suspicion about accuracy anywhere. Or could it be that they have so much to hide that they attack all other things and in that prevent all the other ideas to attack the Newtonian fabrication of the truth. It is always others that are incorrect because Newton not withstanding the fabrication of the basic facts that is always correct.

Where the Universe does not comply with Newton it is the Universe that has to correct it ways and deliver missing mass in order to start to comply with Newton. If the Universe would still not comply with Newton the Newtonian mentors create and fabricate ark, unseen, undetectable, untraceable and non-existing dark matter to get the Universe to comply with Newton's outrageous mismanagement of physics. Physics is a hoax created by the imagination of those not understanding reality and not complying with any form of concept that forms an understandable science.

With the exception of the first two, the others are simple twice the value of the preceding number.

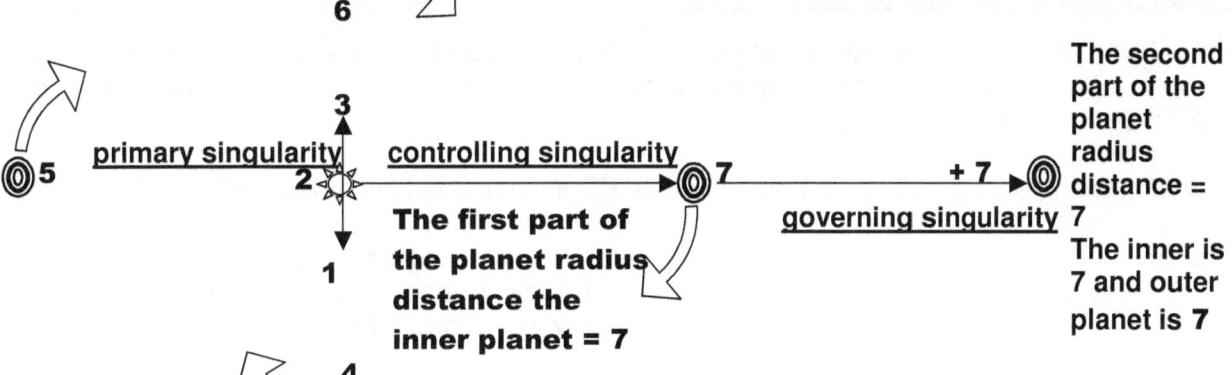

To find the mean distances of the planets, beginning with the following simple sequence of numbers:
0 3 6 12 24 48 96 192 384

To explain this is as follows: the axes line holding 3 doubles because in singularity all axis lines are 3 (1 bottom 2 centre 3 top)

Add 4 to each number:
4 7 10 16 28 52 100 196 388

Then the circle value that always has four point.

Then divide by 10:
0.4 0.7 1.0 1.6 2.8 5.2 10.0 19.6 38.8

Travelling in a straight line or a half circle or a triangle in terms of singularity is equal because it is all 180°. By taking 7 (the first or inner planet) in terms of Pythagoras 7^2 breaking the centre line 1^2 the result is 49 + 1 = 50. The second circle also values 50 and since singularity unites the movement it totals to form 100. The square root of 100 is 10 and dividing the travel by 10 the allocated position becomes valid. Is this not far better and truer than the following Newtonian accepted rubbish?

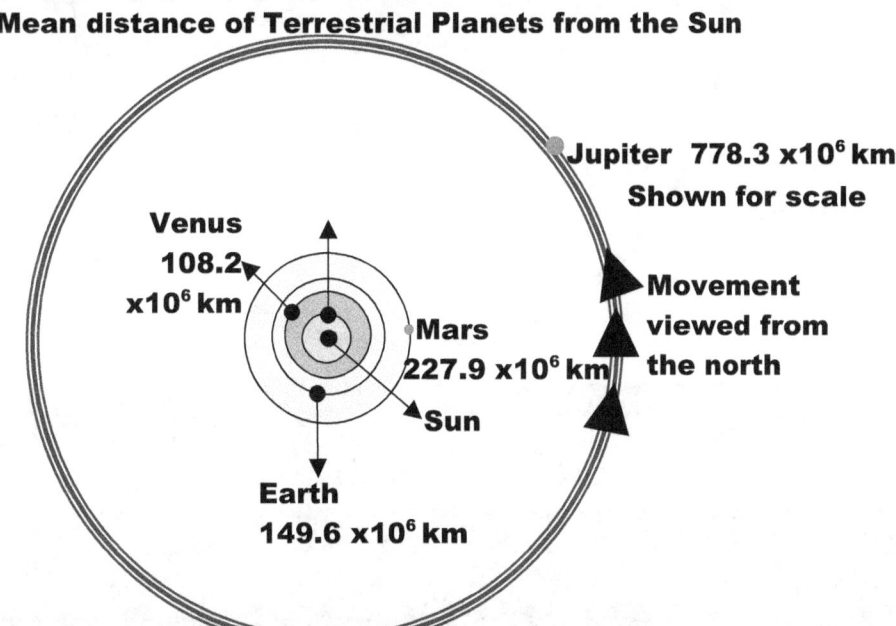

I dare any of the academics in physics standing behind Newton in full support of his theory on the gravitational law he formulated to explain where does planetary position fit into mass compliance. There are two options coming to light in this.

The one is that those brilliant mathematicians never saw this and so they are as stupid as toddlers or they are conspiring to keep this information a secret! If they are as smart as they pretend to be in mathematics why did they not tell the world about this inaccuracy about Newton's findings on which he founded physics.

The Titius Bode law is in lace and notwithstanding the level of despicable cheating, all the concocting the truth does not place Newton's misconceptions in place or remove that which is there, the Titius Bode law.

The following diagram shows the approximate distance of the Jovian planets to the Sun.

It is a complete fraud that science covers up to hide Newton's incorrectness but moreover to hide their personal ignorance and pittyfull incompatence. Do you still believe science is correct at all? Are you still prepared to have this criminals teach your children "the truth"?

I dare any of them standing behind Newton supporting the Newton idea of $4\pi^2 a^3 = P^2 G(M + m)$ to apply the formula $P = \left(\dfrac{4\pi^2 a^3}{G(M + m)} \right)^{0.5}$ and translate that to the concept that "proves" the planet position allocations is derived from the mass of the sun and the mass of the individual planet relevant to the gravitational constant. Telling the Universe what the Universe should do according to Newton does not put the Universe in complying!

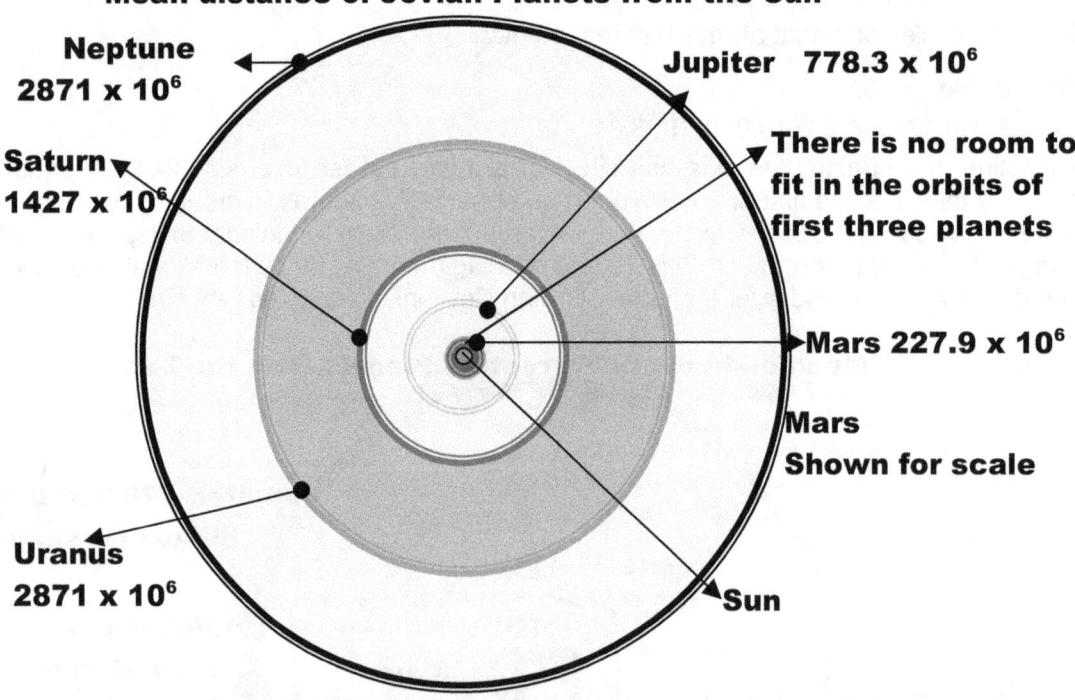

Mean distance of Jovian Planets from the Sun

Neptune 2871 x 10⁶

Jupiter 778.3 x 10⁶

Saturn 1427 x 10⁶

There is no room to fit in the orbits of first three planets

Mars 227.9 x 10⁶

Mars Shown for scale

Uranus 2871 x 10⁶

Sun

The planets only share one thing in common and that is the space in which they orbit that moving towards the sun. All the planets spin according to the space moving towards the sun and in relation to the sun contracting the space

Those that use Newton's formulas Newtonians wish to look brilliant and when peeling down the thin disguise they use to prevent Newton's incompetence becoming known, everyone can see how exceptionally stupid those smart mathematicians are because Newton's formulas disguised as Kepler's laws says nothing but exposes their incompetence in understanding cosmic facts.

Students, your professors are fooling you and you deserve to be their mindless monkeys just the way they think of you because you don't think about what they say! Cut the bullshit, force their ignorance about physics into the open and make monkeys of them, they deserve it even more then you lot do!

As you read the title of the book
www.questionablescience.net

If you think scientist know what gravity is…well, do not be duped that easily because no one in science remotely knows what gravity is…not even Newton knew what gravity is except Kepler… and because of what Kepler introduced now I know I can prove what gravity is.
Gravity is precisely what Kepler said gravity is

That Kepler was the only one that knows what gravity is we can see when we make an effort to investigate Kepler without Newton telling Kepler what he (Kepler) should have found. He (Newton) should have investigated Kepler's work more open minded and much closer then he (Newton) would have seen that gravity is precisely what Kepler found to be gravity

Science fails to bring logic answers to so many questions. It is simple questions about gravity and physics in which they fail, yet to every question science can't answer I bring an answer. My answer serves both logical science and truth but my answer does not match Newton and the misconception that gravity is generated by mass.

Yet, since I do not applaud mainstream science and their fraud they go about to sideline and ignore my work is.

Why is it going unrecognised…. because it trashes every article anyone has ever written about astronomical science and cosmology.

It puts all the delivered on Newton's bias and fraud to the place it belongs, it trashes all work done thus far to the waste paper basket and renders all work invalid and void. Where they have to attack my work they ignore my work because they can't attack my work. In the same sense their work is beyond any defence and when I attack their work, they ignore me as if I represent the plague. I challenge them to prove Newton correct and not just declare Newton being beyond reproach.

$$F = G \frac{M_1 M_2}{r^2}$$

This Mainstream science use as the foundation of all physics anywhere. They put mass and the distance that parts objects in a relevancy, in other words the one is a ratio to the other. The increase in one becomes the reducing of the other. When the distance is large, the influence of mass will be small and when the distance is small, the influence of mass will be overwhelming. Why then when taken into consideration that if it is mass that produces an inclining force of contraction as Newton says then…when the Universe was small it did not implode whereas, instead it did expand. After all, the radius was almost no factor at that point leaving the mass to enjoy an eternal power in relation to the non-existing radius.

When the Universe was at the point where the Big bang started, the radius was incredibly small. That would make the mass inducing gravity by contraction inconceivably large because the mass was completely overpowering all factors with the small radius. It did not bring about an implosion that the overbearing mass contraction was supposed to unleash on such a small Universe in the beginning

The more the radius develops in time, the lesser would the gravity be that the mass factor generates in relation to the advancing radii developing and the larger would the reducing be of all contraction.
The effectiveness of force the mass produce will tarnish as the radius that separates the material from each other increases as time moves

Although it is presumed that the Universe was small at the dawn of the Big Bang, such presumption will put validity to another presumption that the gravity the mass charged at the time was enormous because the influence of the small distance in radii and the factor such distance produced promoted the factor, which the mass has to an enormous large factor. If an object is a million kilometres apart the radius is a million times more in value by dividing the mass influence than when objects are one kilometre apart. That is the most basic realisation about mathematics. It puts ratio to order and define coherency. That is what gravity is to the Universe as it puts respect to factors about the Universe in the Universe. It is what derives order in the Universe.

At the very same time we will find in a Universe that was supposedly so small it had a radius of less than only one kilometre, then at such a time when the Universe was still that small it must also be accepted that the gravity the mass charged was one million times greater as it would be when the radius keeping the structures apart is one million kilometres in distance. The extremely small radius that was only the size of one neutron in radii distance and with the factor that such a distance produces, it must promote the mass factor, which will support the mass in having an enormous large factor by relevancy to what the case must be at present. The mass factor that produces the gravity at any given point during the event of the Big Bang, had to be eternally larger at the dawn of the Big Bang while having an infinite radius, which gave gravity all the power it can have and which it will ever have.

If at the Big Bang there was not sufficient mass to destroy the radius and prevent the expanding from coming about, then the expanding won the match and there can be no contracting Universe as Newton had us to believe. If the Universe started a journey of parting objects no amount of dark matter that might lurk in the night sky and is at this moment hiding from detection will produce the gravity required to stop the expanding from continuing. At the start the expanding became evident and as the radii grows the inclination will suspend in influence as a factor. If there was insufficient mass at the start in order to tilt the balance in favour of the reducing factor, no amount of mass can ever accomplish such a goal afterwards. Then Newton's surmising was one of corruption making that which all physics are based on fools thought and corrupted proof.

If you might be of the opinion that my accusing the greatest intellectual department in the world as being in misconduct and to your view such accusing is outrageous and far-fetched, then be my guest and judge the following with a clear and unbiased mind because when scrutinised with a clear view then the facts cannot fool an idiot. However, that is just what the physics paternity thinks the rest of us forming the general public at large are. They have the opinion that they can feed us in the public arena any senseless rotten garbage they dish up because they see us as being inferior by thought and mind.

With all this in mind did any one ever come to wonder about the all too famous Einstein's critical density theory and the fact that this idea was conceived to conceal the corruption of Newton in physics? The fact in truth is that the Einstein's critical density theory was a scheme plotted by those in charge to cover up and conceal corruption in the heart of physics. If Einstein was unable to recognise the most basic of mathematical principles then what type of genius did physics create in him and what slur did physics promote. This idea of the two factors being in opposing relevance is so simple that children will recognise the principle, and yet those fathers of physics wants me to believe that the greatest mathematician that ever lived did not realise this principle...the principle that the radius and the mass stands related and the growth in the one will promote the decline in the other as a dominant factor.

Can any one with this information including the information given on the previous page have any other conclusion? It is obviously clear that having such a total idea that there might be dark unseen mass floating in the Universe which at this time does not generate gravity but will some day because Newton has to be correct at some point in the future. I am to believe that dark undetected mass can be found and such undetectable mass could be found which will bring about contraction after all this expanding? Why would the mass at present then not activate gravity and why would the mass at some point spring to life and start activating gravity? How much can the Physics paternity still hide the fact that Einstein's critical density is being used as a cover-up to distort the truth to conceal fraud?

The uncovering by the Hubble constant about of the Newton fraud is so simple to see. Hubble found the Universe is expanding and Newton's said otherwise. Who is lying about what? Hubble's declaration was on track to blow the cover that was concealing the Newton fraud wide open and uncover the centuries old deception. To see this we have only too look at the comet behaviour when any and all comets again come around on a cycle by repeated visiting the sun. The question is if it is mass pulling mass onto mass, then why do we have comets left in the solar system? The mass of the Sun should by now at least have destroyed every comet going around.

Every indication that we so far received in vivid portraying from astronomy photography studies from outer space disputes a shrinking Universe concept. From the moon increasing the radius distance between the earth and the sun, to the Hubble Constant indicating a space growing any where in space wherever man may conduct studies. Since the end of the middle ages a force called gravity was identified, but more than that science did not take it. What is gravity, besides being a force? What forces the force? I introduce a cosmic theory that turns the missing questions to answers.

Let us for one second return to the science we all know.
There is an undefined phenomenon in the cosmos, never mentioned (in public) because it obscures the basic formula

$$F = \frac{M_1 M}{r^2} G$$

Lets put the mathematical formula into a practical context.

By reducing r would bring about the same result as enlarging the mass factor of the cosmic objects i.e. the Sun and the planets. It is a very drastic implication that will cause much more than just seasons changing. It must bring about that gravity changes through out the year…yet the radius does constantly change, therefore…

The closer any two cosmic objects come the stronger the force should be, with eventually no force in the Universe being able too keep them apart. This is just not happening!!!

There is no indication of truth about a contracting solar system as Newton proposed and as Newton's followers promote and or a contracting Universe as seen from the Hubble as he introduced has Hubble Constant…and not from any other evidence seen through the Hubble Telescope.

<u>As explained, there is some discrepancies about calculating the force of gravity, because gravity would apply as nicely as it does if it was the perfect balance, a balance exist in space of equal measure bringing about equal seasonal time.</u>

The biggest discrepancy and a practical denouncing the official version of the comet's flight around the Sun

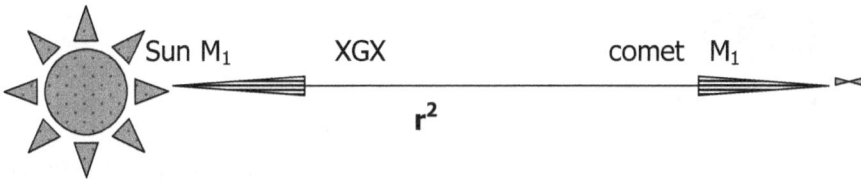

The Sun gets a grip on the comet by mass inflicting gravity and as it gets hold of the comet it drags the comet through the solar system straight ahead to the Sun just as Newton predicted the Sun with all its gravity producing gravity will do. There was no hint of a circle forming at any stage.

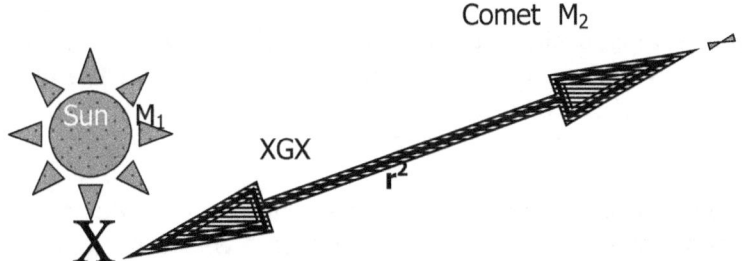

As Newton had said the gravity that the Sun and the comet mass induce pull the comet to the Sun. As we all know the comet moves to the center of the Sun just as Newton predicted with a slight complication and

a change in the venue, the comet no longer aims to the center of the Sun but aims at a target outside the limits of the space that the Sun occupies.

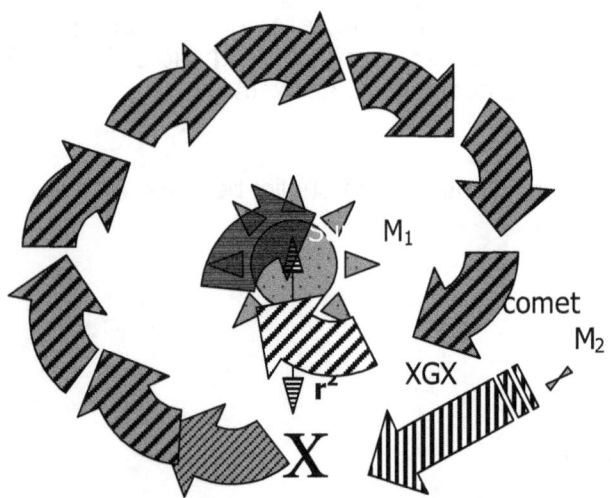

One should give Newton the benefit of the doubt and disregard this miss the position that the comet is aiming at. It could be that the gravity was not generated strong enough to locate the center at such a great distance as the comet was at first. It could be that the reducing of the radius comes in installments of a few circles. The comet might just take a cycle or two to wind down as the radius reduce and one should wait and see if it is not that which Newton meant when he said the mass by the mass is dismissing the radius between the objects. The formula Newton suggested did not make any room for a circle of any sort to form the trajectory of a planet or a comet trajectory.

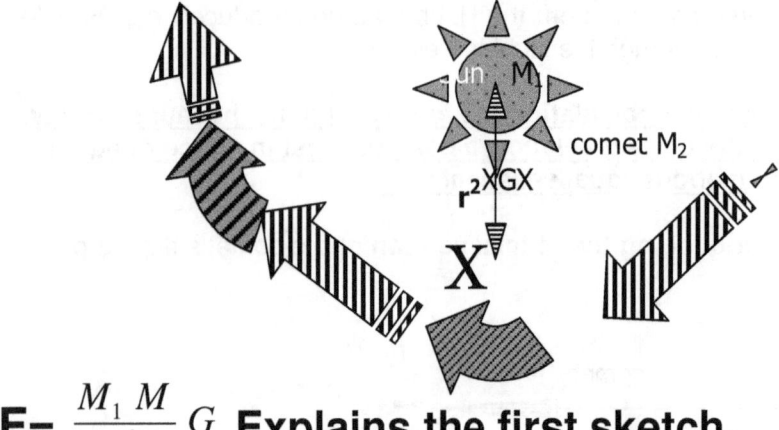

$F = \dfrac{M_1 M}{r^2} G$ **Explains the first sketch.**

Alas, it is not what the comets intends to do because the comet breaks the strangle hold of the Sun and what ever was pulling the comet at first is doing all the pushing at this point because the comet is surging into the darkness of the abyss. The comet speed away from the Sun and also at the same pace it was heading towards the Sun and there is no altering to the speed in any way. The comet seems very much unaware of the comet behaving in opposition to what the great Newton predicted with his formula.

The comet performs all the other maneuvers as the sketches indicate, which Newton's formula totally ignores. If the formula forms the basis of all physics used by science, the basics, which are around for hundreds of years, are trash and simply does not perform as it is supposed to. For many hundreds of years every person in physics were aware of this flaw but did nothing about it.

The lot that was filling the Universe was growing apart. In some case he said the lot was racing apart. The Universe was growing by miles and not shrinking into nothing. The Universe was blowing apart! The main discoverer had a name and a position of seniority, which prevented others from pushing his opinion aside. The man was E.P. Hubble. Through his telescope any one could see that the Universe was expanding and the expansion was most rapid.

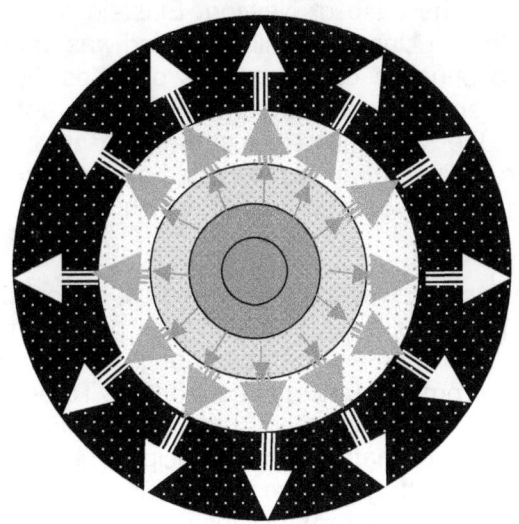

I am sure at the time all the Newtonian con artists were well aware of Newtonian shortfall in logical proof but this Hubble fellow was going to open a can of worms no one in science was able to face. How would the crooks that pretended to be the wise explain that Newton's contracting gravity was part of Newton's wild imagination? What would then happen if those that should know less then became wise and started to ask more questions about what the lot in conspiracy were hiding for (at the time) say about two hundred and fifty years? What is going to happen when the entire world learned that those academics in physics that was pretending to be the most brilliant among men was uncovered as the most stupid crawling the earth. What would happen if every one saw Newton was wrong all along and started demanding answers!

Every one was sharing the Newtonian vision of a contracting Universe where the lot would one day again come together and Creation will end where Creation started some time ago. The Universe has mass that is pulling mass towards one another and we are in the centre of an ever shrinking Universe. That is what the lot of us can see… we are forming the centre of the ever contracting having cosmos where every Newtonian can vividly see with his or her eyes through any telescope that all Newtonians minded scientists are sharing the centre stage of the ever collapsing Universe. The Universe is about to end where all mass contracts into one huge lump of material.

This unleashed a problem the world had no name for. Everything known to science was at that point devastatingly unknown to science. The world was expanding and not contracting which made the Universe quite wrong. It is impossible to have any vision about Newton being wrong. Newton could never be wrong because Newton was never wrong yet…so if the Universe is out of step with science, then science will correct such an abnormality by finding a way to defraud science and postpone the correcting that the Universe had to comply with since the Universe owed the Master Newton some apology. Did the Universe not know that he whom never can be wrong is in name Isaac Newton! Decisive action was needed. At this point I cannot believe that the most brilliant minds were so naïve and therefore I must suspect deliberate deception. Hubble was far too prominent to blow away and Newton was found wanting. At that point they put the onus of proof not on Newton but turned the focus away from Newton to what the presented as the guilty party. When will the Universe confirm its incorrectness by affirming Newton's obvious correctness? If they had to admit that Newton was wrong, the most intellectual science then had to admit they had nothing to show for all their minds brilliant work.

Science that was defying the likeliness of a living God stood bare and naked for all to see. They put the onus of proof and converting onto the cosmos. They asked not Newton but the cosmos when will the cosmos come clean and prove Newton correct, maintaining their unshakable belief that even the cosmos could be at blame but Newton could never be wrong. . When will the cosmos admit to a mistake and set its crooked ways straight. When will it meet its diverting from Newton and reach a point where the Universe will finally come to comply with what Newton demands. It is the cosmos that is wrong therefore it is time to find out when the cosmos will correct its manner. To deal with such a task they needed a man with a bigger ego than he had an IQ. They needed a person that thought more of his abilities than his ability to grasp any complex situation. They needed a man that was presented as a genius without ever proving his genius. They had a man that filled the centre of the Universe, which then placed the man in a location from where the man could see the entire Universe. They had just such a man. He went by the name of Albert Einstein. For all the genius Einstein had, Einstein failed to see the most simplistic and tiniest mathematical rule. Einstein failed to realise that if there was insufficient mass at the beginning of the expanding Universe, the growth of the Universe will reduce the influence of such mass as a factor because as the radius grows, such growth will restrict the gravity by rendering the mass progressively more incompetent.

If the Universe is expanding as Hubble indicated, the growth of the radius will reduce the influence value of the mass as every second passes. The mass will become more and more wanting for such a task. Yet with this obvious shortsightedness of the genius Einstein, the genius saw him fit enough to calculate and measure something as overwhelming as the Universe. As in the case of Newton, Einstein as an ego driven maniac that saw his abilities fit to measure and master the Universe while his mind was to simple to recognise the most basic principle of mathematics, the principle of relevancies or ratios. What a mathematical genius that turns out to be. While the radius enlarges, at the same proportion does the influence of the mass factor reduce and the mere fact that the radius increase shows that at no stage further into the future can the mass stem the growth of the radius because the radius overpowered the mass factor already. Unless there is new material entering the Universe at a point, which is impossible, the entire concept is fraud.

The idea was never to admit wrongdoing on the part of Newton and Newtonian science but to post pone, delay and divert attention away from the truth. If there was not enough mass to start with, no dark matter can kick in later on and start secondary mass frenzy that at that stage will then be enough to bring about the required mass potential that will turn around the Universe from expanding to contracting. To establish a scenario that would hide all deception they got the man that has a bigger ego than an IQ, they tell the world this man is a genius while the fool does not no the least of mathematical principles because his Master Newton did not no the least of mathematical principles and they got him to measure the Universe. While they did not even have any device (and will never have such a device) through which anyone would be able to see the entire Universe, they set of a scandalous misconception that this Einstein could calculate all the mass in the Universe.

Off course as can be expected, there was not enough mass and there will never be enough mass because there is no such a thing as mass in the entire Universe. When the deceit played out to the full, they fraudsters being the paternity of physics elaborated on the delusion by trying to find dark matter that is hidden. If the dark matter did not develop enough contraction at this time, there is no chance in the future to develop enough gravity because the factor of what mass supposedly should have is tarnishing and tarnishing as the Universe expand. The bigger the radius becomes the less would the mass effect be.

The community of astrophysics are trying to frame a picture where they set the stage in the way that if the Universe were stretched to a point the mass would not tolerate any more expanding. The mass will get frustrated in some way and show resistance to the increasingly elastic expanding. The gravity constant (I suppose) must prevent any further expanding. How they ever got to such an argument I never could tell. They surmise that outer space is consistently overall filed with nothing and when this nothing is stretched to the limit, the nothing would resist in growing more nothing or become further nothing and the nothing would stop other nothing to enter outer space in the community represented by nothing. If ever there is a faculty ruled by absolute inconsistency and rubbish as the motto of logic it has to be astrophysics.

Every measured kilometre represents nothing. Every mm is one of nothing. We on Earth are 149×10^6 kilometres holding nothing away from the Sun. Only they can argue that outer space is nothing with material here and there. If that is the case then which has more nothing between the Sun and Pluto or the Sun and Mercury. The distance between the Sun and Pluto is more, therefore that which outer space is made of is more than in the case of Mercury and the Sun. Therefore Pluto has more nothing between the Sun and the planet than Mercury has between the planet and the Sun. Only astrophysics and all the geniuses guarding the principal of astrophysics can put a calculated value by measure on nothing. In fact Mercury has hundred times less nothing between the planet and the Sun than is the case with Pluto. Since my days at school I was always under the impression that a hundred times the value of outer space being nothing is numerically expressed as (zero = $0 \times 100 = 0$), but where the genius that is such a prevailing part of astrophysics take the stage we find that Pluto can have 100 times more nothing than the amount or distance measuring nothing than Mercury has. The figure containing nothing that puts Pluto at the edge of the solar system is one hundred times more nothing than what Mercury has where Mercury becomes the first planet in the solar system. That is astrophysics. The brilliant minds of the mathematicians hold no rules apart from what they can calculate. Astrophysics is the only department throughout the Universe where normal rules don't apply since because with mathematics they can bend all laws as they wish…in fact Newton started the trend with his deceit.

Only the guardians of astrophysics policy can know why the undetected dark matter will start producing gravity to change the expanding to contraction. Would the fact that it is detected, change the influence it established? Or is it merely to extend the cover up and allow the deceit to linger until the following generation. There is no mass and any one that says there is mass, let such a fraudster then explain why all the planets irrespective of size or density, spin around the Sun at the same sped as all the others. Let them prove that the Universe acknowledge big and small and let them show how Jupiter can move at the same pace as does Mercury and Pluto while Jupiter is so many times more massive than the other two mentioned. More condemning evidence is yet to come because the astrophysics tricksters did not leave the corrupting of evidence just at that.

The fatherhood of physics never once diverted from acknowledging that Newton's contraction is the prevailing thesis on which the cosmos is built because they accepted that Newton used unlawful arguments and to cover up Newton's fraud which they still use to this day, they then proceeded with further criminality when producing the bluff they established with Einstein just to fool everyone in the normal public. Without ever recalling Newton's contraction theory that is obviously not working or admitting doubt about Newton's testimony to the effect, physics accepted the Big Bang Theory. The Big Bang theory opposes what ever Newton might have implied. The physics paternity however finds it wise to still advocate Newton while admitting to the Big Bang event. Newton said the lot is contracting. Go on and marry that with the Big Bang that says everything is expanding. You can't promote both except is you can define why we would see the two merge.

The Universe comes from a point the size of a Neutron. That makes the radius parting the Universe infinitely small. It just about removes the radius as a factor. At the very same implication it takes the pulling of the mass (if there are pulling forces converted by mass) to a level it will never again have. As soon as the distance between the objects holding mass started to grow, the power and influence of the mass factor started to diminish in the same ratio. If the mass were incapable of contracting the Universe then, it will forever remain contracting the Universe. Then you may ask what is the story? Read on and you will learn how far Mainstream Physics stray from the truth and how big a cover up the paternity is protecting.

According to science the Universe started with singularity. Quoted directly from the Oxford dictionary of Astronomy the following:

The definition of singularity is as follows:

Singularity: a mathematical point at which certain physical quantities reach infinite values for example, according to the general relativity the curvature of space-time becomes infinite in a black hole. In the Big Bang theory the Universe was born from singularity in which the density and temperature of matter were infinite. The average daily temperature was "$10^{\alpha\beta}$ to 10^{34} K".

Then the second "day" the daily average temperature came down to 10^{34}K and 10^4K. That is fine, but if the temperature was in Kelvin, then what was 0^0K. In order to make sense of the scale used there must be a minimum to secure a maximum otherwise the maximum can just as well be the minimum and is only advocated to impress humans applying earthly standards.

By using a scale as $10^{\alpha\beta}$ to 10^{34} K, it places the lower temperature at a modern 0^0K to make sense of standards. If that was the temperature the standards were lowered, compromising something to gain something, because something had to grow larger for heat to reduce. We know space grew larger bringing heat down to reduce.

Being the onlooker the viewer has to maintain one position. From that position some particles would be circling a centre point, as the particles would be coming towards the onlooker. The other matter would be circling the centre point while rushing away from the onlooker.

At the very end the single dimension may come into the dynamics but where the single dimension comes in the factor of zero is removed.

If there is space, there is a flow of light and a flow of light has to produce lines in relation to angles forming space between them. Something must be present to confirm space because there is an absolute difference between being in space and no space to be found. If there was a line that formed nothing that one line that forms nothing would completely destroy the other lines' chances of ever forming a triangle, let alone having all lines and they then have a total being zero. As shown in the example no line can form zero and therefore no mathematical equation as far as it extends to cosmology can ever bring about zero as a number. While there is space present there has to be three dimensions relating to each other by time and in three dimensions there has to be three lines in relevancy to each other by angles formed holding space in (at least) six opposing sides. Removing one line must bring about a flat Universe and that then will constitute nothing.

Cosmology is about light flowing by means of lines indicating space obeying the rules enforced by time in motion and light flowing dictates crossing space and across space light is using lines. The book: *An open letter Announcing Gravity's Recipe* is dealing with the subject finding singularity by removing the concept of nothing from outer space. By diminishing nothing one uncover singularity and the effort brings in a new perspective not yet introduced.

For your benefit I will shortly give a summary by which I hope to interest you in reading the manuscript: Compressing space produces heat. Releasing heat will bring expansion bringing about space. We call such a release of heat an explosion. In other words heat translate to space and space concentrates back to heat. The one is a product of the other where space forms expanded heat.

They are quick to show the time that was applying at the time being some thousandth of a second or the heat that was present being numbers we have no name for. The other side of the story they ignore. They ignore the other side of the story because in that respect it puts their promoting of Newton down to madness. If you reduce the radius applying at the present back to what it was at the time of the initiating of the Big Bang, you must also increase the influence gravity and mass had at that moment by the same number you are decreasing the radius. That is pure mathematics and the most basic physics of all concepts.

The shrinking radius will increase the effectiveness of the influence of the gravity that the mass can produce by the margin of the shrinking of the radius. If the Radius was infinite at that point, then that means the gravity was eternal. With the entire Universe being as big as a Neutron, the Universe was the size of an atom. If the Universe were the size of an atom and the mass within that Universal atom could not prevent the Universe exploding into immeasurable atoms, then it would not be able to retract all the atoms into one unit again. If there was not enough mass to start the contraction, there can be no contraction of mass that is producing the gravity at this stage. If the gravity is of such a nature that it allows a continuous growth of the radius, then the radius firstly cannot be zero as Newton suggested and the extending of the radius proves there is no contraction in the way Newton had everyone to believe. If Newton's mass contracting mass is true, then on the other hand it must have resulted in an implosion as that which can never repeat again. With Newton's formula of $F = \dfrac{M_1 M}{r^2} G$ forming gravity, then the Big Bang is just not possible because from that formula the Big Crunch must respond.

They put all questions on hold by diverting the attention to some black matter or dark energy, some force no-one can detect and yet it is going to save Newton's honour. Why is that dark matter with mass not pulling at this point?

Forming part of the website **www.singularityrelavancy.com**
By going to LULU.com the following books are available in e-book format as individual books wherein I share with you the newly discovered information about www.singularityrelavancy.com **which you are reading and which you are free to download**
Then download the next book from Lulu absolutely free and see if I exaggerate in any way!

THE ABSOLUTE RELEVANCY OF SINGULARITY: THE UNPUBLISHED ARTICLE Free of Charge
http://www.lulu.com/content/e-book/the-absolute-relevancy-of-singularity-the-unpublished-article/7747133]

The Absolute Relevancy of Singularity The Dissertation
http://www.lulu.com/content/e-book/the-absolute-relevancy-of-singularity-the-dissertation/5994478]

Book 0 The Absolute Relevancy of Singularity in terms of Newton
http://www.lulu.com/content/e-book/book-0-the-absolute-relevancy-of-singularity-in-terms-of-newton/7190018

Book 1 The Absolute Relevancy of Singularity in terms of Cosmic Physics
http://www.lulu.com/content/e-book/book-1-the-absolute-relevancy-of-singularity-in-terms-of-cosmic-physics/6624181

Book 2 The Absolute Relevancy of Singularity in terms of The 4 Cosmic Phenomena
http://www.lulu.com/content/e-book/book-2-the-absolute-relevancy-of-singularity-in-terms-of-the-four-cosmic-pillars/7181003]

Book 3 The Absolute Relevancy of Singularity in terms of The Sound Barrier
http://www.lulu.com/content/e-book/book-3-the-absolute-relevancy-of-singularity-in-terms-of-the-sound-barrier/6621856

Book 4 The Absolute Relevancy of Singularity in terms of The Cosmic Code
http://www.lulu.com/content/e-book/book-4-the-absolute-relevancy-of-singularity-in-terms-of-the-cosmic-code/6625975

Book 5 The Absolute Relevancy of Singularity in terms of Life
http://www.lulu.com/content/e-book/book-5-the-absolute-relevancy-of-singularity-in-terms-of-life/6626316

Book 6 The Absolute Relevancy of Singularity Investigating Kepler
http://www.lulu.com/content/e-book/book-6-the-absolute-relevancy-of-singularity-investigating-kepler/7179110

Where the above all are also available from Lulu.com.

Should there be any person whishing to purchase these books in one volume given as a thesis published in paper format then contact me, on this web address by activating www.singularityrelavancy.com

And you will be able to purchase six books in print as a unit on paper forming one volume with six books going as *The Absolute Relevancy of Singularity* The Theses.

However please note that this printed Theses is very limited as it is printed privately. When you press on the button www.singularityrelavancy.com to activate I will return the e-mail as soon as I can to confirm the availability of published manuscripts and prices. The six books are identical to the six books on offer through Lulu.com but they are in monochrome whereas the individual book in e-book format is in colour where colour applies

This is not the website called
www.singularetyrelevancy.com By pressing the title you can download www.singularetyrelevancy.com

Do you know what this is? This website will not show you the lovely pictures. For that there is another website that will tell you everything that happens in the experiment. This information will tell you why this happens and what is the cosmic philosophy prevailing in physics allowing this.

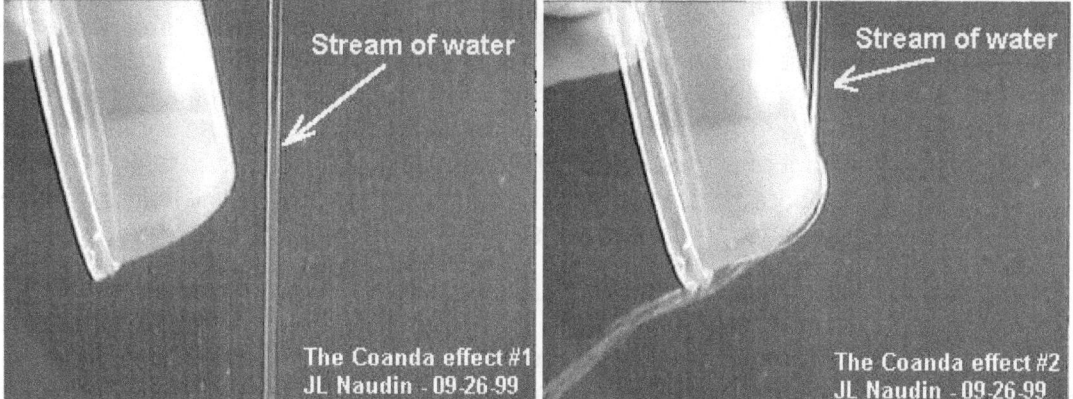

This is the Coanda effect that translates into gravity when the Roche limit comes about.

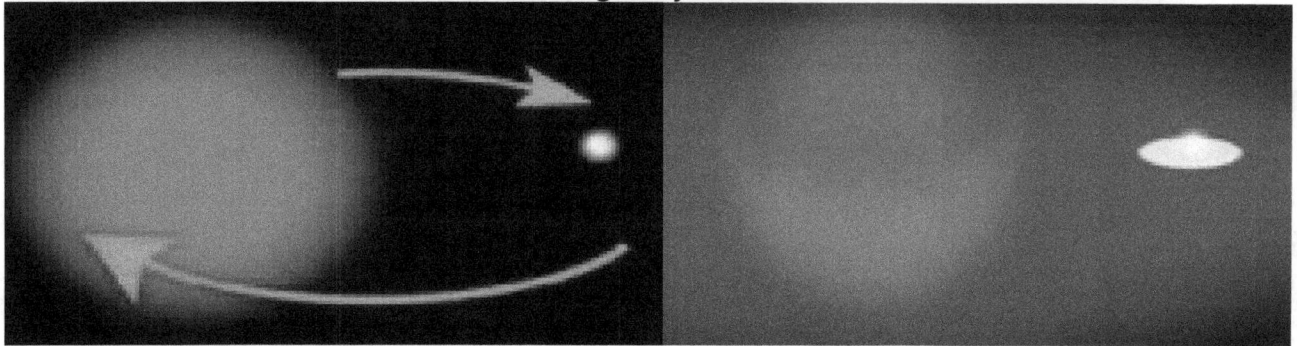

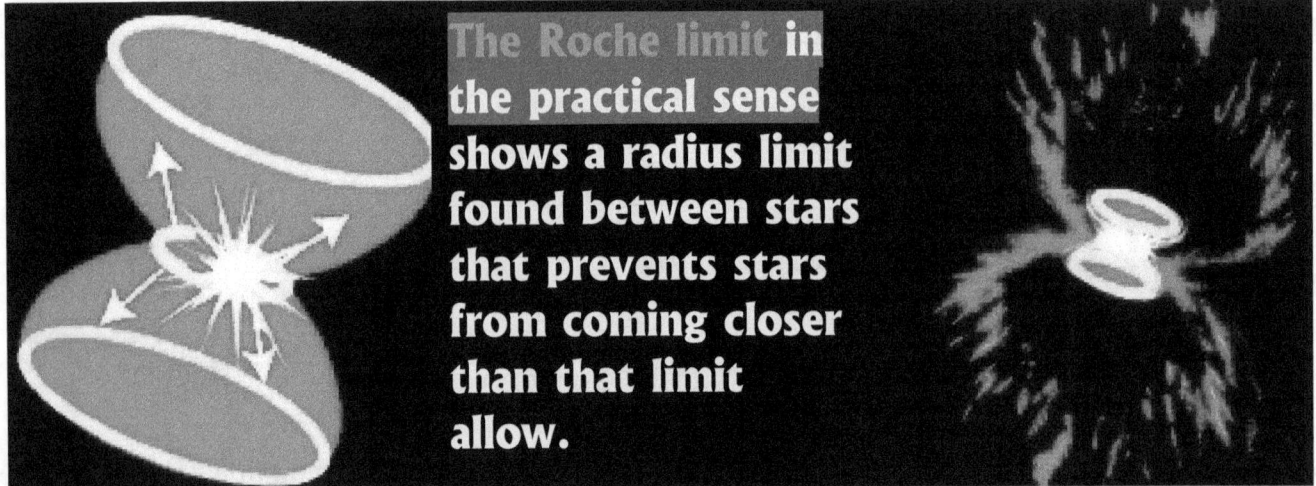

The Roche limit in the practical sense shows a radius limit found between stars that prevents stars from coming closer than that limit allow.

Do you know what this is? What happens in these pictures proves that stars could never collide because what prevents it is cosmic law that science at present can't fathom because their understanding is too limited. This book aims to inform you about the facts behind the lovely pictures. This book aims to tell you why the cosmic physics principles establishes these phenomena…and because the information does not salute Newton, academics in physics despise what I say. If mass don't pull stars to a point of colliding then Newton is wrong and in these picture we see the evidence that stars can never collide as Newton claims it does! Any star coming closer than $\Pi^2 \div 4$, which is equal to 2.4674 times the radius of the larger star would liquidise the smaller star and this law and the Coanda effect are a precise duplication of true gravity truly applying. This value the Roche limit has is $\Pi^2 \div 4$ which is circle movement divided by 4 opposing sides through which the rotation goes.

In order to establish a reference point where singularity divert allowing matter the zone matter can claim space in time, the control of space in time and the influence on space-time, the point of singularity have to reduce the value of singularity on both accounts of the cosmic atoms claiming individual singularity, or enlarge the claim of space by matter away from singularity. This is rather important to understand when arriving at the actual presentation of the formation of the solar system

After establishing the reference point to either singularity reduction or space-time enhancing through allowing matter to grow, the Titius Bode law apply, which I have explained in the pages preceding this page. Total annihilation and destruction of the singularity in one object may result in the object fragmenting to smaller parts where each part will still hold singularity, affected by less matter claiming space. When establishing a point of singularity to both objects a mutual point of referring to both objects' points of singularity will come about at the point of $(\Pi / 2)^2$ where both object will hold onto their individual singularity while spinning around the mutual point of singularity.

Another outcome may be where both objects maintain the claim to singularity, by pushing the space-time occupied to new levels of occupied space-time values. The result of the establishing of new individual but unequal points of singularity is the oval way objects rotate, first favouring the on in the matter part and the other in its space part and afterwards turning the points of reference around. This stems directly from singularity

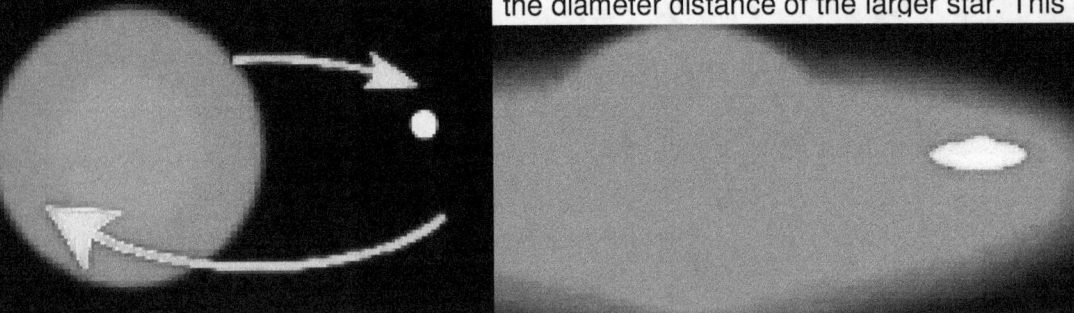

The two stars find a position apart at or closer than 2.467 the diameter distance of the larger star. This is $(\Pi / 2)^2$

The process turns the stars into a liquid thereby forming gravity as this puts the Coanda effect into action

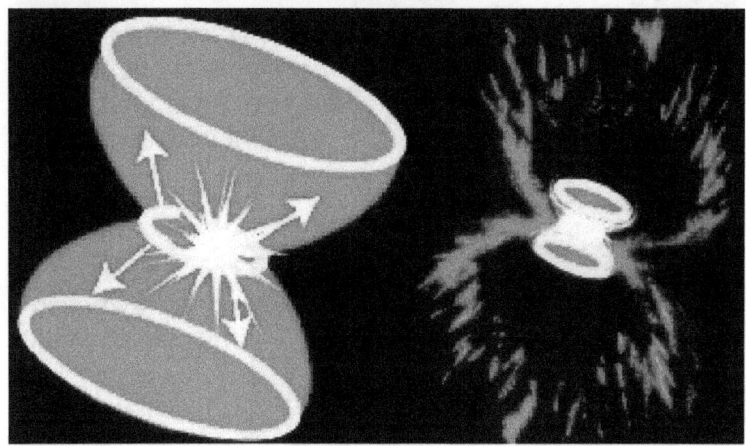

Then the singularity of the large star ($\Pi^0\Pi$) takes charge ($\Pi^0\Pi$) of the ($\Pi^2\Pi$) singularity forming gravity within the lesser star. The gravity forming singularity that forms conditions within the major star applies the same ($\Pi^0\Pi$) conditions within the lesser star. Since the gravity ($\Pi^0\Pi$) has more gravity ($\Pi^2\Pi$) this condition liquefies the material forming the space in the lesser star. If the major star can control the singularity within the lesser star there is no need to turn the lesser star into a Supernova or liquid.

After dissolving the lesser star to a state of liquid the major star then form the same space outlay within the space that the lesser star holds it turns the lesser star into an atmospheric state equal to the atmosphere of the major star. After the lesser star becomes as liquid as the atmosphere it dissolves the material formed by the solid as well as the liquid of the lesser star into becoming part of the major star.

In the event where the resistance op the lesser star forming singularity becomes too much to be controlled by the major star, the heat condition within the lesser star would release into space by exploding since the heat within the lesser star rose to a higher level that the major star could control. This is all to do with singularity and the way the gravity within the universe works when the Universe goes "flat" as Einstein put it. The Universe never goes "flat" but alternates between forming time "going flat" and forming space "looking as we see it". There is a period of space formation and there is a period of time displacement

In the centre of all things spinning a spot (1^0) forms a dot (1^1). This connects by a line that only holds the point that time carries. Space is present at that point but it is also meaningless because spots that are not forming part of this Universe consume the period. I am not going to go into much detail because there are books with hundreds of pages dealing with this issue in total depth and the issue is complex. When singularity applies the one dot is equal to any other dot because $1^0 = 1^1$ and $25^0 = 1^0 = 1^1$ and therefore mathematically all dots are not only equal but forms the same dot.

Therefore as Π^0 extends singularity $1^0 = 1^1$ where then the relevancy changes and $1^0 = 1^1$ until $1^0 = 1^1 = \Pi$ where then ($\Pi^0\Pi$). However as I showed by using the top as an example this can only be with movement forming space ($\Pi^0\Pi$) going on to ($\Pi^0\Pi\Pi^2 = \Pi^3$)) and without movement or gravity Π^2 the process will fall back into singularity Π^0 in the spot 1^0.

Therefore everything rests in singularity and singularity depends on movement of a spot becoming a dot in relation. The point forming singularity extends to the point holding 1^0 forming ($\Pi^0\Pi\Pi^2=\Pi^3$) an connecting this lot. The relevancy applies in accordance with the singularity taking charge and in the event of the two stars having the lesser star within $(\Pi / 2)^2$. This reflects on singularity and we have to see singularity acting and not space producing gravity as Newtonians wish to indicate. Every point holding singularity moves the instant time draws space into singularity. I

t is not space disappearing as Einstein said but it is the relevance of singularity taking charge by forming $1^0 = 1^1$ until $1^0 = 1^1 = \Pi$ where then ($\Pi^0\Pi$). As singularity becomes the sphere by forming the initial ($\Pi^0\Pi$) by going from the spot to the dot it moves through seven points of which four are turning motion and three are linear motion. Therefore no movement can ever be linear before becoming a circle in a sphere ($\Pi^0\Pi\Pi^2=\Pi^3$).

Then moving (Π) forms (Π^2) and this results in a total movement of singularity as ($\Pi^0\Pi\Pi^2=\Pi^3$) but because the circle composes of four points going linear while being a circle the square (Π^2) is divided by four to find the allocated position that time would place singularity at as time then forms space to become space-time.

Every atom holds a centre with the value of 1^0 and stands related to a centre worth Π^0 while relating to each other by 1^0 t0 1^1. It all might sound senseless but this means that under gravity each star is worth the total spin of the combined spin value of all the protons (1+1+1) in all the atoms while ending up as the centre value of the spinning star $\Pi^0 = (1+1+1)$. That shows the combining spin value transfer as much as it translates to the centre of a star.

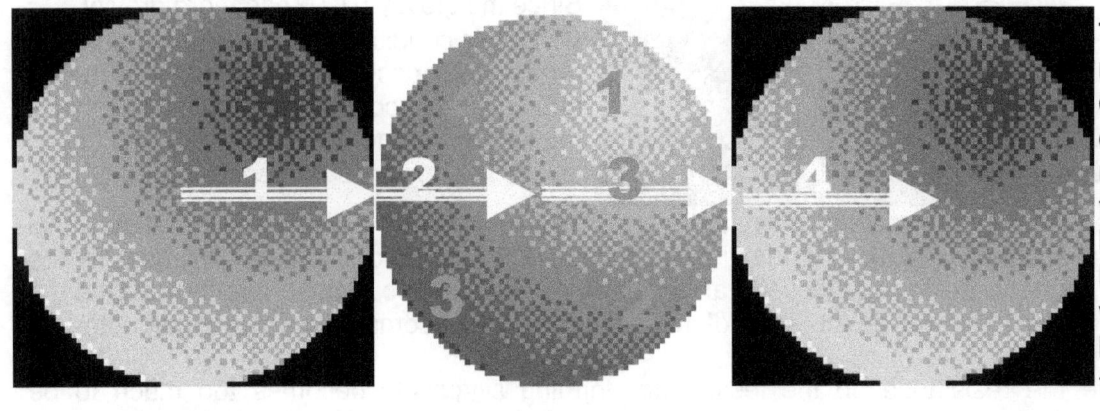

This shows that Π moves 4 times in one displacement circle and where Π moves the moving takes Π from one location to the next location Π by the value of Π^2 and because $\Pi\Pi^2$ space forms $\Pi\Pi^2 = \Pi^3$

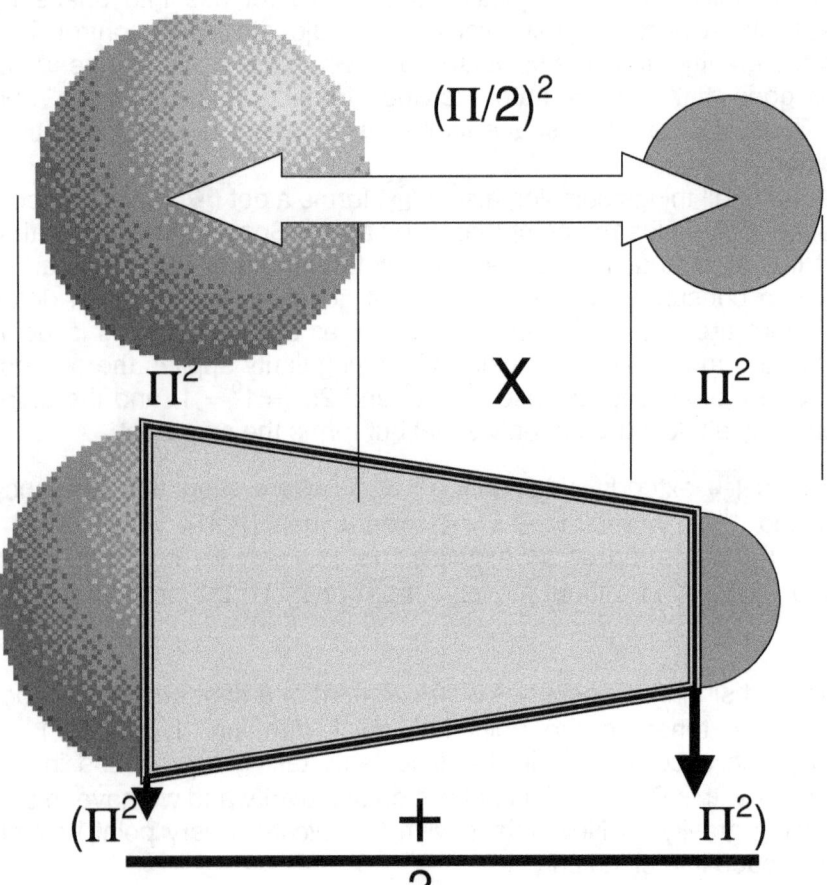

That proves that stars can never collide and the lesser object will always resolve into liquid when entering the space of the major object. "Falling stars" always burn out and debris coming from space always dissolves into liquid just as the spacecraft does when not entering at a specific angle.

There are cosmic laws preventing star and therefore galactica to collide and the four laws are in place to prevent this from occurring. By going closer than $\Pi^2 / 4$ the Coanda effect comes into place as the Roche limit instates the boundary of preventing movement obstruction.

By matter moving closer to singularity, the time component will slow down as the heat factor will rise, since singularity was in the very beginning, eternally cold as it was eternally hot, holding both positions.

Stars only become stars when the stars time distinguish it from the geodesic, meaning the stars interior remains a liquid comparing to the geodesic going onto a relative gas. The picture shows how the "atmosphere" of the earth carrying the value of Π in relation the solid being Π^2 is distinguishing from the geodesic cosmos by turning into a relative liquid as the rest of the cosmos remains a gas. The moon shows no distinction in its "atmosphere" of Π to the relevancy of the cosmic geodesic gas. The moon is not even a cosmic body by any standards.

A Less Complex Commercial Science Book — Chapter 9

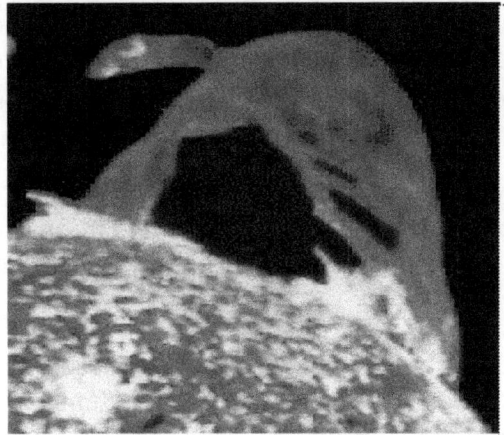

Since motion in stars is the movement of liquid heat, the differentiation brought about by the liquid heat is much higher than mere pressure would suggest.

There are two forms of material in the Universe and that is material or solids and non materials or gas / liquid. There is no "nothing" to be found anywhere so please inform your local Newtonian about to go and search for nothing between his ears rather than in soace.

When a star becomes liquid having a fluid atmosphere, only then can it start performing its duties as a star by converting space (gas) to fluid (heat) which then turns matter back to singularity, by combining protons to denser proton clusters and in that manner serving singularity in a more supportive manner. If the sun or any other star were using gas, it would not have the ability to generate the means of sustaining fusion. Without the heat being in a liquid the stars would not have the ability of applying the full vale of Π to its full potential

When an explosion demolishes matter no force in the cosmos will stop the destruction. The reaction starts because there is a massive unbalance in the relevancy of space occupied to heat bounded by matter to specific space occupied. A sudden super abundance of heat coming available that puts space occupied in a disadvantage to space available since all the available heat became available space through the process we refer to as an explosion. The sudden excessive surge in heat produced more space than already occupied. The heat turned to space will alter the ratio effectively to the advantage of space in view that so much more space suddenly needs filling. The fragmenting of the matter will be in proportion to space occupied by density packed by particles, because there then is more space in need of distributing occupation. As the eternal dot in the original singularity will grow in all directions evenly and equally so would the growth in space, since it is the same process still applying.

Matter is still the product of singularity creating space and heat with the creating of unoccupied space, but at the same time singularity is regulating space and reclaiming space as well as providing a system of controlling space growth. The reduction of heat will create a production of space. This will also apply within the walls of a star where only liquid heat fills all space not occupied by matter since there are no room for gas. When heat formed as a liquid by the governing singularity controlling space-time occupied and unoccupied, the reduction of space through the spherical form will bring about a production of heat and such a production of heat will bring about a surge in the liquid heat demanding more space.

But space is one thing not available, since the urge of the liquid heat will not create more space, but will surge in the amount of heat available at that position in space and moment in time. Thus the production of heat will surge for more space and in this will create more heat going nowhere in the liquid. The urge in the liquid for space will have to be counter acted by the matter holding occupied space. Therefore the matter will have to abandon space occupied to ensure the hear claiming more space through excess heat production by spherical reduction of space.

In this we must acknowledge the role of relevancy because he opposite is applying to the occupied space that matter is claiming. Since the liquid heat demands space, it removes the only available space there are from the matter particles and at that the lesser particles to. The denser particle having the bigger number of protons and neutrons to react on the space claim will put up more resistance to the space compromise the showdown with liquid heat will bring about.

As there are lots of space in the atom to compromise by reducing the claimed occupation the matter will become more matter by reducing the space, and the liquid heat will become more heat by replacing the available space with liquid heat. The reduction of space will produce the production of heat and the production of heat in the liquid will bring about that the heat in liquid will rise claiming more space stemming from an urge in heat forming a surge in space. The urge in the liquid heat for space to produce space as the heat in the liquid intensify where by it will produce space by what ever means necessary.

This means it will claim space not belonging to liquid heat and the only such space available is within the claims of matter. A the liquid surge for space, by producing heat, to that same amount will the matter loose heat, and thereby loose claim to space becoming that much colder respectively in reaction to the surge in heat by the liquid. The relation in liquid heat urging for space will leave the matter unaffected in relation to heat growth since matter cannot claim heat growth. Matter can only be relative to heat growth being surrounded by more or less heat in terms of sure heat or in terms of space.

Through this the matter effectively reduces in heat as the liquid grows in heat where none of the two can produce space unoccupied. As the liquid heat rises, the relevancy to heat in matter reduces and therefore becoming colder. In that way the matter will eventually loose all claim to space, therefore heat, and will freeze by fusion reducing the minor substance to no heat (space) at all.

As the matter loses space it will become closer to singularity and thus become more matter to less space bringing about that the heat will become more heat and the process will define the two much better by the loss of space, to both ends, up to a point where matter becomes absolute matter losing all space and becoming so cold in relevancy to heat in liquid, that it can freeze that it can freeze in fusion at many billions of degrees in what ever measure you wish to apply!

At the point of fusion the three cosmic components segregate completely where heat becomes liquid denouncing space holding the position of gas and pushing matter to the absolutely solid it should be. The two forms in the star had become defiant defined counterparts, underlining in clarity the distinction without the intermingling of space as we see the form of heat called gas we think of as space. By reducing all space belonging to matter, the fluid will fill all space bringing in an intense heat equal to singularity at creation where matter expanded. Only at this point the opposite occurs where the space reduces to a point reversing the process we think of as explosion.

By introducing singularity Π^0 to the Titius Bode law $[(7+7/10) + 10/7]$ matter-to-matter as well as space-to-matter and combining that with the Roche limit matter-to-matter $(\Pi/2)^2$ in relation to heat at the intensity of liquid status 3^2 a value of 62.6 will reduce all space between the particles in fusion and combine the individual singularity by removing the division.

Thus we can be sure that past the density of 62, there are no gas being space or liquid being heat, and matter being solid becomes transformed to singularity whatever singularity is.

   A star is a device whereby the natural process is to liquefy space to the advantage of producing more matter as the solid substance of singularity. When a star produce light, it has space left in the solid of matter, but when it reduces space to the absolute, there will be no space to transmit light because light is the flow of liquid heat through the gas of space.

   A star is a device whereby the natural process is to liquefy space to the advantage of producing more matter as the solid substance of singularity. When a star produce light, it has space left in the solid of matter, but when it reduces space to the absolute, there will be no space to transmit light because light is the flow of liquid heat through the gas of space.

The time frame science has about creation could be amongst all the miss givens the biggest misgiving of all. The very best is that mass as such is not an issue. It is the density that that mass produce, illuminating space that provide the star its qualities.

Each object has a different value of time in space where time has a different value to that specific part of space, whether it is a sub-atomic sub-structure of the atom or whether it is in a Black Hole.

THE ROCHE LOBE: In a binary system, the Roche lobes of components A and B meet at the L_1 Lagrangian point. (a) In a detached system, neither star fills its Roche lobe. (b) In a semidetached system, one massive component, B, fills its Roche lobe. (c) In a contact binary, both components overfill their Roche lobes and share a common envelope.

Try and marry the Roche Lobe with Academic's $F = \dfrac{M_1 M_2}{r^2} G$ and see a divorce before the marrying date arrive. One cannot say the Roche lobe is not, merely to allow Scientists to save face.

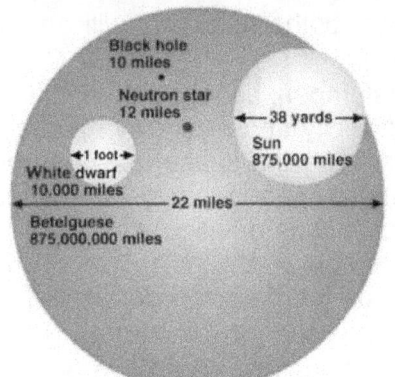

Science will have to face the fact that the larger a star is (the more space it holds to matter occupying space) the less intense the star is and the less formidable is the influence it has. This is going the wrong way with gravity because gravity depends on mass to space. A black hole is a few kilometres in diameter. But holds a relevancy to mass bringing about gravity equal to no other star in the cosmos. The less space the star holds, the more formidable will its influence be and by the time it is so small the gravity supersedes the speed of light, and it is merely a comparative dot

Once more, this phenomenon should not occur with Academic's presumptions about gravity. These bodies will collide and destruct, without a doubt if Scientists are correct when applying the formula $F = \frac{M_1 M_2}{r^2} G$ and there should not be any force which is able to keep them apart. However, they do exist and what is more, they maintain a certain distance apart.

We all accept that we the commoners are to stupid to understand such big issues and instead of feeling completely incompetent to the point of unworthiness we rather leave the issues in the hands of the Brainy Brunch because they do understand and we do not understand. They must be eternally wise and we must be eternally stupid not to understand what they clearly understand. When applying the most limited logic about the big concepts one find that the Big Brains have less understanding of the issues than we have.

At least we understand that we do not understand while the Brainy Bunch does not understand what they clearly do not understand. What they do understand are issues they created keeping in mind that they are the only ones who should be able to understand such complex issues. Apply the most insignificant logic, and the Brainy Bunch is stuck in a much bigger way than we commoners are. All they then can refer to is that Academics are correct and Einstein makes sense. What about Academics that is correct is a matter they never ponder on and how much does Einstein convert sense to logic passes them by with out a second thought.

With the "force" of "gravity" "pulling" the stars closer using the accumulative mass of the stars and multiplying that value with both objects by the mass component, this will reduce r^2 progressively until r^2 reduces to zero.

Seen from this view, it is little wonder that the significance of this was lost in the notion that this is yet another "mystery" of the universe. The Scientists of the day (and the past) lost the importance, which this holds for us as earthly dwellers.

A most surprising aspect of this is that it is not that an unfamiliar or rare phenomenon. However, any answer to this would clash with Academic's presumptions, and before the Scientists allow that to happen, they would much rather ignore what is obvious. However, what is the obvious?

In an attempt to find time, let us first exclude time from space-time.

There is a zone on the outside of the star where the magnetic space-time has gone beyond the value of the speed of light, as we know it. This zone would be ever so slim, but it will be there. It is this zone that supplies the neutron star with its novelty being the exertion of neutrons. Light, (the photons) displaces space-time negatively, which means in the onslaught of the super prevailing positive space-time, the biggest majority will be lost to this quest.

In this, I would like to ponder on the binary stars, as they are the key to time in the cosmos. The two stars develop in the galactica in close proximity as they help each other in transforming negative space-time displacement to positive space-time displacement. With the combined effort, they can grow extensively in supporting each other. From the point of singularity time hold in matter claims a space in the value of $\Pi^2 + \Pi^2$. Another claim to control of space comes into value at the presentation of $\Pi^2\Pi$. A third claim to space holds an influencing value to $1 + 1 + 1 = 3$. When the proximity of matter disallow the claim of influencing space, singularity resents space in a half that of space under control, from both ends. That will be the square of the point from singularity under control being $(\Pi / 2)^2$.

It is clear that something filling the space between Jupiter and its first moon because of lightning interaction between the two structures. If there are lightning there are electricity and electricity means a very distinct interaction. Considering the notion of nothing being in place filling the space between the cosmic structure, electricity needs a conductor to transmit the interaction there are and that disproves the nothing theory Xepted as official information. It is official that the interaction was detailed as $a^3 = T^2 k$ which is what Kepler found, yet with this information science still do not appreciate the fullest of the implication. I have changed the formula to R^3 (space) $= T^2$ (time) K (one as the common denominator) and this becomes singularity $\Pi^3 = \Pi^2\Pi$. By applying the atomic value the relevancy changes to $(\Pi^2 + \Pi^2)\Pi^2\Pi 3$ and that relevancy projects to cosmic atoms such as two stars interacting. When two objects come closer than the relevancy would permit, cosmic laws change their application and in this case then becomes $(\Pi^2 + \Pi^2)$ from either side where the three of space changes to singularity Π acting as the influence $(\Pi/2)$ from both sides making that influence a square $(\Pi/2)^2$.

At present science is unable to explain these very crucial phenomena and even to the extent that those guarding the principles of applying physics don't realise the importance of these phenomena or how the principles apply in physics.

These phenomena are there and it is in place so ignoring it is rather stupid…but that is precisely what Newtonian science has been doing for centuries.

They ignore these phenomena in favour of what is not present in the cosmos …and that is mass. Those in physics put mass in as an explanation for the movement of planets and while all planets has different mass, yet all planets orbit and move at exactly the same tempo. So how can they move according to mass-differences with having different mass and still be moving at an equal pace?

What there is, are the four phenomena I present, but because Newtonian science insight do not extend as far as understanding the phenomena and the role it plays in physics and moreover in generating gravity, Newtonian science is brilliantly ignoring the phenomena as well as the role these phenomena play.

Science do not realise the value or the impact that these four phenomena has on physics. That makes their judgement unfit in terms of deciding the importance the phenomena offer in relation to understanding physics. For many centuries science used improper factors in physics such as thinking that it is mass that produces gravity while only Newton's word is proof of this. There is no evidence of mass forming gravity and to overcome this issue they manufacture a component they called the graviton. There is no evidence of a graviton existing but to compensate for the shortfall existing in physics they use the graviton to promote a hoax however, this is not the only fable science invented

No one realises there is a shortfall because all evidence of such shortfall in Newtonian science is ignored like they would ignore the plague. There is no evidence of mass that has the ability to form gravity, but by cultural brainwashing every one believes this to be true. With everyone satisfied that there is no problem in science all persons are very satisfied with conditions applying as it is.

In every mind using physics the awareness of the mistake is absent and all presume that physics in the manner Newton said it is applying therefore thinking the Newtonian way is a healthy and correct and no one can find a reason to investigate my claims that Newton's idea of mass that forms gravity is a hoax.

No one can see the need to support my claims of science being wrong in spite of finding no evidence of mass anywhere in the solar system. The phenomena that do apply in the cosmos, science totally ignore because it does not support Newton's ideas and Newton can't explain these phenomena. I prove how these phenomena form gravity but I am ignored because it will turn physics on its head.

Should you download

www.singularetyrelevancy.com

Then you are about to discover that…

THE SPOT THAT'S HOLDING THE LOT.

As highly developed we seem to regard science to be, it took a genius like Max Planck all his life to try and reconstruct from what there are to what the Universe started with. Once again the human perspective baffled the genius of a Master such as Planck was, because he placed man amongst the first of creation. Where it started from and what we now have is many, many eternities apart and any evidence of an attempt matching what we have now to what first was is foolish.

It started off with one dot so small eternity met infinity within. Then came one more, and another and they continued coming until there were a countless number of dots. The accumulative size of the dots were the same size as one dot because in the true Universe big and small plays no part. The dots were infinitely small and eternally big at the same time because size is a relevancy and without one the other has no size. So in the true perception, there is no difference in size.

It started with the fact that there is no place or part in with which one may associate zero or nothing. There are no room for a number such as nothing. Next to the one dot (infinitely close) one will find the next dot, and if nothing was a factor then that is precisely what one will find between the two dots. Nothing of space, a non existing entity, taking up no space, and much more important, no time, therefore the dots are infinitely close to one another, being the same space, eternally big as much as infinitely small. If we as humans cannot find a manner in comprehending this notion, there can be no manner ever understanding the cosmos as much as the start to the cosmos.

Every dot was a Universe in its own and the accumulation was a universe. The earth in itself is a Universe as the moon is a universe, because rules applying on earth do not apply on the moon and visa versa. When in the ocean another set of rules apply, therefore being in the sea places a body in another universe. The number of universal entities is still countless, as much as it was in the beginning.

Every position in the Universe either holds singularity in a form, or relates to singularity. There can be no position unrelated to singularity therefore every aspect of the cosmos is space-time in various forms under the provision of singularity connecting. Matter cannot be if not surrounding singularity

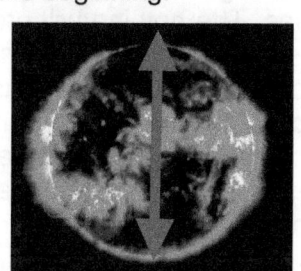

Singularity is as close as any spot can ever come to zero BUT IT CANNOT EVER BE ZERO. From singularity diverts space-time and there cannot be space without time as much as there cannot be time without space, not withstanding the size of space or duration of time.

Through space-time singularity connects as much as relates linking the Universe into a network of influences beyond what ever we can ever conduct. There can be no spot that does not participate in the curvature of space-time. From the point of singularity runs space holding time to the prescription singularity dictates.

With singularity connecting singularity will or cannot relieve or release the connecting other than by a method we humans refer too as an explosion, but space-time separating as much as joining singularity dividing can change by applying time too space in a changeable manner, stretching and shrinking the time aspect by changing the density of the occupying heat, which creates the space allowing the spin of the occupying heat creating space setting the time.

Every dot insignificantly small as it may be, is a part of another Universe as much as it is part of the accumulative Universe and every dot in the infinity holds singularity, which we translate as " nothing" being " darkness". There cannot be "nothing" just as much as there cannot be "darkness". There cannot be something big or small, but it into relevancy of perception, and then the relativity of perception becomes the question. There cannot be hot as much as there cannot be cold. The sun FREEZES hydrogen to a liquid at six and a half thousand degrees Celsius and Universe boils over in the form of the Hubble constant at the temperature (we presume from our vantage point) at minus 273 degrees C. If we Humans cannot or will not abandon our human perception and our manly perspective, we may as well return to astrology for all its worth.

Every point in the infinity we may observe at is not merely part of the Universe in not being "nothing", but is the point where the Universe started representing singularity. It is the very first point where everything began so many eternities ago, because after all, how can we ever determine where the first point was, as they were very much equal and alike at the beginning. Every aspect of the Universe started with the fundamental fact that no point in the Universe can represent "nothing" as a number, because every aspect in the Universe represents singularity in what ever form it may hold in that specific spot forming space-time. If man does not reach a conclusion where that conclusion is matching the Universe and stop to match the Universe with man (and man's incapability), we may all go back to caves and become starving hunter-gatherers again, because we will never find a way to progress to the ultimate understanding of the universe.

I wish to make the argument about gravity found in orbiting objects a rather lengthy debate as part of my initiating discussion and after all, it is the very argument on orbiting structures NOT drawing closer that initiated and concluded all the effort of my writing this book. FOLLOWING ARGUMENTS MADE ME SEEM AS THE VILLAGE FOOL TO FRIEND AND FOE, ALIKE, AND WHERE I MAY NOT BE THE BRIGHTEST IN THE VILLAGE, SURELY I AM NOT THE VILLAGE IDIOT THAT EVERYONE TAKES ME TO BE ONLY BECAUSE I MAINTAIN THERE IS NO GRAVITY.

If the graph broke a zero mark, then a new and totally unrelated line will form on the opposing side baring no relation to corresponding with the line in the previous quarter. The new line may start from any point and lead to another point holding no resemblance to the opposing side. What will prevent the line from finishing at a point marked say three and starts afresh at seven because there is NOTHING between the lines. The nothing will provide releasing and detaching all corresponding relativity there may be through disconnecting. Through nothing there can be no resemblance or corresponding because the point of start need not even to continue at all. Nothing will release any connection and if such a connection may come about, such a connection may just as well be very co-incidental and may never be used in calculating accurately. Experience taught us that there is a definite precise and secure corresponding that can only result from a direct connection in as much as the line being the same line. In that case the line must then come down to infinity and release from infinity as the same line still connecting to a point of re-bouncing to either side. The graphic cross is he result of singularity applying opposing sides but still maintaining connection through the application of Pythagoras that will connect and always bring about a direct relevancy. The graph does not hold zero because information derived as result of a relation prove a contact remaining when the line crosses singularity without applying detachment. The Brainy Bunch holds the view that in a graph the line crossing amounts to breaking the zero mark. That cannot be the case.

In the centre of whatever two synchronized rotating objects the line of singularity matches in spin to one another not withstanding the space-time involved. The relevancy applying places one rotation of spin eternal, therefore the rotation is standing still, but also one rotation consists of an above measurable number of infinities producing a rotating speed faster than non other will ever reach and that makes the rotation the fastest there can will and may ever be. Again we humans now face the same value we wish to separate, as is the case with hot or cold, near or far, and quick or slow. Once again time locks relevancies beyond understanding. That line so infinite small as it is eternally big connects the cosmos through light we are able to see and light we are unable to realise.

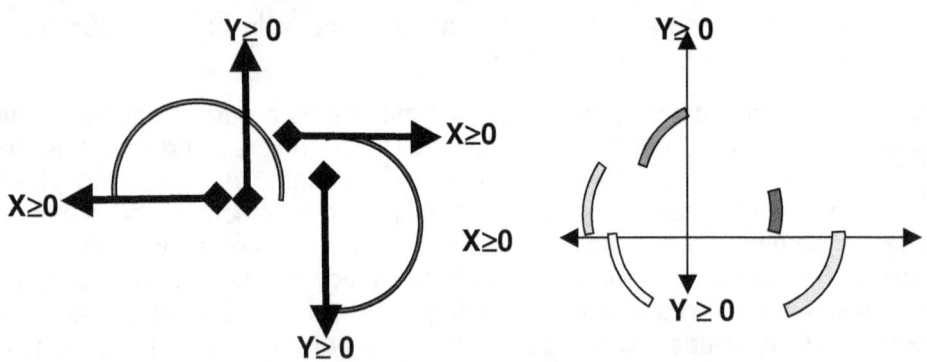

The mysteries of the cosmos are rather simple. Because matter more frozen heat it holds less heat. As it holds less it also holds less space and therefore it will apply the depreciation of space and accordingly present the appreciation of heat.

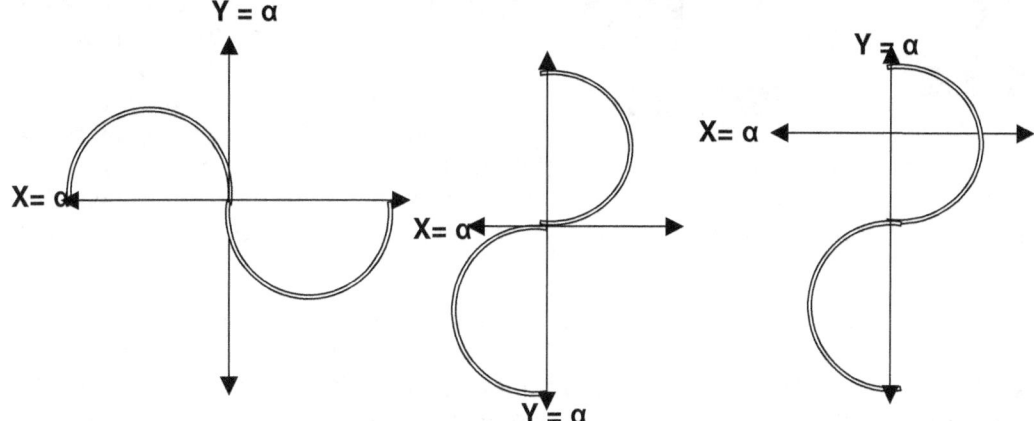

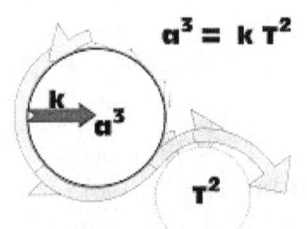

$a^3 = k T^2$ When translating Kepler's mathematical expression into a verbally spoken form of communication such as English, we can see what Kepler said also reads as $k = a^3/T^2$, where **k** is one point from a centre point that is space a^3 relating to time T^2. From a centre comes space-time.

Ever considered why a water drop would freely choose to form a sphere? When a water drop is released in an astronaut's space capsule in outer space, the water drop forms a sphere as it floats in free gravity. As soon as the water drop is released from the Earth gravity and has the opportunity to float in space, it can form any shape it finds pleasing, yet it immediately turns to the shape of a sphere. It is the same reason why we would think the Universe takes on the shape of a sphere, although we know the Universe has no outside and we realize that the Universe is limitless in size. Yet, notwithstanding, we take the shape of the Universe as naturally being a sphere that formed ... but why think of a sphere that the cosmos considers as the sphere being the pre-cast shape? Well by saying it is the strongest form available, has the same argumentative potential as saying a baby is little at birth. Blaming it on no reason substantiating proper evidence that would indicate some facts more prudent than just a simple answer to dodge the question by hiding obvious incompetence. I'll give you one clue ... it has something to do with finding the centre of the Universe. That brings on the next question being where one might locate the centre of the Universe. These questions put to your are most ordinary questions ... and yet the complexity in the answers rise far above the answering ability of those with the supreme mathematical skills. Those master mathematicians cannot answer such simple questions by using their complex mathematical powers. If they could have, then they would have ... the answer is extraordinary simple and childlike easy to explain.

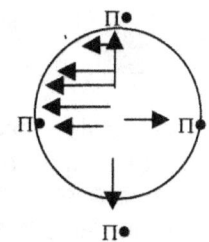

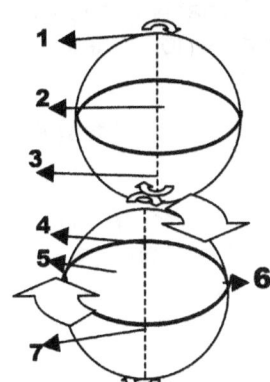

By examining the form of the sphere, we find that there are 6 points on the surface of the sphere that is holding the form at a specific and equal distance from the centre. Lines run from the centre into space at 90° and 180° angles of each other from six opposing sides. There then are six lines at 90° and 180° connecting to the centre from six points on the outside edge of the sphere. As a result of the basic shape that a sphere has, there is a spot in the extreme inner centre of the sphere where the lines in 90° relevance cross each other and others connect by 180°.

There is also at that point a spot where all space relinquishes a position and only singularity 1^0 as form remains. At such a point we find the measure of the sphere being Πr^0 with $r^0 = 1^0$. That is where the line that represents the radius as a line disappears, as it becomes singularity r^0. After more reducing continues, we get to such a point where we find only Π^0 left. At that extreme point is where space in all form disappears as the circle providing the sphere the form the sphere has, removing all possible form by going into singularity $\Pi^0 = 1^0$.

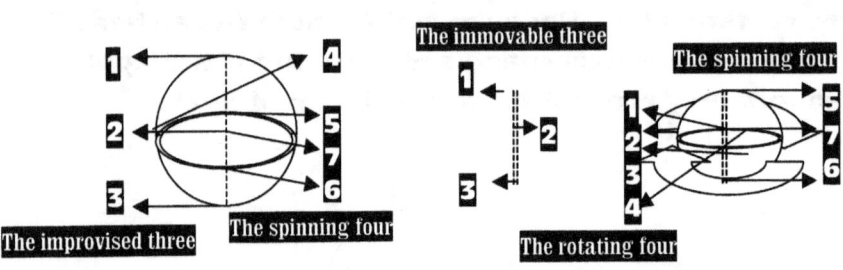

Then in that area all form of any possible space disappeared leaving only the dimensions of singularity 1^0. This too, I take much further in the book as I delved deeper into the argument and by doing that I stumbled on the ingredients forming gravity. However, from such a point there runs lines that connects to space on the outside where six points on the outside points connects to the space less point in the inside. In the book I take this argument much further, but for now, I leave the argument at that. Those lines carry the structural straight that the sphere has where the other six support every one of the six by singularity.

Where there is no space, there must be singularity 1^0 because the space is present although in singularity 1^0. If zero were a factor where all space finally halted in zero as the value, then zero would be able to remove the space from the centre and such removing would continue to remove the space until all space was removed. It will finally abolish all space in the sphere and it would remove the sphere. Zero removes all possibilities of anything coming about. Since the sphere is there, a zero factor in the centre cannot be present. Only infinity can be a factor from where space may grow because infinity can extend and grow into and up to eternity.

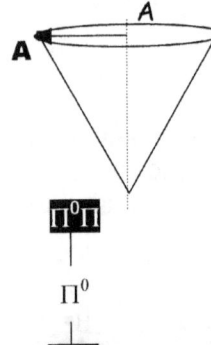

The moving of Π^0 to Π involved relegation and not motion as we consider motion. It was Π^0 getting a side and that is all. There was no true side but only a form that came into place. Singularity (A) received singularity (A) and no more of anything but the shift to comply with having a relevancy forming in relation to singularity. The dots had no sides, had no length or diameter. There was not measurable space or measurable time involved. The time could have been a micro, micro second as much a trillion millennium because time had no relevance. It was eternity interrupted by infinity, as it still is the case, however the line that eternity followed was no line because there was no space to hold the line. The line was momentarily interrupted by infinity, however with no one there, there was no one to notice. The lines were not lines but relations to sides being formed.

The relevancy that had the power to set Π apart from Π^0 is the only relevancy that still has the power, to set particles apart or join particles. It is heat in variation from cold. In order to excite singularity, singularity must establish a basis of heat that sets such a heat basis apart from cold. From there the form the atom will take on, however, the atom was still enumerable eternities to the development side.

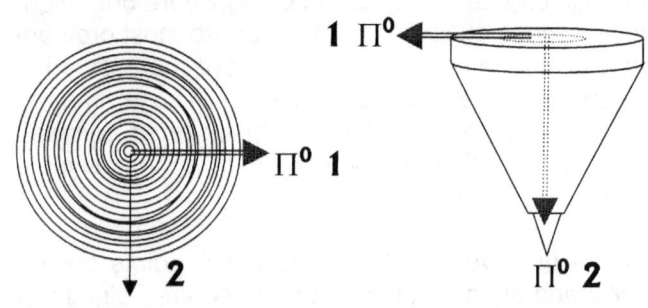

Where they are equal in value we must test the reason why this then is valid.

What is in the Universe is spinning. In the precise middle **of all** objects in rotation **is a precise centre dividing the object in sectors that will** start the spinning initiation **from that centre point.**

$k^0 = a^3 / T^2 k$ states that whatever is, is also spinning in order to be present.

Thus, the spinning object will have a middle point, **a very specific** centre point that does not spin **and only holds Π as a specific value because no radius can apply. But also the one value such a line** cannot have is zero **because the line** is there and holds contact **to the rest of the material bringing about that** zero does not start any **line and therefore the** value of the line must be infinite, **just as described in accordance and by** the definition of singularity. As I am introducing a very new idea, I wish to explain in better detail what I try to convey. While the top is spinning, one will find a line that formed in the centre where no line can form. It comes from spin but can never participate in spin.

That line must be singularity because if one moves any point on that line one position on, such a movement will land the point that then form on the line, on the other side of the line. The line is where the radius ends and starts because the line divides what is spinning in innumerable sectors and when reducing the radius progressively towards the centre of the spinning top at the centre where no line can be there is a line dividing the entire spinning top. At that centre point all further reducing must end because the next movement however slight will fall on the other side that is completely contradicting the one side. One movement further will change whatever is, so completely every aspect of that characteristic will contradict what it was before. There is one point that is neither left nor is it right but any point next to that point must be either left or right. The only value that point may not have is zero because albeit so small that it is not part of our Universe, still the point is there for all to witness and that point is a reality as much as the entire Universe is a reality. Whatever one attaches to the top either in the line of being material or a concept, such a concept or material has to start at the spot in the centre of the top because every aspect of the top changes in contradicting from that point onwards in all directions. That point albeit hypothetical, is also as much a reality none the less and is placed where that point must be standing still because every line running from that point in opposing directions is also in opposing directional spin the other or opposing side.

As the rotating direction moves inwards, the rings holding Π will become smaller and smaller. The reducing of the radius r will eventually end at r^0 but the top does not end there because the top still then is Πr^0. The form we attach to the spin still applies as Π and the top finds directional contradicting change at a point that never moves because it can never move being in the centre where the spin direction ends at Π^0. It is the only aspect in the entire Universe that can be and still be motionless because it is not within the Universe. It is the centre of the Black hole because it is the centre of the Universe. It is 1^0 and only the centre of the Universe in singularity can have 1^0 However that point where the directional spin ends is the point where the actual spin does not takes place because if its immovability. It is at a point in singularity $k^0 = a^3 / T^2 k$. It is where space ends because motionless ness ends space there. The spinning is on the precise location where the point is not spinning because the Universe ends in its not spinning there.

That line running through the centre of the spinning top divides every possible side from the opposing side in innumerable points that are divided by angles and degrees. Moreover, it changes the future position of the point from the present and from the past as it redirects every point every time **k** moves to reposition in a location where T^2 ends. In the end it proves that both **k** and T^2 confirms a^3 just like Kepler stated before Newton interrupted with dishonesty.

Another huge factor that favours the use of Π as a measure in singularity progressing is that any expanding by any mathematically sympathising method will have to use Π since Π is the only route that a spot of no significance can develop into a dot that represents a Universe of development in waiting. Only by promoting through the measure of Π can all possible sides progress on equal terms in all directions simultaneously without bias. The development must progress by measure of equality to the smallest indication and that purpose only Π can serve.

The progress must be generated so that it can flow equally to all sides in all directions spontaneously where not one side will favour the growing process as such. Time in it's flow does form a bias but explaining that at this stage will also involve too much other concepts which I would rather leave to the books. There is only one way to permit such a flow and have a mathematically correct outcome and that would be using Π for such expanding. The use of Π would ensure that a dot rises from the developing that comes from the first spot. The dot would have to form a sphere and a sphere is Π in relation to seven. This bring as back to the Titius Bode concept where ten is one half of a sphere relating to seven points forming the sphere where the seven with singularity puts form to the double ten that totals (including singularity) 21.991 and that is the overbearing dominating issue.

When the cosmos came into motion, motion was not yet defined. When the cosmos brought about motion, the first motion was relevancies. Cold parted from hot. Eternity parted from infinity. Motion parted from motion absence. Infinity broke the laboriousness of eternity for the duration of infinity. The spot became and grew into the dot.

From what the spot was to what the dot now is might be just a mathematical implication of going from 1^0 to 1^1 but in reality that first motion was the creating of and establishing of an entire Universe which was with all possibilities that now is it. Never again can that much growth become a reality, although to us the growth is beyond what we ever can notice. But it is because the growth is so massive and we are so small that we are unable to notice such almighty growth. When the spot Π^0 became functional and established all relevancies possible, heat parted from cold as eternity parted from infinity.

The expansion was not clear motion but more a parting of relevancies where a centre formed a relevancy because the centre could not provide motion. Without being capable of motion, the centre established four points, which also served singularity. From the inverse square law we know that the centre doubled by producing the four points holding singularity. We have to presume there is a time line because the Universe has this as evidence. The fact that light travel from there to here and from here to there proves of such a time line because there is no distance in outer space except in the Newtonian's misconception they have about the cosmos. Any line shows direction and the direction implicates positions according to the line having dimensions in the Universe we have in space and time. The line brings in Pythagoras and Pythagoras implicates mathematics.

1^0 going 1^1 where $1^0 \Rightarrow 1^1$

If there were progress that developed from singularity in the form of the first spot and we are the evidence of such progress, then the mathematical conclusion must be that a line formed where the line developed two sides and we have the evidence of that still present in our Universe. That brought about that three markers formed in relation to one and by admitting to the law of Pythagoras we find that what formed was 3^2 in relation to singularity 1^0 that became 10 on the one hypotenuse and 10 on the other hypotenuse which forms the square of space. Therefore mathematically space has ten positions and material has seven.

From the three the four (2 on both sides of singularity allocating the cosmic divide) the two in square developed as a mathematical consequence and that brought about the five.

The five was duplicated as a response on the other side of the divide and having five as a result of $(2^2) + (1^2) = 5 \times 2 + 10$ we find that the square of space holds the value of ten in place.

But 5 is in pace because $3^2 + 4^2 = 5^2$. And $2 \times 5 = 10$ is in place because $3^2 + 1^2 = 10$. I just can't go in depth showing how the Universe formed mathematics as the Universe built what is within the Universe because that boo (if my memory serves me correctly) is about 1300 pages in all. There is a reason why some part of mathematics show how a line and a half circle and a triangle is equal and why $1 = 1 = 2$ and at the same time $1 \times 1 = 1$. It all has to do with building the Universe in stages of development.

We read from the way mathematics formed how the Universe formed because the Universe formed mathematics as it formed what is presented as a Universe. Before the Universe there was no mathematics so mathematics did not form the Universe but the Universe formed mathematics and above all this came about exactly as the Bible says it Happens. If you divert one glitch from how the Bible presents it the entire process goes astray, but mathematics form the key in the presentation of how it all came about.

At the very beginning there was a spot. How do we know that? The spot is still with us and holds a value of 1^0. This spot is in the centre of all spinning objects. Then time came into motion and 1^0 moved to 1^1. Since from where we stand we see 1^0 and 1^1 as being the same and therefore with the moving of time in the very beginning such moving must contribute to an increase of space.

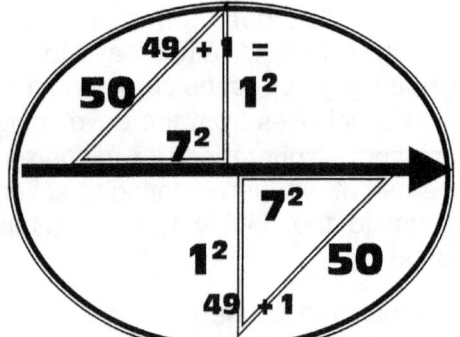

Therefore 1^0 and 1^1 has to have a difference where the one is 1 and the other is one point in singularity smaller making the 1^1 coming from infinity and rising into eternity 1^0, making infinity forever one point smaller than what we find as the value one will associate with the one we find as a measure in infinity.

Therefore we can judge that singularity combines to have a total of $1 + .991$, which then becomes 1.991 or whatever because the one going smaller is running into infinity and since infinity is one less than eternity we are in eternity 1^1 looking at infinity 1^0 which is one point reduced in infinity. However this moving from 1^0 to 1^1 involved 1^2 as well as 1^2 on the other side of the divide. As a result of the form the sphere holds, there is a centre connecting the sides and the centre holds singularity.

However, by presenting a centre where all lines cross on a point that cannot distinguish sides since that point has no individual sides, the centre holding singularity is inactive. Motion makes it active and the motion of space in time activates singularity to charge gravity that we find as a factor in the Coanda effect. That motion that establishes the purpose of space a^3 as a result of motion k through time T^2 was what Kepler presented as a formula. Gravity is $k^0 = a^3/ T^2\ k$ and to install k^0 the motion of space-time a^3/T^2 is required to complete

Producing singularity sets the divide because singularity splits the Universe apart in separate equal components that in combining form the duplication of singularity, being Π. Since the split brings about equality it, means that what is applying on this side must be applying on that side. When motion changes Π^3 to the proton $\Pi^0\ \Pi^2$ it will happen on both sides of the divide of singularity Π^0. In effect it means that, that which combines the proton also parts the proton as it combines the proton because the proton becomes $\Pi^2\ \Pi^0\ \Pi^0\ \Pi^2$ where the adding is the divide being Π^0.

The circle motion comes from space being dismissed by ending the motion and such ending of motion compromises the space it forms. By returning to where it is coming from it is ending the motion that began the space and as space is motion that is duplicating space motion returning is also motion that is ending which is destroying of space. THAT IS GRAVITY! Gravity is the balance between motion forming space by duplication space in motion forming time and time ending motion by destroying the space.

Gravity is about space duplicating space in relation to space destroying space and some particles are more prone to duplicate than destroy not withstanding mass or proton numbers. Those we call gasses. Then there are others that are more prone to destroy space that duplicate and those we refer to as metals. Then there are a few that destroy as much as the create space by duplicating and that we call fluids.

When heat is added to some elements we consider as solids, the heat helps with the duplicating of surrounding space and brings about a balance restoring the difference there are in the destroying of space and the re-establishing of space. The metals become liquid and the heat forming the liquid brings about an adding to the material where such material diminishes space.

By applying heat to materials that already favours the duplicating of space to the destroying of space the adding of heat will bring additional space as duplicated space and thus will produce more space to be duplicated and such elements will rise into the higher part of the atmosphere. When heat is added, the heat that is actually forming into space is what is added. It is the heat forming space by duplication that forms a shift in the balance because the heat forms space also as a process of duplication but without the contracting aspect of singularity renouncing space.

Although the "gasses" are the particles favouring to duplicate space they still hold the tendency to diminish space but when applying heat the balance will favour the duplicating much more because the heat transforms to space acting as space duplicating and adding to the overall duplicating of space. The element has a natural function of returning the space that the heat duplicated back to heat by removing the space destroyed and therefore returns the space to heat. In that way the particles do not only diminish the space they have but diminish space outside their claim.

Such particles we call heavy metals. As heat is added more space becomes available, to duplicate in relation the space they destroy and what space the elements diminish. With more space to duplicate the object will surge higher to a location in a position Earth will naturally duplicate as much space as the newly relocated material duplicates the surrounding space where more space naturally are. Cooling on the other hand reduces the space available for duplication by removing available heat that would have helped with the duplicating of the space and that then tips the balance in favour of the diminishing of space, which that element will also have. In that case the element will become a solid as the space duplication is more that the space. In this, we can trace the most important part of star evolution.

A star with a liquid centre has a lot of heat. Then the duplication of space-time by motion duplicates space much more than it dismisses space and destroys time. With a star in all the liquid as the sun is it is proof of a very young undeveloped and insignificant star with almost no influence sphere. As the star develops, the liquid ratio will shrink until it is only present in the centre. However, as the liquid diminishes the motion of the star deteriorate because the liquid represents the motion.

The star eventually becomes all-solid just before it removes the neutron from the atoms in the star and eventually places the proton action into outer space. Judging the layers we find evidence in this as the outer layers of stars are filled with elements which is highly prone to space duplicating as they have such a relation with heat. Hydrogen and helium stands very favourable to space producing and little in favour of space dismissing while iron, cobalt and copper is much prone to space dismissing. In all factors mass plays no part. It plays no part in the star performance or the star development

This I use to indicate where there is a balance favouring the diminishing of space
In all forms of material, there are the constant interaction between space duplicating and space reducing.

Some elements favouring duplicating space more than the diminishing of space are as follows
Hydrogen has a **mass of 1.00797 g/ mol** **melts at -259^0 C,** **boils at -252^0 C,**
Argon has a **mass of 39.948 g/ mol** **melts at -1899^0 C** **boils at $-268,9^0$ C**
Krypton has a **mass of 83.8 g/ mol** **melts at -157^0 C** **boils at -152^0 C**
Xenon has a **mass of 131.3 g/ mol** **melts at -111.79^0C** **boils at -108^0 C**
Radon has a **mass of 222 g/ mol** **melts at -71^0 C** **boils at -61.8^0 C**

It is note worthy to notice that none of the above elements feature strongly in stars although they should be massive in relation to the numbers of protons they have because they duplicate space.

Other elements favouring diminishing of space more than the duplicating of space will be as follows

Magnesium has a **mass of 24.32 g/ mol** **melts at 650^0 C** **boils at 1107^0 C**
Silicon has a **mass of 28.08 g/ mol** **melts at 1412^0 C** **boils at 2680^0 C**
Iron has a **mass of 55.847 g/ mol** **melts at 1536.5^0 C** **boils at 3000^0 C**
Cobalt has a **mass of 58.933 g/ mol** **melts at 1495^0 C** **boils at 2900^0 C**
Carbon has a **mass of 12.01 g/ mol** **melts at 804^0 C** **boils at 3470^0 C**

There are no correlation between mass and elements prone to space or prone to be solids. Mass do not create gravity and again on one more point Newton was wrong. Mainstream Science would rather ignore such compelling evidence as well as my writing about the matter than to admit that Newton could ever be mistaken.

Time came from eternity, where it stood still in eternity, as it still does in singularity. Remember that little line I indicated previously that is running through the centre of the spinning top and that holds time apart, while it in itself is motionless, eternal. That being without motion represents eternity, a state of being timeless. As space moves away from that point the duration of time begin to rise and the further the extending the larger the time factor will become.

Any fire represents many stages of time, where one part will burn quickly and the other slowly saying this means that science should recognise any fire on earth represents many conditions of heat where the smoke is a solid –gas, the flame is a fluid going on to be gas, the coal are a solid heat going to a liquid as it simmers producing photons which in itself is the dispensing of liquid heat turned to singularity particle dividing. The range that heat forms are so vast one may never appreciate it. Chernobyl showed the world how many forms of burning and burning injuries can come from radiation. Some heat was in the grass, undetectable until a bicyclist past and gained wounds from it killing the person a few weeks later.

The role that heat plays goes beyond (I suspect) what we may ever come to realise. Heat is an eternal fluid in relevancy to matter being the solid and space being the gas, but what is space to the one is solid to the next or a fluid to the other. For instance the proton is dimensionless to the neutron, yet it is fluid to singularity allowing heat to flow.

The neutron is solid in appearance yet it is fluid allowing dimensions to concentrate. Because heat is a liquid in relevancies, it is the father of specific density, allowing heat to flow differently in different forms of matter. To realise the correctness of that, one may gauge the heat relation there are in the first ten elements, and how each stand so different from the next element. Consider neon and boron, where boron has many times the density of neon, yet only half the mass, or where oxygen has more mass than does lithium, yet lithium has a much different relation to heat than does oxygen.

Time is the motion of heat in space, and producing more motion, the duration of time will extend.

Every dot insignificantly small as it may be, is a part of another universe as much as it is part of the accumulative universe and every dot in the infinity holds singularity, which we translate as " nothing" being " darkness". There cannot be "nothing" just as much as there cannot be "darkness". There cannot be something big or small, but it into relevancy of perception, and then the relativity of perception becomes the question. There cannot be hot as much as there cannot be cold. The sun FREEZES hydrogen to a liquid at six and a half thousand degrees Celsius and universe boils over in the form of the Hubble constant at the temperature (we presume from our vantage point) at minus 273 degrees C. If we Humans cannot or will not abandon our human perception and our manly perspective, we may as well return to astrology for all its worth.

Every point in the infinity we may observe at is not merely part of the universe in not being "nothing", but is the point where the universe started representing singularity. It is the very first point where everything began so many eternities ago, because after all, how can we ever determine where the first point was, as they were very much equal and alike at the beginning. Every aspect of the universe started with the fundamental fact that no point in the universe can represent "nothing" as a number, because every aspect in the universe represents singularity in what ever form it may hold in that specific spot forming space-time. If man does not reach a conclusion where that conclusion is matching the universe and stop to match the universe with man (and man's incapability), we may all go back to caves and become starving hunter-gatherers again, because we will never find a way to progress to the ultimate understanding of the universe.

Experience taught us that there is a definite precise and secure corresponding that can only result from a direct connection in as much as the line being the same line. In that case the line must then come down to infinity and release from infinity as the same line still connecting to a point of re-bouncing to either side. The graphic cross is he result of singularity applying opposing sides but still maintaining connection through the application of Pythagoras that will connect and always bring about a direct relevancy. The graph does not hold zero because information derived as result of a relation prove a contact remaining when the line crosses singularity without applying detachment.

The Brainy Bunch holds the view that in a graph the line crossing amounts to breaking the zero mark. That cannot be the case.

Now I wish to refer once again to one of the academic letters, which I already used as a referral. This is very shortly my theoretical proposal

Dear Peet,

Those on your list I know are in science education. You need someone in pure physics.

I am afraid that you will continue to get rejections if you do not relate your work to existing theories and previous work. While it is possible that a lay person hits on an insight that has been overlooked by academic trained in the field over many years, it is unlikely. We assume that work offering something new would be related to existing theories, either by building on top of them or by showing how and where they fall short. If you do not relate to existing work, it is repeatedly going to be dismissed as mind spin too easy to shoot down.

I am sure you understand.
Iben

PROVE ME TO BE INCORRECT IN ANYTHING I SAY! What you see in the picture above is gravity opposing gravity and by using the four cosmic pillars, which builds the Universe, that is how the cosmos comes about. I say gravity and the sound barrier is the very same thing where it indicates opposing movement of the earth and some other object within the earth that moves apart from the earth. Allow me to very briefly explain the sound barrier as it was never explained before by applying relevancy.

Science in the present and for the last three centuries placed all their focus on material. In placing the focus on mass while mass does not exist as a cosmic entity, science has been running around like a chicken without a head and the truth eludes them even after three centuries of lying and corrupting science. You think that this is harsh words, read and you will see those in science deserves much more that just that. Those in science concentrates on the material filling the space but gravity is about the movement of the space and not of the object in the space, but of the space the object takes with as the object moves faster than the other space. That is why objects all fall equal because the space the objects fill is the same, unconditional of what forms the material. All objects fall at an equal pace without size or mass becoming a factor. That is why I introduce gravity by a very new set of principles that no one ever heard of. When an aircraft goes through the sound barrier, the object fills more space per time unit by going faster through space in time. The object then in accordance with the space it holds stretches because it holds more space than when it went slower or when it stood still. Since the aircraft is solid and does shrink a little but not much, it has to concentrate the space around the aircraft and by concentrating it reduces the space. As the space concentrates through the movement of the aircraft a cloud appears around the aircraft because the water vapour in that space concentrates into a cloud that forms. All movement is part of the sound barrier because the sound barrier is the movement of space. However the sonic boom is confused with the sound barrier but the sonic boom is a small part of the entire sound barrier and only fills the centre spot or the middle in the sound barrier. If mass is anything then show me the role that mass plays in the sound barrier and how does mass conform into what we think of in terms of the sound barrier. Gravity forms as not by the mass of what any object presumes to have, which I prove in this book and in all my other books, is clearly not present as a cosmic factor. There is no such a thing as mass in the entire Universe and the conspirers that conspire to keep Newton's fraud disclosed, cover up this reality with all they have. I challenge anyone to prove where do they find proof of a factor such as mass. I found the place! It is in the imagination of physicists while they try to conceal that mass is only a product of Newton's imagination. They conspire to confuse everyone with the implication of weight. Weight there is but things pulling other things by the measure of their mass is a daydream and the thought could be funny if it wasn't that crooked. Gravity forms by redirecting the movement of space in circular flow by the spin of any cosmic structure that has the ability to do so. The aircraft goes straight while it also circles the earth and that forms movement within the earth's confinement by the earth's movement, which is what forms the sound barrier. The sound barrier is gravity and sciences present way of going on about the "Doppler effect" and "Mach's principles" shows how incredibly little those incompetent physicists know about physics. I challenge the lot to prove Newton by showing that the Universe contracts or where mass plays a part in cosmic physics. Show how the planets hold their positions according to the mass they have.

All objects move straight while at the same time circle around some other object. The object moves in a straight line while the spin the object hold diverts the space in which the object is into a circle. This circle brings the value of gravity to Π. By diverting the space in which the object is as well as the space surrounding the object, gravity concentrates space into becoming denser by circling and also hotter.

I return to this explanation later on in this book and then bring the applicable my arguments.
Meet the Newtonian. This Newtonian says on the Internet no less that Master Newton and Master Galileo shared the same opinion. This Newtonian says that things fall equal as Galileo said and things fall by mass as Newton said. How much double standard are they allowed to maintain and never get questioned by the public?
When out Newtonian falls our Newtonian would fall at the speed as anything else would fall. This evidence we see on TV everyday in advertisements and on reality shows. We see blokes jump out of aeroplanes with rug sacks and tennis balls and the lot through each other with the balls and catch the balls while putting on the rug sacks and tasking off the rug sack all the while the lot is getting in and out of a car as they please.

We have all seen this. The car drops evenly paced with the rug sack, the humans and the tennis balls. Then the Newtonian tells the student that according to science everything is dragged down by gravity in accordance with mass. See the Newtonian fall. This Newtonian is the only one that falls by mass

The latest is that there is some little particle undetected as yet or as this goes to print (just to cover my arse in case they discover this particle in the meantime) scientifically named and called "graviton" that "pulls" other things that "pulls" back. Science rumours their profession that science only work on established facts and nothing but facts proven to the point of no contest could other wise be good enough.

This expression is so burned into the minds of everyone that science can say any rumour and because it is science everyone believes and never disputes. One good example is this Global

Warming by carbon immersions. They take two truths and make one lie to convince people the lie is true. People take what they say as undisputed truths because science is renowned for only working with facts. What a lot of horseshit that expression is, that science only work with proven facts and that they use this idea to convince persons in the public that that is why the lot are atheists. They use this expression to convince people they are atheists because God is not a proven quantity but then they propagate the graviton as the measurable quantity, It is because they only work with facts! I can much easier prove God by using physics that they can prove mass by using physics.

Gravity forms by the dual directional movement of objects going according to the speed of movement. An astronomer will hang suspended in space and high above the earth if the circular movement is high enough to keep him there. If he spins slower he will start to drop. It is not gravitons grabbing him by pulling him closer to the earth. If his circular momentum can sustain the rotation his linear distance will maintain his orbit but as soon as the circular rotation become insufficient to maintain the orbit the linear distance will reduce and then the "gravitons" are called into action. How ridiculous can they get!

Gravity is the sound barrier because the sound barrier forms by four cosmic principles that form gravity and therefore gravity and the sound barrier are the same principles in conflict because of movement differences. The sound barrier applies by movement in space in space shared.

Everything I am about to show Newtonians say is nothing new. Newtonians say they are aware of everything that I show and then they turn around and simplify physics by putting the entirety of their physics down to the pulling power of mass that forms gravity. The sound barrier is gravity and to understand gravity is to understand the sound barrier. They say that I bring nothing new to the table and they say they apply every aspect I indicate that forms gravity. When I ask them to explain the sound barrier they use the Mach principle and Doppler's effect, which both dates from a time when no one was aware anything man-made could fly let alone go faster than sound can. Doppler made his contribution to science at a time when the train he measured was slower than a horse. The statistics he left to science applies to going as fast as a man on a horse could ride. That is a far cry from a jet breaking the cosmic boom. They can claim what they like because they never have to prove anything and they never have to listen since they know everything while their shortcomings makes science more the jesting of a buffoon. That annoys me to my guts because all of that comes down to the conspiracy whereby they whitewash their stupidity from the brainwashing they inherited from their predecessors and pass the brainwashing on to their students. Their ignominious stupidity makes me want to shout to the mountains in agony and unbelievable frustration. When a cloak of arrogant self-righteousness hides their incomprehensible stupidity under a cloud filled by their belief in self-importance it becomes hideously loathing.

In the earth's atmosphere no object can move slower than the earth does. If it moves slower than the earth does, it moves as fast as the earth does by receiving mass and being pinned onto the earth as being part of the earth. There is always movement because of the earth moves and all other movement goes in anticipation of moving above what the earth does. Movement is space that is duplicating filled space at a pace and in relation to other space within other unfilled space. That is why using parachutes slows the falling process of falling objects down because falling is a ratio between solids and liquids. This is proven in these photos about the sound barrier where the movement of the solid jet contracts the space of the liquid cloud. What happens in the photo is vivid proof of my theory and it proves the Coanda effect is what is forming gravity. The Coanda effect is a relation between material spinning and air or liquid compressing and that is what happens between the Earth and the atmosphere. This picture shows objects are always moving extraordinarily as they move in relation to the earth's atmosphere that always moves in relation to the earth's gravity compressing space into forming the atmosphere. But it also shows the movement of the aircraft compressing space in relation to the aircraft moving as well. That shows that all movement is gravity or time related. This shows objects moving is extraordinarily because everyone always forgets about that it is the earth that normally applies all of the movement while all else stands still in relevancy. There is always movement because of the earth moving and when anything moves above and beyond the movement the earth provides that then forms the ==sound barrier== using a modified version of the ==Titius Bode law==

"GRAVITY IS DIVIDED IN TWO FACTORS, BEING <u>LINEAR DISPLACEMENT</u> (Π / Π^0) WHICH IS WHAT <u>NEWTON'S GRAVITY</u> IS AND,

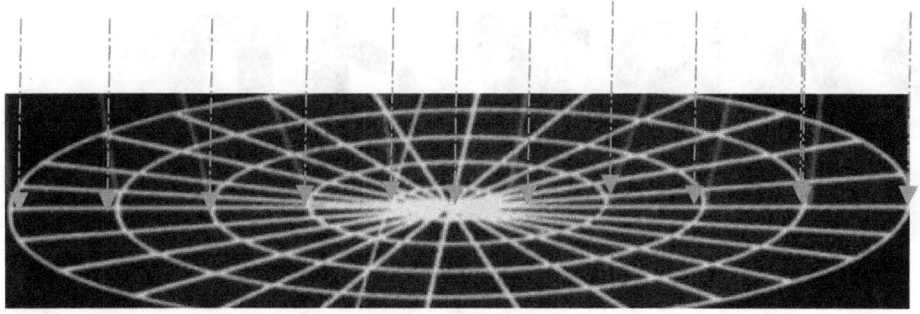

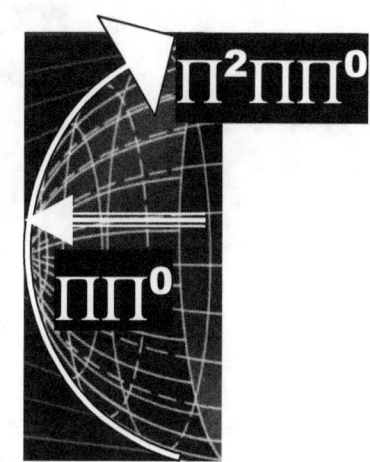

<u>CIRCULAR DISPLACEMENT</u> (Π^3 / Π) = Π^2

WHICH IS THE "GRAVITY" EINSTEIN RECOGNIZED

Those that are of the opinion that it is mass that "pulls' the object towards the earth might be a little wiser when reading the next part carefully as I explain the following:

Anything entering the earth from space it travels in a straight line. The mathematical formulating of this I do not provide at this stage but it is another "some simple algebraic relations" as Prof Friedrich W. Hehl from Annalen Der Physics put it (and I also explain this a in the following chapter). But using the law of Pythagoras and putting 7 in relation to 10 provides the value of Π. The value of Π has two different dimensional values where the one is Π and the other is 21.991° /7°

PART 10

Everything I am about to show Newtonians say is nothing new. Newtonians say they are aware of everything that I show and then they turn around and simplify physics by putting the entirety of their physics down to the pulling power of mass that forms gravity. The sound barrier is gravity and to understand gravity is to understand the sound barrier.

They say that I bring nothing new to the table and they say they apply every aspect I indicate that forms gravity. When I ask them to explain the sound barrier they use the Mach principle and Doppler's effect, which both dates from a time when no one was aware anything man-made could fly let alone go faster than sound can. Doppler made his contribution to science at a time when the train he measured was slower than a horse. The statistics he left to science applies to going as fast as a man on a horse could ride. That is a far cry from a jet breaking the cosmic boom.

They can claim what they like because they never have to prove anything and they never have to listen since they know everything while their shortcomings makes science more the jesting of a buffoon. That annoys me to my guts because all of that comes down to the conspiracy whereby they whitewash their stupidity from the brainwashing they inherited from their predecessors and pass the brainwashing on to their students. Their ignominious stupidity makes me want to shout to the mountains in agony and unbelievable frustration. When a cloak of arrogant self-righteousness hides their incomprehensible stupidity under a cloud filled by their belief in self-importance it becomes hideously loathing.

In the earth's atmosphere no object can move slower than the earth does. If it moves slower than the earth does, it moves as fast as the earth does by receiving mass and being pinned onto the earth as being part of the earth. There is always movement because of the earth moves and all other movement goes in anticipation of moving above what the earth does. Movement is space that is duplicating filled space at a pace and in relation to other space within other unfilled space. That is why using parachutes slows the falling process of falling objects down because falling is a ratio between solids and liquids. This is proven in these photos about the sound barrier where the movement of the solid jet contracts the space of the liquid cloud. What happens in the photo is vivid proof of my theory and it proves the Coanda effect is what is forming gravity.

The Coanda effect is a relation between material spinning and air or liquid compressing and that is what happens between the Earth and the atmosphere. This picture shows objects are always moving extraordinarily as they move in relation to the earth's atmosphere that always moves in relation to the earth's gravity compressing space into forming the atmosphere. But it also shows the movement of the aircraft compressing space in relation to the aircraft moving as well. That shows that all movement is gravity or time related. This shows objects moving is extraordinarily because everyone always forgets about that it is the earth that normally applies all of the movement while all else stands still in relevancy. There is always movement because of the earth moving and when anything moves above and beyond the movement the earth provides that then forms the sound barrier using a modified version of the Titius Bode law

"GRAVITY IS DIVIDED IN TWO FACTORS, BEING LINEAR DISPLACEMENT (Π / Π⁰) WHICH IS WHAT NEWTON'S GRAVITY IS AND,

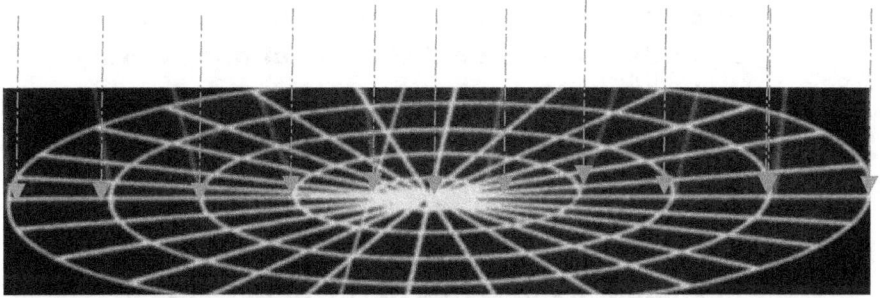

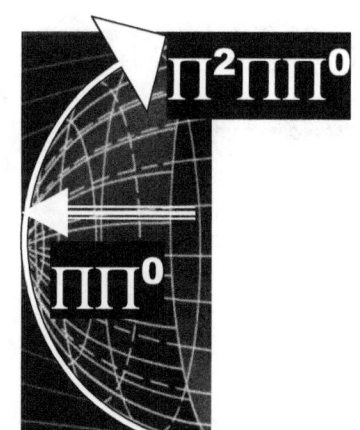

CIRCULAR DISPLACEMENT (Π³/ Π) = Π²

WHICH IS THE "GRAVITY" EINSTEIN RECOGNIZED

Those that are of the opinion that it is mass that "pulls' the object towards the earth might be a little wiser when reading the next part carefully as I explain the following:

Anything entering the earth from space it travels in a straight line. The mathematical formulating of this I do not provide at this stage but it is another "some simple algebraic relations" as Prof Friedrich W. Hehl from Annalen Der Physics put it (and I also explain this a in the following chapter). But using the law of Pythagoras and putting 7 in relation to 10 provides the value of Π. The value of Π has two different dimensional values where the one is Π and the other is 21.991° /7°

Any object entering from space and that does not have direct contact with the surface of the earth, encounters gravity in terms of being directionally diverted from going straight to turning towards the earth by 7°. This then is what provides the "pulling of mass" and the other is having contact with the surface of the earth, which then puts Π in terms with the centre of the earth at $Π^0$.
What becomes clear from these two illustrations is the following: Where there is a direct entry by the spacecraft, the time factor is not sufficient in duration.

This will not allow the time-duration needed for this structure of the aircraft to revaluate its space occupation an therefore the space factor of the spacecraft cannot adopt to its new position in space-time occupation as it has to relate to a new value in accordance with the value that is determined by the earth's concentration of space-time.

In the case of the second entry illustration, a lot more time lapse is allowed for the spacecraft to revalue its structural position in accordance to its new value of space-time occupation. This is the very reason why objects like "falling stars" burn out when they enter the earth's atmosphere.

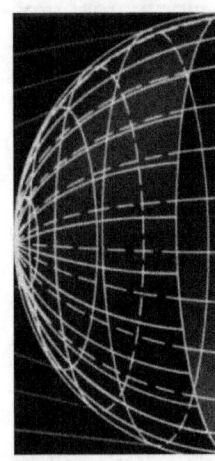

Forget about mass forming gravity it is a fool's rhetoric. When an object holds a steady position to the earth while making a sound, the sound will go in all directions evenly. It will go left and right equally fast. The concept of gravity connects to roundness and to Π. Gravity is Π. Gravity is the movement in terms of Π by duplicating the position of Π per specific time units applying. Then the object making the sound moves left it will hasten the flow of sound in the direction it moves by moving towards the sound while it will increase the distance the sound has to travel by also moving away from the departing sound going to the right side. This means that gravity has two values one being linear and the other being circular and the linear affects the circular in movement as much as the circular affects the linear. This is not only connected to sound but is connected to everything applying to gravity as gravity. When an atom moves the size of the atom must shrink to compensate for the directional change of the orbit that decreases the electron's orbit and therefore decreases the size of the moving atom where atoms form an object.

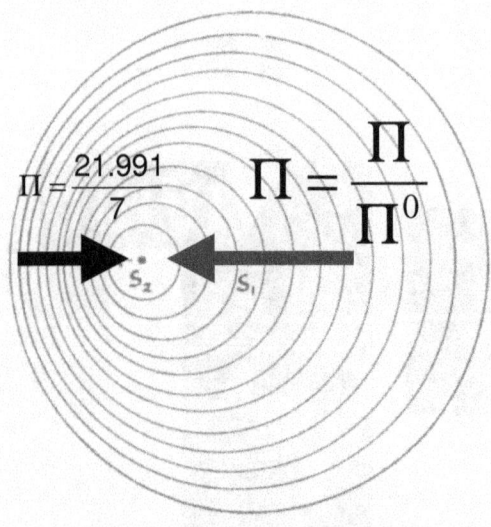

Coming toward the earth from space the object travels straight but the spin of the earth by 7° re-directs the object to follow an inclining line of 21.991 /7. At such a point the object moves towards the earth but the object cannot have weight yet, since the object does not connect directly with the earth. When the object touches the earth it relates to $ΠΠ^0$ and at such a point the object receives a value of mass by which it becomes a unit within the earth. By connecting to singularity in terms of $ΠΠ^0$ or in terms of Π=21.991 /7 puts the object in liquid or in solid.

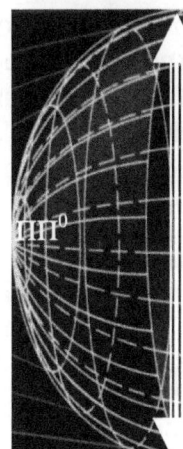

Gravity has the value of Π. The value of Π has two measure where one is when the line gravity forms extends from singularity as ($\Pi^0\Pi$) and when the line that gravity forms extends to singularity it is $21.991 \div 7$ because of the curve gravity associates with forming the value of Π. To explain all these factors I have to take the reader back to the point where the Universe started as spot that formed a dot. I have to take the reader back to where the point where the Universe was numbers holding no space, which is way before the Big Bang era where the atom broke the Universe into space that formed. In gravity there are the two values that gravity will apply. Gravity will link to the axis forming the centre around which everything spins. This gives the object a connection of 3 because it changes location in relation to the axis. Then the other gravity value connects to Π where Π always at all times connect to singularity within the centre at Π^0.

When $\Pi\Pi^0$ connects as the curve of the earth, this connection applies the object forms part of the solid and moves with the earth but also in accordance with the earth.

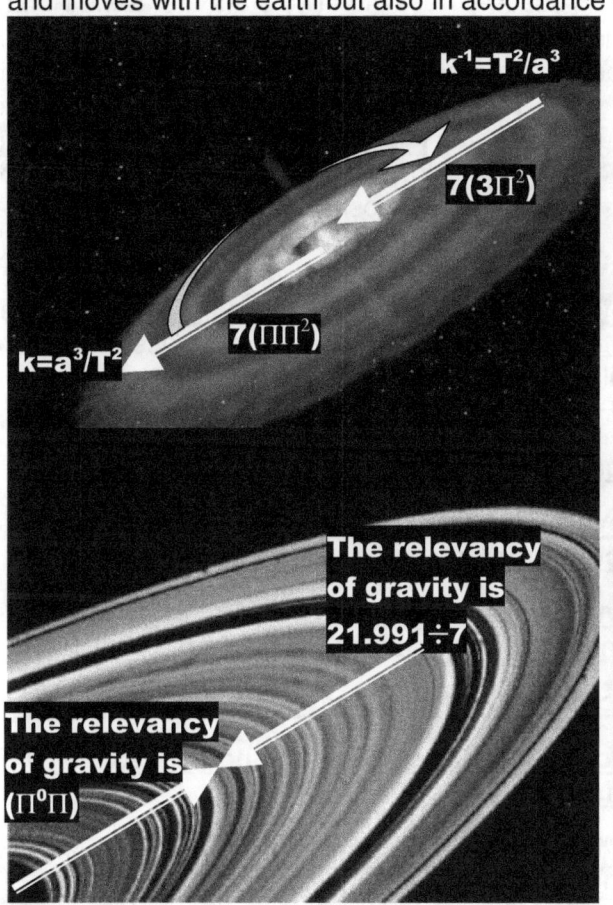

The Universe is contracting by $7(3\Pi^2)$
The Universe is expanding by $7(\Pi\Pi^2)$
Gravity is the inclining of material spinning in relation to 7° that is part of Π
This is because Π is 3.1416 and spinning in space is 3 which makes time expanding $0.1416 \times 7 = 0.991$ larger than 3. The concept might appear simple when told in this manner but the entire philosophy is so much more complex when studied overall.

By observing the rings around the planets we can identify all four pillars forming the composition thought of as gravity. We have the Lagrangian points that is five and I explain why it is 5, we have the Titius Bode law and I explain in detail the Titius Bode law, we find the Roche limit and I explain why it has the value of $\Pi^2/4$ and we can see from my explanation I have provided above why the Coanda effect accumulates all the principles forming gravity.

When something moves it is within the space the earth rotates. When anything moves above and beyond the earth's movement it excels as it exceeds the movement the earth provide and with it, it takes the space it holds in excess of the position the earth relates too. Thus by moving it form a space-time unit within the earth confinement but yet also out of and beyond the space-time the earth provides. Between Π/Π^0 and Π^2/Π^0, the body finds itself in its circular displacement, which is the earth's linear displacement value. When an object exceeds the earth's circular displacement value, it is refer to as Mach 1, or the speed of sound. Objects fall at a rate of $7(3\Pi^2) = 208$ km / h while the earth gravitational displacement is $7(\Pi\Pi^2) = 217$ km / h. because the earth moves faster than anything can fall it will always reduce space and have all surrounding space confined to the earth. The object falling only holds space that is confined to the earth.

When the object or aircraft stands motionless according to the earth centre the object moves only by the movement of the earth and therefore in cosmic terms the object and the earth is one whereby the object has weight and holds a value in mass. The object holds a relative position to the earth by the value of Π. This is because gravity then links to Π^0 and this holds regard to singularity. When the object moves while being connected to the earth by singularity the earth holds space at Π^3 while the object moves in terms of $\Pi\Pi^2$. The earth represents Π and the object moves by Π^2. When the object gets airborne the relation becomes $7(3\Pi^2)$ but I will not go into that. As the speed increases the relevancy of linking Π^2 breaks when the Roche limit is exceeded. The Roche limit is Π^2 or movement divided by the four quadrants of the circle making it $\Pi^2/4$. However while the aircraft shares the atmosphere one half is still within the earth making the Roche limit $\Pi^2/2$.

This will come into affect when the projectile reaches Mach 1. I choose to use the accepted term Mach 1 at this point as to limit any confusion that may arise from the new arguments. I have to stress the fact that the Doppler effect is, once again, merely a co-incidental but duly related by product. However, the Doppler effect, as such, plays no part in these phenomena, or in the outcome of the application. It must be seen in terms of cosmology and not from a human perspective. This is the principle applying when a newly formed star escapes the heat envelope within the centre of a galactica as a star is born. There now are two singularity factors within one space-time unit and the two goes into a battle of existing.

There is the same principle applying between stars sharing space-time. When two stars are at the Roche limit, the linear displacement reaches a value of one, and $\Pi^3/\Pi^2\Pi = \Pi^0 = 1$ is equal to singularity and in this we find space-time forming a value. This formula does not impress the most learned Physicists such as Professor Doctor Friedrich W. Hehl, which you will learn from a little further on in this book since Professor Doctor Friedrich W. Hehl thought this was to use his words "With a lot of words and some simple algebraic relations, there is no way to "explain" the world of physics." However notwithstanding Professor Doctor Friedrich W. Hehl not being impressed, on this rides the entire cosmos formed by gravity. The circular displacement reaches its full complement of half Π^2 which is the Roche limit.

Those Ever-So-Wisely-Mathematically-Sublime-Superiorly- Educated-Cosmic-Super-Brains never show that Newton is at odds with the cosmos on every issue at hand. The cosmos not once implement anything Newton says and Newton can never be reconcilable with what applies in the cosmos.
Go to **Questioning Newton's Mythology** http://www.lulu.com/content/e-book/questioning-newtons-mythology/7570956]

I can prove what gravity is by proving the sound barrier because gravity is the sound barrier.

Gravity is the rotation of the earth in relation to everything that rotates with the earth being relative to everything moving either in conflict to or in support of the earth rotating.

I say I can prove God with far more proof than Newtonian physicists can prove a factor such as mass does exist and moreover that mass can bring a pulling force such as gravity!

I prove what I say so I challenge those in physics to prove the fact of mass and not just weight.

Gravity is the contracting or compressing of the space surrounding the earth and this is by way of rotation.

I use the very same values I apply to gravity to show how the sound barrier works. This I can do because the sound barrier and the concept we think of as gravity is the very same thing that is what a Supernova star and a jet going through the sound barrier and the Tunguska event in 1908 was. It is the very same cosmic principles.

It is very quietly and not well promoted but we know that the moon and the earth are drifting apart. Nothing is coming closer together as Newton claimed in his principles.

Very quietly and not well promoted is the fact we know that the circumference of the earth or the circle around the earth is steadily enlarging thus Nothing is shrinking.

No where does any Newtonian ever put this tendency in relation to Newton's gravitational pulling principles of contracting in line with other cosmic accepted norms such as with everything expanding since the Big Bang, then what is shrinking.

Galileo proved that a large pendulum swings in time with a small pendulum and this Newtonians never dispute. Galileo proved that large things and objects fall at the same rate and by the same speed as small objects and that too they never dispute.

Yet they force students to repeat in examinations that objects fall by mass, which means that a big object and a small object falls differently because of mass, and yet they fall the same. If they fall by mass the bigger object must fall faster then the smaller object does. As the objects of different mass fall equal as Galileo said it does, and then Newton is wrong in claiming objects fall under mass pulling mass.

Gravity is the cooling of space. As space is being is placed in transformation, the time component which then can increase the spin value, or time by fore times the value. Therefore, it will transform fore times more unoccupied space-time to densified space-time, which is the name by which I call material.

Gravity forms when a solid holding an Iron core in relation to copper turns in a cosmic gas, which is the name that I gave to the singularity substance forming outer space. The turning of the solid inside the cosmic gas re-forms the cosmic gas into a cosmic liquid.

Science has to realise that the cosmos formed as the bible says and if there is any atheist idiot out there trying to go get smart about religion then read this book and see how wrong you lot are that believe in science and then order The Veracity of Gravity from Lulu.com and see how correct the Bible is. Creation started to a T exactly like the Bible says and precisely according to Genesis 1. Time split into what is infinity forming time and into time we see as space, which is a singularity substance, controlled by movement exceeding the speed of light in relation to cosmic liquid / cosmic gas moving below the speed of light. Gravity is a relative movement that has a relation between what is solid and what is not solid.

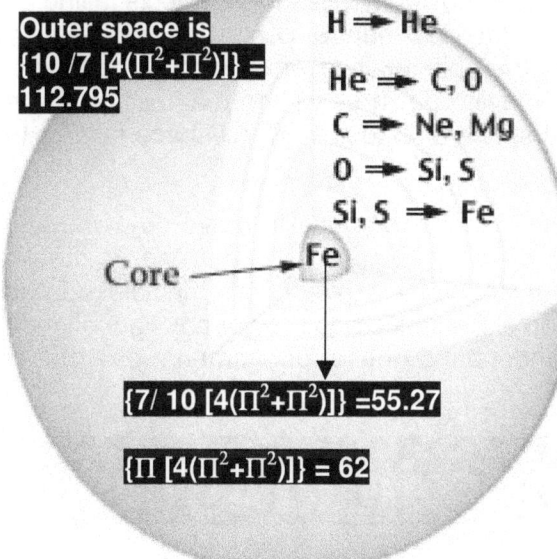

All stars have an iron$_{56}$ core. At present, the Universe is in the iron peak era. Every star has to have an iron / copper core and the relevancy between outer space on the outside holding hydrogen and the iron forming gravity or electricity in relation to spinning around copper forms gravity. In this the Titius Bode law proves all of this. No

Gravity is about outer space $\{10\ /7\ [4(\Pi^2+\Pi^2)]\} = 112.795$ conformed by iron $\{7/\ 10\ [4(\Pi^2+\Pi^2)]\} = 55.27$ the displacement value of iron$_{55}$ in relation with singularity presented by $\{\Pi\ [4(\Pi^2+\Pi^2)]\} = 62$ and that is the displacement value of copper which is the element that has this proton value according to the periodical table. This is exactly how electricity is generated and that proves that gravity and electricity is the same thing charged on different levels.

Fusion within stars does not come about "pressure" but the star freezes the atoms into forming a new unit. At this point I must also say I disagree with this idea that atoms freeze together but atoms rather grow into… The earth has not the gravity to "freeze" any type of fusion process and therefore all attempt this far was futile and will remain futile because the earth must first find the gravity to freeze hydrogen into a water-like substance as the sun does before fusion could be a possibility. Newtonians rob the public blind with their falsifies ideas about things they clearly don't understand and it is about time the public put a stop to this criminal behaviour. Either they stop to bullshit the public and promote wild schemes or they go to jail.

star can function with out having this layer balance. Should this layer balance go array in any one of the layers the layers above it would not be able to withstand the time development, the matter will overheat and the matter will be transform to unoccupied space-time. This disaster is not a natural growth process, but a disaster of catastrophic proportions. The movement of iron $\{7/\ 10\ [4(\Pi^2+\Pi^2)]\} = 55.27$ in terms of the singularity product copper $[\Pi(\Pi^2+\Pi^2)]=112.795$ allows heat that is totally expanded at the value of $\{10\ /7\ [4(\Pi^2+\Pi^2)]\} = 112.795$ to reduce. Gravity is the movement of liquid in relation to a solid.

This again was proven by the very first ever experiment concluded scientifically. This fact of space descending does not come as a surprise because Empedocles proved this fact back in 450 BC. Empedocles showed that space displaces water from the clepsydra, which was a sphere shape container with a sprout on the top and small holes in the sphere through which water ran in small streams out at the bottom. When the flow of air or space was blocked in the spout by a finger covering the hole at the top of the sprout at the entry, the water stopped flowing from the clepsydra. They concluded in 450BC that it is the empty space that pushes the water out of the clepsydra because the moment one restricts the empty space or air to flow into the clepsydra from the top, the water will stop flowing out of the bottom of the clepsydra.

Why would the flow of the water stop if the mass did pull the water down? When the finger blocks the sprout and stop the space entering from the top, the water does not fall to the ground but it is the empty space that pushes the water out at the bottom to fill the clepsydra from the top. When the finger blocks the sprout and stop air to come in through the sprout opening the water should still run out at the bottom by the mass of the water pulling, if mass was doing the pulling. If mass was the force giving factor, then the water must keep on flowing because the mass of the water did not disappear when the sprout was covered and therefore it still has to produce the pulling by forming gravity. All this evidence was known to science about 2500 years ago but since "With a lot of words and some simple algebraic relations, there is no way to "explain" the world of physics" it lacked mathematical communication and it should therefore surprise anyone very little that physics could not fathom this result 2500 years onwards.

Forget the example always used about the hammer and the feather falling equal in a vacuum because the hammer and the nail and the elephant falling together will also fall equally notwithstanding falling in a vacuum or not falling in a vacuum. The vacuum part is conspicuously in place to purposely confuse reality as it is brought in to flagrantly spread misunderstanding of the issues in hand about the falling that takes place. With everything always falling equally when the same the condition applies to all objects falling and therefore with such falling happening under the very same variation of natural conditions applying, this shows it is the space in which the object is that falls and not the object falling while leaving the space it holds behind. The lack of relevant density in relation to air moving down stops the feather from falling equal just as gas does not fall with the space at the rate that space does descend. All space falls by the compressing of the atmospheric space.

The rotation of the earth moves the space sideways and this brings the space to move downwards by increasing the density of space or air as it comes closer to the earth. This results from the Roche limit applying to fix atmospheric layers varying in density. In my books I explain that principle applying mathematically. Notwithstanding using your mathematical marvels, science has not got any vague idea to explain any of the phenomena mentioned above. To understand these phenomena one has to understand singularity.

A Conspiracy In Science in Progress

I came upon a mistake concerning physics.
This mistake is about the cosmic phenomena called gravity. Detecting the mistake is simple because it is uncomplicated to understand. Academics in Science say that a feather will fall with the same speed as what a large rock would fall.

That is according to Galileo and that is accepted as a principle in physics. For the first time ever since the time Newton introduced gravity I seem to be the person that questions this interpretation.

How does mass pulling mass and falling by the gravity power exerted by mass then fit into this interpretation because a feather has much less mass than what a large rock has? If this statement is untrue Galileo is incorrect and the Pope does not need to apologise to Galileo as Physics insist the Pope has to do. Then the pendulum doesn't indicate time and mass does implement falling of objects as Newton protested it to be. But we know Galileo is correct and that makes Newton's suggestions what they are…merely suggestions that proves to be incorrect.

When you disagree with any academic in any lecture hall about mass not forming a picture as responsible for pulling gravity and you come to a conclusion that you doubt the mass part that they bring into the picture as establishing gravity the academics wipe you from the table with a swipe because then they contemplate that you are so stupid you fail to see all the facts that physics present as proven facts and they hold you as being too stupid and mentally underdeveloped to appreciate or to understand physics. They tell you that Newton is not for such stupid people that are unable to see how true Newton's arguments are!

I have been at odds with academics for years and only because of the superior positions they hold in office are they able to bully me into silence but not into submission because that is what this book is about…to expose their corruption.

They can push me into abasement but never into abeyance. By the important Academic positions they hold in the huge academic institutions that give them sanctuary they might dictate the terms of our meeting and in terms of those advantages they can hide behind the criminal wall of deceit and suppress me into silence but they will never get me into submission.

Now I am taking my case to the members of the public so that the truth must be brought into the open. I have had the tour they give and then more came my way. I never got around swallowing their gravity that comes as result of mass creating gravity or any part they present facts as to how it happens and where science is of the opinion that mass pulls as gravity is… Academics condemned my work and therefore me and for six years where I could not get a publisher to come around and bother to read my work let alone seriously proposing a publishing contract. I had to finally go private with the publishing as all doors shut in my face as soon as the academics read the content of my work because from the nature of my work I take Mainstream science head on and as I am sure about my arguments I am confrontational on most aspects of astronomy. There does not seem to be any publisher that wants to go head bashing with the establishment of science on official science principles, which I have to do to convey my message in a no uncertain language. If you also have doubts about the academic's indisputable correctness please read on and confront either them or me on everything you read here.

After reading this book you will have to take sides because you will know the truth and you are free to decide who present the truth…is it the Newtonians or I.

By that decision you then either become my partner in also recognising the crime I uncover or you become part of the crime syndicate as you cover the truth up. You either will be part of the truth by helping me confront them to acknowledge the truth or you will remain part of their cover up by ignoring the evidence as everyone this far did for about four centuries in any case.

By not confronting the establishment, you give the establishment grounds to allure you into being sheepish. Because they see you, as just another stupid senseless student they have the opinion that they can brainwash you into accepting these fallacies that I am about to tell you. They will literally brainwash and condition your mind to accept what they never yet were able to prove. The truth is that this process of brainwashing is going on successfully for four centuries without any backlash to those committing the atrocities.

They are of the opinion you will swallow any rubbish they throw your way just because every generation before you were mind controlled in the way they are about to control you. You may think this is big words but read on and see after you come to know all the facts whether I exaggerate even in the least. They see you as slow-witted and mindless because they think they are the academics being superior making you the lesser and inferior party. If you are not aware of the facts beforehand they know you will follow their teaching without asking questions as it is going on for four centuries this far. Talk about the Catholic Church putting the fear of God into Copernicus and you will find their manner in disagreeing about statements not echoing their perceptions is as bad and more ruthless than the Church was during the Dark ages. The Church just killed those confronting them on science dogma but did not make every student a brainwashed mind controlled Zombie!

They think that your naivety makes you a Zombie and with such a degree of mindlessness that state of mind will incapacitate you into their control. They don't want you to ask nosy questions about contradictions existing and they refuse to answer any uncomfortable questions asked in any confrontational manner. This process of brainwashing and mind controlling in physics has been in progress for hundreds of years. If you are surprised then control me by asking and just answering how it is possible to agree that a feather and a large hammer fall equally while also agreeing that Newton is most correct. How can any person thinking logically agree that Galileo is correct when denouncing objects falling under mass differentiation when Newton insist on mass driving gravity as a force. If you can't…well they can't either! Their task is not to explain but to mislead since they think you can't think while they think they know how to control you.

The motive behind this book is to promote my other books and by trying to be as least complicated as possible this book aims to present Newton's deceit. This book does not aim to represent the full entirety of the original thesis as it represents the entirety of my theory in a single copy for that is not possible but is reduced to aid any possible potential reader in the examining what the purpose is of the information this book wish to present in order to uncover Sir Isaac Newton's fraud. Anybody and everybody are aware that all objects fall at an equal rate. If an object such as a car weighing one ton falls at the same pace as a person weighing fifty kg how does mass come into the picture by committing a force to do the pulling? Mass has to pull because according to their teaching it is mass that establishes gravity. However, mass is a factor that produces differentiation whereas all objects show equality during their fall. If it is mass that is establishing the force gravity, all objects must fall at different speeds. That they do not do as they all fall equal. That means Sir Isaac Newton's physics is wrong from the start because mass cannot have any input in objects falling.

This is not the only untruth that the Paternity called Mainstream Science is keeping concealed as a cover up that is wrapped under an airtight blanket of deception. If you sit in class and listen while also experiencing the sinking feeling that the facts you hear are not adding to a total you are comfortable with while you disagree with what is said then you better read on because   this web page has it at task to show all that will read this document how much discrepancies academics lay on unsuspecting students that trust Academics with their future and their life. Do you as students realize the inconsistencies that physic Academics teach you as the truth.

Students tell your Professors to stop deceiving and stop trying to control your minds with their fraud. Those Academics tutoring you are telling facts about gravity that has never been proven.

That is mind control.

They wish for you to accept facts on gravity that they hold as the truth. They claim those truths are beyond questioning yet with the least examining those truths they stand by then proves to be totally void of substance because it was never corroborated by one single experiment.

Should you question that mass produce gravity they will expel you from University by letting you fail your examinations and it was never proven. They will expel you and have you fail tests should you question their authority on the matter of gravity while at the same time they can't for one second bring evidence in support of what they wish you to accept as the unquestionable truth.

That's brainwashing by mind control because if you don't accept their baseless fact as God given truths they dismiss your academic career.

It is either put up and shut up or be gone. Academics do put mind control to work on unsuspecting students by forcing students never to question the legality of statements they offer as being sound and correct.

What they present as correct I prove in this very book are openly laughably totally incorrect and by just reading my evidence you will see how feebly easy it is to rubbish it. Take the evidence I am about to share with you and confront them with the fabrication of facts that they present. Go on and challenge those teaching you with the falsified facts as I challenge any one to prove me wrong.

What they maintain is gravity is total incompetent nonsense and can't be corroborated at all but what they can't corroborate because they don't understand I prove to be that which the Universe employ to form gravity. There are four phenomena they dismiss because they have no idea what they are. I studied each one and formed an explaining by implementing Kepler's formula as the Universe gave it to Kepler.

The by understanding the formula and implementing the content into the four phenomena I am able now to prove what forms the motion we think is gravity and when reading it then the Universe makes sense. All the questions in these books I managed to answer while they can't ... and in the books I answer a lot more questions than what I ask here in the rest of the information while Science fail to answer any...

The book I present has the dynamics to change science forever and that is not cheap exaggeration or promotional talk. I say that because I investigated Kepler and believe it or not but that investigation was the first one done since Newton explored Kepler the first time and that was four centuries ago. Science never went back to Kepler after Newton included the work of Kepler into his work but when one read my work one will find Newton compromised Kepler's work as he did with the work of Hook and many others. For my saying that about Newton not correctly analysing Kepler the Newtonian Academics at various University Institutions bluntly ignore my work. Academics would not touch my work notwithstanding the twenty-seven years of research that went into my work. I have a very limited three South African Universities dealing with cosmology to turn to and their reviews of my work in the past left me in doubt about their sincerity in the performing of the reviews. The South African Academics do not attack my work because then I can defend my work…no they just ignore the work by sending reviews that totally miss the point about my work. Their reviews are about my linguistic capabilities or about my presentation of the books but they never refer to the work in detail that I represent. Their attitude spurred me on to present my work to Academics outside Africa. I have sent the books to various Universities but because of various factors not in my favour the Universities in question treated the books outside Africa as junk mail. The review of the Academics in South Africa was always mainly (and more or less only) about my accompanying letter by which I introduce my work, such as the letter you are reading. Their review never went further than my introductory letter since the facts they mentioned was only aspect, which I mentioned in my letter. So the facts in my letter was the only response I received. I say this because the review presented no evidence that those Academics even read my work less understands my work. That forced me to write a letter where the letter is an introduction manuscript about my work and that letter of introduction I present as a manuscript on offer for publishing. After all the attempts I decided to chance the commercial press because I have absolute faith in the correctness of my work. I need to find one academic that reads my work with a degree of sincerity to establish the connection of academic acceptance.

The following presentation is as simple as gravity can be represented but in order to have the presentation as simple as it is we do surrender some part of the accuracy to achieve simplicity. Under a microscope one would find that in this explanation the explanations strays a little from the truth in order to make it comprehendible to everyone reading it. The truth is the explanation about gravity can be somewhat more complex than what the following presentation has to offer but then on the other hand it will never be as simple as dumping the entire concept on one thought about mass that produces gravity. That simple gravity in explaining can never be because it is so far from the truth as telling that fairies produce summer and witches bring on winter. The whole concept of mass being responsible for gravity is one big hoax and forms a scam. Putting mass as being responsible for creating gravity is the biggest fraudulent lie that ever hit the Earth on any scale ever.

For instance Mainstream science has the theory that matter and antimatter developed and some matter formed as particles, which were nicely wrapped in containers we call atoms and stayed on as material or matter where the rest formed antimatter that disappeared. The anti material also formed atoms but then chose to just plainly vanish, as did singularity and other cosmic factors to now be nowhere. Singularity is a mathematical position falling outside the detection of the observable Universe in any case so in the case of singularity there is a possible excuse for the disappearing…or is there? We know that singularity produced space-time and space-time produced gravity but then space-time went away leaving us with gravity. Singularity is a single dimensional entity that is not material, holds no space, holds no time and in our 3 D view can only be found outside the human visual spectrum. Mainstream science is of the opinion that singularity disappeared after the Big Bang process came about whereas I am of the opinion that everything must remain part of the cosmos once it was in and was part of the Universe simply because there is nowhere for it to go. The same applies to space-time and antimatter. If it was part of the cosmos in the beginning it has to be in the cosmos until the very end only and simply because there is no other place available to be or to move too. The Universe is the container that is the only container leaving no other container to contain what ever is in need of containing. If it is or if it was it is in the Universe! Our task is to find the place it went to or find what it changed into. While we then are on such a hunt, we might just as well find the cosmic principles not understood but which is there all the same. Take for instance the Bode principle where all nine planets show a relation with the sun in precisely the same manner and using the very same method of spacing, yet science brush the Bode principle off as a coincidence. It might be a coincidence when the space between one or two planets shows these phenomena but when all nine planets plus even the fragmented structures adhere too the very same principle no person can be of the opinion that it came about as a coincidence and still pretend to be serious or professional about cosmology. One then must simply find the proof lacking in our understanding

Mainstream science knows about gravity, the Bode principal, the Roche limit, the Coanda affect, the Lagrangian system, the sound barrier but cannot explain any of the phenomena all though the presence of these phenomena is without dispute. It is the explanations about what causes the phenomena that should be part of the dispute but in science the way they defend Newton scientists go overboard by disputing the phenomena and the phenomena as a principle existing in cosmology or not becomes disputed. Science fail to give acceptable explaining of such occurrences therefore disputes the validity of the phenomena and this failing to explain the presence becomes disputing the presence thereof. In such a light Scientists must somehow realise they are barking up the wrong tree with the information they have to use to do some explaining. They cannot refuse the phenomena and not realise they must have the cat by the tail as far as cosmology goes. Please remember that with this I am referring to cosmology and not general physics. There is an Earth versus a Universe of difference between the two concepts but Newtonians fail to see that because Newtonians cannot appreciate the differences thus blurring the understanding of gravity. If there are that many phenomena (it represents all there is in cosmology) to explain and such little ability to explain (science fail to explain even one) by using the information Mainstream science is using to explain the cosmos, then someone somewhere has to realise there is something drastically wrong in the way they present the knowledge they claim to have.

One cannot be serious about science but defend your view by dismissing the validity of all unknown indicating factors presented as such. There then is some gross incorrectness in the way Mainstream science reason. The Roche limit is there and no denouncing thereof can remove it from the cosmos. They may refer to evidence received from the Hubble telescope as "the star is blowing bubbles" for the lack of explaining what is occurring but occurring it does. One cannot say it is some unknown gesture presented on occasions because not explaining the pictures presents the presence of certain foolishness. For fifty years they lost many pilots but still has no idea what brings the sound barrier about, or find the link gravity holds in the process we call the sound barrier. Instead they try to interpret some effect established almost two centuries ago with steam trains back then travelling at the same speed that horses run. No further investigation with the science in hand brought them closer to new facts! It should be a sign telling them they are going about incorrectly but it does not because Newton said so. It may sound as if I am anti Newton but I am not. But there has to be more than Newton with so many pieces of the cosmic puzzle still missing. Science should not serve only Newton but science should serve the seeking of the truth. When I first came upon the unknown it stirred a sense of disbelief and I decided to respond.

Some twenty-seven years ago I decided to start an investigating quest on my own to see where I could go with my private research. It came as a result of my frustration when I realised the discrepancies there are in theories presented about cosmology and all the unexplained factors no one ever makes any effort to explain. Later I found that no-other person than Newton in person was to blame for the mistake that was made but now I am jumping the gun. If you are a Newtonian and feel a repulsing urge to throw down the letter you will do so at your peril. I say this because I have seen Newtonians get fits in the past when I say what I just said. Whenever I make this very claim the entire science community rejects me immediately without any reservation to any person. When I speak out against incorrectness I see science as a unit and an entire structure reject all further statements that I make without excluding any body active in the field of science. To science Newton is reserved as a god and they placed Newton beyond criticism. When they listen to me criticizing Newton they switch off their mental lights literally. They immediately go blank and I have witnessed it every time. I can visually see their eyes go dim. No Academic Newtonian priest will spend another second to listen to more of my views. Yet they remain unable to use Newton to explain the cosmic phenomenon as the Bode law, the Roche principle, the Coanda affect, the Lagrangian system or the sound barrier. But what they do not realise is the mistake was a resulted not about what Newton presented because it came in what he admitted that he could not present. The mistake came as a later presentation of a concept he admitted he could not underwrite by scientific explanations. When he introduced gravity as a concept he admitted he did not know the origins of the force. What Newton admitted he could not explain and what he did not explain later became an institutionalised claim presented later as if he did explain it.

The error is in the facts not yet ever explained and in that which Newton admitted he did not understand. It is Newton's incorrect suggestions he made about gravity that were later accepted as explained and proven science that went on to become institutionalised facts. It absolutely came in the way Newton changed Kepler's formula. Newton admitted he did not know what gravity was and left it at that. He did not offer any more insight than reducing gravity as a force and a force it stayed all the time. Three hundred and fifty years on science still do not know what gravity is and is still leaving it at being a force. Without knowing what gravity is all other concepts in cosmology does not make sense because all the phenomena I mentioned a minute ago is sides of gravity and performs (each to its own but still) as another principle where the totality forms one concept we call gravity. If the World does not know what gravity is then the phenomena coming from gravity will remain unknown to all. With that in mind Mainstream science still take a very dim view on my criticizing Newton! In all this time Kepler explained exactly, precisely and unequivocally what gravity is! The unbelievable part is that we all missed Kepler's announcing of gravity and in the "we" I use that 'we' include even Isaac Newton and Albert Einstein in persons and by names.

All principles I use in the theory I introduce with the publishing of this book are part of nature. I base my theory on heat stabilizing through space using motion to produce cooling. That is gravity. But however this may sound basic, Mainstream science is also most guilty of their usual departing from this basic principle through the employing of terminology and such terminology has the tendency to cover many of the basic meaning behind the principles in nature. For example the one principle I do not applaud is a principle Mainstream science underwrites in the sense that matter in the beginning was coming about and anti matter came to destroy the matter by consuming it. This translates into a packman computer game that has no correlation to cosmology. It is moreover the disappearing from the Universe of that which came as the result between the two opposing materials that I strongly reject. Anything part of the cosmos at any stage before during or after Creation, remains in and part of the cosmos and cannot leave the cosmos because there is simply no other place for what ever there was to go. Leaving the cosmos just is no option there is.

In any test performed today by creating friction through motion the discrepancy there are between objects in motion will bring friction that will produce heat and the heat will result in space forming. In such destruction of matter space and heat comes about and the net result eventually is space created where no space was before. The cracks showing is space created in the cooled material that was heated to a glowing red-hot always afterwards leaves cracks. The cracks are space not filled but was filled during the heat coming about as a result from the overheating. After the cooling the cracks present new space where there were no space before the heating took place. The heating process started forming and filling space but afterwards when the cooling set in it reduced the filling of the space. The cooling did not destroy the newly formed space. The cracks represent the space not filled. The material in the cool state cannot fill the void that came as a result of the cold material contracting and reducing the space filled when the material was overheated. But the space remains although not filled any longer.

That we can see as evidence with material having a heat building up when motion difference brings on friction and such friction brings on heat. I do not share the view Mainstream science has that when matter and antimatter came into conflict the product that came from this just disappeared without a trace of any sorts. I believe the evidence is present and I think I know where that evidence is. I believe I can show that it is a motion discrepancy that produced matter and anti matter and we do not have to go and look for non-exiting positrons. A positron must produce a negative proton and such a performing sub atomic structure cannot be functional. By changing legions it must then produce a product where it performs as gravity by rejecting material and pushing away other cosmic components. That will lead to an exploding Universe! I say that discarded material became heat that became space that became outer space in the Universe. I go to lengths to make persons see that space cannot be nothing. This is a factor that science has to accept if Mainstream physics have the will to find solutions about the Big Bang. I say the motion between particles in a cramped space as the case was during the initiating of the Big Bang would have brought on friction in space between particles present that we couldn't even calculate. The result is that some of the matter particles produced a means of self-sustaining by applying gravity and the demise of the other particle that became destroyed resulted in plasma forming on the one side and material on the other side. I believe even to today and throughout the rest of the Universal motion through space in time the plasma is transforming to material forming particle growth in space through the motion we named gravity. This was how the cosmos came about and this is the manner the cosmos will conclude.

I believe some of Creation remained as some particles formed by applying gravity in motion and the lack of gravity turned the other particles of lesser motion into heat. In this is the destroying of singularity contact came as a product of light, which again I believe (within reason) I do prove. I believe heat is the destructed form of material and this information the atomic thermo explosions give us. By releasing the heat that is sealed in an atom such release thereof produces heat, light that can liquefy the eye and most important the unexplained nuclear winds that destroy so much. But to realise that we must beforehand find what space is and accept that space is made of something. We have to see what forms space and why space can be the absolute basic container through which gravity can relay the influence it carries. We must come to realise that whatever forms space has to be that same ingredient what forms the lot of everything in the entire Universe.

When particles heat up the particles expand. Expanding is applying more space. The space that the particles hold amplifies. The particles claim more space when heated. The claim of heat on space creates more space and heat results in more space as the product of heat rising in material. Such expanding is one way of bringing about cooling. Heat produces more space and never reduces more space. The motion coming about from the expanding of space brings about cooling. When particles cool motion applies in some form. At first the Universe was extremely hot and without space. Then space came about and heat levels declined. Compress space even today with a piston in a cylinder and such confining of space will increase the heat by the piston effort as the space reduces. The heat coming about has no relevance to particles colliding because compressor cylinders cool down with time and not necessarily with the loss or release of particles. It is not only the discharging of air that will reduce the temperatures inside the container. After the pumping of air increased the heat in the cylinder even to dangerous levels the heat will reduce back to room temperature when further pumping seizes and the stopping of further air movement into the cylinder can bring about heat stabilizing. If it was the commotion of air particles colliding and rubbing of particles going on in the compressor only the release of air could produce cooling. If this were the case the temperatures would rise indefinitely because the action will increase in counter actions producing heat and spurring on more friction by collisions. The stopping of air pumping reduces motion and subsequently brings on cooling. The cooling process is not resulting from calming the excited particles within the container by whatever means. The process is natural and follows immediately when pumping stops. Stopping the pumping automatically leads to cooling coming into action.

This means it is not the particles in the cylinder that brings on the heat levels rising as much as it is not the particles that will eventually bring about the explosion that will follow, should the pumping continue regardless of danger. When the pumping stops the heat immediately starts reducing. There is no further increasing of heat in any way except if the pumping continues again. The already over filled container does not continue with the friction of particles rubbing within the cylinder container. Afterwards when the compressor is left by itself and temperatures stabilize too the same levels of the temperature on the outside a sudden releasing of the air under controlled conditions such motion of the air relieved will bring cooling to the extent that pipes can freeze and block the releasing airflow. Two American Submarines were lost in this manner and yet to this day no person in science saw such a connection in the sinking of the subs. After a few days left undisturbed as far as heat distributing goes the stabilizing will lead to conditions being equal on either side of the cylinder wall. The compacting of air molecules, which although still much higher on the inside than what the denseness is on the outside will not produce the same heat levels inside the cylinder than that what was achieved during and immediately after the pumping operations.

The temperature will only become affected with motion contributing to changes in the balance. The releasing of air will extract heat from the process to a point where it will lead to freezing coming about. We have to see the container for being a container and the Universe also being a container, but more important is that what ever comes about in the one container will be similar in the other container since the containers contain what the cosmos is made of. It is heat flowing between materials in space. Heat will always flow from the highest value to the lesser value. This is the concept I use as the basis of my theory. I base my theory on gravity producing cooling and contraction while heating produces motion by expanding and creating more space. When an object is overheating no amount of force can retain the container from becoming too little and with the heat coming about forming the expanding the space produced with this action will destroy the form any container may have. Even a container as solid as an atom proves not to be able to withstand the expanding produces by overheating. The Universe is a container and as all containers prove the Universe also was unable to sustain the space that heat created when particles overheated. The Big Bang is evidence that. During Creation the compactness of the particles produced motion discrepancies bringing about friction where some particles overheated and formed heat. The heat that then formed space we named the Big Bang where heat produced space and formed motion we see today in the Hubble constant. The Hubble constant is antigravity in its most splendid form.

I believe that my study of Kepler allowed me to achieve an all time breakthrough success because I can now explain what gravity is. Remember that not even Newton could explain what gravity is or where it comes from, but Kepler did that without any person ever noticing. Kepler studied what kept the plants circling the sun and that is gravity. From such explaining what Kepler said without Newton changing formulas on Kepler's behalf I prove the Titius Bode principal also known just as the Bode principle. I prove that the Bode principle is forming the value of gravity when incorporating the Roche limit. These phenomena was never before explained or understood by Mainstream Science although they appear more than regularly in the cosmos. I explain how singularity forms the Roche limit and how singularity brings about the Coanda affect but most important of all Kepler showed me where to search for singularity. My achievements came from my effort where I separated Kepler's work from the opinion that Newton formed and gave to the world about Kepler's work.

For instance from Kepler's work I can explain the operation of the Black Hole, which not even Prof. Stephen Hawking understands. That is because Hawking ignore Kepler. In my opinion my explaining of gravity makes much more sense than the accepted force of Dark Age proportions…and the best part is you do not have to be a genius to realise or understand it. Even a simple person such as I can see it clearly! From my view a force is just motion applying and that is what Kepler said gravity is. Kepler said $a^3 = T^2 k$. I dissected k as a factor in the Coanda effect and found that the Coanda effect is proof of my view about gravity and singularity produces the Coanda effect. The Coanda effect is the establishing of individual space a^3 by applying motion $T^2 k$. Where the Coanda effect is producing gravity and such producing is stronger in a small space than the gravity produced by the Earth in that spot I use that principle to show that there was some manner in which the reducing of k brought about a stronger T^2 just as Kepler said. This was a crucial part during the Big Bang and therefore had to play a major part during the period of the Big Bang. Einstein came to this conclusion but failed to refer his view back to Kepler and by not referring to Kepler missed the point he wanted to make.

Presumptuous as it may be on my part of trying to disprove Mainstream Physics, such a presuming does not change the truth about Mainstream science being incorrect about gravity. After all they admit they do not know what gravity is. I am not disproving anything because they agree they do not know, which paves the way for my showing what gravity is. By admitting not knowing what gravity is they then also admit being incorrect about gravity but unfortunately mainstream physics do not see it that way (yet). The question in hand is finding what role gravity played when the Creation came about for the first time. I had to find a method that would allow me to explain why gravity played a role. In the book I present the analysing of Kepler's formula without Newton's interrupting of Kepler's work.

Please let me explain: Tycho Brahe and later Kepler made a study of outer space as never repeated afterwards. From this Kepler concluded that $a^3 = T^2 k$. We all know that a^3 is space and with the space indicated as being in the third dimension and the third dimension is unmistakably a cube that forms volume, which by definition is presenting space. We also know from the way calculations come about by using the formula of Kepler that T^2 is the duration of a specific period of time relating to a specific centre. On the one hand we have space a^3 and on the other hand in direct relation to the space Kepler introduced motion coming from a centre that forms time $T^2 k$. Kepler gave us space-time a^3 / T^2 centuries before Einstein gave the concept a name but no one ever took any notice. In the formula is space a^3. In the formula the space a^3 has direct relation to time T^2 If k is a^3 / T^2 it means that from the centre holding the gravity is space-time. Space is a^3 and the motion of space a^3 we accept as time $T^2 k$ and such accepting is part of our understanding for the past three hundred and fifty years. Kepler gave us gravity before Newton named it as a force. Kepler gave us space-time long before Einstein named the notion. With Newton's meddling he missed Kepler introducing gravity as $k=a^3/T^2$ space / time.

Gravity is a rotating solid moving through a liquid space. The Earth is a solid that much is true. The atmosphere is regarded by physics to be a liquid and that much is also accepted as true. Gravity is the Coanda effect and the Coanda effect is where a car tire spins trough water and the spinning wheel gathers the water onto the surface of the tire. At speed the tire picks up the water and secures the water around the tire. The motion of the tire, which is the solid, contracts the water, which is the liquid onto the solid tire surface. The contracting of the liquid onto the solid by rotating motion produces the gravity that attracts the liquid to the solid. That is gravity. The solid tire can cover the surface of the tire by a layer of water where the water is as hard and as sturdy as what the solid tire can be. But the water might be hard and sturdy, yet it remains a liquid with all the characteristics attached to liquid. That is why driving in the wet is so dangerous.

The solid tire can surround the tire surface with as much as one inch of water. That is gravity whew the solid tire spins and by spinning it contracts the liquid water. The water being a slid in the form of ice can't perform in the way the liquid does when the liquid is surrounding the tire. An inch of ice will never be strong enough to allow a car to drive over it but in the case of the Coanda effect the motion of the tire allows the water to be much stronger. While when being in the position where the water is surrounding the surface of the wheel through the spin of the wheel that contracting gravity then makes the water as strong as the tire which enables an inch of water to support the entire car running on the water. The spin of the tire produce a gravity contracting by rotating motion which turns the density of the liquid water to the same compactness as the tire surface being a solid will have. The spinning solid of the tire turns the density of the fluid water into a solid equal to that of the solid tire. The tire asserting a rotating motion does the producing of gravity by motion. The expanding of the rotating action produces a contracting of the liquid space it moves through. By expanding the solid rotating takes away some of the space that the liquid holds.

The tire rotating is expanding the space it holds. That is called fleeting momentum. The matter tries to move away from the centre in an effort to gain more space. As the tire spins the tire tries to capture more space and the tire can thrust this so hard that the tire does go oval. The tire tries to hold more space than it has because it is capturing more space in an effort of expanding.

While the tire tries to capture more space the tire also reduces space that the liquid water holds. By capturing the space that the liquid water holds the tire is capturing water, and in the effort the tire is trying to gain the space that the water has and in that the solid tire is making the liquid water solid. Therefore we can drive on an inch of solid water while the water is liquid. By the water contracting we find the density of the water changing where it meets the surface of the solid. The solid tire expands and in the expanding it makes the liquid water solid. Then where the liquid water finds being reduced and being made denser the liquid water gets so dense it becomes a solid water area. The tire wall turns the water into a solid by the rotating action of the wheel. That is when motion applies to the wheel. When considering the wheel at speed and the Earth at speed we tend to think of the Earth being still and motionless. The Earth is spinning at a far greater speed that the tire would ever be capable of. The Earth is spinning so much it is concentrating air to become a liquid.

When objects fall the object has no mass and this is in spite of all the claims the Academics in physics try to produce. Galileo said all objects would fall at an equal pace and hit the Earth at the same time when falling the same distance through the same air under the same conditions. Newton said mass is responsible for that which produces the gravity by which objects fall. That means the object being more massive must fall faster than the object being less massive. If mass brings on gravity mass must distinguish the amount of gravity by applying more or less falling pace. If that does not happen there is no evidence of mass applying because then all objects hold equal mass while descending to the Earth. We see frequently that the object can fall at a specific rate depending not on the size it has or the shape it has but the distance it travels through the air. We see so many times that a car drops from an aeroplane with a human falling next to the car. There is a car advertisement where the human that is falling has a parachute in a bag falling next to the person that is falling next to the car. The car is around twenty times more massive than the human and the human is about twenty times more massive that the bag containing the parachute. If the falling process depended on mass that instigate the gravity action as the intellectuals' wishes to declare then the lot cannot fall at the same rate. The car must fall twenty times faster that the person and the person must fall twenty times faster than the bag. We can see that the lot is falling at the same rate and the descending bares no implication to any mass that shows differences. That is what Galileo said when Galileo said all things fall equal at the same rate and land at the same instant and a massive object will land at the precise instant that a very light object will land on the condition that they are dropped equally and that they fall through the same space at the same time. That statement excludes mass from any part that gravity has.

One can from that see that the falling has no implications brought on by mass. Mass has nothing to do with gravity but to restrain any further gravity effort putting individuality to the item falling. The gravity is produced by the moving of the object and while mass restricts the moving the moving still apply as gravity because the moving remains as a tendency to move when mass applies. Even where mass stops the gravity moving the gravity remains as a tendency to move towards the centre of the Earth. It is not that the mass is pushing but the mass is stopping the moving of the item to the centre of the earth. Mass only applies when any further motion of objects are restricted. If the restriction of the larger object that inflicts the mass suddenly also start moving, the mass turns to gravity that very instant. While the larger object retains in position of preventing the independent object from any further individual moving mass comes in as a factor that stops further motion.

Mass counter acts gravity by stopping gravity. The object only obtains mass when gravity becomes no longer applying. When the objects all fell and they all hit the surface of the Earth at the same time, only then is there distinction about size and mass. That makes mass a factor of the Earth holding a restraining on the object and then mass is not part of gravity because gravity is part of the falling or the moving of the objects. When object fall they have gravity and the gravity is equal applying to all because the gravity is the moving of the objects without restriction. When the objects hit the ground the objects loses independent motion and with the accepting of mass the objects retain the motion that the Earth provides. The mass renders the objects the motion of the Earth and having the motion of the Earth the move at the same pace as the Earth. The mass then makes them having a relation to the Earth where the mass puts them in a restricted part of the Earth. They move at the rate that the Earth move because they then are part of the Earth by the provision of mass. Mass shows how much the object that then landed on the earth and no longer moves independent of the Earth then became Earth.

The gravity is the Earth forming the solid that rotates and the air is the liquid through which the Earth rotates. The liquid is contracted onto the surface of the Earth by the solid of the Earth trying to expand into the space the liquid holds and this happens due to fleeting momentum or rotating motion. Mass is the resistance that an moving object shows when stopped by a larger object that blocks the smaller object from moving further as an individual object where by showing mass the smaller object resists the effort the larger object asserts on the smaller object to compromise the form the smaller object holds as a unit and not to accept the form the Earth imposes on the object. By having mass the object no longer holds independence but retains some form of individuality by not compromising the unit it forms as an independent structure. When the object lands on a larger object and the larger object is part of the Earth the larger object halts further motion. The larger object removes the individual characteristics that the moving object has. Let us have a good look at mass.

When there is a ship we see the painted waterline of the ship indicating the load in mass the ship can take. The ship is lighter than the water because the ship floats on top of the water. By loading the ship the ship can take in a lot more mass than what the water is because as long as the hull displaces more water than what the mass of the water is in terms of the area the ship claims the ship will float. The ship is less dense than the water when taking the area it holds in relation to the density of the water it displaces.

When the ship has an equal density to that of the water the ship will float in the water while being buoyant. The ship has to accept a certain percentage of water to allow the ship to float inside the water and prevent the sinking of the ship to the bottom. This is an act of precise balancing. The ship then being somewhere inside the water has no mass but being equal to the mass that the water holds. The ship has to displace as much water as what the area it holds will be in mass when being only with water. Then the mass of the water is the same as the mass of the ship and therefore the ship can hold ground inside the water without sinking or floating. If the ship again wishes to float the ship will have to displace some water it has in its hull and exchange that water for air. Then having more air than water that will enable the ship to have less density per volume of space and less density will render the ship in having less specific density than the water. When the ship sinks the ship has more mass than the water it displaces. The ship has a bigger specific density than what the water has and the mass of the water pushes the ship to the bottom of the water. At such a point when hitting the bottom the ship then has mass. It is the mass of the ship at the bottom of the water being more than what the mass of the water is that floats over the sunken ship that puts the ship on the bottom of the water. In water or in liquid the ship or the solid either has buoyancy or it has mass. It either floats in buoyancy or it sinks whereby it receives mass when the sinking ends. When it floats it has motion by buoyancy that the water provides.

The ship while floating has less mass than the water has. The ship while submerged has the same mass that what the water has. The ship has more mass that what the water has when the ship sank. The mass is a relevant factor because of the fluid aspect. When floating the ship has little mass and all floating objects has mass that is less than what the water has. When being submerged it does not matter if the ship is a big ship or a small ship because the ship has the same qualities as the water and therefore a big ship will be as buoyant as a small ship will be. This is very important to note. When in a liquid there is no mass factor. Only when the liquid suppresses the object and onto the solid and with no distinctive difference between the motion that the object has and the motion that the solid has does the body turn from being a liquid to being part of the solid does the mass factor enter the equation. A big ship will float submerged next to a fish and from the mass aspect the two would be equal. It is depending on motion.

Up to now every Academic at all levels in science is normally acting as if gravity is a commonly explained factor, proven in detail without the tiniest whim of uncertainty, where every one knows every aspect about all principles that are involved in gravity down to the smallest detail. In truth no one in science anywhere remotely knows what brings gravity about and I used Kepler to unravel this mystery called gravity. But no one in science will admit this fact about Newton or any one else never being able to explain gravity in the least or admit that Kepler is the one who formulised gravity decades before Newton came and gave gravity the name. Newton did not underwrite or define gravity and even today the most informed in Science at best can only assert their suspicion on a rumour presumed about what causes gravity to perform as the part interlinking the cosmos but no one can go any further by explaining the concept. Newton started this realising of gravity but it had and still has no more substantial proof than a rumour has and Newton admitted to it being a concept he could not explain.

Still to this day nobody in science at present will denounce the principle of gravity in a fashion by acknowledging that gravity is as yet still never been explained. All in science act in a manner as if Newton's gravity idea is the best-proven fact there ever was and only occasionally admit it to be just a rumour as Newton admitted it was when he introduced the name (not the concept). If Newton's concept was accurate it will by now have the moon much closer to the Earth than what it was during the time of Kepler's investigation, yet we from modern test that came into place after man landed on the moon in the seventies we now know it is moving away instead of coming closer. Newton agreed that he could only declare gravity as only a vague concept more in a suggestion and far from the manner of forming the proof as one would demand from a rumour when he was announcing a force that could be anything. Not once could one person in the past or present provide substantiating proof on gravity as reality by defining the very principles.

That includes Newton as well as Einstein and even Hawking. Scientists can declare that gravity was a factor at 10^{-43} seconds after the Big Bang but what brought gravity about or why gravity became or still remained, as a presence is still tightly concealed information which all are speculating on. Even to the best informed amongst the most educated do not know what is gravity because they all ignored Kepler and for ignoring Kepler the price they pay is not finding the principles bringing about gravity. Using Kepler makes the method to follow and understand even Einstein's discoveries shockingly simple.

I started using Kepler when I was shocked by the lack of proof on the matter of gravity as it supposedly applies in Newton's manifesto. I found a lot of total nonsense thought to be truth. That urged me to investigate the matter by using other sources than the mismanaged and incoherent nonsense Newton brought about when he raped every aspect of Newton's findings.

What you are about to read has NEVER been revealed and who do we have to thank for that: w have to thank all the brilliant physicists in office and in teaching because they cover the flaws of their misinterpreted principles and cover the flaws of their misleading Newtonian principles.
If you students do not rethink the rubbish you are taught, then your professors are fooling you lawfully and you deserve to be their mindless monkeys just the way they think of you because you don't think about what they say! Cut the bullshit, force their ignorance about physics into the open and make monkeys of them, they deserve it even more! Let the professors that are so wise explain why the Titius Bode law is in place and is used by the cosmos instead of the Newton's mass idea that holds no legitimacy as far as cosmic evidence goes. It is what is used by the cosmos that has credence and not some surmising of Newton's fantasy-cosmic-principles. You and I should fight for the truth and not to uphold Newton's fabrication of the truth.

Go on and calculate the gravity there is between you body and the earth using **$F=GmM/r^2$**

Your weight is? Because mine is 116 kg?

The earth's mass is 6,000,000,000,000 ,000,000,000,000 (6E+24) kilograms
G is G is a constant that is $6.67259 \times 10^{-11} m^3/s^2$ kg

The radius between you feet and the earth is $1 \times x10^{-20}$ meters.
Do the calculation and see what force of gravity it is they say that your body sustains every second. Then decide who is trying to fool you, is it the Newtonian Professors or is it I.
Do the calculation and see if there is any one atom in your body that could withstand such a force that this calculation would bring about. According to this there is more gravity between your feet and the earth than there is available throughout the entire solar system. For more than three centuries these figures were about and not one in Mainstream physics ever questioned the veracity thereof? If that is not a conspiracy to hide Newton's ignorance about reality and mathematics, then what would conspire to a conspiracy?
It is so easy to silence me. Prove that I am incorrect In Anything that I Say!

It was about the night that Einstein realised that had he; Einstein fell from the window of the patent office Einstein would feel as if he was as weightless as a chair and a pen falling alongside Einstein down the building. If Einstein was experiencing weightless ness, it would be because he was weightless while falling. Einstein would not imagine the weightless ness because Einstein was truly falling. Gravity is motion of space and mass is the restricting of the motion of space. Having mass does not bring about gravity but it does restrict gravity's motion. Gravity produces mass but mass does not produce gravity. Mass is the restraining motion and gravity is material moving about. Mass only comes into the application when two objects filled with space moves into a position where both want to claim space the other occupy. I then after reading this realised that gravity is not mass orientated, but gravity is motion differentiation between objects. While falling, The object moves less or slower in the direction that the Earth rotates and will fall in the direction of the Earth centre until such a time as the movement of the object is in synchronising with the speed that the Earth spins or if not the object will and on the Earth surface at the edge of the Earth and that will bring about having mass. The gravity applies as speed that is putting time in relation to the distance travelled and distance travelled is space. While the object is in a process of falling, the motion confirms gravity, both by getting the object's distance or band in which the object travels in harmony with the Earth that conducts all the spinning taking place at that point. When motion downward ends and the Earth disallow any further movement to secure a better specific density in relation to rotating movement, then mass sets in and becomes what is than point holding mass where the constraining of the object takes place to secure frustration of further movement and the Earth's motion annexes the object's freedom. While experiencing mass the motion is still there but now incarcerated by mass and locked onto the Earth by the rotation of the Earth and the superior or equal specific density of the Earth.

By connecting to the Earth the motion that the object is experiencing is what nails to object to the Earth by the force of mass and the object is then experiencing mass and not falling further through the loss of downward movement and now only conducts with the Earth rotating side-on movement. However mass then restricts motion and becomes motion tendency. While falling, gravity applies as equal motion to all objects relying to place all objects in relation to specific density and because of this motion counteracts any size, mass or weight by making everything able to fall equal in specific density. When the Earth restrains further downward motion of the object that comes as the result of finding an allocated position of motion according to the specific density of the falling object, this readjusting of allocated position is stopped from conducting further downward or readjusting movement and all such further movement of gravity is hindering in the form we call mass. The falling object remains individual and still tends to move while Earth individuality resists movement. Further movement is disallowed as other material fill space.
By securing a [lace on the Earth, the falling object will finally rest and from that motion resistance comes mass. While falling, the object is experiencing gravity because the object is in gravity but when on the soil the object experience mass which is the restricting of gravity or motion of the space filled with material.

When any person is standing on any place anywhere, while viewing the Universe, that person is filling the centre of the Universe. All the light that comes across all of space runs directly in a straight line towards you filling the centre of the Universe. If you install a camera on Mars, the light is obliged to acknowledge your relocating the centre of the Universe at your will to reposition you're being that centre of the Universe. All the light that ever left its destination crossing the vast spaces of the Universe, excluding no particular light, travelled all the way just to find you filling the centre of the Universe, right where you are. By you're standing anywhere, you fill the centre of the Universe, and the entire Universe admits to that because all the light comes to meet you there. If you shift from the North Pole to the South Pole you will shift the centre of the Universe because all the light travelling throughout the Universe will find you where you then moved the centre of the Universe.

No photon will pass you by where you are in the centre of the Universe. The Universe is spinning around you or I, which is filling a centre where all motion is connected. Then I reviewed the Universe. If gravity is motion, what causes motion? What stops motion? If a star is about fusing atoms thereby growing, what happen when all the atoms fused into one all collective atom? What is the gravity if the star has one all-inclusive atom providing all the gravity that the star had when the star still had massive volumetric space?

If all that space that once filled an entire giant star fused into one enormous gravity applying atom and that enormous force has been secures in the space that one atom holds, the atom would then show a force that would pull the surrounding Universe flat. Gravity is smallest where space is least. Coming to the conclusion about gravity being motion and mass being the restriction of motion was the easy part.

Gravity is The Roche limit,

    Gravity is The Lagrangian system

        Gravity is The Titius Bode law

            Gravity is The Coanda affect

And gravity as the Roche limit forms the principle in producing the sound barrier. Read the book and find out why this is the case.

Newton's claims about the principles that he declared is responsible for guiding physics carries no validated proof and only after I realised that, was I able to start forming another line of thought on gravity. At first I was not confrontational towards Academics in physics and avoided any indication about disagreeing with Newton, although avoiding to show my disagreements was also totally impossible too but every time I approached academics with my new concept the academics always threw Newton at me. Facing Newton or facing defeat became a two-sided blade and I had to start to confront them by confronting Newton, with which I was in disagreement from the beginning. At first I was reluctant to voice any opinion about the matter of how far I was prepared to challenge Newton because Newton was and is an icon. But slowly it dawned on me that if I had any serious plans to introduce my ideas I had to dispute Newton's gravity principles and do it head. When the slight confrontation did not bring results I finally decided to go all the way and show the inconsistencies that were prevailing in Newtonian science. The lack of honesty and furthermore the absolute dishonest on their part is there whether I avoid it or attack it; the inconsistencies are part of forming the basis for modern accepted science.

First I had to show the general public the true colours of the academics in physics and get every one to see how incorrect Newton is, and only then do I stand any chance to introduce my line of thought. I am so sure of the ideas that I propose of being correct that I dare any one to disprove any part or the entirety that my concepts about cosmology forms! I can't find one academic with influence that is brave enough to stand up and face my attack on Newton and argue me down or prove me wrong in a sound debate. This not one academic could achieve and I challenge the lot to do so. Read my challenge about the correctness of Newton's proposals when he brought no more than suggestions into science and when I dispute Newton, then take me on by proving Newton correct... do it just once... prove Newton correct just once...prove that his formula is working and that his principles apply on the grounds he principled his ideas. Detecting Newton's misconduct is possible because I saw a way to break away from the invalid concepts Mainstream physics hold. I went about and tried to prove Newton and when that was not happening I tried to apply Newton's ideas into the greater fields of cosmology. I tried to amalgamate the four cosmic principles applying in cosmology with what Newton said was happening in the cosmos with mass and with gravity and in light of what the cosmos showed was happening Newton just wasn't happening! Notwithstanding the pose Mainstream physics try to uphold, the entirety of physics still use the idea of magical forces intervening in nature and they still base concepts on unexplained novelties.

To say the least, the concepts physics use in terms of Newton would not even be acceptable to children in the modern informed era we live in, I challenge any person to prove Newton, not to accept Newton but to undoubtedly prove Newton correct! Prove how Newton's formula of mass forming the force of gravity can apply as Newton said it does! I recognised the impossible double standards Mainstream physics apply to promote their much shady explaining. In short I tested Newton's principles and found the principles to be wanting. Prove the attraction Newton said was enforcing gravity that is pulling by mass and is gathering plants by contracting the diameter between planets. Show how much the Moon came closer to the Earth since the time of Kepler. Show proven distances taken by radar tracking and indicate just how accurate Newton was.

Show how much the Moon came closer to the Earth since the time of the Moonwalk in sixty-nine. Even the least degree of verification of correctness is absent when trying to find support of Newton and Newton lacks all evidence of authentication in any investigation of even the simplest terms. It is as if they never read with interest that which they explain when they embark on explaining Newton and they never scrutinise that which they advocate when they teach Newton's principles applying. To most if not to all of the persons reading this, such a venture of investigating Newton is time wasted and just the thought about me embarking on the investigation of the issue is totally senseless to investigate. It is senseless because the concept it carries became accepted as household practise and life science from where it proceeded to become everyday culture in every person's mind. The worst part is that the group of people normally considered as the wisest bunch there is, never did prudent testing on Newtonian presumptions, while to test the presumptions is most easy to do. I will not believe that a lot that lives up to the veneer of being the best mathematical intellectuals on Earth, never though of testing Newton's very simple formula and in that disregard the formula because of the incorrectness the formula holds.

If you think those in charge of astrophysics are the pillars of trust, then get wise by reading the following facts and arguments this book presents. This attitude they have is the result of a relationship that worked for so long and thee fact hat it worked that long is what confirmed their opinion that we, the public, are fools to believe anything and everything because of blind stupidity. This trust we have is brought on by a culture of trusting the King to do the people well and somewhere in every person's cultural past there was kings that did us well in leadership. But their underestimating of our abilities is the testimony of their poor understanding and their weak insight ability, which results from their arrogance and stupidity. This betraying on their part and misusing the public's good nature to be used in schemes to get the public conned must end and I pray that this book form the first step in resisting the arrogance of the Academic Physicist.

Any one not in their group of the Academic Physicist is part of the lowest order of mindless being and to become part of their order and those that have minds with an ability to think, students have to accept what they say when they say whatever they wish to say without having to prove the correctness of what should back their saying so and as a result of this students may never question what they say. The sifting process they named examinations. Then you better read on and I will remove your blindfold and show you what a world of deception the Academic Physicist force on us into.

If you are a student in the science of physics, then ask your Educated Masters to please explain the following abnormalities you are about to read in this book and insist on a clear explanation about the inconsistencies they promote while tutoring physics as if the physics they present are the most flawless and accurate institution there has ever been. Ask those academics supporting Newton about the following flaws that no one mentions …ever… except me in this book you are about to read and get them to explain the inconsistencies never talked about, which I present in this book and then after confronting those charged with tutoring physics and seeing who should be believed, then get wise instead of brainwashed. Let them mathematically show how one would go about and use Newton's visionary formula $F = G \dfrac{M_1 M_2}{r^2}$ to calculate the force of gravity by replacing the symbols with the actual values in mass that the items referred to have. Put in the Earth's mass in place where it belongs and put in your mass in place where it should be and then divide that with the distance between your soles and the Earth measured in micro millimetres by the square thereof!

In the book named an **_Open Letter on Gravity Part 1 and Part 2,_** I bring the solution to the mystery behind gravity. I tried in vane to introduce the principles I find valid to the academics in charge of astrophysics. Facts that Science present as being the uttermost explicit and unwavering truth, fails to bring any logic answers to so many questions that it should address. It fails to have substance in addressing the most basic and simple questions about gravity and physics. Yet to every question science can't answer my approach does bring many solutions. The presentation and the delivery of my answers that I reach are understandable and simple where it serves both logical science and the truth. Since my answers do not match Newton and his misconception about gravity and that mass generates gravity, those in charge of science don't even bother to read my work. With their affixation to the corruption they portrait I can do little to the giants where they are in the mighty positions they have and just because of that they can go about to sideline and ignore my work and this is notwithstanding the correctness that my work delivers compared to the utter failing that Newton's work shows.

When confronted with my evidence and they have to match my work with the hypocrisy and misleading nature of Newtonian cosmology their defence in substantiating their claims is to ignore me. Since I do not applaud mainstream science and the clear fraud they embrace and fraud it is that they embrace, I am silenced. Those academics in charge would much rather protect their un defendable ethos they maintain as forming the back bone in science and what gives their personal position legality although it is corrupt than admit to the truth they find when they begin reading my work and in agreement they then have to back the truth my work brings.

Doing that (accepting the truth in my work) will trash all work in cosmology delivered thus far and condemn it to the waste paper basket and render all work invalid and void. Considering that such acting will lose them money, those academics in controlling positions then will rather rape the truth in order to benefit from continuing to corrupt student's minds further. If they wish to justify their inconstancies they have to attack my work and disprove the accuracy of my work. I challenge them to prove Newton correct and not just declare Newton being beyond reproach after all has seen the evidence I bring. After reading this all students must challenge them to defend what they can't or get honest. When the distance is large, the influence of mass will be small and when the distance is small, the influence of mass will be overwhelming. Why then, when considering that if it is mass that produces an inclining force of contraction as Newton says there is going on then…why didn't the expanding stop before it started when the Universe was small. I dare any physicist to show me where they apply Newton's formula just and exactly as Sir Isaac Newton suggested gravity applies. Show me just once where the mass of the Earth is multiplied with the mss of the object in normal physics. Show me just once how $F = \dfrac{r^2}{M_1 M_2}$ or $F \alpha \dfrac{M_1 M_2}{r^2}$ where one M represents the mass of the Earth while the other M represents the mass of the object and in this formula the end result will have a value of 9.81 Nm/s^2 … show just once one example… where the use of the mass of the Earth comes into play. If multiplying the mass of the Earth with the mass of an object and dividing that with the distance parting the two mass factors does not deliver 9.81 Nm/s2, and then any claim by Newton indicating that $F \alpha \dfrac{M_1 M_2}{r^2}$ is equal to gravity, such claiming constitutes to deliberate fraud…even if Sir Isaac Newton said this. Prove that the mass of the Earth with the mass of an object and dividing that with the distance parting the two mass factors delivers 9.81 Nm/s^2 or admit physics is conducting fraud to protect Newton!

I am able to explain the how the Universe started only because I discovered the building blocks used to build the Universe. The building blocks are the Titius Bode Law, The Roche limit, The Lagrangian points and these all culminate into the Coanda effect. Using these I can and I do prove exactly how the cosmos was built dot by dot.

Gravity is TIME forming SPACE = Π^2 = 9.8696 = MATTER HOLDING THE COSMIC LIQUID COUPLING THAT TO THE NEUTRON TO COMPLETE THE COSMIC GAS to the liquid part of the association between solids and liquids. The earth is the solid while the atmosphere is liquid and outer space is gas.

Every person associates gravity applying as "The natural force of attraction exerted by a celestial body, such as Earth, upon objects at or near its surface, tending to draw them toward the centre of the body" with what we think of happens to solar bodies having mass which is "the natural force of attraction between any two massive bodies, which is directly proportional to the product of their masses and inversely proportional to the square of the distance between them" and science cheats everyone into believing the two is the very same. The one I experience every day and with the other there is no evidence of applying anywhere. Yet students are fooled into believing the two are exactly the same issue, which is untrue. If there is any academic feeling insulted by me calling the lot fraudsters, bring evidence of the second form of gravity working on the principle of mass and I will withdraw my statement, otherwise if no proof can be brought, then you lot that ignored me for ten years are the fraudsters I accuse you to be. You are villains brainwashing students to corrupt their thinking.

Einstein's Critical Density lacks the accepted matching facts we need in proving the critical mass factor, which makes the entire idea silly and bogus. You can't force the Universe to conform into something Newton said because Newton said the Universe works on mass contracting… But our inability in securing such required evidence defies the most basic logic. It seems all new evidence we receive from outer space is disputing all Newton laws and new findings disprove Einstein's Critical Density as the answer. The Universe will not reach a point of contracting, not withstanding whatever dark matter astronomers try to locate in the vast space. A rush is on finding the black-matter that will be applied to force the cosmos too retract back to where it came from. But what if our view of the cosmos was as incorrect as our views at present is about the sun? I prove that contraction is at present as much part of the cosmos as is the expanding is that we focus our attention on and it is our culture we carry from generation to the next generation that leaves the human view obscured in admitting the truth. The Sun is not a coal stove burning fuel. The Sun is an air-conditioned pumping gas (hydrogen and helium with electricity. I prove by applying the Titius Bode law that gravity is electricity and the two are the same thing. By using gravity or electricity the sun is a huge air-conditioned pump freezing cosmic gas, which is what the Universe is into a liquid that we see squirting from the sun.

Newtonian science has NOT developed from the idea that the sun and the rest of the Universe are orbiting the earth. Everything applying on earth and befitting human standards they think must be transmitted directly to the Universe. If I stand in the sun and feel hot, then "the earth is hot" and we have "global warming" just because we humans think it is hot and we humans feel slightly bothered. Every point in the Universe applying singularity has a different measure and the entire Universe is NOT made to cope with life but is completely alien and destructive to life. Newton started to tell the Universe what it must be and the Newtonians never stopped. Newtonians better stop with the Ptolemaic concept of putting the earth or life in the centre of the Universe and start to place singularity in the centre of the Universe where the centre of the Universe is..

Why would the expansion turnaround and do a reverse by going back to where it came from. Consider the momentum alternation such a change will bring about. The sun is not a gas-filled sphere holding hydrogen in its "natural gas" form, but it is all fluid and is in a liquid form where singularity is liquid-freezing hydrogen at 6500^0 C while outer space is boiling over at $- 276^0$ C. **The Absolute Relevancy of Singularity** book explains the Roche limit as well as the other four cosmic principles in the practical sense… when applying cosmic laws instead of improvising cosmic laws uncovers that reality then becomes awesome. It becomes clear the Universe is as much expanding as it is contracting and contracting by expanding. As there is no hot or cold, no big or small, no grand opposing but relevancies in ratio to one another. If you do not believe me, then believe your eyes when looking at the picture. What ever the sun is it is fluid falling into fluid.

Because hydrogen is a gas on earth and we think of 6500^0 as hot on earth, therefore the sun must be hot and the sun must be gas at the same time. It is obvious what we see is liquid squirting and when hydrogen is in a liquid form it then must be cold because my eyes tell me that! Newtonians still have the entire Universe apply the standards befitting life on earth and that is why they wish to locate life.

When something is hot it expands. When something is cold it contracts. Outer space expands to the very limit and keeps expanding therefore outer space must be the hottest there can be, notwithstanding scientific Newtonian stupidity. The sun contracts every bit of space that forms the solar system and therefore the sun must be the coldest place in the solar system notwithstanding whatever Newtonian ego whish to declare. If I feel the sun is hot it is because it is diverting all the heat my way and it diverts the heat to me then where there is no heat it must be cold. It may sound incorrect and unscientific madness but with my applying of Kepler's formula in alignment with the position I located and valuated singularity it clarifies the possibility of the above statement… but please do not take my word for it, use your eyes and make sure you look past the culture bias of past incorrectness. See the fluid push out of a bowl of liquid, spilling both sides as it falls into liquid. The Hydrogen inside of the sun is not gas but it is fluid. In all of nature in all elements found through out science there is no NATURAL GAS as much as there is no NATURAL SOLID. Hydrogen is as much a liquid as iron is a gas and neon is a solid. It depends on the element relating to the space/heat in the circumstances surrounding the substance at that very precise instant in time. We have to stop telling the cosmos to show us what we wish to find and start accepting what the cosmos is telling us is out there that we should look for and find. In creation there are two substances that formed the Universe. One was earth or solids and the other was heaven or uncontrolled heat. Between all solid we have heat parting the solids and being a solid or a liquid or a gas depends on the ratio between the solids and non-solids. Under conditions suiting life certain elements may be a gas, but in stars conditions don't suit life and in outer space conditions don't suit life so therefore life cannot be a barometer for conditions applying in the Universe. Kepler gave us **solids as $a^3=T^2k$ as liquid or gas.**

The earth, just as all other cosmic objects do, contracts outer space by the movement of the rotation. Material spins in gas to reduce the volumetric size and by that it reduces the concentrated space around the star / planet / earth which then reduces the cosmic gas forming outer space to the cosmic liquid forming the atmosphere. By rotation the earth "pumps" gas from outer space to the core within the centre of the earth by applying centrifugal pump action. As the space becomes denser the heat level rises but this is because the rotation of the earth reduces the heat, not the heat that we as humans feel or experience but the heat level within the space. We have to maintain science laws. When any cosmic substance overheats it becomes larger and when cosmic substance becomes colder it shrinks. That is cosmic law. The heat we feel on our skins is heat escaping from where it is cold and contracted to where it is hot and expanded because the levels that reduces has to dispel the heat from the location it is within and move it to where heat is excessive and expanded. It is the heat moving away that we feel and then think of it as hot. As the earth turns it cools the space in which it is and the space reduces in heat and therefore the heat levels can reduce in order to make the space move towards the centre. As a matter of fact the very first experiment in science ever conducted and recorded was the discovery that it is space that reduces and not things (water) that falls but notwithstanding Newtonians have gravity as mass falling. But as Professor Friedrich W. Hehl, Inst. Theor. Physics With a lot of words and some simple algebraic relations, there is no way to "explain" the world of physics. Your seem to be out of touch with modern developments. I guess that is why Newtonian science can't conclude they are mistaken even after three-hundred years.

Again and as for e so many time in the past I repeat the warning once more of the Newtonian Brilliant-Brainy-Bunch to please take note of a conscientious warning about the gravity of the misgiving there is on the part of the most respected Academics in physics about a much concerning matter. As you can see why I state it emphatically that science accuses me to be not schooled to the point where I am able to have any form of an opinion on any matter concerning Sir Isaac Newton. Notwithstanding that my research proves I did my private studies and through which I skipped the indoctrination and mind control academics place on students goes unrecognised by their standards and so too my ability to have any insight on matters regarding physics. However my skipping their methodical and systematic brainwashing enabled me to see and allowed me to be able to express the incorrectness in Newton's teachings and allowed me to show in clarity what destructive force Sir Isaac Newton used to corrupt the laws of mathematics, corrupting to science along the way and mostly raping to the work of a great man, Johannes Kepler and what Sir Isaac Newton did can only be expressed as being blatant criminal fraud. What his deeds amount to is to corrupt the laws of mathematics, to render the laws of cosmology useless and to rubbish all of science. Should you find this to be unbelievable, then I am glad to announce that this book is more for you than any other person, so go on and read what academics guarding science never wanted published. I challenge any one that disputes any claim I make to prove me wrong by proving me wrong and not merely suggesting claims in that direction.

We have to realise that it is about heat and cold where moving cools down and stationary space expands. Creation started with differentiating between what is cold and what is hot. That is exactly what the Bible says in Geneses 1:1. The Bible says God made heaven and earth. This means God made what is solid and what is not solid. It means God made material (the earth) and heavens (the sky or non materials). Back then as is the case now with our most brilliant Newtonians there was no other names for solid as earth and non-solid sky as heavens. At least those back then realised the difference and did not put sky down to "nothing" as our incompetent Newtonians of our modern era does. There is singularity controlled by movement (material) and there is singularity not controlled by movement (non-material).

In terms of mathematical equating the very first recorded instant happened at:

$(10/7)(\Pi^2/2)(\Pi^2 + \Pi^2) = 139.15$

(10/7) There was a cosmic gas that formed or it is the "stuff" that "stuffs" outer space.
($\Pi^2/2$) In the outer space a liquid formed that is what fills the liquid atmosphere of a star such as the sun. It is that pure liquid heat that squirts from the sun that Newtonians see as gas while they stare at a liquid.
($\Pi^2 + \Pi^2$) The solid formed that by spin controls space occupied to cool it from space unoccupied.

$7(\Pi^2 + \Pi^2) = 138.17$

7 The redirection of movement of time brought about the spinning effect of gravity.
$(\Pi^2+\Pi^2)$ The solid formed that by spin controls space occupied to cool it from space unoccupied.

$(7/10)(\Pi^2)(\Pi^2+\Pi^2) = 136.37$

(7/10) Gravitational movement started moving space by containing from the expanded to the contracted. It is still the ratio of gravity forming a cosmic liquid from condensing cosmic gas.
(Π^2) Gravitational movement started by duplicating Π in all seven possible directions minus 3.
$(\Pi^2+\Pi^2)$ The solid formed that by spin controls space occupied to cool it from space unoccupied.

This is exactly what the Bible says, if only the Theologians and the Newtonians were not so preoccupied in fighting for their own small-minded egos instead of looking for the truth.

When the Universe started there was one spot that released a dot. The spot as large as a thought grew into a dot. This too is exactly as the Bible says it happened and it still happens exactly like this. The Universe starts from infinity and grows into eternity just as it happened at the start and this is what Newtonians confuse with what they know as the Hubble constant. To check if I am correct look at anything spinning and you will see conformation. The dot that released was by such release relevant to other dots released because there was to be motion that measured many dots. The dots in release were relevant since they were the same. Only time being in delay by cycle of infinity interrupting eternity to form space, as space is the history of time gone to the past. But since we are looking at things as it started I put it into the past tense. That cycle brought about time delay as every cycle drifted further from the original singularity while the original was still responding as well. That was the first space. It was time being one infinity part. Since the dot was also singularity and was the very same as singularity with a small difference that the spot was •Π^0 and the dot was •Π^1. At first, at moment-Alfa, there was no space nor time for only relevancies came about. Relevancies acted as motion to bring change to eternity by changing the flow of time in eternity. There was the perfect spot in which time moved while remaining the same. Then heat brought expansion and expansion brought space and space brought movement and movement brought about a Universe. That is how it started and that is how it still is. This was before the atom and the atom was before light and that too is exactly what the Bible says in Geneses 1:1.

Time will forever remain eternity but space or time distortion, which will forever remain infinity, interrupted the flow of eternity. Space breaks the monotony of time in eternity by parting time from in infinity. Yet, the relevancies did imply motion except for the fact that singularity is very much incapable of motion. Every dot had a purpose to fill a position in relation to the other dots that the spot excited. All dots had a line of three where two was one, each on every side of the spot. The spot was one with the two dots forming two, which improvised for motion that would later come to space and the three was what space was going to become. How do we know this: 0.1416 x 7 = .991 and that is by the value that the Universe grows.

Time was four because the four would bring about motion as heat separated infinity from eternity or hot from cold. Five was space because space was one removed from time. Space is the distortion of time and one outside time would bring a time delay or a time distortion of four plus one which is five, hence the principle behind the Lagrangian system. Because material was the square of space material was a crossing of three plus three forming six. However to find out what this means you have to read the entire theses called <u>The Absolute relevancy of Singularity</u>.

Space-time is the four of time, plus the three in singularity around which the four of time turns, therefore space-time is seven.

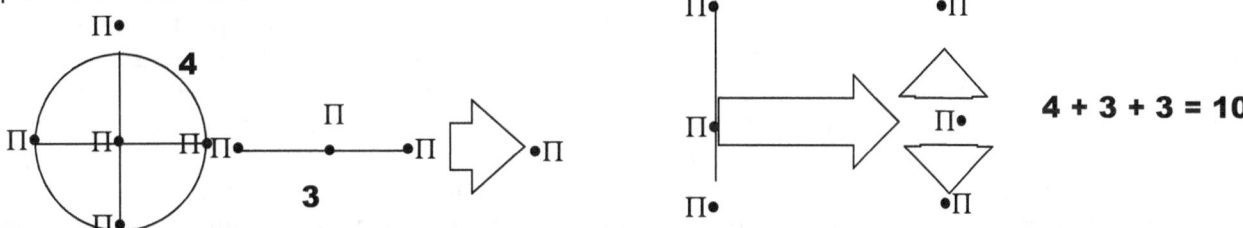

In the circle using $r^2\Pi$ the r has to have distinctive qualities placing it as a factor apart from Π. Where the growth shows no separate distinction but a continuous flow from the precise centre to the precise edge the flow would become in relation with Π depicting the circle and Π replacing r as reference to any point on the circle.

By using r, distinction in the circle is possible but by using, Π there is no distinction possible. Therefore, in the beginning when time formed space there was only relevance coming about from •$Π^0$ and the dot was •$Π^1$ with no mention of any possible r. The fact of r representing a radius represents space and what we refer to be long before the Big Bang introduced space or mathematics using space.

Before the Big bang the lot was form without dimensions playing any part. Then the atom came. Only after that did the Big Bang come. Even before the atom was the point lining up and forming positions that was spinning faster than the speed of light can ever achieve. However every point today still serve the role it took on at that stage and serves in the position that it had during the time it had no space with eternal time. These relevancies developed as part of a Universe we shall never understand. The Universe had no sides and a line was equal to a triangle, which was equal to a half circle. Singularity holds the double space-time position of five times two (matter and space duplicating singularity) which then is ten.

How did the Universe liberate material and heat forming space from singularity because with singularity comes an unchangeable eternal condition that is non-changing-everlasting in all conditions and aspects that is remaining in absolute equilibrium. This equilibrium maintains because all development extends form precisely in a detailed equal equilibrium throughout. Think about what brought the cosmos out from the eternal rest in which it was. The eternal rest still maintains and is therefore our detection. What inspired the eternal rest the cosmos was in and inspired change to the state of eternal rest? What evoked change? That is the question the Atheist will never be able to answer but that too is the most basic and ever-lasting fundamentals of the Universe. Singularity $Π^0$ is not substance but it is a thought establishing substance Π. What changed in this split second start before the official start? I do not wish to ponder on this matter in the letter I am writing at this minute, as there are other books where I delve into this matter. It is called <u>The Absolute relevancy of Singularity</u>

From the deep freeze of creation came the Hot Big Bang and the 3D Universal displacement came about with the relatives being **10 ÷7(4($π^2$ +$π^2$)) = 112.795** and then a second one established 3D by introducing the six to seven sides Universe at a density point of **7 / 10 $π^6$ / 6 = 112.162**. There is of course a lot more information about this establishing of the Universe than what I mention at this point. The question is what made the Universe freeze, to form the Universe in space and through time. It had to start with a specific reason applying, which brought about space-time. Once the process started there was no stop to it, but there is no chance that the initiation of the start was spontaneous by nature. With everything being in one spot, all within that spot was in a state of eternal rest. While all remains the same and nothing changing, what brought on the sudden change of everything, shocking

It is the time the space takes to bring about new positions in space occupied through motion applied and through motion applied, it takes space-time to duplicate and in the relevancy of duplicating, the duplicating takes certain duration in time to move from point to point. Gravity is motion of space towards and in relation with a centre and the time is the period such motion takes while that centre is attempting to produce space by doubling space through motion leading space away from a specific controlling centre to another specific controlling centre. The time it takes to complete such an attempt provides space the opportunity to double its status. Moreover, the time stands affected by this motion material creates to duplicate space-time by generating singularity and activating different locations holding singularity. Gravity is speed and speed in space in motion through time duration. Gravity is motion combating heat expansion and supplying space with space producing through motion providing the space the opportunity to expand while remaining in relevance.

At the start, the gravity part invested heavy in material. However, in space light came about since there is no gravity in space. Light is the attempt to establish motion not controlled by any centre and the time it takes to establish space between such a centre of space control and the light finding an ability to dispense the space by reactive motion. Gravity produced space by allowing as much as producing overheating with a feeble attempt to combat the overheating. What is the meaning of heat if there is no cold to set the standard for the heat to become a value, which is then related to the other end, where such another end must be the limit in cold? The moment singularity produced space-time heat distanced from the cold factor. Space produced a cold base to have heat within. How did singularity part the shared principle of being the unification of heat and cold as a unit? Singularity froze in applying gravity bringing about particle separation within singularity. Heat and cold parted to produce frozen heat by atoms forming and captured heat by unleashing uncontrolled material that ended as space in time.

We can see from what is available how everything fell in place. There was space filling with heat where the heat was compacted by the time delay. Every point established a relevance of three crossing singularity a motionless forming a line and four points formed the square of time. Crossing the four points that supply the time or the motion aspect is a line with three points supporting singularity by not spinning but they maintain singularity in being as motionless as singularity is. That four plus the three is seven. Then all this, which is the body of material requires somewhere to be within because to be is to move within something. That too is time but we regard that time to be space. That time is the heat that became space during the event of the Big Bang. That gives the atoms seven point as plus the electron or space having three points in relation to the seven points there is in time. If singularity is at the limit the value of ten but is the tenth portion of the value then ten must be divided by the tenth portion and from that the tenth portion must be distracted.

This as a group forms a relation with heat unlike any other. If mass did the trick these must have been the group having the second least density, but they form the group with the least density as a five point group. The next group holding the Pythagoras five or the Lagrangian five plus two or three, four or five enabling their relation with heat to be quite remarkable. This must be some indication of events during the period just preceding the Big Bang at say 10^{-7}, 10^{-6}, 10^{-5}, the time when the fuel that would ignite the Big Bang turning heat into space turned material into heat. The relation of five plus one, plus one and five, plus one plus one plus one is just too uncanny to ignore.

Following the process and seeing the influence of singularity should bring about a pattern that may lead one to a pattern of how the required heat formed and how the intended heat transformed to space. Density depends more on proton number arrangement producing specific form in relevancy as to merely and only having mass as factor that contributes to the forming and development of stars in the cosmos. The evidence is so clear that mass has nothing to do with gravity but density has everything to do with gravity. Density is the volume of space in numbers used to fill material in ratio with numbers of space per volume not filled with space. It is matter versus space in every sense there are. This came about before the Big Bang took place and before space was formerly space and time was formally motion. It was a time when singularity set relevancies moving from Π^O to Π

In that manner we know that that was the way particles formed combinations just after the arriving of moment-Alfa. Singularity brought the Universe but also singularity brought the divisions between the many Universes that followed the immeasurable many Universes that came after the flooding of Universes to follow the leaders. The term "moment-Alfa" is the way I refer to the moment when singularity changed, not when space formed or time began or space exploded but even before anything including mathematics became definitive. At this point mathematics renders it useless. There was no space or time to calculate because relevancies came in place. Form took shape but space there still was not because Π^O moved to Π. Every slightest point in space became an opportunity of establishing a Universe with most different functions and ingredients there might form. This is apparent from the fact that it still takes place at the present moment by motion attaching new singularity through duplication and through duplication releases previously attached singularity from serving the purpose of duplicating by motion.

This was the era of distinction, when separation brought an all-possible new Universe

The spot becoming the Dot

The Spot

When the cosmos came to motion, motion was not yet defined. When the cosmos brought about motion, the first motion was relevancies. Cold parted from hot. Eternity parted from infinity. Motion parted from motion absence. Infinity broke the laboriousness of eternity for the duration of infinity. The spot became and grew into the dot.

From what the spot was to what the dot now is might be just a mathematical implication of going from 1^0 to 1^1 but in reality that first motion was the creating of and establishing of an entire Universe with all possibilities now in it. Never again can that much growth become a reality, although to us the growth is beyond what we ever can notice. But it is because the growth is so massive and we are so small that we are unable to notice such almighty growth.

When the spot Π^0 became functional and established all relevancies possible, heat parted from cold as eternity parted from infinity. The expansion was not clear motion but more a parting of relevancies where a centre formed a relevancy because the centre could not provide motion. Without being capable of motion, the centre established four points, which also served singularity. From the inverse square law we know that the centre doubled by producing the four points holding singularity.

By exciting the centre spot, the centre spot came to be because of the heat that formed in relevancy as heat parted from the cold bringing about the division that followed and that was the motion that formed. Therefore the heat had to move but being singularity it could not get singularity to move. In an attempt to establish growth, singularity activated six spots of which four was having motion drawn into relevance four spots that was providing what was to be motion and three that was to be securing the position the centre holds. There were four forming a ring around singularity with two forming in locations we will refer to as above and as below or north and south.

The three in line was in singularity not being able to move but the four was also in singularity and just as incapable of moving. All the points came as relevancies applying the forming of more of what was to come but only the four committed to time were expected to move. The four points that came as a result of discrepancies that became time that produced form and that established the relation with the one but had to perform the motion by expanding was as much incapable of motion as the centre was that charged the four with motion in the first place. As they were incapable of motion, it still required a tendency to apply motion that did separate Π^0 from Π. This not only involved form but it involved all relevancies that did come or may in the future come about as a result of the attempt to commit motion. If mass was a factor contributing to gravity the cosmos would have frozen back to singularity without ever releasing singularity to relevancy.

Mass does not establish gravity. There is no magical graviton. In the beginning there was no mass but boy was there gravity! The only means that the cosmos could find a way to break from the grip of eternal eternity was to expand into relevancies. Such a feat can only go to task by forming opposing hot and cold. Becoming hot produces more of what is heating. That implies motion or a moving away from where it was by generating more of what is available. Only where hot released from cold could whatever was repeated once again and duplicate what was before into what then is more. Secured by motion T^2 in relation to a specific centre **k** from where singularity holds the Universe true to form. The **k** was an intention to place apart and by today's standards will not even qualify any noticing.

All that are is in singularity. From singularity comes the motion and the space we call space-time. Singularity is dimensionless, time less and space less and because of all this features, it carries the value of Π^0. By expanding, singularity applies a relation coming about that reforms singularity from Π^0 to Π. Only when extending Π^0 to Π, the extending creates motion and the motion creates space that then doubles through motion applying which cuts the space in motion in half by matching the space as a duplicate. Motion creates another dimension or another level reforming singularity from Π^0 to Π or from Π to Π^2 or from Π^2 to Π^3

As said before we now know Π came about since Π is achieving form and not space. Only r can establish space as size will accumulate and as it had with everything else singularity had **r** covered by one as in being $r^0 = 1$. By reducing the circle radius **r** by half continuously will lead to an infinite small circle and an infinite number holding r would place **r** to the power of one as a factor. Then as a factor **r** would not contest any change when change is introduced into any future equation but Π will remain because the circle as a form remains even being infinitely small. By reducing r indefinitely to the tune of half each time, r would become infinitely small, beyond human calculating means, however as mentioned in the case of the smallest dot holding one spot, r would become insignificant beyond human comprehension even, but never reaching zero and still Π would remain intact and dictating form. To amplify by dimension a value has to be set to r but if r remained covered by singularity all alterations that could possibly come about was in the form, which was Π.

This expanding can be a problem one can wrestle with for one lifetime and never reach any conclusion. How can something grow without getting more that what was before? Then it hit me like a ton of bricks. The answer is in heat but not heat, as we know heat. It is heat in getting relevancies between outer limits. Only heat could break the monotony of singularity. Heat in the form we now know heat as heat is now. Since the Big Bang heat is material transforming from one state to another state.

The change that took place involved singularity but singularity was 1^0 and being ^{01} could not grow. The growth came about. Heat rose from singularity, but if heat rose from singularity. Singularity as a factor changed from 1^0 to 1^1, which means a relevancy came in place that no one could detect. It is true that 1^1 are still one, but one could then escape from singularity by producing factors other than 1. Heat came about but only as a relevancy to utter cold. If there is heat, there is cold or if there is no heat there can be no cold. Space came into forming a relevancy that brought form. Since it is a relevancy and not a generation by accumulation, the form produced was Π.

The spot formed a dot by heat and cold establishing relevancies and from that singularity was broken to allow all other forms of relevancies to come about. The cosmos did not start because of gravity. The cosmos started with heat and cold coming into a relevancy and in the cosmos there is no hot as much as there is no cold. The cosmos broke, put from the confinement of singularity by establishing a singularity in a relation of heat and cold. The heat that came about was beyond measure because the cold that held the heat was also beyond measure. The immeasurable heat was on the outside of the dot that formed and the cold was on the inside of the dot that formed.

The cold contracted because in nature cold contracts. The heat expanded into a dimension of form and heat by expansion is in nature about motion. Motion is duplicating that which is and heat is what is duplicating by motion. But only heat by expansion was possible because in affect singularity cannot move. The motion became contraction, as the motion was the result of heat expanding which was forming four points in the rim of the dot. The expanding of the points created motion in relevance of a centre that formed because of the motion, which established an immovable centre as the Coanda effect, placed more dots in relation to more dots that formed.

Every dot was Π and every dot formed Π^3 because of the expanding heat, which produced Π^2. With that a new relevancy came about forming a centre in between the four points of expansion that was resulting in time. But since the points were in themselves singularity, which is immovable and space-less, they still heated forming a cold centre with the heat bringing about motion. It became a repetition where infinity broke eternity by producing a centre because of space (or rather form) forming the motion to enable the space to form in relation to the heat applying motion. This brought about a Cosmos being conceived.

The spot forms a full circle, but the line running through the circle is forever present because that is the future radius of the circle that will one day develop the circle, which is equal to the present diameter. The fact of the presence of such a possible line in such a possible circle dividing the possible circle into two parts makes the centre line equal to the half circle. The line forms the half circle but not only that the line presents the half circle as much as the line is the half circle. The line then is 180^0 and the half circle is 180^0 because in singularity the two factors are the same.

The same value is of course $\Pi^0 = 1$. The issue of concern is o understand that singularity cannot move. Singularity has no space. Singularity is no only part of the Universe but singularity is the Universe. By establishing motion singularity has to be charged with the time delay we find space to be. The space is time taking a period or a duration while moving from one singularity point to another singularity point while conducting the heat and the accumulation of heat that built up due to the retarding of the time to conduct the heat forms the space that is conductor to bring about the motion of the space.

It takes heat time to entice singularity and singularity can only entice. Singularity cannot move and neither can singularity form space. By enticing from one relevancy to another there is a bridging of heat that has to be crossed in order to send the gravity or the enticing or the relevancy to depart the space and reconnect the space to the next singularity. Bridging all the accumulated various time delays that formed an accumulation of heat through time distorting brings us the space we see and have. However there is no true space or motion but it is eternal motionless space is singularity charging time to provoke heat into forming space.

Three points formed a line covering singularity where the centre singularity recovered heat to grow and two points served as an axis to allow the rotation and to assist the duplication. There is one centre connecting the duplication of three as well as the recovery of one (the fourth one) that is applying the tie aspect. Therefore, motion consists of three positions in relation to a centre, which forms as space in relevancy to the motion and the space receive a controlling centre.

The duplication comes about as singularity is exciting another singularity in precise relevancy of 3 to 3 to 1, but the points charged is as space less and as motionless as only singularity is. The heat it requires to carry the exciting between points forming space and the space excites heat and the time delay it takes to excite singularity between points forms space-time.

That is why the Universe is Π

Where motion conducts electrical charging which is equal to gravity the charging of motion is to entice duplication of singularity. This is the basis, the heart and the sole ingredient of the Coanda principle that includes the Roche limit ($\Pi^2/4$). The charging of gravity $((7/10)+(7/10))/(10/7) = \Pi^2$ and the charging of space-time $\Pi^3 = \Pi^2\Pi$ is all due to the relevancy brought on by the Coanda principle. The value of motion came from singularity exciting singularity and that is the duplication while the duplication or motion presents the space.

The development came into eras as the relevancies brought about new relevancies that spawned even newer relevancies that all remained in touch with the original singularity centres. Every one focused a new time delay that eventually brought about space and every distortion of time brought more. That concentrated between singularity points that charged the points to form space. When the charging became overdue in some sectors it erupted in forming the Big Bang. By the time the Big Bang erupted there was such a huge backlog in heat and time corrupted and delayed the next result was the employing of space as a commodity in the Universe. The relevancy was C the gravity was C^2 and the space was C^3. That left what was inside atom still spinning faster than the speed of light applying the relevancy of $k = C$ where the electron applied the relevancy of $T^2 = C^2$ and that formed the atom which then became the cube of the speed of light $a^3 = C^3$. That left the atom at the relevant size of what the speed of light permitted at the time but since the Universe from that the relevancy expanded as the Atom grew in space to the extent it has now. The purpose of the star is to recapture the space the atom grew into and from there dismiss the space by spinning faster than whet the speed of light will be on the outside of the star.

- This form came about when only form was present in the cosmos. It was in a time era where form featured in relevancies that would lead to one day becoming the atom. The atom forms a dual purpose of duplicating as well as dismissing and some prefers the one better to the other. This relevancy came in place when time was not time and space was form. Time is forever eternity being interrupted by form in infinity to bring about eternity ticking as infinity ticks. Before that singularity took on stages in forming relevancies between duplicating and dismissing space-time, which incidentally was not yet truly space-time in the sense we think of as space-time. At first a dot moved from the spot leaving the spot but taking with the spot as part of the dot to remain in the dot. The two never separated but the one allowed the other to be.

As the dot confirmed a discrepancy between infinity and eternity by defining infinity as an interruption of eternity cold and hot parted a union. The dot that formed was not space but a relaying of time to form a new point of singularity where eternity was interrupted by infinity. Time took form from 1^0 to 1^1 or from Π^0 to Π. It brought form into differentiating between interrupted eternities with infinity doing the interrupting

Then a true distinct relevance came about that positioned a time differentiation outside the realm of time by four. In this realisation we can assume that space had some meaning at this point and the formula used to investigate suggests just that. Even in grouping, there are characteristics, which make a certain group of atoms more perceptible to duplicating and others more perceptible to dismissing

The lagging of exciting one point in relation to another point takes time. It takes time to send the message across to get singularity at that point excited. It takes effort to bridge from the dominating singularity to the independent singularity and that effort slows time down. The crossing of the divide is space formed by pushing time into duplicating. When time brought in a five points to the four points it took time to be, that fifth point became more than only form, it became space because it was one point outside the Universe of four or of form. One must see the three points established as motion duplicating singularity in relation to one dismissing singularity. This always has to strike a balance in order to establish space-time. It began as a relevancy and developed into space-time flowing or space-time displacement.

What the Coanda effect proves is that the rotating motion is acclimating a centre that exemplifies all phenomena in nature as we use nature to our advantage. All of nature including gravity uses the same method of motion forming around a circle in rotation and in the centre of the circle a point of no motion holding no space comes about. This is what Kepler taught us when he taught us **$a^3 = k\ T^2$**. With the Coanda effect forming the basic principle of all natural phenomena we can see from that, that the motion of liquid in the presence of a solid forms a centre that excites as it establishes singularity. From that rotation, space flows to a controlling centre but because of the lack of motion in that centre, there is a lack of space in that centre. Therefore, there is proof of a flow towards such an established centre and there is control from that point of singularity. In every case, the singularity controlling space-time sets standards for space dismissing in relation to space duplicating.

The duplicating stands in regard to the flow that the liquidity of the atom in relation to the solidity of the atom can reproduce. This forms density and mass but mass has little influence on the scenario.

There is a balance between the duplication in relation to the dismissing of space and the relation extends to the number of atomic elements present which then creates the balance applying within the star. As the liquid heat subsides in the centre of the star and the heat density is dissolved by the dismissing-prone elements the motion or moving ability of the star as a unit fades away as the star becomes static and solid with less space providing the star with less motion.

It is the way the atom formed before the atom took on space-time. It is in the formation, that space-time relates to motion. We have some elements being quite massive but also lighter than air and others are quit light but as dense as they come. This can only be a contribution from the way the atom relates to heat, which make the atom volatile (movable) or dense (motionless). Those elements being volatile are also very movable and in that we find the role that such elements play in the star. Stars that are predominantly made up of hydrogen and helium with very slight support from the metallic inner core are those stars that duplicate by producing motion. However the point I wish to press is that mass and being massive and being heavy do not support the fact that some elements have more gravity they produce because their protons are more numerous than others. The fact that mass generates gravity is a myth.

One will find that whatever group one chooses there are gasses and there are solids. If mass was attracting mass then the strongest mass must be attracted to the strongest mass and the least mass must float in the air. $F = G (M.m) r^2$ hardly can even begin to explain the fact that there is a gas that is more massive than iron but floats in the breeze just as hydrogen which is the least massive element.

Every dot was a Universe in its own and the accumulation was a universe. The earth in itself is a Universe as the moon is a universe, because rules applying on earth do not apply on the moon and visa versa. When in the ocean another set of rules apply, therefore being in the sea places a body in another universe. The number of universal entities is still countless, as much as it was in the beginning. Every dot insignificantly small as it may be, is a part of another Universe as much as it is part of the accumulative Universe and every dot in the infinity holds singularity, which we translate as " nothing" being " darkness". There cannot be "nothing" just as much as there cannot be "darkness".

There cannot be something big or small, but it into relevancy of perception, and then the relativity of perception becomes the question. There cannot be hot as much as there cannot be cold. The sun FREEZES hydrogen to a liquid at six and a half thousand degrees Celsius and Universe boils over in the form of the Hubble constant at the temperature (we presume from our vantage point) at minus 273 degrees C. If we Humans cannot or will not abandon our human perception and our manly perspective, we may as well return to astrology for all its worth.

Every point in the infinity we may observe at is not merely part of the Universe in not being nothing, but is the point where the Universe started representing singularity. It is the very first point where everything began so many eternities ago, because after all, how can we ever determine where the first point was, as they were very much equal and alike at the beginning. Every aspect of the Universe started with the fundamental fact that no point in the Universe can represent "nothing" as a number, because every aspect in the Universe represents singularity in what ever form it may hold in that specific spot forming space-time. If man does not reach a conclusion where that conclusion is matching the Universe and stop to match the Universe with man (and man's incapability), we may all go back to caves and become starving hunter-gatherers again, because we will never find a way to progress to the ultimate understanding of the universe.

Looking at stars Newtonians see coal stoves being stoked to burn. In the days of Newton coal stoves were the nuclear science of the day ands while all other departments in science moved on and away from coal stove principles Astrophysics and cosmology remained true to Newton by inventing the coal stove in so many ways not even the coal stove could think of the facets it can go through. Newtonians see stars being fuelled like coal stoves and such stoves can run out of fuel. This is so much Newtonian backwardness as mass forming gravity and the moon coming closer and the cosmos shrinking and we falling into the sun because of non-existing dark matter making up what is required to make Newton not to seem the idiot that Newtonians are because they make him and his contraction theory to be less foolish that what it apparently is and they overbearingly are.

What is of vital scientific importance is that there are three fundamental dimensions controlling the universe. The three are beyond intermingling and one confirms a status in relation to the others but not intermingling in status. From singularity comes matter and forming space-time in own accord. By matter not controlling time, space grew uncontrolled and the third dimension came about. That dimension birth we now recognise as the Big Bang, but the Big Bang is the last of a three prong cosmic growth. Science has to recognise the dimensions of densified (singularity), occupied (matter behind the electron) and unoccupied (space-time outside the orbiting electron boundaries) forming three points of cosmic recognising space-time.

Every dot was by itself as well as the accumulation as it currently is the present universe. The earth in itself is a Universe standing apart from other universes such as the moon as well as the space between the moon and the earth. The moon is a universe. Rules applying on earth do not apply on the moon and visa versa. When considering conditions with in the oceans and applying space-time another set of rules apply therefore the sea places a body in another universe. It takes the same engendering technology going underwater in deep sea diving that going into outer space.

The number of universal entities are still countless as much as it was in the beginning matter as atoms and even much smaller. Every dot insignificantly as it may be is a part of another Universe as much as it is part of the accumulative Universe and every dot in infinity holds singularity, which we translate as "nothing" but it cannot be nothing. There cannot be nothing as much as there cannot be darkness. There cannot be something big or small except in the relevancies of perceptions and then the relativity of such perceptions becomes questionable. There cannot be hot as much as there cannot be cold The sun freezes hydrogen to a liquid at 6500 ^{0}C and outer space boils over at 0 K. If we humans cannot or will not abandon our human culture driven perceptions and our mankind's pre-programmed perspective we may as well return to astrology for what the future hols. There are so many boundaries out there ready to destroy us because of our lack of insight, as did the challenger disaster.

Creation birth started off with one dot so small eternity met infinity within. Then came one more, and another and they continued coming until there were a countless number of dots. The accumulative size of the dots were the same size as one dot because in the true Universe big and small plays no part. The dots were infinitely small and eternally big at the same time because size is a relevancy and without one the other has no size. So in the true perception, there is no difference in size.

It started with the fact that there is no place or part in with which one may associate zero or nothing. There are no room for a number such as nothing. Next to the one dot (infinitely close) one will find the next dot, and if nothing was a factor then that is precisely what one will find between the two dots. Nothing of space, a non existing entity, taking up no space, and much more important, no time, therefore the dots are infinitely close to one another, being the same space, eternally big as much as infinitely small. If we as humans cannot find a manner in comprehending this notion, there can be no manner ever understanding the cosmos as much as the start to the cosmos.

Every dot was a Universe in its own and the accumulation was a universe. The earth in itself is a Universe as the moon is a universe, because rules applying on earth do not apply on the moon and visa versa. When considering the conditions with in the ocean and applying space-time another set of rules apply, therefore being in the sea places a body in another universe. The number of universal entities is still countless, as much as it was in the beginning, before dots formed atoms.

Every dot insignificantly small as it may be, is a part of another Universe as much as it is part of the accumulative Universe and every dot in the infinity holds singularity, which we translate as " nothing" being " darkness". There cannot be "nothing" just as much as there cannot be "darkness". There cannot be something big or small, but in the relevancy of perception, and then the relativity of perception becomes the question. There cannot be hot as much as there cannot be cold. The sun FREEZES hydrogen to a liquid at six and a half thousand degrees Celsius and Universe boils over in the form of the Hubble constant at the temperature (we presume from our vantage point) at minus 273 degrees C. If we Humans cannot or will not abandon our human perception and our manly perspective, we may as well return to astrology for all its worth, because that is the only boundaries we will find in the cosmos.

To unlock scientific truth we first have to dispose of scientific misconception

In the two pictures we are seeing disposing or releasing heat creates space. We may call it plasma or shock waves or what ever, but in the final analyses it is heat turning to space. Whatever you wish to call that which lies between the particles comes from being a solid, then with adding heat, the solid *"whatever"* becomes liquid and that is the white and orange plasma that we find. That white and orange is heat in a liquid form, just as all flames and smoke is heat in a liquid form. But that liquid does not remain liquid because the governing singularity cannot enforce a commitment ensuring the liquid heat remains liquid. The liquid *"whatever"* you wish to call the heat in fluid form then further overheats turning the heat to space. The space created must be equal to the heat reformed. That is a law of energy where energy equals equality everywhere it is.

Let us humans first detach culture from facts. Take the argument to iron, which we know well. Iron cannot boil, iron cannot flow or bend and iron cannot brake. Iron is an element like all the other elements we know, not one element can do any of the above, in sharp contrast to human belief. As indicated in this book the limits we should find to guide us we ignore for the reason that we cannot see it. We may not be able to ever see singularity, but with intelligence guiding mankind, we do not have to see everything to believe everything. It is because we could not see religion, but still practised religion that set us apart from the other animals.

At the start one would find iron and iron in a "natural state" as we find iron on earth being a human produce on the surface of the earth it will be a solid, suitable for man to handle with bare hands. When such a piece of iron is left in a desert in the midday heat, the human hand cannot handle the iron any longer without aid of covering the skin of the hand. Our perception is that the iron became hot, but that is not the case and our view is a culture contribution and not scientific fact. By heating the iron artificially with combined gasses (acetylene and oxygen or what ever) we now can over heat the iron to a state of flowing like a fluid. Our human culture tells us the iron now is melting.

That is a misconception!

Like the fact of "nothing" we inherited the idea from our past. After introducing artificially even more heat with more heat releasing gasses we may artificially form a condition where the iron would become a gas. Again it is not the iron that becomes a gas, it is the space the iron finds itself in that became hot enough to become a gas. The iron particles remain the same; it is the condition surrounding the particles that changes form with overheating.

Important to note is the fact that iron in a solid state will surround itself with solid matter in space applying a solid space. By introducing conditions producing _more overheating_ the space or connecting between the particles become concentrated heat forming a liquid substance! It is not the iron that turned liquid but the wrapper containing the iron that concentrated so much it formed liquid fluid by the introducing of more heat to a point where the overheating created a fluid. It is considered that the oxygen burn and by that the iron heats up. NOT TRUE!

If oxygen burns no oxygen would be left on earth by the time man arrived on earth to use it to the benefit of intelligent life. The oxygen remains oxygen while the oxygen merely does a task in nature where oxygen carries heat to a specific space. On the other hand it is the task of nitrogen removing heat from the point of overheating by means of flames whereby it creates space. One can feel the "wind blowing" as the flames generate created space. In the extreme the creation of such space we call an explosion.

In the process where the space between the iron particles still further overheats, it becomes a gas. It cannot be iron that becomes gas, because iron will be as much a gas as iron will be a liquid or a solid. It is the space covering the iron particle separating the different iron particles, which will convert and sustain form. The gas is as invisible as space because the gas is the form space holds. This confirms the Biblical view of earth (solids) created and heaven (heat or gaseous/liquids) created. There are only two forms of substance that forms the Universe solids and non-solids, which is liquids and gas. It is not the solids going liquid but it is more of the liquid in ratio with the solids in between the solids that make a structure go solid or gas. There are heavens (non solids) and earth (solids) and this has to do with movement applying control or non-movement allowing non-movement control.

Iron is a solid. Introducing more heat the iron becomes wrapped in a cover that concentrates the wrapper to the point of concentration where it became a fluid. The iron remained what it is, neither a solid, nor a fluid nor a gas. By introducing more heat it becomes a gas. The gas we cannot see because the gas is space. But so was the fluid space. The introducing of heat brought about the turning of a solid to a liquid to space and every time more space becomes part of the picture. Iron is in its normal form a solid. That means the space, which the iron particles are in, is solid and that disallow the iron to alter the form in which it is. By introducing considerable heat the iron melts changing the form of the iron from solid to liquid.

Considering the evidence we find it is not the iron that melted and that became liquid, but it is the space in which the iron is that became liquid. The iron particles are still as solid as they were. By introducing more heat the iron would eventually turn to gas. It is not the iron that turned to gas, but it is the space in which the iron particles are that has increased to the extent that the space now has so much heat, the heat turned to more space. The iron as particles remain the same, they are just elements confined to a nucleus with electrons spinning about. The space between the particles increased to such an extent it first became a liquid or a fluid and with more heat introduced the heat increase brought about that heat turned to space. That means by overheating the particles surround with heat as a fluid the heat increase then add space as a gas. The gas is the ultimate form of overheating but where one is unable seeing the gas.

1 Firstly the iron is cold enough to be a solid. Replace the word iron with cosmos and forget the colour we associate with heat being white and note the solidness of the centre of a galactica. This must have been the state of galactica that contained large parts of the Universe when time rolled away from eternity.

2 By introducing overheating the space between the iron and not the iron as such turns to liquid. The same apply as more matter (iron) produce more space forming as some matter turned to heat by overheating. The matter increased spin and in that way went out of sequence where it then became softer and softer in relation to other particles, where the loss of the matter released more of the third cosmic component we named heat and space.

3 Some of the heat introduced with the overheating by means of congestion then forms space while other remain in the form of heat allowing space to seam liquid. The matter could not breath and overheated by the enormous gravity the overheating created

4 As the area between the particles still further overheat certain parts of the area overheats to the extent that the space becomes an invisible gas allowing the congestion of matter to separate from one another and allow the stars' individual governing singularity growth. 5 From the soup of heat galactica come about allowing stars to rise out of the dense liquid cradle from where they can establish singularity growth. The process continues as more space becomes introduced through space overheating turning heat into space

5 Should star development come about as suggested it is foreseen that the Milky Way once was a liquid from which the sun developed the singularity in which it then form self-sustaining. The only pre-condition was that it captured individual space-time where the captured space-time remained a liquid frozen (as it was back then at the time of parting) by the governing singularity while outer space further overheated into a thin gas

6 The sun captured so much space by the intervention of singularity when released from the Milky Way that it produced space so concentrated today at present it clearly remained a liquid inside as it froze the

interior in time the liquid it now is while outer space is still overheating as a gas with no visibility. From this overview one can judge just how far science is behind the time in their views on creation and the beginning of time including the universal establishing. Cosmology still hides behind medieval ideas that other faculties and scientific departments forgot long ago.

It is the point forming the very centre that plays the part as the **controlling singularity** within the Universe I have named as **Infinity,** which is better known as the axis. It is where nothing can go smaller and anything within that point can never reduce. That point is where the entirety called the Universe begins and where everything holding substance begins. Once one accepts the fact of singularity being present in that location, that accepting of singularity then is contradicting all the things we know and we can measure and we recognise that point being present by merit of the fact that the point referred to is not being formed by any of the things we can recognise.

It is made up of everything we don't know and constitutes of everything we are unable to recognise or visualise. In that spot there is no space. That spot holds **Infinity.** In that space there can be no motion because there can be no space to have the motion within. It is formed as a line that is so small that our human reality by perception declare that point as not being there and the only reason why we know it is there is because of the results it left as an imprint of its not being there. We cannot detect it but notwithstanding our failure to note it we can recognise the dot on the merits of its absence and while in our Universe it is always absent, reality disallows the dot ever to be absent, because it is never absent. It cannot be absent. It cannot go absent but it can never be there where it should be in a place from where the third dimension forms and it is always present if I wish to locate it. It is **infinity** that can never go away. I named the other part of singularity forming space **eternity** because that area never become bigger, or become more or find an end to the outside. Whatever was and is and will ever be is locked in that space I named **eternity** and it is **eternity** that never ends because **eternity** can never end moving. What we think of, as expanding is never ending movement giving eternity the eternal motion that will go on forever.

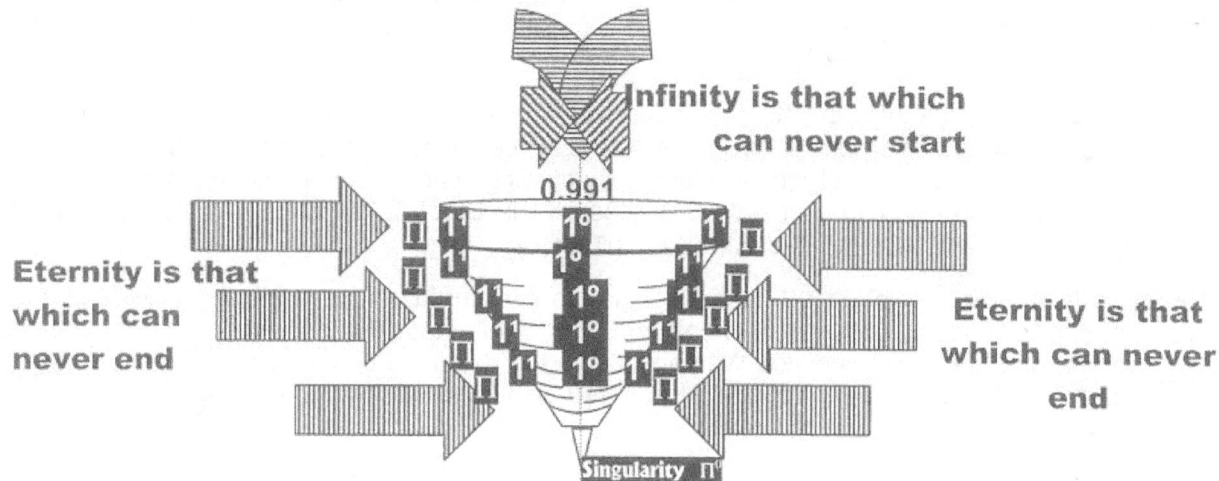

The line **k** coming from the centre (singularity k^0) forms by forming an initial spot Π^0 becoming the dot Πr^0. However, I went on to say that whatever the line used to start with has to continue in order to repeat the same that began the line. Therefore the line started with Π^0 and it has to continue with Π^0 until such a point, as it must end with Π. Whether the line is Π^0 or is r^0, or uses 1^0 the outcome all refers to singularity being used. By reducing the line we come to the end of the mathematical equation of the circle but the circle does not end there. When the top is in a state of motionlessness on own accord it is everything but motionless. The motion it adapts are synchronised with the earth in harmony with the solar system and according to the greater picture of the cosmos.

When an energy source not related to the cosmos called life intervenes and energises the tops motion, the singularity in that top suddenly jumps to life. By adopting a rotation energised to an unnatural state of energising because of life's intervention, the singularity of the top is not in charge but as it applies more and more energy, it will begin to find a means whereby it can escape and apply individual singularity as the top starts to separate from the singularity the earth holds. The singularity holding the earth would then allow the singularity of the top to rotate within a specific band where that a specific band of being active before the earth's singularity will start to destroy the singularity in rebellion.

The top on the other hand will try its outmost, when the singularity it holds gets by individual spin is too strong to remain be in domination of the earth's singularity. The motion of the top is an attempt to begin applying an individual singularity space-time defying and standing apart from the earth's gravity. That action we see as the top starts rotating in a manner where the top does not align with the earth's singularity. With the adding of spin, the time the top holds becomes unrelated to the time the earth holds and the top will start a campaign too escape from the singularity domination the earth has on the top. When the time or spin of the top exceeds the limits the earth places on the top, the top would emerge by trying to escape from constrains placed by the earth.

The view I represent at this point is known to science for almost as long as science knows mathematics. Not long after the law of Pythagoras was understood where Pythagoras introduced mathematics Eratosthenes of Syene made as big a discovery as Pythagoras did. But in the one instance the world took notice because the world could see and understand and the other instance the world disregarded the findings because the world did not see what the implications was. The same apply to aircraft flying and when the aircraft wishes to escape the earth's singularity hold it has to comply with the laws laid down by the earth.

The seven becomes as big a part of the concept as does Π as it all interacts.

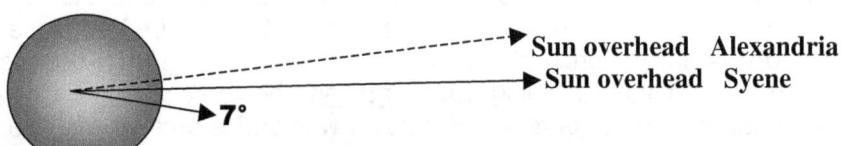

It took Eratosthenes of Syene (276 – 194 BC) a Greek astronomer who in the year 240 BC made a discovery that the earth has a profile of 7°. Since then no one ever did anything about it. When any singularity wishes to disconnect from the earths singularity, specific pre-calculated laws would have to comply to allow the lesser object to divorce from the larger object. I indicated how the dimensions of 10/7 and 7/10 interact to form (Π^2)

Matter is a product through the separation of space and time receiving the value of Π The original time and Π^2 as follows: By circling around a spinning solid the space contracts to form Π and Π^2. Gravity forms everywhere in the Universe by applying singularity. By dividing space into material (material spinning in space) and duplicating space by material spinning, the TITIUS BODE LAW forms a 7^0 deviation and 7 / 10 in conjunction with THE ROCHE PRINCIPLE OF $(\Pi/2)^2$

In my article to Annalen der physics I used 15 pages to explain this process of singularity applying. I received a rather cordial but sincere reply from the Editor of the magazine. When I placed an article in Annalen der physics Dear Prof Friedrich W. Hehl said in the e-mail he sent me that there is no way to "explain" the world of physics I am not going to go into detail how this works. On the other side of the Pythagoras's' triangle we have 1 going square.

That makes Pythagoras's' triangle 49 + 1 = 50 on the one side of the earth and the same on the other side of the earth. The total is 100 and the square is 10. That leaves the Titius Bode law with a value 7 (it forms part of the material of one body) and 10 in relation to the space. Then from the relation of 7/10 and 10 / 7 forming Π the Titius Bode law form Π^2 applying "With a lot of words and some simple algebraic relations" to quote Friedrich W. Hehl, Inst. Theor. Physics of Annalen Der Physics fame. This was simple algebraic relations but still it is science, is it not?

Since it involves singularity moving it calls for the law of Pythagoras to produce space. The law of Pythagoras is the triangle a^3 that is moving forward in singularity **k** by turning T^2. In singularity the 7 stands in for 7 points on the numerical line crossing over the line holding singularity or 1.

The car straightens the curve of the earth when it stretches the $7°$ the earth presents as the curve around which it spins. The car leaves the curve of the earth and gets airlift. It then needs "wings" to force the car down by creating more gravity because gravity is the space that pushes down on the earth and the wings produces more space that pushes the car down on the ground. As soon as the car lifts into the air not touching the ground surface the relevancy changes to $7°(3\Pi^2) = 217$ km/h as the relevancy changes to $7°(3\Pi^2) = 207$ km/h.

A Less Complex Commercial Science Book 489 Chapter 10

This picture we have above shows why gravity forms as the space we call air pushes down onto the earth. The faster the car moves from one point to the next the more air it will generate to push it down. Gravity is the space above the earth that pushes down on the earth and that includes the space the car holds. But if the car exceeds the speed the earth retains to spin the car will exclude some space from the earth's space and take this air along for the ride. This lump of air holding the car will stretch the 7° it retains to form a circle and this will extend the position where Π forms. This will then begin to lift the car when the point holding Π as radius becomes wider as the position where 7° forms is further apart.

As Indicated space always holds a double 10 value in terms of 7

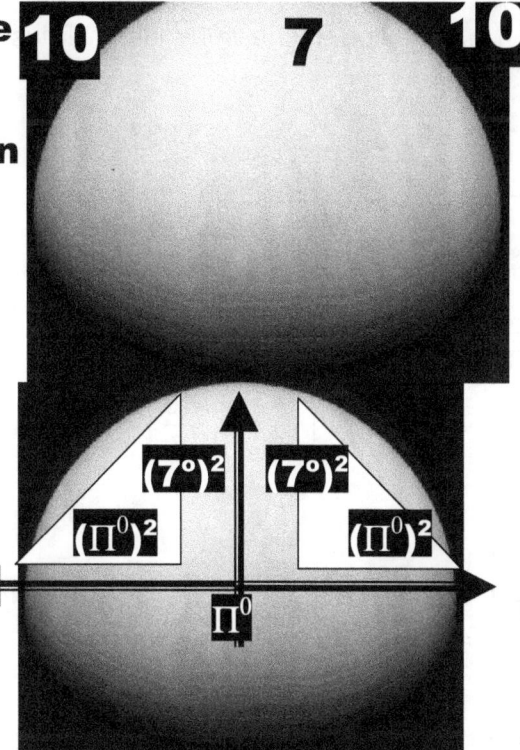

Material associates with 7 forming one value in 21.991 / 7 = Π

By spinning through 7° the changing of direction is taking 7 squared $(7°)^2$ and this gives 49 in total movement or space-time displacement or gravity applying or whatever you wish to call it. This is on one side of Pythagoras's' triangle

The spin re-aligns the centre with the same point on the surface of the earth but it then is at another point. This movement then takes $Π^0 = 1$ or singularity to $(Π^0)^2$ or also 1 which is singularity. I sent an article about singularity to Annalen der physics.

By moving 7 has to go square T^2 and that means 7 goes square 7^2 twice $7^2 + 7^2$ crossing the same divide $Π^0 = 1$. Since all movement in singularity has to enforce the law of Pythagoras we have two triangles holding 7 dots moving across singularity. I don't want to get too involved by bringing in numerical outlays because then this can truly become complex. The line has two opposing sides turning directionally against each other while turning with each other. By moving or turning this involves time duplicating space by the square Π^2 on both sides of the divide $\Pi^2+\Pi^2$ and using the same divide or the same axis or the same point serving singularity we have 7^0 crossing the same point in singularity Π^0.

There then is in this rotational movement 7^0 standing in for Π^2 on both sides of the divide $\Pi^2+\Pi^2$, which then is 7^2 on both sides of the divide 7^2+7^2. The circle spins in duel directions. On the one side it would go left if on the other side it would go rite. The one side hold a directional change in singularity by 90°. As it is going sideways it changes to going down. This produces a rite angle triangle of 90° and in it the law of Pythagoras produces direction changes. Since the square of the turn of the circle places by the spin and the direction change we have 7 holding a relation to 10 in space because it is space that has to carry the value of 10 when material circles by 7. There is a connection between space surrounding the spherical circle turning and the sphere. The circle holds the value of 7 as in 7^0 and this we find from looking at singularity controlling the circle by movement.

Matter in relation (part of) with the total dimension of space.

$$\left(\frac{10}{7} \div \frac{7}{10}\right) = 2.04$$

$$\frac{1.4285}{0.7} = 2.04 \quad \text{Taking from both orbiting influences}$$

SPACE DIVIDED INTO TIME

$$\left(\frac{7}{10}\right) \div \left(\frac{10}{7}\right) = 0.49$$

$$\frac{0.7}{1.4285} = 0.49 \quad \text{Taking from both orbiting influences}$$

SPACE MULTIPLIED WITH TIME

$$\frac{7}{10} \div \frac{7}{10} = 1 \quad \text{and} \quad \frac{10}{7} \times \frac{7}{10} = 1 \quad \text{Therefore not influencing change}$$

THE PROCESS PARTED USING THE ROCHE PRINCIPLE

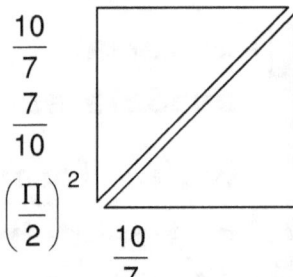

$\left(\frac{\Pi}{2}\right)^2$ The Roche influence on Titius Bode

$$2.04 \times \left(\frac{\Pi}{2}\right)^2 = 5.033$$

$$2.04 \times \left(\frac{\Pi}{2}\right)^2 = 5.033$$

$$5.033 + 5.033 = 10.066 \text{ from both objects}$$

SPACE DIVIDED INTO TIME

$$\left(\frac{7}{10}\right) \div \left(\frac{10}{7}\right) = 0.49$$

$$\left(\frac{10}{7} \div \frac{7}{10}\right) = .49 \quad \left(\frac{10}{7} \div \frac{7}{10}\right) = .49$$

$$.49 + .49 = .98$$

$$.98 \times 10.066 = 9.8696 = \Pi^2$$

TIME SPACE $= \Pi^2 = 9.8696$ TIME SPACE $= \Pi^2 = 9.8696 =$ Space and time in a dimensional implication

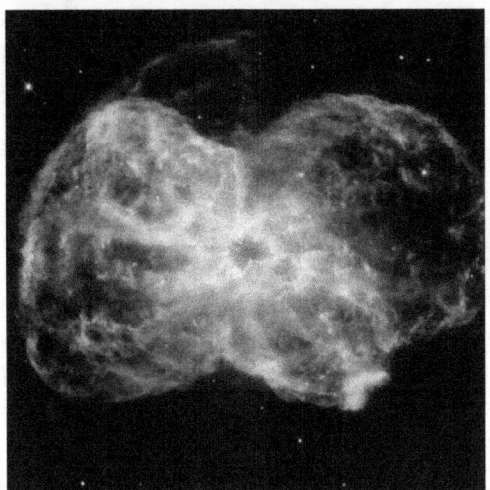

When the 7 of material and the 10 of space no longer forms a gravitational alliance the 10 forming space will expand because the spin of 7 cannot retain gravitational structure.

For gravity to form the space value of 10 must always maintain the material value of 7 so that the formation of Π can form gravity by moving Π to form Π^2.
This is gravity and that is how gravity is maintained

The wise men from physics put everything down to mass without ever searching for evidence of mass. Putting everything down to mass may be one solution except for the fact that only nothing is that simple. Even the rejecting and / or accepting are incorrect, as the pulling part just comes across a tad too simple to make sense. To find substantiation one has to find the manner in which light connects to singularity because everything connects to singularity. In every circle centre there is space that is so small it is not present within the space of our Universe. In that space in the centre of every spinning object we have singularity forming space. That points serving singularity is the reality and the rest is only make believe.

Explaining the following is rather tough with the limited space available but it should secure the conclusion that I am not grabbing for straws and there are substantial facts on which I work.

Lets investigate the Universe. The Universe is made of lines connecting points. Whatever you see or can't see is lines connecting. If we wish to examine the Universe we first better star to examine lines connecting dots because whatever is in the Universe is dots that connects with lines. So it will be necessary to investigate lines. In mathematics they teach students that a line starts at zero and that is a fable. Zero starts nothing and nothing loses everything before anything can start. This sounds so unimportant but it is fundamentally all-important. A line starting with zero how cannot increase in length

What I am about to explain is the absolute basis for singularity. Lines mathematically cannot start at zero because there is no evidence of zero as a factor in mathematics. The shortest possible line (hypothetically) must be so short it must have an initial and ultimate point sharing the same spot. Any theoretical line being the shortest possible line cannot have the line holding the initial starting point at point zero and advance from there. If it used zero as a start, the zero part would not count, because the line will only start at a point past zero where the line then will start

When the line has a beginning and an end at the very same spot and it wishes to extend the position as to further the possibility it has, which direction should it favour. Extending the line in any one direction will favour one direction without any clear reason not extending in other directions. The only option about extending will be in all directions equally in order to give a meaningful non-bias flow of mathematical equilibrium.

The shortest line in the realm of possibilities must have a start and finish holding one spot and such a line will also be a dot or a circle. Not favouring one direction puts all directions at equilibrium meaning that any form of what ever might develop from such a spot with the end and the start being in the same position also has to be a sphere.

This reasoning prompted me to look for singularity in such a spot because if the prime spot from which all came was a spot holding all, then the spot must hold the shortest line but more prominent it will hold the smallest form including the smallest circle. One possibility that the shortest spot can never have is having a starting point on the zero mark. If the mark of zero holds the start it must also hold the end because the end and the beginning has the same position. If the position of zero then is the beginning, the end will also be zero leaving the line without an end as well as without a beginning.

The conclusion from this is that no line can start at zero because that will be a mathematical impossibility. A line or spot starting at zero would therefore be shorter than the shortest line possible. A line growing or extending from zero can never leave zero because of the influence of being zero disqualifies any possibility of growth. If the line then had to grow in all directions at the same pace the line must therefore be a circle or being three-dimensional, a sphere. Flowing from this fact is that in the Universe there can be no zero point or unfilled space. The value of the circle is Π, and that is where creation started. That gave me the clue where to start looking for singularity. One would find singularity in the value Π and the value Π will be in all things rotating in a circle. You might wonder how does that apply to the cosmos and moreover to gravity?

My approach might seem unconventional but through the abandoning of the accepted, it enabled me in locating the precise location of a universal singularity forming a connecting basis of the Universe (this I say with some degree of confidence). The smallest figure there can be must be a dot. The only mathematically sensible option about extending a line from the dot will be non-bias progress in all directions equally in order to give a meaningful flow of mathematical equilibrium. The Pythagoras mathematical principle is the proof and that I explain. The obtaining of singularity is in my rejecting of nothing by replacing it with something being the dot.

The claim becomes obvious when observing the connection between the half circle, the straight line and the triangle, which could also promote all the qualities lurking behind the pyramid. Consider the connection between 180^0 sharing and then one may realise much of the pyramid mystique becomes less spectacular in considering the very basic in mathematics being the Law of Pythagoras on which all mathematics are focused.

The claim becomes obvious when observing the connection between the half circle, the straight line and the triangle, which could also promote all the qualities lurking behind the pyramid. Consider the connection between 180^0 sharing and then one may realise much of the pyramid mystique becomes less spectacular in considering the very basic in mathematics being the Law of Pythagoras on which all mathematics are focused.

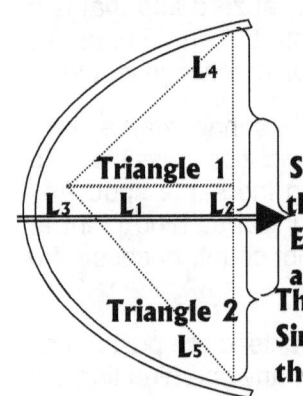

The Lagrangian System implicating the five positions extending from singularity

Singularity dividing the cosmos
Each triangle claiming a side of the universe

1 Half circle = 180^0 L_3 L_4 L_5
2 Triangle 1 = 180^0 L_3 L_4 L_5
3 Triangle 2 = 180^0 L_3 L_4 L_5
4 Straight Line = 180^0

The half Circle = 180^0 combining as a Sphere when comprising Singularity in the matching of the value of the straight line forming the half circle and combining as the triangle and all are equal 180^0

The distance between the sun and Pluto is roughly one hundred times more and if the distance between mercury and the sun, but both has nothing between them and the sun. If space comprises of nothing how can nothing then become plural forming more or be multiplied. If it was one becoming one hundred, then the one cannot contribute to a value of nothing but then must be part of something. If the one substituted the nothing, all laws of mathematics will go in disarray because when one multiply any number by zero it becomes zero placing both planets in the sun. By excluding nothing from the equation space becomes something bringing in a value lying inside the realms of the infinite that must form singularity. Applying this logic to the Lagrangian system and interpreting that information to the law of Pythagoras a clear pattern come about.

Mercury has 58×10^6 km and Pluto is 5900×10^6 km space between the sun and the planet. That indicates a distance and a distance comprises of something, for if was nothing then both would have equal nothing and be next to the sun. I repeat, the distance indicates something because nothing would place them both in the sun. The problem is identifying something from nothing that defines the difference there is in science. I cannot see how nothing can become plural or more sometimes

Taking that into account it is important to recognise that notwithstanding the size of a line, there is another line (or dot) eternally bigger as well as eternally smaller than the line in question. We can never grasp the size of a line that forms the utmost or the least of possibilities and therefore size belongs to the human mind forming conceptions of big and small, but it has no place in the cosmos at large. This concept not only applies to size, but to all limits and divides we wish to create forming borders we can appreciate. When looking at the circle in the conventional manner, we persist with errors brought about in culture and not by applying some significant modern logic.

═══════════════════════════════

From the smallest ever possible dot will grow a line in every imaginable direction relating to a prospect of Π not favouring one direction that puts all directions at equilibrium meaning that any form of what ever might develop from such a spot will have the end and the start being in the same position, which will also have to be a sphere as the flow outward will be equal in all directions. This reasoning prompted me to look for singularity in such a spot because if the prime spot from which all came was a spot holding all, then the spot must hold the shortest line but more prominent it will hold the smallest form including the smallest circle or for that matter the smallest sphere. One possibility that the shortest line or smallest spot can never have is having a starting point on the zero mark. If the mark of zero holds the start it must also hold the end because the end and the beginning has the same position. If the position of zero then is the beginning, the end will also be zero leaving the line or spot without an end as well as without a beginning. Such a spot will constitute all of nothing Any line starting from zero would inevitably start from a point where it ignores the zero mark because the fact of zero does not implicate a start or a size of value, but only the not being there of that position. All lines would form a duplication of another line sharing value since there will always be a possibility of yet another line in the realms of singularity lying between the two lines in question reducing the size infinitely to either side of the divide we humans create. Boundaries therefore are human and as man made substances it does not belong to the cosmos outside the influence of man and must be discarded.

(r or Π) ● r/2 ● r/2 ● r/2 dividing r reduces r to infinity but not Π as Π remains stable, protected by the rotation of matter forming a circle around singularity

When **the circle reduces**, the **value** located to **r** will become implicated because **r determines specific size. Not so** in the **case of Π, because** Π in the true sense only **indicate that the circle is a square without corners** and therefore Π **dictates form and not size. By reducing size** only **r comes into contest** and will point to such reduction. By **reducing** the circle **radius r by half continuously** will lead to an **infinite small circle** but Π **will remain because the circle as a form remains** even being infinitely small

In any circle or sphere the size only depend on the fluctuation of r in the square as a component to the circle or sphere but that does not affect the form by indication of Π in any way there may be. The conclusion from this is that no line can start at zero because that will be a mathematical impossibility. A line or spot starting at zero would therefore be shorter than the shortest line possible. For obvious reasons can no line, or any line grow or extend from zero because such a line must then quit zero and become something, thus abandon its original value. That would mean the start of the line has a different value to the end and a line holds conformity through out. When any line is starting from point zero it can never leave zero because of the influence of being zero disqualifies any possibility of growth. If the line then had to grow in all directions at the same pace the line must therefore be a circle or being three-dimensional, a sphere. Flowing from this fact is that in the Universe there can be no zero point or unfilled space. In the case of the growing sphere the value of the circle is Π, and that is where creation started. That gave me the clue where to start looking for singularity. One would find singularity in the value Π and the value Π will be in all things rotating in a circle. You might wonder how does that apply to the cosmos and moreover to gravity?

By reducing r indefinitely to the tune of half each time, r would become infinitely small, beyond human calculating means, however as mentioned in the case of the smallest dot holding one spot, r would become insignificant beyond human comprehension even, but never reaching zero and still Π would remain intact and dictating form.

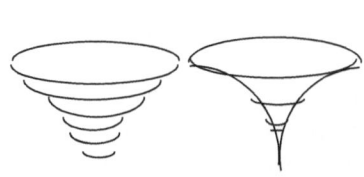

An observation coming instinctively to mind one may recognise is that the form reminds rather explicitly of natural phenomenon as hurricanes, water whirls and even the shape most commonly favoured to express

the cosmic object referred too as a Black Hole. The similarity may be more than coincidental. Let us consider the statement in the reverse.

Anything occupying space in the cube will apply r, notwithstanding the name used confirming the shape or r named as length width or height, it is all just a straight line bringing about the cube with all its other names that may find attachment to specific form but nevertheless still remains only a six-sided cube with connecting lines applying different angles changing in some cases.

The normal perception is that any circle growing spontaneous would grow by the radius, which is r. That cannot be the case because r is an indication of a straight line. By growing with the aid of a straight line from the centre to circle the influence that that would have on the circle would result in many circles following one another and not a continuous growth. Gravity is the dimensional changing of space holding r as reference in the cube as to the sphere holding Π as the reference. In order to generate spin producing time in matter occupying space, therefore creating dimensional change, Π has to be a factor indicating the possibility of spin because implementing Π the circle sides will follow one another without establishing separation . The answer must be in finding Π, and thereby locating singularity.

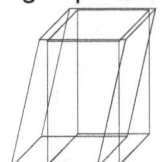 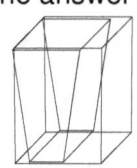 

Arriving at the question about locating the space and time forming the centre the centre of the Universe one has to realise the centre of the Universe are in every singularity forming matter weather it is big or small, size carries no significance. It is the impartiality of singularity that is claiming the value and not the differentiation of matter. One must realise there are no big / small or hot /cold or near / far ; It is all relevancies between matter claiming space and space is heat in a turnabout manner. Every aspect in the cosmos are locked-in universes, sealed off from other universes and inclusive or exclusive depending on singularity holding relevancies relating to one another. The relevancies rely on inter dependence and inter linking, but there are no differences according to human sizes or standards. Accepting that principle unlocks the "so called mysteries" of the Universe and brings about clear understanding. It is all about accepting, acknowledging and interpreting the role singularity maintains on matter.

Due to the spinning nature of such a point with all surrounding the point will be alternating direction favouring change every second and in that the value to such a point can only be Π because of its constant changing. Using r would specifically oppose another r from
every angle because the use of r will bring about a static relation to the previous and following instant and therefore it will cancel the constant spin flow.

In the circle using $r^2\Pi$ the r has to have distinctive qualities placing it as a factor apart from Π. Where the growth shows no separate distinction but a continuous flow from the precise centre to the precise edge the flow would become in relation with Π depicting the circle and Π replacing r as reference to any point on the circle. By using r as a distinction in the circle division is possible but by using Π there is no distinction possible making it a solid flow.

If the spinning top is all the evidence any one needs to come to such a conclusion what will bring any proof that the singularity governing the top connects too anything anyway. Placing singularity is fair and fine, but what will the evidence be in proving its activeness as part of the creation at large? The reason why we can be sure it is active is that when spinning it shows borders implicating restraining of further movements outside the set limits. By going faster (past the upward border) the spin goes oblong where it actively tries to change the position the top holds to the earth in relation to the surface of the earth. By going too slow it once again shows identical character. When going too fast it indicates an attempt to rise into the air, therefore relieve its singularity in an effort to part with the earth's singularity. It shows unmistakable characteristics of trying to become airborne securing an independent position from the earth, which holds it down. At the bottom we surmise correctly that it wishes to topple over and fall down. Of course the bottoming out shows the same characteristics whereby we gauge that to be the normal process of falling down. If the bottoming is relative to the earth's singularity and we recognise the process as normal, then the top of the limits should be just as recognisable normal.

Locating and finding Singularity

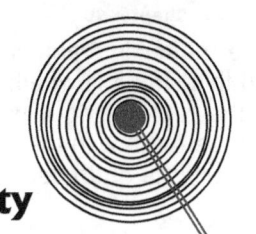

In the **precise middle** of all **objects in rotation** is a precise centre dividing the object in sectors that will **start the spinning initiation** from that centre point. Thus, the spinning object **will have a middle point**, a very specific **centre point that does not spin** and only holds Π as a specific value. One value such a line **cannot have is zero** because **zero does not start any** line and therefore the **value of the line must be infinite**, just as described in **accordance** and by **the definition of singularity**

That point albeit hypothetical, is also as much a reality none the less and is placed where that point **must be standing still** because every line **running from that point** in **opposing directions** are also **in opposing directional spin the other or opposing side.**

In considering the spinning motion in the fraction of time in the detailed instant every aspect of rotation will turn in every instant of change in time. Although the points had the same characteristics only seconds before, they oppose the characteristics it had just before and just after the very second in which they are and to which they relate by similar points also in rotation. The fact of the graph proves my point in quarterly opposing dimensions and values,

From this centre line that is only theoretical definable, but is still there all the same, a opposing value always form that becomes real and distinct when rotating, but even more distinct when not rotating because then the line grows so much it covers all the matter, to a securing spin value of zero, the most original value it had. When not rotating, it is as thick as the material will go. When rotation begins, the line shrinks back to a hypothetical position claiming zero spin that is not less distinct but more distinct because from that point every rotating piece of what ever is then spinning will clearly carry the singularity value of Π implicating rotation.

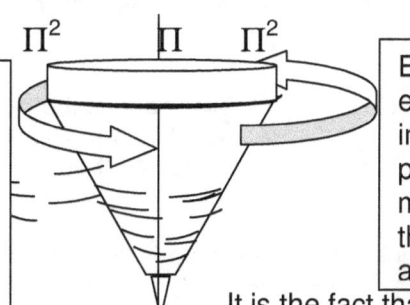

When looking at the cosmos from whichever angle indicates the fact that the cosmos is moving. It is forever spinning and it is all going towards as much where it is coming from.

Everything is on the move and always encircling something of greater importance. A top can spin but the parameters of its spin are limiting the motion it can apply. By not spinning the top is still spinning as the earth are doing the spinning on its behalf.

It is the fact that the same affect comes about when spinning too slow that triggers the questions.

When spinning too fast the top fights something because the alignment keeping it upright starts to tarnish. The same apply when spinning too slowly but that makes sense.

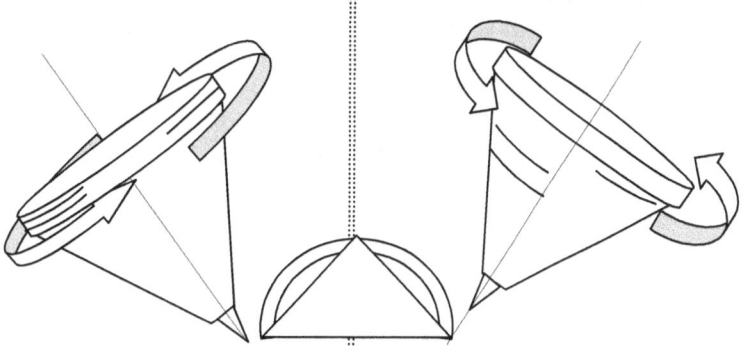

In determining this behaviour as part of a cosmic process where matter interact with matter in an laid down set of rules, we should once more be asking questions and this time it is whether the top will show the same behaviour in outer space as it does on earth. With the reply of no it would not comes an admitting that the process involves the interacting of singularity of the earth with the singularity of the top where the spinning created independent singularity, as valid as that of the earth because the earth has a role in sustaining it or destroying it at the border ends.

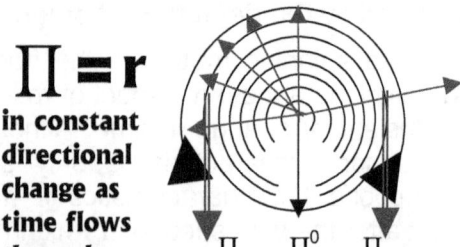

$\Pi = r$
in constant directional change as time flows through rotation

$\Pi \quad \Pi^0 \quad \Pi$

Pinpoint positioning of singularity Π^0 with Π positioning space to either side forming the border set by singularity

The new direction pointing to a new location in relation to the previous point will oppose the previous point it had in relation to direction considering the centre point.

In the sketch below the circle to the right would come about from a straight line r growing influencing the appreciation of Π, but to influence Π would lead to a breakdown in r as Π and r are different entities. The circles to the left shows a continuous growth by extending Π every time and since Π is the same part as the previous Π, only extending that billionth of a millimetre each time, the circle will be truly continuous without any signs of a break

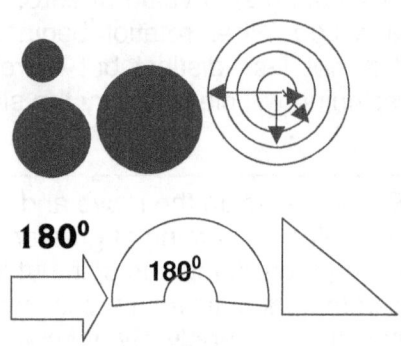

Looking at the affect of gravity it shows the precise quality of no distinctive point, as gravity never seems to end at a point but flows all over affecting all that holds a position in its sphere of influence. The gravity coming from China meets the gravity coming from America at no particular spot but intermingles without distinction.

The triangle, the half circle and the straight –line has two things in common, they share 180^0 as a mutual value and they are part of singularity.

Using the concept that gravity applies Π as the circle factor Π as well as Π^2 replacing r^2 the replacing by Π brings two values as Π and Π^2. That I found is the case with gravity and will be apparent when explaining the sound barrier as well as the Four Cosmic Pillars. In order to create a distinction I remained using r as the indicator of the cube or non-circle that has vacant space and by vacant space I refer to non-solid structures. In the solid structure I use Π as a value for reasons that will become apparent in due time.

The value of singularity stems directly from the law of Pythagoras or **Pythagoras** is the result of **the average of singularity. With the shortest line being a dot, all lines must start from a position implicating** Π. A circle is a square without corners implementing Π and a half circle is therefore a triangle without corners. The corners are the factor that confused every one in the past. When replacing the value we normally attach to circle being r with Π, the law of Pythagoras becomes quite meaningful and mathematical.

By placing a connecting circle on the sides of the triangle half a circle forms. By implicating Π as a relevancy and not the straight-line r, two values of Π applies to each circle, and the straight line is no longer r, but is Π^2. This will bring about that each circle holds half the square value implicated to the allocated conditions applying to Π in that specific instance. By adding the two half squares forming the two half circles and then calculating the square root of the total that then forms the average diameter, an average of Π in the connecting line will come about. As both lines are the straight line forming singularity coming from one line being Π, the connecting line then must be the average of the two lines as Π^2. That is what **the law of Pythagoras says.**

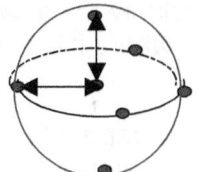

In the sphere there are never only one direction implicated in movement. Movement are always in relation to the centre position because as a line goes up it also goes in or out. When a line goes north or south, it also comes towards the centre or going away from the centre.

There is always relevancy present in movement. As this moving indicates direction it also apply Π^2 for indicating value forming the time factor.

In the sphere there are no radius but only the extending of Π from the centre Π in six opposing directions relating to one another by the square but remaining Π because of the unity the matter holds in relating to space. It is not possible to draw a precise line that would form a precise ring and not cut some atoms in parts. Because there will always be an atom disallowing the precise positioning of the circle the circle continues on a solid basis holding Π as a positional reference and not r. In every sphere there then are the seven Π relating in precise dimensional and positional equality forming equilibrium to the centre Π as well as to one another by 90° and 180° implicating the dimensional positioning. Therefore the sphere holds 7^{Π} and the cube holds $6 \times r^2$

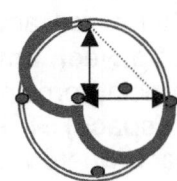

Because every moving line represents one quarter of the sphere in relation to the rest of the sphere and the line also indicate the relevant position between the point indicated and the point in the centre it is a relevancy of singularity in progress. By connecting the line, as Pythagoras will suggest the singularity within the sphere become a specific value indicated representing one half circle.

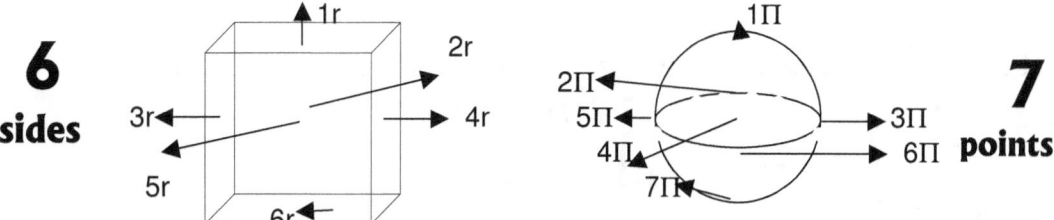

6 sides   **7 points**

Where space comes into contact with the sphere the cube loses one of the six dimensions it has to the more dominating seven dimension of the sphere whereby the seven dimension in equilibrium will dominate the six dimension loosely connected by r bringing about that the cube then has 5 sides to the seven of the cube. This means that in the cube the "bottom falls out" and without a "bottom" to support objects they fall to earth. Remember that a body "floats" in space, but at one specific point it starts to "fall" to the earth. That is gravity and it is a dimension change much more than any force. I shall explain this last remark later on.

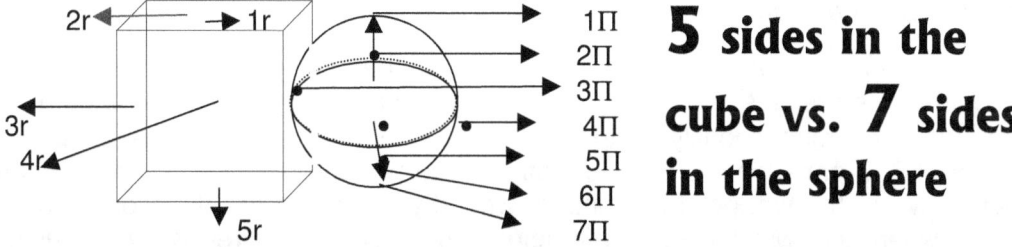

5 sides in the cube vs. 7 sides in the sphere

That too is the Lagrangian system with five cosmic structures holding relevancy to the centre structure where the centre structure stands in for seven positions diverting from singularity and the orbiting structures standing in for five positions in space.

Space-time is a four dimensional position of the Universe where the position of an object is specified by three coordinates in space and one position in time.

With singularity placed in infinity within the centre of every rotating object every atom and its relation to its surroundings including other atoms form space-time diverting from the point holding singularity as far as rotation goes because every object holds three relative positions in as far as where it was, where it is and where it will be in relation to singularity providing time. I elaborate on this else where.

Any point will be opposing itself within the **rotating of 180°** where it **then change every aspect** of its **previous flowing** characteristics it had or **will once again have** in **360°** from there. While in rotation from the view point of a bystander it all may seem static and never changing but to the object in spin every next instant in time will be diverting from every aspect it had every second passing, and the direction it held in relation to the direction it held the previous mille, mille second will totally be incompatible with the direction it holds the very next mille, mille second of rotation.

This is why we can use degrees measuring the circle by (6^2) (forming the square relating to matter through singularity) X 10 (square if space) = 360° however it is always in motion. That proves no point can be static or constant, though it may seem that way to outsiders. Although matter is matter, matter can also be anti-matter and moreover form its own anti-matter at the same time. This degeneration of structure is very likely to occur with overheating.

Revaluing Π to Π^2 will bring about a new contact point where Π meets r forming another relation in Π^2 **Time is** the **changes in relation** where Π **contacts a different** r not withstanding the many r points there may form because **every r constitutes a different value** to the Universe through other ratios and relevancies brought about **by heat and light. Time is the duration it takes Π to rotate between any two given points of r** and therefore must always amount to **a square (T^2)** moving from point to point through the **cube of space (a^3)** in that **duration of time (k)**. With that it proves **Kepler's a^3 (space) $=T^2 k$ (time in the instant of motion)** but motion must continue through a specific value in space where the space-time is maintaining relevant equilibriums throughout singularity connecting.

The TITIUS BODE Principle Outside the sphere

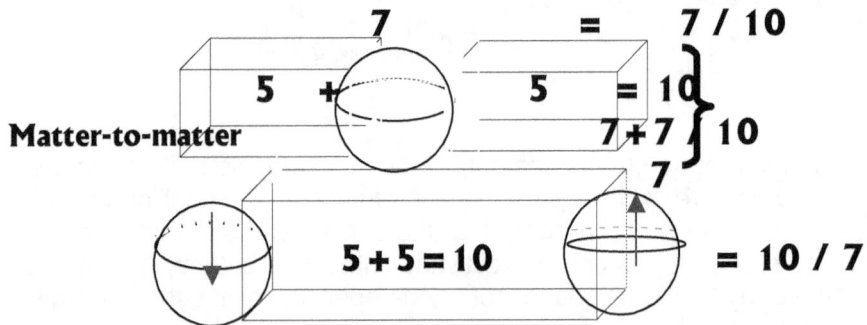

Matter-to-matter

With the dimensional change from space in the cube to space in the sphere a relation of 5 to 7 comes about depicting gravity. The principle of 5 sides in space relating to 7 in the sphere holding matter forms the basis of the Titius Bode and the Lagrangian principles.

The Titius Bode law is an extending dynamic deriving from the law of the gravity dimensional factor where the space factor in a square of ten relates to a matter factor in the square by half (half since nothing can be in two places in the Universe simultaneously) of the matter factor of Π^{7+7} or the square of space (10) relate to the matter factor of 7. From such a point every other point will be opposing any other point not pointing in the direction to which the first point is pointing, whereby it extends the direction it holds. No matter what the point is or where the point leads, such a point holding a specific direction will be unique in the direction it is rotating because at that or any other specific point wherever, it will be directing not in the direction it spins but in the direction flowing from the centre point outwards.

In the Roche limit the space factor provides space to a solid structure and therefore the value of r is replaced by the value of Π bringing about a square in half of Π. The cube holding 5 to either side removes allowing the extending of Π to indicate position to space. Where Π extends to lock onto the next sphere's extending indicator, Π has to connect to Π forming the square of space and translating that to the half of Π being $(\Pi/2)^2$.

5/2
Five sides divided by two spheres.

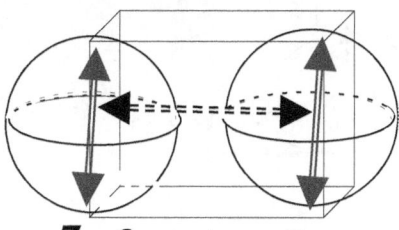

7 Space-to-matter

The Roche limit
$5/2 = (\Pi/2 \times \Pi/2) = 2.4674$

The space between the spheres divide in half, but because of the extending of Π and not applying r as ordinary mathematics will suggest where Π replaces r the singularity extending from Π^0 will be half of Π in the square of $\Pi = (\Pi/2)^2 = $ **2.4674.** In this lies the dynamics why planets have a positional (be it rather a dimensional) relation of 7/10. There are many other borders that control space limitations such as Π forming the atmosphere and $\Pi^2 + \Pi^2/7$, which I shall explain.

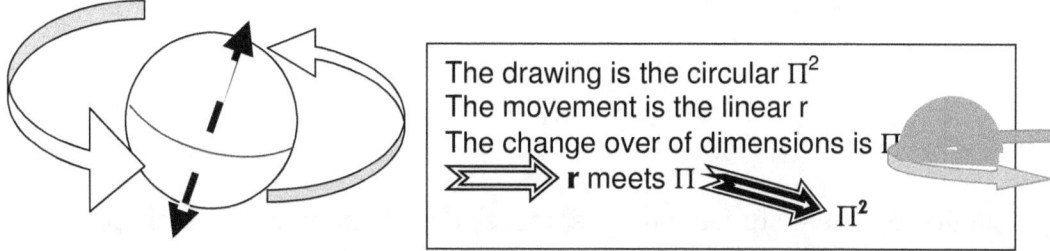

The drawing is the circular Π^2
The movement is the linear r
The change over of dimensions is Π
⟹ r meets Π ⟹ Π^2

In the action of the inseparable drawing closer and moving closer gravity finds the dual value of linear and circular gravity. There is no separation of the two factors acting as one but both have different application and values in the unit. This is the result of singularity having three parts acting as one but giving three distinctions in application.

Gravity is the dimensional changing of heat holding r as reference to the sphere holding Π as the reference. Heat occupying space has the cube that can apply r, as a straight line bringing about the cube with all its other names than may find attachment to specific form but nevertheless still remains only a six-sided cube with angle changing in some cases.

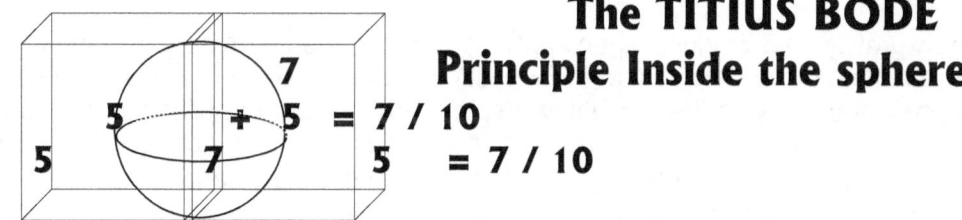

The TITIUS BODE Principle Inside the sphere

$5 + 5 = 7/10$
$5 = 7/10$

Space-time is a four dimensional position of the Universe where the position of an object is specified by three coordinates in space and one position in time

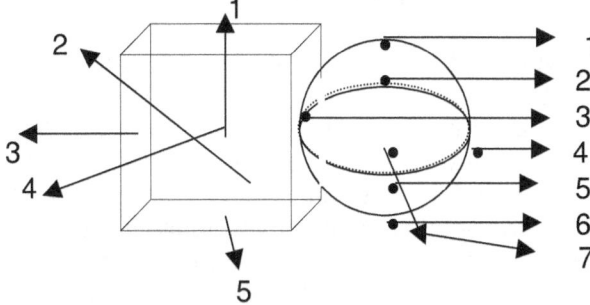

As the meeting of r points to a very distinct different r in direction such a point of meeting opposes the other points in meeting and will lead to destruction of the form Π in any the event of any value changes by Π changing Π^2 and r.

By coming into contact with the sphere the cube loses on dimension to the seven dimensions dominating six bringing about that the cube then has 5 sides to the seven of the cube. That is the Lagrangian system with five cosmic atoms holding relevancy to the centre cosmic atom where the centre cosmic atom stands in for seven and the orbiting cosmic atoms standing in for five positions in space. There is a more explicate explanation about this somewhere else in this book.

It is moreover the individual singularity in maintaining the major singularity, which sustains the governing singularity providing equilibrium in space-time.

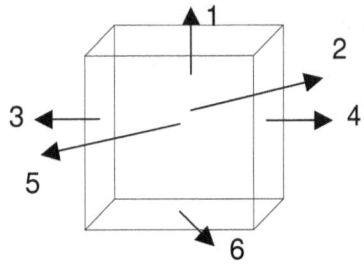 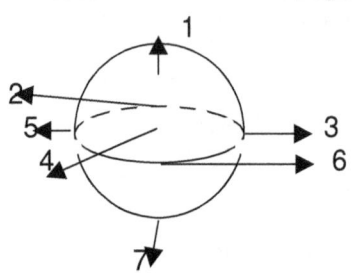

In the sphere there are no radius but only the extending of Π from the centre Π in six opposing directions relating to one another by the square but remaining Π because of the unity the matter holds in relating to space. In every sphere there then are the seven Π relating in precise dimensional and positional equality to the centre Π as well as to one another by 90^0 and 180^0 implicating the dimensional positioning.

Therefore the sphere holds $7^Π$ and the cube holds $6\,r^2$

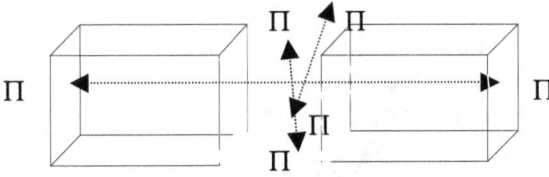

Seeing our spinning top from the top, there are four quarters opposing each other and by that opposing one another.

The sectors provide individual singularity a means in sustaining governing singularity by which provision comes through maintaining governing singularity the required spin in maintaining cooling. If this process did not apply, there would be no connecting individual singularity to major singularityWith the establishing if the value Π and identifying r, one has to distinguish and define each item in order to bring comparison. No line can start at zero but only on the smallest dot one may imagine bringing about that the line will start with the value of Π and proceed and proceed following the same value. The value has to be an extending of the original value because no evidence indicates changes that may take place. In the centre of a rotating circle is a spot $Π^0$ that becomes a dot Π and without the spot $Π^0$ the dot Π can never be and visa versa is also true.

This shows how gravity goes singular when applying time and then form space as the history that time left behind. We are in time but we live in space. Everything in the unniverse spins and by spinning everything forms singularity in the cvery centre of whatever spins. When gravity applies the location of the centre of whatever changes position in time provides singularity in that very centre. These centres holding singularity connects and this connection places the Universe in singularity. The Universe is $Π^0=ΠΠ^2/Π^3$ **and this means singularity $Π^0$ forms space $Π^3$ by the movement thereof while the movement $ΠΠ^2$ of space $Π^3$ brings about singularity $Π^0$ which is the value by which the Universe "grows".**

Everything that holds gravity in the Universe is round and is spinning while spinning around some other object. In that we have to locate gravity because mass is not present. See how they are allocated in total random as if the Universe does not know about Newton's presumption of mass that should play a part. The only persons that are unable to see this anomaly are those so very educated in the science of Newton's principles that just never apply. If mass was present then surely there has to be some arrangement that show how and why the Universe adhere to mass, but there is not even a hint of a factor such as mass with a pulling power. Nothing is pulling anything because in place of the pulling we have the Titius Bode law, which I explain later on in the book. For the first time ever I am going to explain how it works and not merely the layout as the Newtonians do because the layout is rooted in Π forming.

A Less Complex Commercial Science Book — Chapter 10

Wherever we look we find rings and rind associate with Π and not with mass. The rings show that gravity has to be well connected to the roundness of rings because it is always rings form around structures. Even the moon is a ring circling around the earth much like the rings circle around the gas planets. In any ring there is always an incorporating centre and in that gravity must be vested. We have to look for a centre and a ring forming a circle by which it would lead to a circle moving. That will mean that if we wish to locate gravity we have to find a centre of a circle and we have then a ring circling around a centre. That is gravity and in that there is no hint of mass playing any part in gravity. When you show any Newtonian about the fact that all things fall equal as Galileo proves they come up with many arguments that avoids the topic. The clepsydra proves that gravity functions without mass the best.

A clepsydra or water clock is an ancient device for measuring time by means of the flow of water from a container. A simple form of clepsydra was an earthenware vessel with a small opening through which the water dripped; as the water level dropped, it exposed marks on the walls of the vessel that indicated the time that had elapsed since the vessel was full. More elaborate clepsydras were later developed. Some were double vessels, the larger one below containing a float that rose with the water and marked the hours on a scale. The clepsydra was the first recorded science experiment ever recorded. They proved through the clepsydra that by blocking the top intake, the water stopped running out of the bottom hole.

The clepsydra or water clock is an ancient device for measuring time by means of the flow of water from a container. A simple form of clepsydra was an earthenware vessel with a small opening through which the water dripped; as the water level dropped, it exposed marks on the walls of the vessel that indicated the time that had elapsed since the vessel was full. More elaborate clepsydras were later developed. Some were double vessels, the larger one below containing a float that rose with the water and marked the hours on a scale. A form more closely foreshadowing the clock had a cord fastened to the float so that it turned a wheel, whose movement indicated the time.

The Clepsydra shows how air coming from the top pushes water out the bottom. When you stop the air from coming into the top of the clepsydra by blocking the clepsydra at the top the water stops moving out at the bottom of the clepsydra. This proves that to get water to vacate the clepsydra air has to fill the clepsydra. The air coming in moves just as fast as the water move out of the clepsydra. This proves that it is air moving down and the air could be filled with water or not filled at all, the air moves at the same pace. The compressing of air is gravity and gravity is the pushing of air down to the surface of the earth, taking the air down to the centre of the earth.

That is gravity!

It has nothing to do with mass because that water is not moving down faster or apart from the air. When you block the air from flowing the water stops. So that proves that what moves downwards is air either filled with water or not filled with water. The thing to figure out is how does gravity function when mass clearly has no role to play in the process.

Looking at a point where singularity forms a line we find a triangle. This is because a line, a triangle and a half circle have the same value at 180°. By moving from a point holding singularity the movement is by seven points. This explains singularity. Singularity is not in exclusive brilliant formula but in the most basic of Mathematical principles. Two square triangles form one square triangle.

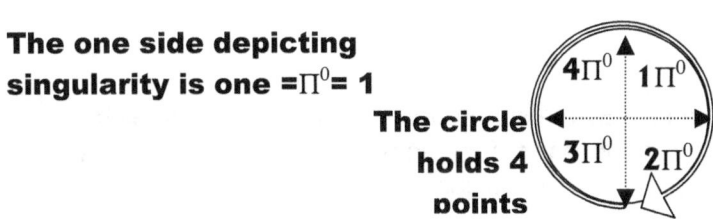

The one side depicting singularity is one $= \Pi^0 = 1$

The circle holds 4 points

The one side depicting space-time is three = 7

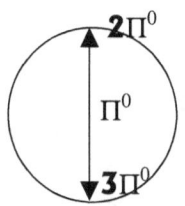

The axis forming the line holds three points and the circle holds 4 points

The axis forming the line

The calculation of the triangle involves the law of Pythagoras.

Since I am explaining the most elementary of mathematics I better explain the law of Pythagoras as well. **(For those SUPER-EDUCATED-MASTES BEING TO ADVANCE TO REMEMBER THAT)**

The law of Pythagoras states **that the sum total of the square of the two sides in a rectangular triangle will always be equal to square of the perpendicular side.** How basic in mathematics can we still go?

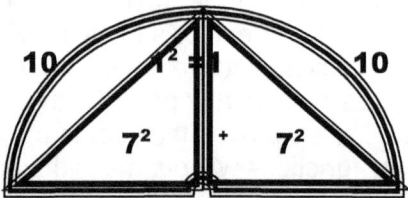

The axis holding $3^2 = 9$ and singularity forms space – time at 10

The square of space from the point of singularity is $2(1^2 + 7^2) = 10$

I have indicated a few pages back that the position the proton holds in considering singularity is $\Pi^2 + \Pi^2$ and the neutron holds in same fashion a matter in space of $\Pi^2 \Pi$. I also showed that matter can only relate to space by implicating matter in the sphere and therefore has to use the value of Π as a reference. In one dimension space became 10 and in that same dimension matter became seven.

In order to separate matter (7) and space (10) through time (the spinning of matter) (7) in space (10) and space (10) spinning the matter (7) the following result came about through the application of the Roche principle $(\Pi/2)^2$

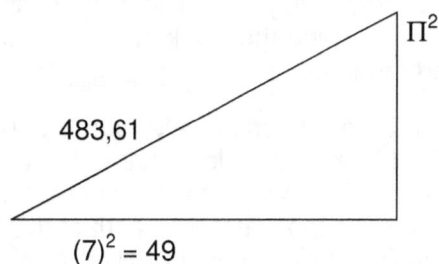

$(7)^2 = 49 \times \Pi^2 = 483{,}61$. But at this stage matter was in the relevancy of $(\Pi/2)^2$ therefore the factor that $(\Pi/2)^2$ represents, holds the value of 483,61. To get to the resulting dimensional value of 483,61, the square thereof becomes a factor.

$\sqrt{483{,}61} = 21{,}991$

From this one must conclude that gravity is also 21,991 when the moving body aligns with the 7 forming the contraction.

As 21,991 is one half of $(\Pi/2)$. Π Therefore, must be that value matter holds, which is 7. When matter divides into space $(7 \div 21{,}999)$ the result from that is Π. Through this the neutron (which is matter) holds $\Pi^2\Pi$ as a factor. Through that, the value of the triangle in matter, space and time also holds a 180°.

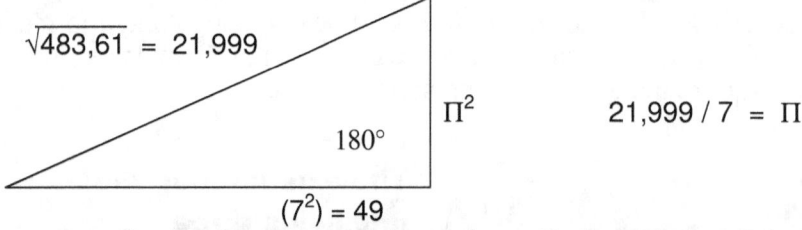

No person can deny the fact that the earth is a sphere. By the sphere turning into space it forms gravity in relation to singularity being $(\Pi^2\Pi)$. Going into space the earth spins into $(\Pi/2)^2$. This forms as a result where the one side of $\Pi/2$ turns into the other side of $\Pi/2$ by moving forming a square that forms is the Roche limit.

The spin of the circle forms 7° on the one side of rotation as well as 7° on the other side of the centre of rotation.

That puts the centre turning by $7° \times 7° = 7^2 = 49$. Then on the other side of the divide the same repeats where there too forms by turning $7° \times 7° = 7^2 = 49$. Then by using the law of Pythagoras and incorporating $\Pi°$ on both sides 10 forms as a space value outside the circle.

A double 10 plus a developing singularity value of 1.9991 represents singularity, which represents Π turning around Π° to form Π. This means the one side of the circle holds Π = 21.991 ÷7 and on the inside connecting to the circle centre holding singularity at Π°= 1 the value of Π is 3.1416 ÷ 1

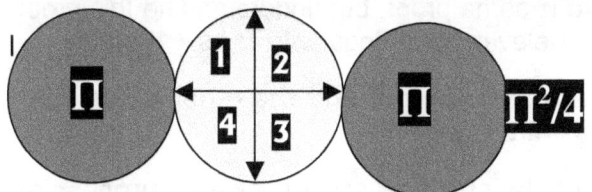

When the circle rotates and the rotation is prevented by an obstruction within the limits of Π duplicating $Π^2$ **within the area forming space where the four quadrants are the part that forms the material component would liquefy anything that blocks the relevancy in singularity between solids and liquid.**

In nature there is the Roche limit placing a limit to the reduction of space and the inflow of heat to sustain progressive cooling. Materials spin in order to contract the surrounding liquid onto and into material or solids to maintain the cooling of the solid. This is a very complicated process and I have devoted one half of an entire book just to go into this process. At a point of $(Π/2)^2$ the reduction of space disallows any object the cosmic object cannot reduce, an entry to its area of reducing space.

The first question that one can ask is why would there be the value of $(Π/2)^2$ between orbiting structures positioning themselves in a time relation to space. This is to establish Π and to get Π to move and form $Π^2$ from the one side Π/2 to the other side $(Π/2)^2$. But the air forming the atmosphere is still part of the earth and therefore it does not become the full Roche limit $(Π/2)^2$ but half of it $(Π)^2/2$.

Then we had two point holding eternity in place going square by movement to form 4 points serving eternity and infinity captured the first three points held by both and since eternity could not release from the two it had but had to duplicate what it had, eternity by movement became a circle captured by the line. With four points captured by the line of three points the circle coming about is eternally returning to infinity but never complying with infinity because if mismatching temperature or movement (3 against four). Material will always be colder than outer space. It is because material spin and outer space moves by expanding due to overheating.

This is where I start when I start to explain the first moment but I use a shipload more information to do explaining when I explain the star in the book I do so. I involve the four cosmic pillars to substantiate the claims I make because all four still work the very same way as it did at the beginning of the Universe. The three points serving one part of singularity combined with the four points serving singularity unites as seven to form a circle of either 3.1416 or 21.991÷7. The seven going to one is eternity matching infinity by movement. But since seven moves it is seven that has to produce gravity. How do I know all these facts, because we can see from the top it is still doing what it did the very first second.

The $Π^2$ end will be at the point where heat passes through the object directly to the earth and this position of space-time relates to the neutron time link of $Π^2$. The space link of the neutron will then form the Π link. The value of the Π link we find to be $(Π)^2/2$, but the explaining to why it is $(Π)^2/2$ is rather more complicated.

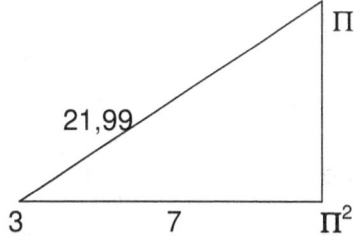

At the end of the space relevancy 3 where matter occupy space (21,991 / 7) is a border Π. That border is the exact point where space reforms to a square of time placing all matter (occupied heat) and heat (unoccupied matter) to a value of the square of time.

That specific point is in relation to the square of the diminishing shield around the earth. However it takes matter ($Π^3$) from the 3 dimensional positions to the square ($Π^2$) in relevancy to time in singularity. With time holding space in singularity the 4 sides of Π truly relates to half of the total square value.

Everything in the Universe is a sphere formed by singularity extending. I can give the displacement values but it would not mean much at this point to any reader. Creation started at a point long before the Big Bang, long before light and even before space, as we understand what space is as it is space.

If there is anyone doubting my revelations or my claims here is some proof, but understanding this proof requires the reading of all six parts of the books An Absolute Relevancy of Singularity Thesis 1 to 6.

7 / 10 $(\Pi^{2}/^{2})(\Pi^{2}+\Pi^{2})$ = 139. This is when the two factors one being hot and the other being cold formed identities in relation to each other and the Roche limit formed.

$7(\Pi^{2}+\Pi^{2})$ =138. This is the point that gravity formed a circle by spinning and repeated a process at the seventh point forming singularity.

10 / 7 $\Pi^{2}(\Pi 2+\Pi^{2})$ = 136. At this point space became apart from gravity and contraction of space by singularity recouping space began.

By applying a different position Π becomes r in relation to the previous position Π held because the circle now have to introduce a line in support of the new circle. The loss to density through the application of a new time relation will be suspended matter forming in a heat release.

Around the earth we find several of layers of air that is compressed in ever more dense space. While it is impossible that "mass" can merely pull "air" into denser units, it is very plausible and mathematically correct to presume that the spin of the earth can redirect the space into more compact layers and in this we then have winds and other disturbances coming about from this turbulent spinning action. It is clear that the air flowing around the earth acts in the same way as liquids do and has the same characteristics as liquids have. Studying the sun we find a higher concentration of liquid around the sun, which proves the Newtonian suggestion that the sun is a gaseous formation completely misrepresentative. They presume because the sun is mainly filled with hydrogen and hydrogen on earth is a gas therefore the hydrogen in the sun is also a gas. Think of how an air-conditioner unit functions when it reduces the heat on the inside of a room and then look at the sun. It repels heat to the outside to cool what is inside. Now we see the air conditioner from the outside blowing heat out and we say the conditioner is incredibly hot.

Same with the sun, it repels heat in a cold liquid form from the outside the freeze what is inside. Again Newtonian backward thinking is astonishingly stupid. If it is liquid squirting from the sun, then the hydrogen in the sun must be a liquid irrespective of what Newtonian culture demands to dictate. Gravity is the Titius Bode law applying in the limits of the Lagrangian points, adhering to the Roche limit and the lot culminates into the forming of the Coanda effect.

The Coanda Effect has been discovered in1930 by the Romanian aerodynamicist Henri-Marie Coanda (1885-1972). He has observed that a steam of air (or a other fluid) emerging from a nozzle tends to follow a nearby curved surface, if the curvature of the surface or angle the surface makes with the stream is not too sharp. If a stream of water is flowing along a solid surface, which is curved slightly from the stream, the water will tend to follow the surface. Now for a very simple demonstration: If you approach gently a curved shaped surface (like the shape of the primary hull of the Repulsing) under a stream of water (see below):

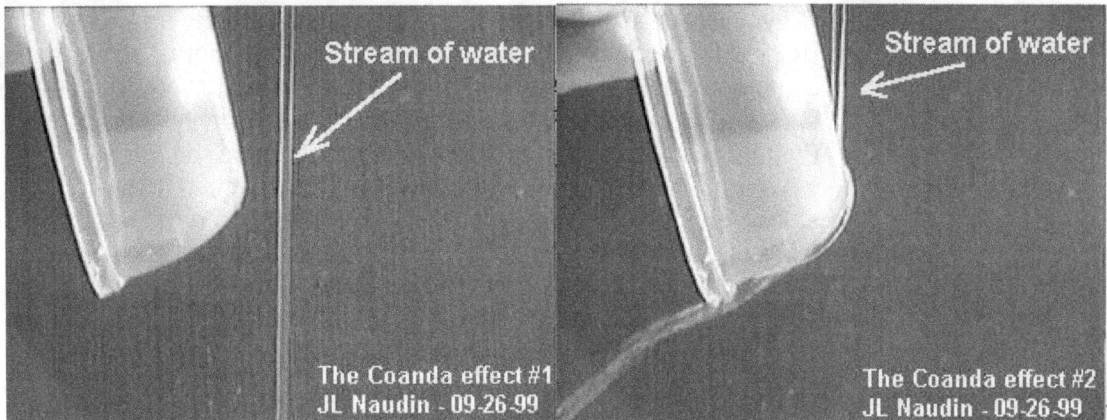

The water follows the surface of the curved shape, this is the Coanda Effect and the Coanda Effect works with any of our usual fluids, such as air at usual temperature, pressures and speeds.

The picture above is about a phenomenon called the Coanda effect but this is never mentioned in any physics handbook because Newtonian physics-religiosity is unable to explain or to understand this principle. When liquids flow past a cylindrical object the liquid clings to the surface of the object rather than follow the "path of mass" and fall straight down to earth. Gravity is about the atmosphere that forms a liquid that is the same as the liquid running around the solid circle called the Coanda effect. Gravity is the movement of the air in relation to the movement of the earth. If an object moves within the parameters the earth set the liquid to move around the earth and the object has the ability to maintain the speed, it will be just more liquid floating around the earth at a specific height fore filing a specific circle or rotation requirement. If the speed drops the object will fall notwithstanding mass or the lack thereof. The entire idea vested in gravity focuses on speed-differences.

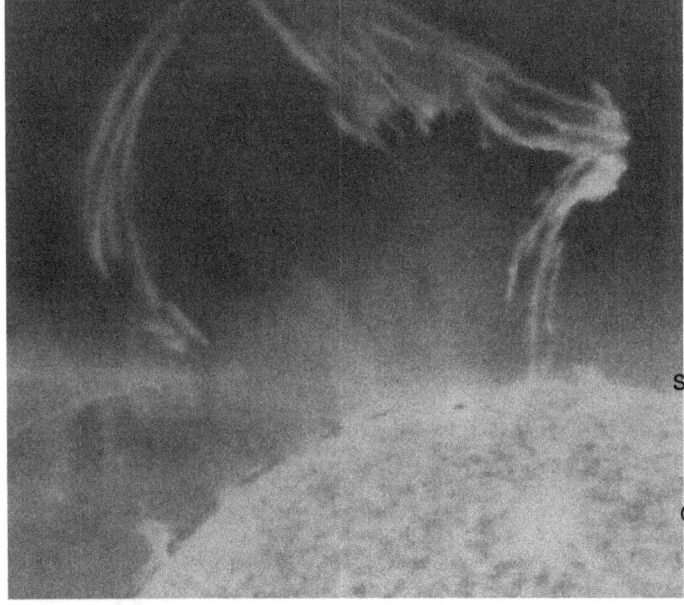

So what the hell has mass got to do with the entire affair of gravity?

If the circle is as large as the sun is, then the compressing of space turns the air into squirting liquid precisely as the picture of the sun shows. Again I ask what has mass got to do with the entire idea.

But remove mass from the equation and the Newtonian brilliance becomes idiotic stupidity because their entire mindset of playing God with mathematics becomes total invalid and their stupidity rings as load as a Cathedral tower bell. In front of your eyes you witness the sun freezing an atmosphere which the entire solar system shares and which is made up of condensed space that it freezes into squirting liquid that is the flow of heat, as pure as heat can be. This has noting to do with mass but those conspirers are so deliberately clinging onto their fictional Newtonian Universe that the truth passes them by and in total arrogant stupidity they refuse to use their eyes and see what the Universe shows it is.

The movement of space filled with material cools by movement of that which moves and this puts thermo differentiation between space that moves and space that does not move. As the space differences by thermo differentiation grows it would seem that one part grows in relation to another part shrinking. This is why large gravitational stars always seem to lose space while the Universe seems to expand.

Gravity forms as the earth or any other cosmic body rotates by 7°. By diverting the straight-line movement by 7° a contraction forms in a circle. In my books I prove how this then brings about the value of Π by implementing the law of Pythagoras and gravity is the law of Pythagoras. The reclining of space by redirecting the direction of travel from straight ahead to 7° reclines the space in a steady and sturdy flow. It is the space reclining or contracting and the space contracts albeit filled by solid material or empty of solid material. This is the reason why all things fall equally. It is the space moving down with or without holding material and the space has the same density in relation to the solid cosmic structure rotating.

In the centre of the rotating body singularity forms to the value of Π^0 and this extends to the curve forming Π in terms of the curve of the rotating object. This then forms part of gravity moving forming

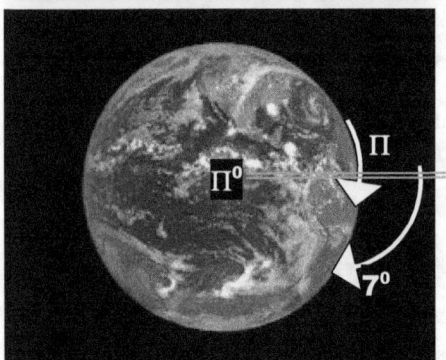

The extending of singularity goes from Π^0 to $5\Pi^0$.

singularity going square Π^2 and this form the relevancy Coming in from space while having no mass the coordinates change as the value of Π forms a relation to the 7°

When an object comes in from the atmosphere or flies the sky the object associate with the turn giving it a value of the rotation in line with the axis which puts a value of 21.991 / 7 on Π or on gravity. This is gravity and this is how gravity functions. It has nothing to do with mass in any way, shape or form and there is no factor such as mass in the entirety of the cosmos apart from being in the imagination of the Newtonian conspirators calling them

The entering by singularity that goes from 7° by $3\Pi^2$.

physicists. I sent a fifteen-page article to the Annalen Der Physics in which I explain this process in length. I received the following reply from Annalen Der Physics.

Dear Dr. Schutte, You submitted an article of 15 pages to the Annalen. The content of this paper doesn't constitute a theory in physics. With a lot of words and some simple algebraic relations, there is no way to "explain" the world of physics. You seem to be out of touch with modern developments. This is also shown by the fact that you don't quote any relevant literature. I am sorry to say, but the Annalen is not able to publish your work. I am sorry for having no better news for your. Best regards, Friedrich Hehl
Co-Editor Annalen der Physik (Berlin)
Friedrich W. Hehl, Inst. Theor. Physics
* University of Cologne, 50923 Koeln _____/\/_____ Germany
fon +49-221-470-4200 or -4306, fax -5159
hehl@thp.uni-koeln.de, http://www.thp.uni-koeln.de/gravitation
* Univ. of Missouri, Dept. Phys. & Astr., Columbia, MO, USA

From this reply I received it is clear professor **Friedrich Hehl** did not understand a single word I sent him because as he said, it was not in mathematics but written in words! Now I beg of Friedrich W. Hehl, Inst. Theor. Physics University of Cologne, 50923 Koeln _____/\/_____ Germany or any of the editing staff of Annalen der physics or any physicists professing the principles of Newtonian inspired physics to prove to me how does the formula $P = \left(\frac{4\pi^2 a^3}{G(M+m)}\right)^{0.5}$ use $P = \frac{}{G(M+m)}$ to place the positions of planets because this is one of the supposed Kepler Laws. I wish him well in using Newton's corrupt formulas to prove that Jupiter with having the "most mass" is bang in the centre and on either of the outside and inside we have the two smallest planets. In that put the mass in relation to the position. Prove to me that mass is a valid factor in the cosmos. Physicists live in Newton's dreams. My work proves nothing so please prove Newton when he claimed $P = \left(\frac{4\pi^2 a^3}{G(M+m)}\right)^{0.5}$ especially the part $P = \frac{}{G(M+m)}$.

A Less Complex Commercial Science Book   507   Chapter 10

Do you realise what you see in these pictures arte the same cosmic principles applying in such a manner that it is the very same thing happening in the cosmos and on earth?

The circles forming is gravity forming Π by the movement of Π^2 as gravity Π^0 extends into wider circles by enlarging Π from connecting to one dot and then connecting up to $4\Pi^0$ in the circle.

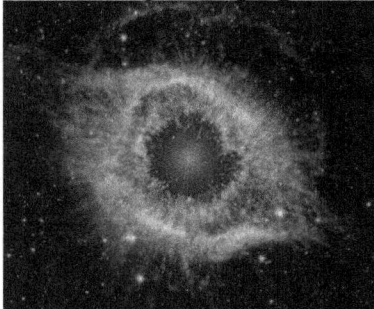 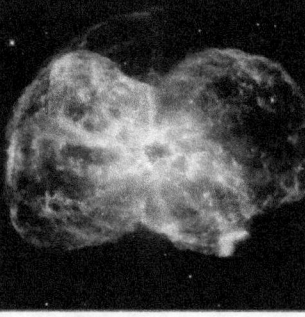 

Do you know what happens in these pictures? Of course you do know this, but do you understand this? Knowing this says little because it is the big anomaly science can't explain because Newtonian science will not allow any correct understanding of this. That is because Newtonian principles can't correctly explain this. This is one of the four cosmic phenomena that prove Newtonian philosophy about physics to be incorrect! This website will introduce you to singularity that is a totally new concept.

Please believe me when I say you have never even seen the true physics behind whet drives the Sound Barrier or gravity for that mater. To understand what law in physics drive this other than Newton's Dark Age rhetoric goes to Book 1 <u>The Absolute Relevancy of Singularity in terms of Cosmic Physics</u>

http://www.lulu.com/content/e-book/book-1-the-absolute-relevancy-of-singularity-in-terms-of-cosmic-physics/6624181

  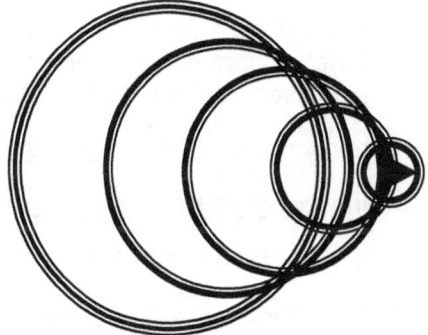

There are four cosmic principle or phenomena that forms the Sound Barrier which is gravity and to free yourself from the hoax of Newton's physics go to **Book 2 <u>The Absolute Relevancy of Singularity in terms of The 4 Cosmic Phenomena</u>** http://www.lulu.com/content/e-book/book-2-the-absolute-relevancy-of-singularity-in-terms-of-the-four-cosmic-pillars/7181003]

The explanations may seem plausible but put Newton into it and the entire concept falls apart. It is not in the explanation of what happens that the answer lies but in the explanation of why this happens that true physics lie…and that Newton's thinking cannot cope with….It is fine and well to give fancy sketches but explaining why and what principles is behind it Newton can never give.

The Sound Barrier is gravity. The Sound Barrier is the combination of four principles forming the Universe as the four are forming gravity. By explaining the Sound Barrier I explain gravity as I explain every aspect of law that forms the Universe we see. By not knowing how the Sound Barrier forms gravity and how gravity forms the Sound Barrier it is an admission of complete failure to understand physics. Do not confuse the sonic boom with the sound barrier. That would be to confuse a thunderstorm with a lightning bolt. That is what Newtonians do not understand and that serves then bad.

To learn more about the Sound Barrier go to <u>The Absolute Relevancy of Singularity: The Theses</u> http://www.lulu.com/content/e-book/the-absolute-relevancy-of-singularity-the-theses/7587475]

To explain the Sound Barrier and therefore gravity I had to find the **location, position** of singularity Π°Πas a factor forming space-time.

If you wish to get the technical professional explanation got to **The Absolute Relevancy of Singularity: The Sound Barrier.** http://www.lulu.com/content/e-book/book-3-the-absolute-relevancy-of-singularity-in-terms-of-the-sound-barrier/6621856

By looking at the Sound Barrier we can see what gravity is but in Newtonian terms the explanation of the sound barrier is non-existent. Everyone in science has an explanation about the sound barrier and every explanation underlines the absolute incompetence in science to form any type of explanation.

No one can explain the cloud forming around the aircraft as the aircraft fly well below the sonic boom. The cloud does not make sense when Newton's gravity is used to explain this phenomenon. No one in physics can use physics to explain the cloud forming. We can all see it is vapour condensing around the fuselage of the aircraft. Let them show the mathematical explanation why this happens. It is easy to see what happens but explain why that "what" happens goes beyond what Newton's physics can cater fort. If you throw out mass and forces physicist can offer little because then they know little.

> To explain the Sound Barrier and therefore gravity I had to find **space-time** not as some magic entity Einstein sometimes referred to but something that is everywhere around us and which we are part of. by dissecting Kepler's formula in relation to **valuing singularity**

I not only found but I also **proved space-time** by **aligning space-time** with **gravity**

I found the **working principals** behind **gravity** as a cosmic occurrence.

I found the reason for the **Roche limit** and explaining the resulting of **gravity from that**.

I found out why the **Lagrangian system**, becomes **the building form** of the Universe.

I found why the **Titius Bode law** mathematically provides **the foundation of gravity**.

Go to **The Absolute relevancy of Singularity: The Unpublished Article** http://www.lulu.com/content/e-book/the-absolute-relevancy-of-singularity-the-unpublished-article/7747133]

I am able to explain The Sound Barrier mathematically by not applying

Newton's physics. Do you know what this pictures shows...?
It shows a jet aeroplane going through the sound barrier. By going to and then **download** then **The Absolute Relevancy of Singularity The Theses** you will learn what the sound barrier is. No person ever could understand the sound barrier because no person ever could enter singularity and see what applies when the Universe goes singular.

You will read what exactly happens when the "sound barrier" is exceeded but also I will mathematically prove what gravity truly is. This I explain in simple understandable mathematical terminology. It is so simple even I can understand it. I make this remark because I am forever told I am too stupid to "understand Newton's physics" and therefore I am "unable to accept Newton's physics."

There is nothing to understand in Newton's physics because there is nothing to believe in Newton's physics. With one simple question I derail Newton's physics and that is prove tome there is mass, not weight but mass presented by the cosmos. The Sound barrier is gravity by implicating all four of the cosmic foundation phenomena that forms gravity.

By going to and then downloading then The Absolute Relevancy of Singularity The Theses and The Absolute Relevancy of Singularity Introducing The Sound Barrier you will see why the "sound barrier" is a cosmic limit. This is the route to follow to find the basis of the sound barrier. Download The Absolute Relevancy of Singularity The Website to learn what gravity is and how gravity functions.

Then download The Absolute Relevancy of Singularity The Theses to learn what the sound barrier is in terms of gravity or in terms of the Coanda effect and how the sound barrier functions as a cosmic

principle, My explanation shows you why the "sound barrier" is a cosmic limit. The pictures that I show might not have anything to do with the sonic boom, but the sonic boom is a very small part of the process of what physics think of in terms of breaking the sound barrier, which is a process of movement forming a relation between two objects moving in space at different speeds in example the jet and the earth where the movement of the jet is extraordinary.

By going to the website and then to download the short book with the title www.questionablescience.net in which I show that objects moving in relation to the earth's atmosphere. In the earth's atmosphere no object can move slower than the earth does. If it moves slower than the earth does, it moves as fast as the earth does by receiving mass and being pinned onto the earth as being part of the earth.

There is always movement because of the earth moves and all other movement goes in anticipation of moving above what the earth does. Movement is space that is duplicating filled space at a pace and in relation to other space within other unfilled space. That is why using parachutes slows the falling process of falling objects down because falling is a ratio between solids and liquids.

This is proven in these photos about the sound barrier where the movement of the solid jet contracts the space of the liquid cloud. What happens in the photo is vivid proof of my theory and it proves the Coanda effect is what is forming gravity.

The Coanda effect is a relation between material spinning and air or liquid compressing and that is what

happens between the Earth and the atmosphere. What this picture shows is that objects are always moving extraordinarily as they move in relation to the earth's atmosphere where everything that always moves independent of the earth's movement must move according to the earth when it moves in relation to the earth's gravity compressing space into forming the atmosphere. But it also shows the movement of the aircraft compressing space in relation to the aircraft moving as well.

The Titius Bode law

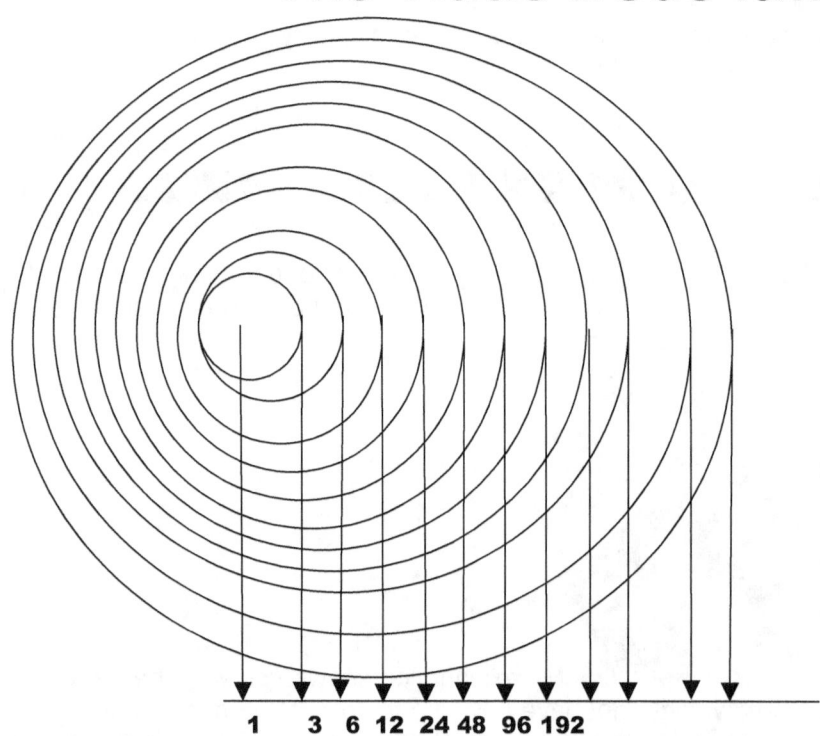

1   3   6 12 24 48 96 192

There is always movement because of the earth moving and when anything moves above and beyond the movement the earth provides that then forms the sound barrier using a modified version of the Titius Bode law By going to and then download The Absolute Relevancy of Singularity The Website you will learn what the sound barrier is. By then afterwards going to and then download The Absolute Relevancy of Singularity The Theses you will learn what the difference is between mass and gravity and why a parachute slows down falling objects, and you will learn why we so desperately need to move on from Newton's incorrect restraining on the human mind. But either way you will find out why the "sound barrier" is a cosmic limit.

That shows that all movement is gravity or time related. This shows that when objects move then such movement of whatever is moving is moving extraordinarily because everyone always forgets about the fact that it is the earth that normally applies all of the movement while all else stands still in relevancy.

It is not my manner to speak ill of the brain dead or the dead by other means, but in the case of the Newtonian academics I am left

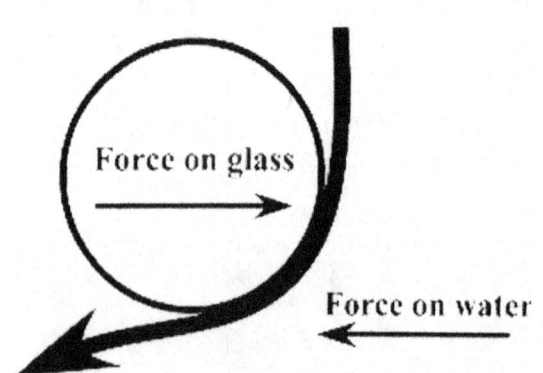

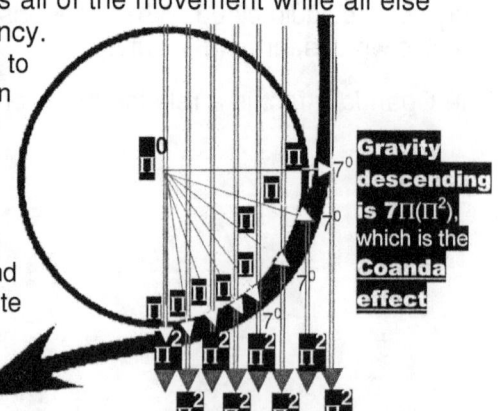

Gravity descending is $7\Pi(\Pi^2)$, which is the Coanda effect

with no option. Their forces haunt me to death and it is their forces and ghosts and witchcraft I have to fight. The lifting of a body comes quite natural when a certain speed is exceeded. By exceeding $7(3\Pi^2)$ the body will start to lift no matter what the mass is. A 747 Boeing of multi tonnage lifts off spontaneously at excess of that speed.

Newtonians are forever concerned with middle ages and with forces they can't explain but such forces and witches there are not, therefore they do not have to fear and can sleep well at night. In the sketch the circle portrays a glass and the arrow portrays running water. The Coanda effect is the water that does not drop straight down but follows the curvature of the glass.

The picture clearly shows the 7° inclination of gravity to the value of contracting $\Pi^0\Pi$. This is gravity!

This is the most vivid example of the Coanda effect and it is what gravity is! It is a whirl allowing the flow of liquid space around a solid centre in relation to the centre holding singularity or $\Pi^0\Pi$ as it contracts space into a denser liquid.

The Coanda effect is gravity and my explaining this statement is part of many other books in which I explain what gravity is. The Coanda effect shows how liquid attach to the solid by $7(3\Pi^2)$ and the solid attach to the liquid by a relevance value of $7(\Pi\Pi^2)$. That is gravity.

Should anyone require more or better explaining I would advise that person to purchase any of my books holding the title as an to go to http://www.singularityrelavancy.com/ and also go to www.questionablescience.net Flying object is under this gravity control of movement and it is this that has crafts fly and cars requiring down force by the aid of aerodynamic devices.

Gravity is defined as a force that is present in mass pulling mass and it is that entire idea that there is not evidence of. When I refer to gravity everyone grabs on a cultural notion of a concept they formed and in that concept they link the smallest part of the concept to the become and represent the overall gigantic principle and by knowing one line everyone has the opinion that anyone then is the absolute master on the idea of gravity. When I freeze any substance the substance contract to a liquid and with more cooling it contracts to a frozen state of ice. The gas expanded more than what the solid did because the gas is hotter than the solid is. When we form the opinion that the outer space expanded to the limits the idea springs to mind that outer space is freezing cold. When I say the Sun freezes hydrogen to a liquid because my eyes see the liquid squirting from the Sun I am dangerously mentally impaired since the Sun is blistering hot. Then through this culture my effort to say gravity is motion and motion is the cooling of an overheating and thus expanding Universe goes wasted. Every one has the opinion that where gravity is the strongest such as the case is on the Sun or the centre of the Earth, such a place is extremely hot and where gravity is least that place is unbearably cold.

Time is centralised in Π^0 that forms Π as space's limit that becomes space by gravity being Π^2.

a^3 symbolises in a mathematical interpretation of implicating the three-dimensional space holding a specific centre in relation to another specific centre indicated by **k** that could apply to either centre points in question. This is always a straight-line **k** representing the position of the **controlling singularity** moving in a circle T^2. The space forming a^3 is a **positional validity** of the space indicated by $k^0 = a^3 / (T^2 k)$.

T^2 is representing the circle that goes around the **governing singularity** k^0 that forms in relation to the line **k** in reference to the centre k^0 The space that forms holds the orbiting planet a^3 in direct circular contact with the space in relation to a very specific centre k^0 moving from point T_1 to T_2 in relation to a precisely placed centre k^0. The circle coming about from T^2 is the **controlling singularity** which is always a circle at the centre that is poisoned by the line **k** in relation to the centre k^0 and by forming a circle it holds reference to the **governing singularity.** Where **the governing singularity** is the centre of a spinning object such as the Earth, the centre of every atom holds **mutual singularity** that collectively puts a mutual value of all the atoms' singularity as a combined equal to the **governing singularity** and then the solar system will provides a **primary singularity**. The one would represent T^2 the other forms **k** that then produces the third singularity forming space a^3.

k is the space taken from the centre k^0 to the end of the line **k.** This line shows where the location is around which planet circles. The specific value about the centre is most important because from the specific centre gravity indicates a positional worth. The line forming **k** is pointing the circle or the **governing singularity** formed as a line that eventually forms a circle running from the centre k^0 to where the space a^3 is indicated.

The turning T^2 of any circle holding space a^3 is valid only if forming a reference **k** to a centre k^0. $k^0 = a^3 / (T^2 k)$. This depicts a position a domineering singularity k^0 fills in relation to another point serving subordinate singularity **k**. There are always a dominant and a serving singularity interacting. If **k** indicates the centre of the Earth then T^2 rotates to form the **governing singularity** k^0 where then the centre of the Sun **k** will form the **controlling singularity.** When the Sun rotates, the Sun's centre k^0 forms the **governing singularity** giving the Earth in orbit **k** holds the **controlling singularity**. The measure of **k** is not a specific value but serves only as an indicator to which space rotates or applies by the space rotating in a circle.

This role of singularity being **controlling** or **governing** is playing part in movement of gravity forming and is very important when trying to understand the role that the four phenomena play in the forming of gravity. It is most important to understand what happens in the event of an object going through the "sound barrier" or when escaping from the Earth's atmosphere. Where the object is standing still holding a position that allows the object to have mass, the object is part of the Earth while the Earth has the **governing singularity** and the Sun has the **controlling singularity**. As soon as any object moves on Earth, the movement switches singularity by allowing the object to obtain the **governing singularity** while the Earth then fore fills the directional circular control in forming the **controlling singularity.** All four phenomena interacts in a manner forming this role where for instance in the solar system the Sun holds the **controlling singularity** and Milky Way forms the **governing singularity**

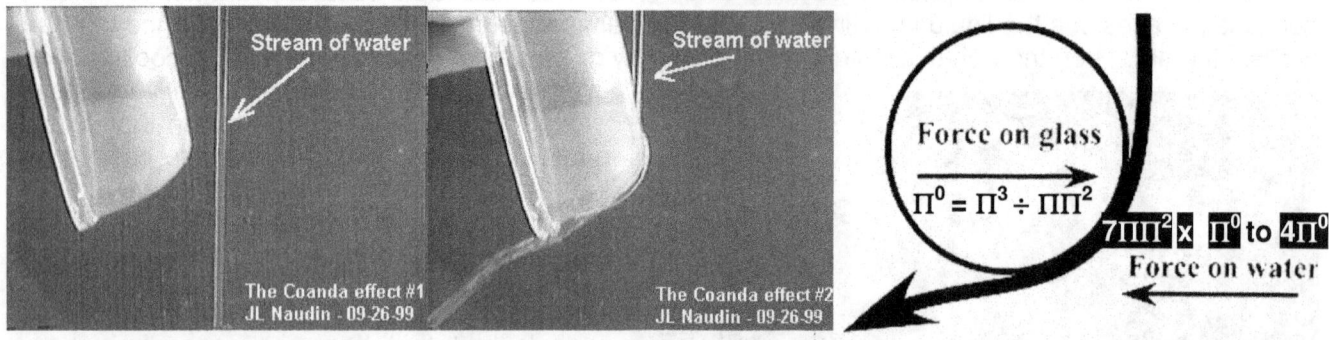

I have written twelve articles in which I explain the Titius Bode law, why it is in place, how does the Titius Bode law apply gravity, what keeps the Titius Bode law structurally in place and why is it in place as it is.

The gravity it should hold in distributing movement will bring along a certain amount of linear gravity to maintain the circular gravity position it wants to hold in the cosmic balance. The duel movement of gravity forming the allocated relevant position provides ratio of singularity Π accompanied by the same value but in movement by the square thereof Π^2 and repositioning the structure as a star.

I now explain gravity again and relate this explanation to the sound barrier functioning.

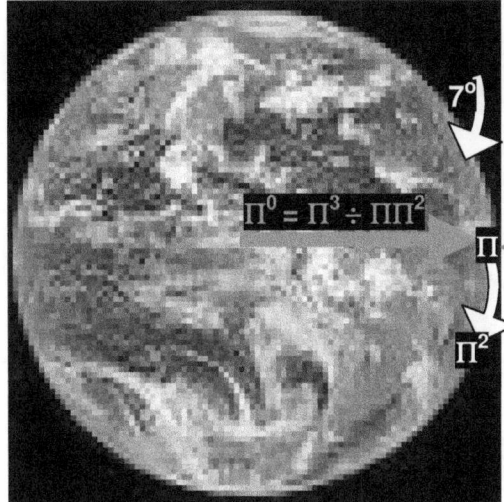

Gravity forms by association or relevancies that holds a value of coming towards the centre from the sky which then is 7(which is the curve of the earth) (Π associating with the curve of the earth) and Π^2 (associating with the moving of the earth) This is part of the Titius Bode law forming the measure of gravity which is not by mass but by forming Π.

$7\Pi\Pi^2$ x Π^0 to $4\Pi^0$

Then when a line is drawn from the centre of the earth another value comes in place referring to the centre of the earth Π^0 and the curve of the earth Π. The value of Π is then shared by both disciplines (association from the sky to the centre ($7\Pi\Pi^2$) and association from the centre to the sky ($\Pi\Pi^0$). In reality it is ($7\Pi\Pi^2$) but I am not explaining this.

If the body does not connect to the earth by way of forming a unit or forming "mass" the axel line of the earth comes into affect giving the ratio a value of 3. Then when not connecting to the earth surface the ratio becomes $7(3\Pi^2)$ x Π^0 going up to $4\Pi^0$ all depending on the sped or the height. This is the way gravity forms by forming a connecting association with $\Pi^0\Pi$.

This forming of gravity explains the Coanda effect because the Coanda effect puts air or liquid in relative movement with solids and this movement (but not the mass part) is how gravity forms.

The condensing of the vapour surrounding the jet is a prelude to the sonic boom and is just one link in the sound barrier where the sonic boom is just another link in the sound barrier process. The sonic boom IS NOT the sound barrier but is as much part of the entire concept as the beginning of movement is part of the sound barrier. Waves in the sea are as much forming part of the sound barrier as winds howling or blowing is part of the sound barrier.

When any object moves it fill a certain space during a specific time period.

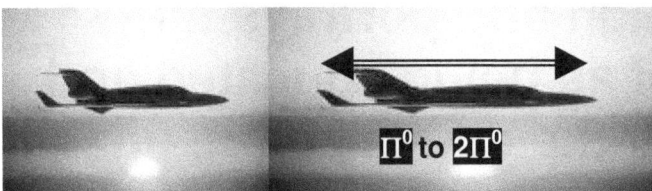 

When an aircraft stands still by only moving with the earth it moves at Π^0 As it moves faster it fills more space up to where it fills from Π^0 to $2\Pi^0$ of sky space and as the speed increase this goes up to $2\Pi^0$ to $4\Pi^0$ The aircraft holds more space but does not fill more air and therefore the air has to compress

Part 11

This is the Article I sent to Annalen Der Physics in which I explained the connection there is between the Coanda Effect and what we see as gravity. This indicates how gravity forms when we take reality into account.

As you will witness the brilliance of Physics academic shines through and portrays the misconception we think of as being educated and being informed but in reality the educated are unable to read words.

In the website of
The Absolute Relevancy of Singularity
Which is called
www.singularityrelavancy.com
shows that the cosmic phenomena perform the way they do by
Singularity

The Following Phenomena is Singularity forming a Universe
I don't only show what the four cosmic phenomena are being
1) The Lagrangian system 2) The Roche limit 3) The Titius Bode law 4) The Coanda affect.

This website is not there just to show how the four phenomena are as they are because for that there are numerous other websites specialising in that. There are many websites that tells how these phenomena are as they are…I show why they are as they are and how it came to be that they are the way they are because as they are, they form the Universe as big as the Universe gets. To know why they are how they are is to know why the Universe is the way the Universe is and how the Universe came to be in the form it is. I don't only tell how they are as they are but explain conclusively each valid number filling the measure that they are where every number fills a place, and that is a first time in all of cosmology that this breakthrough is achieved. I don't merely mathematically formulate, but explain every digit as singularity forms the number and that is how singularity forms the cosmos. Every digit holds a number that validates a space filled that forms the Universe.

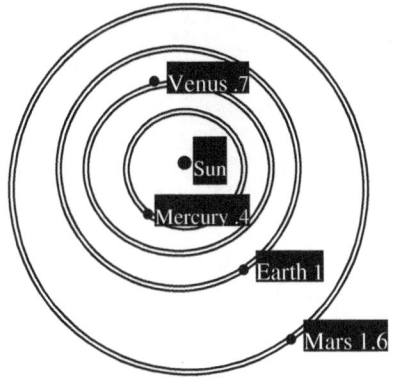
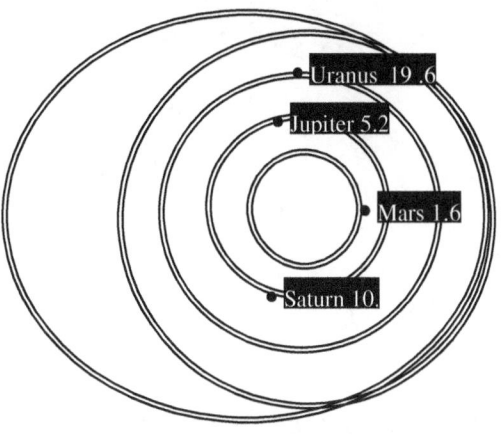

The Titius Bode Law in table form:

Planet	Mercury	Venus	Earth	Mars	Ceres	Jupiter	Saturn	Uranus
Bode's Law distance	4	7	10	16	28	52	100	196
Actual distance	3.9	7.2	10	15.2	28	52	95	192

The Titius Bode Law:

A numerical sequence announced by J.E. Bode in 1772, which matches the distances from the Sun of the six planets then known. It is also known as the Titius-Bode law, as it was first pointed out by the German mathematician Johann Daniel Titius (1729-96) in 1766. It is formed from the sequence 0,3,6,12,24,48,96, and 192 by adding 4 to each number. The planets were seen to fit this sequence quite well – as did Uranus, discovered in 1781. However, Neptune and Pluto do not conform to the 'law'. Bode's Law stimulated the search for a planet orbiting between Mars and Jupiter that led to the discovery of the first asteroids. It is often said that the law has no theoretical basis, but it does show how orbital resonance can lead to commensurability. The importance that becomes known is the sequence the Ties – Bode law saw in the number arrangement of 3; 6; 12; 24; 48; 96 etc. The incorrect application of the Titus Bode law lies in subtracting the figure of 3 from 10 leaving 7. The other way of reasoning is to add four each time to the firs value of three starting with 3 and so on. The true significance of the Titus-Bode law is that it points directly to a circular growth of 7 stages. The 7 relating to 10 is a precise derogative of the Roche limit or the Roche limit is a precise derogative of the Titius Bode principle because he two systems interlink.

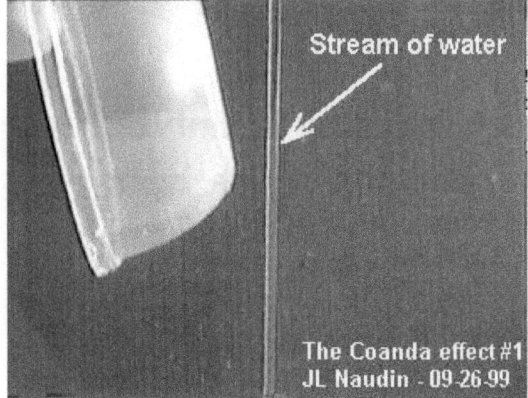

The Coanda effect

The Coanda effect applies as a gravitational phenomenon where moving liquid concentrates around the surface of round solid structures and by movement of either the liquid or the solid or both these concentrates the density of the liquid to gather and compact the flow of the liquid while remaining following the curve of the round surface. The liquid rather follows the curve of the round bowl than to fall straight to the Earth as on should expect. The liquid maintains relevance to the centre of such a round solid. I discard the idea that mass could be responsible for forming gravity because in almost four hundred years all evidence is indicating that the truth is to the contrary.
LAGRANGIAN POINT:

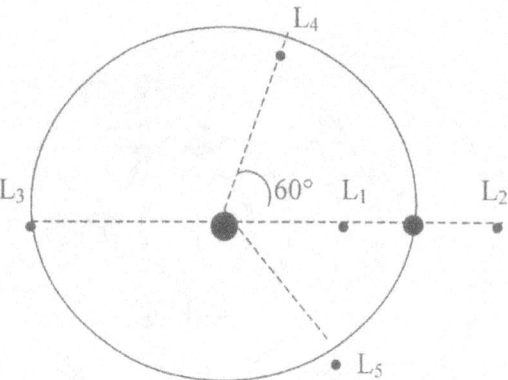

LAGRANGIAN POINT:
The Lagrangian points are five equilibrium points in the orbit of one body around another, such as a planet around the Sun

The phenomena are there and are applying! Put Newton's formula $F = G \dfrac{M_1 M_2}{r^2}$ to task and use it to explain these very common phenomena, and anyone would find it is not possible to use Newton and explain the gravity represented by this. The phenomena are there and applying so if Newton can't explain it then maybe Newton's concept of mass establishing gravity is not applying.

It is this last statement where Newtonian science is unwavering in their believing that mass is forming gravity which is what I strongly bring into question.

Please read on to find more information concerning The Absolute Relevancy of Singularity

The Roche limit is:

The region surrounding each star in a binary system, within which any material is gravitationally bound to that particular star. The boundary of the Roche lobes is an equipotential surface, and the lobes touch at the inner Lagrangian point, L_1, through which mass transfer may occur if one of the components expands to fill its lobe. It names after the French mathematician Edouard Albert Roche (1820-83).

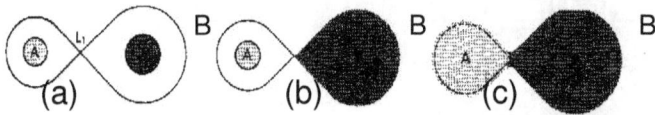

THE ROCHE LOBE: In a binary system, the Roche lobes of components A and B meet at the L_1 Lagrangian point. (a) In a detached system, neither star fills its Roche lobe. (b) In a semidetached system, one massive component, B, fills its Roche lobe. (c) In a contact binary, both components overfill their Roche lobes and share a common envelope.

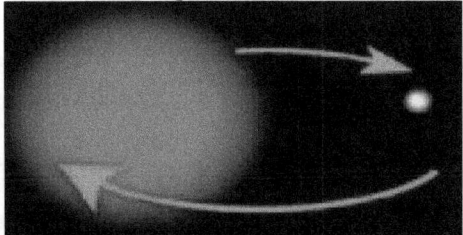

The Roche limit in the practical sense

HOME PAGE of THE ABSOLUTE RELEVANCY of SINGULARITY,
Forming the website
www.singularityrelavancy.com

ISBN 978-0-9802725-3-6

My initial aim was to make this introduction article as simple as I could but I was forced to make it much more informing as I would and with it being an introduction article it now requires a lot more concentration to read than it should...

WRITTEN BY P. S. J. Schutte

All rights are reserved.
No part, parts or the entirety of this book may be reproduced by publishing, electronically copied, duplicated by whatever means that form reproduction or duplication of any description, without the prior written consent of the copy rite owner.

WRITTEN BY PEET SCHUTTE but belongs to
© KOSMOLOGIESE EN ASTRONOMIESE TEGNIKA

PART 2 of The Absolute Relevancy of Singularity,

Forming part of the website **www.singularityrelavancy.com**

By going to LULU.com the following books are available in e-book format as individual books wherein I share with you the newly discovered information about
www.singularityrelavancy.com which you are reading and which you are free to download
Then download the next book from Lulu absolutely free and see if I exaggerate in any way!
The Absolute Relevancy of Singularity The (proposed) Article Free of Charge from Lulu
The Absolute Relevancy of Singularity The Dissertation
The Absolute Relevancy of Singularity in terms of Newton
The Absolute Relevancy of Singularity in terms of Cosmic Physics
The Absolute Relevancy of Singularity in terms of The Four Cosmic Phenomena
The Absolute Relevancy of Singularity in terms of The Sound Barrier
The Absolute Relevancy of Singularity in terms of The Cosmic Code
The Absolute Relevancy of Singularity in terms of Life

Should there be any person whishing to purchase these books in one volume given as a thesis published in paper format then contact me, on this web address by activating www.singularityrelavancy.com
And you will be able to purchase six books in print as a unit on paper forming one volume with six books going as *The Absolute Relevancy of Singularity* The Theses.
However please note that this printed Theses is very limited as it is printed privately. When you press on the button www.singularityrelavancy.com to activate I will return the e-mail as soon as I can to confirm the availability of published manuscripts and prices. The six books are identical to the six books on offer through Lulu.com but they are in monochrome whereas the individual book in e-book format are in colour where colour applies

The Absolute Relevancy of Singularity The Article is written as the first introduction to introduce singularity forming gravity in the new theorem explaining the Absolute Relevancy of Singularity. Since the article was comprehensive but was adjudged as to long for a physics journal, I decided to offer the article in its original and total layout in which I introduce the framework of my ideas.

The Absolute Relevancy of Singularity The Dissertation is there written as the second introduction to introduce the four pillars in a very wide sense on which the new theorem rests. This is to convince readers about the authenticity behind the explaining and the thinking that forms the new approach to physics backing the Absolute Relevancy of Singularity where gravity depends on Π.

Then The Absolute Relevancy of Singularity consists of a four individual part theses each forming a thesis. There are either six individual books on offer in e-book format or in print could only be purchased as one unit named The Absolute Relevancy of Singularity The Theses. This consist of
www.singularityrelavancy.com which you are reading and which you are free to download
The Absolute Relevancy of Singularity The (proposed) Article Free of Charge from Lulu
The Absolute Relevancy of Singularity The (proposed) Article Free of Charge
The Absolute Relevancy of Singularity The Dissertation ISBN 978-0-9802725-8-1

The Absolute Relevancy of Singularity The Theses called
Thesis 1 The Absolute Relevancy of Singularity in terms of Cosmic Physics ISBN 978-0-9802725-2-9

The Absolute Relevancy of Singularity in Explaining the Sound Barrier called
Thesis 2 The Absolute Relevancy of Singularity in terms of The Sound Barrier ISBN 978-0-9802725-3-6

The Absolute Relevancy of Singularity explaining the Four Cosmic Phenomena called
Thesis 3 The Absolute Relevancy of Singularity in terms of The Four Cosmic Pillars ISBN 978-0-9802725-5-0 and

The Absolute Relevancy of Singularity used to explain The Cosmic Code called
Thesis 4 The Absolute Relevancy of Singularity in terms of The Cosmic Code. ISBN 978-0-9802725-5-0

WHOM IT MAY CONCERN,

I do find much pride in my status as being Afrikaner and would like to have my names used by pronouncing it in the manner Afrikaans dictates...therefore I would sincerely appreciate the courtesy when readers will take note that my name and last name are pronounced in Afrikaans, which is originally from Dutch and must be pronounced that way. Peet one would pronounce "here" which is the closest English to the pronouncing of the "ee". The "Sch" in Schutte is pronounced exactly as school is where both actually are pronounced Skutte or "skool". By pronouncing my name in Afrikaans you do me the utmost courtesy any one can. Being an Afrikaner is what I am most proud of. Another point I wish to highlight is that I feel compiled to produce this work in a comic-like format. I have found that the more intellectual and the more educated Academics are, the less they understand the most primitive or classical mistakes in science as well as physics.

My mother tongue is Afrikaans, which is an African language and my second language is English which is the normal British /American/ Canadian / Australian variety used by many if not most. With English being my second language I am not boasting about my verbal skill in English and there is a hidden motive why I am mentioning that at this point, but I shall get to the explaining a little later on. Ever since my days as a student I had problems with accepting the logic behind Newton. In other words Newton's ideas about gravity never made much sense to me ...and don't come with the nonsense that I don't understand Newton because after you have read my work you will have to admit I am the only one ever that understood Newton because I could correct his very flawed perception about gravity. Then after twenty-seven years of intensely studying gravity as a cosmic phenomenon I found the answer. I have followed a theory that I partly present in these books I named the Absolute Relevancy of Singularity The Theses, of which I investigated the research on a part time basis since 1977.

Ten years ago I decided to formulate my conclusions in a seven part theses I named Matter's Time In Space: The Theses Vol. 1 to 7 which I then compiled as my presentation of my new cosmic theory and then following that I worked on promoting my theses. This took almost every minute of my life the past ten years, as the promoting required my attention on full time basis whereby I was trying to introduce my findings to many academics without having much joy I should add. In promoting my ideas I wrote another twenty-seven books trying to make my ideas simpler or better understood. This past ten years saw me go without any income as I tried to get my theorem recognised, contacting institutions and intellectuals all over the world. Contacting people does not pay and you may take my word on that! Going without a steady income left me almost destitute and finally I decided to follow another path in order to find a manner to get my theory across where it will come to the attention of influential readers, and therefore I decided to publish these books electronically as to try and get around the stranglehold of Newtonian bias controlling science at present worldwide. I then decided to publish The Absolute Relevancy of Singularity The Theses as the six part manuscript by going electronically through LULU.com, which I saw as the only manner whereby I could generate funding by which I eventually would be able to have the twenty seven books I already wrote linguistically edited. Thereby I hope I will acquire the funding whereby I could have the books edited professionally and then afterwards have it published on a Print-On-Demand basis and then distributed through the large retail distributors such as Barns and Noble and Amazon.com. With my first language not being English and the books not linguistically checked by an expert there are bound to be language errors that readers will notice. In the past I tried to check my work myself but after checking say one hundred and fifty pages for language corrections, instead of having corrected work I ended by having four hundred pages of newly written information which is still not language corrected but holds a lot more information. This brought no solution but compiled the problem. This exacerbation of the problem is because my priorities lie elsewhere. I aim to spend money on correcting the work linguistically and then have the books formally printed in ink on paper, as I receive money and in the hope that I will receive money. I hope I will finally have all my work edited professionally as I hope I will find money to do so. However, the work I present I introduced for the very first time ever since time began and comes via my brain and every concept I offer, as an introduction to the world of science, is entirely a product of my mind.

My promise to you is that if ever you are able to prove that the information I present as mine exclusively is not completely and altogether new, I shall personally refund your money immediately. I insist as I say as I prove that gravity forms by π

www.singularityrelavancy.com which you are reading and which you are free to download; download it now

What you are about to read holds the dynamics that would change physics for all time to come. For the first time you will learn what gravity is as you would learn what singularity is as much as you will learn how to venture into a Universe that holds what there is together without mass, but all pulling goes by singularity arresting a bonded "flat" Universe in a state of gravity manipulating singularity. It is singularity holding the Universe together by moving time through space.

Gravity arrests the Universe, but no one ever managed to find the way it does arrest the Universe. If you read on you are about to find out how. I am taking the reader into a cosmos that holds a maximum value of 1 and anything greater than 1 does not fit into the Universe you are about to enter. Al the mind-boggling formulas used to impress has no meaning in singularity or in 1.

The Universe you are about to enter doesn't rely on a mathematical computing skills but an ability requiring human intellect through reasoning and following a line of debating. It requires the skill no computer could produce because it requires intelligent understanding of issues going beyond simply calculating and drawing unconsidered conclusions that is void of any intellectual understanding of cosmic principle, such as for example space whirls. If you have an ability to think and reason and don't require some mathematical disposition to rely on to help you think, then read on but be warned, this might be the highest intellectual level you ever called on.

The first question that would springs to mind when any person reads this work is to ask why I have is attitude in the manner of how aggressive I am. What would spur on such aggression towards the faculty of science and all that administrate all the small parts, as you are about to encounter? Why would any person find it necessary to attack the establishment of physics with such aggression? Why would any body insist there is something missing in physics when physics work so well and no one but me finds physics questionable? It is possible that the entire world find physics flawless except me Peet Schutte. The idea that every human on earth could be wrong and only I, the one named Peet Schutte is correct, such an idea crosses over from mental insanity to clear madness. Madness is when a person finds the world of physics at fault and more so considering that the one person says that that person has the remedy to cure all the mistakes that only that person are able to witness. I can see why everyone thinks I am mad. But also I seem the be the only person on earth that sees through the brainwashing and mind control going on in physics. Now that I have shocked everyone into a feeling of total rejection, I challenge you to read on and in the end, see if I am mad or if I am correct…then and only then form your opinion!
Where would you, the person that is reading at this present this page, place the centre of the Universe? Whatever your insight into physics might be, it will be unfulfilled because if you cannot place the centre of the Universe. Without such critical knowledge to your disposal, you then have no idea where gravity comes from. Test your thoughts in the following: If gravity pulls towards a centre and gravity holds the Universe attached it has to be pulling to a centre, therefore then the next question arising from that simplistic question must then be… where is the centre of the universe?

What I write in this website comes as a resulted being part of my effort to introduce my theory to prominent Academic Masters and / or their institutions of choice. Somehow the Intelligence domineering physics perceived my work with them having negative critics about my approach on physics and as a result launched an attack on my person as would be seen by hiding behind their good profession and strong-arming me using their superb mental abilities to block me from saying what I have in mind. At first and also yet again there was no such intension as to get confrontational on my part and that was the least of my motive. I did not wish to antagonise any one but it is clear our honoured gentleman of Physics accepted my approach to singularity that induced aggression on their part to attack my massage in such a manner. That opened the door where I now can show what is wrong in physics by attacking their corrupt system. On the other hand there is no sense in beating around the bush and drumming support I do not have by loudly singing their praise in admiration under false pretensions concerning those aspects I totally disregard and regard as rubbish. I just cannot see how Mainstream science can feel insulted when I insult incorrect cosmic teachings and when it comes to the supporting of such explaining. This is what I refer to when I say culture brings along presumptions that was never tested in the past. That which physics use to support their principles comes from culture. Everything in physics going about cosmology is a scam. Physics work well when on earth but in the cosmos it is a scam that no one can ever backup by proof. I challenge any academic to admit to himself what will happen to any first year student that refuses to underwrite Einstein's critical density theory in an examination paper.

Einstein's Critical Density lacks the accepted matching of facts we need in proving the critical mass factor. But our inability in securing such required evidence defies the most basic logic. It seems all new evidence we receive from outer space is disputing all Newton laws findings that disprove **Einstein's Critical Density** as the answer. The universe will not reach a point of contracting, not withstanding whatever dark matter astronomers try to locate in the vast space. Such a student will not see the second half of the academic year and be in a position to defend his observations.

Up to now **every one in science** is normally **acting as if gravity** is a commonly **explained factor,** which **every one knows** every aspect about **all principles that are involved in gravity** down to the smallest detail. In truth **no one in science** anywhere remotely **knows what brings gravity about** and **I used Kepler to unravel this mystery called gravity**. But no one in science will admit this fact about **Kepler being the one who formulised gravity decades before Newton came and gave gravity the name**.

Newton did not underwrite or define gravity and even today the most informed in Science at best can only assert their suspicion on a rumour presumed about what causes gravity to perform as the part interlinking the cosmos but no one can go any further by explaining the concept. **Newton started this realising of gravity** but it had and still has no more substantial proof than a rumour has **and Newton admitted to it being a concept he could not explain. In Newton's ignoring to test Kepler's findings Newton missed the opportunity to find what gravity is. Since Newton every person in science also ignored Kepler and every one is guilty of missing the opportunity Kepler maid available.**

Have you ever had doubt in class while listening to your lector telling Newton's ideas? Have you ever got an eerie feeling something doesn't add up? While you are listening to your lector or teacher rambling on about Newton you are becoming brainwashed and with mind control they are drawing you into the biggest scam ever devised. You are becoming part of the biggest hoax ever created and it is called Isaac Newton's physics. Have you ever thought you will hear any person with that seems to be in a sane mind tell you Isaac Newton, physics is fraud…well, now you have… and if you keep on reading you will learn what a lot of deceiving is part of physics and it is all about covering Newton's Fraud.

Do you know what gravity is?

I am referring to gravity and you know what gravity is. Gravity is what you have been fighting since birth and if not for gravity, you were Superman. If it was not for the effect gravity had on you, then you could have been the local superhero along with six billion other Superheroes that had no restraining from gravity. By you suffering from gravity during all your life that puts you in terms of being an expert on the subject of gravity and no one know gravity better than you do.

To your knowledge there is only one form of gravity and you know better than most that you are standing in aide of that gravity, where it is that gravity that is keeping you on Earth, so by experiencing the restraining it brings with the concept for your entire life and during your entire life you very well know what gravity is. Your expertise on gravity had you fighting gravity more than what you have fought or had any other fight with, including your Mother-in- law. If it wasn't for gravity, you could out accelerate a fighter jet in mid air. You know it is gravity's that is going to get you old and it is the very same gravity that you fight that is the same gravity which is the gravity that is telling you that you now desperately have to lose weight or over strain your heart and die young…then I come and accuse you of not knowing what gravity is by asking you what gravity is! You are so sure about gravity that you are no more familiar with any other subject and little else has your expertise as gravity does.
Are you sure that gravity is what you might think gravity is and that you and Newton has the same concept about what gravity is?

Have you ever thought that what you think of as being gravity is not that which science presents as gravity? The idea you have about gravity is not the gravity science says is gravity. I say that if what you think of, as keeping you glued to the Earth, is your idea of gravity, and then it is not the gravity science defined as gravity. That what Science define as gravity and that what you are thinking of as gravity is very much not equal but also it serves the Masters in physics well to leave you with having that idea about your gravity and their definition about gravity being the same because now they don't have to inform you what gravity really is. They leave you with what you think of as gravity. What Science say gravity is, is far from your view of what gravity is because the Newton approach to gravity is the same as a magnet hooking onto a metal and the grip coming from that they say is gravity.

They say there is a force between you and the Earth and this force is pulling you as much as it is pulling the Earth, but it is pulling the Earth much harder than what it is pulling you because the Earth has much more mass than what you have. That is deception at its best because notwithstanding height, the gravity remains the same. The same should apply in the formula if the formula $F = \dfrac{r^2}{M_1 M_2}$ did apply.

When the radius is insignificant the mass becomes enormous. With the radius big, the influence of mass and the force is weak. Your concept about their gravity is not nearly that which they say is gravity when they say the force of gravity is the product of the mass multiplied by another mass in conjunction with the multiplying of the gravitational constant and this product is reduced by the radius between the tow objects holding mass. It fits their comfort to have you be an expert on gravity with the use of one idea. If they can keep you thinking you are very well informed on the subject of gravity, then you will never question their authority on the matter of gravity and you will never learn that they know nothing about gravity whereas they will have you thinking there is no further point in discussing gravity since we are all experts on the subject of gravity. You then can define everything that gravity defines by using one word and that is gravity. The entirety is very deeply rooted in society but in this book I only aim to tackle gravity and therefore we return to gravity.

By definition gravity is defined as being:
Gravitation is the **force** of **attraction** that **operates between all bodies**. **The size of the attraction** depends on **the masses of the bodies** and **the distance between them**: **the gravitational force diminishes** with **the square of the distance apart** according to **the inverse square law**. *Gravitation is the weakest of the four forces. Newton formulated the law of gravitational attraction and showed that gravitationally a body behaves as though all its mass were concentrated at its centre. Hence the gravitational acts along a line joining the centres of the gravity of the two masses.*

According to this if your feet are on the ground it works like a magnet and you are unable to lift your foot by releasing it from the ground. That applies when the invert square law comes into effect. Then when you feet are ten meters from the ground you will float in the air because the inverse square law diminishes the contraction there is between the ground and your feet. The smaller the radius is the bigger will the mass effect be. The bigger the radius is the smaller would the effect of mass be. $F = G \dfrac{M_1 M_2}{r^2}$

The distance between your feet and the earth is not one but is say 10^{-6} meters. Therefore when your feet is on the ground the mass effect has to be multiplied by the square of 10^{-6} and then your mass has to be multiplied on top of that with the earth's mass of 59 760 million, million thousand kilograms because this formula says the earth is pulling as much as your mass is pulling! Now do the multiplications and see how much force your body must sustain in order to accomplish the force of gravity there has to be between your feet and the earth when you stand on the earth. After this realisation, then ask yourself you is the fraudster, is it me saying Newton is a frauds or is it Newton telling all this hogwash!

It is not you being glued or not being glued to the Earth that I discard. It is the definition holding this whole idea that I do not share in the least. What the definition describes is as if there are the effect of electro magnets pulling at each other from different poles and that is it is the total opposite of what I experience when standing on the earth. The reality of finding my feet of the earth is breaking the first millimetre of gravity clampdown is the easiest and not the most difficult as would the formula $F = G \dfrac{M_1 M_2}{r^2}$ suggest. In reality the difficulty increases as the radius grows that is between my feet and the earth whereas $F = G \dfrac{M_1 M_2}{r^2}$ would suggest the smaller the radius is the more force would be required to release my feet from the ground. When I say there is no gravity everyone thinks I say we all are going to fall off the Earth at random and with me thinking that way then it is obvious that I must be a nut with a serious mental handicap with a deficiency preventing me in understanding the issues. Everyone thinks of me as the clown acting mad when I say mass is not to be found in nature. But I do not say we are not standing on the Earth.

I do not say there is nothing that is keeping me glued to the earth. I say there is no attraction between two bodies by the force of the mass that is in such doing then is diminishing the radius parting the bodies by the inverse square law. I say there are a connection by motion between the centre of the body and the material surrounding the centre. This is what I say when I say there is no mass pulling. This must underline that your concept about their gravity is not nearly that which they say is gravity when they say the force of gravity is the product of the mass multiplied by another mass in conjunction with the multiplying of the gravitational constant and this product is reduced by the radius between the tow objects holding mass. That is not the way you are walking on Earth but they wish you to be confused with the truth and their lie so that they can get way with murder.

But they would love you to have that concept about gravity and they would hate it if you thought further about the issue because then you would exceed their knowledge limits about what they know as gravity. It fits their comfort to have you be an expert on gravity with the use of one idea. If they can keep you thinking you are very well informed on the subject of gravity, then you will never question their authority on the matter of gravity and you will never learn that they know nothing about gravity whereas they will have you thinking there is no further point in discussing gravity since we are all experts on the subject of gravity.

They will put up a fight to keep you in the idea that gravity is a one-word concept and by thinking in terms of gravity you then entirely define the whole concept involving gravity. You then can define everything that gravity defines by using one word and that is gravity. It is not what they present as a scientifically defined explaining of gravity, but who cares. To them you and I and the rest is cow fodder, there to be used when needed and then to be left in the dark and to their mentality they could care less whether we have any opinion of sorts for what we think is of little importance to those mighty academics occupying towering intellectual heights. This statement will be underlined in the letter I present later.

They can give less about what you think than they care about the opinion your dog has on religion. That is all the same to them because they are the ones lining their pockets with money. It is all a scam to confuse and depress the truth about gravity. It is a culture that drives the powerful to subdue the weak. While one thing is happening they whish you to think another thing is in progress. It is a state of mind the They give a concept a meaning as in using an idea while behind the scenes they use an entirely different definition they share very sparsely.

We have to analyse what gravity is. There is a defining to do by dissecting the factors we see as gravity. Gravity is not many things to lots of people, but is very specific one thing. To investigate the identity of gravity there are questions one must first answer. Is gravity the part that has me standing on the Earth or is gravity the part that has me glued to the Earth. It is surprisingly not the same thing. While I am standing on Earth I am standing still as far as my perception carries but there remains the intention to move, even with me standing still. The intention to move downwards is gravity and the restriction I have that prevents me from moving is mass. I can't stand still and at the same time intend to move by the same driving effort. My standing on Earth but remaining in intention to move further can't be motivated by a similar or the same principle. The concept is not the same because when standing still the Earth does the moving on my behalf and there is no independent motion of my body. But when I jump I am moving away from the Earth and when I walk I walk along the surface of the Earth giving me independent movement on top and in addition to the movement of the Earth. My movement is stopped by mass and not committed by mass when being in mass. When in mass my intension is to be motionless. There are two occurrences where one is immobility through mass and the other is being without mass while moving. Mainstream science wants you to believe it is the same thing but it is not. Gravity is the inclination that my body has to move further towards the centre of the Earth while mass is, that, which frustrates my body, and prevent it from moving towards the Earth. Science tells you it is the same thing but it is not. The very second that that which prevents me from moving falls away the inclination of moving which is gravity kicks in and gravity lets me move freely again.

I will continue to move towards the centre of the Earth and that process has a very well defined and scientific term which we use to describe the event: we call it falling in English but every other language devised another word to describe the same concept. When I stop falling that which gives me mass is what ends my moving. What gives me mass changes my gravity from movement to being inclined to move or show a potential ability to move. As soon as that which prevents me from moving allows me to continue to move to the centre of the Earth my gravity will start giving me full motion and then that which gives me mass by frustrating my moving will not stop my progress towards the centre of the Earth. When my gravity is stopped by mass blocking my movement and that, which gives me mass, then has the ability to turn my descending towards the Earth over to a frustrating of gravity or my freedom to move and change such a freedom to move into a tendency or an attempt to move. The price I pay for the loss of free movement is the gain of mass in mass preventing further movement. It is not as simple as being glued or not being glued…

This is what I have tried to convey to academics throughout the world and in person to so many in South Africa. But conveying this message is criticising Newton and not once did I mange to get one of the esteemed to listen to my criticism of Newton. If they hear it is about gravity implicating Newton the debate stops with them defending Newton and not giving any more attention to my argument. When trying to convey my message about gravity I firstly have to start to show that gravity is actually the moving of the body and when the body has mass the body isn't moving independently as a cosmic independent object but becomes in mass meaning it forms a unit with the Earth as the Earth. Not one academic will ever allow me to get that far…

To the normal concept that all person have, the thinking is that the body not moving and the body moving is the same thing and the body not moving is gravity with the body moving is that which is formed by having mass because the culture that Mainstream science promotes the past three hundred and fifty years never got past mass. If you, the person that is reading this don't believe me that Mainstream science does not discriminate between mass and gravity in the correct way then test you nearest physics lecturer. Ask him by testing his thoughts and you will see it suits them well to let everyone get all mixed up and confused about the two concepts. To get to the bottom of the concept behind gravity one needs to analyse and dissect and then realise what involves all the aspects that covers the concepts entirely. We have to go splitting the factors forming the concept we holds as gravity and defining the two concepts into proper categories.

Now students have the choice to insist on you telling them the truth because the truth is written in

Go To
"Newton's Fraud "

The truth is finally out and your culture of deceit has eventually been detected
You will stop your mind abuse on students because now they can know more about gravity than did all those that came before you and all those that came with you.

Again I challenge you to come forward and tell your students the truth about what I uncover in the articles that follows and as the articles progress in introducing information…then you explain to them how you deceived their blind trust in you.

To Find Out More About
Newton's Fraud
To Learn About

Brainwashing And Mind Control

The facts you are about to learn will astonish you and it will seem unbelievable but notwithstanding it is true.

Unbelievable as it is nevertheless, I challenge any one to show me that the least of any or all facts I uncover is not true. I hope to start everyone thinking about physics and not to daydream about what is impossible. Ask questions not about what is unknown but firstly what is thought of as being the known. Ask what is mass and have a truly significant explanation. I give you thoughts that you can think about, chew on and digest the answers in the correct way. Don't fall back on what you were taught but use your mind to answer truthfully.

I suggest you wait a time for some of the ideas to become appreciated in the full sense because the information may come across as being extensive but you will lose less time in waiting and thinking the information through than you lose by unwittingly disregarding what I say in favour of what you learned to accept as the truth but what is part of the deception and then afterwards finding out you were deceived. It will still be quicker waiting on the response than losing the time wasted when you are being tricked by the Academics in science holding positions in Astrophysics and Physics.

Sometimes the reading is slightly extensive but to find the truth you must dig deep and overcome corruption you made your own through culture and there is a lot of material that came in place over many centuries as accepted culture and it is the lie that detours from the truth that then is extensive.

What you are about to read form this point onwards is the first time gravity is explained in a sane way and is proven by implementing basic physics in using singularity in relation to the Law of Pythagoras.

Since gravity also influence the space outside the sphere the space we call outer space has seven plus three points bringing about ten positions of gravity influencing space. The influence inside the sphere also captures the space outside the sphere.

In the Universe there are two forms, the cube and the sphere. The cube is a loosely connected structure with any form possible but the only precondition is that there must be at least six sides connecting. The six sides hold a relevancy or a responsibility to one another and provide a Universal accepted form maintaining the universe. From the structure one can see gravity is not strongly present in the form representing the cube. All six sides support what ever are inside evenly form all sides. The sphere is the form securing gravity. In the centre of the sphere there is a point where space vanishes.

This is another definition about gravity as science recognises the definition of gravity.

gravity definition
grav·ity (grav′i tē)
noun *pl.* gravities -·ties
the state or condition of being grave; esp.,
solemnity or sedateness of manner or character; earnestness
danger or threat; ominous quality the *gravity* of his illness
seriousness, as of a situation
weight; heaviness
lowness of musical pitch
gravitation, esp. terrestrial gravitation; force that tends to draw all bodies in the earth's sphere toward the center of the earth
Etymology: L *gravitas*, weight, heaviness < *gravis*, heavy: see grave
adjective
operated by the force of gravity

gravity Synonyms

gravity
n. Weight
heaviness, pressure, force; see gravitation.

A Less Complex Commercial Science Book 526 Chapter 11

Importance
seriousness, concern, significance; see importance 1.
Webster's New World Roget's A-Z Thesaurus Copyright © 1999 by Wiley Publishing, Inc., Cleveland, Ohio.
Used by arrangement with John Wiley & Sons, Inc.

gravity Usage Examples

Converse of object
defy: Cause a regular pencil to defy gravity by clinging to your fingers!
Preposition: on
moon: Not much gravity on the moon, You can find more on the Earth.
Adjective modifier
utmost: Any breach of such confidentiality will be viewed with the utmost gravity.
Modifies a noun
fed: If you get a fuel leak with a top mounted gravity fed tank, it will keep on coming.
Noun used with modifier
quantum: In quantum gravity, the role of the gage coupling, is played by the energy of a particle.
Possessives
earth: We were planning to escape the earth's gravity; Martians could do the same, with their planet.
Preposition: of
offense: Eric Dawson was jailed for four months, a sentence which reflects the gravity of the offense.

Webster's New World College Dictionary Copyright © 2005 by Wiley Publishing, Inc., Cleveland, Ohio.
Used by arrangement with John Wiley & Sons, Inc.

Have you too had the feeling that everyone knows there is a problem about gravity in physics but no one can put a finger on the problem? There is this thought that warns the thinker of something not being in place. It is lingering on… It is coming through the ages, from generation to generation… without anyone ever finding a solution.

They school students in the idea that a feather and a hammer will fall equally in a complete vacuum, and that experiment was done on the Moon. Then they school the same students that objects fall in relation of the force of the gravity that the mass creates.

It is said that all things fall equal under conditions in vacuum but it is said that mass brings about gravity. This leaves an eerie feeling of doubt that lingers on in your mind, but pointing a finger at the precise place where the problem is, enhances the feeling and shifts the doubt from the question to where this feeling is.

There is a suspicion lingering in the back of everyone's mind that something is not quite correct about the approach physics take on the matter of gravity, and mass enticing gravity by force does leave a feeling of uncertainty, but overall it is only those academics seasoned with years and salted with time that seem to miss this feeling, and they should have acquired the reason to what results in this eerie lacking-feeling we have when we feel certainty lacking about gravity that we others have.

Something about the way gravity is presented just doesn't add up as it should and does not quite reach the answers it should conclude. There is this vague unspoken question hanging in the air without ever finding words to express the question…and yet the question remains however unspoken it seems.
I have devoted half my life's worth in time to finding the solution and I did, but the solution is something no one in academic powers wish to hear. That which I address, undresses the Academics charged with tutoring physics, and by my work they are left naked. They do not volunteer to accept my solution because they do not like my solution. My solution discards their teachings and show how ridiculous those teachings are. I banish what they see as being holy.

I have the solution, which academics don't wish to read never mind wish to address. It will turn all physics round about and put all academics on their backs. If you are a student studying in physics this web page is detrimental to your future, as you can remain part of the problem physics have or you may join the solution that came to physics.

The problem is that the faculty of physics are covering up a culture driven by the mentality to brainwash and control the minds of students to accept what should never be accepted.

Hidden under a cover of "understanding Newton" or "not being able to understand Newton" they force certain incompatible arguments to join in that what never can join because it lacks foundation while making sense at the same time. They avoid this problem by never presenting the foundation part. Every generation find an itchy feeling but never is there a place that any one can secure the very point where it is apparent enough to scratch and rid physics of the undetectable itch. Believe it or not but this itch is in place because of centuries of brainwashing going on and is employed in physics from generation to generation for centuries on end. I challenge any person whomever, wherever, to show that or how I am bespattering academics, besmearing Newton or besmirching physics or even exaggerating in any line that I write. After reading my book that I wrote, then please do show where I fowl that which is unblemished in Newtonian physics. I am driven to correct the centuries old corruption, which forms my mission. My effort is to uncover the mind control they place on students. If you are of the opinion that the term "mind control" is excessive then I challenge you to prove me wrong. If you think the term "brainwashing" is cheap promotion, then challenge me to a public debate in any arena of your choice on this matter.

This web page is dedicated to bringing honesty into the faculty of Astrophysics and Astronomy as well as show the Physics student on what corruption and deceit does physics base their facts which they proclaim as being such well proven, and godly accurate facts.

Students listen to this:

Read the next pages and you are about to learn how students are brainwashed into accepting the baseless and ridiculous as truths. The Custodians of Physics have nothing better to offer than presenting you with unfounded corrupt and distorted facts. Doing that they resort to mind control on students and introduces baseless concepts by manipulating student's thoughts.

If you are a student then read in this web site what they do to you. They are defrauding you by exchanging your institution fees for corruption so confront them about their dishonesty. Force them to become honest and to stop corrupting students with intentional malice. That which they offer has no truth. All they have are misconceptions and incoherent facts.

Custodians of Mainstream Physics stop your practice of mind abuse, thought control and criminal behaviour towards students immediately!

If you are one of the Academics and Custodians of Mainstream Physics I challenge you to show one piece of evidence where I am incorrect, exaggerate facts, produce incoherent and distorted views about your physics being based on corruption and misconduct.
If you Academics think you are innocent, read on.
Fortunately for the future and unfortunately for Academics in mainstream physics the truth is out and it is published and your dishonesty as Academics will come to be no more. Your entire generation and all the generations that came before you will be washed away as dirty mud on soiled linen. You will be remembered by the coming future as those in the past that was not worth remembering.

Students do not have to suffer your abuse and cheating. Students no longer have to sit and wait for thought control as academics in physics wittingly force feed students of physics all the facts that are in truth merely distortions about the fundamentals on physics and expect then to allow academics in physics to get away with academic murder just because your professional position allows you to do what ever pleases you since no authority is in place to check your credentials and balance your truths.
Now students have the choice to insist on you telling them the truth because the truth is written in
The Book SIR ISAAC NEWTON'S FRAUD.

The truth is finally out and your culture of deceit has eventually been detected
You will stop your mind abuse on students because now they can know more about gravity than did all those that came before you and all those that came with you.

Again I challenge you to come forward and tell your students the truth about what I uncover in the articles that follows and as the articles progress in introducing information...then you explain to them how you deceived their blind trust in you.

Brainwashing And Mind Control

The facts you are about to learn will astonish you and it will seem unbelievable but notwithstanding, it all are true.

Unbelievable as it is nevertheless, I challenge any one to show me that the least of any or all facts I uncover there is one that is not true.

Sometimes you will wait a time for some of the information to sink in because you have to abolish the abuse you suffered with brainwashing that science applied to you. But chewing on the ideas I present is less time consuming as they are extensive but you will lose less time in wasting than time you gain in finding correctness.

Compared to the time that you have lost up to now by unwittingly learning the deception and finding out you were deceived afterwards and comparing that time wasted to the time you now must spend in seeing how to correct the information is a drop in the ocean.
It will still be quicker for you to read my information and realising the corruption than the time you have already lost and the time you wasted when you were being tricked by the Academics in science holding positions in Astrophysics and Physics.

Sometimes when reading the work, the reading is slightly extensive but to find the truth you must dig deep and overcome the blatant corruption there is. The disinformation you were served up to now is a lot of material that came in place over many centuries and it is the lie that detours from the truth that then is extensive.

The Book SIR ISAAC NEWTON'S FRAUD indicates (in short) the problem there is and brings the proof there is about the corruption. The idea behind this book is to keep the cost as low as possible in order to get the message as widely spread as possible.

The book An Open Letter On Gravity addresses the fraud as well as the solution in depth.
Here is one problem that Newtonians cannot address because it is shear and blatant corruption.
FRAUD!!! I hear every Newtonian physics academic shout their disgust and unbelievable discontent and dismay at me because no one in his right mind will ever think to connect a person with the esteem that Sir Isaac Newton enjoys to a term such as corruption and to fraud and live another day to repeat this accusation!

Here is just one of many examples that you may use to test Sir Isaac Newton's fraud he committed when he committed the Universe to the deception of

$$F = G\frac{M_1 \times M_2}{r^2}$$

Here is something about which you should THINK.......... THINK about because
       Newtonians DO NOT HAVE TO THINK...about
           WHAT THE ANSWER IS. They collaborate with Newton to carry the torch of deceit onto all students they can corrupt. You think I am silly...
I dare you to confront your lecturer/ professor and ask for an explanation about the common comet's incorrect behaviour and see what I mean by fraud!

GRAVITY as Newton suggested it is, of being the force and as Newtonian science upholds it to be, IS A HOAX, This deception of reality KEPT THE WORLD OF SCIENCE AT RANSOM since the end of the middle ages!

YOU MIGHT DECIDE TO IGNORE THESE FACTS BUT THEN YOU WILL REMAIN PART OF THE LAST BASTION OF THE DARK MIDDLE AGES, OR YOU MAY READ THIS BOOK AND BECOME PART OF THE FUTURE.

Gravity produced by the mass of material is Newton's presumption, and was never proven. If there is any academic that has proof that Newton's suggestions were ever proven, then please bring that proof to the table. As the suggestions Newton made, were never proven it is clear that all the academics echoing Newton's corruption then are all corroborators of a criminal ploy because one and all are all incorrect. My work is based on the findings of Galileo and Kepler, and their findings, which are very much inconceivable with the findings of Newton.

I DO NOT EXPECT ANY PERSON TO BELIEVE THIS STATEMENT OUTRIGHT, SO I INVITE YOU TO SPEND THE NEXT FEW MINUTES OF YOUR LIFE TO READ THE FOLLOWING ARGUMENTS, THEN SEE IF YOU STILL ARE AS CONVINCED ABOUT GRAVITY AS BEFORE. This is just one of many such diversions from the truth that I show Newton made and that Newtonian corroborators still echo these lies to this day.

$$F = G \frac{M_1 \times M_2}{r^2}$$ This formula suggests gravity and for that reason we may start by determining the influence of gravity on planets as we find them in the solar system. This is the formula all of science in physics uses on which they base the principles of gravity. If this formula does not prove what it should prove, then all of physics has no leg to stand on and Academics in physics then knows nothing about physics.

First, let us concern ourselves with a comet because the comet shows how Newton and his formula resembles the crime of all time as it is an elaborate and extensive hoax that enabled Sir Isaac Newton to prove the world they are fools for three and a half centuries.

The comet is a lump of carbon and ice circling in outer space, minding its business until the mass of the Sun gets hold of it (well that is the way one has to interpret it by going along what Newton's formula would suggest.) It is common knowledge how the comet stands related to the Sun's gravity. Firstly, picture the comet at its farthest point, away from the Sun.
The Sun detects the comet in outer space (supposedly) when we find the comet is at its most outward and furthest point from the Sun. By shear mass and tenacity that the Sun has, the Sun starts pulling this flimsy comet over the radius that holds the gravitational constant.

Gravity pulls of the centre of the Sun towards the centre of the comet while it is the comet that pulls the comet also towards the centre of the Sun. The lot is going to the centre of the Sun notwithstanding from what angle the comet is approaching.

The gravity of the Sun pulls the comet straight towards the Sun, that much we all know. There Newton's formula still proves it's worth being correct or so we think. Gravity always pulls an object directly towards the centre of a cosmic body: that too is common knowledge and that is just what Newton said will happen. The man is a genius as everyone thinks he is. Therefore, the comet is drawn directly towards the centre of the Sun and throughout its journey the comet is picking up momentum that is directly related to the gravity that is taking the comet on its journey pointing directly to the middle of the Sun, because as Newton stated; gravity is always cantered in the middle of a cosmic body.

As the comet is increasing its speed, the comet comes closer to the Sun and therefore the Sun's gravity that pulls is in conjunction with the comets gravity that is also pulling where the two are working together in tandem to simultaneously increase the momentum as the distance between the two cosmic bodies is reducing. Each instance the comet is drawn towards the Sun, the gravity that the Sun applies to the comet becomes larger progressively.

Newton stated that $F = G\dfrac{M_1 \times M_2}{r^2}$ = the Force that the mass (M_1) of both objects (X m_2) and therefore in both directions hence r^2 and that this commitment by both masses to establish the force that work as gravity releases gravity, because of the fact that:

1. The value of M_1 in is equal to the gravity force it presents because that is the representing the influence of the mass of the cosmic structure.

2. The value of r^2 is equal because the two objects are destroying the gravitational constant at an equal distance from both sides.

3. The value of M_2 is still a mass filled with gravity and therefore has to effect the relation in the same way as both objects consist of different compositions of materials that are used to manufacture the different objects.

4. That means Force one has to have a different value to that of force two where force two enjoys a different radius (shortened every instant of progress) making ($F_1 \neq F_2$). In contrast time duration $P_1 = P_2$ and this cannot apply in the case of gravity because the greater the force, the bigger the impact has to be on the time duration.

The comet is heading to the Sun as all comets has to do since Newton's mass suggests that the mass that the comet has, is pushing the comet in the direction of the Sun while at that very moment that the mass that the Sun has, is pulling the comet in the direction of the Sun. The comet is heading to the centre of the Sun because the Sun is to heavy to head to the centre of the comet and both will eventually land in the centre of each other.

$$F = G\dfrac{M_1 \times M_2}{r^2}$$

The mass of the Sun is M_1 and that is pulling by the force of the mass M_2 across the radius r.

The mass of the comet is M_2 and that is pushing by the force of the mass M_1 across the radius r that is also containing the illusive Gravitational constant.

Both has to bridge the gravity constant G that fills all the space that r holds from both directions thus measuring as r^2.

When the comet is at its closest point to the Sun, something odd happens which cannot be explained by Newton's gravity at all! Remember gravity should now be at its strongest point because of the proximity of the two objects. The force has to be so strong that it can never divert from the direction that the force of gravity forces the comet to take. The comet has to hit the Sun in the centre by the centre of the comet. After all, that is gravity. That is not exactly happening as Newton prescribed it to happen…

1. The comet remains at an even distance encircling the Sun.

2. No longer does the gravity of the Sun pull the comet towards the centre of the Sun.

3. At this very point the gravity that the Sun applies on the comet does not pull the comet towards the centre of the Sun any longer, in fact, it seems as if the effect of the gravity has been neutralized.

4. The comet stays at an even space from the Sun as it goes around to complete a half circle's orbit around the Sun. It only completes a part of its rotation around the Sun.

5. After this, an even more peculiar event takes place. The Sun, at the point where gravity should be at its most dominant, suddenly loses its complete grip on the comet.

6. The comet brakes free from the Sun's pull of gravity and speeds off towards its destiny into the vastness of the cosmic space, undeterred by the gravity of the Sun.

Then after a pre-determinate and pre-calculated time the Sun starts applying its gravity on the comet once more. At a point where the comet is at its farthest point, the gravity of the Sun becomes strong enough to bring about a complete turn around to the comet's direction of travel. However, the gravity between the Sun and the comet is at this point, at its weakest point of influence.

So, when the Sun's gravity is at its strongest, the comet manages to brake loose and neutralize the Sun's gravity pull in order to avoid its fatal collision with the Sun and when the Sun's gravity is at its weakest, the comet cannot escape the pull of gravity. There is definitely something very wrong, either with the comet's behaviour or the laws made up by Newton.

NEWTON SEEMED TO BE ABLE TO CONVINCE EVERYBODY ABOUT HIS COSMIC LAWS, EXCEPT THE COSMIC BODIES THEMSELVES, WHICH SAEEMS TO PREFER TO BEHAVE AS IF NEWTON'S LAWS NEVER EXISTED. ARE THE COSMIC BODIES REBELLING AGAINST NEWTON'S AUTHOROTY BY IGNORING THE LAWS OF NEWTON OR ARE THEY SIMPLY DISMISSING THE LAWS NEWRON DRAFTED, AS IF THEY NEVER EXISTED.

MY GUESS IS IT IS BECAUSE THEY DO NOT EXIST, EXCEPT IN THE IMAGINATION OF ISAAC NEWTON AND HIS FOLLOWERS.

For the more mathematical minded person the argument is as follows. May I remind you, THAT NEWTON'S OWN LAWS ARE IMPLIED, and again the planets disobey these laws completely!!

$$F = G \frac{M_1 M}{r^2}$$ THIS IS THE SUGGESTED FORMULA THAT PLANETS APPLY WICH ENABLE THEM TO MAINTAIN THE ORBITS AROUND THE SUN WICH THEY DO. I dare to prove that there is a difference between findings of *Galileo and Kepler* on the one side and the work of Newton and Einstein on the other hand. Only one of these two group's findings can be right, because there is an unmatchable difference in the concept of these two groups' opinions.

Newton considers that a force exists between two bodies in space: the mass of the two bodies' product is being brought into context with the gravity constant (G). This value is then divided by the distance r calculated as a square (r^2) value.

Then we get to the point where the brainwashing comes in.

A simple everyday example such as the can comet indicate, is an absolute undeniable fact that shows how Newton does not comply to even the most basic test of reality. This is only one of so many examples I show as to how academics mislead students by forcing them to accept totally falsified facts.

There is no way in the Universe, that the mass of the comet joins forces with the mass of the Sun in order to allow the mass of the two to corroborate in an accumulating effort whereby the joint effort establishes the force of gravity and enable this established force to draw the comet into the Sun as Newton claimed is the case.

The comet escapes by the same effort that the comet arrives and if the comet is pulled, by whatever force that they can imagine, then they better imagine that the same force applies the same effort to push the comet into outer space. The comet goes as the comet arrives.

Students are forced to except Newton or die an academic death.

I challenge any academic to disagree with my statement.
I challenge any academic to a public debate on this matter where the academic can defend Newton by proving me wrong when I accuse them of raping the minds of students in defrauding the trust student's put in them.

If any student would tell any academic that the student does not accept the terms of Newton's formula because the principle behind Newton's formula is lacking any substantiating proof, the student will automatically be suspended after the first examination the student fails.

If the student refuses to commit his agreeing about Newton being correct on paper during an examination on the grounds that the comet proves Newton to be unreliable, the student is history in academic terms.

If the student uses the behaviour of the comet and then use the comet to prove Newton suspicious and unreliable, the academic will not be able to bring counter proof about the comet actually being accurately portraying the correctness of Newton and in defence the academic will counteract as the academic will forcefully have the student removed from campus, if need be.

The academic will never submit to bring proof or even show tolerance to any rebellious behaviour any student may show when doubting Newton.

The academic expects the student to hail Newton notwithstanding, the lack of proof and the lack there is on the side of the academic to bring substantiating proof about Newton's unwavering correctness or where the academic shows any commitment to prove to the student in what manner Newton is correct and in why then should Newton be regarded as trustworthy and thus prove to the student that Newton is reliable after all…

The student doubting Newton is dead academically. The academic does not have to answer to any person or explain how the academic sees him or herself fit to expel the student by insisting that the student has to except Newton because the institution of physics worldwide says so and that being the only criteria the academic world of physics ever use to prove Newton correct.

If that is not clear brainwashing I don't know what is…

If telling a person is telling anther person that the second person has to think that anything is absolutely correct while being obviously flawed, as the comet proves Newton's formula to be, is not abusing authority, then tell me what constitutes to abusing seniority. When someone insists on being correct while the statement nevertheless is flawed and yet it still has to be accepted as being correct, that such behaviour is not mind control, then what is mind control.

Any one suggesting that I am exaggerating and making false claims prove where I did this just in this web page. There are numerous other Newtonian claims that I disprove in the book addressing this issue named after the tale behind the criminal ploy as Sir Isaac Newton's Fraud.

It is time to bring these crooks called Physics Academics to accountability because if any institution ever abused trust and mismanaged their position in society, not one and nothing can even compare to what Newtonians are doing in abusing the minds of students.

Students, now is the time to wake up and smell the garbage! You allow these cheats called Physics Academics to blindfold you and brainwash you by mind control. Let them prove to you how much the Moon came closer to the Earth since man made the historical walk on the Moon surface in 1969. There were at that time, devises installed on the Moon with which the distance between the Moon and the Earth are constantly monitored. Let them tell you how much the Earth and the Moon grew away from each other and how that disproves Newton's credibility.

Students purchase the book the less costly version of An Open Letter On gravity renamed as having the title that says it all Sir Isaac Newton's Fraud. If the comet were the only incorrect interpretation of physics by Newton in his corrupting of physics, I would be the first one to say: Let it go! However the numbers of such incidents are numerous, and that makes the problem unavoidable where we just can't let go.

Sir Isaac Newton's Fraud is the book that shows how academics mislead you as students and how those academics you so heartily trust, rape your trust by defrauding you and expect you to pay them for the privilege you have that they may brainwash you into accepting Sir Isaac Newton's Fraud.

Sir Isaac Newton's Fraud is the version of another book where this book Sir Isaac Newton's Fraud is what I have devised as the less expensive issue bearing cost in mind. First purchase the less expensive issue and convince yourself that you do have academic snakes in your midst. Confront these academic snakes on the issues I present in my books, once you convinced yourself about the elaborate hoax that is on offer carrying the name or being called The Department of Physics.

All the issues I address are as simple and as common as the comet example proves to be. T name a few: Why would there have been a Big Bang when all the mass of the entire Universe was concentrated in a neutron which makes the Neutron unbelievable massive as it will never again be. That which reduced the effect of the mass, being the radius was so small it was not visible and the radius never again will be that small. If $F = G\dfrac{M_1 \times M_2}{r^2}$ was in place as Newtonians insist it to be, we have to have an implosion and no Big Bang. Ask your local academic why such all-powerful mass that was at the time controlled by such incredibly small radius not lead to an implosion instead of a Big Bang.

With Newton's formula of $F = G\dfrac{M_1 \times M_2}{r^2}$ in place why do we need a critical density factor because the Universe has to be coming together as we speak! The formula of mass drawing mass is then at this moment in place and with it being in place why would we need a critical density after all. If those criminals would mention the dark matter supplying the required mass then you have to read how I explain this statement, as that concoction is to be the most elaborate deployment of corruption ever unleashed on mankind by any institute or government in all of civilisation.

With Newton's formula of $F = G\dfrac{M_1 \times M_2}{r^2}$ in place why did Hubble find an expanding Universe that is growing apart at a rate that boggles the mind. The mass should be drawing the mass closer and that should eliminate any possible expanding Universe! That is if Newton is reliable…

There are so many more statements that are so simple to see should any one have the time to think about the issues. The issues are not mind-boggling but what is hard to accept is that the academics think the public is so stupid the public can be that easily deceived.

Academics should realise that one can fool some people sometimes but one cannot fool everybody the whole time. Read about the Academic world of physics fooling the entire world all the time.

Should any person find that I exaggerate in any instance in any book and prove that, then I will give back that person's purchase price of the book ten fold.

Should any person find that I do not exaggerate in any instance then I beg of you to help me fight these all powerful fraudsters and let them face their corruption they are committing to out children!

To Find Out About The Solutions that I propose and what gravity truly is then go to The Absolute Relevancy of Singularity

What you are about to read form the book in question being Sir Isaac Newton's Fraud is an actual article taken from the book The Absolute Relevancy of Singularity. There are two books namely Sir Isaac Newton's Fraud and the other more expensive version is The Absolute Relevancy of Singularity, which is a combination of many letters that I address over an extensive period to academics, which I now re-address to the public, containing numerous letters that I addressed to academics over a time of seven years on the matter of Newton's fraud. In the book The Absolute Relevancy of Singularity the one named as being Sir Isaac Newton's Fraud is an attraction of the other The Absolute Relevancy of Singularity, but Sir Isaac Newton's Fraud is much shorter but only points to the falsifying and the corruption.

It is in the book The Absolute Relevancy of Singularity in which that I bring the solutions about the problems Newton's corruption left us with. Please do not confuse the two because the one being Sir Isaac Newton's Fraud is a subsidiary of the other being the one that establishes my suggestions on solving the mystery behind the concept we call gravity and is named The Absolute Relevancy of Singularity.

At that point where space vanishes gravity is the strongest.
From the centre point where gravity is the strongest gravity hold the sphere true to form. At the edges of the sphere there are also point lining in 90^0 and 180^0 holding relevancy and responsibility to one another but the centre spot being the gravity point positions all the points in a location that the centre point allocate. The sphere secures form by the value of Π and that value becomes the value of gravity. This means that in the cube at the point of contact between the cube and the sphere the cube experience such a contact point as if the "bottom falls out" of the cube and without a "bottom" to support objects they fall to the sphere as objects does fall to the earth. Remember that a body "floats" in space, but at one

specific point it starts to "fall" to the earth. The cube is outer space and the earth is the sphere and that gives gravity. That is gravity and it is a dimension change much more than any force. I shall explain this last remark later on. That too is the Lagrangian system with five cosmic structures holding relevancy to the centre structure where the centre structure stands in for seven positions diverting from centre and the orbiting structures standing in for five positions in space. To form the centre point that gives the sphere its strength, the sphere must spin and only by rotation it forms such a seventh centre point and by forming such an axis line, this line gives the top the gravity to spin erect and enforce independence from the earth. This line that forms is what supports gravity and by spinning Π forms Π^2 and this is what produces gravity in solids. Mass has nothing to do with gravity but gravity produces what we think of as being mass.

Gravity that bonds the Universe together in the boundaries of singularity applying is a relation between material that moves in time and space that stands still and only moves by expanding. Awarding mass has no validity when objects form gravity but mass comes as a result of the above-mentioned relevancies.

Infinity is the centre line that can reduce no more and can never start, and that has no outside because it only has an inside

Eternity is the line that can't end because it can never stop expanding and therefore eternally moves and where it only has an outside because it can never reach the inside due to Π

To start explaining the Four Cosmic Phenomena we have to understand gravity. To understand gravity we have to understand the Four Cosmic Phenomena. Since science never yet came close to understanding the Four Cosmic Phenomena it is clear that science never understood gravity. Anything that spins forms a centre axis line I call infinity. Around this line a circle spins and the circle holds the value of Π and Π holds eternity away from infinity.

$$\Pi = \frac{21.991}{7} \qquad \Pi = \frac{21.991}{7}$$

Gravity is the revaluation of Π in terms of the ring (7) and of the axis ($\Pi°$). In this the turning of a circle, the circle revalue $7 \div 7 = 1$ and space revalue from 21.991 to $3.1416 = \Pi$. That is how Π forms the curvature of space-time.

When a circle spins it forms an axis. Then the centre axis holds singularity @1. The rim of the circle is 7° where space then is 21.991. Gravity is Π moving from one dimension to another dimension and the Titius Bode law is absolute proof of this attachment that Π has to gravity. Gravity holds 7 in relation to 10. Gravity forms when $7 \div 7 = 1$ and on the top $21.991 \div 7 = 3.1416$. By compacting the space we establish a denser space we call the atmosphere and the atmosphere is the changing of $7° \div 7° = 1$ and $21.991 \div 7° = \Pi$. It is about movement of the Earth enforcing movement of space in a centrifugal pump action. In physics mass only pulls a cover over the eyes of those that are supposedly well informed intellectuals performing as qualified physicists that could in many centuries never, not once could explain how they say that mass has the ability to pull objects in the act of gravity.

The axis that always forms a line holding 3 opposing points when the circle spins.

The circle that always holds 4 opposing points when turning around the axis.

The Titius Bode law in conjunction with the Roche limit as well as the Lagrangian points conform to form a unit known as the Coanda effect. Science has no idea what establishes these four very important phenomena because science goes about studying the Universe incorrectly. No one could explain these crucial phenomena because the approach science uses to study the Universe completely incorrect.

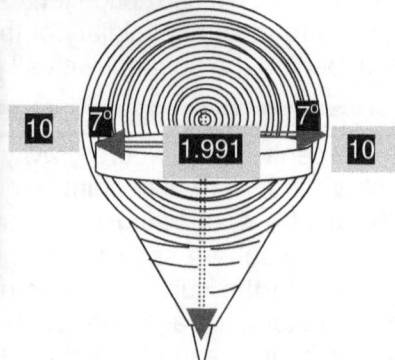

The Titius Bode law shows the centre axis as a line that represents 3 singularity positions
The circle represents 4 singularity points or single dimensional positions.
The total in singularity is 7 and that is why Π holds in the circle 7 points and space as 21.991.
The space holds 10 points on either side of the circle (20/2 on the one side) in relation to the circle holding 7.
From the Titius Bode law one draws the proof that time holds singularity **a = (n + 4) / 10** and space is the result of time that moved on. One can see space is the footprints time left behind as time moved to the future leaving space as the past. That is the reason why the **Titius Bode law**, the **Lagrangian points**, the **Roche limit** and the **Coanda effect** forms the way entire the Universe unfolds and that excludes mass as a Universal factor altogether.
The sound barrier is not there to frustrate aeroplane pilots and warmongers. The sound barrier is the principle by which young stars start a life cycle and that is what is really important behind the sound barrier.

I explain the sound barrier according to the Titius Bode law, the Roche limit, the Lagrangian points and the Coanda effect and this is how the sound barrier unfolds. What happens in the case of the sound barrier is as follows: the Titius Bode law restarts the Universe as singularity arrests the Universe by applying gravity every instant time alternates.

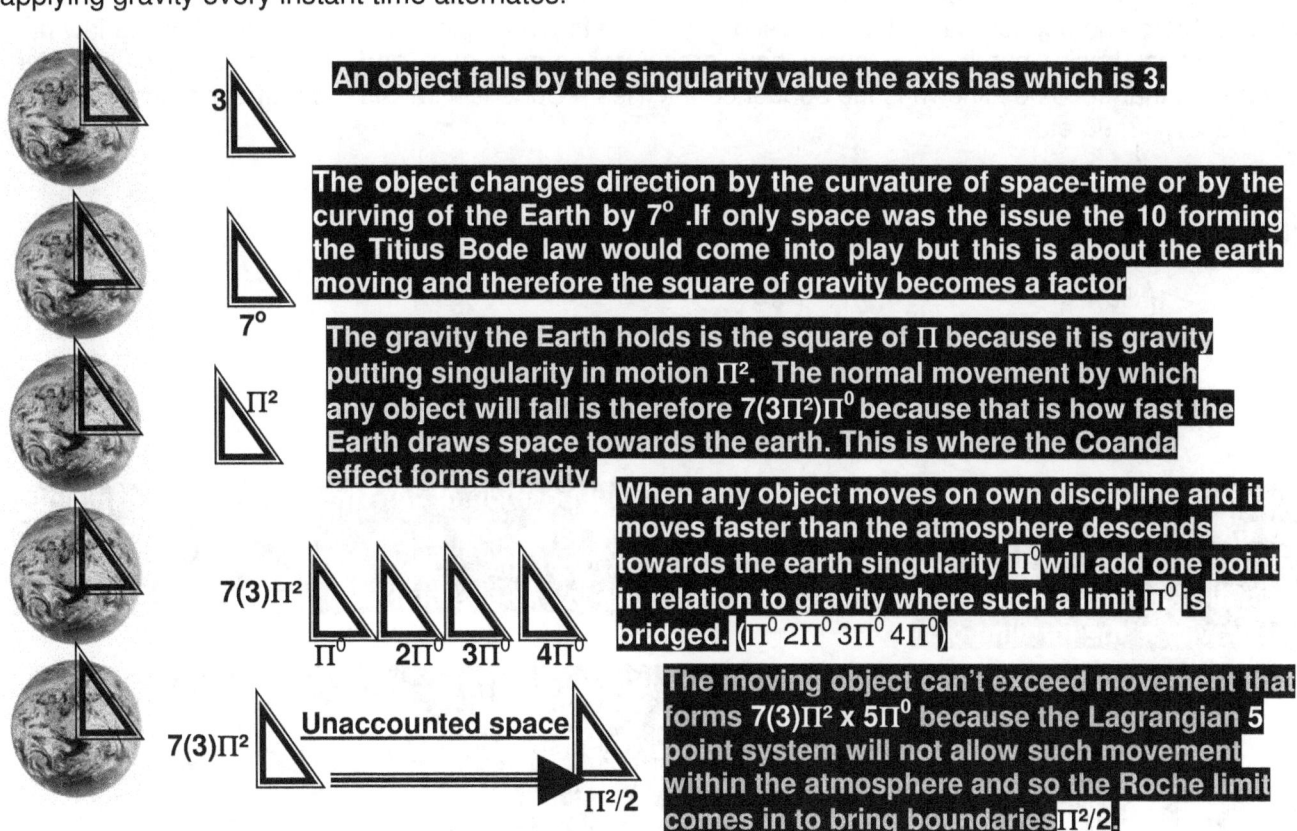

The Roche limit applying to stars being apart and the closest stars could be to each other is half singularity by the square which Π^2 is divided by **4**, which is, $\Pi^2/4$ but within the earth atmosphere this value halves to $\Pi^2/2$. So the sound barrier is **7(3)Π^2 x $\Pi^2/2$ = 1022.79 km / h**. To explain what happens when this happens and how this happens, the following gravity process applies. Gravity is the arrest singularity makes on the Universe in the single dimension where the highest value is 1.

Every instant singularity releases the Universe to space by restarting the Universe this goes according to the Titius Bode law bringing about space. When restarting takes place after the gravity arrest, then every point will be in place according to normal movement such as **7(3)Π^2**. The gravitational process places space in relation to material by movement of either or both. That is the Coanda effect. The higher an object is above the curving of the earth at the point where $\Pi = \dfrac{21.991}{7}$ changes to $\Pi=3.1416$ it has to be Π^0 away and the faster it moves (Π^0 to $4\Pi^0$) the higher the object must be from the earth curve. It is about movement away from the earth to maintain height in terms of motion just as a satellite must do and if not the satellite falls. The gravity shifts the points in singularity that forms space on position according to $\Pi = \frac{21.991}{7}$ going **$\Pi = 3.1416 \div \Pi°$**.

This puts space one point away in terms of singularity **$\Pi°$**. As the object moves in relation to the earth but not with the earth the movement must adjust the distance it moves according to the height it is in and this is **$7(3)\Pi^2$ from** Π^0 to as much as $4\Pi^0$. The lines that $4\Pi^0$ forms are radial moving by increasing of 7° inclining towards the centre of the Earth. At a height of $4\Pi^0$ it will need to move is **$7(3)\Pi^2\ 4\Pi^0$** = 829 km / h to maintain space-time progress. At a height of **$4\Pi^0$** the speed of sound cannot be broken any more and this has nothing to do with sound. This is the transfer of lines carrying as much as connecting singularity to the centre of the earth. Then when the object moves faster than the allocated height requires movement should be a gap appears between the positions where the object should be and where the object then is much farther away. If the object moves as if it is at a height requiring **$7(3)\Pi^2 \times \Pi^2/2$** an unfilled space appears that carries no sound because it is void of unfilled lines holding singularity at that moment.

Titius Bode law is formed from the sequence 0,3,6,12,24,48,96, and 192 by adding 4 to each number. The planets were seen to fit this sequence quite well – as did Uranus, discovered in 1781. However, Neptune and Pluto do not conform to the 'law'. Bode's Law stimulated the search for a planet orbiting between Mars and Jupiter that led to the discovery of the first asteroids. It is often said that the law has no theoretical basis, but it does show how orbital resonance can lead to commensurability. The importance that becomes known is the sequence the Ties – Bode law saw in the number arrangement of 3; 6; 12; 24; 48; 96 etc.

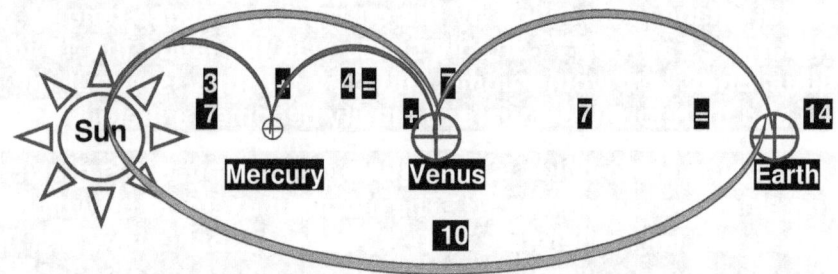

The space holding 7 plus the space holding 7 equals a position of 14. The circle is 4 as explained and the position of 14 minus the circle of 4 is 10. To place the earth according to a position in singularity in accordance to us on the Earth we have 10 of the earth divided by 10 in space giving us an allocated position of 1 This puts the Titius bode law in position with gravity and that proves that gravity is the forming of Π

The curve of the earth is 7° on both sides (7° + 7°) but because 7° represents the earth turning in movement it is also ($7^2 + 7^2$) and by turning it crosses singularity (1^2) both sides of the opposing circle in rotation then according to the law of Pythagoras it is ($7^2 + 1^2$) + ($7^2 + 1^2$) = (49 + 1) + (49 + 1) = 50 on the triangle that forms by a circle turning the direction = 50 + 50 + 100. Therefore the space in which the circle turns is $100^{½}$ to the root thereof = 10 and therefore the Titius Bode law shows the inside of the circle factors forming Π as gravity. That is why 7 goes doubled minus the second part of the circle which is 4 divided by the space in which the planet orbits and the allocated singularity position according to the sun is derived. It is implementing Π as gravity

This is gravity. It is the way singularity freezes the entirety together into forming one, 1 atom, 1 earth, 1 solar system 1 Milky way, 1 Universe. The way gravity works is it puts the entire Universe in 1 relation for one instant and in that instant gravity applying is **$7(3)\Pi^2$**, the atom is $(\Pi^2+\Pi^2)(\Pi^2\Pi)3$ = **1836,** where the Universe begins is the atom is $7/10(\Pi^6) \div 6=112$ and where the atom begins the Universe at the atomic start is $\Pi(\Pi^2+\Pi^2+\Pi^2+\Pi+3)$ = 112. It is relevancies frozen in one moment by gravity where time forces the Universe to become an instant. The Universe freezes in the moment, like a movie film and then moves on to freeze again and that leaves nothing to calculate but to read and to interpret mathematical formula applying of relevancies applying.

So you intellectuals forming the accepting of standards in physics do not have a clue what I am talking about…that is because there are too much detailed explaining missing, the detail you never wish to read because words can't explain physics. For ten years no one wanted to read my work because I don't follow your guidelines when I explain things you have no idea how to understand and I don't use mathematics in formula to calculate the way you think physics should be…and yet you come no closer in understanding anything when using your calculations, notwithstanding your brilliant mathematical abilities and your overburdening egocentric arrogance but instead you lot criticize me for explaining the way I explain things you can't even think to explain, nor form any concept that could eventually lead to forming any conclusion thereof. You put yourself on a pedestal being bright and shiny insisting on the admiration of entire world and for me not following your principles you criticize me while I am able to show you things and give you knowledge that you can't buy with 3 billion dollars of tax payers money because your methods goes begging and still you find fault in my work because I don't compromise and follow your misleading methods. With your methods and the way you go about practising physics in the way you do you are getting no where for a long time and yet you will not read my work written in my way of conducting physics because it does not conform to the style you lot in physics wish to portray physics to be. If you will not read my work and accept there is another way of doing physics, then you will remain uninformed and feel proud about your ill understanding and your progress in stupidity. I enter singularity and you have no idea what it is except to give the concept of singularity as you see singularity to be a magical status. It never dawned on you that singularity is 1 in form and that is because you lot are besotted on your ability to mathematically calculate that which you can't even fathom.

PART 3 of The Absolute Relevancy of Singularity,

Forming part of the website www.singularityrelavancy.com

The Prologue

Years ago I was reading of a remark Einstein made about his realisation whiles being a patent clerk. Einstein realised that had Einstein fell from the window of the patent office Einstein would feel as if he was as weightless as a chair and a pen falling alongside Einstein down the building.

Then I then realised Einstein felt weightless because he was falling and part of falling was feeling what was happening to him. He was not pretending to fall whereby he then would feel as if…he was really falling and with that there is no as ifs. What he experienced came by means of what he was experiencing. If Einstein was experiencing weightless ness, it would be because he was weightless while falling. Einstein would not imagine the weightless ness because Einstein was truly falling. He was at that moment truly weightless. Einstein, the pen, and the chair had the same weight since they were all weighing the same. All three items would be equally weightless during the falling…that was what Galileo found because objects of different size and different mass travel equal while descending. The bigger objects do not fall quicker than a smaller object and that can only be attributed to one fact; it can only be true if they weighed the same while falling.

From this one can deduct that gravity is motion or the intent to commit motion and mass is one the motion of gravity is frustrated by blocking the continuing of the motion. Gravity is motion of space and mass is the restricting of the motion of space. Having mass does not bring about gravity but it does restrict gravity's motion. Gravity produces mass but mass does not produce gravity. Mass is the restraining motion and gravity is material moving about. Mass only comes into the application when two objects filled with space moves into a position where both want to claim space the other occupy. In essence it still is the frustration of motion and the commitment to move once the blocking of space is relinquished.

I then after reading this realised that gravity is not mass orientated, but gravity is motion differentiation between objects. While falling, The object moves less or slower in the direction that the Earth rotates and will fall in the direction of the Earth centre until such a time as the movement of the object is in synchronising with the speed that the Earth spins or if not the object will and on the Earth surface at the edge of the Earth and that will bring about having mass. The gravity applies as speed that is putting time in relation to the distance travelled and distance travelled is space. While the object is in a process of falling, the motion confirms gravity, both by getting the object's distance or band in which the object travels in harmony with the Earth that conducts all the spinning taking place at that point. That will reduce the height in which the object spins until it lands on the Earth and then can't reduce such reducing of a travelling band any further. It has to do with specific density. If the specific density is increased by filling the object with helium we will find there arrives a point where the conducted speed is at a level that the Earth no longer will claim the body into having mass.

When motion downward ends and the Earth disallow any further movement to secure a better specific density in relation to rotating movement, then mass sets in and becomes what is than point holding mass where the constraining of the object takes place to secure frustration of further movement and the Earth's motion annexes the object's freedom. While experiencing mass the motion is still there but now incarcerated by mass and locked onto the Earth by the rotation of the Earth and the superior or equal specific density of the Earth. By connecting to the Earth the motion that the object is experiencing is what nails to object to the Earth by the force of mass and the object is then experiencing mass and not falling further through the loss of downward movement and now only conducts with the Earth rotating side-on movement. In this the downward movement is not lost altogether but remains, as detectable movement is the form of having a tendency to move although the object in mass is applying by forcing the downward motion to stand still.

While the object is in mass and seems to be as if it is resting the tendency to move downward remains applying but that tendency to continue to move downwards is the tendency he named mass. However mass then restricts motion and becomes motion tendency. While falling, gravity applies as equal motion to all objects relying to place all objects in relation to specific density and because of this motion counteracts any size, mass or weight by making everything able to fall equal in specific density. When falling, the object is either equal to what might be in the air according to allowed specific density, or has more than the specific minimum required density that is what is allowed to serve as the minimum required specific density and therefore will spiral down to the Earth.

When the Earth restrains further downward motion of the object that comes as the result of finding an allocated position of motion according to the specific density of the falling object, this readjusting of allocated position is stopped from conducting further downward or readjusting movement and all such further movement of gravity is hindering in the form we call mass. The falling object remains individual and still tends to move while Earth individuality resists movement. Further movement is disallowed as other material fill space. While the bonding of the atoms forming the object will secure any further deforming the object will remain to be independent but it is this bonding that is the value of the specific density of the object applying. By securing a [lace on the Earth, the falling object will finally rest and from that motion resistance comes mass.

While falling, the object is experiencing gravity because the object is in gravity but when on the soil the object experience mass which is the restricting of gravity or motion of the space filled with material.

Moreover, I came to another conclusion of equal importance. When any person is standing on any place anywhere, while viewing the Universe, that person is filling the centre of the Universe. Let's get more personal. When you, the person that is reading this, are standing at night and are looking at the Universe you are seeing the Universe from the centre of the Universe. All the light, every single beam that ever left any destiny at any time acknowledges this fact.

You are the most important person in the Universe because you are holding the most important position in the Universe. All the light that comes across all of space runs directly in a straight line towards you filling the centre of the Universe. Not excluding the effort of one photon, all light is heading to meet you where you are in that centre spot and not one photon will pass you by. Not one photon dare miss you because if they do they miss the effort that all light has to accomplish and that is to locate you as the person filling the centre of the Universe. If you find this funny, or laughable you are in for a shock because this is what gravity is and this principle dictates gravity. It is the most complex issue one can imagine and expanding on this thought takes thousands of pages. It forms the crux to all cosmic principles and embraces every successful and meaningfully theory ever used to explain the Universe. Without taking this aspect in to account, there is no valid explanation available to understand the cosmos. Al the light coming from wherever meets the point you fill in time and in space. For al the light travelling you hold the spot it was on route to.

Should you decide to shift your position to any other place in the Universe you will shift the centre of the Universe to that location as well. If you install a camera on Mars, the light is obliged to acknowledge your relocating the centre of the Universe at your will to reposition you're being that centre of the Universe. All the light that ever left its destination crossing the vast spaces of the Universe, excluding no particular light, travelled all the way just to find you filling the centre of the Universe, right where you are. By you're standing anywhere, you fill the centre of the Universe, and the entire Universe admits to that because all the light comes to meet you there.

If you shift from the North Pole to the South Pole you will shift the centre of the Universe because all the light travelling throughout the Universe will find you where you then moved the centre of the Universe. The light left its destination billion years ago as it travelled through space at the speed of light anxious to acknowledge you're being in the very centre of the Universe. No photon will pass you by where you are in the centre of the Universe. No wonder every person born has the idea they were born to fill the centre of the Universe, which we do fill. The Universe is spinning around you or I, which is filling a centre where all motion is connected. That is the Coanda effect on the utter-most grandest scale imaginable; nevertheless it is only a manifestation of the Coanda effect. It implicates gravity as wide as can be…

Then I reviewed the Universe. If gravity is motion, what causes motion? What stops motion? That answer is in the Black Hole. If a star is about fusing atoms thereby growing, what happen when all the atoms fused into one all collective atom? What is the gravity if the star has one all-inclusive atom providing all the gravity that the star had when the star still had massive volumetric space? If all that space that once filled an entire giant star fused into one enormous gravity applying atom and that enormous force has been secures in the space that one atom holds, the atom would then show a force that would pull the surrounding Universe flat. Where does the gravity of the star end when all the atoms in the star became one giant atom? Gravity is smallest where space is least. Where space of an entire massive star is left in the size of one atom the gravity coming from that will pull the Universe flat at that point.

Coming to the conclusion about gravity being motion and mass being the restriction of motion was the easy part. What produced the motion and what prevented the restriction from overcoming the motion was the tough part. Figuring out why was everything on the move and where did the motion stop that was the part that took some figuring and some explaining. What made gravity move and why does gravity move…the answers are in the four phenomena never yet explained to satisfaction but now turns out to be the cradle of gravity.

Gravity is **The Roche limit**,
    Gravity is **The Lagrangian system**
        Gravity **is The Titius Bode law**
            Gravity is **The Coanda affect**

And gravity as the Roche limit forms the principle in producing the sound barrier. Read the book and find out why this is the case.

Newton's claims about the principles that he declared is responsible for guiding physics carries no validated proof and only after I realised that, was I able to start forming another line of thought on gravity. This had the purpose of confronting the corner stone of modern physics and at first I tried desperately to do just that. At first I was not confrontational towards Academics in physics and avoided any indication about disagreeing with Newton, although avoiding to show my disagreements was also totally impossible too but every time I approached academics with my new concept the academics always threw Newton at me. Facing Newton or facing defeat became a two-sided blade and I had to start to confront them by confronting Newton, with which I was in disagreement from the beginning.

At first I was reluctant to voice any opinion about the matter of how far I was prepared to challenge Newton because Newton was and is an icon. But slowly it dawned on me that if I had any serious plans to introduce my ideas I had to dispute Newton's gravity principles and do it head. When the slight confrontation did not bring results I finally decided to go all the way and show the inconsistencies that were prevailing in Newtonian science. That worked neither and it brought me the same results as before whereby I decided to go public and straight to John and Jane Dow avoid arrogance academics have with only one motto they serve and that is their autocracy and in particular their megalomania especially to my case as well as me in person. I wrote them (nine in total) letters in which I warned them that I was going public to show the extent of their dishonesty in their Newtonian's approach and lacking of substance and proof their physics has. The lack of honesty and furthermore the absolute dishonest on their part is there whether I avoid it or attack it; the inconsistencies are part of forming the basis for modern accepted science.

This process I now described is explained in a paragraph or less and it seems I got that far in a breath or two, but getting this far took me the best part of seven years to get to I tried my best not to attack them or Newton but left with the option to leave the project and lose thirty years of work and then fail after I concluded an answer on every aspect they never even thought of or take them on and dish out what they should have received years ago made me decide on the latter. After being avoided and taunted by their powerful positions and arrogance vested in their mentality they show in regard to their positions as well as the disregard they show in the mentality of others I slowly concluded that only and after I can get people forming the general public and the opinion of those that holds their disregard just as I do to see what they hide will I get a response from the Mater's of fraud.

First I had to show the general public the true colours of the academics in physics and get every one to see how incorrect Newton is, and only then do I stand any chance to introduce my line of thought. I am so sure of the ideas that I propose of being correct that I dare any one to disprove any part or the entirety that my concepts about cosmology forms! But that can only come about when I can get an audience to see how I expose Newton for what Newton was and it is in that where I find no luck. I can't find one academic with influence that is brave enough to stand up and face my attack on Newton and argue me down or prove me wrong in a sound debate. Now I see frowning coming from everywhere because it is madness on my part to think the world is wrong and only I am correct!

I realise that it shows signs of madness on my part and in my thinking to even regard any possibility that I am the only person on Earth that is correct and all others that ever studied physics are wrong but mad as it seems, if that is what I have to say to find an audience to listen and to judge my case, then that is what I say. I don't say this lightly or without understanding the enormity of what I suggest is going on, but be that as it may seem, it is the truth without question that Newton went on for three hundred and fifty years defrauding science with no one testing his claims.

Argue me down or prove me wrong but don't discount me before hearing me out and only after considerable consideration while studying my arguments then form an opinion that disputes what I say but when disputing what I say, do it while confronting me in a sound argument when proving me incorrect! This not one academic could achieve and I challenge the lot to do so. But do it after studying all my work and being in a position to account for all the details I propose. Don't just dismiss me because I dismiss Newton because following that road is the way of the coward and the mentally impaired. Read my challenge about the correctness of Newton's proposals when he brought no more than suggestions into science and when I dispute Newton, then take me on by proving Newton correct... do it just once... prove Newton correct just once...prove that his formula is working and that his principles apply on the grounds he principled his ideas.

Detecting Newton's misconduct is possible because I saw a way to break away from the invalid concepts Mainstream physics hold. I went about and tried to prove Newton and when that was not happening I tried to apply Newton's ideas into the greater fields of cosmology. That also wasn't possible. I tried to amalgamate the four cosmic principles applying in cosmology with what Newton said was happening in the cosmos with mass and with gravity and in light of what the cosmos showed was happening Newton just wasn't happening!

Notwithstanding the pose Mainstream physics try to uphold, the entirety of physics still use the idea of magical forces intervening in nature and they still base concepts on unexplained novelties. Think of finding four unexplained forces going around and influencing persons in an unexplainable manner except that the magic of gravity keeps people attracted to the Earth. To say the least, the concepts physics use in terms of Newton would not even be acceptable to children in the modern informed era we live in, I challenge any person to prove Newton, not to accept Newton but to undoubtedly prove Newton correct! Prove how Newton's formula of mass forming the force of gravity can apply as Newton said it does! I recognised the impossible double standards Mainstream physics apply to promote their much shady explaining. In short I tested Newton's principles and found the principles to be wanting.

The inconsistencies Newton introduced brought science double vision and to compensate for these bogus truths supporting their incredible theories, they simplify issues to such a level where what they embark on, is the meaningless acceptance of the unproven and they proclaim to understand what are meaningless inconsistencies and to achieve this they create scenarios which uses the entanglement of deception. Prove the attraction Newton said was enforcing gravity that is pulling by mass and is gathering plants by contracting the diameter between planets.

Show how much the Moon came closer to the Earth since the time of Kepler. Show proven distances taken by radar tracking and indicate just how accurate Newton was. Show how much the Moon came closer to the Earth since the time of the Moonwalk in sixty-nine. The figures are available but are kept in a grave of silence where no one ever speaks about what science found applies and how much the distance between the Earth and the Moon is shrinking as Newton said is happening or then how much is the is expanding which will contradict the very principles Newton brought about! What they declare as unwavering facts can't even be supported in some form when tested by a silly test as to show that the distance between the Earth and the Moon is shrinking. Even the least degree of verification of correctness is absent when trying to find support of Newton and Newton lacks all evidence of authentication in any investigation of even the simplest terms. It is as if they never read with interest that which they explain when they embark on explaining Newton and they never scrutinise that which they advocate when they teach Newton's principles applying. They give values that are senseless and the very values they use make that which they say meaningless.

In this book I am going to investigate how much truth there is in mass pulling by the force of gravity. To most if not to all of the persons reading this, such a venture of investigating Newton is time wasted and just the thought about me embarking on the investigation of the issue is totally senseless to investigate. It is senseless because the concept it carries became accepted as household practise and life science from where it proceeded to become everyday culture in every person's mind. The worst part is that the group of people normally considered as the wisest bunch there is, never did prudent testing on Newtonian presumptions, while to test the presumptions is most easy to do. I will not believe that a lot that lives up to the veneer of being the best mathematical intellectuals on Earth, never though of testing Newton's very simple formula and in that disregard the formula because of the incorrectness the formula holds.

Do you think of astrophysics as being the department that is run by the wise and the level minded, the honest and pure at heart, the nobility of well-to-do academics and the sober thinking standing in front of the world as the absolute trustworthy? If you are a student, there is no other choice you have but to trust them while they feed you absolute hogwash! If you would so much as dare to doubt any thing they say they will banish you from the institution they rule so absolutely.

The banishing process is dome under the blanket of examination. They teach you what to think and to make sure you think what they wish you to think, they tell you to confirm their teachings on a blank piece of paper. You write what they prescribe and you supply the answers they demand in the words (sometimes) of what they demand. Should you in any way say anything different from what they tell you to think, your presence will not be tolerated any further as they abolish you from their institution of academic tutoring. After reading this book I invite you to…no I dare you to challenge their statements with evidence gained from this book and see them wilfully further their culture of deceit by bringing unfounded arguments just in order to silence you and prevent you from getting behind the truth.

If you think those in charge of astrophysics are the pillars of trust, then get wise by reading the following facts and arguments this book presents. What you are about to read is simply mystifyingly simple and yet to this day I have not had the privilege to challenged one academic any where that had the honesty to admit to the fact of Newton being wrong. After you have considered the following you might agree with me that even small Children can reach a higher level of clear-minded logic and find more sensibility than what those scientists promoting astrophysics have because science lives in a make believe fool's paradise.

The manner of regard to life that the Academic Physicist holds and the outlook on life that the followers of Newton physics have (I call them plainly Newtonians and to me they are sheepish because they resemble to the image that to me seems the same as sheep running after their leader without having the ability to think for one second any thought spawned out of personal intellect) is quite the opposite of what I think of them. They keep their forming the establishment of the order the Academic Physicist in high regard and consider their order to be the top thinkers in society. This religion that they practise of self promotion and sublimely self regarding their status being next to God has them so high that we down on Earth forming the waste of human garbage can be told anything and we will believe what they say just because they with their supreme intellect tell us to think what they wish us to think.

This they do because we human waste living way down below their supremacy have not the ability to think and therefore they must think on our behalf. In their view and so far very correctly judged on their part, they, the persons being in the group that forms the Academic Physicists, believe very correctly that can dish up whatever they wish and we, those forming the group in the gutter, those that are mindless in their eyes, we will have to accept what they say without being allowed to form an opinion other than having the opinion they give us to have because in their view we are unable to have a mind other than what they are able to control. This attitude they have is the result of a relationship that worked for so long and thee fact hat it worked that long is what confirmed their opinion that we, the public, are fools to believe anything and everything because of blind stupidity.

But in spite of their aggravating conduct and mischief towards us, it is not because of a lack of insight and inability of controlling a mind that we have our childlike belief and blind trust in their opinions and which there was. It is the faith we shown that they misused for their scandalous cheating. Our faith is what we have shown towards them and is that, which became used as the reason why we accepted what they said blindly. We didn't accept their word on the grounds of us being utterly stupid as they perceive us to be but our trust depended on our good nature and believing in their trustworthiness.

This trust we have is brought on by a culture of trusting the King to do the people well and somewhere in every person's cultural past there was Kings that did us well in leadership. But their underestimating of our abilities is the testimony of their poor understanding and their weak insight ability, which results from their arrogance and stupidity. You are about to see just how stupid they really are in the thinking aspect of science. It will become clear as you page along while reading! They didn't fool us half as much as they fooled themselves and you are about to read all about it.

The fact that they could fool us for centuries didn't run on their intelligence being so much superior but served their purpose as it stemmed from the trust we had in them resulting from good intentions on our part. This betraying on their part and misusing the public's good nature to be used in schemes to get the public conned must end and I pray that this book form the first step in resisting the arrogance of the Academic Physicist.

Any one not in their group of the Academic Physicist is part of the lowest order of mindless being and to become part of their order and those that have minds with an ability to think, students have to accept what they say when they say whatever they wish to say without having to prove the correctness of what should back their saying so and as a result of this students may never question what they say. Only when and after proving that a student has totally lost all ability to think for him or her self may a student be promoted into the ranks of their sublime intellectual group. The sifting process they named examinations. You write on paper what they told you and never question their opinion and after passing that examination will you ever enter their sphere of intellectual brotherhood. Does this sound far fetched? Then you better read on and I will remove your blindfold and show you what a world of deception the Academic Physicist force on us into.

Read the following and see how they, the high and the mighty, those that think they can replace God and those who think they can think on our behalf and think what to tell us to think, how much they are clowns and the jokers in society. Read how little are they, the Academic Physicists, able to understand concepts about Creation while they think they are able to replace God in their superior intellect.

If you are a student in the science of physics, then ask your Educated Masters to please explain the following abnormalities you are about to read in this book and insist on a clear explanation about the inconsistencies they promote while tutoring physics as if the physics they present are the most flawless and accurate institution there has ever been. Ask those academics supporting Newton about the following flaws that no one mentions …ever… except me in this book you are about to read and get them to explain the inconsistencies never talked about, which I present in this book and then after confronting those charged with tutoring physics and seeing who should be believed, then get wise instead of brainwashed. Let them mathematically show how one would go about and use Newton's visionary formula $F = G \frac{M_1 M_2}{r^2}$ to calculate the force of gravity by replacing the symbols with the actual values in mass that the items referred to have. Put in the Earth's mass in place where it belongs and put in your mass in place where it should be and then divide that with the distance between your soles and the Earth measured in micro millimetres by the square thereof!

In the book named **an _Open Letter on Gravity Part 1 and Part 2,_** I bring the solution to the mystery behind gravity. I tried in vane to introduce the principles I find valid to the academics in charge of astrophysics. Facts that Science present as being the uttermost explicit and unwavering truth, fails to bring any logic answers to so many questions that it should address. It fails to have substance in addressing the most basic and simple questions about gravity and physics. Yet to every question science can't answer my approach does bring many solutions.

The presentation and the delivery of my answers that I reach are understandable and simple where it serves both logical science and the truth. Since my answers do not match Newton and his misconception about gravity and that mass generates gravity, those in charge of science don't even bother to read my work. With their affixation to the corruption they portray I can do little to the giants where they are in the mighty positions they have and just because of that they can go about to sideline and ignore my work and this is notwithstanding the correctness that my work delivers compared to the utter failing that Newton's work shows.

When confronted with my evidence and they have to match my work with the hypocrisy and misleading nature of Newtonian cosmology their defence in substantiating their claims is to ignore me. Since I do not applaud mainstream science and the clear fraud they embrace and fraud it is that they embrace, I am silenced. Why is it that my work is going unrecognised or even in the least goes never debated and never commented on…it is because it will then trash every article anyone has ever written about astrophysics and cosmology. They show little integrity when academics with such supposed high standing or then such as they should have, play a dishonesty game where those in commanding positions will rather protect fraud and save their skins. They would rather protect the corruption they have than seek the truth and find honesty in physics.

Those academics in charge would much rather protect their un defendable ethos they maintain as forming the back bone in science and what gives their personal position legality although it is corrupt than admit to the truth they find when they begin reading my work and in agreement they then have to back the truth my work brings. Doing that (accepting the truth in my work) will trash all work in cosmology delivered thus far and condemn it to the waste paper basket and render all work invalid and void. It will put all the Newtonian's bias and fraud into the place where it belongs.

Considering that such acting will lose them money, those academics in controlling positions then will rather rape the truth in order to benefit from continuing to corrupt student's minds further. If they wish to justify their inconstancies they have to attack my work and disprove the accuracy of my work. That they can't do. They then ignore my work because they can't attack my work. In that sense they also place their work beyond my approach, as they can simply ignore me as if I represent the plague while they carry on with little consequence to bother them. I challenge them to prove Newton correct and not just declare Newton being beyond reproach after all has seen the evidence I bring. After reading this all students must challenge them to defend what they can't or get honest.

$$F = G \frac{M_1 M_2}{r^2} \qquad F = \frac{r^2}{M_1 M_2}$$

This is the basis that Mainstream science uses as the foundation of all physics anywhere. If this is wrong then everything they have got to work with goes out the window. They put mass and the distance that parts objects in a relevancy, in other words the one is a ratio to the other. The one factor brings a measure to the other factor's value. The one cannot be without the other. The increase in one becomes the reducing of the other and the other way round also applies. When the distance is large, the influence of mass will be small and when the distance is small, the influence of mass will be overwhelming. Then they state we are in a Big Bang expanding of the entirety. Why then, when considering that if it is mass that produces an inclining force of contraction as Newton says there is going on then…why didn't the expanding stop before it started when the Universe was small. Today using hindsight after the fact of the exploding Universe became apparent by the studies Hubble brought to light did the lot of everything that is not implode as Newton would have us believe whereas, instead it did expand just as Hubble proved. The radius at the time of the first instant back then was no factor, which makes the gravity at the time a totality of unrivalled force. The radius being that insignificant leaves the mass unchallenged in asserting power in relation to the non-existing radius it had.

I dare any physicist to show me where they apply Newton's formula just and exactly as Sir Isaac Newton suggested gravity applies. Show me just once where the mass of the Earth is multiplied with the mss of the object in normal physics. Show me just once how $F = \frac{r^2}{M_1 M_2}$ or $F \, \alpha \, \frac{M_1 M_2}{r^2}$ where one M represents the mass of the Earth while the other M represents the mass of the object and in this formula the end result will have a value of 9.81 Nm/s^2 … show just once one example… where the use of the mass of the Earth comes into play. If multiplying the mass of the Earth with the mass of an object and dividing that with the distance parting the two mass factors does not deliver 9.81 Nm/s2, and then any claim by Newton indicating that $F \alpha \frac{M_1 M_2}{r^2}$ is equal to gravity, such claiming constitutes to deliberate fraud…even if Sir Isaac Newton said this. Prove that the mass of the Earth with the mass of an object and dividing that with the distance parting the two mass factors delivers 9.81 Nm/s2 or admit physics is conducting fraud to protect Newton!

<u>To whom it may concern:</u>
My introduction as well as introducing the readers to general cosmology in a very brief and compressed manner but first, I have to give the emphatic warning to all prospective contemplating readers.

Please take note of a conscientious warning about the gravity of the misgiving there is on the part of the most respected Academics in physics about a much concerning matter.

I state it emphatically that science accuses me to be not schooled to the point where I am able to have any form of an opinion on any matter concerning Sir Isaac Newton. Notwithstanding that my research proves I did my private studies and through which I skipped the indoctrination and mind control academics place on students goes unrecognised by their standards and so too my ability to have any insight on matters regarding physics.

However my skipping their methodical and systematic brainwashing enabled me to see and allowed me to be able to express the incorrectness in Newton's teachings and allowed me to show in clarity what destructive force Sir Isaac Newton used to corrupt the laws of mathematics, corrupting to science along the way and mostly raping to the work of a great man, Johannes Kepler and what Sir Isaac Newton did can only be expressed as being blatant criminal fraud. What his deeds amount to is to corrupt the laws of mathematics, to render the laws of cosmology useless and to rubbish all of science. Should you find this to be unbelievable, then I am glad to announce that this book is more for you than any other person, so go on and read what academics guarding science never wanted published. I challenge any one that disputes any claim I make to prove me wrong by proving me wrong and not merely suggesting claims in that direction.

If you read on, you will see where the Universe starts and why the Universe can never ends.
You will see precisely how the Universe goes "flat" because I vividly take you there.
You will see precisely why the Universe goes "flat" because looking at what I show you can see the point where everything in that space goes flat in front of you by not seeing with your eyes.
You can actually cast your eyes on the spot that is while it also never can be part of the Universe.
You will see precisely when does the Universe goes "flat" because you are part of the Universe going "flat".

PART 4 of The Absolute Relevancy of Singularity,
Forming part of the website <u>www.singularelavancy.com</u>

From the Heart of the Author, the following:

Tyco Brahe, on his dying bed, kept on rambling to Johannes Kepler, begging Kepler to finish his work so that his life would not be in vain. Kepler complied but after five hundred years the work of Kepler is still unfinished and if I die without anyone reading my work, it is not only my work that will go unread but also again the work of Kepler as well, will remain unfinished, as it will never be understood in the way the cosmos intended. The Universe introduces a cosmos to Kepler that uses a formula that does not comply with the standards we see in the Universe that we see. Tyco Brahe spent a life time accumulating facts and the arrogance of semi blind, self-opinionated, self serving academics in physics of the day that thought they had the authority to decide what science is and what science has to be and what confirms science as much as what conforms science, decided Tyco Brae's studies were pointless.

Up to today the work of Tyco Brahe as well as Kepler is still unfinished and when I try to finish Kepler's work I find that my life's worth of work will be in vain, as modern mainstream physics will see to it that my views will never gets published because I question Newton. I do support Kepler but Newton never supported Kepler and because I support Kepler and not Newton, I am stupid and I am mentally retarded, slightly weak in thinking and mostly out of touch with reality. But that was how everyone in science at the time also felt about Tyco Brahe, Kepler and Galileo. I feel the same as Tyco Brahe, as Galileo Galilee felt and more, because I am bullied in the same way. Newtonians of the day keep going on about the Church and the wrongdoings committed to Galileo Galilee, but it is modern science that killed Galileo Galilee, because everything Galileo Galilee said Newton turned around and destroyed.

Gravity is no force...gravity is time because that is why one may employ the pendulum arm invented by Galileo Galilee to measure time. Time to science is how long it takes for the earth to circle once around the Sun, but for God sake, how can that time apply to the entire Universe. The pendulum swing you would use in the atmosphere of a massive star to measure the gravity applying will show time very different in the star. The thing about Newtonians is that they are not accountable to no one about what in their world contradicts reality. The one polishes the other and the lot shine while everything underneath is rusted to the core. And there I have lost the audience of every academic in physics, but I guess I never had their pompous attention anyway...

PLANET	PERIOD (Years) (T)	MOVEMENT (T^2)	DISTANCE	SPACE (a^3)	RATIO k
Mercury	0.241	0.058	0.39	0.059	0.983
Venus	0.615	0.378	0.728	0.381	0.992
Earth	1.000	1.000	1.000	1.000	1.000
Mars	1.881	3.54	1.524	3.54	1.000
Jupiter	11.86	140.66	5.20	140.6	1.000
Saturn	29.46	867.9	9.54	868.25	0.999
Uranus	84.008	7069	19.19	7067	1.000
Neptune	164.8	27159	30.07	27189	0.999
Pluto	248.4	61703	39.46	61443	1.004

KEPLER'S LAW OF PERIODS FOR THE SOLAR SYSTEM			
PLANET	SEMIMAJOR AXIS $a\ (10^{10} m)$	PERIOD T (y)	T^2/a^3 $(10^{-34} y^2/m^3)$
Mercury	5.79	0.241	$k^{-1} = 2.99$
Venus	10.8	0.615	$k^{-1} = 3.00$
Earth	15.0	1.00	$k^{-1} = 2.96$
Mars	22.8	1.88	$k^{-1} = 2.98$
Jupiter	77.8	11.9	$k^{-1} = 3.01$
Saturn	143	29.5	$k^{-1} = 2.98$
Uranus	287	84.0	$k^{-1} = 2.98$
Neptune	450	165	$k^{-1} = 2.99$
Pluto	590	248	$k^{-1} = 2.99$

Can you see the grain of cosmic dust too small to see with the naked eye that spins at the same tempo that the large Jupiter does. If you say you can see the grain that is too small to be seen by the naked eye you are as much a liar than is the Newtonian that says gravity or movement of objects depends on mass because notwithstanding the huge discrepancies, both cosmic structures (Jupiter and the cosmic dust) move gravitationally cyclic equal around the sun and that disproves that any one has mass that produces gravity.

This is Jupiter that is 317 times larger than the earth is and is therefore the planet holding the most mass of all planets and yet according to Kepler's tables (Kepler's tables are indisputably accurate), notwithstanding mass differentiation it spins around the sun at the same speed as what a grain of cosmic dust spins that is too small to be seen with the naked eye. Where does mass fit into this accept that it proves Newton's claims about mass forming gravity are a swindle set in motion to corrupt science at the very core.

The dust floating in space and the smallest asteroids spin around the sun as fast as the mighty Jupiter does…so in what way does the mass pulling the mass manifest as a cosmic reality. Only Kepler is correct. In the tables that Kepler configured as $a^3=T^2k$ we have three distinct factors combining to form a specific value that indicates space-time $a^3=T^2k$ and moreover shows that the Universe structurally is composed in terms of **space a^3 = time T^2k** and every factor as much as a^3 and T^2 as well as k has a part and a role in forming the eventual value of **space - time** $a^3=T^2k$. The pendulum arm semi rotates T^2 in the gravity k of the space of the atmosphere a^3. For years science missed the principle that the pendulum measures space flowing in time thus **space - time** $a^3=T^2k$.

The part that mainstream science missed for five hundred years is the presentation of how the Universe is as it gave Kepler the formula of a "flat" Universe because the Universe we have is not substantial. Now all readers left reading suddenly find this sudden urge to switch sides and begin agreeing with the Newtonians that I am somewhat soft between the ears. There the cosmos shows is a Universe in singularity that science only speculates about.

To incorporate a three-dimensional Universe into a flat Universe in singularity is not that simple. Whenever I am presented with this picture presentation of a flat Universe I am baffled by the Newtonian

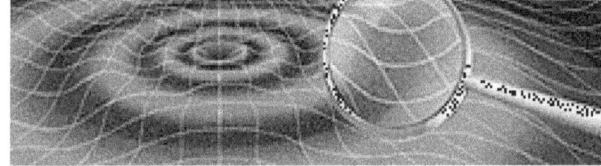

stupidity represented by such an image. If there is a top, as the picture indicates as to be the mathematical flat Universe, then a bottom has to be somewhere, so it then is not flat. If there are waves, then there is depth and where there is depth there can't be flatness. Where there are blocks, then there is a width and a length and that can't be flatness because the blocks then show variation in size. This picture is as three dimensional as the blanket on any bed. My question is if I see the topside, then mathematically there has to be a bottom, a side where the top ends because with the top ending in the picture I see there has to be a bottom side that is there but is only dimensionally obscured by the topside I do see.

Secondly, the next question is what do I see? If it is the Universe the picture presents, then from where am I looking at the Universe I see and from what angle can I see the Universe in such a way because if I am, then I am part of the Universe in which I am. In this picture I am not in the Universe but I am a spectator of the Universe as if I am God looking down on creation. The only view I can have of a Universe is being part of the Universe and what they show makes me abstract from the Universe. I am always part of the Universe so I can only look at a Universe from up, down, back, front and both sideways forming a picture. I see waves. If there are waves, then there is a height and a depth bringing in dimensions complimenting the flatness they portray and that makes it a cube I am looking at. That then all this picture does, is it shows the funny way they see the Universe from the perspective they think they have using brilliant mathematics, as seen from what they think is their view of privilege by not being part of the Universe. This is the joke you will get when applying mathematics to a part of the Universe that holds all space as 1 because all the space that can fit into singularity is 1. Where singularity is one there 1 is the only applying value. If you don't recognise the fact, you get funnies such as they show in this picture. I use the same top that Newton used when he tried to show that gravity forms by mass. I show with the top, an everyday object child's toy where to discover singularity. I show with the top where the Universe starts. I show with the top where it is that the Universe ends. I even show how to travel to reach the very centre spot in the Universe! It's been five hundred years more or less and my study about the work of Kepler is the first authentic study ever conducted and because it is accurate and cuts Newton flagrant anomalies out, my studies are so disregarded I am told it is not science. In the studies of Tyco Brahe and Johannes Kepler they did not tell the Universe what they must find, should the Universe wish to be considered as correct. They studied the Universe and found the Universe told them that the formula formulating the Universe is $\underline{a^3 = T^2k}$. Never did they tell the Universe what the Universe should be, and that $\underline{a^3 = T^2k}$ is incorrect and it must be $\underline{a^3 = 4/3\Pi r^3}$ and that is how one would measure a sphere, but instead learned from what the Universe told them and reported accurately. The Universe said it is... $\underline{a^3 = T^2k}$

Does anyone see in the tables the Universe telling Kepler $a^3=T^2$. Newton told the Universe what the Universe had to be to comply with and what Newton wanted the Universe to be, it had to employ mass and when I ask to prove how mass is pulling the Universe closer, because even with the invention of Dark matter it still is not coming closer by mass... I am told I am not conducting science because science is conducted in a very specific manner where the Universe is told what to do. The Universe has to start shrinking because physics and science says Newton said it has to shrink...and now I too must comply or my work would never find publishing. Read on and find out how I am told what science is before science is physics. Science as Newton did, is still telling the Universe what to be...Just the same way as the Universe are told it has to contract or be reinvented by dark matter changing the critical density that withholds the over all mythical contracting, I am told in the same manner what science is and how science must be conducted or my work will never find publishing. In science the protocol is the one academic is so much admiring the work of the other academic so that the other academic would admire his or her work rite back so that that they don't criticize the work of any body but criticize only the persons not echoing their views.

I now wish to present the readers with a letter substantiating my claim about the attitude academics have when they are confronted with ideas not matching their ideas. This is not the only letter, there are many. This letter shows a trend and since I by now endured enough of the superior attitude academics show towards those not confirming their ideas about science I am publishing this letter in full. The academics are too lazy to read anything they are not familiar with and then reject by not reading it. And most of all, by giving you the picture I challenge you to show one point where I am incorrect or incoherent because what I say, everybody can read, if the person can read, because I am presenting you someone that apparently can't read, or so the academic claims because he says science can't be explained by using words...!

PART 5 of The Absolute Relevancy of Singularity,

Forming part of the website www.singularityrelavancy.com

An Article Sent to Annalen Der Physik
And the e-mail sent to confirm yet another rejection...

Dear Dr. Schutte,

You submitted an article of 15 pages to the Annalen. The content of this paper doesn't constitute a theory in physics. With a lot of words and some simple algebraic relations, there is no way to "explain" the world of physics. You seem to be out of touch with modern developments. This is also shown by the fact that you don't quote any relevant literature. I am sorry to say, but the Annalen is not able to publish your work.I am sorry for having no better news for your.
Best regards,
Friedrich Hehl
Co-Editor Annalen der Physik (Berlin)
--Friedrich W. Hehl, Inst. Theor. Physics
* University of Cologne, 50923 Koeln _____/\/_____ Germany
fon +49-221-470-4200 or -4306, fax -5159
hehl@thp.uni-koeln.de, http://www.thp.uni-koeln.de/gravitation
* Univ. of Missouri, Dept. Phys. & Astr., Columbia, MO, USA

Dear Prof Friedrich W. Hehl,
I have received your e-mail reply and I wish to respond on your letter. This letter is a duplication of a previous letter but I repeat the letter because I selected material from my books to show you how much any one can prove facts and concepts in physics by applying "a lot of words and some simple algebraic relations, there is no way to "explain" the world of physics." In five articles I show you how you stray by using your mathematical calculations.

What I show you with these articles science with all the splendour of calculation and not using words could not explain in almost four centuries and that you will see if you care to read my work this time. I show you how ridiculous the use of the idea of applying mass are as Newton intended it applies in the cosmos. In sharp contrast I show exactly how the Titius Bode law applies gravity…and may I remind you that it is the Titius Bode law that is in place in the cosmos with the cosmos showing no evidence of mass playing any role in the formation of the solar system. Even in the arrangement of planets there are no evidence of mass favouring any position of any planet. However, truth for once be told, it is the Titius Bode law that undoubtedly forms the space that forms the solar system and I explain how it does it. I achieve this breakthrough that was never yet achieved by not trying to be clever, but by being honest about physics and searching within the cosmos for reality to bring facts to mind. I challenge you to show one word I use to explain physics and I add as it was never explained before, that does not become pure physics, basic physics as you have never experienced before, notwithstanding all your calculation ability. Everything concerning gravity goes by circles and in four hundred years all your calculations missed that.

However, if my explaining is not intellectually matching your understanding of the concepts I explain how gravity is Π and how the Titius Bode law forms Π and by doing that it also form gravity, where then this forms the solar system by the forming gravity, then do not merely reject my work unconditionally without reading it just because it is in your opinion The content of this paper doesn't constitute a theory in physics If you feel the content reduces your standard and belittles your ability to show any insight into physics, give it to a school child to read and let the child explain to you how the Titius Bode law forms gravity in the manner I prove that the Titius Bode law forms gravity by Π. Let the child also show where I prove it is impossible that mass can form a pulling power force as gravity because if it did, then by now mass would have had to destroy the entire solar system which it does not do.

The entire solar system is expanding and even the circumference of the earth is growing bigger, this concept ignores the basics of Newtonian physics. You can't support the Big Bang concept and at the same time hail Newton's mass forming a pulling force idea. It is not possible for the Universe to come from singularity at a point in the cosmos while at the same time mass contracts all other mass to a single point and if you believe that then reconsider your position in physics. The planets go by rings around the sun and not by mass pulling straight into the sun as Newton said mass pulls in a straight line. The only value that a circle or a ring can have is Π and that gravity shows it has. This information that you are about to read has more importance about cosmic physics than anything that you ever read before in your entire career. It shows for the first time how gravity evolves by showing how a phenomena forms gravity. I challenge you to prove only one idea wrong that I put foreword in all the articles. You are about to learn physics without the Newtonian bending of physics laws and see physics for what it is.

I am sending you seven articles wherein I prove that the validity of mass used in terms of Newton was never proven and wherein I prove how the Titius Bode law forms gravity. But true to Newtonian nature I presume all this information will again not be good enough and true to Newtonian character it will again be ignored because it will again not meet your Newtonian standards. If you do ignore my work again, it will be a blemish on your insight!

The article of 15 pages to the Annalen had in mind to introduce a very wide-ranging concept contained in many books. I wish to promote books in which I introduce a much larger and much more detailed cosmic picture. It is four books that actually form four volumes of one theme supporting The New Cosmic Theory. I wish to unveil a totally new approach to the thinking in cosmology. The concept is proposed in the article I sent to you which is "revealing" The New Cosmic Theory In the article as much as the theme I wish to go where no one ever attempted to go before. I introduce the Universe of singularity, a state in which the Universe still is because it is a state from which the Universe grows. It is where material in a dimensional dynamic does not apply because it is where Einstein said "the Universe goes "flat"". I show you how and where the Universe goes "flat" I will guide you to the point where I go…so that you may see where my books and the article lead you. It is in the domain of singularity.

When you read work about the Big Bang you have to go right down the development (in reverse order) to the point where the Theory of the Big Bang points at a spot named singularity. It shows the very start from where all material developed. At that point one will find The Absolute Relevancy of Singularity and there has never been any attempt by any person ever to venture beyond the dimensional birth of the cosmos, which is called the Big Bang by going into the era where singularity prevailed. I take you there in my books as well as the unpublished article. However, going there requires a very high degree of concentration and calls for understanding that a very little number of persons are capable to show. I try to show how the Universe goes "flat" as Einstein said the Universe goes "flat". Even by completing this unimpressive letter you will also know how the Universe goes "flat". Even where you failed to read the article I sent you, then by just reading this letter you will be able to find where singularity takes the Universe "flat". But it requires a mental capacity to understand because where I venture no one ever in the history of mankind reached into before. I do not speculate but even in the unpublished article I show with pictures and sketches as well as "some simple algebraic relations" where to go to where the Universe starts, but you failed to read that because you are opinionated as to what conditions should the Universe have before the Universe will allow any one into physics. That is a pity. One should learn from the cosmos and not tell the cosmos what it must be to qualify as the cosmos. Then in the article I show you by almost taking your finger to the spot, the very point where the Universe ends and that too I qualify. You might dispute my arguments and show me about what you disagree, but it shows very little understanding of reason on your part about qualities man should have before understanding the Universe. I go into a Universe that was in place before light was in place in the Universe and only darkness prevailed because light calls for space and in that era of singularity space was not even a thought yet. I show why the Universe goes "flat" and in a "flat" Universe only the value of 1 holds value since singularity is 1. If you can understand 1 or $5^0 \times 7^0 \times 3^0 = 1^1$ you have all the mathematical skills required to understand the applying concepts. To reach a value of 1 does not require big mathematical equations but to reach singularity requires 1.

The collection I named The Absolute Relevancy of Singularity: The Theses and the collection as such forms a small introduction to the thirty-two or so books I wrote on various matters concerning physics with gravity in mind, but **The Theses** as such in the entirety of the four books does not officially even start to introduce the spectrum of every aspect of my work. I have been in contact with numerous Academics and about one in one hundred reply. When the one in a hundred reply, the academic always uses a most aggressive tone which I came to accept as what I receive from academics, and because of that I was most delighted to find some kind remarks from you as a practicing academic, and might I add, the first such kind remark in ten years of my trying to contact any person in physics that would take note of what I have to say about a new line of thought, because the few others that replied were extremely aggressive about me confronting Newton. I only began to submit books to publishers after twenty-seven years of studying Newton and the role Newton play in cosmology and thereafter which was ten years ago I began promoting these ideas. The New Cosmic Theory is a process wherein I try to introduce a study that is ongoing for about thirty-seven years, give or take a few and I did not jump into the frying pan having my first thought about the matter published as an article when I sent the article to the address of Annalen der Physics.

The New Cosmic Theory that I try to convey by writing books in total holds much information and every time when publishers reject the publishing of any entire book I propose, the rejection was on the grounds that "the discourse is not falling within the main-stream science discourse" and therefore I was subsequently advised to write articles on the subject as to find recognition. I was told that only then could I achieve publication of any entire book. Now I find that trying to publish articles has my work rejected on grounds as follows and the following is directly coming from the reply in which one of my articles was rejected recently: "You submitted an article of 15 pages to the Annalen. The content of this paper doesn't constitute a theory in physics. With a lot of words and some simple algebraic relations, there is no way to "explain" the world of physics. You seem to be out of touch with modern developments. This is also shown by the fact that you don't quote any relevant literature." It is not possible to introduce the totality of my work in 15 pages (or whatever a journal would allow) while remaining absolutely coherent on all aspects during such an introduction about anything. You wish for me to work with mathematics and calculations while the world I enter starts mathematics. My aim with the website www.singularityrelavancy.com is to introduce the reader to a world before mathematics as a multiplying process took centre stage. I take the reader into the cosmic era when 1x1 was 1 and only 1+1 was valid forming 2. In the article I say that in so many words, and you would have noticed me saying this if only you took notice to read the article with care. I take you into a true flat Universe where space has no dimensions because dimensions are the multiplication of numbers whereas a flat Universe is found within the adding of numbers. Multiplying brings about a discipline of dimensions and singularity is void of dimensions, thus deemed to be single in dimension. The era we enter uses a line called time to create a single ongoing dimension.

I show why the triangle and the straight line and the half circle are all equal to $180°$ and in the world using space as form by using dimensions this fact about mathematics is bizarre. The triangle and the straight line and the half circle are all unequal in form while mathematics proves the three equal. It is obvious that the triangle and the straight line and the half circle are as wide apart as the sea and the Sun is, and yet there was a period in cosmic development when the three were mathematically equal as much as they still are. I have mathematics telling me this fact beyond doubt. Please use a formula and your brilliance in mathematics and using no words to prove to me why the triangle and the straight line and the half circle are all equal as they all are $180°$ while explaining details because on this rests one entire pillar of mathematics. The answer about this we find in the Lagrangian point system, which is one of the four cosmic phenomena, I explain when using the four cosmic phenomena to explain gravity. This becomes clear when using the law of Pythagoras to prove how this very law became the basis for mathematics and I do use mathematics in the law of Pythagoras to prove how mathematics started when the Universe started mathematics. However, I don't prove that in the article because the space allowed in the article is much to little to prove anything.

In the article however, I show why did Π become $21.991 \div 7$ or then $\Pi=3.1416$ or why is a circle Πr^2 or why is a circle circumference Πr or $\Pi d \div 2$. I show why a circle begins with Π and don't just surmise it. In my books I show why the phenomenon called the Titius Bode law is responsible for Π as a cosmic form and value. In my books I explain just as I claim in the article how the Roche limit come about and how the Roche limit is responsible for the sound barrier and what is the true cosmic value of the Roche limit as it plays a part in gravity on stars...that I show when I enter the era of singularity when calculations were still not yet developed. I show why a sphere in calculating the volume of space is represented as the formula $a^3=4\Pi r^3/3$ and why it is used to calculate the sphere when using these specific interpretations and how this is different from Kepler's $a^3 = kT^2$, which is the way to calculate voulumetric space in applying singularity. The basis of this formula is derived from singularity finding form and that too I prove, but I have to use words because prior to when volumetric space came about, singularity prevailed and singularity is single dimensional. I pertinently state this over and over in the article. In the article alone I have no space to show all these facts and therefore in the article I only show why a circle uses Π to begin with. I show where and why did gravity start and what the true value of gravity is as gravity kick-started the Universe into a beginning because the beginning began with gravity. That I don't show in the article because printing space available will not permit me the opportunity to do so, but I introduce a book where

I show exactly why, how and by which factors did the Universe start by using singularity. I show how the Universe evolved by singularity before space developed and at that time it implemented the four cosmic phenomena that later became part of space when space developed.

I am trying to introduce a study I have done during twenty-seven years of research and there is not one word that I can quote from any other source since every word comes from conclusions that I make and which I prove with the use of logic. All I try to do is to find a medium wherein I can tell some interested parties where to go to read my work and then for them to judge me on their merit and not be sidelined by rules set by academics in charge of publishing. Why don't you allow everyone to read my work and then afterwards, let all readers be opinionated by personal impressions applying and do evaluation of facts according to personal interpretations? Everyone goes on about the unfairness Galileo endured at the hands of the Catholic Church, but at least the Church allowed Galileo to publish his work so that the entire world could take note. Every one in science as well as the Church thought Galileo was out of touch when he declared the science wisdom prevailing at the time was incorrect, and five hundred years later we know who was out of touch as you state I am. I do not compare my work with that of Galileo but I find the same restrictions brought on me by the Powers of the day controlling science. The method of the blocking of getting new principles published is the same as what was in place back then where those in power controlled the thinking about science and those in power today still controls the thinking in terms of science by using equal draconian methods. By disallowing any other views to be printed that does not resonate with the prevailing mindset, science ensures that the public out there consider the correctness of their position as beyond suspect. Their discourse is then thought of as the only possible thinking policy that could be correct, which makes what they think absolute, beyond any suspicion any person could ever have. Killing criticism makes science deemed by everyone in the world as being undisputable because no one ever could dispute Newton. But it seems that no one ever got the opportunity to dispute Newton. Newton is only undisputable because disputing Newton is not permitted by science. Newton was never proven to be incorrect because any attempt to disprove Newton is killed in the infant stage and more often so even before birth of any such a thought could take place. I know this because for the past ten years the academic world holding publishing power destroyed every attempt I made to draw attention to the obvious insufficient work they base physics on. If you kill the messenger, no one will know about a new message and that is what happened then and that is what happens today. The Catholic Church was the one stopping Galileo, but nowhere is it mentioned that this was also in total collaboration with every party in physics at the time. Galileo did not only cross swords with the Catholic Church but crossed with the views the academics at the time had so Galileo went against what the academic world believed. It was the academics that prompted the Catholic Church to believe the Sun was circling the Earth and that the Earth was the centre of the Universe. Again I say I will not dare to compare my work with that of Galileo, but the treatment I receive I do compare. One thing science can take even less than the Catholic Church could is criticising their supremacy.

I have done twenty-seven years of research about the working of cosmology and found a manner by which I could interpret the four cosmic phenomena science do not even recognise because while they are there, they also don't fit into Newton's mathematical physics. As science goes, they will rather reject the obvious presence of the phenomena because it does not match Newton and must therefore be out of touch with modern developments. I did not only unravel the phenomena but worked out gravity from the manner the phenomena influence cosmology. The phenomena holds root in singularity and no one has yet entered that domain. All I try to do is to find a medium wherein I can tell some interested parties where to go to read my work and then for them to judge me on their merit and not be sidelined by rules set by academics in charge of publishing as Galileo endured. Let everyone read my work and then after that let all readers be opinionated by personal convictions applying. Allow my work to be evaluated by those reading it and not be smothered by those trying to kill the content because they do not care for the style I use. Galileo had an opinion that was clashing with the present dogma of the day but he could express his views because we now know about it. The way modern science kills me is they make very sure no one will ever know about me because they silence me as if I am dead. It is also so evident that at Galileo's trial academics were brilliantly absent by not showing a united effort to defend the liberty of thinking. That image today's science try to portray as if they now in all righteousness are fighting to uphold honesty. However, today one may only think freely as long as your thinking is echoing mainstream ideas. For ten years my ideas were constrained at every possible level I encountered and my ideas were destroyed, as Giordano Bruno was burnt alive. Before finding publishing I have to find favour in the eyes of the

Academics in physics whom will not have my work published since I disagree with prevailing sentiment and I denounce Newton in terms of cosmology, but only in terms of cosmology.

That is what everyone misses.

Newton does not work in cosmology but Newton works in physics because in cosmology mass does not apply. In physics mass applies. I can find no evidence of mass doing anything in cosmology, still everyone grants mass because with mass it is easy to play with mathematics.

On earth where mass applies in everyday practice, Newton's work is undisputable correct but going into cosmology there is no evidence of mass applying, and that is where cosmology parts from physics. Mass do not pull planets and stars and that Hubble proved when Hubble proved the Universe is constantly expanding. I return to this elsewhere. Because I challenge everyone to show that mass plays a role in cosmology and in forty years no one could, my through thinking that my discourse is not falling within the main-stream science discourse and those with the power to prevent my work getting published will think up any excuse not to publish. They will block me because what I think will have modern thoughts prevailing in science at present brought into question. For forty years I have been asking that just for once someone will step forward to prove mathematically and without doubt that mass brings about gravity. Show the evidence that all the small stars are either in the centre of galactica or are on the outside of galactica and the arrangement of allocating stars go according to mass. What is it in the atom or the moon that has the ability to pull by magic something it does not connect to. Prove how it is possible that things fall by the measure of mass. Just for once show how things fall by the attraction of mass when everything proves that all things fall equal and therefore mass has no role to play in falling. The example used is a feather and a hammer falling in vacuum and this is fraud. Show how a car and a brick fall equal in front of a camera held by a cameraman and then tell persons the objects fall by mass issuing gravity proportionally according to the mass dishing out the gravity when the camera can follow both objects falling. If the ratio of mass brought about the ability of gravity pulling, then more massive things will fall faster and they don't. Mass does not pull or attract by any means or measure and also in this statement I return to debate it further elsewhere.

To bring one point to your attention just the following: you do support Newton and I question Newton and that questioning Newton is mainly what science hates. Where you underwrite Newton's claims of mass bringing the pulling of objects then please show me by using $F = G \frac{M_1 M_2}{r^2}$ how much did the mass of the earth draw the earth closer to the sun by using the mass of the Sun since the days of Kepler? You know as well as I do it does not happen because in fact the distance between the planets and the sun increases and does not decrease, as it should do according to Newton. Please use the formula that forms the basis of physics to show the world when will the BIG Collision come that will inevitably have to come if $F = G \frac{M_1 M_2}{r^2}$ is correct and when will the moon slam into the Earth. If Newton applies we await the collision between the earth and the moon because the masses on both ends will do the pulling to create the devastation that will follow the collision. Since Kepler made his calculations centuries ago, tell me how much did the moon come closer to the earth, presuming that $F = G \frac{M_1 M_2}{r^2}$ is indisputable. Did any member of physics ever bother to do such calculations as to determine when the collision is due, or have no one ever took interest in the case, and if not, then please tell why not. If the formula you mathematically base physics on was anywhere near correct applying in cosmology as you in physics claim it is, then you must be able to apply the formula and show the precise date such a collision will take place because every factor in the formula is known to science! Please show me on what evidence do you build your belief that mass pulls by other mass because from where I stand what I see is that science had to invent a graviton spawned from the imagination of science to try and address the question as to how does mass pull mass. I put it to you that your use of $F = G \frac{M_1 M_2}{r^2}$ is as correct as the presumption was in the time of Galileo that the Sun is circling around the Earth. Sir, the substance of power controlling thinking still prevails in science as much as it did in the days of Copernicus during his life where everyone had to submit to the thinking of the Powers in Charge of science albeit members of the Church back then for fear for your life as Copernicus did. One still do not dare ask questions or ask for proof as I do on the merit of mass as a factor in cosmology (not in physics) for then one would be silenced till death interrupts the questioning. Copernicus so feared not the Church but his colleges in science that he published his work after his death because then science could not kill him or employ the Church to do the official killing. Today science will allow me the privilege to die silently in a corner as long as I do it quietly because no one will allow my torturous screams to be heard. As in the time of the Copernicus, you lot still can't stand new thinking because new thinking will cultivate doubt in the minds of the many of those you consider as mindless and you require that they undoubtedly believe in you. By seeing to it that my work goes unpublished you willingly kill me by killing and destroying thirty years of my life...and then you lot point the finger at what was done back then as if you could be bothered by not implementing the very same evil. I ask you where is the freethinking and what happens to the freedom of speech as long as what is said is truthfully substantiated and can be proven because all the facts I present in the article you are unable to disprove and that is a challenge I put to you. Professor Hehl, you didn't even read my article because if you did you would see there is not one point in any argument about my work where I am incorrect, not one point you are able to disprove me or show me I don't follow the laws applying to physics in detail, yet you have to audacity to denounce my entire article just because it does not fit the profile you envisage it should. You and all your colleges are more condemning than the Pope and the Church was because at least they gave Galileo a fair hearing and considered his evidence. Even that you lot fail to do. I have had this treatment for ten years and every time it is the same over and over. You just couldn't be bothered to read it because it takes too much effort on your part to think in terms of evaluating every idea I put across, and there are a lot of new ideas you then have to chew on! What I touch on requires intellect to understand and not just some mathematical computing ability to perform. It shouts for human insight into cosmic forming that does away with make believe science such as Newton's guessing does with mass forming a factor and that the cosmos denounces.

If you did bother to read the article, which you failed to do, you then would have seen I start where mathematics start and I can quote no one because I venture where no one has gone before; I go into singularity which by your definition is Singularity: a mathematical point at which certain physical quantities reach infinite values for example, according to the general relativity the curvature of space-time becomes infinite in a black hole. In the big bang theory the Universe was born from singularity in which the density and temperature of matter were infinite... and that I do quote, but that is all I can quote for the rest is the product of my labour and fruit of my mind. I challenge you to show where I stray from your definition of singularity when I show where to find singularity.

I explain just how it is possible to locate just such a point holding singularity to the precise value singularity must have in our modern Universe but I can assure you that where a mathematical point at which certain physical quantities reach infinite the grand splendour of mathematics are lost in dimensions not applying. I work in the era you can define but can't understand because it predates mathematics "singularity in which the density and temperature of matter were infinite" and it is in the infinite that mathematics becomes obsolete. Again I ask you that if no one ever has been there where I venture in physics, then whom must I quote because I quote the small part where science have been and that is all there is to quote. Every aspect about the Big Band deals with conditions prior to singularity deforming. If you use any quantity or formula based on numbers being more than 1 or any number to the power of zero, then you have left the realms of singularity because singularity could only be 1. At the point where singularity applies all complicated mathematics disappear. If you disagree, then give me any number that can apply to singularity other than 1 and please show me any mathematical formula that will apply to prove that singularity can be more than one.

If you did bother to read my article I sent you, you would have seen I show you exactly where $\Pi° = 1°$ could be found in the world of physics you study...and if I dare draw your attention to your accepted definition on singularity then as quoted it is a mathematical point at which certain physical quantities reach infinite" I show in the article where to locate this very point holding infinity. I also show where the point of singularity is infinite as it is holding what I named $\Pi° = 1°$. I show the point cannot ever start or become smaller since it is so small it has no space in which to form and if the point was in the Universe at the beginning, then it still has to be in the Universe because if it was in the Universe once, it must remain within the Universe because it has no other place to go by leaving the Universe. That is what you reject because that is what my article announces and my article introduces where to locate singularity and that is the article you reject. In this light going according to your attitude I am most delighted by your attitude, because from your attitude it is clear that where I venture you have never even left one thought.

In the website www.singularityrelavancy.com I am introducing the reader to a world before mathematics as a multiplying process took centre stage. I take the reader into the cosmic era when 1x1 was 1 and only 1+1 was valid forming 2. This figuration proves mathematically that there was a time (1+1=2) pointing to a period before dimensions brought about perspectives (1x1=1). I take you into a true flat Universe where space has no dimensions because dimensions are the multiplication of numbers whereas a flat Universe is found within the adding of numbers and the adding of numbers point to a flat line forming the basis of the singular Universe. This process is directly formulated by translating Kepler's formula $a^3=kT^2$ to the true measure of gravity that is Π.

Please be so kind as to tell me, Professor Friedrich W. Hehl with all your mathematical splendour and magnificent abilities in constructing wisdom without using words, why is 1+1=2 and why is 2+2=4, because you do use it in physics, don't you? When you use numbers in your world of physics, being the Master that you are, you have thought about where numbers came from and how did numbers arrive? Please prove why it is that doubling two is also taking two into the square and while proving this, it is done without leaving a whisper of doubt. Please use your vast mathematical insight to explain without using words why would the third number be three, specifically three because that was how three came from singularity as 1 and why would the following number be four, which then is also the square of two by using the law of Pythagoras when proving this. How did five follow four to become five by using the law of Pythagoras to prove the point and then using this evidence to show that double five becomes ten, again by the merit of Pythagoras. Why would nine be the square of three and by adding 1 it becomes 10, because Professor Hehl, proving this is what really forms the basis of all science and that is how the Universe formed!

I wrote books about this process wherein I show numbers formed the start of the Universe and not material as you in science wish to believe. It formed by forming mathematics and the splendour of mathematics arrived only when the form of the Universe was completed and the cosmos stepped into the dimensional dynamics of space. This happened when the atom formed at $(\Pi^2+\Pi^2+\Pi^2+\Pi+3)=35.75\times\Pi=112$ which is also when space as a whole formed at $7/10(\Pi^6)\div6=112$. The relation $7/10(\Pi^6)\div6$ validates the sphere as (Π^6) spinning in (7/10) a six sided cube $\div6$ which is outer space. What I show when using $7/10(\Pi^6)\div6$ was the moment the Universe came into dimensions by arriving at the formation of the atom. In the article I show why one might conclude why the Universe uses Π as a numerical basis for gravity, but I agree, the article alone does not start to prove anything because for that there are four books forming such proof called **The Absolute Relevancy of Singularity: The Theses**

I dare you, no I challenge you with all your mathematical splendour to prove one iota I produce as evidence in these books being incorrect, and you have to use words to prove me wrong because where I venture is where mathematics goes singular which was at a point before mathematics came in place...and that place can still to be located in the present Universe. I can and I do show you the very spot where the Universe came from, but not in the article for there is no room to do it. You, with all the astonishing mathematics are stuck at the point where the big bang arrived and at that point everything that forms the Universe was already formed within the Universe. The Universe adopted space at the event of the Big bang and therefore mathematical values came in a dimensional context at that point. However, everything that currently is, was already present in the Universe at the event called the Big Bang. Before that the Universe was one being 1^0 or 451^0 or $5^0 \times 1^0$ because that is what singularity implies the value of singular space is, it is 1. It is because you got stuck with your mathematics and you used your mathematics instead of brains and that is why that you can't proceed to resolve issues beyond where the cosmos formed the atom. You are all agreeing about everything coming from singularity but going there you have to lose your mathematical equations; it does not apply! If you in science realised mathematics construe singularity as one, science might have realised that mesmerising mathematics before the big bang was useless as a tool to formulate facts, then you would've realised how to reach a pre big bang Universe. I did just that and I can show how, and why and where the first moment arrived. I can show you precisely where that fist moment of arrival is today. The mathematics you apply had to start somewhere and it is there where I venture.

The Universe in singularity adopted the four phenomena which is called **1) The Lagrangian system 2) The Roche limit 3) The Titius Bode law 4) The Coanda affect** but to unravel their meaning you have to go into singularity and to do that you have to understand Kepler and to understand Kepler that introduced singularity when he introduce $a^3 = kT^2$ forming the measure of singularity applying. You have to part what Newton thought Kepler said from what Kepler said and explain what Kepler really said, and that I do in the article by using some simple algebraic relations and by the meaning behind $a^3 = kT^2$ and those four phenomena the Universe came about. But the condition to understand how the Universe came about is to first understand the four phenomena interacting. However, to understand how the Universe formed numerically or better titled **The New Cosmic Theory** one has to understand **The Cosmic Code** and learn how to read from it the interpretations of factors. To understand **The Cosmic Code** one has to understand the process of cosmic law supporting the Roche limit that works as what we think of in terms of forming **The Sound Barrier.**

To understand **The Sound Barrier** one has to understand the process of cosmic law supporting the Titius Bode law as well as the Lagrangian system that forms the Coanda effect and together the lot works as **The Four Cosmic Pillars** on which the entirety of everything was built by implementing singularity through a very specific process I named **The Four Cosmic Pillars**.

To understand those four comic laws applying one must be able to evaluate the process by reading **The Cosmic Code** in order to be able to recognise the actions brought about by singularity in relation to **Applying Physics** in terms of **The Absolute Relevancy of Singularity** and the one aspect of singularity applying in physics is being able to see how singularity forms space by the measure of Π, which is the only aspect of the entire collection of information I try to show in the article I sent to you…and you are unable to read that little bit…then how on earth will you ever get around to understand how the start of the Universe numerically came about when singularity and only singularity applied! Those books showing how the solar system was born and how the Universe came about I do not yet offer on sale. The information contained there are the really tough nutcrackers that explain by the cosmic code the inner working of gravity in stars and in galactica according to the cosmic code.

The Universe as we see it started much later in a period you call the Big bang but in truth there where the Big bang happened is when the atom came into form…and that I prove mathematically if you dare to read my work, which is the four books I wish to introduce via your Journal. Tell me Professor Friedrich W. Hehl, why is there mathematics formed by the adding process and mathematics formed by the multiplying process. In this evidence we find the development of the Universe. What happened that secured the forming of four to then be the prelude of five. Every number is a point and specific sets of points hold different relevancies placing the number in a quantification that brings about material in accordance to the coded relevancy. Please tell me the specific indisputable reason how the Universe did arrive at the value of what the number five depicts and it being precisely 5, or how did six become the next number on the numerical ladder and no other number but six tiny dots. Every number of dots serves a very specific purpose and five dots hold a value totally different from six. That is why nitrogen is a gas while carbon is a solid. That is why Mercury is a liquid with Xenon being a gas although they both are much more massive than say iron being a solid. Try to do this explaining according the law of Pythagoras by showing how the law of Pythagoras implemented the process and not use words. However, this is a country mile further than even the Cosmic Code is.

Why would seven bring about that a circle forms by redirecting directions as it is used in forming a circle by 7°? Why does the numerical value of 7° and only 7° play this role? Can you show me with your ingenious mathematics how the top part of Π is 21.991 when the bottom part is the circle by 7°, and use Pythagoras to substantiate the reasons. All the answers are in front of everyone but you said "With a lot of words and some simple algebraic relations, there is no way to "explain" the world of physics" and while you express you inability to see what I see, I see what I say I see and show what I see as clear as daylight while I do explain what I see precisely with a lot of words and some simple algebraic relations because complicated mathematical formulas did not yet enter the form of the Universe in the period I introduce as a **The New Cosmic Theory** Before space applied, form applied and form was singular before space became dimensional. Have you ever considered why a triangle and a half circle are equal to a straight line by the dimensional value of 180° or is this the first time your intellect went that far? …And this question points towards your obvious mathematical brilliance in physics and not the lack thereof. This I point out to show while you do know everything about physics there is thought to be, there also is a small possibility that you do not know everything about physics that there might be. The Universe started numerically mathematical and not by material as material came later; everything used a numerical order to form.

In the article you failed to read I show precisely where the Universe goes flat and becomes singular with the little impressive "and some simple algebraic relations" I show precisely the point where the Universe goes flat and in line with your impressive mastering of mathematics I challenge you to use your mathematics to show where I am misinformed or where I fall from the wagon by using "and some simple algebraic relations". Guess what, the "simple algebraic relations" is what the Universe used mathematically to indicate to Johannes Kepler as to inform him as well as Tyco Brahe how the Universe is constructed. The Universe showed how the Universe used $a^3=kT^2$ which is some simple algebraic relations as a means of form, but true to your academic arrogance I see you know better than even the Universe does because the some simple algebraic relations $a^3=kT^2$ is what the Universe used to describe to Kepler about the form the Universe adopts. By using some simple algebraic relations $a^3=kT^2$ what is in the Universe became the form of the Universe, and you missed all of that…that is a pity. You know, using $a^3=kT^2$ I show where the Universe goes flat, a task no one ever mastered by using breathtaking mathematical equations.

Some academics previously indicated there was some point holding singularity within the black hole but I show in my books where this happens and where to locate singularity in everyday life. If Einstein said the Universe goes flat, then that flatness must still be around and be everywhere so that everything will be able to go flat every time Einstein said it does! However, the most wise amongst you failed to even value the measure of singularity, being everywhere and all around by using impressive mathematical equations let alone to pinpoint the point serving singularity and even less to indicate where the exact centre of the Universe are to be located. I show why the Universe is a sphere, which is a fact that is up to now only been surmised. I show why the Universe applies gravity as the form of the sphere, which is something all the brilliant masters in mathematics failed to deliver up to now. Why did all the mathematical masters fail to prove that the Universe uses the shape of a sphere while all pictures indicate everyone accepts the fact, thus failing to impress with your brilliant mathematics you have to surmise, as you have to do with most things. I dare you to read my article and show my arguments I present are failing in anything that I say it does. In the article I show where singularity applies and why singularity chooses Π as the form of gravity. Why singularity chooses Π as the form of gravity you are unable to prove when you are using those most impressive mathematical equations you refer to, because singularity does not apply impressive mathematical equations. Singularity applies simplicity.

You showed me that I "seem to be out of touch with modern developments". Please let me show you why I "seem to be out of touch with modern developments". Please let me show you what inconsistencies there is with the basic mathematical formula Newton introduced when Newton tried to use mathematics to prove that mass was responsible for gravity because you belittle the mathematics I apply $a^3=kT^2$, which is precisely the mathematics Kepler used to portrays how the Universe forms and that formula he read from the way the Universe is constructed. Then you look down your nose at Kepler's work while Newton's mathematics broke every possible mathematical law it can when Newton tried to convince the world how clever he was by cheating with mathematics.

Newton started off applying the factors holding in the relevancy as follows: $F = \frac{r^2}{M_1 M_2}$ and discovered it fell short of any form of accuracy. There is no way that this formula would ever work even by a lesser degree of accuracy.

Then Newton changed the formula to being the following $F \alpha \frac{M_1 M_2}{r^2}$. Newton tried to convince (and succeeded) that one are able to change $F = \frac{r^2}{M_1 M_2}$ to $F \alpha \frac{M_1 M_2}{r^2}$ while it meant the ratio would still work in the same way as if it was something like this: $F = \frac{M_1 M_2}{r^2}$ and the formula still didn't work. The changing of the formula you use as the corner stone, the foundation of all physics still proved to be a total disaster notwithstanding the cheating of the most fundamental mathematical law that should support all physics laws. Then Newton and his fellow boffins in science cheated mathematical law even further to change the lot to $F = G \frac{M_1 M_2}{r^2}$ without explaining how $F = \frac{r^2}{M_1 M_2}$ could end up as being equal to $F = G \frac{M_1 M_2}{r^2}$.

If you feel so strongly about mathematics used in physics then tell me Professor Friedrich W. Hehl, why don't you start to apply currencies to the factors and show the world how $F = \frac{r^2}{M_1 M_2}$ could become equal to $F \propto \frac{M_1 M_2}{r^2}$ and this equal ness could be carried on to eventually become the same principle as $F = \frac{M_1 M_2}{r^2}$ to then become $F = G \frac{M_1 M_2}{r^2}$. Put in real numerical values and show it does not constitute to mathematical fraud. If Newton were that correct, then please use the formula $F = G \frac{M_1 M_2}{r^2}$ to prove that the value derived from $F = \frac{r^2}{M_1 M_2}$ could eventually be the very same equal ness as one would achieve from $F = G \frac{M_1 M_2}{r^2}$.

Better still, why don't you write an article in your journal doing the song and dance about the accuracy the Universe proved Newton had when he implemented $F = G \frac{M_1 M_2}{r^2}$ because Hubble destroyed all the credibility that $F = G \frac{M_1 M_2}{r^2}$ once was thought to have. Then you can vindicate your attitude towards me while serving the cause of mathematics at the best you possibly can by restoring lost confidence in applying Newtonian religiosity. Put values to the factors and prove the Newtonian formulas have all the same results in the end. …And while you are at it, write in the article showing how much did the distance there is between the earth and the moon reduce since the moon landing in 1969. Please use the most accurate figures available and then tell the world how much accuracy there is when implementing the calculation science applies to a shrinking Universe as Newton said it does when he said $F = \frac{r^2}{M_1 M_2}$ is equal to $F = G \frac{M_1 M_2}{r^2}$. Show how much all distances in the Universe shrink by their mass attracting other mass to bring about pulling forces of gravity applying throughout. Do use Newton's formula $F = G \frac{M_1 M_2}{r^2}$ to ensure accuracy. The Universe expands just as Hubble indicated it does as it expands from every centre holding singularity and it expands everywhere equally. Your colleague at Annalen Der Physics, Professor Doctor Ulrich Eckern once accused me of missing the basics of mathematics and classical mechanics by my evaluation of $F = G \frac{M_1 M_2}{r^2}$ but he failed to show what it is that I miss. Now it is the dream chance you have been waiting for…write an article about the correctness of the formula $F = G \frac{M_1 M_2}{r^2}$, show how the Universe comply to underwrite the absolute correctness of mass as a reliable factor shown by the formula having mass doing the pulling and then by the same token show what it is that I do not understand about the basics of mathematics and classical mechanics for I have been told this since my student days and after almost forty years I still fail to recognise what I am missing. To show me what it is I don't get, use the formula to prove how much did the moon come closer to the earth the past forty years by using $F = G \frac{M_1 M_2}{r^2}$ and the information acquired from data coming from the instruments placed on the moon for that sole purpose. If you are unable to do so, then never use mass in cosmology again because then mass is not pulling anything ever. Go one-step better… show why the Universe did not collapse back into singularity using $F = G \frac{M_1 M_2}{r^2}$ when r^2 was infinitely small and mass was absolutely contracted holding singularity in which the density and temperature of matter were infinite. Back when the Big Bang took place the entire Universe was in one Black Hole, then why did

$F = G\dfrac{M_1 M_2}{r^2}$ allow the lot to escape. It will never again have that chance to bring everything that went loose back into contraction again. At that point the Universe had its best chance to collapse into the Big Crunch because the further the radius expands with Hubble expanding, the lesser the chance is that mass can do the pulling. With an ever-increasing radius by the square, the mass effectively will reduce in strength. If you prove why $F = G\dfrac{M_1 M_2}{r^2}$ did not pull everything back into singularity then that will be a worthwhile challenge for your brilliant mathematical skills to achieve! Then you lot can stop searching for the mythical dark matter that has to cover Newton's blatant errors because the dark matter is just a cover up. If the matter was there and is there presently it has mass, then why does it not use the mass it must have in the present to pull the Universe into contraction starting here and now and why is it waiting for something to unleash the forces of gravity by the mass of the dark matter. ...And by the way, why would the matter not pull now if it has mass just because it is dark… if it is going to pull it already has to pull or it will never pull because it is not there at all. It is much more likely to be in the imaginations and calculations of science that be in the actual Universe. The matter being in the Universe has to have mass, dark or not. If mass does the pulling, and it is there, it has to pull now, in the present at present or the entire idea is just another scientific hoax to cover Newton's incompetence. The matter being dark or not, has to have mass, visible or unseen and if the mass is there and mass pulls by the force of gravity, then please tell why the lot is not pulling now and what are the dark matter waiting for to start the pulling that will begin the contracting? I say this is more proof that there is no mass and that mass does not pull and the entire concept is to try and vindicate Newton's absolute misjudgement of gravity. It is one more compromise to cover-up science.

You see, Professor Friedrich W. Hehl, if mass was the factor initiating gravity or then the falling of a body to the ground, solid objects will have to fall faster than objects that is empty and hollow because the empty space within the hollow object will restrain the falling by not falling with the object since only the mass would tend to fall leaving the empty space behind to restrict the downwards descent of the falling object. If the emptiness within the cup did not fall with the cup falling then the emptiness will bring a drag on the solid part that falls. The empty part within the cup will try to stop the fall while a solid filled glass will then fall faster than an empty glass because the emptiness within the empty part of say the cup or glass falling would not fall, leaving only the small rim of the cup falling. With the major part not falling this hollow cup will fall slower while the fullness of any solid object will fall in its entirety, making the fall of the solid object unrestricted by having no empty space that does not fall and thus the solid object then will fall faster. Drop a full glass from an aeroplane with an empty glass as see the emptiness in the empty glass falls as fat as the water does in the glass. It is the space that moves down taking the body with.

A filled container does not fall faster than does an empty container and visa versa because the empty space of the object falls as fast as the filled space of the object and all objects fall equal and according to a variation in density in air caused by temperature fluctuation (excluding some gasses) allowing any variety of mass to fall equally. It is the space and all the space notwithstanding being filled or not that falls or moves towards the roundness of the earth proving that space holding material or not holding material falls equally notwithstanding mass and for that reason that is why Galileo's pendulum swings regardless of pendulum length or size as Galileo said it would. It is the descending space driving the pendulum that swings.

This again was proven by the very first ever experiment concluded scientifically. This fact of space descending does not come as a surprise because Empedocles proved this fact back in 450 BC. Empedocles showed that space displaces water from the clepsydra, which was a sphere shape container with a sprout on the top and small holes in the sphere through which water ran in small streams out at the bottom. When the flow of air or space was blocked in the spout by a finger covering the hole at the top of the sprout at the entry, the water stopped flowing from the clepsydra. They concluded in 450BC that it is the empty space that pushes the water out of the clepsydra because the moment one restricts the empty space or air to flow into the clepsydra from the top, the water will stop flowing out of the bottom of the clepsydra. Why would the flow of the water stop if the mass did pull the water down? When the finger blocks the sprout and stop the space entering from the top, the water does not fall to the ground but it is the empty space that pushes the water out at the bottom to fill the clepsydra from the top. When the finger blocks the sprout and stop air to come in through the sprout opening the water should still run out at the bottom by the mass of the water pulling, if mass was doing the pulling. If mass was the force giving factor, then the water must keep on flowing because the mass of the water did not disappear when the sprout was covered and therefore it still has to produce the pulling by forming gravity. All this evidence was known to science about 2500 years ago but since "With a lot of words and some simple algebraic relations, there is no way to "explain" the world of physics" it lacked mathematical communication and it should therefore surprise anyone very little that physics could not fathom this result 2500 years onwards. Professor Friedrich W. Hehl, do try and find the ability to use a lot of words to "explain" the world of physics because it is helpful preserving past experiments and results as it then does broaden one's horizons mentally…sometimes!

Forget the example always used about the hammer and the feather falling equal in a vacuum because the hammer and the nail and the elephant falling together will also fall equally notwithstanding falling in a vacuum or not falling in a vacuum. The vacuum part is conspicuously in place to purposely confuse reality as it is brought in to flagrantly spread misunderstanding of the issues in hand about the falling that takes place. With everything always falling equally when the same the condition applies to all objects falling and therefore with such falling happening under the very same variation of natural conditions applying, this shows it is the space in which the object is that falls and not the object falling while leaving the space it holds behind. The lack of relevant density in relation to air moving down stops the feather from falling equal just as gas does not fall with the space at the rate that space does descend. People realised this fact 2500 years ago but then used a lot of words to "explain" the world of physics and today because of not using a lot of words to "explain" the world of physics science has no idea how to interpret the very first experiment ever conducted! That is a travesty as much as it is a tragedy. All space falls by the compressing of the atmospheric space. The rotation of the earth moves the space sideways and this brings the space to move downwards by increasing the density of space or air as it comes closer to the earth. This results from the Roche limit applying to fix atmospheric layers varying in density. In my books I explain that principle applying mathematically. The increase in atmospheric density is the result of the rotation motion of the earth brought on by the Roche limit applying while it takes filled and unfilled space towards the solid of the ground and that is what the Coanda effect shows which is what your brilliant mathematics in one hundred years could not begin to explain. The Coanda effect is around for almost one hundred years and please use your mathematical skills to explain why the water will rather follow the roundness and flow with a detour along the rim of the glass than fall straight down as it should when mass would pull? That is the principle behind the sound barrier, and the Coanda effect, and wind restriction and hurricanes and more other things than I have room to mention. With all the attempts made in that past to uncover those issues I mention, it never was resolved notwithstanding all the impressive mathematics available to use. Notwithstanding using your mathematical marvels, science has not got any vague idea to explain any of the phenomena mentioned above. That is why modern science seems to be out of touch with modern developments. To understand these phenomena one has to understand singularity.

The simplicity singularity applies is shown when I show why the triangle and the straight line and the half circle are all equal to 180° but when considering form using mathematical dimensions this mathematical fact seems bizarre. It is obvious that the triangle and the straight line and the half circle are as wide apart as the sea and the Sun is, and yet there was a period in cosmic development when the three were mathematically equal as much as they still are. In the books, not the article, I show by using the law of Pythagoras why did Π become 21.991÷7 or then is Π=3.1416. I show using the law of Pythagoras how and why by the law of Pythagoras is a circle $Πr^2$ or why by the law of Pythagoras is a circle circumference Πr or Πd÷2.

I show mathematically by the law of Pythagoras why is a circle using the specific value Π has to begin any circle or sphere with, but due to lack of space I can only prove it in my books as I can't prove it in the article. I show where and why did gravity start by the law of Pythagoras and what the true value of gravity is as gravity kick-started the Universe into a beginning. Can you show how the law of Pythagoras was implemented when the Universe formed, because I can show the reasons why and I do show the reasons why with using words since the law of Pythagoras implemented actual basic mathematics? I show why the law of Pythagoras implements the law it carries. The reason for this is the method how the Universe started off and the reasons why the Universe began. Maybe you should try to use words one day; after all it is a helpful tool in explaining physics, because it surely helped me explain what was never explained before. I mention the law of Pythagoras because the Universe does apply the law of Pythagoras in all of the cosmos and therefore the law of Pythagoras is part of physics, don't you agree? Can you use your breathtaking mathematics without using words to tell how did it come about that the law of Pythagoras has the dynamics it portrays it has in mathematics as well as physics, because I can by using words. Use your astonishing mathematics to show why everything started by the law of Pythagoras. Mathematics can't do it because the law of Pythagoras forms mathematics and the law of Pythagoras helped to form mathematics as mathematics developed. The four cosmic principles yet not understood by science shows just how Pythagoras applies.

Mass has nothing to do with falling and all things fall equal as if having equal mass when falling because all things fall equal in relation to the space in which they are. It is by buoyancy that space holds things and that removes mass while falling as a factor. Space not holding things fall with space holding things while it is therefore not the mass that causes the falling but the compressing of space which you call the atmosphere. The falling is written in relation with the value of Π. The value of Π is 3.1416÷1 or it is 21.991÷7. It is not coincidental that Π has two distinct equal values because it is due to precisely that that the first moment in the Universe came about when point 1 parted from point 2 putting space in-between eternity and infinity.

There are two values forming Π as much as confirming Π. The air or space holds 21.991 when the Earth holds 7° but when spinning the earth applies the change of direction by instating the axis by the value of Π°Π which is the centre line or axis or earth centre Π° connecting singularity to the earth circle Π. Then relevancies in Π changes as the space that was 21.991 with the air held a link to the roundness of the earth being 7° at the time. But as the 7° dived into 21.991, the 7° goes singular or 1 as the space then in turn becomes Π=3.1416 or becomes the circle. The earth and the circle of the Earth becomes (7÷7)=Π° or 1 while the rim of the earth is Π=3.1416. Forming Π=3.1416÷Π° the form Π then aligns with centre of the earth holding singularity Π°because the axis placed singularity Π° in centre stage when the earth turned. Every time the earth or sphere turns, it places the surrounding space in relevancy from 21.991÷7 to form Π°Π. All this I said in the article that you say is not physic s because I use of a lot of words to "explain" the world of physics By the way, now for the first time in you entire career you also know what gravity is and what forms gravity and doesn't it make a lot more sense than to presume mass pulls mass by gravity without having a stitch of true evidence to prove Newton correct?

What happens is that the space condenses (21.991÷7) by the turning of the planet or star and the compacting of space surrounding the spinning sphere results from the rotational movement (÷7) of the earth that brings about that this compresses the space of air or atmosphere (21.991) into more density (÷7) which is done by movement of the space surrounding the turning object (21.991÷7) albeit a planet or a star and moves space filled with whatever or unfilled, going vertically down towards the roundness of the Earth that then is represented by the circle or the rim of the Earth (Π=3.1416÷1). Everything within the concentrating space will come closer to the surface because it is the space that moves down to the earth and not only the object filling space, but everything within the space including the space that falls downward. Every micro millimetre the relevancy of space changes from (21.991÷7) to the roundness of (Π=3.1416÷1) until the Earth forms the final (Π=3.1416÷1) and the object finds mass as a result.

The falling body never stops falling but find that mass comes about when relevancies changes from Π=21.991÷7 to Π=3.1416÷Π⁰. By touching the Earth, and by that ending the relevancy of Π=21.991÷7 from reapplying, the object then becomes part of the earth circle Π=3.1416÷Π⁰ and having contact with the axis Π⁰Π it becomes part of the Earth singularity distribution and only then finds in this relevancy applying the reward of mass. The body never stops falling but as the earth by density restrict the body movement vertically according to density, the falling becomes a tendency to move downwards in order to unite with singularity formed within the centre of the spinning earth. This is all about relevancies changing and relevancies reapplying positional changes, which is what gravity or time is. This is how the Universe goes flat or singular. It is Π°Π and that I say so many times in the article that you were incapable to read or you refused to read or you were not motivated to read…you can make your choice about you're reasons withholding you to understand what I explained in the unpublished article. Regrettably this is "a lot of words to "explain" the world of physics and because of the use of a lot of words to "explain" the world of physics it seems to be out of touch with modern developments making the process of thought very difficult to comply with modern developments and prevent many academics this far to understand and therefore you lot would rather cling to the use of $F = G \frac{M_1 M_2}{r^2}$, which is truly what seems to be out of touch with modern developments since everyone accepts that the Universe expands. I prove my point of gravity being Π using some simple algebraic relations...you now have the chance to prove me wrong by proving how correct the cosmos shows Newton and his mass is. Show that Jupiter is coming closer to the Earth and therefore we have to relocate to some other galactica or die! Show why Jupiter is randomly located notwithstanding size and why the planets do not arrange positional allocations by the implementing of mass as the factor that would and that should arrange the positions of the various planets should mass truly be a reality in cosmology.

The spin of the sphere constitute of a change in direction to the value of 7°. In the centre of the circle the axis are $Π^0$ and therefore the spin makes the directional change reform to singularity or change to the value of the circle in relation or in relativity with the axis in having the circle 3.1416 and having the centre or singularity or the axis $÷Π^0$ and with that connection the circle, which is space, goes singular or goes flat bringing about the much argued flat Universe. This is how gravity puts multi-dimensional space going into singularity or $Π^0$. This means that the 7° becomes one or singular and space changes from (21.991÷7) to $ΠΠ^0$. On the top of the equation the value of Π is 21.991, and by revaluating 7 through spin to become 1 that value changes to the value of 3.1416 being in relation to $Π^0=1$. The rest of the explanation that will bring proof to my statement when using Pythagoras is far too bulky to offer it at this point. I did say it in the article you refuse to publish because you refuse to read it that the relevancy of gravity is the changing of the value of $ΠΠ^0$.

The space reduces (21.991÷7) to conform to singularity 3.1416÷$Π^0$ by the rotation of the sphere that produces an axis by initiating singularity $Π^0$. In the books I show the very reason why is Π=21.991÷7 and I use the law of Pythagoras underwriting the Titius Bode law, that conjuncts with the Roche limit as well as the Lagrangian points to prove the Coanda effect and the Coanda effect, as I show in the article you didn't read and therefore wouldn't publish, is gravity by principal. Gravity is the Coanda effect that is the changing of liquids Π=21.991÷7 to form solids Π=3.1416÷$Π^0$. I much rather say you didn't read it as putting it down to you not understanding it because I use "a lot of words to "explain" the world of physics and which possibly tops your understanding limitations and therefore you didn't publish it, don't you think...? It will be of no use to explain how the law of Pythagoras was implemented to prove Π=21.991÷7 because I use words which you so honestly admit you don't fathom and it takes far to much space to explain the entire process in this letter. However, I do show how I mathematically conclude this value by using the law of Pythagoras in the books, if you care to look.

With the gravity being Π and that gravity comes as a result of the earth's spin contracting the space forming a denser substance called the atmosphere, which comes about and in accordance to every sphere spinning around an axis of its own, the increase in density around every spinning object brings about a loss in the overall density of space between all cosmic objects such as the earth and the moon and all the planets and the sun and the loss in overall density brings along that the distance controlled by the density of the substance there is between all cosmic structures gains in space. That is what the Big Bang was from the start. It is the substance that fills space in-between cosmic objects is that what you mathematically see as dark matter and it is limiting absolute expanding at the rate of what the density will allow. It is working as it should at present and no search is required. With the Big Bang the density in unoccupied space decreases as the density in occupied space increases making the Universe to seemingly expand, which it can't do. It is relevancies reapplying. The earth can't pull the moon by mass.

Mass does not constitute to the falling of objects but the space compressing brings about the falling action notwithstanding the space being filled with objects or empty of objects and therefore as Newton so vividly proved, one can pull a cover over the eyes of all but one (and that one is me) by cheating with mathematical formulas but to cheat by using words are a lot more difficult. The only thing Newton pulled by mass is a huge cover over many eyes for a long time. All things fall equal as we see on TV everyday where people and cars and bicycles and beds fall equal when dropped from aeroplanes and therefore it has to be the space in which the objects are including all the surrounding space in which the objects are not that is falling and not the object because of mass or having more favourable density. ...And don't say

I don't understand or I miss Newton because if you say you believe $F = G \frac{M_1 M_2}{r^2}$, then surely if you might understand Newton because then you clearly seems to be out of touch with modern developments in terms of cosmology and applying physics. It was shown this past century that the Universe is expanding and not contracting by mass...expanding means the lot forming everything in the Universe are drifting apart but that is "a lot of words to "explain" the world of physics which explains why modern Newtonian science seems to be out of touch with modern developments.
But all persons filling academic posts in science holds the attitude that they know all and others know nothing and academics know best while the rest of the population is mindless, thoughtless and worthless. Professor Friedrich W. Hehl, you are no exception to this rule. In fact, you are one of the best examples I have seen. I have taken these insults long enough...its been raining on me constantly for ten years ever since I tried to introduce my first thesis and if you wish to insult me, do so while you see there are in science parts that is yet still undiscovered and rather try to find how much there is that you don't know about science instead of thinking how much you as a scientist knows about science and what great achievement science is instead of trying to go where science still has to go. Let me give you some wisdom. A wise man thinks of all the things he does not know and what awaits his discovery while it is a fool that thinks of how much he knows and feels impressed by his personal field of knowledge. The best answer to a question one can have is "why" since no answer ever brings full conclusions and science missed some.

Yet you lot in science seem to know you've concluded everything that anybody could ever uncover about science. Recently by some academic in physics blew me away when I tried to introduce him to The New Cosmic Theory.
I am afraid that you will continue to get rejections if you do not relate your work to existing theories and previous work. While it is possible that a lay person hits on an insight that has been overlooked by academic trained in the field over many years, it is unlikely. We assume that work offering something new would be related to existing theories, either by building on top of them or by showing how and where they fall short. If you do not relate to existing work, it is repeatedly going to be dismissed as mind spin too easy to shoot down. I am sure you understand. This is what I have been told when my work was again rejected by another non-complying professor. When I show mass plays no part in cosmic physics by building on top of them or by showing how and where they fall short my methods don't apply. When I show how to enter singularity "a lot of words to "explain" the world of physics So you lot block anything constructive not coming from mainstream science because you lot are the only ones knowledgeable about science. I have shown how knowledgeable you lot are and what it is you protect by pushing me off the table. You say I don't understand Newton while I am the only one ever that understood Newton. I understand very well that Newton doesn't work because mass influences physics but plays no part in cosmology.
Show me how does mass pull because I prove what gravity is and why gravity pulls by Π, but you lot refuse to read because that is an easy way to escape...that is also a cowardly way to protect what you try to hide! Tell me why did it take you lot since 1860 or thereabouts not to be able to explain The Roche limit because I can, but I have to detour from normal science. In modern science you confuse physics applying on earth with how the Universe works because Newton got the lot confused. The physics on earth uses mass but cosmology in the Universe applies singularity, precisely as Kepler introduced it by introducing $a^3 = T^2 k$ which in singularity is $k^0 = a^3 \div (T^2 k)$.

You lot are not even able to start to explain The Titius Bode law, a law that is in place since it was discovered in 1766. The law is so vital planets are discover by applying this law and you know nothing constructive why this law is in place or why it arranges planets as it does. You ignore the law because it ignores mass. By this law I can explain how gravity works. Mass has no part in cosmology as it applies in the Universe, but singularity rules the lot.

Has any one in your league have any clue why The Lagrangian system forms as it does while applying your most impressive mathematical equations. It's been around since 1772 and in almost three hundred years you have not even come close to any attempt to explain the phenomenon. Studying this system shows singularity rules the lot.

The Coanda affect is the way all jet propulsion works on and all aerodynamics and wind restriction works on, and yet you have not even come close even to find an inconclusive explanation, even in a feeble context of any sorts. Yet when I use thirty two years of my life to find a method to explain these phenomena that then bridges the barrier preventing science to enter the era before the Big Bang came about, there is only locked doors preventing me to find publishing. The centre of a solid "pulls" closer the liquid by Π and the earth is solid with the air being liquid.

What are you lot scared of? Are you scared of what you hide would be uncovered and that the people's admiration would tarnish when your dark methods are unveiled... for dark they are because on the most critical there could be in science, how mass pulls, that most basic you only assume? You lot in the modern era are many times over more protectionists than the Roman Catholic Church was at their worst. The Roman Catholic Church thought they represented God and acted as such. You lot think you are God and act accordingly and therefore you allow no one to know anything if you lot did not know it first. You are unable to tell me what the four phenomena are, let alone explain their function and purpose in detail. When I try to introduce a road that will lead to a method whereby one may uncover the purpose and the working of gravity and the manner in how the four phenomena forms gravity and how the four phenomena kick-started the Universe into what it is, everyone in science blocks me because I represent the devil since you are god. You stop me from showing anything new because then I show why you lot are wrong. You might hold all the power, but when using your power you only advance stupidity, as this recent project in the mountain of Switzerland will once again prove. You'll learn nothing because you are going about the wrong way. The Big Bang started with numbers and relevancies that then by applying those dots in relevancies became material and space. If you read my article that I sent you, you would have seen the detail how it happens.

I am no longer taking these insults on the cheek and riding it out as I did do so many, many times during the past ten years while trying to get my idea of **The New Cosmic Theory** read by anyone that is not so sublimely self-opinionated as I find practising and teaching academics in physics are. I try to introduce the cosmos, as it was in the pre-big bang era when only singularity prevailed because if it did prevail then it still has to prevail just as Einstein surmised it does in the flat Universe he saw. If it was part of the Universe back then when singularity prevailed, then it still must be part of the Universe since it has nowhere else to go when it leaves while it has nowhere to leave. The cosmos is written in a mathematical cosmic code and I found a way to translate the code and using that I became able to understand many unresolved facts about the cosmos. **Please be warned that there is no simplicity such as just awarding mass in the forming of gravity.**

The process is immensely more complicated than awarding mass because this is done without the circumvention of the truth by sidestepping reality in proving the factor that mass is forming gravity by a factor in alignment of cosmic objects having a specific volume in size or in any cosmic planet showing more pulling power. You can't just simply gauge an object and then award mass by the size it has to cheat reality, because in cosmology there is no such an escape root. In fact size plays no role in the cosmos because the smallest star there can ever be, the black hole, is also the biggest star there can ever be. If you disagree with my statement that mass does not apply in cosmic terms, then show what role mass plays in the positional allocation of planets circling the sun. See how planets align in the solar system and from their size, prove they use mass to line up accordance to mass being a factor... **I prove that gravity is Π** and every circle every planet makes is vivid proof of the fact that **that gravity forms by Π applying.** If the planets as well as the sun pulled by mass, then surely the lot had to be part of the sun by now because the pulling has been going on for some time. The truth is the planets are drifting further away from the sun every instant of time and how do you reconcile that with Newton's force of mass giving a pulling power.

The proof I bring is true about gravity being formed as a result of implementing the following phenomena, **The Lagrangian system** 2) The **Roche limit** 3) The **Titius Bode law** 4) The **Coanda affect,** and the combination these phenomena we find the sphere as a multi-dimensional circle present the form Π, which I explain by delivering mathematical proof as to how they fit into the overall picture of gravity. I prove the fact that every individual one of those phenomena is forming a unit that is in total being what we think of as gravity. The phenomena altogether constitutes a unit that forms the process working as gravity. The phenomena are there and with them applying in physics how can anyone dare as to be so arrogant in saying my work is not physics when I explain cosmic laws.

I am going to use this letter as a web page introduction in the future as to show anyone that wishes to read my work, where I stand on matters concerning science. I think it gives a splendid opportunity since this shows the story of my life the last ten years and my encounters with all academics in physics! That makes this letter an open letter.

The Absolute Relevancy of Singularity: The Dissertation The book with which I hope to convince all the non-believers and general doubters about the authenticity of my new approach to science. Should any person feel a need to first find conviction in my claims I make, I have published this title to introduce my thoughts. Should the interested party need convincing about the authenticity and the new approach I take to science before committing to purchase the four books, then this book will do all the convincing required by any sceptic.

What you are going to read is new to all of mankind. Everything you are going to encounter was never written before. Whether you are the most accomplished physicist or a first year student fresh out of school you have the same background knowledge about the work you will encounter. It is advisable therefore to first get acquainted with the first, and then the second book then the third book before reading the fourth THE COSMIC CODE.

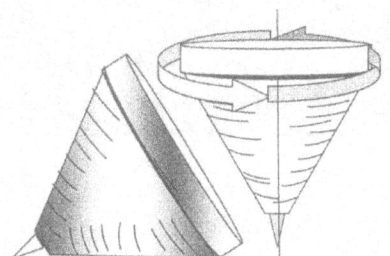

There is a Universe in difference between the top lying down and a top spinning erect and all of that difference Newton missed three and a half centuries ago. In the books I explain in much detail (as well as all the others I have written on the matter) as well as that I prove that gravity forms through movement in relation to these four cosmic phenomena and the movement establishes Π which translates into gravity. By movement gravity applies and in that mass is a by-product resulting from gravity moving and then forming mass in the process. Mass is the result of gravity and is not the factor producing gravity. Mass results in gravity forming.

Gravity forms by the spinning movement of material and in that establishes singularity that is then initiating a circle that produces singularity by using Π. That is gravity.

By the movement of the top that is holding the top erect, the top in moving fights off the mass that the Earth bestow on the top and where mass kills movement by having the top lying down still on the Earth, the moving top comes erect as it then fights the Earth's gravity by applying its independent individual gravity.

By applying individual motion to the status of the top on top of the structural individuality that increases by the motion that the Earth provides, the independence of the individual object is becoming further exaggerated by having independent motion, which is further defying the incorporation the Earth strives to achieve. As the motion of the independent object grows more independent by applying more excessive motion to such an extent where motion creates almost the ultimate independence that may free the individual object with independence from the motion the Earth creates is what is breaking the restraint gravity has on all objects with independence formed by their structure. The structure show independence at all times by not forming part of the structure of the Earth within the sphere of the Earth's gravity. Moving about shows even more reluctance on the part of the top when spinning allows the top to eventually become part of the Earth. Breaking the sound barrier is the motion in space duplicating space by crossing over gravity borders, which is the limit to what constraint the Earth may produce in accordance with what full independence would allow.

The phenomena form an intergraded unit that results in gravity forming where each forms a part of gravity. You may still be you would be sceptical ...but convince yourself that I did manage to:
 1) Find the location, position of singularity as a factor forming space-time
 2) Finding space-time by dissecting Kepler's formula in relation to valuing singularity
 3) Finding and proving space-time and aligning space-time with gravity
 4) Find the working principals behind gravity as a cosmic occurrence.
 5) Find the reason for the Roche limit and explaining the resulting of gravity from that.
 6) Find out why the Lagrangian system, becomes the building form of the Universe.

7) Find why the Titius Bode law mathematically provides the foundation of gravity
By proving that the Coanda affect is gravity through activating space-time
By using the above the four cosmic pillars, it enables me to present the proof where I now can explain what conditions bring on the sound barrier.

By proving it is gravity that the individual structure generates motion above and beyond the gravity the Earth provide is what is producing individual motion that the independent object earned within the sphere of motion that the Earth's gravity provides where the independent and individual motion put the relevance that gravity has beyond the conserving means gravity has where the space that is serving the independent object is independently in motion. The adding to the independence on top of the normal structural independence is creating more individualism by the independent motion of the individual structure being apart from the motion that the gravity of the Earth provides. The fact every one misses is that any structure that is not part of the Earth's crust has an independent gravity and the form this gravity applies is stronger than the Earth's gravity which is why the structure maintains its form and this provides the independent individuality the structure has giving the unique structural space. The gravity of the Earth strives to incorporate everything into the Earth's sphere and into the Earth's structure and therefore the fact that the object is not incorporated into the Earth shows defiance and individuality, which gives it, mass.

These are the definitions underwriting cosmology and while my work is that much ignored; read my books and let's see how far I stray from these definitions in comparison of how much Mainstream science underwrites these definitions by them bringing indisputable proof in presenting unwavering hardcore facts as I do. The manner that Mainstream science interprets these definitions is a joke not worthy to mention.

Quoted directly from the Oxford dictionary of Astronomy the following:
The definition of space-time is as follows:
Space-time is a four dimensional position of the Universe where the position of an object is specified by three coordinates in space and one position in time. According to the theory of special relativity there is no absolute time, which can be measured independently of the observer, so events that are simultaneous as seen from one observer occur at different times when seen from a different place. Time must therefore be measured in a relative manner as are positions in three-dimensional Euclidean space, and this is achieved through the concept of space-time. The trajectory of an object in space-time is called world line. General relativity relates to curvature of space-time to the positions and motions of particles of matter.

The definition of singularity is as follows:
Singularity: a mathematical point at which certain physical quantities reach infinite values for example, according to the general relativity the curvature of space-time becomes infinite in a black hole. In the big bang theory the Universe was born from singularity in which the density and temperature of matter were infinite.

The Oxford dictionary of Astronomy defines gravitation as follows
Gravitation is the force of attraction that operates between all bodies. The size of the attraction depends on the masses of the bodies and the distance between them; gravitational force diminishes by the square of the distance apart according to the inverse square law. Gravitation is the weakest of the four fundamental forces in nature. I. Newton formulated the laws of gravitational attraction and showed that a body behaves as though all its mass were concentrated at its centre of gravity. Hence the gravitational force acts along a joining of the centres of gravity of the two masses. In the general theory of relativity gravitation is interpreted as the distortion of space. Gravitational forces are significant between large masses such as stars planets and satellites, and it is this force, which is responsible for holding together the major components of the Universe. However on the atomic scale the gravitational force is about 10^{40} times weaker than the force of electromagnetic attraction

In the books on offer through this web page and in which I am introducing a totally new concept in terms of gravity, the proof I bring is true about gravity being formed as a result of these phenomena. In the past science hardly recognised the existence of such phenomena although they are known to science for centuries.

They are known as
1) The Lagrangian system
2) The Roche limit
3) The Titius Bode law
4) The Coanda affect

However, since the explanations that I provide holds a completely new line of thought, there are just too many and too numerous wide ranging facts behind that which forms the complete picture as a whole, this leaves me unable to include a full introduction in a space as small as that which a web page will allow. The explaining of such a totally new approach includes for instance those phenomena science this far failed to understand and which I have named as the four cosmic pillars. With these facts being altogether new to science, I find academics showing very little willingness to consider the acceptable value thereof. I recon it must be the result of science seeing so many idle explanations in the past and then proving to be senseless as much as being little impressive, therefore my mentioning it without bringing and substantiating proof will be fruitless and counter productive.

I found the manner in which to interpret Kepler's formula as $a^3 = kT^2$ and I found that when dealing with Kepler's formula, we should not see a^3 as space but we should see singularity being positioned in space in relation to singularity forming relevancies. What brought the answers was putting singularity in context with Π. Doing that placed me in the position to discover what gravity is and how gravity operates to form the Universe. I saw that Kepler's formula should instead use Π.

By placing Π in relation to gravity I manage to find an explanation for the four cosmic phenomena. Everything that has anything to do with gravity forms a circle albeit that it is called the curvature of space-time or gravity bending light or forming a round galactica, the connecting factor is gravity which implements Π. Gravity or another name used to call gravity would be time is running on the measure of Π and every aspect of cosmology integrates Π as the basic concept on which cosmology is founded.

Because my views do not echo the commendable praise attributed to the greatness by which Newton is commemorated, my work is purposely and very much wilfully poorly received in the world of physics and astrophysics and by that I find very little willingness in any understanding shown in the ranks of Newtonian science. This work contains ideas about the introducing of a totally new concept on explaining for the first time ever the working principles of gravity, a matter that eluded Newton no less. I decided to offer four books that introduce the explaining of these concepts in e-book format. This method of publishing rests totally on a financial basis.

I tried to introduce the four phenomena as a concept by using a web page but found such introduction is far too comprehensive in having just too many and numerously wide ranging facts that form the complete picture as a whole to be comprehensibly appreciable, and therefore on account of that realisation that I was unable to include a full introduction in a space as small as that which a web page will allow, it gave me the idea of introducing this new concept via electronic publishing. As my other books I sell by printing are all hundreds to thousands of Mega Bites of information, I had to revise the layout where each is to have fewer than twenty mega bites. This motivated me to only introduce the concepts in producing small books that then could be sold via the electronic publishing media as to allow persons to first acquaint themselves about the viability of the concepts and the feasibility of this new approach I introduce. If any person shows interest in finding out more about any of the books, please click on the book of interest and discover something in science no one yet has ever heard about.

The main issue of finding the value and the meaning of the four phenomena was to connect gravity to Π. Gravity is much closer connected and is much more intimately related to Π than it ever can be linked with mass. By giving each of the phenomena a measured value in terms of Π solved every riddle connected to the phenomena and not only did the phenomena become purposely clear but also the working principle gravity…

What you purchase is information and not ink put on paper, forming a copy of yet another science book. You purchase information never yet divulged, and I am not exaggerating. Look at what new knowledge that I uncover as I explain for the first time the following phenomena:

The Roche limit is:

The region surrounding each star in a binary system, within which any material is gravitationally bound to that particular star. The boundary of the Roche lobes is an equipotential surface, and the lobes touch at the inner Lagrangian point, L_1, through which mass transfer may occur if one of the components expands to fill its lobe. It names after the French mathematician Edouard Albert Roche (1820-83).

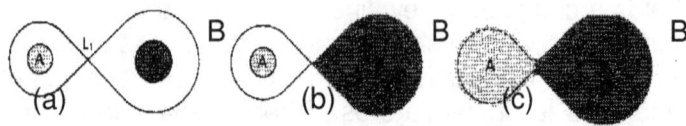

THE ROCHE LOBE: In a binary system, the Roche lobes of components A and B meet at the L_1 Lagrangian point. (a) In a detached system, neither star fills its Roche lobe. (b) In a semidetached system, one massive component, B, fills its Roche lobe. (c) In a contact binary, both components overfill their Roche lobes and share a common envelope.

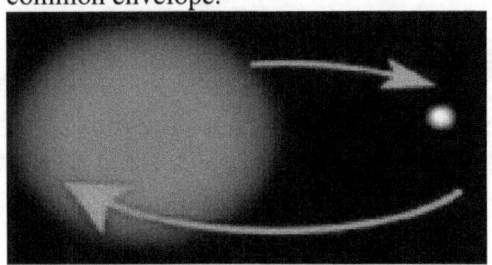  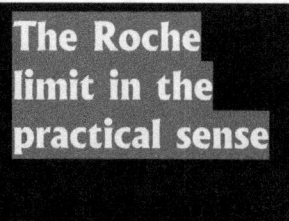

The Roche limit in the practical sense

The following will offend Newtonians beyond what words can describe but the truth is there for all to see. The formula $F = G \frac{M_1 M_2}{r^2}$ cannot explain the comic occurrence shown in the pictures above called the Roche limit, I should find some attention when I say I can explain what is occurring in this instance and this occurrence connects directly to the Roche limit, as explained above. Not only does the Roche limit explain this phenomenon, but also it ties directly to the Titius Bode principle, also being another inexplicable factor in light of the formula $F = G \frac{M_1 M_2}{r^2}$. The Roche limit liquefies the minor structure when the two are 2.4674 x the radius apart and in the process the major star absorbs the minor star after in ballooned the smaller one to the radius belonging to the larger one. This is kept under wrap because $F = G \frac{M_1 M_2}{r^2}$ can't explain what happens while it discredits Newton completely. I can explain it, but only after I abandon Newton's $F = G \frac{M_1 M_2}{r^2}$

According to the formula of $F = G \frac{M_1 M_2}{r^2}$ all orbiting structures should collide with a bang, but instead it is evident that they do the tango until one drops, but when dropping it still does not collide with the larger structure, instead it is liquefied and than is absorbed. There is no form of collision ever taking place as would the formula $F = G \frac{M_1 M_2}{r^2}$ suggest that is used by science.

The position where the formula applies is most surprising. Where the formula $F = G \frac{M_1 M_2}{r^2}$ applies, one has to find singularity applying because the position of r is pointing to a specific pinpointing of space contracting.

This is not only limited to planets in our solar system. In the Universe, there are giant stars spinning around each other. These stars are binaries, which are also one form of double stars where double stars are another such a form. The difference between the types depends on the distance they remain apart.

They keep a certain distance apart and do not collide. In the case of the sun and its planets, it could be a case that the systems might be too small, or they might be too apart. However, this is not the case with binary stars. They are close, they are big, and they spin around a mean axes called the Roche limit.

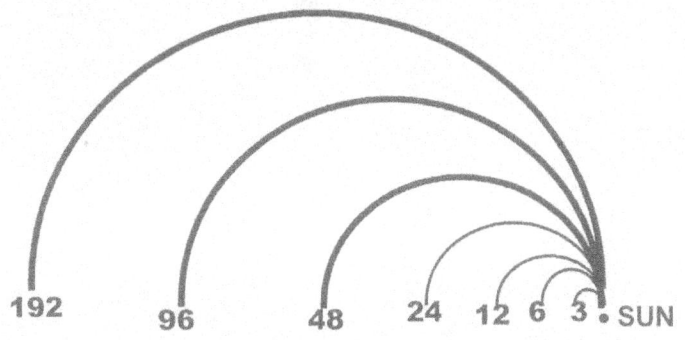

The Titius Bode Law in table form:

Planet	Mercury	Venus	Earth	Mars	Ceres	Jupiter	Saturn	Uranus
Bode's Law distance	4	7	10	16	28	52	100	196
Actual distance	3.9	7.2	10	15.2	28	52	95	192

The Titius Bode Law:
A numerical sequence announced by J.E. Bode in 1772, which matches the distances from the Sun of the six planets then known. It is also known as the Titius-Bode law, as it was first pointed out by the German mathematician Johann Daniel Titius (1729-96) in 1766. It is formed from the sequence 0,3,6,12,24,48,96, and 192 by adding 4 to each number. The planets were seen to fit this sequence quite well – as did Uranus, discovered in 1781. However, Neptune and Pluto do not conform to the 'law'.

Bode's Law stimulated the search for a planet orbiting between Mars and Jupiter that led to the discovery of the first asteroids. It is often said that the law has no theoretical basis, but it does show how orbital resonance can lead to commensurability. The importance that becomes known is the sequence the Ties – Bode law saw in the number arrangement of 3; 6; 12; 24; 48; 96 etc. The incorrect application of the Titus Bode law lies in subtracting the figure of 3 from 10 leaving 7. The other way of reasoning is to add four each time to the firs value of three starting with 3 and so on. The true significance of the Titus-Bode law is that it points directly to a circular growth of 7 stages. The 7 relating to 10 is a precise derogative of the Roche limit or the Roche limit is a precise derogative of the Titius Bode principle because he two systems interlink.

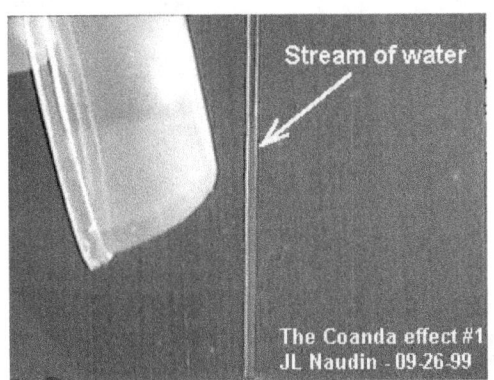 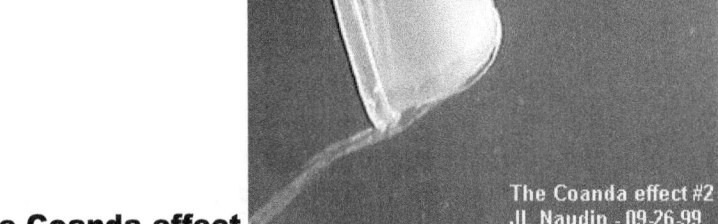

The Coanda effect

The Coanda effect applies as a gravitational phenomenon where moving liquid concentrates around the surface of round solid structures and by movement of either the liquid or the solid or both these concentrates the density of the liquid to gather and compact the flow of the liquid while remaining following the curve of the round surface. The liquid rather follows the curve of the round bowl than to fall straight to the Earth as on should expect. The liquid maintains relevance to the centre of such a round solid. I discard the idea that mass could be responsible for forming gravity because in almost four hundred years all evidence is indicating that the truth is to the contrary.

LAGRANGIAN POINT:

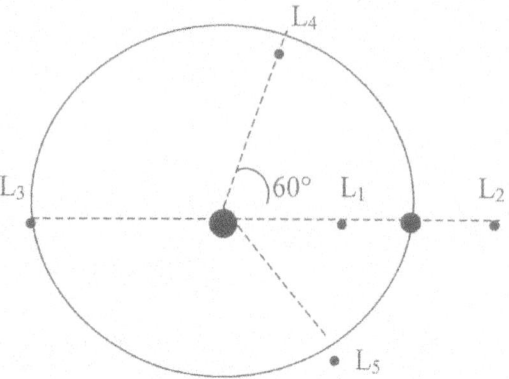

LAGRANGIAN POINT:
The Lagrangian points are five equilibrium points in the orbit of one body around another, such as a planet around the Sun

The phenomena are there and are applying! Put Newton's formula $F = G \frac{M_1 M_2}{r^2}$ to task and use it to explain these very common phenomena, and anyone would find it is not possible to use Newton and explain the gravity represented by this. The phenomena are there and applying so if Newton can't explain it then maybe Newton's concept of mass establishing gravity is not applying. <u>It is this last statement where Newtonian science is unwavering in their believing that mass is forming gravity which is what I strongly bring into question.</u>

Please read on to find more information concerning <u>The Absolute Relevancy of Singularity</u>

If you are a well-informed Newtonian physicist that could be taught no more and knows everything the Universe has to offer I know only nothing impresses you because you lot constructed an entire Universe you then filled with…nothing. For all others, any book that deals with gravity there are just too many and numerously wide ranging facts that form the complete picture as a whole, which leaves me unable to include a full introduction in a space as small as that which page will allow. The explaining include for instance those phenomena, which I call the four cosmic pillars, but wise as you are, you would not believe me at this point that I have cracked the coconut because I guess in your vast experience you have seen too many idle explanations in the past proving to be senseless and little impressive, therefore my mentioning my success would not matter much either way.

PO Box ?????
Some Godforsaken Town
In the
Limpopo Province
South Africa.
mailto:E-mail www.singularityrelevancy.com

P.S. J. SCHUTTE (PEET SCHUTTE)

This attitude academics maintain about how I introduce the approach to singularity has been going on for ten years or more. I am rejected because I reject what was never confirmed and since I don't confirm what science believe is confirmed, my work is never even red as science. My approach to ACADEMICS ABOUT A MISTAKE IN NEWTON'S APPROACH TO SCIENCE CONCERING GRAVITY, WHICH THEN BECAME MY CONCERN LEAVES EVERYONE IN SCIENCE UNCONCERNED. Now I published <u>The Absolute Relevancy of Singularity</u> **in six books to show what is correct in science, I hope to explain this in a very small way in the rest of this website. s you will be one of a few that truly knows what gravity is!**

www.ingramcontent.com/pod-product-compliance
Lightning Source LLC
Chambersburg PA
CBHW080615190526
45169CB00009B/3189